21st Century Astronomy

Stars and Galaxies

SEVENTH EDITION

21st Century Astronomy

Stars and Galaxies

SEVENTH EDITION

Stacy Palen

WEBER STATE UNIVERSITY

George Blumenthal

UNIVERSITY OF CALIFORNIA, SANTA CRUZ

W. W. NORTON & COMPANY

Independent Publishers Since 1923

W. W. Norton & Company has been independent since its founding in 1923, when William Warder Norton and Mary D. Herter Norton first published lectures delivered at the People's Institute, the adult education division of New York City's Cooper Union. The firm soon expanded its program beyond the Institute, publishing books by celebrated academics from America and abroad. By midcentury, the two major pillars of Norton's publishing program—trade books and college texts—were firmly established. In the 1950s, the Norton family transferred control of the company to its employees, and today—with a staff of five hundred and hundreds of trade, college, and professional titles published each year—W. W. Norton & Company stands as the largest and oldest publishing house owned wholly by its employees.

EDITOR: Rob Bellinger
PROJECT EDITOR: Laura Dragonette
DEVELOPMENTAL EDITOR: John Murdzek
ASSISTANT EDITOR: Selin Tekgurler
MANAGING EDITOR, COLLEGE: Marian Johnson
MANAGING EDITOR, COLLEGE DIGITAL MEDIA: Kim Yi
PRODUCTION MANAGERS: Eric Pier-Hocking, Sean Mintus, and Richard Bretan
MEDIA EDITORS: Miryam Chandler and Meg Leary
ASSOCIATE MEDIA EDITOR: Arielle Holstein
MEDIA PROJECT EDITOR: Danielle Belfiore
MEDIA EDITORIAL ASSISTANTS: Aly Grindall and Lily Edgerton
MARKETING MANAGER: Ruth Bolster
DESIGN DIRECTOR: Rubina Yeh
DESIGNER: Anne DeMarinis
PHOTO EDITOR: Amla Sanghvi
PHOTO RESEARCHER: Jane Sanders Miller
DIRECTOR OF COLLEGE PERMISSIONS: Megan Schindel
COLLEGE PERMISSIONS MANAGER: Bethany Salminen
PHOTO DEPARTMENT MANAGER: Stacey Stambaugh
PERMISSIONS ASSOCIATE: Elizabeth Trammell
COMPOSITION: Graphic World
MANUFACTURING: Transcontinental

Library of Congress Cataloging-in-Publication Data
Names: Palen, Stacy, author. | Blumenthal, George (George Ray), author.
Title: 21st century astronomy / Stacy Palen, Weber State University, George
    Blumenthal, University of California—Santa Cruz.
Other titles: Twenty-first century astronomy
Description: Seventh edition. | New York : W. W. Norton & Company, [2022] |
    Includes index.
Identifiers: LCCN 2021032451 | ISBN 9780393877021 (paperback) |
    ISBN 9780393877113 (epub)
Subjects: LCSH: Astronomy—Textbooks. | LCGFT: Textbooks.
Classification: LCC QB45.2 .P35 2022 | DDC 520—dc23
LC record available at https://lccn.loc.gov/2021032451

ISBN 978-0-393-53915-8 (pbk)

W. W. Norton & Company, Inc., 500 Fifth Avenue, New York, NY 10110
wwnorton.com
W. W. Norton & Company Ltd., 15 Carlisle Street, London W1D 3BS
1 2 3 4 5 6 7 8 9 0

Stacy Palen thanks everyone at Bellwether Farm for their support during this project that spanned the challenges of the COVID-19 pandemic.

George Blumenthal gratefully thanks his wife, Kelly Weisberg, and his children, Aaron and Sarah Blumenthal, for their support during this project. He also wants to thank Professor Robert Greenler for stimulating his interest in all things related to physics.

BRIEF CONTENTS

Chapters 8–12 are not included in this edition.

CONTENTS

Chapters 8–12 are not included in this edition.

PART I INTRODUCTION TO ASTRONOMY

6 The Tools of the Astronomer   142

PART II   THE SOLAR SYSTEM

7 The Formation of Planetary Systems   174

PART III   STARS AND STELLAR EVOLUTION

16 Evolution of Low-Mass Stars   456

17 Evolution of High-Mass Stars   486

PART IV   GALAXIES, THE UNIVERSE, AND COSMOLOGY

21 The Expanding Universe 604

22 Cosmology 626

WORKING IT OUT

Chapters 8–12 are not included in this edition.

ASTROTOURS

INTERACTIVE SIMULATIONS

ASTRONOMY IN ACTION VIDEOS

Chapters 8–12 are not included in this edition.

AstroTour animations, Interactive Simulations, and Astronomy in Action videos are all available from the free Student Site at the Digital Resources Site, and they are also integrated into assignable Smartwork exercises. **digital.wwnorton.com/astro7.**

Dear Student,

Why is it a good idea to take a science course, and in particular, why is astronomy a course worth taking? Many people choose to learn about astronomy because they are curious about the universe. Your instructor likely has two basic goals in mind for you as you take this course. The first is to understand some basic physical concepts and how they apply to the universe around us. The second is to think like a scientist and learn to use the scientific method not only to answer questions in this course but also to be able to evaluate and understand new information you encounter in your life. We have written the Seventh Edition of *21st Century Astronomy* with these two goals in mind.

Throughout this book, we emphasize not only the content of astronomy (for example, the differences among the planets, the formation of chemical elements) but also *how* we know what we know. The scientific method is a valuable tool that you can carry with you and use for the rest of your life. One way we highlight how science works is in the **Process of Science Figures**. In each chapter, we have chosen one discovery and provided a visual representation illustrating the discovery or a principle of the process of science. In these figures, we develop the idea that science is not a tidy process that proceeds in a straight line from one idea to the next. Discoveries are made because people are trying to answer a question and show why or how we think something is the way it is. This is often surprising.

The most effective way to learn something is to "do" it. Whether playing an instrument or a sport or becoming a good cook, reading or listening can only take you so far. The same is true of learning astronomy. We have written this book to help you "do" as you learn. Sometimes this means doing a hands-on activity to apply a concept and sometimes we provide a tool to make reading a more active process.

The following tools in each chapter help you "do" as you learn:

- **Active Learning Figures** open each chapter and ask you to "do" science by setting up an experiment and then making either a prediction or an observation and recording the results. Keep in mind that the "answer" isn't the most important part of the activity. Rather, we want the experience of thinking about a physical phenomenon and predicting what will happen next to become a natural way for you to apply your knowledge and understand new concepts.

- **What if** questions throughout the chapter prompt you to apply what you have learned to situations both real and imagined. Many of these are based on questions that students have asked, when they wondered, "What if the universe were different than it is?" These questions are open-ended, so think of them as guides for how to wonder about scientific ideas, rather than tests of your preexisting knowledge.

- Each chapter's **What an Astronomer Sees** feature helps you understand how astronomers interpret astronomical images, obtaining enormous amounts of information from a single picture. This feature is accompanied by an end-of-chapter question to teach you to interpret astronomical images yourself. Similarly, we have identified all of the figure-based questions at the end of the

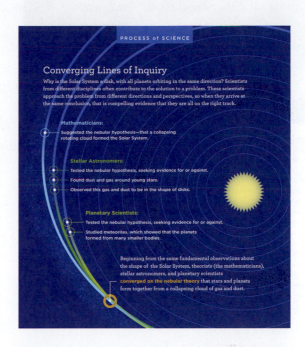

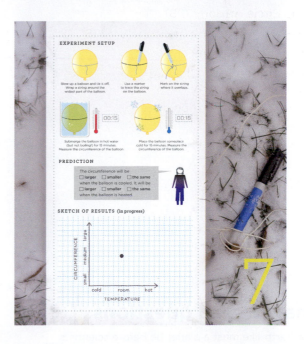

what if . . .
What if you observe a close pair of stars forming a binary system? How would you expect the disk around such a pair of stars to differ from the disk around a single star?

Figure 7.8 ★ WHAT AN ASTRONOMER SEES This gorgeous image from the Hubble Space Telescope shows a portion of the Carina Nebula. An astronomer looking at this image will immediately notice the colors, because color often indicates where different atoms or molecules are present. She will not necessarily know which atoms or molecules are represented by these colors because different astronomers will use a different palette. But even without reading any background on the image, she will know that there is something different about the hazy blue areas and the hazy pink or green ones. An astronomer will recognize and then mostly ignore the diffraction spikes (mentioned in Chapter 6) that form an X around each bright star. She will notice the brown clumps of material that are too dense to see through. These high-density regions indicate that star formation might be happening in this nebula, and this will be confirmed by the small oval protostar in the upper right corner and by the dense "fingers" that stick out in various places. These fingers point the way toward a source of interstellar wind outside the image to the upper right. That wind has eroded away the less dense material around these denser regions and may have triggered star formation by compressing the material at the top of each finger. Each finger is a dense blob of material that creates a wind "shadow" behind it. A new star has just formed at the top of the finger near the middle of the image. An astronomer will identify this new star because of the thin jets of material that are being ejected in opposite directions; new stars sometimes create such jets.

chapter, which should also help you develop your ability to interpret images and graphs.

- There are **Check Your Understanding** questions at the end of each chapter section. These questions are designed to be answered quickly if you have understood the previous section. The answers are provided in the back of the book so you can check your answer and decide whether further review is necessary.

CHECK YOUR UNDERSTANDING 7.1

Which of the following pieces of evidence support the nebular hypothesis? (Choose all that apply.) (a) Planets orbit the Sun in the same direction. (b) The Solar System is relatively flat. (c) Earth has a large Moon. (d) We observe disks of gas and dust around other stars.

Answers to Check Your Understanding questions are in the back of the book.

- **Reading Astronomy News** sections in each chapter include a news article or press release with questions to focus your attention on how the science is presented. As a citizen of the world, recognizing what is credible and questioning what is not are important skills. You make judgments about science, distinguishing between good science and pseudoscience, in order to make decisions in the grocery store, pharmacy, car dealership, and voting booth. You base these decisions on the presentation of information you receive through the media, which is very different from the presentation of information in class. The goal of Reading Astronomy News is to help you build your scientific literacy and your ability to challenge what you hear elsewhere.

- While we know a lot about the universe, science is an ongoing process, and we continue to search for new answers. To give you a glimpse of what we don't know, we provide **Unanswered Questions** features in each chapter. Most of these questions represent topics that scientists are currently studying.

unanswered questions

How Earth-like must a planet be before scientists declare it to be "another Earth"? An editorial in the science journal *Nature* cautioned that scientists should define "Earth-like" in advance—before multiple discoveries of planets "similar" to Earth are announced and a media frenzy ensues. Must a planet be of similar size and mass, be located in the habitable zone, and have spectroscopic evidence of liquid water before we call it "Earth 2.0"?

Reading Astronomy News

NASA's TESS Mission Uncovers Its 1st World With Two Stars
NASA

Discovering more planets means that we discover more unusual ones; like this planet that orbits two stars at once

The TOI 1338 system lies 1,300 light-years away in the constellation Pictor. The two stars orbit each other every 15 days. One is about 10% more massive than our Sun, while the other is cooler, dimmer and only one-third the Sun's mass.

Newly discovered TOI 1338 b is the only known planet in this binary star system. It's around 6.9 times larger than Earth, or between the sizes of Neptune and Saturn. The planet orbits in almost exactly the same plane as the stars, so it experiences regular stellar eclipses.

Scientists use the observations from TESS to generate graphs of how the brightness of stars change over time. When a planet transits in front of its star from our perspective, its passage causes a dip in the star's brightness.

Planets orbiting two stars are more difficult to detect than those orbiting one. TOI 1338 b's transits are irregular, between every 93 and 95 days, and vary in depth and duration thanks to the orbital motion of its stars. TESS only sees the transits crossing the larger star; the transits of the smaller star are too faint to detect.

"These are the types of signals that algorithms really struggle with," said lead author Veselin Kostov, a research scientist at the SETI Institute and Goddard. "The human eye is extremely good at finding patterns in data, especially non-periodic patterns like those we see in transits from these systems."

After identifying TOI 1338 b, the research team used a software package called eleanor, named after Eleanor Arroway, the central character in Carl Sagan's novel "Contact," to confirm the transits were real.

TOI 1338 had already been studied from the ground by radial velocity surveys. Kostov's team used this archival data to analyze the system and confirm the planet.

QUESTIONS

1. The period of the two stars in the binary pair is 15 days. Are these two stars very close to each other, or very far apart?

2. The planet's period is 95 days. Is the planet's orbit larger or smaller than the orbit of the two stars around one another?

3. Make a sketch of this system, including the two stars and the planet. Draw circles to show how the orbits are related, and whether or not they cross.

4. The article states that the discovery was confirmed using archival data from ground-based telescopes. Speculate: Why was the planet not discovered prior to this, using that ground-based data?

5. This system provides a window into the naming conventions for stars and planets. Study the names of the stars and planet. How are the two stars of a binary star system designated? How is the first planet in a planetary system designated?

Source: https://www.nasa.gov/feature/goddard/2020/nasa-s-tess-mission-uncovers-its-1st-world-with-two-stars.

working it out 7.1

Angular Momentum

In its simplest form, the angular momentum (L) of a system is given by

$$L = m \times v \times r$$

where m is the mass, v is the speed at which the mass is moving, and r represents how spread out the mass is.

Let's apply this relationship to the angular momentum of Jupiter in its orbit about the Sun. The angular momentum from one body orbiting another is called *orbital* angular momentum, $L_{orbital}$. The mass (m) of Jupiter is 1.90×10^{27} kilograms (kg), the speed of Jupiter in orbit (v) is 1.31×10^4 meters per second (m/s), and the radius of Jupiter's orbit (r) is 7.79×10^{11} meters. Putting all that together gives

$$L_{orbital} = (1.90 \times 10^{27} \text{ kg}) \times (1.31 \times 10^4 \text{ m/s}) \times (7.79 \times 10^{11} \text{ m})$$

$$L_{orbital} = 1.94 \times 10^{43} \text{ kg m}^2/\text{s}$$

Calculating the *spin* angular momentum of a spinning object, such as a skater, a planet, a star, or an interstellar cloud, is more complicated. Here, we must add up the individual angular momenta of *every tiny mass element* within the object. For a uniform sphere, the spin angular momentum is

$$L_{spin} = \frac{4\pi m R^2}{5P}$$

where R is the radius of the sphere and P is the rotation period of its spin.

Let's compare Jupiter's orbital angular momentum with the Sun's spin angular momentum to investigate the distribution of angular momentum in the Solar System. The Sun's radius is 6.96×10^8 meters, its mass is 1.99×10^{30} kg, and its rotation period is 24.5 days $= 2.12 \times 10^6$ seconds. If we assume that the Sun is a uniform sphere, the spin angular momentum of the Sun is

$$L_{spin} = \frac{4 \times \pi \times (1.99 \times 10^{30} \text{ kg}) \times (6.96 \times 10^8 \text{ m})^2}{5 \times (2.12 \times 10^6 \text{ s})}$$

$$L_{spin} = 1.14 \times 10^{42} \text{ kg m}^2/\text{s}$$

$L_{orbital}$ of Jupiter is about 17 times greater than L_{spin} of the Sun. Thus, most of the angular momentum of the Solar System now resides in the orbits of its major planets.

For a collapsing sphere to conserve L_{spin}, its rotation period P must be proportional to R^2. As with the skater, when a sphere decreases in radius, its rotation period decreases; that is, it spins faster.

- The language of science is mathematics, and it can be as challenging to learn as any other language. The choice to use mathematics as the language of science is not arbitrary; nature "speaks" math. To learn about nature, you will need to speak its language. We don't want the language of math to obscure the concepts, however, so we have placed this book's mathematics in **Working It Out** boxes to make it clear when we are beginning and ending a mathematical argument, so that you can spend time with the concepts in the chapter text and then revisit the mathematics of the concept to study the formal language of the argument. You will also learn to work with data and identify when data aren't quite right. We want you to be comfortable reading, hearing, and speaking the language of science, and we provide you with tools to make it easier. At the end of each chapter, there are several mathematical problems. For each Working It Out box, there is a "scaffolded" end-of-chapter problem that guides you to predict the answers, calculate them, and then check your own work. Predicting the answer helps you develop intuition about mathematics, and checking your own work is a skill you will find useful later in life every time you need to calculate costs or estimate budgets.

- Each chapter concludes with an **Origins** section, which relates material or subjects found in the chapter to the origin of the universe and the origin of life. In recent years, astrobiologists have made much progress understanding these issues.

- At the end of each chapter, we have provided several types of questions, problems, and activities for you to practice your skills. The **Test Your Understanding**

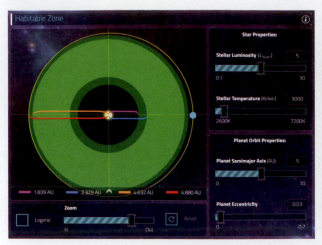

questions focus on more detailed facts and concepts from the chapter. **Thinking about the Concepts** questions ask you to synthesize information and explain the "how" or "why" of a situation. **Applying the Concepts** problems give you a chance to practice the quantitative skills you learned in the chapter and to work through a situation mathematically.

- At the very end of each chapter, an **Exploration** activity shows you how to use the concepts and skills you learned in an interactive way. About half of the book's Explorations ask you to use animations and simulations found on the Student Site, while the others are hands-on, paper-and-pencil activities that use everyday objects such as ice cubes or balloons.

Digital resources linked to the book can help you understand and visualize many of the physical concepts described in the book. **AstroTours** and **Interactive Simulations** are represented by icons in the margins of the book (which are links in the ebook) and in Smartwork online homework. There is also a series of short **Astronomy in Action** videos embedded in the ebook (and represented by icons in the margins of the print book). The videos are available at the Student Site and in Smartwork online homework questions that your instructor might assign. These videos feature one of the authors (and several students) demonstrating physical concepts at work. Your instructor might assign these videos to you, or you might choose to watch them on your own to create a better picture of each concept in your mind.

Astronomy gives you a sense of perspective that no other field of study offers. The universe is vast, fascinating, and beautiful—filled with a wealth of objects that, surprisingly, can be understood using only a handful of principles. By the end of this book, you will have gained a sense of your place in the universe—both how incredibly small and insignificant you are and how incredibly unique and important you are.

Sincerely,
Stacy Palen
George Blumenthal

▶▶ **Interactive Simulation:** Habitable Zone

Astronomy in Action: Angular Momentum

Dear Instructor,

We wrote this book with a few overarching goals: to inspire students, to use active learning to build scientific literacy, and to create a useful and flexible tool that offers diverse approaches to the content.

As scientists and as teachers, we are passionate about the work we do. We hope to share that passion with students and inspire them to engage in science on their own. Through our own experience, familiarity with education research, and surveys of instructors, we have come to know a great deal about how students learn and what goals teachers have for their students. We have explicitly addressed many of these goals and learning styles in this book, sometimes in large, immediately visible ways, such as the inclusion of features, but also through less obvious efforts, such as in questions and problems that relate astronomical concepts to everyday situations or a fresh approach to organizing material.

One way we do this is through the *new* **Active Learning Figures** at the beginning of each chapter. These figures model student engagement and provide an opportunity for students to do an experiment or make an observation on their own using only everyday objects they can find around their house or dorm room. These are particularly useful in online classes, to encourage students to engage with the universe from their own home.

Many instructors state that they would like their students to become "educated scientific consumers" and "critical thinkers" or that their students should "be able to read a news story about science and understand its significance." We have specifically addressed these goals in our **Reading Astronomy News** feature, which presents a news article and a series of questions that guide a student's critical thinking about the article, the data presented, and the sources. Questions based on the Reading Astronomy News feature are available in Smartwork online homework.

Education research shows that the most effective way to learn is by doing. **Exploration** activities at the end of each chapter are hands-on, asking students to take the concepts they've learned in the chapter and apply them as they interact with animations and simulations on the Student Site or work through pencil-and-paper activities. Many of these Explorations incorporate everyday objects and can be used either in your classroom or as activities at home, and nearly all are assignable through Smartwork online homework.

We also believe students should be exposed to the more formal language of science—mathematics. We have placed the math in **Working It Out** boxes, so it does not interrupt the flow of the text or get in the way of students' understanding of conceptual material. But we've gone further by beginning with fundamental ideas in early Working It Out boxes and slowly building in complexity through the book. We've also worked to remove some of the stumbling blocks that affect student confidence by providing calculator hints, references to earlier Working It Out boxes, and detailed, fully worked examples. Many chapters include problems on reading and interpreting graphs. Appendix 1, "Mathematical Tools," summarizes some math concepts for students. Each Working It Out box is accompanied by a "scaffolded" problem at the end of the chapter that guides students to predict, solve, and check their work. Even students who are mathematically adept often forget to think about the problem before they attack it, or forget to check that their answer is sensible. Both of these skills are necessary for developing a mathematical intuition about the universe.

Discussion of basic physics is contained in Part I to accommodate courses that use the *Solar System* or *Stars and Galaxies* volumes. A "just-in-time" approach to introducing the physics is still possible by bringing in material from Chapters 2–6 as needed. For example, the sections on tidal forces in Chapter 4 can be taught along with the moons of the Solar System in Part II, or with mass transfer in binary stars in Part III, or with galaxy interactions in Part IV. Spectral lines in Chapter 5 can be taught with planetary atmospheres in Part II or with stellar spectral types in Part III, and so on.

In our overall organization, we have made several efforts to encourage students to engage with the material and build confidence in their scientific skills as they proceed through the book. For planets, stars, and galaxies, we have organized the material to cover the general case first and then delve into more details with specific examples. Thus, you will find "planetary systems" before our own Solar System, "stars" before the Sun, and "galaxies" before the Milky Way. This allows us to avoid frustrating students by making assumptions about what they know about stars or galaxies or forward-referencing to basic definitions and overarching concepts. This organization also implicitly helps students understand their place in the universe: our galaxy and our star are each one of many. They are specific examples of a physical universe in which the same laws apply everywhere. Planets have been organized comparatively to emphasize that science is a process of studying individual examples that lead to collective conclusions. All of these organizational choices were made with the student perspective in mind and a clear sense of the logical hierarchy of the material.

For example, we begin in Chapter 19 by introducing galaxies as a whole and our measurements of them, including recession velocities. Then we address the Milky Way in Chapter 20—a specific example of a galaxy that we can discuss in detail. This follows the repeating motif of moving from the general to the specific that exists throughout the text and gives students a basic grounding in the concepts of spiral galaxies, supermassive black holes, and dark matter before they need to apply those concepts to the specific example of our own galaxy. Chapter 21, "The Expanding Universe," covers the cosmological principle, the Hubble expansion, and the observational evidence for the Big Bang.

The **Origins** sections illustrate how astrobiologists and other scientists approach the study of a scientific question from the chapter related to the origin of the universe and of life. Material about exoplanets is introduced in Chapter 7, incorporated into other chapters when appropriate, and continued in Chapter 24. We revised each chapter, streamlining some topics and updating the science to reflect the progress in the field. This includes

- updates on new telescopes and space missions;
- new images and results from Solar System exploration missions to the Moon, Mars, Jupiter, and Venus;
- updated graphical data on climate change;
- results from Kepler and other missions on populations of exoplanets;
- results from the *Gaia* mission on stellar distances, the H-R diagram, and the Milky Way;
- results from LIGO/VIRGO on merging compact objects;
- many new articles for Reading Astronomy News; and
- revised simulations for some Exploration exercises, as well as new simulations.

Other items new to the Seventh Edition include the following:

- *New* **Active Learning Figures** at the start of each chapter, described previously.

- Each chapter contains a **What an Astronomer Sees** figure that demonstrates for students how an astronomer can derive a wealth of information from a single image. Each figure is accompanied by an end-of-chapter question and questions in Smartwork that further guide students in developing the skill of interpreting astronomical imagery.

- Multiple times in each chapter, we ask students "**What if**?" These "What if" questions often ask students to imagine the universe other than it is, or to extrapolate from the information they have learned. These advanced comprehension questions give students the opportunity to experiment with developing and testing hypotheses by asking, "If the universe were not as it is, what else would have to be true?" This feature is accompanied by a "guiding" question in Smartwork that helps students organize their thinking. The in-text questions can be particularly useful for sparking discussion, either in class or in an LMS.

- Teaching Astronomy by Doing Astronomy (**tada101.com**) is a blog for introductory astronomy instructors. Stacy Palen writes regular posts and provides suggestions for adding active learning to any size of class. Stacy also posts suggested Reading Astronomy News articles, discusses ways to integrate math into a course, and hosts discussions about recent developments in practical applications of astronomy education research. Join the community at tada101.com.

Many professors find themselves under pressure from accrediting bodies or internal assessment offices to assess their courses in terms of learning goals. Each chapter has Learning Goals and an end-of-chapter Summary to correspond to the chapter's Learning Goals. In Smartwork, questions and problems are tagged by type and can be sorted by Learning Goal. Smartwork contains more than 2200 questions and problems that are tied directly to this text, including the Check Your Understanding questions, versions of the Reading Astronomy News and Exploration questions, questions based on the Chapter-opening experiments, and many ranking, sorting, and labeling tasks. Smartwork now contains pre- and post-activity questions based on *Learning Astronomy By Doing Astronomy*, Second Edition, so you can help students prepare for each activity and assess learning afterwards.

Every question in Smartwork has hints and answer-specific feedback so that students are coached to work toward the correct answer. Smartwork is flexible: premade assignments based on each chapter make it easy to assign, and instructors can easily modify any of the provided questions, answers, and feedback or can create their own questions.

We've also created a series of 23 Astronomy in Action videos explaining and demonstrating concepts from the text, accompanied by questions in Smartwork. You might assign these videos prior to lecture—either as part of a flipped modality or as a "reading quiz." In either case, you can use the diagnostic feedback from the questions in Smartwork to tailor your in-class discussions. Or you might show them in class, to stimulate discussion. Or you might simply use them as a jumping-off point—to get ideas for activities to do with your own students.

We continue to look for better ways to engage students, so please let us know how these features work for your students!

Supporting Resources for Students

digital.wwnorton.com/astro7

Smartwork Online Homework

smartw⊕rk

Steven Desch, Guilford Technical Community College
Shimonee Kadakia, El Camino College
Ana M. Larson, Emerita, University of Washington
Violet Mager, Penn State Wilkes-Barre
Doug Stuffle, Tidewater Community College
Dave Wood, San Antonio College
Todd Young, Wayne State College
William Younger, Tidewater Community College

More than 2200 questions support *21st Century Astronomy*, Seventh Edition—all with answer-specific feedback, hints, and ebook links. Questions include Summary Self-Tests, versions of the Explorations (based on AstroTours and new Interactive Simulations), and questions based on the Reading Astronomy News features. Astronomy in Action video questions focus on getting students to come to class prepared and on overcoming common misconceptions. Process of Science Guided Inquiry Assignments help students apply the scientific method to important questions in astronomy, challenging them to think like scientists. The course also contains ranking, sorting, and labeling exercises based on art from the text. Questions span Bloom's Taxonomy, engaging and challenging students. Smartwork can be set up to work right in your LMS, with student scores flowing directly to your LMS gradebook. Your local Norton representative can help you set up LMS integration.

Norton Ebook 🄴

The *21st Century Astronomy*, Seventh Edition ebook provides students an enhanced reading experience at a fraction of the cost of a print textbook. Students are able to have an active reading experience and can take notes, bookmark, search, highlight, and even read offline. Instructors can add notes for students to see as they read the text. Norton Ebooks can be viewed on—and synced among—all computers and mobile devices. Every question in Smartwork provides a reference link to the ebook so that students have easy access to their textbook when completing homework assignments.

Twice per year, the Seventh Edition ebook will be updated with exciting new research from the field of astronomy, ensuring that the book remains timely and relevant. New Smartwork questions will support these features.

Student Site

W. W. Norton's student website features the following:

- Thirty AstroTour animations. These animations, some of which are interactive, use art from the text to help students visualize important physical and astronomical concepts. All are now tablet compatible.

- Eight Interactive Simulations, authored by Stacy Palen and paired with in-text Exploration activities, allow students to explore topics such as Moon phases, Kepler's laws, and the Hertzsprung-Russell diagram.
- Twenty-three Astronomy in Action videos feature author Stacy Palen demonstrating the most important concepts in a visual, easy-to-understand, and memorable way.

Learning Astronomy by Doing Astronomy, Second Edition: Collaborative Lecture Activities

Stacy Palen, Weber State University
Ana M. Larson, Emerita, University of Washington

Students learn best by doing. Devising, writing, testing, and revising suitable in-class activities that use real astronomical data, illuminate astronomical concepts, and pose probing questions that ask students to confront misconceptions can be challenging and time consuming. In this workbook, the authors draw on their experience teaching thousands of students in many different types of courses (large in-class, small in-class, hybrid, online, flipped, and so forth) to bring 36 field-tested activities that can be used in any classroom today. The activities have been designed to require no special software, materials, or equipment and to take no more than 50 minutes to do. Pre- and post-activity questions are now assignable within Smartwork.

If you are interested in packaging the workbook with this text, please contact your local Norton representative.

Starry Night Planetarium Software (College Version) and Workbook

Steven Desch, Guilford Technical Community College
Michael Marks, Bristol Community College

Starry Night is a realistic, user-friendly planetarium simulation program designed to allow students in urban areas to perform observational activities on a computer screen. Norton's unique accompanying workbook offers observation assignments that guide students' virtual explorations and help them apply what they've learned from the text's reading assignments. If you are interested in packaging the Starry Night software and workbook with this text, please contact your local Norton representative.

For Instructors

Instructor's Manual

Ana M. Larson, Emerita, University of Washington

This resource includes brief chapter overviews; suggested discussion points; notes on the Process of Science figures, AstroTour animations, Interactive Simulations, Astronomy in Action videos, Reading Astronomy News, and Explorations; and worked solutions to Check Your Understanding and end-of-chapter Questions and Problems. Also included are notes on teaching with *Learning Astronomy by Doing Astronomy: Collaborative Lecture Activities* and answers to the *Starry Night Workbook* exercises.

Teaching Astronomy by Doing Astronomy Blog, tada101.com

Teaching Astronomy by Doing Astronomy is a community-focused blog that features posts by Stacy Palen and contributing astronomy instructors on teaching ideas, classroom activities, and current events. Instructors looking for additional ways to incorporate active learning will find this blog to be an invaluable resource.

PowerPoint Lecture Slides

Allison Kirkpatrick, University of Kansas

These ready-made lecture slides integrate selected textbook art, all Check Your Understanding and Working It Out questions from the text, classroom response questions, and links to the AstroTour animations, Interactive Simulations, and Astronomy in Action videos. These lecture slides are fully editable and are available in Microsoft PowerPoint format.

Test Bank

Matthew Newby, Temple University

The Test Bank has been revised using Bloom's Taxonomy and provides over 2400 multiple-choice and short-answer problems. Each chapter of the Test Bank consists of five question levels classified according to Bloom's Taxonomy:

Remembering

Understanding

Applying

Analyzing

Evaluating

Problems are further classified by section and difficulty level, making it easy to construct tests and quizzes that are meaningful and diagnostic. The Test Bank assesses a common set of Learning Objectives consistent with the textbook and Smartwork online homework.

Resources for Your LMS

These files, available for use in various Learning Management Systems (LMSs), add high-quality Norton digital resources to your course, including direct links to the ebook, Smartwork, AstroTours, Interactive Simulations, and Astronomy in Action videos. They also contain Flashcards for students' self-study. Resources for Your LMS are available in Blackboard, Canvas, Desire2Learn, and Moodle formats.

Acknowledgments

The authors would like to acknowledge the extraordinary efforts of the staff at W. W. Norton: Selin Tekgurler, Assistant Editor, who kept everything flowing smoothly while never losing sight of a single detail; Amla Sanghvi for managing the complex photo program; Elizabeth Trammell for managing text permissions; and the copyeditor, Carla Barnwell, who made sure that all the grammar and punctuation survived the multiple rounds of the editing process. We would especially like to thank John Murdzek for the developmental editing process, and Laura Dragonette, who shepherded the manuscript through production.

Rob Bellinger was the editor. Eric Pier-Hocking, Sean Mintus, and Richard Bretan managed the production; Anne DeMarinis designed the book; and Rubina Yeh was the design director. Miryam Chandler, Meg Leary, Arielle Holstein, Aly Grindall, and Lily Edgerton worked on the assessment and visualization resources, and Ruth Bolster will help get this book into the hands of people who can use it.

We gratefully acknowledge the contributions of the authors who worked on previous editions of *21st Century Astronomy*: Jeff Hester, Gary Wegner, the late Dave

Burstein, Ron Greeley, Brad Smith, and Howard Voss, with special thanks to Dave for starting the project, to Jeff for leading the original authors through the First Edition, and to Brad for leading the Second and Third Editions. Laura Kay led the team through the Fourth, Fifth, and Sixth Editions, and we miss her insightful and critical eye.

Stacy Palen
George Blumenthal

We would also like to thank the reviewers, whose input at every stage improved the book:

Seventh Edition Reviewers

Amy Abe, Lake Forest College
Paulo Afonso, American River College
Marilyn Akins, Bluegrass Community Technical College–Newtown
Heather Appleby, Richland College
Scott Armel, Trident Technical College
Brandon Bear, Virginia Polytechnic Institute
Raymond Benge, Tarrant County College
Philip Bennett, Dalhousie University
Ryan Bennett, University of North Texas
Katie Berryhill, Los Medanos College
Brett Bochner, Hofstra University
Julie Bray-Ali, Mt. San Antonio College–Walnut
Robert Brandenberger, McGill University
Michael Briley, Appalachian State University
Meade Brooks, College of Mainland
Shea Brown, University of Iowa
Spencer Buckner, Austin Peay State University
Philip Choi, Claremont College–Pomona
Micol Christopher, Mount San Antonio College
Asantha R Cooray, University of California–Irvine
Jonathan Craig, Horry Georgetown Technical College–Conway
James Dickinson, Clackamas Community College–Oregon City
Edmund Douglass, Farmingdale State College
Donald D. Driscoll, Kent State University
Jay Dunn, Georgia State University–Dunwoody
Robert Egler, North Carolina State University
Emmanuel Fonseca, West Virginia University
Douglas Gobeille, University of Rhode Island
Jose J. Gomez, The Citadel
Dirk Grupe, Morehead State University
George Hassel, Sienna College
David John Helfand, Columbia University
Jerry Horne, College of Southern Nevada–West Charleston
Manish Jadhav, Dutchess Community College–Poughkeepsie
Jennifer Jones, Arapahoe Community College
Charles Kerton, Iowa State University
John LaBrasca, Clark College
Peter Lanagan, College of Southern Nevada–Henderson
Denis Leahy, University of Calgary
Jane H. MacGibbon, University of North Florida
James Maxin, Louisiana State University–Shreveport
John McClain, Temple College

Marcio B. Melendez, Anne Arundel Community College
Sharon Morsink, University of Alberta
Irina Mullins, Houston Community College
Alexis Nduwimana, Georgia State University–Decatur
Jisun Park, Kingsborough Community College
Victoria Plaveti, Manhattan College
Valerie A. Rapson, Schenectady County Community College
Elliot Richmond, Austin Community College
Edward Rhoads, Indiana University–Purdue
Allen Rogel, Bowling Green State University
Ajani Luister Walden Ross, Wright State University–Dayton
Takashi Sato, Kwantlen Polytechnic University
Eric Schlegel, University of Texas at San Antonio
Arthur Schneider, Amarillo College
Parandis Tajbakhsh, University of Toronto–Mississauga
William K. Teets, Vanderbilt University
Jose Vazquez, Washington State University–Vancouver
Jordan Watkins, Brookhaven College
Casey Watson, Millikin University
Melinda Weil, City College of San Francisco
Lee Widmer, Xavier University
Nelka Wijesinghe, Long Star College–Montgomery
Jeffrey Wilson, Texas A&M University–Commerce
David Wood, San Antonio College
William Younger, Tidewater Community College–Portsmouth/ Virginia Beach

Reviewers of Previous Editions

Manuel Alvarado, El Paso Community College
Gagandeep Anand, Boston University
Scott Atkins, University of South Dakota
Simon Balm, Santa Monica College
Timothy Barker, Wheaton College
Peter A. Becker, George Mason University
Timothy C. Beers, Michigan State University
Raymond Benge, Tarrant College
Ryan Bennett, University of North Texas
David Bennum, University of Nevada–Reno
Edwin Bergin, University of Pittsburgh
William Blass, University of Tennessee
Steve Bloom, Hampden Sydney College
Brett Bochner, Hofstra University
Daniel Boice, University of Texas at San Antonio

Bram Boroson, Clayton State University

David Branning, Trinity College

Julie Bray-Ali, Mt. San Antonio College

Jack Brockway, Radford University

Shea Brown, University of Iowa

Spencer Buckner, Austin Peay State University

Suzanne Bushnell, McNeese State University

Paul Butterworth, George Washington University

Juan E. Cabanela, Minnesota State University–Moorhead

Amy Campbell, Louisiana State University

C. Austin Campbell, El Paso Community College

Michael Carini, West Kentucky University

Christina Cavalli, Austin Community College

Gerald Cecil, University of North Carolina–Chapel Hill

Supriya Chakrabarti, Boston University

Robert Cicerone, Bridgewater State College

David Cinabro, Wayne State University

Scott Cochran, Central Michigan University

Judith Cohen, California Institute of Technology

Eric M. Collins, California State University–Northridge

Tara Cotton, University of Georgia

John Cowan, University of Oklahoma–Norman

Robert Coyne, Texas Tech University

Debashis Dasgupta, University of Wisconsin–Milwaukee

Steve Desch, Arizona State University

Robert Dick, Carleton University

Gregory Dolise, Harrisburg Area Community College

Michael Endl, Austin Community College

Tom English, Guilford Technical Community College

David Ennis, The Ohio State University

Duncan Farrah, University of Hawaii

Sunil Fernandes, University of Texas at San Antonio

John Finley, Purdue University

Matthew Francis, Lambuth University

Keigo Fukumura, James Madison University

Kevin Gannon, College of Saint Rose

Todd Gary, O'More College of Design

Christopher Gay, Santa Fe College

Ken Gayley, University of Iowa

Christopher Gerardy, University of North Carolina–Charlotte

Parviz Ghavamian, Towson University

Martha Gilmore, Wesleyan University

Guillermo Gonzalez, Ball State University

Bill Gutsch, St. Peter's College

Karl Haisch, Utah Valley University

Ioannis Haranas, Wilfrid Laurier College

Javier Hasbun, University of West Georgia

Charles Hawkins, Northern Kentucky University

Sebastian Heinz, University of Wisconsin–Madison

Jim Higdon, Georgia Southern University

Scott Hildreth, Chabot College

Barry Hillard, Baldwin Wallace College

Paul Hintzen, California State University–Long Beach

Paul Hodge, University of Washington

Christopher Hodges, Sacramento State University

William A. Hollerman, University of Louisiana at Lafayette

Hal Hollingsworth, Florida International University

Mike Hood, Mt. San Antonio College

Olencka Hubickyj-Cabot, San Jose State University

Kevin M. Huffenberger, University of Miami

James Imamura, University of Oregon

Adam Johnston, Weber State University

Jennifer Jones, Arapahoe Community College

Bruno Jungweirt, North Carolina State University

Steven Kawaler, Iowa State University

Patrick Kelly, Dalhousie University

Bethuel Khamala, El Paso Community College

Monika Kress, San Jose State University

Jessica Lair, Eastern Kentucky University

Rafael Lang, Purdue University

Kenneth Lanzetta, Stony Brook University

Lee LaRue, Paris Junior College

Alex Lazarian, University of Wisconsin–Madison

Denis Leahy, University of Calgary

Hector Leal, University of Texas–Rio Grande Valley

Hyun-chul Lee, University of Texas–Rio Grande Valley

Kevin Lee, University of Nebraska–Lincoln

Ludwik Lembryk, University of Toledo

Matthew Lister, Purdue University

M. A. K. Lodhi, Texas Tech University

Leslie Looney, University of Illinois at Urbana–Champaign

Isaac Lopez, Boston University

Jack MacConnell, Case Western Reserve University

Kevin Mackay, University of South Florida

Dale Mais, Indiana University–South Bend

Michael Marks, Bristol Community College

Norm Markworth, Stephen F. Austin State University

Kevin Marshall, Bucknell University

Petrus Martens, Georgia State University

Stephan Martin, Bristol Community College

Justin Mason, Old Dominion University

Amanda Maxham, University of Nevada–Las Vegas

Chris McCarthy, San Francisco State University

Douglas McElroy, Chaffey College

Ben McGimsey, Georgia State University

Charles McGruder, West Kentucky University

Janet E. McLarty-Schroeder, Cerritos College

Stanimir Metchev, Stony Brook University

Chris Mihos, Case Western Reserve University

Milan Mijic, California State University–Los Angeles

J. Scott Miller, University of Louisville

Scott Miller, Sam Houston State University

Kent Montgomery, Texas A&M University–Commerce

Robert Morehead, Texas Tech University

Andrew Morrison, Illinois Wesleyan University

Edward M. Murphy, University of Virginia

Kentaro Nagamine, University of Nevada–Las Vegas

Hon-kie Ng, Florida State University

Merav Opher, Boston University

Robert Parks, Louisiana State University

Jon Pedicino, College of the Redwoods

Nicolas Pereyra, University of Texas–Rio Grande Valley

Ylva Pihlström, University of New Mexico

Jascha Polet, California State Polytechnic University

Bob Powell, University of West Georgia

Dora Preminger, California State University–Northridge

Daniel Proga, University of Nevada–Las Vegas

Laurie Reed, Saginaw Valley State University

Barry Rice, Sierra College

Judit Györgyey Ries, University of Texas at Austin

Allen Rogel, Bowling Green State University

Kenneth Rumstay, Valdosta State University

Masao Sako, University of Pennsylvania

Samir Salim, Indiana University–Bloomington

Ruben Sandapen, Acadia University

Ata Sarajedini, University of Florida

Eric Schlegel, University of Texas at San Antonio

Paul Schmidtke, Arizona State University

Ann Schmiedekamp, Pennsylvania State University

Michael Schwartz, Santa Monica College

Jonathan Secaur, Kent State University

Ohad Shemmer, University of North Texas

Caroline Simpson, Florida International University

Parampreet Singh, Louisiana State University

Paul P. Sipiera, William Rainey Harper College

Ian Skilling, University of Pittsburgh

Tammy Smecker-Hane, University of California–Irvine

Allyn Smith, Austin Peay State University

Jason Smolinski, State University of New York at Oneonta

Roger Stanley, San Antonio College

Ben Sugerman, Goucher College

Neal Sumerlin, Lynchburg College

James Sy, El Paso Community College

Catherine Tabor, El Paso Community College

Angelle Tanner, Mississippi State University

Christopher Taylor, California State University–Sacramento

Benjamin Team, University of West Georgia

Fiorella Terenzi, Florida International University

Donald Terndrup, The Ohio State University

Todd Thompson, The Ohio State University

Glenn Tiede, Bowling Green State University

Frances Timmes, Arizona State University

Trina Van Ausdal, Salt Lake Community College

Lennart Van Haaften, Texas Tech University

Walter Van Hamme, Florida International University

Karen Vanlandingham, West Chester University

Nilakshi Veerabathina, University of Texas at Arlington

Paul Voytas, Wittenberg University

Ezekiel Walker, University of North Texas

Jasper Wall, University of British Columbia

Colin Wallace, University of North Carolina–Chapel Hill

G. Scott Watson, Syracuse University

James Webb, Florida International University

Julia Wickett, Collin College

Paul Wiita, Georgia State University

Richard Williamon, Emory University

Kurt Williams, Texas A&M University

Fred Wilson, Collin College

David Wittman, University of California–Davis

David Wood, San Antonio College

ABOUT THE AUTHORS

Stacy Palen is an award-winning professor in the Department of Physics and Astronomy at Weber State University. She received her BS in physics from Rutgers University and her PhD in physics from the University of Iowa. As a lecturer and postdoc at the University of Washington, she taught Introductory Astronomy more than 20 times over four years. Since joining Weber State, she has been very active in science outreach activities ranging from star parties to running the state Science Olympiad. Stacy does research in formal and informal astronomy education and the death of Sun-like stars. She spends much of her time thinking, teaching, and writing about the applications of science in everyday life. She then puts that science to use on her small farm in Ogden, Utah.

George Blumenthal has been a professor of astronomy and astrophysics at the University of California, Santa Cruz, since 1972, and served as the university's Chancellor from 2006 to 2019. He received his BS degree from the University of Wisconsin–Milwaukee and his PhD in physics from the University of California, San Diego. As a theoretical astrophysicist, George's research encompasses several broad areas, including the nature of the dark matter that constitutes most of the mass in the universe, the origin of galaxies and other large structures in the universe, the earliest moments in the universe, astrophysical radiation processes, and the structure of active galactic nuclei such as quasars. Besides teaching and conducting research, he has served as chair of the UC Santa Cruz Astronomy and Astrophysics Department, has chaired the Academic Senate for both the UC Santa Cruz campus and the entire University of California system, and has served as the faculty representative to the UC Board of Regents. He is currently the director of UC Berkeley's Center for Studies in Higher Education and serves on several nonprofit boards, including the California Institute for Regenerative Medicine and the California Association for Research in Astronomy, which he chairs.

21st Century Astronomy

Stars and Galaxies

SEVENTH EDITION

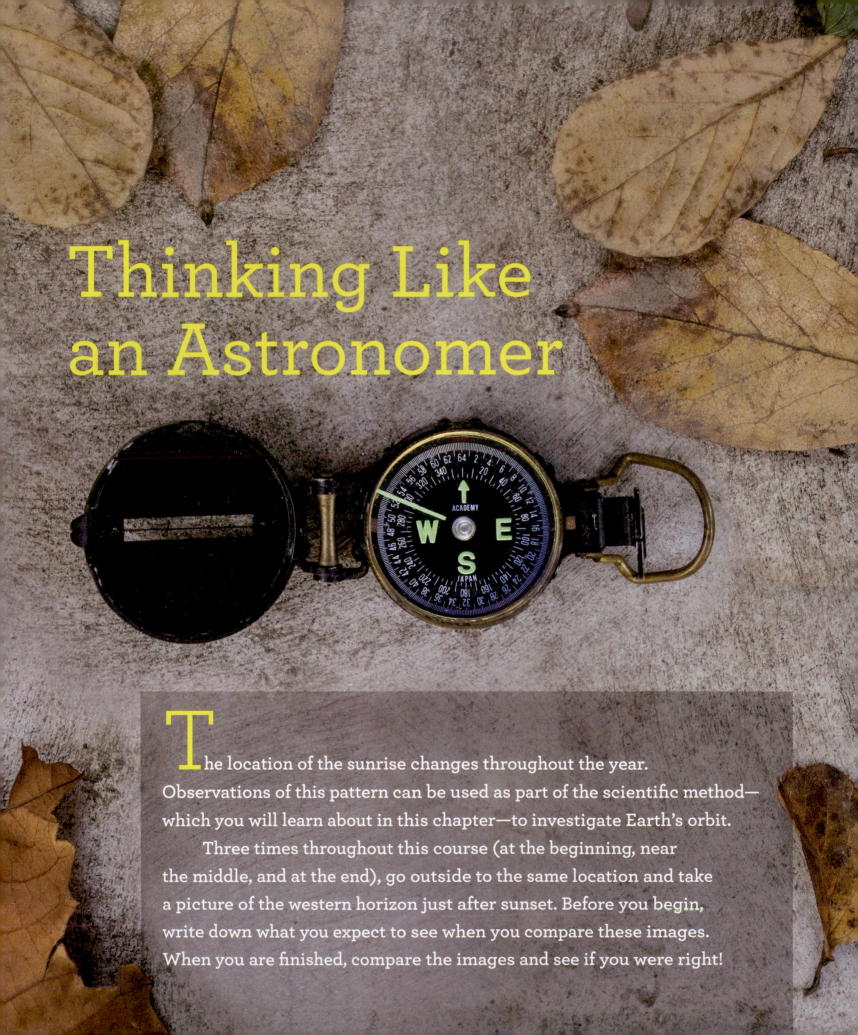

Thinking Like an Astronomer

The location of the sunrise changes throughout the year. Observations of this pattern can be used as part of the scientific method—which you will learn about in this chapter—to investigate Earth's orbit.

Three times throughout this course (at the beginning, near the middle, and at the end), go outside to the same location and take a picture of the western horizon just after sunset. Before you begin, write down what you expect to see when you compare these images. When you are finished, compare the images and see if you were right!

EXPERIMENT SETUP

PHOTO 1

PHOTO 2

PHOTO 3

Go to the same location and take three photos of the horizon at a time of day when the Sun is just below the horizon. Be sure that you have a stationary object (like a tree) in each photo. Take a photo at:

1 beginning of semester
2 middle of semester
3 end of semester

PREDICTION

1
2
3

When I compare the three images, I expect to see:

_____

_____

SKETCH OF RESULTS

1

Loosely translated, the word **astronomy** means "patterns among the stars." But modern astronomy—the astronomy we talk about in this book—is about far more than looking at the sky and cataloging the visible stars. The contents of the universe, its origin and fate, and the nature of space and time have become the subjects of rigorous scientific investigation. All of these subjects are related to questions humans have long had about our *origins*. How and when did the Sun, Earth, and Moon form? Are other galaxies, stars, planets, and moons similar to our own? The answers that scientists are finding to these questions are changing our view of both the cosmos and ourselves.

LEARNING GOALS

In this chapter, we will begin studying astronomy by exploring our place in the universe and the methods of science. By the end of Chapter 1, you should be able to:

1. Describe the size and age of today's universe and Earth's place in it.

2. Explain how astronomers use the scientific method to study the universe.

3. Show how scientists use mathematics, including graphs, to find patterns in nature.

4. Summarize our astronomical origins.

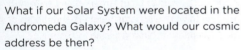

what if . . .

What if our Solar System were located in the Andromeda Galaxy? What would our cosmic address be then?

1.1 Earth Occupies a Small Place in the Universe

Locating Earth in the larger universe is the first step in learning the science of astronomy. In this section, you will get a feel for the neighborhood in which Earth is located, and begin to develop a framework to organize your growing knowledge of the universe by both size and distance from Earth. You will also begin to explore the scale of the universe in both space and time.

Our Place in the Universe

Many people receive their postal mail at an address—house or building number, street, city, state, and country. If we expand our view to include the enormously vast universe, our "cosmic address" might include our planet, star, galaxy, galaxy group or cluster, and galaxy supercluster. Follow along in **Figure 1.1** as we describe your cosmic address.

We reside on a planet called Earth, which is orbiting under the influence of gravity around a star called the **Sun**. The Sun is an ordinary middle-aged star. It is more massive and luminous than some stars but less massive and luminous than others. The Sun is extraordinary only because of its importance to us within our own **Solar System**. Our Solar System consists of eight planets (listed in order of distance from the Sun): Mercury, Venus, Earth, Mars, Jupiter, Saturn, Uranus, and Neptune. It also contains many smaller bodies, such as dwarf planets (for example, Pluto, Ceres, and Eris), asteroids (for example, Ida and Eros), and comets (for example, Halley). All those objects are gravitationally bound to the Sun.

The Sun is one of several hundred billion stars in the **Milky Way Galaxy**, a pancake-shaped disk of stars, gas, and dust. The Sun is located about halfway

Figure 1.1 Our place in the universe is given by our cosmic address: Earth, Solar System, Milky Way Galaxy, Local Group, Virgo Supercluster, and Laniakea Supercluster.

★ **WHAT AN ASTRONOMER SEES** In this type of figure, an astronomer will be especially sensitive to the arrows, which show that the figure "zooms out" from panel to panel. While each panel is the same size on the page, they represent dramatically different sizes in space. This figure is representative, without precision, but an astronomer will know that the Laniakea structure in the last panel is much larger—more than 100,000,000,000,000,000 times larger—than Earth, in the first panel. Learning to work with large numbers and ranges of size is one of the challenges of thinking like an astronomer.

out from the center of this disk. Like the Sun, most stars in the Milky Way Galaxy have planets in orbit around them.

The Milky Way is a member of a collection of a few dozen galaxies, including the Andromeda and Triangulum Galaxies, similar to the Milky Way, and about 20 smaller galaxies, called the **Local Group**. Most galaxies in that group are much smaller than the Milky Way. The Local Group, in turn, is part of a vastly larger collection of thousands of galaxies—a **supercluster**—called the Virgo Supercluster, which astronomers have more recently learned is part of an even larger grouping called the Laniakea Supercluster. (Unusually for astronomy, there are not special names that distinguish various sizes of superclusters.) The observable universe has millions of superclusters.

We can now define our cosmic address—Earth, Solar System, Milky Way Galaxy, Local Group, Virgo Supercluster, Laniakea Supercluster. Yet even that address is incomplete because it encompasses only the *local universe*. The part of the universe that we can see—the *observable universe*—extends to many times the size of Laniakea in every direction. This observable part of the universe contains roughly 2 trillion (two thousand billion) galaxies. The entire universe is much larger than the local universe and contains much more than the observed planets, stars, and galaxies. Astronomers estimate that about 95 percent of the universe is made up of matter that does not interact with light, known as *dark matter,* and a form of energy that permeates all space, known as *dark energy.* Dark matter and dark energy aren't well understood, and they are among the many exciting areas of research in astronomy.

The Scale of the Universe

As shown in Figure 1.1, the size of the universe completely dwarfs our human experience. We can start trying to understand astronomical size scales by comparing astronomical sizes and distances to something more familiar. For example, the diameter of our Moon (3474 kilometers [km]) is slightly greater than the distance between New York, New York, and Ogden, Utah (**Figure 1.2a**), about 80 percent of the way across the United States. The distance from Earth to the Moon is about 100 times that distance, which is comparable

unanswered questions

What makes up the universe? We have listed planets, stars, and galaxies as components of the universe, but astronomers now have evidence that 95 percent of the universe is in the form of dark matter and dark energy, which we do not yet understand. Scientists are using the largest telescopes and particle colliders on Earth, as well as telescopes and experiments in space, to explore what makes up dark matter and what constitutes dark energy.

a.

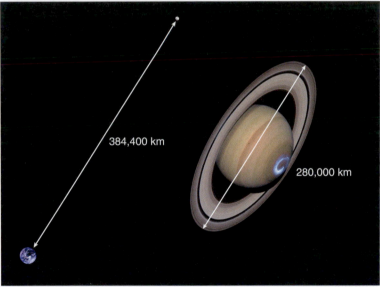

b.

Figure 1.2 **a.** The diameter of the Moon is about the same as the distance between New York, New York, and Ogden, Utah. **b.** The size of Saturn, including the rings, is about 70 percent of the distance between Earth and the Moon.
Credit (part b.): NASA, ESA, J. Clarke (Boston University, USA), and Z. Levay (STScI). https://esahubble.org/products/calendars/cal200604/. https://creativecommons.org/licenses/by/4.0/.

to the diameter of the planet Saturn with its majestic rings (**Figure 1.2b**). The distance from Earth to the Sun is about 400 times the Earth–Moon distance, and the distance to the planet Neptune is about 30 times the Earth–Sun distance. We could multiply all those relationships together, to find that the distance from the Sun to Neptune is 1.2 million times greater than the distance from New York, New York to Ogden, Utah.

The enormous distances beyond the edge of the Solar System are even more difficult to comprehend, but astronomers have a useful "trick" to help with this problem. In common language, we often use distance and time interchangeably; for example, if someone asks you how far it is to the nearest city, you might say 100 km or you might say 1 hour. In either case, you will have given that person an idea of how far away the city is. It would be unusual to say that the city is one "car-hour" away, but inventing that term would be one way to be more precise in your language. The city is as far away as a car can travel in one hour; it is one car-hour away.

In astronomy, the speed of a car on the highway is far too slow to be useful. Instead, astronomers use the fastest speed in the universe—the speed of light—to express the vast distances. Light travels at 300,000 kilometers per second (km/s). Traveling at the speed of light, you could travel all the way around Earth—a distance of 40,000 km—in just under $\frac{1}{7}$ of a second. Astronomers would then say that the circumference of Earth is $\frac{1}{7}$ of a "light-second." Most distances in astronomy are so vast that they are measured in units of **light-years (ly)**: the distance light travels in 1 year—about 9.5 trillion km or 6 trillion miles. Pause for a moment to think about how much longer a year is than a second; this is how much bigger a light-year is than a light-second.

Because light takes time to reach us, we see astronomical objects not as they are now, but as they used to be. We see them as they were in the past, when the light left them to begin its travel toward Earth. Because light from the Moon takes $1\frac{1}{4}$ seconds to reach us, we see the Moon as it was $1\frac{1}{4}$ seconds ago. Because light from the Sun takes $8\frac{1}{3}$ minutes to reach us, we see the Sun as it was $8\frac{1}{3}$ minutes ago. We see the nearest star as it was more than 4 years ago and we see the Andromeda Galaxy as it was 2.5 million years ago. The light from the most distant observable objects has been traveling for almost the age of the universe—nearly 13.8 billion years. **Figure 1.3** shows distances ranging from the size of Earth to the size of the observable universe, expressed as the time it takes light to travel that far. This is often called the "light travel time."

These vast distances show that we occupy a very small part of the space in the universe. We also occupy only a very small part of time. Imagine that the entire history of the universe took place within a single day. The universe begins the cosmic day at midnight, and hydrogen and helium cool enough to combine with electrons within the first 2 seconds. The first stars and galaxies appear within the first 10 minutes. Generations of stars are born and die before our Solar System forms from recycled gas and dust at about 4 P.M. The first bacterial

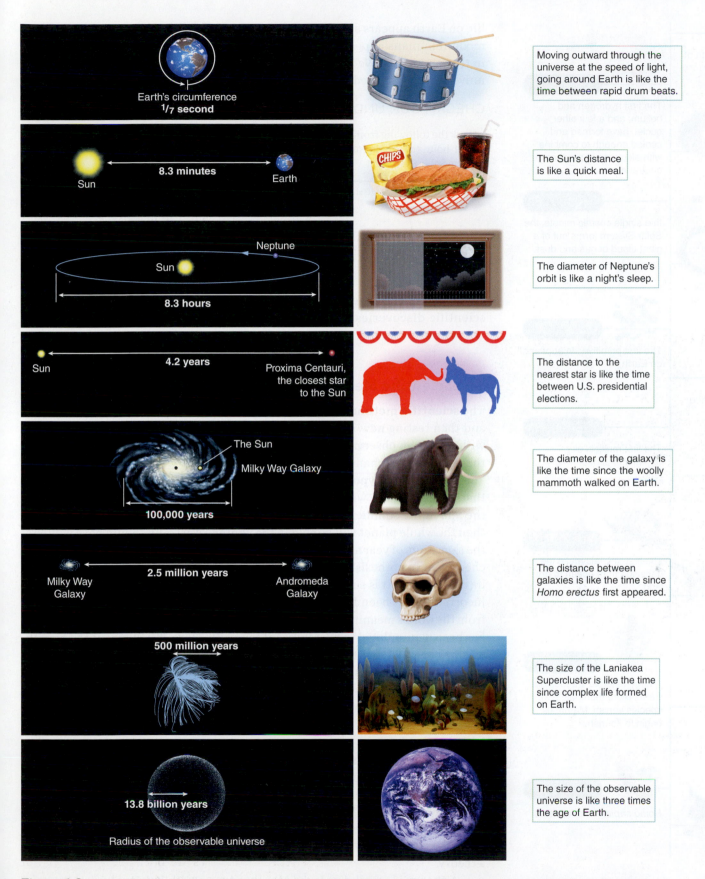

Figure 1.3 Thinking about the time light takes to travel between objects helps us comprehend the vast distances in the observable universe.

2:00:00 AM
The universe is a hot bath of photons and elementary particles.

2:10:00 AM
Stars appear and then galaxies. The Milky Way Galaxy becomes visible as star formation begins.

4:05:00 PM
A Mars-sized planetesimal crashes into Earth, forming the Moon.

3:40:00 PM
More complex single-celled organisms appear.

1:00:00 PM
Multicellular organisms become abundant.

1:35:00 PM
The first dinosaurs appear.

1:59:20 PM
The earliest human ancestors appear on the plains of Africa.

1:59:59.8 PM
Modern humans appear

12:00:02 AM
The first hydrogen and helium, and a few other nuclei, have formed and cooled enough to combine with electrons to produce neutral atoms.

4:00:00 PM
In a single cosmic minute, the Solar System forms out of a giant cloud of gas and dust.

5:20:00 PM
The first primitive life appears on Earth.

9:30:00 PM
The first multicellular organisms appear on dry land.

11:20:00 PM
The first animals make the transition from ocean to dry land.

11:53:10 PM
A large asteroid crashes on Earth. More than half of all species vanish. Mammals begin to flourish.

11:59:58.5 PM
Homo sapiens first appears.

Figure 1.4 This cosmic timeline presents the history of the universe as a 24-hour day.

life on Earth appears at 5:20 P.M., the first land animals appear at 11:20 P.M., and modern humans appear at 11:59:59.8 P.M.—that is, with just $\frac{1}{5}$ of a second left in the cosmic day. We humans occupy only a sliver of time in the history of the universe, as illustrated in **Figure 1.4**.

CHECK YOUR UNDERSTANDING 1.1

Rank the following from smallest to largest: (a) a light-minute, (b) a light-year, (c) a light-hour, (d) the radius of Earth, (e) the distance from Earth to the Sun, (f) the radius of the Solar System.

Answers to Check Your Understanding questions are in the back of the book.

1.2 Science Is a Way of Viewing the Universe

To view the universe through the eyes of an astronomer, you need to understand how science itself works. Throughout this book, we emphasize not only scientific discoveries but also the *process* of science. This section outlines the scientific method.

The Scientific Method

The **scientific method** is a systematic way of exploring the world by developing and then testing new ideas or explanations. You might begin a scientific study with a fact—an observation or a measurement. For example, you might observe that the weather changes predictably each year and wonder why that happens. You then create a **hypothesis**, a testable explanation of the observation: "I think that it is cold in the winter and warm in the summer because Earth is closer to the Sun in the summer." You come up with a test: if your hypothesis is correct, then the whole planet will be cold in the winter—Australia should have winter at the same time of year as the United States. This is a prediction that you can use to check your hypothesis! In January, you travel from the United States to Australia and find that it is summer in Australia. Your hypothesis has just been proved incorrect; it has been **falsified**. (Notice that this usage of "falsified" is different from the word's meaning in common usage. In common usage, "falsified" evidence has been manipulated to misrepresent the truth. Here, it just means the hypothesis has been shown to be incorrect.) Your test has two important elements that all scientific tests share. Your observation is reproducible: anyone who goes to Australia will find the same result. And your result is repeatable: if you conducted a similar test next year or the year after, you would get the same result. Because you have falsified your hypothesis, you must revise or replace it to be consistent with the new data.

Any idea that is not testable—that is, not falsifiable—must be accepted or rejected based on intuition alone, so it is not a scientific idea. A scientific idea does not have to be testable using current technology, but we must be able to imagine an experiment or observation that could prove the idea wrong. These tests must be repeatable over time and reproducible by everyone. As continuing tests support a hypothesis by failing to disprove it, scientists come to accept the hypothesis as a theory.

A **theory** is a well-developed idea that agrees with known physical laws and makes testable predictions. As with "falsified," the scientific meaning of "theory" is different from the meaning in common usage. In everyday language, theory may mean a guess: "Do you have a theory about who did it?" In everyday language, a theory can be something we don't take very seriously. "After all," people say, "it's only a theory." In stark contrast, scientists use the word *theory* to mean a carefully constructed proposition that accounts for every piece of relevant data as well as our entire understanding of how the world works. A theory has been used to make testable predictions, and all those predictions have come true. Sometimes, competing theories exist to explain a phenomenon. The success or failure of the predictions is the deciding factor between competing theories.

Einstein's theory of general relativity, which underlies the modern understanding of gravity, is an example of a scientific theory. For more than a century, scientists have tested the predictions of Einstein's theory of general relativity and have not been able to falsify it. As Einstein himself noted, a theory that fails only one test is proved false. Even after 100 years of verification, if a prediction of the theory of general relativity failed tomorrow, the theory would require revision or replacement. In that sense, all scientific knowledge is subject to challenge. This openness to challenge is one of the greatest strengths of science.

Let's pause to summarize this science-specific vocabulary. An *idea* is a notion about how something might be. A *fact* is an observation or measurement—for example, the measured value of Earth's radius is a fact. A *hypothesis* is an idea that leads to testable predictions. A hypothesis may lead to a scientific theory, may be based on an existing theory, or both. A *theory* is an idea that has been examined carefully, is consistent with all existing theoretical and observational knowledge, and makes testable predictions. Scientists also often use the term "law." A scientific *law* is a series of observations that can be used to predict a phenomenon but has no underlying explanation of why the phenomenon occurs. A *law* of daytime might say that the Sun rises and sets once each day, whereas a *theory* of daytime might say that the Sun rises and sets once each day because Earth spins on its axis. A **model** describes the properties of a particular object or system in terms of known physical laws or theories. These models are often computational, and use computers to predict the behavior of a complex system, like a system of multiple stars or planets. Scientists themselves can be sloppy about how they use these words, so you may sometimes see them used differently from how we have defined them here.

As the **Process of Science Figure** shows, the steps of the scientific method are interrelated. Scientists often begin with an observation or idea, followed by analysis, followed by a hypothesis, followed by a prediction, followed by further observations or experiments to test the prediction. Typically, the process of testing a theory is never completely finished; there is always another test to be performed under another set of conditions.

Scientific Principles

Scientific **principles** are general rules about the universe that provide guidelines for the formulation of scientific theories. Two important principles used in astronomy are the *cosmological principle* and *Occam's razor*.

The **cosmological principle** assumes that matter and energy behave throughout space and time as they do today on Earth. This means that the same physical laws and theories that we observe and apply in laboratories on Earth can be used to understand what goes on in the centers of stars or in distant galaxies. The principle also implies that the universe has no special locations or directions. In a

what if . . .

Would a theory still be a scientific theory if it were not likely to be falsifiable in many human lifetimes?

The Scientific Method

The scientific method is a formal procedure used to test the validity of scientific hypotheses and theories.

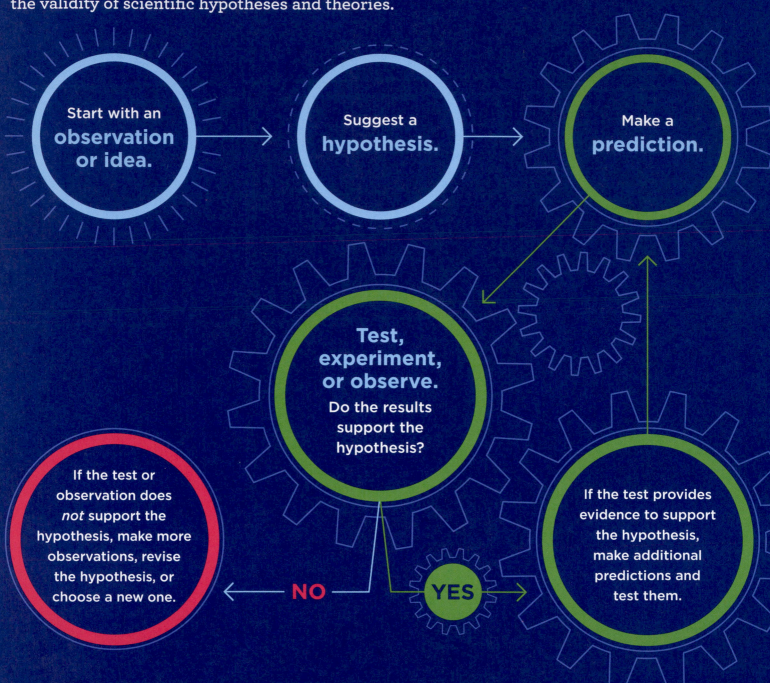

Start with an **observation or idea.**

Suggest a **hypothesis.**

Make a **prediction.**

Test, experiment, or observe. Do the results support the hypothesis?

If the test or observation does *not* support the hypothesis, make more observations, revise the hypothesis, or choose a new one.

NO

YES

If the test provides evidence to support the hypothesis, make additional predictions and test them.

An idea or observation leads to a falsifiable hypothesis. This hypothesis is tested by observation or experiment, and is either accepted as a tested theory or rejected based on the results. The **green loop** repeats indefinitely as scientists continue to test the theory.

sense, the cosmological principle is tested each time we apply our theories to new observations of astronomical objects. Thus far, observations of the universe around us have only added to scientists' confidence in the validity of the cosmological principle. If reasonable experimental evidence ever challenges the validity of the cosmological principle, scientists will construct a new description of the universe that takes those new data into account.

How should you choose between two hypotheses if both explain all of the observations equally well? According to **Occam's razor**, you should choose to adopt the hypothesis that requires the fewest assumptions until there is evidence to the contrary. For example, the nucleus of an atom is known to be positively charged, here in the Milky Way. Suppose you have two competing hypotheses about the charge of the nucleus of an atom in the Andromeda Galaxy: one hypothesis is that the nucleus is positively charged, like nuclei in the Milky Way; the second hypothesis is that the nucleus is negatively charged, opposite to nuclei in the Milky Way. The negative-nucleus hypothesis would require you to make additional assumptions about the location of the boundary between Andromeda-like matter and Milky Way–like matter and about why atoms on the boundary between the two regions did not destroy each other. You also would need an assumption about how the atoms in the two regions came to be constructed so differently. You also would need an assumption about why the Andromeda Galaxy is "special," which violates the cosmological principle. It is far simpler, and requires fewer assumptions, to hypothesize that atoms in the Andromeda Galaxy are the same as atoms in the Milky Way Galaxy.

In many sciences, researchers can conduct controlled experiments to test hypotheses. That experimental method is typically unavailable to astronomers. Astronomers cannot change the tilt of Earth or the temperature of a star to see what happens. Instead, astronomy is an observational science: astronomers apply physical theories based on Earth-bound experiments to observations of astronomical objects. Astronomers typically make multiple observations using as many methods as possible, and then create mathematical and physical models based on established science to explain the observations.

One example of observation leading to new theories comes from the study of planets orbiting stars other than the Sun. Planets orbiting other stars are called *exoplanets*. The first exoplanets to be discovered were giant planets (similar to Jupiter) orbiting very close to their star. However, planets like these are not found in our own Solar System, where the giant planets are all far from the Sun. Those discoveries challenged existing ideas about how our Solar System formed. As different observers using multiple telescopes found more and more of these unexpected planets, astronomers realized they needed new ideas of planet formation to explain how such large planets could wind up so close to their star. Astronomers could not actually build different systems of stars and planets to run a controlled experiment, but they could use the known laws of physics to create computer simulations of planetary systems. When researchers did that, they discovered that the orbits of planets can migrate, becoming closer or farther from the central star. Planetary scientists are searching for evidence that planets migrated early in the history of our own Solar System. The new theory of star and planet formation, which includes planetary migration, has successfully explained thousands of planetary systems, and is at present the best theory available to explain all of the known facts.

Sometimes, science proceeds from theory to observation, instead of from observation to theory. One example is the discovery of black holes. In the late 18th century, John Mitchell (1724–1793) and Pierre-Simon Laplace (1749–1827) independently hypothesized the existence of "dark stars": massive objects having

what if . . .

What if we discovered a region in the universe where the laws of physics are different from the ones on Earth (that is, the cosmological principle is wrong): How should we then think about the laws of physics in yet other parts of the universe?

Reading Astronomy News

Astronomers Find Massive Black Hole in the Early Universe

Brooks Hays

Even though you may not yet know much about black holes, you can begin to make sense of some of the numbers astronomers use to talk about them.

June 25 (UPI)—With the help of a trio of Hawai'ian telescopes, astronomers have imaged the 13-billion-year-old light of a distant quasar—the second-most distant quasar ever found.

Scientists gave the new quasar an indigenous Hawaiian name, Pōniuā'ena, which means "unseen spinning source of creation, surrounded with brilliance." Researchers described the brilliant object in a new paper, which is available in preprint format online and will soon be published in the *Astrophysical Journal Letters*.

Quasars are like lighthouses, their beams hailing from far away in the ancient universe. Powered by supermassive black holes at the center of galaxies, quasars are some of the brightest objects in the universe.

As astronomers peer deeper into the cosmos, they're able to see what the universe was like during its earliest days. In this instance, the Pōniuā'ena's lighthouse-like beacon hails from a period when the universe was still in its infancy—just 700 million years after the Big Bang.

The light of J1342+0928, a quasar spotted in 2018, is older and more distant, but the power and size of Pōniuā'ena is unmatched in the early universe. Spectroscopic observations of Pōniuā'ena, recorded by the Keck and Gemini observatories, revealed a supermassive black hole with a mass 1.5 billion times that of the sun.

"Pōniuā'ena is the most distant object known in the universe hosting a black hole exceeding one billion solar masses," lead study author Jinyi Yang, postdoctoral research associate at the University of Arizona's Steward Observatory, said in a news release.

According to Yang and colleagues, for a black hole to grow to such a tremendous size so early in the history of the universe, it would have needed to start out as a 10,000-solar-mass "seed" black hole, born no later than 100 million years after the Big Bang.

"How can the universe produce such a massive black hole so early in its history?" said Xiaohui Fan, associate head of the astronomy department at the University of Arizona. "This discovery presents the biggest challenge yet for the theory of black hole formation and growth in the early universe."

The light of distant objects, including quasars and massive galaxies in the early universe, can help scientists pinpoint the reionization of the universe. Astrophysicists estimate reionization occurred between 300 million years and one billion years after the Big Bang, but astronomers haven't been able to determine exactly when and how quickly it happened.

The phenomenon describes the ionization of hydrogen gas as the first stars, quasars, galaxies, and black holes came into existence. Prior to the reionization, the universe was without distinct light sources. Diffuse light dominated, and most radiation was absorbed by neutral hydrogen gas.

"Pōniuā'ena acts like a cosmic lighthouse," said study coauthor Joseph Hennawi, a cosmologist and an associate professor in the department of physics at the University of California, Santa Barbara. "As its light travels the long journey towards Earth, its spectrum is altered by diffuse gas in the intergalactic medium which allowed us to pinpoint when the Epoch of Reionization occurred."

Pōniuā'ena was initially spotted by a deep universe survey using the observations of the University of Hawai'i Institute for Astronomy's Pan-STARRS1 telescope on the Island of Maui. Later, scientists used the Gemini Observatory's GNIRS instrument, as well as the Keck Observatory's Near Infrared Echellette Spectrograph, to confirm the identify of Pōniuā'ena.

"The preliminary data from Gemini suggested this was likely to be an important discovery," said study coauthor Aaron Barth, a professor in the physics and astronomy department at the University of California, Irvine. "Our team had observing time scheduled at Keck just a few weeks later, perfectly timed to observe the new quasar using Keck's NIRES spectrograph in order to confirm its extremely high redshift and measure the mass of its black hole."

QUESTIONS

1. The article describes the light from this object as being 13 billion years old. Is this object inside or outside of the Milky Way Galaxy?

2. Which panel of Figure 1.3 would be most useful for showing someone else how far away this object is?

3. This supermassive black hole has a mass of 1.5 billion times that of the Sun. Write this number in scientific notation.

4. This black hole must have been born from a "seed" that formed no later than 100 million years after the Big Bang. In an astronomical context, is that "soon" after the Big Bang, or a "long time" after the Big Bang? How do you know?

5. What is the new question that Xiaohui Fan is asking, now that this observation has been made? Where do questions like this fit into the scientific method shown in the Process of Science Figure and discussed in Section 1.2?

Source: https://www.upi.com/Science_News/2020/06/25/Astronomers-find-massive-black-hole-in-the-early-universe/7781593108062/.

such strong gravity that light could not escape. At that time, there was no way to test that hypothesis. More than 100 years later, in the early 20th century, Karl Schwarzschild (1873–1916) studied Einstein's relativity equations and calculated that, despite their mass, those dark stars would be very small, with a radius of only a few kilometers. Fifty years after that, those objects were named *black holes*. No evidence of their existence was available until the 1970s and 1980s, when the new technology of space-based X-ray telescopes made possible the observations needed to test the hypothesis. In this case, the theory preceded the observation by nearly 200 years.

The scientific method provides a mechanism for testing new scientific ideas, but it offers no insight into where the idea came from in the first place or how an experiment was designed. Scientists discussing the creation of new ideas and experiments use words such as *insight*, *intuition*, and *creativity*. Scientists speak of a beautiful theory in the same way that an artist speaks of a beautiful painting or a musician speaks of a beautiful song. Science has an aesthetic that is as human and as profound as any found in the arts.

Scientific Revolutions

Scientific inquiry is necessarily dynamic. Scientists must constantly refine their ideas in response to new data and new insights. This vulnerability of scientific knowledge may seem like a weakness. "Gee, you really don't know anything," the cynical person might say. But that apparent vulnerability is actually a great strength, because it means that science self-corrects. New information eventually overturns incorrect ideas. In science, even our most cherished ideas about the nature of the physical world remain subject to challenge by new evidence. Many of history's best scientists earned their status by falsifying a universally accepted idea. That is a powerful motivation for scientists to challenge old ideas constantly—to formulate and test new explanations for their observations.

For example, the classical physics that Sir Isaac Newton developed in the 17th century to explain motion, forces, and gravity withstood the scrutiny of scientists for more than 200 years. During the late 19th and early 20th centuries, however, a series of scientific revolutions completely changed our understanding of the nature of reality. The work of Albert Einstein (**Figure 1.5**) is representative of those scientific revolutions. Einstein's special and general theories of relativity replaced Newton's mechanics. Einstein showed that Newton's theories were a special case of a far more general and powerful set of physical theories. Einstein's revolutionary new ideas unified the concepts of mass and energy and destroyed the conventional notion of space and time.

Throughout this text, you will encounter many other discoveries that forced scientists to abandon accepted theories. Einstein himself never embraced the view of the world offered by *quantum mechanics*—a second revolution he helped start. Yet quantum mechanics, a statistical description of the behavior of particles smaller than atoms, has held up for more than 100 years. In science, all authorities are subject to challenge, even Einstein.

Science is a way of thinking about the world. It is a search for the relationships that make our world what it is. A scientist assumes that order exists in the universe and that the human mind can grasp the essence of the rules underlying that order. Scientists build on those assumptions to make and then test predictions, finding the underlying rules that allow humanity to solve problems, invent new technologies, or find a new appreciation for the natural world. Scientific knowledge

Figure 1.5 Albert Einstein is perhaps the most famous scientist of the 20th century, and he was *Time* magazine's selection for Person of the Century. Einstein helped usher in two scientific revolutions.

working it out 1.1

Mathematical Tools

Scientists use mathematics to understand the patterns they observe and to communicate that understanding to others. The following mathematical tools will be useful in our study of astronomy:

Units. Scientists use the metric system of units because most metric units are related to each other by multiplying or dividing by 10. Thus, converting from one unit to another means simply moving the decimal point to the right or to the left. Metric measurements also have prefixes that identify the relationship of the units. There are 100 *centi*meters (cm) in a meter, 1000 meters in a *kilo*meter (km), and 1000 grams in a *kilo*gram (kg). (A more complete listing of metric prefixes is located inside the front cover of this book.) One way to check your own work is to think about whether the answer should be larger or smaller than the original number. The number of centimeters, for example, should always be larger than the number of meters because there is more than 1 cm in a meter. Often, a quick estimate can also help. For example, 1 ft is about an "order of magnitude" (a power of 10) larger than a centimeter; in other words, there are a few tens—not hundreds—of centimeters in a foot. If you have converted 1.2 ft to centimeters and calculated an answer of 3200 cm, you should try again. You can also include the units in every step of the calculation, and at each step, strike out the ones that cancel. For example, if you are multiplying 10 minutes by 60 seconds per minute, the unit *minutes* appears in the numerator and the denominator and divides out.

$$10 \ \text{minutes} \ \times \ \frac{60 \ \text{seconds}}{1 \ \text{minute}} = 600 \ \text{seconds}$$

This can help you remember if you should multiply or divide by the conversion factor; you want the answer in the correct units.

Scientific notation. Scientific notation is how we handle numbers of vastly different sizes. Writing out 7,540,000,000,000,000,000,000 in standard notation is inefficient. Scientific notation uses the first few digits (the *significant* ones) and counts the number of decimal places to create the condensed form 7.54×10^{21}. Similarly, instead of writing out 0.000000000005, we write 5×10^{-12}. The exponent on the 10 is positive or negative depending on whether the decimal point moves left or right, respectively. For example, the average distance to the Sun is 149,600,000 km, but astronomers usually express that value as 1.496×10^8 km.

Ratios. Ratios are a useful way to compare objects. A star may be "10 times as massive as the Sun" or "10,000 times as luminous as the Sun." Those expressions are ratios.

Proportionality. Often, understanding a concept amounts to understanding the *sense* of the relationships that it predicts or describes. "If you have twice as far to go, getting there will take you twice as long." "If you have half as much money, you will be able to buy only half as much gas." Those are examples of proportionalities.

Appendix 1 further explains the mathematical tools used in this book.

---

is an accumulated collection of ideas about how the universe works, yet scientists are always aware that what is known today may be superseded tomorrow. Science has found such a central place in our civilization because science makes the most accurate predictions about how natural systems will behave.

CHECK YOUR UNDERSTANDING **1.2**

The scientific method is a process by which scientists: (a) prove theories to be known facts; (b) gain confidence in theories by failing to prove them wrong; (c) turn theories into laws; (d) survey what most scientists think about a theory.

1.3 Astronomers Use Mathematics to Find Patterns

Scientific thinking allows scientists to make predictions. Once a pattern has been observed, such as the daily rising and setting of the Sun, scientists can predict what will happen next. Making an accurate prediction typically means making a numerical prediction. To do that, scientists need math.

The rhythms of nature produce patterns in our lives, and those patterns give us clues about the nature of the physical world. The Sun rises, sets, and then rises again at predictable times and in predictable locations. Spring turns into summer, summer turns into autumn, autumn turns into winter, and winter turns into spring again. Astronomers identify and characterize those patterns and use them to understand the world around us. As shown in **Figure 1.6**, the visible star patterns in the sky change predictably with the seasons. You will learn many other examples of patterns in the sky in Chapter 2.

Astronomers use mathematics to analyze patterns of all kinds and to communicate complex material compactly and accurately. Many people find mathematics to be a major obstacle that prevents them from appreciating the beauty and elegance of the world as seen through the eyes of a scientist. In this book, we strive to explain any necessary math in everyday language. We describe what equations mean and help you use them in a way that allows you to connect scientific concepts to the world. You will learn to appreciate the beauty of mathematics if you accept the challenge and make an honest effort to understand the mathematical material. **Working It Out 1.1** introduces units and scientific notation. Units are not only necessary to understand relationships between numbers (how large is an object a meter in diameter compared to one that is a kilometer in diameter?), but they also provide a convenient way to check your own work! Numbers in astronomy are typically so large that we must express them in scientific notation. **Working It Out 1.2** reviews some basics of mathematical tools and graphs. In modern society, the ability to read and interpret graphs is indispensable, whether they are graphs of natural systems or graphs of systems like the stock market. You will need the tools in these two boxes as you learn astronomy, whether or not your course is quantitative. But more importantly, you will need these tools in your later life, whether or not you ever pick up another astronomy book. In addition to these tools, Appendix 1 at the back of the book gathers together some essential mathematics, and Appendix 2 collects the physical constants of nature. Other appendixes contain data tables with key information about planets, moons, and stars.

Figure 1.6 Since ancient times, people recognized that patterns in the sky change with the seasons. Those and other patterns shape our lives. These star maps show the sky in the Northern Hemisphere during each season. Find the constellation Ursa Major in each star map to see a pattern over the course of the year.

CHECK YOUR UNDERSTANDING 1.3

When you see a pattern in nature, it is usually evidence of: (a) a theory's being displayed; (b) a breakdown of random clustering; (c) an underlying physical law; (d) a coincidence.

working it out 1.2

Reading a Graph

Scientists often convey complex information and mathematical patterns in graphical form. Graphs typically have two axes: a horizontal axis (the *x*-axis) and a vertical axis (the *y*-axis). The *x*-axis usually shows an independent variable, the one a researcher might have control over in an experiment, whereas the *y*-axis shows the dependent variable, which is typically the variable a researcher is studying.

Graphs can take different shapes. Suppose we plot the distance a car travels over time, as shown in **Figure 1.7a**. In a linear graph, each interval on an axis represents the same-sized step. Each step on the horizontal axis of the graph in Figure 1.7a represents 5 minutes. Each step on the vertical axis represents a distance of 5 km traveled by the car. Data are plotted on the graph, with one dot for each observation; for example, after 20 minutes, the car has traveled 20 km.

Drawing a line through those data indicates the trend (or relationship) of the data. To understand what the trend means, scientists often find the slope of the line, which is the relationship of the line's rise along the *y*-axis to its movement along the *x*-axis. To find the slope, we look at the change between two points on the vertical axis divided by the change between the same two points on the horizontal axis. For example, finding the slope of the line gives

$$\text{Slope} = \frac{\text{Change in vertical axis}}{\text{Change in horizontal axis}}$$

$$\text{Slope} = \frac{(15 - 10)\,\text{km}}{(15 - 10)\,\text{min}}$$

$$= 1\,\text{km/min}$$

There, the trend tells us that the car is traveling at 1 kilometer per minute (km/min), or 60 kilometers per hour (km/h).

Many observations of natural processes do not yield a straight line on a graph. When you catch a cold, for example, you feel fine when you get up in the morning at 7 A.M. At 9 A.M. you feel a little tired. By 11 A.M. you have a bit of a sore throat or a sniffle and think, "I wonder whether I'm getting sick," and by 1 P.M., you have a runny nose, congestion, fever, and chills. The illness progresses faster with each hour as time goes by. That process is exponential because the virus that has infected you reproduces exponentially.

For the sake of this discussion, suppose that the virus produces one copy of itself each time it invades a cell. (Viruses actually produce 1000–10,000 copies each time they invade a cell, so the exponential curve is much steeper.) One virus infects a cell and multiplies, so now two viruses exist—the original and a copy. Those viruses invade two new cells, and each one produces a copy. Now four viruses are there. After the next cell invasion, we have eight. Then 16, 32, 64, 128, 256, 512, 1024, 2048, and so on. That behavior is plotted in **Figure 1.7b**.

Seeing what's happening in the early stages of an exponential curve can be difficult because the later numbers are so much larger than the earlier ones. That's why we sometimes plot that type of data *logarithmically*, by putting the logarithm (roughly the exponent of the 10 in scientific notation) of the data on the vertical axis, as shown in **Figure 1.7c**. Now each step on the axis represents 10 times as many viruses as the previous step. Even though we draw all the steps the same size on the page, they represent different-sized steps in the data (for example, the number of viruses). We often use that technique in astronomy because it has a second, related advantage: very large variations in the data can easily fit on the same graph.

Each time you see a graph, you should first understand the axes—what data are plotted on the graph? Then you should check whether the axes are linear or logarithmic. Finally, you can look at the actual data or lines in the graph to understand how the system behaves.

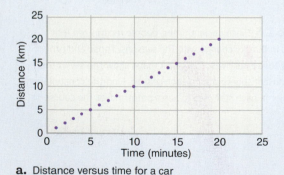

a. Distance versus time for a car

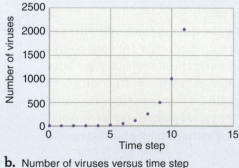

b. Number of viruses versus time step

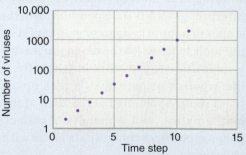

c. Number of viruses versus time step (log plot)

Figure 1.7 Graphs such as these show relationships between quantities. **a.** The relationship between time and distance traveled. **b.** The relationship between time and number of viruses. **c.** The data in part b. plotted logarithmically.

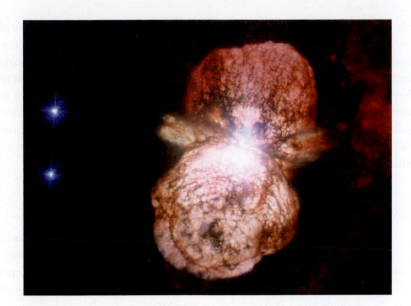

Figure 1.8 You and everything around you are composed of atoms forged in the interiors of stars that lived and died before the Sun and Earth formed. The supermassive star Eta Carinae, shown here, is ejecting a cloud of chemically enriched material just as earlier generations of stars once did to enrich the gas that would become our Solar System.
Credit: ESA/Hubble & NASA. https://esahubble.org/images/potw1208a/. https://creativecommons.org/licenses/by/4.0/.

Origins: An Introduction

How and when did the universe begin? What combination of events led to the existence of humans as sentient beings living on a small, rocky planet orbiting a typical middle-aged star? Are others like us scattered throughout the galaxy?

In these "Origins" sections, which conclude each chapter, we look into the origin of the universe and the origin of life on Earth. We also examine the possibilities of life elsewhere in the Solar System and beyond—a subject called **astrobiology**. This origins theme includes the discovery of planets around other stars and how they compare with the planets of our own Solar System.

Later in the book, we present observational evidence that supports the **Big Bang** theory, which states that the universe started expanding from an infinitesimal size about 13.8 billion years ago. In the early universe, only the lightest chemical elements were found in substantial amounts: hydrogen and helium, and tiny amounts of lithium and beryllium. However, we live on a planet with a central core consisting mostly of heavy elements such as iron and nickel, surrounded by outer layers made up of rocks containing large amounts of silicon and various other elements—all heavier than the original elements. The human body contains carbon, nitrogen, oxygen, calcium, phosphorus, and a host of other chemical elements—all of which are heavier than hydrogen and helium. If those heavier elements that make up Earth and our bodies were not present in the early universe, then where did they come from?

The answer to that question lies within the stars themselves (**Figure 1.8**). Earlier "generations" of stars supplied the building blocks for the chemical processes that we see in the universe, including life. In the core of a star, light elements, such as hydrogen, combine to form more massive atoms, which eventually leads to atoms such as carbon. When a star nears the end of its life, it often loses much of its material—including some of the new atoms formed in its interior—by ejecting it back into interstellar space. That material combines with material lost from other stars—some of which produced even more massive atoms as they exploded—to form large clouds of dust and gas. Those clouds go on to make new stars and planets, similar to our Sun and Solar System. The phrase "we are stardust" is not just poetry; we are literally made of recycled stardust.

what if . . .

What if Earth formed 4 billion years after the Big Bang, rather than 9.3 billion years after it: How might Earth be different?

unanswered questions

Does life as we know it exist elsewhere in the universe? At the time of this writing, no scientific evidence indicates that life exists on any other planet.

Studying origins also provides examples of the process of science. Many physical processes in chemistry, planetary science, physics, and astronomy that are seen on Earth or in the Solar System are observed across the Milky Way and throughout the universe. As of this writing, though, the only biology we know about is that which exists on Earth. At this point in human history, then, much of what scientists can say about the origin of life on Earth and the possibility of life elsewhere is reasoned extrapolation and educated speculation. In these "Origins" sections, we address some of those hypotheses and try to be clear about which are speculative and which have been tested.

SUMMARY

Astronomy seeks answers to many compelling questions about the universe. Astronomy uses all available tools to follow the scientific method. The process of science is based on objective reality, physical evidence, and testable hypotheses. Scientists continually strive to improve their understanding of the natural world and must be willing to challenge accepted truths as new information becomes available. Understanding numerical measurements is an important part of understanding science.

(1) Describe the size and age of today's universe and Earth's place in it. We reside on a planet orbiting a star at the center of a solar system in a vast galaxy that is one of many in an ancient universe that is roughly three times as old as the Sun. We occupy a very tiny part of the universe in space and time.

(2) Explain how astronomers use the scientific method to study the universe. The scientific method is an approach to learning about the physical universe. The method includes observing phenomena, forming hypotheses, making predictions to test and refine those hypotheses, and repeated testing of theories. All scientific knowledge is subject to challenge. Like art, literature, and music, science is a creative human activity; it is also a remarkably powerful, successful, and aesthetically beautiful way of viewing the world.

(3) Show how scientists use mathematics, including graphs, to find patterns in nature. Mathematics provides many of the tools that astronomers need to understand the patterns we see and to communicate that understanding to others.

(4) Summarize our astronomical origins. We are a product of the universe: Except for hydrogen, the very atoms we're made of formed in stars that died long before the Sun and Earth formed. The atoms that make up ordinary matter come from particles that have been around since the Big Bang, processed through stars, and collected into the Sun and the Solar System.

QUESTIONS AND PROBLEMS

TEST YOUR UNDERSTANDING

1. Rank the following in order from smallest to largest.
   - **a.** Local Group
   - **b.** Milky Way
   - **c.** Solar System
   - **d.** universe
   - **e.** Sun
   - **f.** Earth
   - **g.** Laniakea Supercluster
   - **h.** Virgo Supercluster

2. If an event were to take place on the Sun, approximately how long would the light it generates take to reach us?
   - **a.** 8 minutes
   - **b.** 11 hours
   - **c.** 1 second
   - **d.** 1 day
   - **e.** It would reach us instantaneously.

3. *Understanding* in science means that
   - **a.** we have accumulated lots of facts.
   - **b.** we can connect facts through an underlying idea.
   - **c.** we can predict events on the basis of accumulated facts.
   - **d.** we can predict events on the basis of an underlying idea.

4. The cosmological principle states that
   - **a.** on a large scale, the universe is the same everywhere at a given time.
   - **b.** the universe is the same at all times.
   - **c.** our location is special.
   - **d.** every location in the universe has its own laws and theories.

5. The Sun is part of
   - **a.** the Solar System.
   - **b.** the Milky Way Galaxy.
   - **c.** the universe.
   - **d.** all of the above.

6. A light-year is a measure of
   - **a.** distance.
   - **b.** time.
   - **c.** speed.
   - **d.** mass.

7. Occam's razor states that
   a. the universe is expanding in all directions.
   b. the laws of nature are the same everywhere in the universe.
   c. if two hypotheses fit the facts equally well, the one with fewer assumptions is preferred.
   d. patterns in nature are really manifestations of random occurrences.

8. The circumference of Earth is $1/7$ of a light-second. Therefore, if you were traveling at the speed of light,
   a. you would travel around Earth's equator 7 times in 1 second.
   b. you would travel around Earth's equator once in 7 seconds.
   c. you would travel across Earth's diameter 7 times in 1 second.
   d. you would travel across Earth's radius once in 7 seconds.

9. According to the graphs in Figures 1.7b and c, by how much did the number of viruses increase in four time steps? ★
   a. It doubled.
   b. It tripled.
   c. It quadrupled.
   d. It went up more than 10 times.

10. Any explanation of a phenomenon that includes a supernatural influence is not scientific because
   a. it is not based on a hypothesis.
   b. it is wrong.
   c. people who believe in the supernatural are not credible.
   d. science is the study of the natural world.

11. "All scientific knowledge is provisional." This means:
   a. Everything we know will be proved wrong, eventually.
   b. Scientists don't really know anything.
   c. Science is no better at predicting the natural world than any other way of knowing.
   d. Science is always improving through challenge and further experimentation.

12. When we observe a star that is 10 light-years away, we are seeing that star
   a. as it is today.
   b. as it was 10 days ago.
   c. as it was 10 years ago.
   d. as it was 20 years ago.

13. Which of the following elements formed in the Big Bang? (Choose all that apply.)
   a. hydrogen   **c.** beryllium
   b. lithium   **d.** carbon

14. "We are stardust" means that
   a. Earth exists because two stars collided.
   b. the atoms in our bodies have passed through (and often formed in) stars.
   c. Earth is formed primarily of material that used to be in the Sun.
   d. Earth and the other planets will eventually form a star.

15. According to our current understanding of the universe, the following astronomical events led to the formation of you. Place them in order of their occurrence over astronomical time.
   a. Stars die and distribute heavy elements into the space between the stars.
   b. Hydrogen and helium are made in the Big Bang.
   c. Enriched dust and gas gather into clouds in interstellar space.
   d. Stars are born and process light elements into heavier ones.
   e. The Sun and planets form from a cloud of interstellar dust and gas.

THINKING ABOUT THE CONCEPTS

16. ★ **WHAT AN ASTRONOMER SEES**   Figure 1.1 repeatedly "zooms out" to show larger and larger segments of space. Compare the steps shown in this figure with the size of the steps shown in Figure 1.3. Each step of "zoom" shown in Figure 1.1 does not show the same increase in size. Why was this figure drawn this way? 👁

17. Suppose you lived on the planet named "Tau Ceti e" that orbits Tau Ceti, a nearby star in our galaxy. How would you write your cosmic address?

18. Imagine living on a planet orbiting a star in a very distant galaxy. What does the cosmological principle tell you about the physical laws at that distant location?

19. Consider Figure 1.3. If the Sun suddenly exploded, how soon after the explosion would we know about it? 👁

20. Consider Figure 1.3. If a star exploded in the Andromeda Galaxy, how long would that information take to reach Earth? 👁

21. Give an example of a scientific theory that a newer theory has superseded. As scientists developed that new theory, where on the Process of Science Figure did a change occur so that the old theory became invalid and the new theory was accepted? 👁

22. Some people have proposed the theory that extraterrestrials (aliens) visited Earth in the remote past. Can you think of any tests that could support or refute that theory? Is it falsifiable?

23. Explain how the word *theory* is used differently in the context of science than it is in everyday language.

24. What is the difference between a *hypothesis* and a *theory* in science?

25. Suppose the tabloid newspaper at your local supermarket claimed that children born under a full Moon become better students than children born at other times.
   a. Is that theory falsifiable?
   b. If so, how could it be tested?

26. A textbook published in 1945 stated that light from the Andromeda Galaxy takes 800,000 years to reach Earth. In this book, we assert that it takes 2,500,000 years. What does that difference tell you about a scientific "fact" and how our knowledge evolves?

27. Astrology makes testable predictions. For example, it predicts that the horoscope for your star sign on any day should fit you better than horoscopes for other star signs. Read the daily horoscopes for all the astrological signs in a newspaper or online. How many of them might fit the day you had yesterday? Repeat the experiment every day for a week and record which horoscopes fit your day each day. Did your horoscope sign consistently best describe your experiences?

28. A scientist on television states that it is a known fact that life does not exist beyond Earth. Would you consider that scientist's statement reputable? Explain your answer.

29. Some astrologers use elaborate mathematical formulas and procedures to predict the future. Does doing so show that astrology is a science? Why or why not?

30. Why can it be said that we are made of stardust? Explain why current theory supports that statement.

APPLYING THE CONCEPTS

31. The distance to the Sun is 8.3 light-minutes.
   a. If you converted this to light-hours, should you get a larger or a smaller number?
   b. Convert the distance to the Sun from light-minutes to light-hours.
   c. Check your work by comparing the value in light-hours to the value in light-minutes.

32. Suppose that you want to travel 100 km at 30 km/h, and you want to know how much time it will take.
   a. Your answer should have units of "hours." Why?
   b. If you multiply the two numbers together, what units will the result have? Are these the units of time?
   c. If you divide the speed by the distance, what units will the result have? Are these the units of time?
   d. If you divide the distance by the speed, what units will the result have? Are these the units of time?
   e. Carry out the operation (multiplication or division) that will give you the units of time that you want for your answer. How long will it take you to drive 100 km at 30 km/h?

33. a. Write 86,400 (the number of seconds in a day) and 0.0123 (the Moon's mass compared to Earth's) in scientific notation.
   b. In each case, check your work by carefully thinking about the exponent of 10, which should be negative if the number is less than 1 and positive if the number is greater than 1.

34. The average distance from Earth to the Moon is 384,000 km. In the late 1960s, astronauts reached the Moon in about 3 days. How fast (on average) must they have been traveling (in kilometers per hour) to cover this distance in this time? Compare this speed to that of a jet aircraft (800 km/h). ●–●–●
   a. The problem asks for a speed in kilometers per hour, but you have been given a time in days. How many hours are in a day?
   b. Should the number of hours be larger or smaller than the number of days? Do you want to multiply by the number of hours per day, or divide by the number of hours per day to convert 3 days into hours?
   c. The problem is asking for a speed. What are the units of speed?
   d. How should you combine the units of these numbers to find a speed? Multiply or divide them? If divide, which way should you divide them?
   e. Combine the distance in kilometers and the time in hours to find a speed in kilometers per hour.
   f. Compare your answer to the speed of a jet aircraft.

35. Light takes about 8⅓ minutes to travel from the Sun to Earth, and, when it is closest, Neptune is 30 times farther from Earth than the Sun is. How long does light from Neptune take to reach Earth? ●–●–●
   a. Make a prediction: Should light take more or less time to reach Earth from Neptune than from the Sun?
   b. How should you combine the numbers 30 and 8⅓ to get a number consistent with your answer to a? That is, should you multiply or divide them? If you should divide them, which way should you divide them so that the units come out in minutes?
   c. Calculate the time it takes for light to travel from Neptune to Earth.
   d. Check your work by verifying that your units are units of time (minutes or hours), and your answer is consistent with your prediction.

36. Review Working It Out 1.1., and then convert the following numbers to scientific notation:
   a. 7,000,000,000   **c.** 1238
   b. 0.00346

37. Review Working It Out 1.1, and then convert the following numbers to standard notation:
   a. 5.34×10^8 **c.** 6.24×10^{-5}
   b. 4.1×10^3

38. If a car is traveling at 35 km/h, how far does it travel in
   a. 1 hour?   **c.** 1 minute?
   b. half an hour?

39. Review Appendix 1. The surface area of a sphere is proportional to the square of its radius. How many times larger is the surface area if the radius is
   a. doubled?   **c.** halved (divided by 2)?
   b. tripled?   **d.** divided by 3?

40. The average distance from Earth to the Moon is 384,400 km. How many days would it take you, traveling at 800 km/h—the typical speed of jet aircraft—to reach the Moon?

41. The distance from Earth to Mars varies from 56 million km to 400 million km. How long does a radio signal traveling at the speed of light take to reach a spacecraft on Mars when Mars is closest and when Mars is farthest away?

42. The surface area of a sphere is proportional to the square of its radius. The radius of the Moon is only about one-quarter that of Earth. How does the surface area of the Moon compare with that of Earth?

43. A remote Web page may sometimes reach your computer by going through a satellite orbiting approximately 3.6×10^4 km above Earth's surface. What is the minimum delay, in seconds, before the Web page shows up on your computer?

44. Imagine that you have become a biologist, studying rats in Indonesia. Usually, Indonesian rats maintain a constant population. Every half century, however, those rats suddenly begin to multiply exponentially. Then the population crashes back to the constant level. Sketch a graph that shows the rat population over two of those episodes.

45. The circumference of a circle is given by $C = 2\pi r$, where r is the radius of the circle.
   a. Calculate the approximate circumference of Earth's orbit around the Sun, assuming that the orbit is a circle with a radius of 1.5×10^8 km.
   b. A year has 8766 hours, so how fast, in kilometers per hour, does Earth move in its orbit?
   c. How far along in its orbit does Earth move in 1 day?

EXPLORATION The Scale of the Universe

Neptune is about 30 times farther from the Sun than Earth is. The nearest star is about 9000 times farther from the Sun than Neptune is. The diameter of our Milky Way Galaxy is 30,000 times the distance to that nearest star. The Andromeda Galaxy, the nearest large galaxy similar to the Milky Way, is about 600,000 times farther than that nearest star. The diameter of the Local Group of galaxies is about 4 times the distance to Andromeda, and the diameter of the Laniakea Supercluster, which includes the Local Group and many other galaxy groups and clusters, is 50 times larger than the Local Group. That supercluster is just one of millions in the observable universe. The radius of the observable universe is about 110 times larger than the Laniakea Supercluster.

These distances are incredibly difficult to visualize. In order to better picture the relative sizes and distances of things, astronomers often "scale" them. This means that they imagine them much smaller, in the same way that we imagine our town is much smaller when we make a map of it. To get a better sense of the problem in visualizing these distances, imagine a model in which the Sun is 1 centimeter away from Earth—about the size of a peanut M&M. In this model, Earth would be so small as to be barely visible. Calculate the distances to the other objects in the universe, on this scale. We've done the distance to Neptune and the distance to Andromeda for you, as an example. Practice changing units and switching to scientific notation when the numbers get large.

1. At this scale, how far is Neptune from the Sun?

   Neptune is 30 times farther, so 30 cm. _____

   _____

2. How far is the nearest star?

   _____

   _____

3. What is the diameter of the Milky Way?

   _____

   _____

4. How far away is the Andromeda Galaxy, on this scale?

   Andromeda is 600,000 times farther than the nearest star, so

   1,620,000,000 km, or 1.6×10^9 km, or 1.6 billion km.

5. What is the diameter of the Local Group, on this scale?

   _____

   _____

6. What is the diameter of the Laniakea Supercluster, on this scale?

   _____

   _____

7. What is the radius of the observable universe, on this scale?

   _____

   _____

As you can see, even when we imagine Earth so small as to be barely visible, the sizes of structures in the universe are unimaginably large.

Patterns in the Sky—Motions of Earth and the Moon

Each month, the Moon changes its appearance in a predictable way. For 1 month, go outside on 10 different dates to the same location and take a picture of the Moon. You will have to think about what time of day to go outside to take these photos. Label each photo with the date and time of the observation. For each photograph, make a sketch of the relative positions of the Sun, Earth, and the Moon when you took the photograph. Think about where you must have been standing on Earth to see that phase of the Moon at that location in your sky, and add a stick figure to your Earth sketch to show where you must have been standing.

EXPERIMENT SETUP

For 1 month, go outside on 10 different dates to the same location, and take a picture of the moon. You will have to think about what time of day to go outside to take these photos.

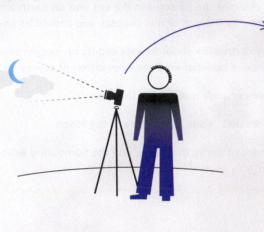

DAY 14

DAY 5

Label each photo with the date and time of the observation.

SKETCH OF RESULTS

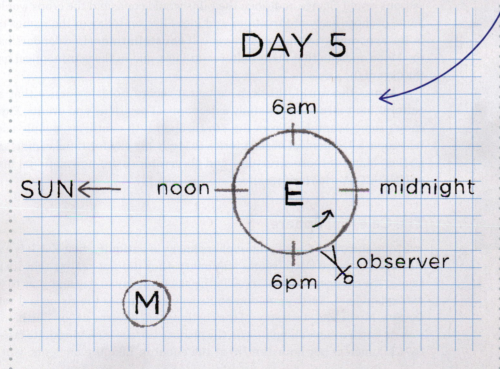

DAY 5

6am

SUN ← noon E midnight

6pm observer

M

2

Ancient peoples learned to use the patterns they observed in the sky to predict how the length of day, the seasons, and the appearance of the Moon would change. Some people understood those patterns well enough to create complicated calendars and tables of predictions of rare eclipses. Now, we see those patterns with the perspective of modern science, and we know that those changes are caused by the relative motions of Earth and the Moon. Discovering the causes of those patterns has shown humankind the way outward into the universe.

LEARNING GOALS

Here in Chapter 2, we examine the patterns in the sky and on Earth and the underlying motions that cause them. By the end of this chapter, you should be able to:

1. Describe how Earth's rotation about its axis and its revolution around the Sun affect how we perceive celestial motions from different places on Earth.

2. Explain why seasons change throughout the year.

3. Describe the factors that create the phases of the Moon.

4. Sketch the alignment of Earth, the Moon, and the Sun during eclipses of the Sun and the Moon.

a.

b.

Figure 2.1 a. One suspected use of Stonehenge 4000 years ago was to keep track of celestial events. **b.** The Mayan El Caracol at Chichén Itzá in Mexico (906 CE) is believed to have been designed to align with Venus.

2.1 Earth Spins on Its Axis

Ancient humans may not have known that they were "stardust," but they did sense a connection between their lives on Earth and the sky above. By watching the repeating patterns of the Sun, Moon, and stars in the sky, people could predict when the seasons would change and when the rains would come and the crops would grow. Some of those early observations and ideas live on today in the names of stars, in the apparent grouping of stars we call **constellations**, in calendars based on the Moon and Sun that many cultures still use, and in the astronomical names of the days of the week.

Across the world, the archaeological record holds evidence that ancient cultures built structures to study astronomical positions and events. **Figure 2.1** shows some examples. From the 8th through 17th centuries, people used pretelescopic astronomical observatories to study the sky for timekeeping and navigation. Many of those structures and observatories are now national historical or UNESCO World Heritage sites.

The Celestial Sphere

Aristotle and other Greek philosophers knew, in antiquity, that Earth is a sphere. However, because Earth seems stationary, they did not realize that its motion caused the changes they observed in the sky from day to day and year to year. As we explain here, Earth's rotation on its axis determines the rising and setting of the Sun, Moon, and stars—one of the rhythms that governs life on Earth.

As Earth rotates about its axis, the planet's surface is moving quite fast—about 1674 kilometers per hour (km/h) at the equator. Like the ancients, you do not feel that motion. Because the motion is uniform, you do not feel it any more than you would feel the motion of a car with a perfectly smooth ride cruising down a straight highway. Nor do you feel the direction of Earth's spin, although the apparent daily motion of the Sun, Moon, and stars across the sky reveals that direction. Imagine that you are in space far above Earth's **North Pole**, which is located at the north end of Earth's rotation axis. From there you would see Earth complete a counterclockwise rotation, once each 24-hour period, as **Figure 2.2** shows. As the rotating Earth carries an observer on the surface from west to east, objects in the sky appear to move in the opposite direction, from east to west. As seen from Earth's surface, each object seems to move across the sky in a path called its **apparent daily motion**.

Several special terms are needed to describe the sky that you can see above you while standing on Earth. The **zenith** is the point in the sky directly above you (**Figure 2.3a**). Wherever you are on Earth, you can see only half the sky. The **horizon** is the boundary between the part of the sky you can see and the other half of the sky that is blocked by Earth (**Figure 2.3b**). You can find the horizon by standing up and pointing your right hand at the zenith and your left hand straight out from your side. Turn in a complete circle. Your left hand has traced out the entire horizon. In common language, the "horizon" is thought of as where the sky meets the ground. Objects like mountains or tall buildings can make that complicated. So astronomers ignore those complications, and consider the horizon to be perfectly straight, as if you were on a boat in the middle of a calm ocean. The **meridian** is an imaginary line that runs from north to south, through the zenith. It divides the sky into an eastern half and a western half. The meridian is shown as a dashed line in Figure 2.3a. The meridian line continues around the far side of Earth, through the **nadir** (the point directly below you), and back to the starting point due north.

To help visualize the apparent daily motions of the Sun and stars, think of the sky as a huge imaginary sphere, called the **celestial sphere**, with the stars painted on its surface and Earth at its center. From ancient Greek times to the Renaissance, many people believed that to be a true representation of the heavens. The celestial sphere is a useful concept because it is easy to visualize, but never forget that it is, in fact, imaginary.

For an Earth-bound observer, each point on the celestial sphere indicates a direction in space (**Figure 2.4**). An observer looking straight up from Earth's North Pole is looking in the direction of the **north celestial pole (NCP)**. Directly above Earth's **South Pole**, which is at the south end of Earth's rotation axis, is the **south celestial pole (SCP)**. Directly above Earth's **equator** is the **celestial equator**, an imaginary circle that divides the sky into a northern half and a southern half. Just as Earth's North Pole is 90° away from Earth's equator, the north celestial pole is 90° away from the celestial equator. If you are in the Northern Hemisphere and you point one arm toward the celestial equator and one arm toward the north celestial pole, your arms will always form a right angle. The same holds true for the Southern Hemisphere: the angle between the celestial equator and the south celestial pole is always 90° as well.

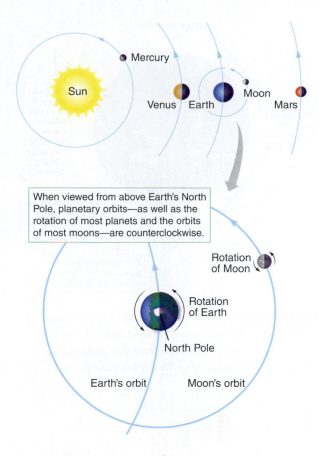

When viewed from above Earth's North Pole, planetary orbits—as well as the rotation of most planets and the orbits of most moons—are counterclockwise.

Figure 2.2 Motions in the Solar System, as viewed from above Earth's North Pole (not drawn to scale).

▶❚❚ **AstroTour:** The Celestial Sphere and the Ecliptic

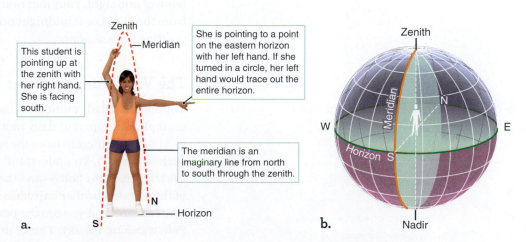

This student is pointing up at the zenith with her right hand. She is facing south.

She is pointing to a point on the eastern horizon with her left hand. If she turned in a circle, her left hand would trace out the entire horizon.

The meridian is an imaginary line from north to south through the zenith.

Figure 2.3 **a.** The meridian is a line on the celestial sphere that runs from north to south, dividing the sky into an east half and a west half. **b.** At any location on Earth, the part of the sky above the horizon is divided into an east half and a west half by an imaginary meridian projected onto the celestial sphere.

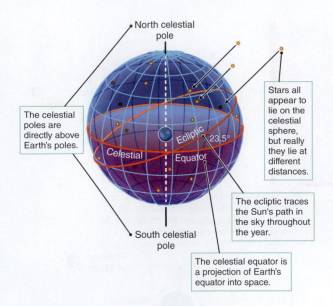

The celestial poles are directly above Earth's poles.

North celestial pole

Stars all appear to lie on the celestial sphere, but really they lie at different distances.

The ecliptic traces the Sun's path in the sky throughout the year.

South celestial pole

The celestial equator is a projection of Earth's equator into space.

Figure 2.4 The celestial sphere is a useful fiction in thinking about the stars' appearance and apparent motion in the sky. Because Earth's axis is tilted 23.5°, there is an angle of 23.5° between the ecliptic and the celestial equator.

Astronomy in Action: Vocabulary of the Celestial Sphere

AstroTour: The View from the Poles

Between the celestial poles and the equator, objects have positions on the celestial sphere with coordinates analogous to latitude and longitude on Earth. **Latitude** indicates distance north or south from Earth's equator. You find your latitude by imagining one line from the center of Earth to your location on the surface, and a second line from Earth's center to the point on the equator closest to you. The angle between those two lines is your latitude. At the North Pole, for example, those two imaginary lines form a 90° angle. At the equator, they form a 0° angle. So the latitude of the North Pole is 90° north, and the latitude of the equator is 0°. The South Pole is at latitude 90° south. Similarly, on the celestial sphere, **declination** indicates the distance of an object north or south of the celestial equator (from 0° to ±90°).

On Earth, **longitude** measures position east or west from the Royal Observatory in Greenwich, England. Similarly, on the celestial sphere, **right ascension** is an eastward and westward measure. The **ecliptic** is the Sun's path in the sky throughout the year. Notice in Figure 2.4 that the ecliptic is tilted relative to the celestial equator. The two circles in the sky intersect in two locations. Right ascension is equal to zero at one of these crossing points, and increases towards the east. Right ascension and declination are used to locate objects in the sky as precisely as anyone might require. In Appendix 7, we discuss detailed descriptions and illustrations of latitude and longitude as well as coordinates used with the celestial sphere.

Take a moment to visualize all those locations in space. To see how to use the celestial sphere, consider the Sun at noon and at midnight. Local noon occurs when the Sun crosses the meridian at your location. That is the highest point above the horizon that the Sun will reach on any given day. The highest point is almost never the zenith. You have to be in a specific place on a specific day for the Sun to be directly overhead at noon. One such place and time is at latitude 23.5° north of the equator on June 20.

From our perspective on Earth, the celestial sphere appears to rotate, carrying the Sun across the sky to its highest point at noon, over toward the west to set in the evening. In *reality*, the Sun remains in the same place in space through the entire 24-hour period, and Earth rotates so that any given location on Earth faces a different direction at every moment. When it is noon where you live, Earth has rotated so that you face most directly toward the Sun. Half a day later, at midnight, your location on Earth has rotated to face most directly away from the Sun. Local midnight occurs when the Sun is precisely opposite from its position at local noon.

The View from the Poles

The apparent daily motions of the stars and the Sun depend on where you live. For example, the apparent daily motions of celestial objects in a northern place such as Alaska are different from the apparent daily motions seen from a tropical island such as Hawai'i. To understand why location matters, let's examine the view of the stars from the poles—and then use them to guide our thinking about the view of the stars from other latitudes.

Imagine that you are the person in **Figure 2.5a** who is standing on the North Pole watching the sky. There, the north celestial pole is directly overhead at the zenith. You are standing where Earth's axis of rotation intersects its surface, which is like standing at the center of a rotating carousel. As Earth rotates, the spot directly above you remains fixed over your head while everything else in the sky appears

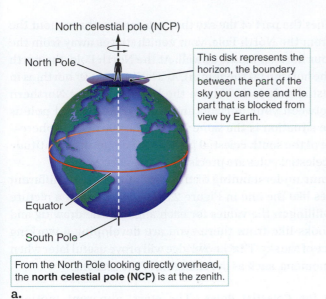

North celestial pole (NCP)

North Pole

This disk represents the horizon, the boundary between the part of the sky you can see and the part that is blocked from view by Earth.

Equator

South Pole

From the North Pole looking directly overhead, the **north celestial pole (NCP)** is at the zenith.

a.

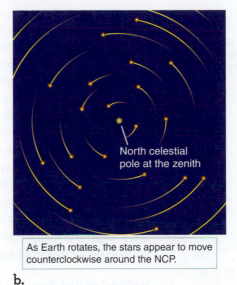

North celestial pole at the zenith

As Earth rotates, the stars appear to move counterclockwise around the NCP.

b.

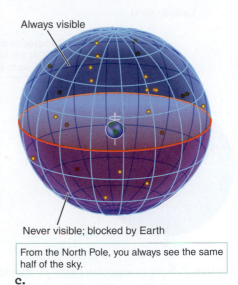

Always visible

Never visible; blocked by Earth

From the North Pole, you always see the same half of the sky.

c.

Figure 2.5 As viewed from **a.** Earth's North Pole, **b.** stars move throughout the night in counterclockwise, circular paths about the zenith. **c.** The same half of the sky is always visible from the North Pole.

to revolve counterclockwise around that spot. **Figure 2.5b** depicts that view of the sky from the ground, when the Sun is below the horizon.

If you are standing at the North Pole, the zenith and the horizon are always indicating the same locations in space. Objects visible from the North Pole follow circular paths that always have the same **altitude**, or angle above the horizon. Objects close to the zenith appear to follow small circles, whereas objects near the horizon follow the largest circles (Figure 2.5b). The view from the North Pole is special: you will always see the same half of the celestial sphere from there because nothing rises or sets each day as Earth turns (**Figure 2.5c**).

The view from Earth's South Pole is much the same—with two differences. First, the South Pole is on the opposite side of Earth from the North Pole, so the visible half of the sky at the South Pole is precisely the half hidden from view at the North Pole. Second, stars appear to move clockwise around the south celestial pole rather than counterclockwise as they do at the north celestial pole. To visualize why those motions are different, stand up and spin around from right to left. As you look at the ceiling, things appear to move counterclockwise, but as you look at the floor, they appear to be moving clockwise.

CHECK YOUR UNDERSTANDING 2.1a

No matter where you are on Earth, stars appear to rotate about a point called the: (a) zenith; (b) celestial pole; (c) nadir; (d) meridian.

Answers to Check Your Understanding questions are in the back of the book.

The View Away from the Poles

Away from the poles, the visible half of the sky changes constantly as Earth carries you around; your horizon cuts across different points in space, and your zenith moves across the sky as Earth rotates. Suppose that you leave the North Pole to travel south to lower latitudes. Your latitude is equal to the angle of your position from the equator, as shown for the person standing at 60° north latitude in **Figure 2.6**.

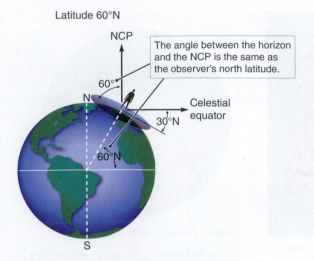

Latitude 60°N

The angle between the horizon and the NCP is the same as the observer's north latitude.

Figure 2.6 Your perspective on the sky depends on your location on Earth. The locations of the celestial poles and the celestial equator in an observer's sky depend on the observer's latitude. Here, an observer at latitude 60° north sees the north celestial pole at an altitude of 60° above the northern horizon and the celestial equator 30° above the southern horizon.

Figure 2.7 Time exposures of the sky showing the apparent motions of stars through the night. Note the difference in the circumpolar portion of the sky as seen from the two northern latitudes.

From a location in the Canadian woods, the north celestial pole appears high in the sky…

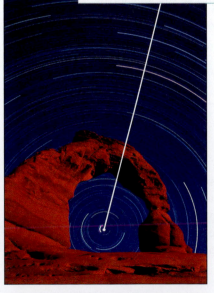

…but at a lower latitude in Utah, the north celestial pole appears closer to the horizon.

Your latitude determines the part of the sky that you can see throughout the year. As you move south from the North Pole, your zenith moves away from the north celestial pole, and your horizon moves as well. At the North Pole, the north celestial pole is 90° above the horizon (at the zenith). At a latitude of 60° north, as in Figure 2.6, the north celestial pole is 60° above the horizon. In the Northern Hemisphere, the angle between your horizon and the north celestial pole is equal to your latitude. The situation is the same in the Southern Hemisphere— your latitude is the altitude of the south celestial pole. At the equator, at a latitude of 0°, the north and south celestial poles are precisely on the horizon.

One way to solidify your understanding of the view of the sky at different latitudes is to draw pictures like the one in Figure 2.6. If you can draw a picture like that for any latitude—filling in the values for each angle in the drawing and imagining what the sky looks like from there—you are developing a working knowledge of the appearance of the sky. That knowledge will prove useful later when we discuss a variety of phenomena, such as the changing of the seasons.

Motions of the Stars and the Celestial Poles The stars' apparent motions about the celestial poles also differ from latitude to latitude. At the poles, all the stars are **circumpolar**; they go "around the pole" and never rise or set. As an observer moves farther from the poles, they will see fewer circumpolar stars. Two time-lapse views of the sky from different latitudes are shown in **Figure 2.7**. More stars are circumpolar for the more northern latitude, in Canada. The image taken from Utah is farther south, so it has fewer circumpolar stars. The Utah image also demonstrates the added complication that objects on the horizon often block low altitudes in the sky. Even stars that ought to be visible all the time, because they are circumpolar, might drop behind nearby objects that block the horizon.

From the Canada and Utah latitudes, if we focus our attention on stars near the north celestial pole, we see much the same thing we saw from Earth's North Pole. All the stars appear to move throughout the night in counterclockwise, circular paths around the north celestial pole. The north celestial pole is no longer directly overhead as it was at the North Pole, however, so the apparent circular paths of the stars are now tipped with respect to the horizon. (More correctly, your horizon is now tipped with respect to the apparent circular paths of the stars.)

Recall from Figure 2.6 that your latitude is equal to the altitude of the north celestial pole. Stars closer to the north celestial pole than that angle are above the horizon all the time, so they are circumpolar. Another group of stars, near the south celestial pole, are below the horizon all the time; they are never visible at all from your location. Stars between those that never rise and those that are circumpolar can be seen for only part of each 24-hour day; they appear to rise and set as Earth rotates.

Figure 2.8 shows the orientation of the sky as seen by observers at the North Pole, the equator, 30° north, and 30° south. The view from the North Pole (Figure 2.8a) was previously shown in Figure 2.5, but it is reproduced here for ease of comparison. For an observer at the equator (Figure 2.8b), the celestial poles are both at the horizon, and all the stars are visible in a 24-hour period, rising straight up and setting straight

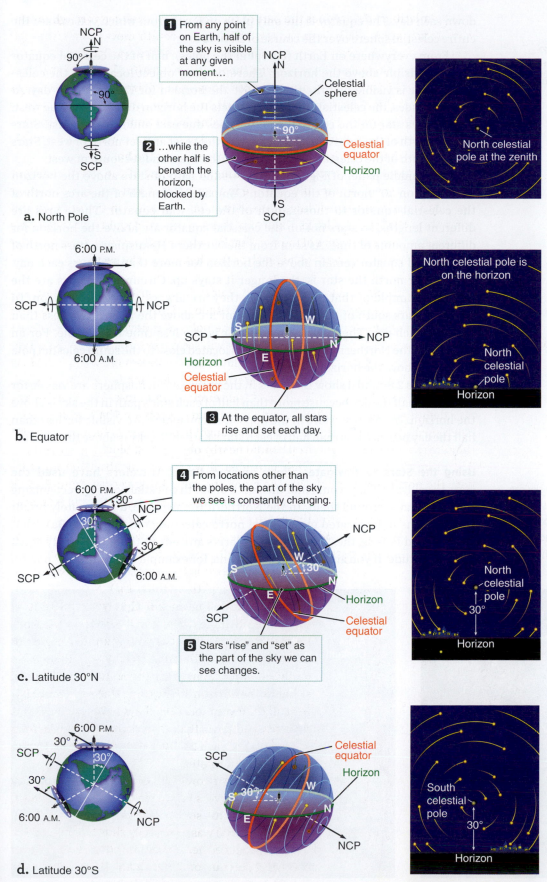

1 From any point on Earth, half of the sky is visible at any given moment…

2 …while the other half is beneath the horizon, blocked by Earth.

a. North Pole

NCP
N
90°
90°
S
SCP

NCP
N
Celestial sphere
90°
Celestial equator
Horizon
S
SCP

North celestial pole at the zenith

b. Equator

6:00 P.M.
SCP ← ← NCP
6:00 A.M.

SCP ← S W NCP
N
E
Horizon
Celestial equator

3 At the equator, all stars rise and set each day.

North celestial pole is on the horizon

North celestial pole

Horizon

c. Latitude 30°N

4 From locations other than the poles, the part of the sky we see is constantly changing.

6:00 P.M.
30°
NCP
30°
30°
SCP
6:00 A.M.

NCP
S W 30° N
E
SCP
Horizon
Celestial equator

5 Stars "rise" and "set" as the part of the sky we can see changes.

North celestial pole
30°
Horizon

d. Latitude 30°S

6:00 P.M.
30°
SCP
30°
30°
6:00 A.M.
NCP

Celestial equator
SCP
Horizon
30° W
S
N
E
NCP

South celestial pole
30°
Horizon

Figure 2.8 The celestial sphere is shown here as viewed by observers at four latitudes. At all locations other than the poles, stars rise and set as the part of the celestial sphere that we see changes during the day.

what if . . .

What if Earth did not rotate at all relative to the stars as it orbited the Sun? What effect would that have on the science of astronomy?

down each day. The equator is the only place on Earth from which you can see the entire celestial sphere over the course of 24 hours.

From everywhere on Earth (except at the poles), half of the celestial equator is always visible above the horizon. Therefore, any object located on the celestial equator is visible half the time—above the horizon for 12 hours each day. At other latitudes, the celestial equator intersects the horizon due east and due west. Therefore, a star on the celestial equator rises due east and sets due west. Stars located north of the celestial equator rise north of east and set north of west. Stars located south of the celestial equator rise south of east and set south of west.

The middle panel of Figure 2.8c shows the path of stars above the horizon for a location 30° north of the equator. Compare the length of the arcs north of the celestial equator to those south of the celestial equator. These arcs are different lengths, so stars not on the celestial equator are above the horizon for different amounts of time. As seen from the Northern Hemisphere, stars north of the celestial equator remain above the horizon for more than 12 hours each day. The farther north the star is, the longer it stays up. Circumpolar stars are the extreme example of that phenomenon; they are always above the horizon. In contrast, stars south of the celestial equator are above the horizon for less than 12 hours each day. The farther south a star is, the less time it is visible. For an observer in the Northern Hemisphere, stars located close to the south celestial pole never rise above the horizon.

Figures 2.8c and d show that stars in the observer's hemisphere are visible for more than half the day because more than half of each star's path in the sky is above the horizon. In contrast, stars in the opposite hemisphere are visible for less than half the day because less than half of each star's path in the sky is above the horizon.

Using the Stars to Navigate   Since ancient times, travelers have used the stars to navigate. They would find the north or south celestial poles by recognizing the stars that surround them. In the Northern Hemisphere, a moderately bright star happens to be located close to the north celestial pole (**Figure 2.9a**). That star is called Polaris, the "North Star." Polaris's altitude in the sky is nearly equal to your latitude. If you are in Phoenix, Arizona, for example (latitude 33.5° north),

Figure 2.9 Groups of stars near the pole stars in the sky can be used to locate a pole star. **a.** Two bright "pointer stars" in the cup of the Big Dipper point toward Polaris, the "North Star." **b.** In the Southern Hemisphere, the constellation Crux and two of its bright pointer stars can be used to locate the relatively faint southern pole star.

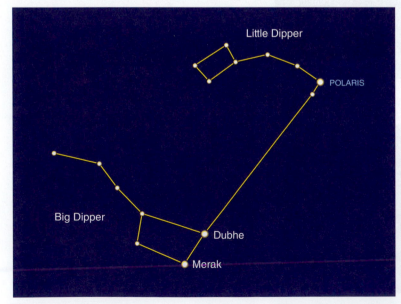

a.

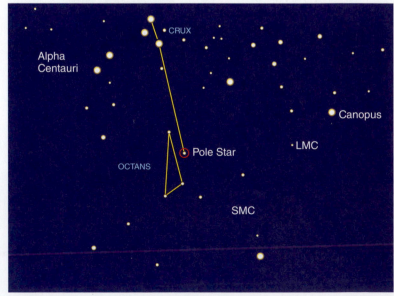

b.

working it out 2.1

How to Estimate Earth's Size

We can use the location of the north celestial pole in the sky to estimate Earth's size. Suppose we start out in Phoenix, Arizona, and we observe the north celestial pole to be 33.5° above the horizon. If we head north about 290 km to the Grand Canyon, the north celestial pole has risen to about 36° above the horizon. That difference between 33.5° and 36° (2.5°) is 1/144 of the way around a circle. (A circle is 360°, and 2.5°/360° = 1/144.)

That means we must have traveled 1/144 of the way around Earth's circumference, C, by traveling the 290 km between Phoenix and the Grand Canyon. In other words,

$$\frac{1}{144} \times C = 290 \text{ km}$$

When we rearrange the expression, Earth's circumference is given by

$$C = 144 \times 290 \text{ km} \approx 42,000 \text{ km}$$

Earth's actual circumference is just over 40,000 km, so our simple calculation was close. The circumference of a circle is equal to $2\pi r$, where r is its radius. Earth's radius, then, is given by

$$\text{Radius} = \frac{C}{2\pi} = \frac{40,000 \text{ km}}{2\pi} = 6400 \text{ km}$$

In about 230 BCE, using much the same method, the Greek astronomer Eratosthenes (276–194 BCE) was the first to accurately measure Earth's size. As **Figure 2.10** illustrates, Eratosthenes used the distance between his home city of Alexandria, Egypt, and the city of Syene (now Aswân, also in Egypt), which was 5000 "stadia." He noticed that on the first day of summer in Syene, the sunlight reflected directly off the water in a deep well, so the Sun must have been nearly at the zenith. By measuring the shadow of the Sun from an upright stick in Alexandria, he saw that the Sun was about 7.2° south of the zenith on the same date. Assuming that Earth was spherical and that Syene was directly south of Alexandria, he determined the distance between the two cities to be 7.2° divided by 360, or 1/50 of Earth's circumference.

Although historians know Eratosthenes concluded that the cities were 5000 stadia apart, they are still not sure of the value of his stadion unit. Other sources indicate that it may have been about 185 meters. We can use this conversion factor to compare Eratosthenes' value to modern values:

$$\frac{1}{50} \times C = 5000 \text{ stadia} \times 185 \text{ meters/stadion}$$

$$= 925,000 \text{ meters} = 925 \text{ km}$$

$$C = 50 \times 925 \text{ km} = 46,250 \text{ km}$$

That result is only about 16 percent higher than the modern value.

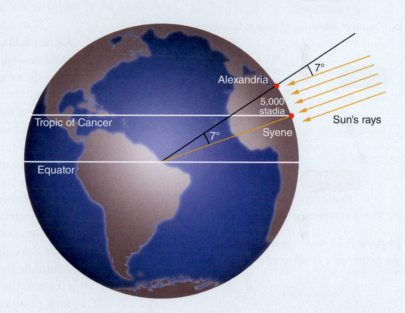

Figure 2.10 Eratosthenes used observations and basic calculations to estimate Earth's size.

Polaris has an altitude very close to 33.5°. In Fairbanks, Alaska (latitude 64.6° north), Polaris sits much higher, with an altitude very close to 64.6°. In the Southern Hemisphere, the constellation Crux (commonly called the Southern Cross) points to a faint star near the south celestial pole (**Figure 2.9b**). A navigator who has located a pole star can identify north and south and therefore east, west, and her latitude. Doing so enables the navigator to determine which direction to travel. The location of the north celestial pole in the sky was also used to measure Earth's

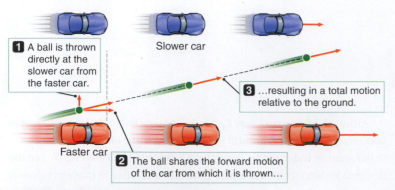

1 A ball is thrown directly at the slower car from the faster car.

Slower car

3 …resulting in a total motion relative to the ground.

Faster car

2 The ball shares the forward motion of the car from which it is thrown…

a. Frame of reference: Viewer on the street

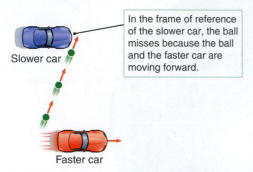

In the frame of reference of the faster car, the ball misses because the slower car is moving backward.

Slower car

Faster car

b. Frame of reference: Viewer in faster car

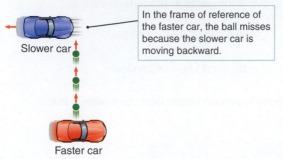

In the frame of reference of the slower car, the ball misses because the ball and the faster car are moving forward.

Slower car

Faster car

c. Frame of reference: Viewer in slower car

Figure 2.11 The motion of an object is always measured relative to the frame of reference of the observer.

size, as described in **Working It Out 2.1**. Because of Earth's rotation, determining longitude by astronomical methods is much more complicated. Longitude cannot easily be determined from astronomical observation alone.

Relative Motions and Frame of Reference

Why don't we feel motion as Earth spins on its axis and moves through space in its orbit around the Sun? The answer lies in the concept of a **frame of reference**, a coordinate system within which an observer measures positions and motions. The difference in motion between two individual frames of reference is called the **relative motion**. For example, imagine that you are riding in a car traveling down a straight section of highway at a constant speed. Under those conditions, without looking out the window or feeling road vibrations, you cannot easily tell the difference between being in the moving car and sitting in a parked car. Because everything inside the car is moving together, you can measure only the relative motions between objects in the car, and they are all zero—no motion is observed. Similarly, the resulting relative motions between objects that are near each other on Earth are zero because they have the same rotational motion. That is why we do not notice Earth's motion.

Now imagine two cars driving down the road at different speeds, as **Figure 2.11a** shows. For the moment, ignore any real-world complications, such as wind resistance. If you were to throw a ball from the faster-moving car directly out the side window at the slower-moving car as the two cars passed, you would miss. The ball shares the forward motion of the faster car, so the ball outruns the forward motion of the slower car. From your perspective in the faster car (**Figure 2.11b**), the slower car lagged behind the ball. From the slower car's perspective (**Figure 2.11c**), your car and the ball sped on ahead.

Although you cannot feel Earth's rotation, it influences motions around you as diverse as weather patterns and the flight of artillery shells toward a distant target. Any object that travels a significant distance northward or southward above Earth's surface will be affected. Here's why: All points on Earth's surface move in a circle each day around the planet's rotation axis. That circle is larger for points near the equator and smaller for points closer to one of the poles, but all points complete their circular motion in the same amount of time: exactly 1 day. A point closer to the equator has farther to go each day than one nearer a pole does. Therefore, points nearer the equator must be moving faster than ones at higher latitude. If an object starts out at one latitude and then moves to another, that difference in the speed of the surface influences the apparent direction of the object's motion.

Imagine a cannonball launched directly north from a point in the Northern Hemisphere, as in **Figure 2.12a**. Because the cannon is closer to the equator than the target, the cannon is moving toward the east faster than its target. Even though the cannonball is fired toward the north, it shares the eastward velocity of the cannon itself, just like the ball thrown from the faster car in Figure 2.11. Therefore, the cannonball is *also* moving toward the east faster than its target. As the cannonball flies north, it moves toward the east faster than the ground underneath it does. To an observer on the ground, the cannonball appears to curve toward the east as it outruns the eastward motion of the ground it is crossing. The farther north the cannonball flies, the greater the difference between its eastward velocity and the ground's. The cannonball follows a path that appears to curve more and more

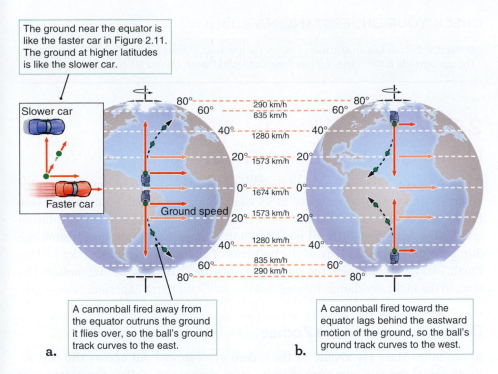

The ground near the equator is like the faster car in Figure 2.11. The ground at higher latitudes is like the slower car.

Slower car

Faster car

Ground speed

80°
60°   290 km/h
        835 km/h
40°   1280 km/h
20°   1573 km/h
0°   1674 km/h
20°   1573 km/h
40°   1280 km/h
60°   835 km/h
        290 km/h
80°

80°
60°
40°
20°
0°
20°
40°
60°
80°

A cannonball fired away from the equator outruns the ground it flies over, so the ball's ground track curves to the east.

a.

A cannonball fired toward the equator lags behind the eastward motion of the ground, so the ball's ground track curves to the west.

b.

Figure 2.12 The Coriolis effect causes objects to be deflected as they move across Earth's surface. The green dashed curve shows the cannonball's path as measured by a local observer.

to the east the farther north it goes. The cannonball also curves east if shot toward the south in the Southern Hemisphere, as shown in Figure 2.12a.

If you are located in the Northern Hemisphere and fire a cannonball *south* toward the equator, as in **Figure 2.12b**, the opposite effect will occur. Now the cannon is moving toward the east more slowly than its target. As the cannonball flies toward the south, its eastward motion lags behind that of the ground underneath it, so the cannonball appears to curve toward the west.

That curving motion of objects as a result of the difference in Earth's rotation speeds at different latitudes is called the **Coriolis effect**. In the Northern Hemisphere, the Coriolis effect causes a cannonball fired north to drift to the east as seen from the ground. In other words, the cannonball appears to curve to the right, as viewed by the cannon operator. A cannonball fired south appears to curve to the west, which is also to the right as observed by the cannon operator. In the Northern Hemisphere, the Coriolis effect seems to deflect things to the *right*. Viewed from space, this deflection will produce a clockwise motion of weather patterns. In the Southern Hemisphere, it seems to deflect things to the *left*. Viewed from space, this will produce a counterclockwise motion of weather patterns. In between, at the equator itself, the Coriolis effect vanishes.

On Earth, the Coriolis effect is large enough to deflect a fly ball hit north or south deep into the outfield by about a half a centimeter. At some time or other, then, the Coriolis effect has probably affected the outcome of a baseball game. On the other hand, the story about water spiraling in different directions down a drain in different hemispheres is just a myth. On the scale of sinks or toilets, the Coriolis effect is far too small to overcome other effects, such as the shape of the basin itself. However, this story often helps people remember that the Coriolis effect works oppositely in each hemisphere, so it may be worth remembering, even if it isn't true.

2.2 Revolution about the Sun Leads to Changes during the Year

Earth orbits (or **revolves**) around the Sun in the same direction that Earth spins about its axis—counterclockwise as viewed from above Earth's North Pole (see Figure 2.2). A **year** is defined as the time Earth takes to complete one revolution around the Sun—it's a little bit less than 365.25 days. Earth's motion around the Sun is responsible for many of the patterns of change we see in the sky and on Earth, including changes in the stars we see overhead. Because of that motion, the stars in the night sky change throughout the year, and Earth experiences seasons.

Constellations and the Zodiac

About 9000 stars are visible to the naked eye—spread out across the sky. As Earth orbits the Sun, our view of these stars changes. To follow the patterns of the Sun and the stars, early humans grouped together stars that formed recognizable patterns, called constellations. People from different cultures saw different patterns and projected ideas from their own cultures onto what they saw in the sky. Constellations are creations of the human imagination.

If you noted the position of the Sun in relation to the stars each day for a year, you would find that the Sun traces out a great circle against the background of the stars (**Figure 2.13**). On September 1, the Sun appears to be in the direction of the constellation Leo. Six months later, on March 1, Earth is on the other side of the Sun, and the Sun appears from our perspective on Earth to be in the

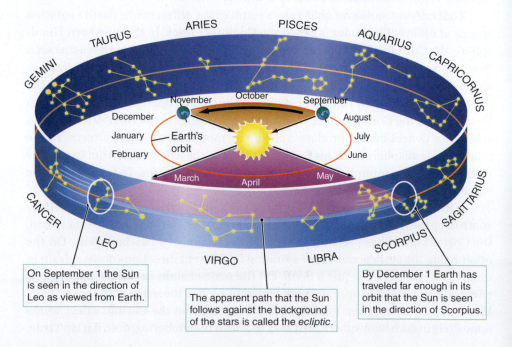

Figure 2.13 As Earth orbits the Sun, different stars appear in the night sky, and the Sun's apparent position against the background of stars changes, blocking our view of stars in that direction. The imaginary circle that the Sun's annual path traces is called the ecliptic. Constellations along the ecliptic form the zodiac.

On September 1 the Sun is seen in the direction of Leo as viewed from Earth.

The apparent path that the Sun follows against the background of the stars is called the *ecliptic*.

By December 1 Earth has traveled far enough in its orbit that the Sun is seen in the direction of Scorpius.

direction of the constellation Aquarius. Recall that the Sun's apparent path against the background of the stars is called the ecliptic (the blue band in Figure 2.13). The constellations that lie along the ecliptic through which the Sun appears to move are called the constellations of the **zodiac**.

Today, astronomers use an officially sanctioned set of 88 constellations to serve as a road map of the sky. Constellations visible from the Northern Hemisphere draw their names primarily from the list of constellations that the Alexandrian astronomer Ptolemy compiled 2000 years ago. Constellation names in the Southern Hemisphere come from European explorers who visited the Southern Hemisphere during the 17th and 18th centuries. Every star in the sky lies within the borders of a single constellation. This makes it possible to catalog all the stars using the names of their parent constellation. Following the Greek alphabet, astronomers call the brightest star in a constellation *alpha*; the second brightest, *beta*; and so forth. For example, Sirius, the brightest star in the sky, lies within the constellation Canis Major (meaning "big dog"). So Sirius is also called "Alpha Canis Majoris," indicating that it is the brightest star in Canis Major. Appendix 7 includes sky maps showing the constellations.

Earth's Axial Tilt and the Seasons

To understand why the seasons change, we need to consider the combined effects of Earth's rotation on its axis and its revolution around the Sun. Many people believe that seasons change because Earth is closer to the Sun in summer and farther away in winter. How can we test that hypothesis? We might predict: if the distance between Earth and the Sun caused the seasons, then all of Earth should experience summer at the same time of year. But the United States has summer in June, whereas Australia has summer in December. In modern times, we can directly measure the distance between Earth and the Sun, and we find that Earth is actually closest to the Sun at the beginning of January, by about 5 million km. Yet, in the Northern Hemisphere, we have winter in January. We have just falsified the hypothesis tying seasonal change to distance in two ways, so we need to look for a different explanation that accounts for all the available facts.

Earthbound observers can see that as the year passes, the Sun is above the horizon longer in summer than in winter. We often say "the days are longer in the summer", meaning there are more daytime hours, and fewer nighttime ones. The Sun is also higher in the sky as it crosses the meridian in summer than in winter. Observing more closely, the Earthbound observer finds that the Sun appears to move along the ecliptic, which is tilted 23.5° with respect to the celestial equator.

Looking at Earth from space, **Figure 2.14** shows that as Earth moves around the Sun over the course of the year, its rotation axis always points in the same direction, toward distant Polaris. This axis is tilted 23.5° from the perpendicular to Earth's orbital plane, causing the ecliptic to likewise be tilted 23.5° from the celestial equator. As it orbits, Earth is sometimes on one side of the Sun and sometimes on the other. As a result, Earth's North Pole is sometimes tilted more toward the Sun, and other times the South Pole is tilted more toward the Sun. This accounts for the changing altitude of the Sun as it crosses the meridian, and the changing length of daytime and nighttime. When Earth's North Pole is tilted toward the Sun, an observer on Earth views the Sun north of the celestial equator. Therefore, for observers in the Northern Hemisphere, the Sun is above the horizon more than

▶ǁ **AstroTour:** The Earth Spins and Revolves

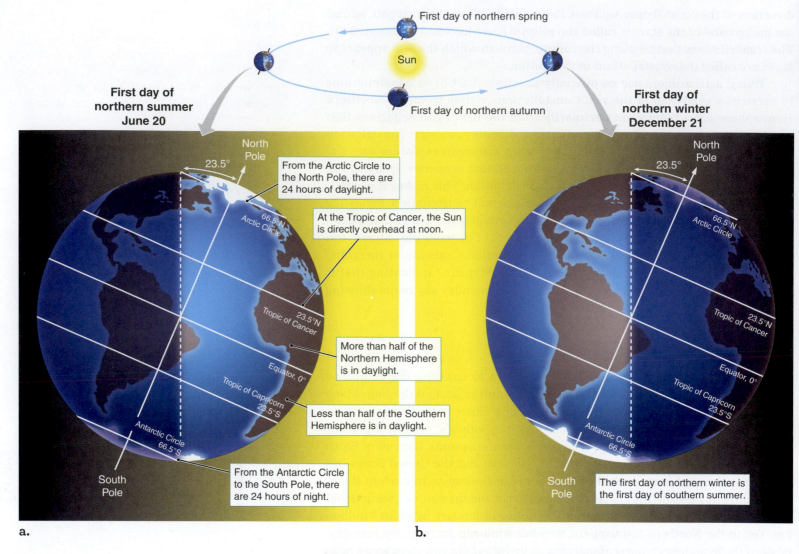

First day of northern spring

Sun

First day of northern autumn

First day of northern summer June 20

First day of northern winter December 21

a.

North Pole

23.5°

From the Arctic Circle to the North Pole, there are 24 hours of daylight.

66.5°N Arctic Circle

At the Tropic of Cancer, the Sun is directly overhead at noon.

23.5°N Tropic of Cancer

More than half of the Northern Hemisphere is in daylight.

Equator, 0°

Tropic of Capricorn 23.5°S

Less than half of the Southern Hemisphere is in daylight.

Antarctic Circle 66.5°S

South Pole

From the Antarctic Circle to the South Pole, there are 24 hours of night.

b.

North Pole

23.5°

66.5°N Arctic Circle

23.5°N Tropic of Cancer

Equator, 0°

Tropic of Capricorn 23.5°S

Antarctic Circle 66.5°S

South Pole

The first day of northern winter is the first day of southern summer.

Figure 2.14 a. On the first day of the northern summer (around June 20, the summer solstice), the northern end of Earth's axis is tilted most nearly toward the Sun, whereas the Southern Hemisphere is tipped away. **b.** Six months later, on the first day of the northern winter (around December 21, the winter solstice), the situation is reversed. Seasons are opposite in the two hemispheres.

12 hours each day, thus making the daytime longer than 12 hours. Six months later, when Earth's North Pole is tilted away from the Sun, an observer in the same place views the Sun south of the celestial equator and the daytime is less than 12 hours.

In the preceding paragraph, we specified the *Northern* Hemisphere because seasons are opposite in the Southern Hemisphere. Look again at Figure 2.14. Around June 20, while the Northern Hemisphere is enjoying the long days and short nights of summer, Earth's South Pole is tilted away from the Sun. It is winter in the Southern Hemisphere; less than half of the Southern Hemisphere is illuminated by the Sun, and the days are shorter than 12 hours. On December 21, Earth's South Pole is tilted toward the Sun. It is summer in the Southern Hemisphere, so its days are long and its nights are short.

To understand how the combination of Earth's axial tilt and its path around the Sun creates seasons, consider what would happen if Earth's spin axis were exactly perpendicular to the plane of Earth's orbit (the **ecliptic plane**). First, the Sun would

Astronomy in Action: The Cause of Earth's Seasons

always be on the celestial equator. Then, at every latitude (except the poles), the Sun would follow the same path through the sky every day, rising due east each morning and setting due west each evening. The Sun would be above the horizon exactly half the time, so days and nights would always be exactly 12 hours long everywhere on Earth. In short, Earth would have no seasons.

The changing length of days through the year only partly explains seasonal temperature changes. A more important effect relates to the angle at which the Sun's rays strike Earth. The Sun is higher in the sky during summer than during winter, and sunlight strikes the ground *more directly* during summer than during winter. To see why that is important, study **Figure 2.15**. During summer, Earth's surface is more nearly face-on to the incoming sunlight. The light is concentrated and bright so more energy falls on each square meter of ground each second. During winter, the surface is more tilted with respect to the sunlight, so the light is more diffuse. Less energy falls on each square meter of the ground each second. For example, at latitude 40° north, which stretches across the United States from northern California to New Jersey, more than twice as much solar energy falls on each square meter of ground per second at noon on June 20 as falls there at noon on December 21. That variation is the main reason why summer is hotter and winter is colder.

The **Process of Science Figure** demonstrates how determining the causes of seasonal change requires accounting for all the known facts. The two effects of Earth's tilted axis—the directness of sunlight and the different lengths of the night—mean that the Sun heats a hemisphere more at one time of year than another, causing summer and winter.

The Solstices and the Equinoxes

Four days during Earth's orbit (two solstices and two equinoxes) mark special moments in the year. The day when the Sun is highest in the sky as it crosses the meridian is called the **summer solstice**. On that day, the Sun rises farthest north of east and sets farthest north of west. In the Northern Hemisphere, that occurs each year near June 20, the first day of summer. Figure 2.14a shows that orientation of Earth and Sun.

Six months after the summer solstice, around December 21, the North Pole is tilted away from the Sun (Figure 2.14b). That day is the **winter solstice** in the Northern Hemisphere—the shortest day of the year and the first day of winter in the Northern Hemisphere. Almost all cultural traditions in the Northern Hemisphere include some sort of major celebration in late December. Those winter festivals celebrate the return of the source of Earth's light and warmth. The days have stopped growing shorter and are beginning to get longer, and spring will come again.

Between the two solstices, the ecliptic crosses the celestial equator on two days. On those days, the Sun lies directly above Earth's equator. We call those days *equinoxes*, which means "equal night," because the entire Earth experiences 12 hours of daylight and 12 hours of darkness. Halfway between summer solstice and winter solstice, the **autumnal equinox** marks the beginning of fall in the Northern Hemisphere; it occurs around September 22. Halfway between winter solstice and summer solstice, the **vernal equinox** marks the beginning of spring in the Northern Hemisphere; it occurs around March 20. In the Southern Hemisphere, the dates for the summer and winter solstices, and the autumnal and vernal equinoxes, are reversed from those in the Northern Hemisphere.

what if . . .

What if humans establish a colony on Mars, where the axial tilt is 25.19°? Considering only this single piece of information, would you expect seasonal variations to be larger, smaller, or about the same as seasonal variations on Earth? Explain.

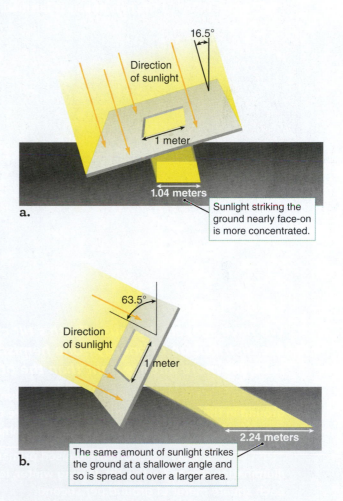

16.5°

Direction of sunlight

1 meter

1.04 meters

a.

Sunlight striking the ground nearly face-on is more concentrated.

63.5°

Direction of sunlight

1 meter

2.24 meters

b.

The same amount of sunlight strikes the ground at a shallower angle and so is spread out over a larger area.

Figure 2.15 Local noon at latitude 40° north. **a.** On the first day of northern summer, sunlight strikes the ground almost face-on. **b.** On the first day of northern winter, sunlight strikes the ground more obliquely, and less than half as much sunlight falls on each square meter of ground each second.

Theories Must Fit All the Known Facts

Many people misunderstand the phenomenon of changing seasons because they do not account for all the relevant facts.

HYPOTHESIS 1

We have seasons because Earth is closer to the Sun in the summer and farther away in winter.

THE TEST   If this were true, both the Northern and Southern Hemispheres would have summer in July. However, the Northern and Southern Hemispheres experience opposite seasons.

THE CONCLUSION   The hypothesis is falsified.

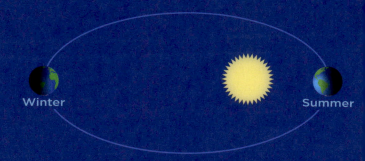

Winter

Summer

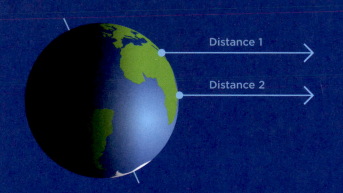

Distance 1

Distance 2

HYPOTHESIS 2

We have seasons because the tilt of Earth's axis causes one hemisphere to be significantly closer to the Sun than the other.

THE TEST   If this were true, the distances would have to be very different to cause such a large effect. Earth is tiny compared to its distance from the Sun: The distances between each hemisphere and the Sun differ by less than 0.004 percent.

THE CONCLUSION   The hypothesis is falsified.

HYPOTHESIS 3

We have seasons because Earth's tilt changes the distribution of energy—one hemisphere receives more intense light than the other.

THE TEST   If this were true, the amount of sunlight striking the ground in the summer would have to be more than in the winter, and the days would have to be longer in summer.

THE CONCLUSION   Seasons are caused primarily by a change in illumination due to Earth's tilt. During winter, less energy falls on each square meter of ground per second.

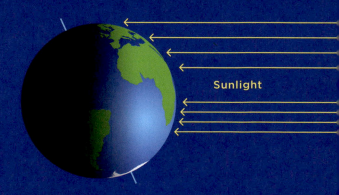

Sunlight

New information often challenges misconceptions. Incorporating this new information may alter current theories to improve the explanation of observed phenomena.

Figure 2.16 shows the solstices and equinoxes from two perspectives, oriented so that Earth's North Pole points straight up. Figure 2.16a is in the reference frame of the Sun. It shows Earth in orbit around a stationary Sun, whereas Figure 2.16b is in the reference frame of Earth. It shows the Sun's apparent motion along the celestial sphere, around a stationary Earth. Practice shifting between those perspectives. You will know that you understand them when you can look at a position in either panel and predict the corresponding positions of the Sun and Earth in the other panel.

Just as a pot of water on a stove takes time to heat up when the burner is turned on and to cool when the burner is turned off, Earth takes time to respond to changes in heating from the Sun. The hottest months of northern summer are usually July and August, which come after the summer solstice, when days are growing shorter. Similarly, the coldest months of northern winter are usually January and February, which occur after the winter solstice, when days are growing longer. Temperature changes on Earth lag behind changes in the amount of heating we receive from the Sun.

That picture of the seasons must be modified somewhat near Earth's poles. At latitudes north of 66.5° north and south of 66.5° south, the Sun is circumpolar for a part of the year surrounding the first day of summer. Those lines of latitude are the **Arctic Circle** and the **Antarctic Circle**, respectively (see Figure 2.14). When the Sun is circumpolar, it is above the horizon 24 hours per day, earning the polar regions the nickname "land of the midnight Sun" (**Figure 2.17**). An equally long period surrounds the first day of winter, when the Sun never rises and the nights are 24 hours long. The Sun never rises very high in the Arctic or Antarctic sky, so the sunlight is never very direct. Even with the long days at the height of summer, the Arctic and Antarctic regions remain relatively cool.

In contrast, on the equator, *all* stars, including the Sun, are above the horizon approximately 12 hours per day. Days and nights there are 12 hours long throughout the year. The Sun passes directly overhead on the first day of spring and the first day of autumn because on those days the Sun is on the celestial equator. Sunlight is most direct, perpendicular to the ground, at the equator on those days. On the summer solstice, the Sun is at its northernmost point along the ecliptic. On that day, and on the winter solstice, the Sun is farthest from the zenith at noon, and therefore sunlight is least direct at the equator.

Latitude 23.5° north is called the Tropic of Cancer, and latitude 23.5° south is called the Tropic of Capricorn (Figure 2.14). The band between those latitudes is called the **tropics**. If you live in the tropics—in Rio de Janeiro or Honolulu, for example—the Sun will be directly overhead at noon twice during the year.

Precession of the Equinoxes

Two thousand years ago, when Ptolemy and his associates were formalizing their knowledge of the positions and motions of objects in the sky, the Sun appeared in the constellation Cancer on the first day of northern summer and in the constellation

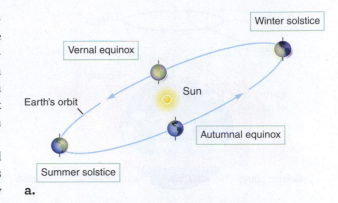

Winter solstice

Vernal equinox

Sun

Earth's orbit

Autumnal equinox

Summer solstice

a.

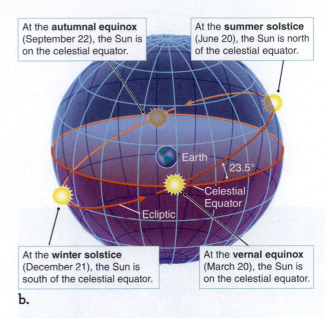

At the **autumnal equinox** (September 22), the Sun is on the celestial equator.

At the **summer solstice** (June 20), the Sun is north of the celestial equator.

Earth

23.5°

Celestial Equator

Ecliptic

At the **winter solstice** (December 21), the Sun is south of the celestial equator.

At the **vernal equinox** (March 20), the Sun is on the celestial equator.

b.

Figure 2.16 Earth's motion around the Sun from the frame of reference of **a.** the Sun and **b.** Earth.

🎥 **Astronomy in Action:** The Earth-Moon-Sun System

Figure 2.17 This composite photo shows the midnight Sun, which is visible in latitudes above 66.5° north (or south). In the 360-degree panoramic view, the Sun moves 15° each hour.

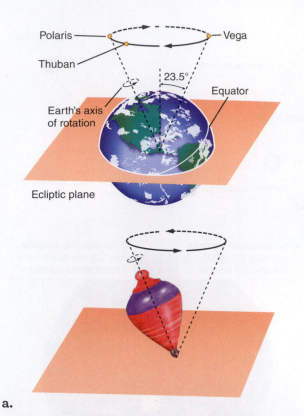

a.

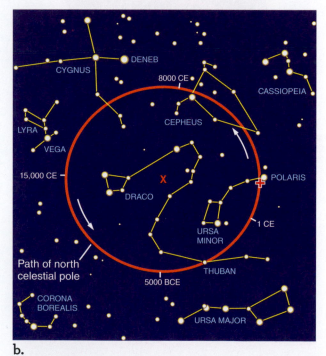

b.

Figure 2.18 **a.** Earth's axis of rotation changes orientation in the same way that the axis of a spinning top changes orientation. Polaris is the current North Pole star, Thuban was the North Pole star in 3000 BCE, and Vega will be the pole star in 14,000 CE. **b.** That precession causes the projection of Earth's rotation axis to move in a circle, centered on the north ecliptic pole (orange *X* in the center). The red cross on the circle marks the projection of Earth's axis on the sky in the early 21st century.

Capricornus on the first day of northern winter. Today, the Sun is in Taurus on the first day of northern summer and in Sagittarius on the first day of northern winter. Why have the constellations in which solstices appear changed? Two motions are actually associated with Earth and its axis: Earth spins on its axis, but its axis also wobbles like the axis of a spinning top slowing down (**Figure 2.18a**). The wobble is very slow: the north celestial pole takes about 26,000 years to complete one circle in the sky. Polaris is the star we now see near the north celestial pole. However, if you could travel several thousand years into the past or the future, the point about which the northern sky appears to rotate would no longer be near Polaris. Instead, the stars would rotate about another point on the path shown in **Figure 2.18b**. That figure shows the path of the north celestial pole through the sky during one cycle of that wobble.

The celestial equator is perpendicular to Earth's axis. Therefore, as Earth's axis wobbles, the celestial equator also must wobble. As it does so, the locations where the celestial equator crosses the ecliptic—the equinoxes—change as well. During each 26,000-year wobble of Earth's axis, the locations of the equinoxes make one complete circuit around the celestial equator. That change of the position of the equinox, due to the wobble of Earth's axis, is called the **precession of the equinoxes**.

CHECK YOUR UNDERSTANDING 2.2

If Earth's axis were tilted by 45°, instead of its actual tilt of 23.5°, how (if at all) would the seasons be different from what they are now? (a) The seasons would remain the same. (b) Summers would be colder. (c) Winters would be shorter. (d) Winters would be colder.

2.3 The Moon's Appearance Changes as It Orbits Earth

After the Sun, the most prominent object in our sky is the Moon. Just as Earth orbits around the Sun, the Moon orbits around Earth once every 27.32 days. In this section, we discuss the phases of the Moon as seen from Earth.

The Changing Phases of the Moon

The Moon and its changing aspects have been the frequent subject of mythology, art, literature, and music. In mythology, the Moon was the Greek goddess Artemis, the Roman goddess Diana, and the Inuit god Igaluk. We speak of the "man in the Moon," the "harvest Moon," and sometimes a "blue Moon."

Unlike the Sun, the Moon has no light source of its own; instead, the Moon shines by reflected sunlight. As the Moon orbits Earth, our view of the illuminated portion of the Moon constantly changes. Those different appearances of the Moon are called **phases**. During a new Moon, when the Moon is between Earth and the Sun, the side facing away from us is illuminated, and during a full Moon, when Earth is between the Sun and the Moon, the side facing toward us is illuminated. The rest of the time, we can see only part of the illuminated portion from Earth. Sometimes the Moon appears as a circular disk in the sky. Other times it is nothing more than a thin sliver, or its face appears dark.

To help you visualize the changing phases of the Moon, use a sphere (such as an orange or a softball), a lamp, and your head (**Figure 2.19**). Your head is Earth, the sphere is the Moon, and the lamp is the Sun. Turn off all the other lights in the room, and step back as far from the lamp as you can. Hold up the sphere slightly above your head so that the lamp illuminates it from one side. Move the sphere clockwise around your head and watch how the appearance of the sphere changes. When you are between the sphere and the lamp, the face of the sphere that is toward you is fully illuminated. The sphere appears to be a bright, circular disk. As the sphere moves around its circle, you will see a progression of lighted shapes, depending on how much of the bright side and how much of the dark side of the sphere you can see. That progression of shapes mimics the changing phases of the Moon.

Figure 2.20 shows the changing phases of the Moon, from two perspectives. The orbit is shown as viewed from a point above Earth's North Pole; the outer circle of Moon images then shows how the Moon appears from Earth at that phase. The **new Moon** occurs when the Moon is between Earth and the Sun. The far side of the

Figure 2.19 A sphere (such as an orange) and a lamp can help you visualize the changing phases of the Moon.

Astronomy in Action: Phases of the Moon

AstroTour: The Moon's Orbit: Eclipses and Phases

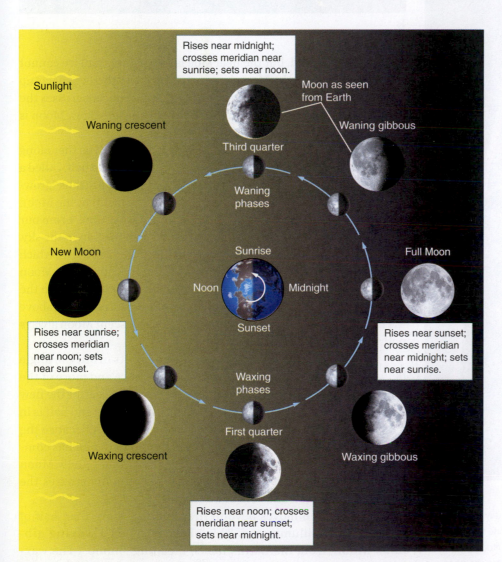

Figure 2.20 The inner circle of images (connected by blue arrows) shows the Moon as it orbits Earth, as seen by an observer far above Earth's North Pole. The Sun is on the left. The outer ring of images shows the corresponding phases of the Moon as seen from the Northern Hemisphere of Earth.

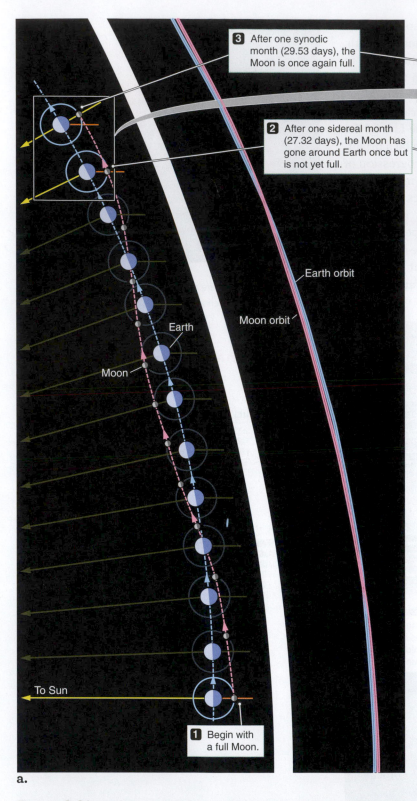

3 After one synodic month (29.53 days), the Moon is once again full.

2 After one sidereal month (27.32 days), the Moon has gone around Earth once but is not yet full.

Earth orbit

Moon orbit

Earth

Moon

To Sun

1 Begin with a full Moon.

a.

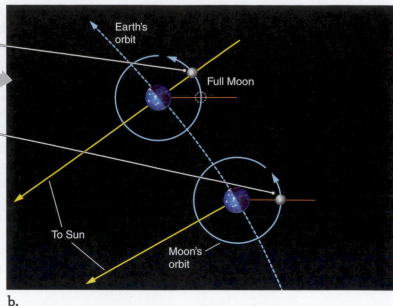

Earth's orbit

Full Moon

To Sun

Moon's orbit

b.

Figure 2.21 **a.** The Moon completes one sidereal orbit in 27.32 days, but the synodic period (the period between phases seen from Earth) from one full Moon to the next is 29.53 days. The horizontal orange line to the right of the Moon indicates a fixed direction in space. **b.** Here the orbits of Earth and the Moon are shown to scale, but the sizes of Earth and the Moon are not.

Moon is illuminated, but the near side is in darkness and we cannot see it. At new phase, the Moon is close to the Sun in the sky, so it is up in the daytime with the Sun: it rises in the east at sunrise, crosses the meridian near noon, and sets in the west near sunset. A new Moon is never above the horizon in the nighttime sky.

A few days after a new Moon, the Moon has moved farther along its orbit of Earth. At this point, a sliver of its illuminated half, called a **waxing crescent Moon**, becomes visible. *Waxing* here means "growing in size and brilliance"; the name refers to the fact that the Moon appears to be "filling out" from night to night at that time. From our perspective, the Moon has also moved away from the Sun in the sky. Because the Moon travels around Earth in the same direction in which Earth rotates, we now see the Moon following the Sun, so the Moon is east of the Sun in the sky. A waxing crescent Moon is visible in the western sky in the evening, near the setting Sun but remaining above the horizon after the Sun sets. The "horns" of a crescent Moon always point directly away from the Sun.

As the Moon moves even farther along in its orbit, the angle between the Sun and Moon grows larger, so more and more of the Moon's near side becomes illuminated. About a week after the new Moon, half of the near side of the Moon is illuminated and half is in darkness. This phase is called a **first quarter Moon** because the Moon has moved a quarter of the way around Earth and has completed the first quarter of its cycle from new Moon to new Moon (Figure 2.20). The first quarter Moon rises at noon, crosses the meridian at sunset, and sets at midnight.

As the Moon moves beyond first quarter, more than half of its near side is illuminated. This phase is called a **waxing gibbous Moon**, from the Latin *gibbus*, meaning "hump." The waxing gibbous Moon continues nightly to "grow" until finally we see the entire near side of the Moon illuminated—a **full Moon**. Earth is now between the Sun and the Moon, so they appear opposite each other in the sky when viewed from Earth. As a result, the full Moon rises as the Sun sets, crosses the meridian at midnight, and sets in the morning as the Sun rises.

The second half of the Moon's orbit is the reverse of the first half. The Moon continues in its orbit, again appearing gibbous but now becoming smaller each night. This phase is called a **waning gibbous Moon**, where *waning* means "becoming smaller." When the Moon is waning, the left side—as viewed from the Northern Hemisphere—appears illuminated. A **third quarter Moon** occurs when half of the near side is illuminated by sunlight and half is in darkness. A third quarter Moon rises at midnight, crosses the meridian near sunrise, and sets at noon. The cycle continues with a **waning crescent Moon** in the morning sky, until the new Moon once again rises and sets with the Sun, and the cycle begins again. Notice that when the Moon is farther from the Sun than Earth is, it is in gibbous (or full) phases, whereas when the Moon is closer to the Sun than Earth is, it is in crescent (or new) phases.

You can always tell a waxing Moon from a waning Moon because the illuminated side is always the side facing the Sun. When the Moon is waxing, it appears in the evening sky, so its western side is illuminated. That is the right side as viewed from the Northern Hemisphere. Conversely, when the Moon is waning, it appears in the morning sky, so the eastern side appears bright. That is the left side as viewed from the Northern Hemisphere.

Figure 2.21 illustrates two types of lunar periods. The first one is based on the Moon's orbit in space, and the second is based on the alignment of the moving Moon, Earth, and Sun. The Moon completes one orbit around Earth in 27.32 days. That **sidereal period** is how long the Moon takes to return to the same location in its orbit. However, the relationships among Earth, the Moon, and the Sun change as a result of Earth's orbital motion, so going from one full Moon to the next takes 29.53 days. That cycle is known as the Moon's **synodic period** and is the basis for our "month," because it is what we can easily observe from Earth.

Do not try to memorize all possible combinations of where the Moon is in the sky at what phase and at what time of day. Instead, work on understanding the motion and phases of the Moon, and then use your understanding to figure out the specifics of any given case. To study the phases of the Moon, draw a picture like Figure 2.20, and use it to follow the Moon around its orbit. From your drawing, figure out what phase you would see and where it would appear in the sky at a given time of day. You might also try the simulations described in "Exploration: Phases of the Moon" at the end of the chapter.

The Moon's Visible Face

Although the Moon's illumination varies as it orbits, one aspect of the Moon's appearance does *not* change: the face that we see. If we were to go outside next week, next month, 20 years from now, or 200 centuries from now, we would still see the same side of the Moon that we see tonight. That is because the Moon rotates on its axis exactly once for each revolution that it makes around Earth.

Imagine walking around a tree while always keeping your face toward the tree. By the time you complete one circle around the tree, your head has turned completely around once. When you were south of the tree, you were facing north; when you were east of the tree, you were facing west; and so on. But someone looking at you from the tree would always see your face. The Moon does the

what if . . .

What if the Moon were a cube rather than a sphere? How would this change the appearance of the Moon's phases?

▶▶ **Interactive Simulation:** Phases of the Moon

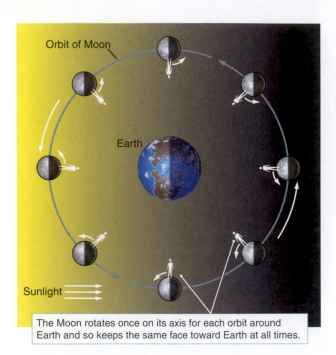

The Moon rotates once on its axis for each orbit around Earth and so keeps the same face toward Earth at all times.

Figure 2.22 The Moon rotates once on its axis for each orbit around Earth—an effect called synchronous rotation. Here, the Sun is far to the left of the Earth–Moon system.

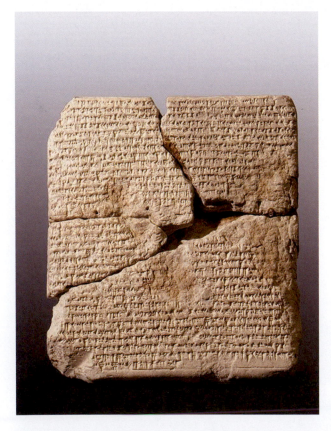

Figure 2.23 Hieroglyphic calendar at the Kom Ombo Temple. The calendar used a system of 12 months, plus festival days.

same thing, rotating on its axis once per revolution around Earth, always keeping the same face toward Earth (**Figure 2.22**). Objects whose revolution and rotation are synchronized (or "in sync") with each other are said to be in **synchronous rotation**. It occurs as a result of the gravitational and tidal forces between Earth and the Moon (Section 4.4). In later chapters we discuss other examples of synchronous rotation.

The Moon's *far side*, facing away from Earth, is often called the "dark side of the Moon." In fact, no side of the Moon is always dark. At any given time, half of the Moon is in sunlight and half is in darkness—just as for Earth. The side of the Moon that faces away from Earth, the "far side," spends just as much time in sunlight as the side of the Moon that faces Earth.

CHECK YOUR UNDERSTANDING 2.3

If you see the Moon rising just as the Sun is setting, what is the phase of the Moon? What is the phase if the Moon is setting at noon?

2.4 Calendars Are Based on the Day, Month, and Year

Archaeologists tell us that developing agriculture was crucial for the rise of human civilization, and keeping track of the seasons and best times of the year to plant and harvest was critical to successful farming. Records going back to the dawn of humanity suggest that people kept track of time by following the patterns in the sky, especially those of the Sun, Moon, and stars. Some anthropologists have speculated that notches on fragments of bone found in southern France represent a 33,000-year-old lunar calendar. In this section, we examine some different calendars.

Lunar and Lunisolar Calendars

As civilizations developed around the globe, different cultures tried to solve the problem of keeping track of time. The rates of rotation of Earth and the revolutions of the Moon around Earth and of Earth around the Sun are not integer multiples of one another. A lunar cycle (from full Moon to full Moon) is 29.5 days, and a solar cycle—the time the Sun takes to appear to move from and return to its highest possible point in the sky at noon on the summer solstice—is 365.24 days. One solar cycle, then, has 12.38 lunar cycles. Those fractions of days and months make calendars complicated.

Some of the oldest known calendars come from the Egyptians (**Figure 2.23**), the Babylonians, and the Chinese. The ancient Egyptians used a system of 12 months of 30 days each—which added up to 360 days—and then added five "festival days" to the end of the year. Without leap years, the seasons in that 365-day year drifted, so an extra month was added when necessary. When we consider how some Western countries celebrate the days between the modern December holidays and the new year, an end-of-year calendar break for festivals seems like a good solution!

The Babylonians started the 24-hour day and 7-day week—7 for the Sun, the Moon, and the 5 planets visible with the naked eye. They created the first **lunisolar calendar**, in which a month began with the first sighting of the lunar crescent, and a 13th month was added when needed to catch up to the solar year.

As the Babylonians developed mathematics, they discovered that 235 lunar months equals 19 solar years (and 6940 days). Then they created a calendar cycle that consisted of 19 years, 12 of which have 12 months and 7 of which have 13 months, and then the cycle repeats. The ancient Hebrew calendar adopted that lunisolar cycle, and the Jewish calendar still uses it today. That type of calendar keeps holidays in the same season from year to year, even though the dates change.

The ancient Chinese calendar, dating back several thousand years, is lunisolar, too. By about 500 BCE, the Chinese were using a year of 365.25 days and a system similar to that of the Babylonians of adding a 13th month to some years. To mark out a year, a few other cultures used stellar calendars, which followed the position of a bright star such as Sirius or certain prominent groups of stars in their sky, such as the Pleiades or the Big Dipper.

The Islamic calendar is purely lunar, with no added 13th lunar month. Their 12 months of 29 or 30 days each add up to 354 days—11.24 days short of the solar year. For that reason, the Islamic New Year and all other holidays drift earlier in each solar year. In the Islamic calendar, a holiday may fall in the winter in some years, and then a few years later it will have moved back to autumn.

The Modern Civil Calendar

The international civil calendar used today is a solar calendar known as the **Gregorian calendar**. It is based on the **tropical year**, which measures the 365.242 solar days between vernal (spring) equinoxes. A **solar day** is the 24-hour period of Earth's rotation that brings the Sun back to the same local meridian. That measure is in contrast to the **sidereal day**, the time Earth takes to make one rotation and face the same star on the meridian. The sidereal day is about 23 hours 56 minutes and is slightly shorter than the solar day because of Earth's motion around the Sun. This is similar to the two measures of the time it takes the Moon to orbit (Figure 2.21).

The Gregorian calendar includes a system of **leap years**—years in which a 29th day is added to February—decreed by Julius Caesar in 45 BCE to make up for the extra fraction of a day. Leap years prevent the seasons from slowly sliding through the year to become increasingly out of sync with the months, so that we don't end up experiencing winter in December one year and in August in later years.

The Gregorian calendar is named for Pope Gregory XIII. He was concerned that the Easter holiday—the first Sunday after the first full Moon after March 21—was drifting away from the vernal equinox. Julius Caesar's rule of one leap year every 4 years resulted in an average year of 365.25 days, but the actual year is 365.242 days. That difference of 0.008 day is about 11.5 minutes per year, or about 3 days every 400 years, and by Gregory's time it had caused the date of the vernal equinox to drift in the Julian calendar by about 10 days. In 1582, therefore, Pope Gregory decreed that 10 days would be deleted from the calendar to move the vernal equinox back to March 21. To make that system work out better in the future, he declared that only century years divisible by 400 are leap years, thereby deleting 3 leap years (and the 3 days) every 400 years. Catholic countries followed that system immediately, but Protestant countries did not adopt it until the 1700s. Eastern Orthodox countries, including Russia, did not switch from the Julian to the Gregorian calendar until the 1900s. One slight further revision—making years divisible by 4000 into common 365-day years—has been proposed so the modern Gregorian calendar will slip by about only 1 day in 20,000 years.

what if . . .

What if the astronomical cycles were even multiples of each other, so that a cycle of lunar phases was precisely 30 days and Earth's orbital period was precisely 12 months? In such a system, how would the dates of "wandering" holidays, such as Chinese New Year or Ramadan, change?

Despite international adoption of the Gregorian calendar, billions of people still celebrate holidays and festivals according to a lunisolar or lunar calendar. Chinese New Year, Passover, Easter, Ramadan, Rosh Hashanah, and Diwali, among others, have dates that change from one year to the next because they are based on lunar months from lunisolar or lunar calendars. The astronomy of people from long ago is still in use today.

CHECK YOUR UNDERSTANDING 2.4

Suppose Earth orbited the Sun twice as fast as it currently does. Would the solar day be longer or shorter than it currently is, in that case?

2.5 Eclipses Result from the Alignment of Earth, the Moon, and the Sun

Ancient peoples must have been terrified to see an eclipse, as they watched the Sun or the Moon being eaten away as if by a giant dragon. An **eclipse** is the total or partial obscuration of one celestial body, or the light from that body, by another celestial body. In this section, we describe the types of eclipses and their frequency.

Solar Eclipses

A **solar eclipse** occurs when the Moon passes between Earth and the Sun; observers on Earth in the shadow of the Moon will see the eclipse. Three types of solar eclipses exist: *total*, *partial*, and *annular*. Consider the structure of the shadow of the Sun cast by a round object such as the Moon, as shown in **Figure 2.24**. An observer at

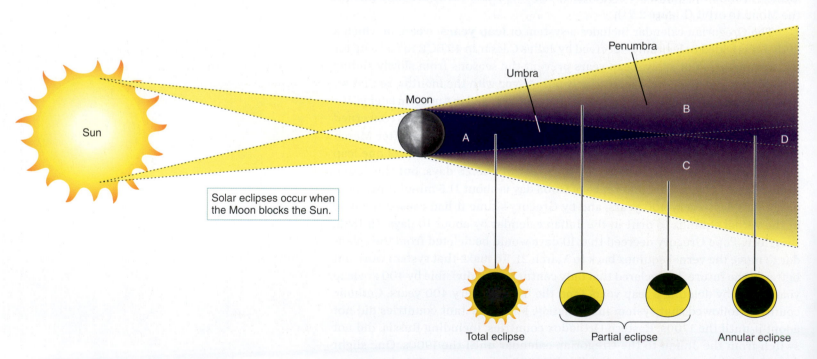

Solar eclipses occur when the Moon blocks the Sun.

Total eclipse   Partial eclipse   Annular eclipse

Figure 2.24 Different parts of the Sun are blocked at different places within the Moon's shadow. An observer on Earth in the umbra (point A) sees a total solar eclipse, observers in the penumbra (points B and C) see a partially eclipsed Sun, and observers at point D see an annular solar eclipse. Not to scale.

point A would see the Sun blocked. That darkest, inner part of the shadow is called the **umbra**. If a location on Earth passes through the Moon's umbra, the Moon blocks all the Sun's light, and a **total solar eclipse** will be observed (**Figures 2.25** and **2.26a**). At points B and C in Figure 2.24, an observer can see one side of the disk of the Sun but not the other. Only partially in shadow, that outer region is the **penumbra**. If a location on Earth passes through the Moon's penumbra, viewers there will observe a **partial solar eclipse**, in which the Moon's disk blocks the light from a portion of the Sun's disk.

In the third type of solar eclipse, called an **annular solar eclipse**, the Sun appears as a bright ring surrounding the dark disk of the Moon (**Figure 2.26b**). An observer at point D in Figure 2.24 is far enough from the Moon that it appears smaller than the Sun. An object's apparent size in the sky depends on the object's actual size and its distance from us. The Sun is about 400 times the diameter of the Moon, and the distance between the Sun and Earth is about 400 times more than the distance between the Moon and Earth. As a result, the Moon and Sun have almost the same apparent size in the sky. In addition, the Moon's orbit is not a perfect circle. When the Moon and Earth are a bit closer together than average, the Moon appears slightly larger in the sky than the Sun. An eclipse occurring then will be total for some observers. When the Moon and Earth are farther apart than average, the Moon appears smaller than the Sun, so eclipses occurring then will be annular for some observers. Among all solar eclipses, one-third are total at some location on Earth, one-third are annular, and one-third are only partial.

Figure 2.27 shows the geometry of a solar eclipse when the Moon's shadow falls on Earth's surface. Figures like this one usually show Earth and the Moon much closer together than they really are. The page is too small to draw them to scale and still show the critical details. The relative sizes and distances between Earth and the Moon are roughly equivalent to the difference between a basketball and a tennis ball, respectively, placed 7 meters apart. Figure 2.27b shows the geometry of a solar eclipse with Earth, the Moon, and the separation between them

Figure 2.25 The full spectacle of a total solar eclipse.

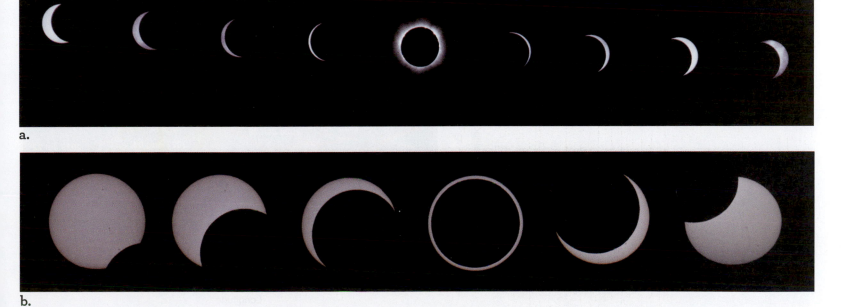

a.

b.

Figure 2.26 Time-lapse images of the Sun taken **a.** during a total solar eclipse and **b.** during the annular solar eclipse of May 20, 2012. The Sun set during the ending phases.

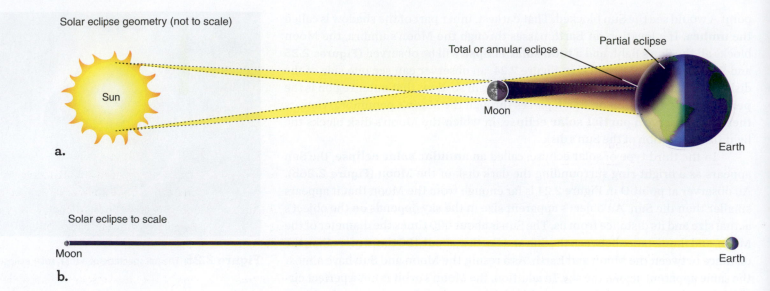

Figure 2.27 **a.** A solar eclipse occurs when the shadow of the Moon falls on the surface of Earth. **b.** The same as part a., but now the Moon and Earth are drawn to scale.

drawn to scale. Compare that drawing with Figure 2.27a to see why drawings of Earth and the Moon are rarely drawn to scale. If the Sun were drawn to scale in Figure 2.27a, it would be bigger than your head and located almost 64 meters off the left side of the page.

From any particular location, you are more likely to observe a partial solar eclipse than a total solar eclipse. Where the Moon's penumbra touches Earth, it has a diameter of almost 7000 km—large enough to cover a substantial fraction of Earth. Thus, a partial solar eclipse is often visible from many locations on Earth. In contrast, the path along which a total solar eclipse is visible, shown in **Figure 2.28**,

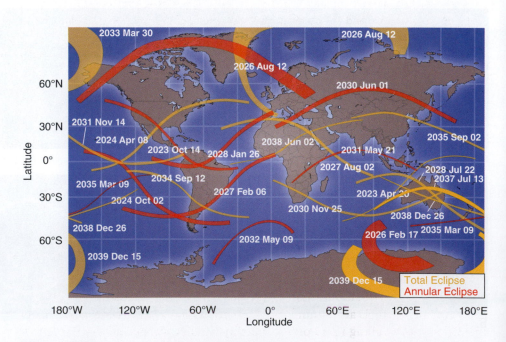

Figure 2.28 The paths of total solar eclipses through 2032. Solar eclipses occurring in Earth's polar regions cover more territory because the Moon's shadow hits the ground obliquely.

Reading Astronomy News

2500 Miles of Citizen Scientists

Laura Reed, sciencenode.org

Citizen scientists across the nation captured images of the Sun's corona to create the largest dataset of its kind.

On August 21 [2017] many of us stood outside and marveled at the solar eclipse.

But for some, the eclipse was more than just a show. It was an opportunity to participate in the National Solar Observatory (NSO) Citizen Continental-America Telescopic Eclipse (CATE) Experiment.

The Citizen CATE Experiment is funded by federal, corporate, and private sources, including several companies that donated equipment to the project. Volunteer observers and schools get to keep the equipment used in the study.

A grand experiment

Scientists, students, and volunteers assembled along the 2500-mile path of eclipse totality tracked the Sun using 68 identical telescopes.

Using specialized software and instrument packages, they produced more than 1000 images from the start of the partial solar eclipse until totality.

The goal was to capture 90 minutes of continuous, high-resolution, rapid-cadence images detailing a difficult-to-capture region of the solar atmosphere: the Sun's inner corona.

This is the first time scientists have collected research-quality observations of the corona during the eclipse's entire transit across the United States.

"This dataset is extraordinary," says Matt Penn, principal investigator for the project. "Normally during a solar eclipse, we get about 2 minutes of data in the region closest to the photosphere. But Citizen CATE allows us to get an hour and a half of data."

My corona

The corona is the Sun's outer atmosphere. It is difficult for scientists to study because the photosphere, or solar surface, is so bright that it overpowers the faint corona. We can only see the corona when something obscures the photosphere.

(Think of a light so bright that it makes it difficult to see an object close to you. Blocking the light with your hand allows a better view of the object.)

Scientists can create artificial eclipses using an instrument called a coronagraph that covers the Sun's bright disk. The proximity of the instrument to the observer distorts the Sun's edge, making precise observation and measurement difficult.

During a real eclipse, the Moon blocks the Sun. The Moon's great distance lets scientists measure and study the corona in greater detail.

The process

On the day of the event, skies were clear for 58 of the 68 observation sites.

Observers, spaced about 50 miles from each other, started observations when the Moon's shadow appeared on their horizon. They captured images every 10 seconds during totality.

The rapid cadence of imagery along with a 2-arcsecond pixel resolution should help scientists understand the intensity of the corona over an extended period as well as the motions of prominences, coronal inflows, coronal mass ejections, and other active regions.

The next total solar eclipse in the United States will be April 8, 2024. The path of totality will begin in Texas, moving north from Mexico, and will exit the U.S. via Maine.

Start making plans to join the next team of citizen scientists.

QUESTIONS

1. Why does a solar eclipse move across the surface of the Earth?

2. Why did scientists want data from along the whole track of the eclipse?

3. Why can the corona be seen only during an eclipse?

4. Would you have expected a lunar eclipse near the date of the solar eclipse? Explain.

5. See whether any results are available on the project website (http://citizencate.org/). What did participants learn from this experiment?

Source: https://sciencenode.org/feature/2,500-miles-of-citizen-scientists.php.

covers only a tiny fraction of Earth's surface. Even when the distance between Earth and the Moon is at a minimum, the umbra is only 270 km wide on Earth. As the Moon moves along in its orbit, that tiny shadow sweeps across Earth at a few thousand kilometers per hour. In addition, the Moon's shadow falls on Earth's curved surface, causing the region shaded by the Moon during a solar eclipse to

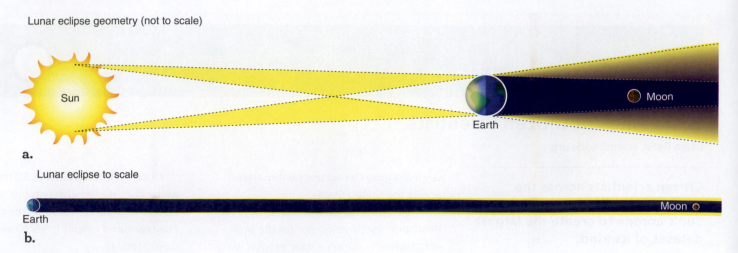

Lunar eclipse geometry (not to scale)

Sun

Earth

Moon

a.

Lunar eclipse to scale

Earth

Moon

b.

Figure 2.29 **a.** A lunar eclipse occurs when the Moon passes through Earth's shadow. **b.** The same as part a., but now Earth and the Moon are drawn to scale.

be elongated by various amounts. The curvature can even cause an eclipse that started out as annular to become total.

As a result of those factors, a total solar eclipse can never last longer than $7\frac{1}{2}$ minutes and is usually significantly shorter. Even so, it is one of the most amazing sights in nature. People all over the world flock to the most remote corners of Earth to witness the fleeting spectacle of the bright disk of the Sun blotted out of the daytime sky. Perhaps you saw the total eclipse visible from much of the United States in August 2017. The next total solar eclipse visible in the continental United States will take place in 2024. Annular eclipses will be visible from parts of the United States in 2021 and 2023. Viewing a solar eclipse should be on your lifetime to-do list!

unanswered questions

How long will Earth continue to have total solar eclipses? Those occur because the Moon and the Sun are coincidentally the same size in our sky, but will that always be the case? An object's observed size in the sky depends on its actual diameter and its distance from us. One or both of those can change. The Moon is slowly moving away from Earth by about 4 meters per century. Over time, the Moon will appear smaller in the sky, and it won't be able to cover the full disk of the Sun. Although we can measure how fast the Moon is moving away from Earth now, we are less certain of how that rate may change with time. A lesser and more uncertain effect comes from the Sun—which will continue to brighten slowly, as it has throughout its history. With that brightening, the Sun's actual diameter will slightly increase, and it will appear larger in our sky. A more distant Moon and a larger Sun will eventually (in hundreds of millions of years) end total eclipses on Earth.

Lunar Eclipses

Lunar eclipses occur when the Moon moves through Earth's shadow. **Figure 2.29a** shows the geometry of a lunar eclipse and is drawn to scale in **Figure 2.29b**. During a lunar eclipse, Earth is between the Sun and the Moon. Because Earth is much larger than the Moon, the dark umbra of Earth's shadow at the distance of the Moon is more than 2 times the Moon's diameter. A **total lunar eclipse**, in which the Moon is entirely within Earth's shadow, lasts as long as 1 hour 40 minutes. In a total lunar eclipse, the Moon often appears red (**Figure 2.30a**). That "blood-red Moon," as literature and poetry have called it, occurs because red light from the Sun is bent as it travels through Earth's atmosphere and then illuminates the Moon. Earth's atmosphere absorbs or scatters other colors of sunlight away from the Moon and therefore does not illuminate it.

A **penumbral lunar eclipse** occurs when the Moon passes through the penumbra of Earth's shadow; those are noticeable only from a very dark location or when the Moon passes within about 1000 km of the umbra. If Earth's shadow incompletely covers the Moon, some of the Moon's disk remains bright and some of it is in shadow. The result is a **partial lunar eclipse**. **Figure 2.30b** shows a composite of images taken at different times during a partial lunar eclipse. In the middle frame, Earth's shadow nearly completely eclipses the Moon.

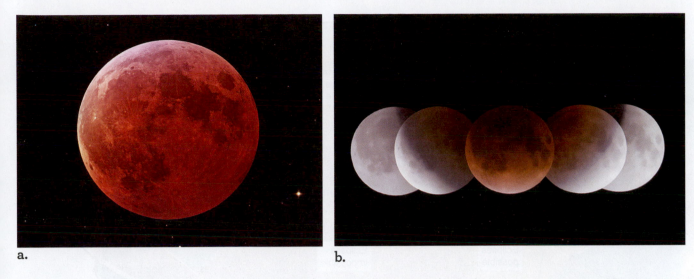

a.   b.

Figure 2.30 a. During a total lunar eclipse, the Moon often appears blood red. **b.** A time-lapse series of photographs of a partial lunar eclipse clearly shows Earth's shadow.

★ **WHAT AN ASTRONOMER SEES** Astronomers learn to be sensitive to color variations. An astronomer looking at these two images for the first time would immediately conclude from the difference in color between the images that the Moon in a. is in full lunar eclipse. An astronomer would also notice in b. that the images of the Moon were carefully aligned, so that in each image Earth's shadow remained stationary. Combining the images in this way makes it possible to see (or measure) the size of Earth's shadow relative to the Moon.

Many more people have observed a total lunar eclipse than have observed a total solar eclipse. To see a total solar eclipse, you must be located within that very narrow band of the Moon's shadow as it moves across Earth's surface. In contrast, when the Moon is immersed in Earth's shadow, anyone located in the hemisphere facing the Moon can see it. As a result, total eclipses of the Moon are relatively common from any location, so you may have seen at least one.

Frequency of Eclipse Seasons

How could some people in ancient cultures predict eclipses? From their understanding of lunar and solar cycles for making calendars, they computed cycles of eclipses. Imagine Earth, the Moon, and the Sun all sitting on the same flat tabletop. If the Moon's orbit were in the same plane as Earth's orbit, the Moon would pass directly between Earth and the Sun at every new Moon. The Moon's shadow would pass across the face of Earth, and we would see a solar eclipse. Similarly, Earth would pass directly between the Sun and the Moon every synodic month, and a lunar eclipse would occur at each full Moon.

Solar and lunar eclipses do *not* happen every month, however, because the Moon's orbit does *not* lie in the same plane as Earth's orbit. As **Figure 2.31** shows, the plane of the Moon's orbit around Earth is inclined by about 5.2° with respect to the plane of Earth's orbit around the Sun. The line along which the orbital planes of the Sun and the Moon intersect is called the **line of nodes**. For part of the year, the line of nodes points generally toward the Sun. During those times, called **eclipse seasons**, a new Moon passes directly between the Sun and Earth, casting its shadow on Earth's surface and causing a solar eclipse. Similarly, a full Moon occurring during an eclipse season passes through Earth's shadow, causing a lunar eclipse. An eclipse season lasts only 38 days. That's how long the Sun is close enough to the line of nodes for eclipses to occur. Usually, the line of nodes

what if . . .

What if Earth had two moons, each having the same size and the same orbit, but located 120° apart in that orbit? Would we expect to see more solar eclipses in this case? Would the eclipses for both moons occur during the same eclipse seasons?

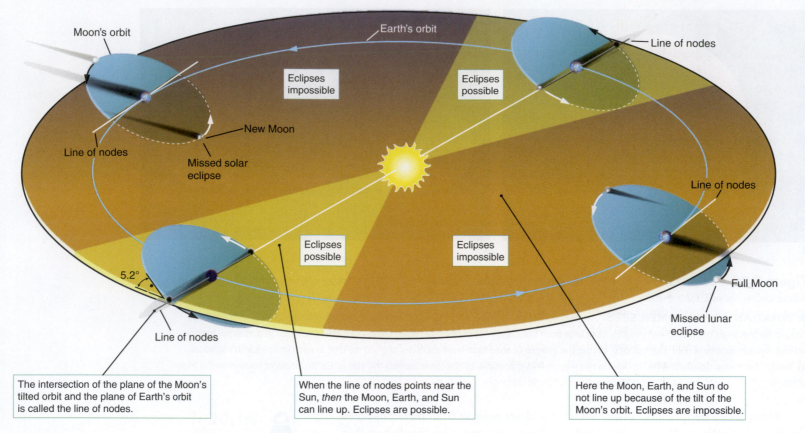

Figure 2.31 Eclipses are possible only when the Sun, the Moon, and Earth lie along (or very close to) an imaginary line known as the line of nodes. When the Sun does not lie along the line of nodes, Earth passes under or over the shadow of a new Moon, and a full Moon passes under or over the shadow of Earth.

points farther away from the Sun, so Earth, the Moon, and the Sun cannot line up closely enough for an eclipse to occur. At those times, a solar eclipse cannot take place because the shadow of a new Moon passes "above" or "below" Earth. Similarly, no lunar eclipse can occur because a full Moon passes "above" or "below" Earth's shadow.

If the plane of the Moon's orbit always had the same orientation, eclipse seasons would occur twice a year, as suggested in Figure 2.31. In actuality, eclipse seasons occur about every 5 months 20 days. The roughly 10-day difference occurs because the plane of the Moon's orbit slowly wobbles, much like the wobble of a spinning plate balanced on the end of a circus performer's stick. As the Moon's orbital plane wobbles, the line of nodes changes direction. That wobble rotates in the direction opposite the direction of the Moon's motion in its orbit. That is, the line of nodes moves clockwise as viewed from above Earth's orbital plane. One wobble of the Moon's orbit takes 18.6 years, so we say that the line of nodes regresses by 360° every 18.6 years, or 19.4° per year. That rate amounts to about a 20-day regression each year. If January 1 marks the middle of an eclipse season, the next eclipse season will be centered around June 20, and the one after that around December 10.

CHECK YOUR UNDERSTANDING 2.5

If the Moon were in its same orbital plane but twice as far from Earth, which of the following would happen? (a) The Moon would not go through phases. (b) Total eclipses of the Sun would not be possible. (c) Annular eclipses of the Sun would not be possible. (d) Total eclipses of the Moon would not be possible.

Origins: The Obliquity of Earth

Earth's various motions give rise to the most basic of patterns faced by life on Earth. Earth's rotation is responsible for the cycle of night and day. Earth's axial tilt and its passage around the Sun bring the change of the seasons. As life evolved on Earth, it had to adapt to those patterns.

Earth's range of climate, based on distance from the equator, probably has contributed to the broad diversity of life on our planet. Earth's biodiversity includes life that adapted to the long, cold polar nights and to the much higher temperatures of the tropics. Earth's life adapted to seasonal patterns in rain and drought, leading to acquired seasonal patterns of migration and reproduction. If Earth had no axial tilt, the poles would continuously be in winter and probably too cold for humans. Latitudes near the equator would be consistently even warmer than they are now.

Periodic changes in Earth's axial tilt might have affected life. If Earth's tilt were larger than 23.5°, the seasonal variation would be even stronger. If the tilt were smaller, the seasonal variation would be weaker. Chinese, Indian, Greek, and Arabic records going back 3000 years indicate that people estimated the tilt by measuring the length of the shadow from a vertical pole on the day of the solstice. Earth's axial tilt actually varies from 22.1° to 24.5° over a 41,000-year cycle. Currently, the tilt is about midway between the two extremes and getting smaller. It will reach its minimum value of 22.1° in about 10,000 years. Scientists are studying whether that variation in tilt correlates with periods of temperature change on Earth, especially the times of ice ages.

SUMMARY

The motions of Earth and the Moon are responsible for many of the repeating patterns that can be observed in the sky. Calendars keep track of time by using those patterns. Earth's rotation about its axis causes daily patterns of rising and setting, and Earth's revolution around the Sun causes yearly patterns of the stars in the sky and marks the passage of the seasons. Earth's tilt on its axis changes both the length of daytime and the intensity of sunlight, causing the seasons. Life on Earth adapted to those seasonal variations. Earth's axial tilt varies slightly over tens of thousands of years. The Moon's revolution around Earth causes the monthlong pattern of the phases of the Moon. Occasionally, alignments of Earth, the Moon, and the Sun cause eclipses.

(1) **Describe how Earth's rotation about its axis and its revolution around the Sun affect how we perceive celestial motions from different places on Earth.** Earth's daily rotation on its axis causes the apparent daily motion of the Sun, Moon, and stars. Our location on Earth and Earth's location in its orbit around the Sun determine which stars we see at night. You can determine your latitude from the altitude of the pole star. When observing objects in our sky, we need to consider how Earth moves in relation to other objects. The ecliptic is the path that the Sun appears to take through the stars.

(2) **Explain why seasons change throughout the year.** A year is the time Earth takes to complete one revolution around the Sun.

Constellations are patterns of stars that reappear in the same place in the sky at the same time of night each year. The tilt of Earth's axis determines the seasons by changing the angle at which sunlight strikes Earth's surface in different locations. The changing angle of sunlight and the differing length of the day cause the seasonal variations on Earth. Equinoxes and solstices mark the changing seasons.

(3) **Describe the factors that create the phases of the Moon.** The relative locations of the Sun, Earth, and the Moon determine the phases of the Moon. The Moon takes one sidereal month to complete one revolution around Earth and one synodic month to go through a cycle of phases. The Moon's motion around Earth causes it to be illuminated differently at different times. When farther from the Sun than Earth is, the Moon is in gibbous phases. When closer to the Sun than Earth is, the Moon is in crescent phases.

(4) **Sketch the alignment of Earth, the Moon, and the Sun during eclipses of the Sun and the Moon.** A solar eclipse occurs when the new Moon is in the plane of Earth and the Sun, and the shadow of the Moon falls on Earth. A lunar eclipse occurs when the full Moon is in the plane of Earth and the Sun, and the shadow of Earth falls on the Moon. Twice a year, at new or at full Moon, the Moon is exactly in line between Earth and the Sun. At those times, called eclipse seasons, eclipses can occur.

QUESTIONS AND PROBLEMS

TEST YOUR UNDERSTANDING

1. Constellations are groups of stars that
   a. are close to one another in space.
   b. are bound to each other by gravity.
   c. are relatively close to one another in Earth's sky.
   d. all have the same composition.

2. Where on Earth can you stand and, over the course of a year, see the entire sky?
   a. only at the North Pole
   b. at either pole
   c. at the equator
   d. anywhere

3. Day and night are caused by
   a. the tilt of Earth on its axis.
   b. the rotation of Earth on its axis.
   c. the revolution of Earth around the Sun.
   d. the revolution of the Sun around Earth.

4. Polaris, the North Star, is unique because
   a. it is the brightest star in the night sky.
   b. it is the only star in the sky that doesn't move throughout the night.
   c. it is always located at the zenith, for any observer.
   d. it has a longer path above the horizon than any other star.

5. An angle exists between the ecliptic and the celestial equator because
   a. Earth's axis is tilted with respect to its orbit.
   b. Earth's orbit is tilted with respect to the orbits of other planets.
   c. the Sun follows a rising and falling path through space.
   d. the Sun's orbit is tilted with respect to Earth's.

6. Earth's axial tilt causes the seasons because
   a. one hemisphere of Earth is closer to the Sun in summer.
   b. the days are longer in summer.
   c. the rays of light strike the ground more directly in summer.
   d. both a and b
   e. both b and c

7. Which of the following are true on the vernal and autumnal equinoxes? (Choose all that apply.)
   a. Every nonpolar place on Earth has 12 hours of daylight and 12 hours of darkness.
   b. The Sun rises due east and sets due west.
   c. The Sun is located on the celestial equator.
   d. The motion of the stars in the sky is different from that on other days.

8. We always see the same side of the Moon because
   a. the Moon does not rotate on its axis.
   b. the Moon rotates on its axis once for each revolution around Earth.
   c. the other side of the Moon is unlit when it is facing Earth.
   d. the other side of the Moon is on the opposite side of Earth when it is facing Earth.

9. If you see the Moon on the meridian at sunrise, the phase of the Moon is
   a. waxing gibbous.
   b. full.
   c. first quarter.
   d. third quarter.

10. A lunar eclipse occurs when _____ shadow falls on _____ .
   a. Earth's; the Moon
   b. the Moon's; Earth
   c. the Sun's; the Moon
   d. the Sun's; Earth

11. Different cultures created different calendars because
   a. they had measured different lengths of the day, month, and year.
   b. they used different definitions of the day, month, and year.
   c. the number of days in a month and the number of days and months in a year are not integers.
   d. calendars are completely arbitrary.

12. Which stars we see at night depends on
   a. our location on Earth.
   b. Earth's location in its orbit.
   c. the time of the observation.
   d. all of the above.

13. On the summer solstice in June, the Sun will be directly above _____ and all locations north of _____ will experience daylight all day.
   a. the Tropic of Cancer; the Antarctic Circle
   b. the Tropic of Capricorn; the Arctic Circle
   c. the Tropic of Cancer; the Arctic Circle
   d. the Tropic of Capricorn; the Antarctic Circle

14. The Sun, Moon, and stars
   a. appear to move each day because the celestial sphere rotates about Earth.
   b. change their relative positions over time.
   c. rise north or south of west and set north or south of east, depending on their location on the celestial sphere.
   d. always remain in the same position with respect to one another.

15. If you see the first quarter Moon on the meridian, the Sun must be
   a. on the western horizon.
   b. on the eastern horizon.
   c. below the horizon.
   d. on the meridian.

THINKING ABOUT THE CONCEPTS

16. Seafarers such as Columbus used Polaris to navigate as they sailed from Europe to North America. When Magellan sailed the South Seas, he could not use Polaris. Explain why.

17. If you were standing at Earth's North Pole, where would you see the north celestial pole in relation to your zenith?

18. Figure 2.13 shows that observers in the Northern Hemisphere see the zodiacal constellation Gemini in the winter. Why don't they see it in the summer? 👁

19. Imagine that you are flying along in a jetliner.
   a. Describe ways to tell that you are moving.
   b. If you look down at a building, which way is it moving relative to you?

20. Astronomers are sometimes asked to serve as expert witnesses in court cases. Suppose you are called as an expert witness, and the defendant states that he could not see the pedestrian because the full Moon was casting long shadows across the street at midnight. Is that claim credible? Why or why not?

21. Imagine that one person was developing a theory of seasons as described in the three hypotheses in this chapter's Process of Science Figure. Compare that process with the flowchart of Chapter 1's Process of Science Figure. How would the development of that theory look on the diagram? 👁

22. Why is the winter solstice *not* the coldest time of year?

23. Earth spins on its axis and wobbles like a top.
    a. How long does it take to complete one spin?
    b. How long does it take to complete one wobble?

24. At what approximate time of day do you see the full Moon near the meridian? At what time is the first quarter (waxing) Moon on the eastern horizon? Use a sketch to help explain your answers.

25. Assume that the Moon's orbit is circular, and imagine that you are standing on the side of the Moon that faces Earth.
    a. How would Earth appear to move in the sky as the Moon made one revolution around Earth?
    b. How would the "phases of Earth" appear to you compared with the phases of the Moon as seen from Earth?

26. If people on Earth were observing a lunar eclipse, what would you see from the Moon?

27. From any given location, why are you more likely to witness a partial eclipse of the Sun than a total eclipse?

28. Why don't we see a lunar eclipse each time the Moon is full or witness a solar eclipse each time the Moon is new?

29. ★ **WHAT AN ASTRONOMER SEES**   Figure 2.30b shows the shadow of Earth, projected onto the Moon over the course of a partial lunar eclipse. From this shadow, what can you determine about the shape of Earth? Many observations like this, taken from many different locations on Earth in different years, show the same shadow shape. From those observations, what can you determine about the shape of Earth? 👁

30. How would Earth's temperature variation be different if its axis were tilted 90° to the plane of its orbit (like the planet Uranus)?

APPLYING THE CONCEPTS

31. If, as some historians believe, Eratosthenes used the Egyptian stadion, about 157.5 meters, what would he have computed for the size of Earth?
    a. In Working It Out 2.1, the circumference of Earth is calculated using 1 stadion = 185 m. Make a prediction: Do you expect your answer to be larger or smaller than the answer from Working It Out 2.1? Will it differ by a lot or a little?
    b. Follow Working It Out 2.1 to calculate the circumference of Earth.
    c. Check your work by comparing your answer to your prediction.

32. A point on the surface of Earth spins along at 1674 km/h at the equator. Use that fact, along with the length of the day, to calculate Earth's equatorial diameter. ●–●–●
    a. Study Working It Out 2.1 to make a prediction about how large Earth's diameter should be. Is it about 1000 km, about 10,000 km, or about 100,000 km?
    b. The units of speed given here are km/h. You have the hint about using the length of the day to calculate diameter. In what units should you write the length of the day (seconds, minutes, or hours)? Should you multiply or divide the speed by the length of the day to get a distance in km for an answer?
    c. Calculate Earth's diameter.
    d. Check your work by comparing the diameter you calculated with the radius from Working It Out 2.1.

33. The Moon's apparent diameter in the sky is approximately ½°. About how long does the Moon take to move a distance equal to its own diameter across the sky? (Hint: How long does it take the Moon to move 360°—that is, through one orbit?)
    a. Make a prediction: Compare ½° to 360°; do you expect it to take the Moon more or less time to move through ½° than 360°? Do you expect the difference in time to be large or small?
    b. Reason: The ratio of the angles should equal the ratio of the times, because the Moon is moving at constant speed through those angles.
    c. Calculate: Set up the ratio and calculate the time for the Moon to move through ½°.
    d. Check your work by comparing your answer to your prediction.

34. An object's apparent size in the sky is proportional to its actual diameter divided by its distance. The Moon has a radius of 1737 km, with an average distance of 3.780×10^5 km from Earth. The Sun has a radius of 696,000 km, with an average distance of 1.496×10^8 km from Earth. Show that the apparent sizes of the Moon and the Sun in our sky are approximately the same. ●–●–●
    a. Make a prediction: If the apparent sizes of the Moon and the Sun in the sky are approximately the same, what value would you expect to get if you divided one apparent size by the other?
    b. Calculate the apparent size of the Moon and the Sun, and divide the apparent size of the Moon by the apparent size of the Sun to find the ratio of sizes.
    c. Check your work by comparing your answer to your prediction.

35. Earth has an average radius of 6371 km, while the radius of the Moon is 1737 km. An object's apparent size in the sky is proportional to its actual diameter divided by its distance. If you were standing on the Moon, how much larger would Earth appear in the lunar sky than the Moon appears in our sky?
    a. Reason: How does the distance from Earth to the Moon compare to the distance from the Moon to Earth? In the calculation of the apparent size of Earth as seen from the Moon, then, what is the difference in the calculation of the apparent size of the Moon as seen from Earth? Set up an algebraic equation for the ratio of the apparent size of Earth (as seen from the Moon) to the apparent size of the Moon (as seen from Earth).
    b. Make a prediction: Compare the radius of Earth to the radius of the Moon. Do you expect the apparent size of Earth to be a little larger, a few times larger, or hundreds of times larger than the apparent size of the Moon?
    c. Calculate the ratio of the apparent size of Earth (as seen from the Moon) to the apparent size of the Moon (as seen from Earth).
    d. Check your work by comparing your result to your prediction.

36. Determine the latitude where you live. Draw and label a diagram showing that your latitude is the same as (a) the altitude of the north celestial pole and (b) the angle (along the meridian) between the celestial equator and your local zenith. What is the Sun's altitude at noon as seen from your home at the winter solstice and at the summer solstice?

37. If the angle between your zenith and Polaris is 40°, what is the altitude of Polaris? What is your latitude?

38. The southernmost star in a group of stars known as the Southern Cross lies approximately 65° south of the celestial equator. What is the farthest-north latitude from which the entire Southern Cross is visible? Can it be seen in any of the U.S. states? If so, which ones?

39. Imagine that you are standing on the South Pole at the time of the southern summer solstice.
    a. How far above the horizon will the Sun be at noon?
    b. How far above (or below) the horizon will the Sun be at midnight?

40. Suppose the tilt of Earth's equator with respect to its orbit were 10° instead of 23.5°. At what latitudes would the Arctic and Antarctic circles and the Tropics of Cancer and Capricorn be located?

41. The Moon's orbit is tilted by about 5° with respect to Earth's orbit around the Sun. What is the highest altitude in the sky that the Moon can reach, as seen in Philadelphia (latitude 40° north)?

42. Suppose you would like to witness the midnight Sun (when the Sun appears just above the northern horizon at midnight), but you don't want to travel any farther north than necessary.
    a. How far north (that is, to which latitude) would you have to go?
    b. At what time of year would you make that trip?

43. ⊙
    a. The vernal equinox is now in the zodiacal constellation Pisces. The precession of Earth's axis will eventually cause the vernal equinox to move into Aquarius. How long, on average, does the vernal equinox spend in each zodiacal constellation?
    b. Stonehenge was erected roughly 4000 years ago. Referring to the zodiacal constellations shown in Figure 2.13, identify the constellation in which those ancient builders saw the vernal equinox.

44. Use Figure 2.18 to estimate when Vega, the fifth-brightest star in our sky (excluding the Sun), will once again be the northern pole star. ⊙

45. How would the length of the eclipse season change if the plane of the Moon's orbit were inclined less than its current 5.2° to the plane of Earth's orbit? Explain your answer.

EXPLORATION Phases of the Moon

digital.wwnorton.com/astro7

In this Exploration, we will be examining the phases of the Moon. Visit the Student Site at the Digital Resources page and open the Phases of the Moon Interactive Simulation in Chapter 2. This simulator animates the orbit of the Moon around Earth, allowing you to control the simulation speed and a number of other parameters.

Begin by starting the animation to explore how it works. Examine all three image frames. The large frame shows the Earth-Moon system, as looking down from far above Earth's North Pole. The upper right frame shows what the Moon looks like to the person on the ground. The lower right frame shows where the Moon appears in the person's sky. Stop the animation, and press "Reset Animation."

1. What time of day is this for the person shown on Earth?

_____

_____

_____

_____

2. What phase is the Moon in?

_____

_____

_____

_____

3. Where is the Moon in this person's sky?

_____

_____

_____

_____

Run the animation until the Moon reaches waxing crescent phase.

4. As viewed from Earth, which side of the Moon is illuminated (the left or the right)?

_____

_____

_____

5. The person shown on Earth will observe this waxing crescent Moon either after sunset or before sunrise. At which of these times can the person see the waxing crescent Moon?

_____

_____

_____

_____

Run the animation until the Moon reaches first quarter and the Sun is setting for the person on Earth. (Hint: You may want to slow the animation rate!)

6. How many full days have passed since new Moon?

_____

_____

_____

_____

7. At this instant, where is the first quarter Moon in the person's sky?

_____

_____

_____

_____

8. If an astronaut were standing on the near side of the Moon at this time, what phase of Earth would he see?

_____

_____

_____

_____

Three observations about the phases of the Moon are connected: the location of the Moon in the sky, the time for the observer, and the phase of the Moon. If you know two of these, you can figure out the third. Use the animation to fill in the missing pieces in the following situations:

9. An observer sees the Moon in _____ phase, overhead, at midnight.

10. An observer sees the Moon in third quarter phase, rising in the east, at _____.

11. An observer sees the Moon in full phase, _____, at 6:00 A.M.

Motion of Astronomical Bodies

Like the Moon, Venus has phases as a result of the relative positions of Venus, Earth, and the Sun. By studying these phases carefully, Galileo figured out that Venus orbits the Sun rather than orbiting Earth. To see why his observations show this, set up a light source and four roundish objects of the same size to represent Venus at different locations in its orbit. If your desk were a clock, the objects might be at 7, 9, 10, and 11. Take a picture from a vantage point like the one shown, which is at about the position of the 6. Keep the light source out of the frame. Compare the sizes of the objects in your picture. Each represents a different phase of Venus, because each is illuminated differently from your vantage point.

EXPERIMENT SETUP

Use a desk lamp and a few roundish objects to set up a model of the orbit of Venus. Make a prediction: How will the size of the crescent phase "Venus" compare to the size of "Venus" showing nearly full phase? Take a picture, keeping the light source out of the frame.

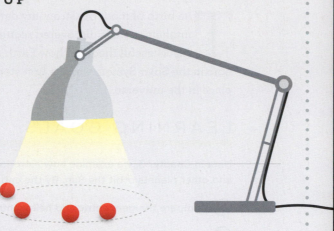

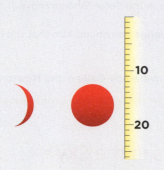

Using your picture, measure the size of "Venus" at crescent and full phase. What does this tell you about the relative distances of these objects from your camera? The real Venus also behaves this way, so what can you conclude about the orbit of Venus? Specifically, is it centered on the Sun or centered on Earth?

PREDICTION

I predict that when the object is in "full" phase, it will appear ☐ larger ☐ smaller than when it is at "crescent" phase.

SKETCH OF RESULTS (in progress)

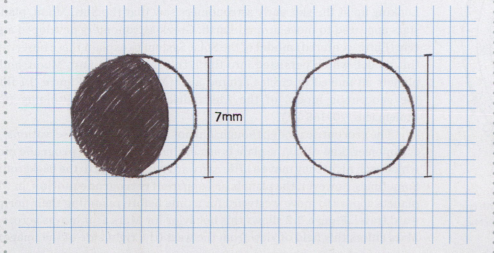

7mm

3

The birth of modern astronomy dates back to when astronomers and mathematicians discovered regular patterns in the motions of the planets. A successful theory of how Earth moved and how it fit in with its neighbors in the Solar System was the first step toward understanding our planet's place in the universe.

LEARNING GOALS

Here in Chapter 3, we examine how astronomers came to understand that Earth and other planets orbit the Sun. By the end of this chapter, you should be able to:

1. Compare the geocentric and heliocentric models of the Solar System.

2. Use Kepler's laws to describe how objects in the Solar System move.

3. Explain how Galileo's astronomical discoveries convinced him that the heliocentric model was correct.

4. Describe the physical laws of motion discovered by Galileo and Newton.

3.1 The Motions of Planets in the Sky

When people in ancient times looked up at the sky, they saw the Sun, Moon, and stars rise in the east and set in the west and *appear* to move around Earth. The ancient peoples were aware of five planets (*planet* means "wandering star") because from one night to the next they moved in a generally eastward direction among the stars, whose positions appeared fixed on the celestial sphere. One thing the ancients did *not* know, however, was that Earth was similar to those planets. Developing a successful theory of how Earth and the planets move and how Earth fits in with its neighbors in the Solar System was the first step to understanding Earth's place in the universe. The history of how those ideas evolved—from Earth at the center of all things to Earth as just an ordinary planet—is a good example of how science is self-correcting.

The Geocentric Model

Looking up at the sky, early astronomers saw that the Sun, Moon, planets, and stars appeared to move around Earth. Greek astronomers developed a **geocentric** (Earth-centered) **model** of the Solar System to explain these observations. In this model, which persisted into the 17th century, the Sun, Moon, and known planets all moved in circles around a stationary Earth. **Figure 3.1** illustrates the geocentric model of the Alexandrian astronomer Ptolemy (Claudius Ptolemaeus, 90–168 CE).

The geocentric model, though, did not account for all observations. Ancient astronomers knew that the planets usually have an eastward **prograde motion**, in which each night they move a little eastward with respect to the background stars. Sometimes, however, the planets had apparent **retrograde motion**, in which the planets appear to move westward for a time before resuming their normal eastward travel. **Figure 3.2** shows this behavior for Mars as it moves across the sky. Mars enters this field of view from the west (the right side of the image), makes

Figure 3.1 In the Ptolemaic view of the heavens, Earth is at the center, orbited by the Moon, Mercury, Venus, the Sun, Mars, Jupiter, and Saturn.

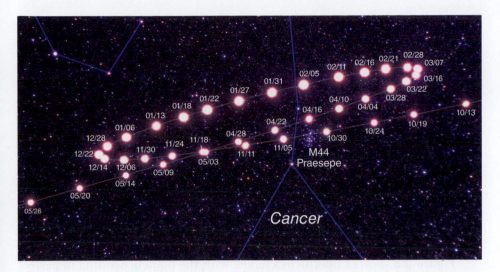

Figure 3.2 This time-lapse photographic series shows Mars as it moves in apparent retrograde motion.

★ **WHAT AN ASTRONOMER SEES** An astronomer will notice from the dates shown that the photographic series begins with Mars on the right side of the image. As time passes, Mars moves toward the left, then completes the loop, and leaves the image on the left, about 7.5 months later. An astronomer will also notice that Mars is larger and brighter on the image in the retrograde part of the loop, indicating that Earth and Mars are closer together at that time. This color image clearly shows that some stars are red and some stars (like the ones in the group of stars called M44) are blue.

a loop, and leaves the field of view to the east (the left side of the image) about seven months later.

The apparent retrograde motion of the five "naked eye" planets—Mercury, Venus, Mars, Jupiter, and Saturn—created a puzzling problem for Ptolemy's geocentric model, as it stood in 150 CE. Because the geocentric model in its simplest form failed to explain the apparent retrograde motion of the planets, Ptolemy added an embellishment called an *epicycle*—a small circle superimposed on each planet's larger circle (**Figure 3.3**). As the planet travels along its larger circle around Earth, it also moves along its epicycle. When moving along the smaller circle in the opposite direction of the forward motion of the larger circle, the planet's forward motion would be reversed and it would appear to move backward in the sky. This embellishment made the model reasonably successful at predicting the positions of planets in the sky. For nearly 1500 years, Ptolemy's model, in which the Sun, Moon, and planets all moved in perfect circles around a stationary Earth, with the "fixed stars" located far beyond the planets, was the accepted paradigm in the Western world.

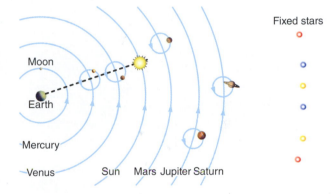

Figure 3.3 To reconcile retrograde motion with the geocentric model of the Solar System, Ptolemy added loops called epicycles to each planet's circular orbit around Earth.

CHECK YOUR UNDERSTANDING 3.1a

How did the ancients know the planets were different from the stars?

Answers to Check Your Understanding questions are in the back of the book.

Copernicus Proposes a Heliocentric Model

Nicolaus Copernicus (1473–1543—**Figure 3.4**) was not the first person to suggest that the Sun might be at the center of the Solar System. A few ancient Greek and medieval Arab astronomers, such as Aristarchus of Samos (310–230 BCE), had briefly considered the idea when they saw problems with the Earth-centered model. These astronomers lacked the observational or mathematical tools to test the Sun-centered hypothesis, and most astronomers found the fact that they could not feel Earth's motion around the Sun to be a powerful argument in favor of the Earth-centered model.

Copernicus, however, was the first to develop a comprehensive mathematical model with the Sun at the center and that later astronomers could test. That **heliocentric** (Sun-centered) **model** was the beginning of the Copernican

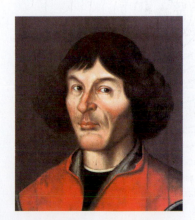

Figure 3.4 Nicolaus Copernicus rejected the ancient Greek model of an Earth-centered universe and replaced it with a model that centered on the Sun.

what if . . .

What if we launch an Earth satellite that orbits at half the distance to the Moon? Would astronauts aboard that satellite observe retrograde motion of the Moon?

Revolution. Through the work of 16th- and 17th-century scientists such as Tycho Brahe, Galileo Galilei, Johannes Kepler, and Isaac Newton, the heliocentric model of the Solar System became one of the best-corroborated theories in all of science.

Copernicus was multilingual and highly educated: he studied philosophy, canon (Catholic) law, medicine, economics, mathematics, and astronomy in his native Poland and in Italy. He conducted astronomical observations from a small tower, and sometime around 1514 he started writing about a heliocentric model. Eighteen years later, he completed his manuscript. He did not publish the book because he knew his ideas would be controversial: philosophical and religious views of the time held that humanity, and thus Earth, must be the center of the universe. Late in his life, Copernicus was finally persuaded to publish his ideas, and his great work *De revolutionibus orbium coelestium* ("On the Revolutions of the Heavenly Spheres") appeared in 1543, the year he died.

Figure 3.5 shows Copernicus's model with the planets orbiting in perfect circles around the Sun. That model explained the observed motions of Earth, the Moon, and the planets, including apparent retrograde motion, much more simply than the geocentric model did. In this model, relative motion is important; when you are in a car or train and you pass a slower-moving car or train, the other vehicle seems as though it is moving backward.

Similarly, in the Copernican model, planets farther from the Sun appear to have retrograde motion when Earth overtakes them in their orbits. **Figure 3.6** illustrates that effect for Mars; compare this diagram to the observations shown in Figure 3.2. Conversely, planets closer to the Sun than Earth is—Mercury and Venus—appear to have retrograde motion when overtaking Earth. Except for the Sun and the Moon, all Solar System objects exhibit apparent retrograde motion. The effect diminishes with increasing distance from Earth. Apparent retrograde motion is caused by the relative motion between Earth and the other planets.

Although Copernicus correctly placed the Sun at the center of the Solar System, he still conceived of the planets as moving in perfectly circular orbits at constant speeds, so he needed to use some epicycles to match the observations. His model made testable predictions of the location of each planet on a given night. Those predictions were at least as accurate as those of the geocentric model. Overall, however, the heliocentric model was simpler than the geocentric model and became the basis for further refinements in understanding how Earth moved. As copies of *De Revolutionibus* and Copernicus's ideas slowly spread across Europe, other scientists began to consider and then accept the heliocentric model, and a scientific revolution began.

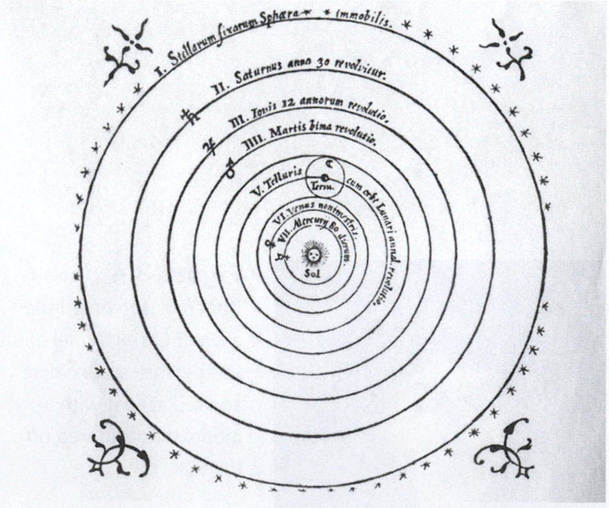

Figure 3.5 The Copernican heliocentric view of the Solar System (II–VII) and the fixed stars (I). The Sun is at the center, orbited by Mercury, Venus, Earth, Mars, Jupiter, and Saturn. The Moon orbits Earth.

Scaling the Solar System

From his observations, Copernicus deduced the correct order of the planets and concluded that planets closer to the Sun travel faster than planets farther out. He also realized that he needed to consider two categories of planets: **inferior planets**, which are closer to the Sun than Earth is, and **superior planets**, which are farther from the Sun than Earth is.

In Copernicus's model, Earth, another planet, and the Sun periodically align in space to form either a line or a right triangle. There are special words for each of these alignments, as shown in **Figure 3.7**. When a superior planet is in line with the Sun and Earth but on the other side of the Sun from Earth, the planet is in *conjunction* (Figure 3.7a) A superior planet in conjunction rises and sets in the sky with the Sun. When a superior planet is in conjunction, it is at its farthest from Earth, so it is at its faintest, too. Exactly at conjunction, however, you won't see the superior planet at all because it's behind the Sun in the sky.

In contrast, when a superior planet is in line with the Sun and Earth on the same side of the Sun as Earth, the configuration is an *opposition* (Figure 3.7a). At opposition, the superior planet is "opposite" the Sun in the sky. Like a full Moon, the planet rises when the Sun sets and sets when the Sun rises. When the superior planet is in opposition, it is at its closest to Earth during that orbit and thus at its brightest; therefore, opposition is the best time to observe the planet in the sky. Opposition occurs during the time when the planet exhibits retrograde motion because that is exactly when Earth is overtaking the planet in its orbit. *Quadrature* occurs when Earth, the Sun, and a superior planet form a right triangle in space (Figure 3.7a).

For an inferior planet, the configurations are slightly different (Figure 3.7b). When the inferior planet is between Earth and the Sun, it is closest to Earth and the configuration is called *inferior conjunction*. If the inferior planet is on the other side of the Sun from Earth, it is farthest from Earth, and the configuration is called a *superior conjunction*. *Greatest elongation* occurs when the inferior planet forms a right triangle with Earth and the Sun, and thus is the farthest it gets from the Sun in the sky. The inner planets are always close to the Sun in the sky, so Mercury and Venus are visible only within a few hours of sunrise or sunset. The best time to observe those planets is at greatest elongation because they will have the greatest separation from the Sun in the sky.

Copernicus realized that two types of orbital periods existed. We name these periods with terms similar to those used for lunar orbits. A planet's *sidereal period* is how long the planet takes to make one orbit around the Sun with respect to the stars and return to the same point in space. A planet's *synodic period* is how long the planet takes to return to the same configuration with the Sun and Earth, such as from inferior conjunction to inferior conjunction or from opposition to opposition.

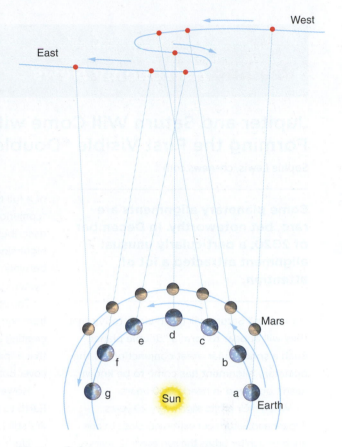

Figure 3.6 The Copernican model explains the apparent retrograde motion of Mars (see Figure 3.2) as seen in Earth's sky when Earth passes Mars in its orbit. (Not drawn to scale.)

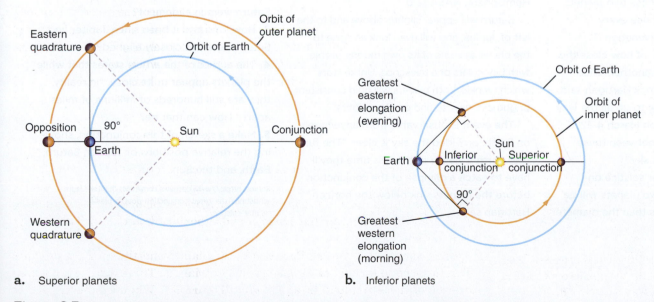

a. Superior planets

b. Inferior planets

Figure 3.7 Planetary configurations for **a.** superior (outer) planets and **b.** inferior (inner) planets.

Reading Astronomy News

Jupiter and Saturn Will Come within 0.1 Degrees of Each Other, Forming the First Visible "Double Planet" in 800 Years

Sophie Lewis, cbsnews.com

Some planetary alignments are rare, but noteworthy. In December of 2020, a particularly unusual alignment attracted a lot of attention.

Jupiter and Saturn will be so close today that they will appear to form a "double planet." Such a spectacular great conjunction, as the planetary alignment has come to be known, hasn't occurred in nearly 800 years.

When their orbits align every 20 years, Jupiter and Saturn get extremely close to one another. Jupiter orbits the sun every 12 years, while Saturn's orbit takes 30 years, so every few decades Jupiter laps Saturn, according to NASA.

The 2020 great conjunction is especially rare—the planets haven't been this close together in nearly 400 years, and haven't been observable this close together at night since medieval times, in 1226.

"Alignments between these two planets are rather rare, occurring once every 20 years or so, but this conjunction is exceptionally rare because of how close the planets will appear to one another," Rice University astronomer Patrick Hartigan said in a statement. "You'd have to go all the way back to just before dawn on March 4, 1226, to see a closer alignment between these objects visible in the night sky."

Aligning with the winter solstice on December 21, 2020, the two planets will be just 0.1 degrees apart—less than the diameter of a full Moon, EarthSky said. The word "conjunction" is used by astronomers to describe the meeting of objects in our night sky, and the great conjunction occurs between the two largest planets in our Solar System: Jupiter and Saturn.

The planets will be so close, they will appear, from some perspectives, to overlap completely, creating a rare "double planet" effect. So close, that a "pinkie finger at arm's length will easily cover both planets in the sky," NASA said.

However, while they may appear from Earth to be very, very close, in reality, they are still hundreds of millions of miles apart.

During the last great conjunction in 2000, Jupiter and Saturn were so close to the Sun that the event was difficult to observe. But skywatchers should have a clearer view of the celestial event this time around. The great conjunction will be shining bright shortly after sunset, low in the southwestern sky, as viewed from the Northern Hemisphere, NASA said.

Saturn will appear slightly above and to the left of Jupiter, and will even look as close to the planet as some of its own moons, visible with binoculars or a telescope. Unlike stars, which twinkle, both planets will hold consistent brightness, easy to find on clear nights.

The event is observable from anywhere on Earth, provided the sky is clear. "The further north a viewer is, the less time they'll have to catch a glimpse of the conjunction before the planets sink below the horizon," Hartigan said.

The planets will appear extremely close for about a month, giving skywatchers plenty of time to witness the spectacular alignment throughout the holiday season. The event coincidentally aligns with the December solstice, marking the shortest day of the year in the Northern Hemisphere.

This will be the "greatest" great conjunction for the next 60 years, until 2080. Hartigan said that, following that conjunction, the duo won't make such a close approach until sometime after the year 2400.

QUESTIONS

1. The use of the word "conjunction" in this context is more general than the formal usage introduced in the text of this chapter. In the chapter, we use "conjunction" to mean "solar conjunction." Compare that usage with the usage in this article. What does the term mean more generally?

2. How long had it been since Jupiter and Saturn were in alignment?

3. How long had it been since Jupiter and Saturn were this closely aligned?

4. The author of the article states that while the planets appear quite close, "in reality, they are still hundreds of millions of miles apart." How can that be?

5. Make a sketch of this conjunction, showing the relative positions of Jupiter, Saturn, Earth, and the Sun.

Source: https://www.cbsnews.com/news/jupiter-saturn -christmas-star-great-conjunction-double-planet -winter-solstice/.

The synodic period is what can be observed directly from Earth. In **Figure 3.8a**, Earth and the superior planet are in opposition at point A. Superior planets move around the Sun more slowly than Earth does, so Earth orbits the Sun once and then catches up to the superior planet to form the next opposition at point B. In **Figure 3.8b**, Earth and the inferior planet are in inferior conjunction at point A. An inferior planet moves around the Sun faster than Earth does, so it completes one sidereal period and then must continue in its orbit to catch up to Earth for the next inferior conjunction at point B.

Copernicus used the geometry of these alignments along with his observations of the positions of the planets in the sky, including their altitudes and the times they rose and set, to estimate the planet–Sun distances. He could not figure this out in terms of meters or feet, but instead determined these distances in multiples of the Earth–Sun distance. The average distance from Earth to the Sun is so useful that it has its own unit, the **astronomical unit (AU)**. **Table 3.1** shows that the relative distances Copernicus calculated are remarkably close to the distances modern methods yield. **Working It Out 3.1** describes the numerical details. Copernicus's model not only predicted planetary positions in the sky but also could be used to accurately compare the distances between the planets and the Sun.

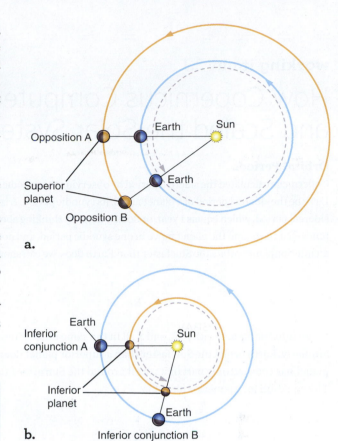

a.

b.

Figure 3.8 The synodic periods of planets indicate how long they take to return to the same configuration with Earth and the Sun. **a.** Earth completes one orbit around the Sun first and then catches up to the superior planet. **b.** Inferior planets complete a full orbit around the Sun first and then catch up to Earth.

CHECK YOUR UNDERSTANDING 3.1b

The planet Uranus will be observed in retrograde motion when: (a) Uranus is closest to the Sun; (b) Uranus is farthest from the Sun; (c) Earth overtakes Uranus in its orbit; (d) Uranus overtakes Earth in its orbit.

3.2 Kepler's Laws Describe Planetary Motion

Copernicus did not understand *why* the planets move about the Sun, but he realized that his heliocentric model offered a way to compute the planets' relative distances. His theory is an example of **empirical science**, which seeks to describe patterns in nature as accurately as possible even if it's not yet possible to explain why those patterns exist. Copernicus's work was revolutionary because he challenged the accepted geocentric model and proposed that Earth is just one planet among many. His conclusions paved the way for other great empiricists, including Tycho Brahe and Johannes Kepler.

Tycho Brahe's Observations

Tycho Brahe (1546–1601—**Figure 3.9**) was a Danish astronomer of noble birth who entered university at age 13 to study philosophy and law. After seeing a partial solar eclipse in 1560, Tycho (conventionally referred to by his first name) became interested in astronomy. A few years later, he observed Jupiter and Saturn near each other in the sky, though not exactly where they were expected to be from the astronomical tables based on Ptolemy's model. Tycho gave up studying law and devoted himself to making better tables of the planets' positions in the sky.

The king of Denmark granted Tycho the island of Hven, between Sweden and Denmark, to build an observatory. Tycho designed and built new instruments,

Table 3.1	Copernicus's Scale of the Solar System	
Planet	Copernicus's Value (AU)	Modern Value (AU)
Mercury	0.38	0.39
Venus	0.72	0.72
Earth	1.00	1.00
Mars	1.52	1.52
Jupiter	5.22	5.20
Saturn	9.17	9.58

AU = Astronomical Unit.

working it out 3.1

How Copernicus Computed Orbital Periods and Scaled the Solar System

Orbital Periods

Copernicus calculated the sidereal period by observing the synodic period. Let P be the sidereal period of a planet and S, its synodic period. E is Earth's sidereal period, which equals 1 year, or 365.25 days. By thinking about the distance that Earth and the planet move in one synodic period, and noting that an inferior planet orbits the Sun faster than Earth does, we can show that

$$\frac{1}{P} = \frac{1}{E} + \frac{1}{S}$$

for an inferior planet, with P, E, and S all in the same units of days or years. Similarly, Earth orbits the Sun faster than a superior planet does, so the planet has traveled only part of its orbit around the Sun after 1 Earth year. The equation for a superior planet is

$$\frac{1}{P} = \frac{1}{E} - \frac{1}{S}$$

For Saturn, the time that passes between oppositions—the date of maximum brightness—shows that Saturn's synodic period (S) is 378 days, or $378 \div 365.25 = 1.035$ years. Then, to compute Saturn's sidereal period (P) in years, we use $S = 1.035$ yr and $E = 1$ yr in the equation for a superior planet:

$$\frac{1}{P} = \frac{1}{1\,\text{yr}} - \frac{1}{1.035\,\text{yr}} = 1 - 0.966\,\text{yr}^{-1} = 0.034\,\text{yr}^{-1}$$

Thus,

$$P = \frac{1}{0.034\,\text{yr}^{-1}} = 29.4\,\text{yr}$$

Saturn's sidereal period is 29.4 years, which means that Saturn takes 29.4 years to travel around the Sun and return to where it started in space.

Scaling the Solar System

Copernicus used the configurations of the planets shown in Figure 3.7 along with their sidereal periods to compute the relative distances of the planets. For the superior planets, he measured the fraction of the circular orbit that the planet completed in the time between opposition and quadrature, and then he used trigonometry to solve for the planet–Sun distance in astronomical units (see Figure 3.7a). For the inferior planets, he had a right triangle at the point of greatest elongation, and then he used right-triangle trigonometry to solve for the planet–Sun distance in astronomical units (see Figure 3.7b). Copernicus's values are impressively similar to modern values (see Table 3.1). Copernicus still did not know the actual value of the astronomical unit in miles or kilometers, but he was the first to accurately compute the relative distances of the planets from the Sun.

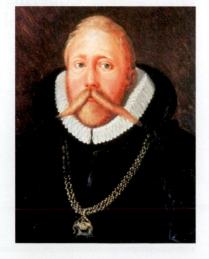

Figure 3.9 Tycho Brahe, known commonly as Tycho, was one of the greatest astronomical observers before the invention of the telescope.

operated a printing press, and taught students and others how to conduct observations. With the help of his sister Sophie, Tycho carefully measured the precise positions of planets in the sky over several decades, developing the most comprehensive set of planetary data then available. He created his own geo-heliocentric model, shown in **Figure 3.10**. In Tycho's model, the planets orbit the Sun, and the Sun and Moon orbit Earth. His model gained some acceptance among people who preferred to keep Earth at the center for philosophical or religious reasons. Tycho lost his financial support when the king died, and in 1600 he relocated to Prague.

Kepler's Laws

In 1600, Tycho hired a more mathematically inclined astronomer, Johannes Kepler (1571–1630—**Figure 3.11**), as his assistant. Kepler, who had studied Copernicus's ideas, was responsible for the next major step toward understanding the planets' motions. When Tycho died, Kepler inherited the records of his observations. Working first with Tycho's observations of Mars, Kepler deduced three rules that accurately describe how the planets move. These are now generally referred to as **Kepler's laws**. Kepler's laws are empirical: they use existing data to make predictions about future behavior but do not include an underlying theory of why the objects behave as they do.

Kepler's First Law Comparing Tycho's extensive planetary observations with predictions from Copernicus's heliocentric model, Kepler expected the data to confirm circular orbits for planets orbiting the Sun. Instead, he found disagreements between his predictions and Tycho's observations. He was not the first to notice such discrepancies. Rather than discard Copernicus's model, Kepler revised it.

By replacing Copernicus's circular orbits with *elliptical* orbits, Kepler could predict the positions of planets for any day, and his predictions fit Tycho's observations almost perfectly. An **ellipse** is an oval that is symmetric from right to left and from top to bottom. As shown in **Figure 3.12a**, you can draw an ellipse by attaching the two ends of a piece of string to a piece of paper, stretching the string tight with the tip of a pencil, and then drawing around those two points while keeping the string tight. Each point at which the string is attached to the paper is a **focus** (plural: **foci**) of the ellipse. If the two foci are close together, then the ellipse is more circular (**Figure 3.12b**). If the two foci are farther apart, the ellipse is more elongated. The long axis of the ellipse is called the **major axis**, and the short axis is called the **minor axis**. The **eccentricity** (*e*) of an ellipse measures that elongation; it is determined by the distance from the center of the ellipse to a focus, divided by half the length of the major axis. A circle has an eccentricity of 0 because the two foci coincide at the center. The more elongated the ellipse becomes, the closer its eccentricity gets to 1.

Kepler's first law of planetary motion states that the orbit of each planet is an ellipse with the Sun located at one focus. The other focus, as shown in **Figure 3.13**, is nothing but empty space. The ellipse in Figure 3.13 has very high eccentricity, compared to actual planetary orbits, in order to better distinguish its features. The dashed lines represent the major and minor axes of the ellipse. Half the length of the major axis of the ellipse is called the **semimajor axis**, *A*. The average distance between the Sun and a planet is equal to the semimajor axis of the planet's orbit.

Figure 3.10 Tycho's geo-heliocentric model, showing the Moon and Sun orbiting Earth, with the other planets orbiting the Sun.

Figure 3.11 Johannes Kepler explained the motions of the planets with three empirically determined laws.

▶❚❚ **AstroTour:** Kepler's Laws

▶▶ **Interactive Simulation:** Planetary Orbits

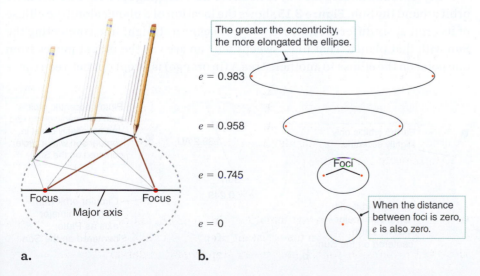

The greater the eccentricity, the more elongated the ellipse.

$e = 0.983$

$e = 0.958$

$e = 0.745$

Foci

$e = 0$

When the distance between foci is zero, *e* is also zero.

Focus   Focus
Major axis

a.   b.

Figure 3.12 **a.** We can draw an ellipse by attaching a length of string to a piece of paper at two points (called foci) and then pulling the string around as shown. The long axis is called the major axis. **b.** Ellipses range from circles (*e* = 0) to elongated eccentric shapes. *e* = eccentricity.

Figure 3.13 According to Kepler's first law, planets move in elliptical orbits with the Sun at one focus. (Nothing is at the other focus.) The orbit's eccentricity is the center-to-focus distance divided by the semimajor axis.

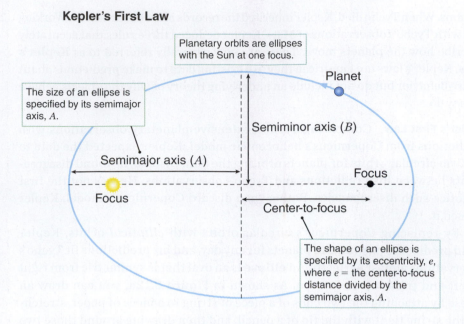

Kepler's First Law

Planetary orbits are ellipses with the Sun at one focus.

The size of an ellipse is specified by its semimajor axis, A.

Planet

Semiminor axis (B)

Semimajor axis (A)

Focus

Focus

Center-to-focus

The shape of an ellipse is specified by its eccentricity, e, where e = the center-to-focus distance divided by the semimajor axis, A.

Most planetary objects in our Solar System have nearly circular orbits (that is, they are ellipses with very low eccentricities). As **Figure 3.14a** shows, Earth's orbit, with an eccentricity of 0.017, is very nearly a circle centered on the Sun; therefore, the variation in distance between Earth and the Sun is small. In contrast, dwarf planet Pluto's orbit (**Figure 3.14b**) has an eccentricity of 0.249. The Sun is noticeably offset from center, and the orbit is elongated.

The **Process of Science Figure** traces the steps that led to the development of Kepler's first law. Along the way, several theories were falsified. Recall from Chapter 1 that in order for a theory to be scientific, it must be falsifiable.

Kepler's Second Law   By analyzing Tycho's observational data of changes in the planets' positions, Kepler found that a planet moves fastest when closest to the Sun and slowest when farthest from the Sun. We now know that Earth's average speed in its orbit around the Sun is 29.8 kilometers per second (km/s). When closest to the Sun, Earth travels at 30.3 km/s. When farthest from the Sun, Earth travels at 29.3 km/s.

Kepler found an elegant way to describe the changing speed of a planet in its orbit around the Sun. **Figure 3.15** shows the location of a planet along the ellipse of its orbit at six different times (t_1 to t_6). Imagine a straight line connecting the Sun with that planet. That line "sweeps out" an area as the planet moves from one point on the ellipse to another. Area A (in orange) is swept out between times

Figure 3.14 The orbits of **a.** Earth and **b.** Pluto in comparison with circles around the Sun. e = eccentricity.

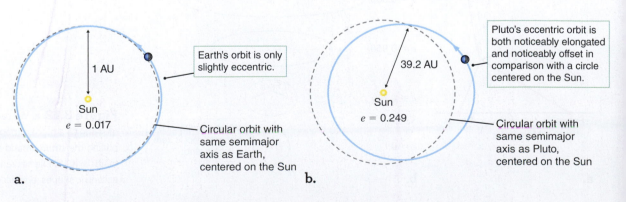

1 AU

Sun

$e = 0.017$

Earth's orbit is only slightly eccentric.

Circular orbit with same semimajor axis as Earth, centered on the Sun

39.2 AU

Sun

$e = 0.249$

Pluto's eccentric orbit is both noticeably elongated and noticeably offset in comparison with a circle centered on the Sun.

Circular orbit with same semimajor axis as Pluto, centered on the Sun

a.   **b.**

Theories Are Falsifiable

Early astronomers studied the motions of the planets,
but did not understand why the planets behaved as they do.

Copernicus
1473–1543

● **The Hypothesis**
Copernicus proposed that planets moved in
circular orbits (with epicycles) around the Sun.

Tycho
1546–1601

The Observation
Tycho, born three years after Copernicus died, observed and
collected enormous amounts of data about planet positions.

Kepler
1571–1630

The Prediction
Kepler, 25 years younger than Tycho, used Copernicus's model
and Tycho's data to predict where the planets should be.

The Test
When Kepler compared the predictions with actual data,
they disagreed. Copernicus's idea was falsified!

The New Hypothesis
Planet orbits are not circular. They are elliptical.

Newton
1643–1726

● This new hypothesis gained acceptance a century later
when Newton showed that planetary orbits are an
inevitable consequence of gravitational attraction.

In order for a theory to be scientific, it must be falsifiable, even if the test can't
be carried out until decades or centuries later. Disproving an old theory is part of
the self-correcting nature of science: It always leads to deeper understanding.

Kepler's Second Law

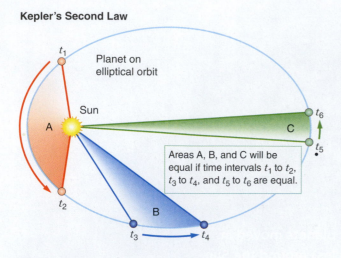

Figure 3.15 An imaginary line between a planet and the Sun sweeps out an area as the planet orbits. According to Kepler's second law, the three areas A, B, and C will be the same if the three time intervals shown are equal.

t_1 and t_2, area B (in blue) is swept out between times t_3 and t_4, and area C (in green) is swept out between times t_5 and t_6. Kepler realized that if the three time intervals in the figure are equal (that is, $t_1 \rightarrow t_2 = t_3 \rightarrow t_4 = t_5 \rightarrow t_6$), then the three areas A, B, and C must be equal as well. In order for this to be true, the planet must move fastest when it is closest to the Sun (area A), because it must cover the largest distance in its orbit. When the planet is farthest from the Sun (Area C), it must move more slowly because it covers a smaller distance in its orbit. This change in speed balances the change in distance so the area remains the same.

Kepler's second law, also called Kepler's **law of equal areas**, states that the imaginary line connecting a planet to the Sun sweeps out equal areas in equal times, regardless of where the planet is along the ellipse of its orbit. That law applies to only one planet at a time. The area swept out by Earth in a given time interval is always the same. Likewise, the area swept out by Mars in a given time is always the same. But the area swept out by Earth and the area swept out by Mars in a given time are *not* the same. Kepler's second law can be used to find the speed of a planet anywhere in its orbit.

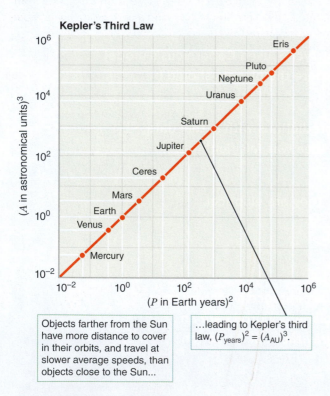

Figure 3.16 A plot of A^3 versus P^2 for objects in our Solar System shows that they obey Kepler's third law. (By plotting powers of 10 on each axis, we can fit both large and small values on the same plot. We will use that approach often.)

Table 3.2	Kepler's Third Law: $P^2 = A^3$		
Planet	**Period P (years)**	**Semimajor Axis A (AU)**	$\dfrac{P^2}{A^3}$
Mercury	0.241	0.387	$\dfrac{0.241^2}{0.387^3} = 1.00$
Venus	0.615	0.723	$\dfrac{0.615^2}{0.723^3} = 1.00$
Earth	1.000	1.000	$\dfrac{1.000^2}{1.000^3} = 1.00$
Mars	1.881	1.524	$\dfrac{1.881^2}{1.524^3} = 1.00$
Ceres	4.599	2.765	$\dfrac{4.599^2}{2.765^3} = 1.00$
Jupiter	11.86	5.204	$\dfrac{11.86^2}{5.204^3} = 1.00$
Saturn	29.46	9.582	$\dfrac{29.46^2}{9.582^3} = 0.99^*$
Uranus	84.01	19.201	$\dfrac{84.01^2}{19.201^3} = 1.00$
Neptune	164.79	30.047	$\dfrac{164.79^2}{30.047^3} = 1.00$
Pluto	247.92	39.482	$\dfrac{247.92^2}{39.482^3} = 1.00$
Eris	557.00	67.696	$\dfrac{557.00^2}{67.696^3} = 1.00$

*Ratios are not exactly 1.00 because of slight perturbations from the gravity of other planets.

working it out 3.2

Kepler's Third Law

Kepler's third law states that the square of the period of a planet's orbit measured in years, P_{years}, is equal to the cube of the semimajor axis of the planet's orbit measured in astronomical units, A_{AU}. As an equation, the law says

$$(P_{years})^2 = (A_{AU})^3$$

For mathematical convenience, Kepler used units based on Earth's orbit—astronomical units and years. If other units were used, such as kilometers and days, P^2 would still be proportional to A^3, but the constant of proportionality would not be 1.

To calculate the average size of Neptune's orbit in astronomical units, you first need to find out how long Neptune's period is in Earth years,

which you can determine by observing the synodic period and from that computing its sidereal period (using Working It Out 3.1). Neptune's sidereal period is 165 years. Plugging that number into Kepler's third law gives the following result:

$$(P_{years})^2 = (165)^2 = 27{,}225 = (A_{AU})^3$$

To solve that equation, you must first square 165 to get 27,225 and then take its cube root (see Appendix 1 for calculator hints). Then

$$A_{AU} = \sqrt[3]{27{,}225} = 30.1$$

The semimajor axis of Neptune's orbit—that is, the average distance between Neptune and the Sun—is 30.1 AU.

Kepler's Third Law   Kepler looked for patterns in the planets' orbital periods. Compared with planets closer to the Sun, he found that planets farther from the Sun have longer orbits *and* move more slowly in those orbits. Kepler discovered a mathematical relationship between a planet's sidereal period—how many years the planet takes to go around the Sun and return to the same position in the Solar System—and its average distance from the Sun in astronomical units. **Kepler's third law** states that the square of the sidereal period (P) is equal to the cube of the semimajor axis (A). This is true in the Solar System, where the period is measured in Earth years, and the semimajor axis is measured in AU.

Table 3.2 lists the periods and semimajor axes of the orbits of the eight classical planets and three of the dwarf planets, along with the values of the ratio P^2 divided by A^3. Those data are also plotted in **Figure 3.16**. Kepler referred to that relationship as his **harmonic law** or, more poetically, as the "Harmony of the Worlds." Kepler's third law is explored further in **Working It Out 3.2**. Kepler's laws enhanced the heliocentric mathematical model of Copernicus and led to its greater acceptance.

○ **what if . . .**

What if you read online that "experts have discovered a new planet with a distance from the Sun of 2 AU and a period of 4 years"? Use Kepler's third law to argue that this is impossible.

CHECK YOUR UNDERSTANDING **3.2**

Order the following from largest to smallest semimajor axis: (a) a planet with a period of 84 Earth days; (b) a planet with a period of 1 Earth year; (c) a planet with a period of 2 Earth years; (d) a planet with a period of 0.5 Earth years.

3.3 Galileo's Observations Supported the Heliocentric Model

Galileo Galilei (1564–1642—**Figure 3.17**) was the first to use a telescope to make and report significant discoveries about astronomical objects. Galileo's telescopes were small, yet they were sufficient for him to observe spots on the Sun, the uneven

Figure 3.17
Galileo Galilei laid the physical framework for Newton's laws.

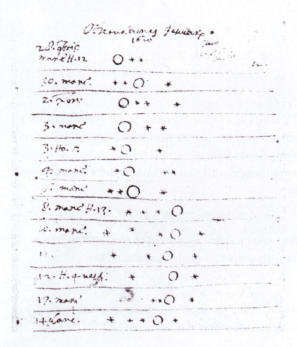

Figure 3.18 A page from Galileo's notebook shows his observations of Jupiter's four largest moons.

surface and craters of the Moon, and the many stars in the band of light in the sky called the Milky Way.

Galileo's Observations

Galileo provided the first observational evidence that some objects in the sky do not orbit Earth. When Galileo turned his telescope on Jupiter, he observed four "stars" in a line near the planet. Over time he observed that the objects remained near Jupiter, but changed position from night to night (**Figure 3.18**). Galileo hypothesized that those objects were moons orbiting Jupiter. Those four "Galilean moons" are the largest of Jupiter's many moons. Galileo also estimated the relative distance of each moon from Jupiter and the periods of their orbits, and he showed that the square of the period was proportional to the cube of the radius of the orbit for each moon, consistent with Kepler's third law.

Galileo also observed that Venus went through an entire set of phases like the Moon. He noticed that as the phases of Venus changed, so did the size of Venus in his telescope. In a geocentric model in which Venus orbits Earth like the Moon does, the apparent size of Venus would change only slightly, when it looped an epicycle. In the heliocentric model, however, the Earth–Venus distance varies by a lot, and Venus's size changes accordingly. When Venus is in its gibbous to full phases, it is farther away, on the other side of the Sun from Earth, and smaller in the sky. When Venus is in its crescent to new phases, it is closer, on the same side of the Sun as Earth, and larger in the sky (**Figure 3.19**). In the experiment that opens this chapter, you can build a model that demonstrates this relationship. Galileo's observations of Jupiter's moons and the phases of Venus in particular convinced Galileo that Copernicus was correct to place the Sun at the center of the Solar System.

In addition to his astronomical observations, Galileo did important work on the motion of objects. Unlike natural philosophers, who thought about objects in motion but did not actually experiment with them, Galileo conducted experiments with falling and rolling objects. As with his telescopes, Galileo improved or developed new technology to enable him to conduct those experiments. For example, by carefully rolling balls down an inclined plane and by dropping various objects from a height, he found that a falling object travels a distance proportional to the square of the time it has been falling. If he simultaneously dropped two objects of different masses, they reached the ground at the same time, showing that all objects falling to Earth accelerate at the same rate, independent of their mass.

Galileo's observations and experiments with many types of moving objects, such as carts and balls, led him to disagree with the Greek philosophers about when and why objects continue to move or come to rest. Before Galileo, it was thought that an object's natural state was to be at rest. But he found that an object naturally does what it was doing until a force acts on it. That is, an object in motion continues moving along a straight line with a constant speed until a force acts on it to change that state of motion. That idea of *inertia*, which Newton later adopted as his first law of motion, has implications for not only the motion of carts and balls but also the orbits of planets.

Figure 3.19 Modern photographs of the phases of Venus show that when we see Venus more illuminated, it also appears smaller, implying that Venus is farther away then.

Dialogue Concerning the Two Chief World Systems

Galileo faced considerable danger because of his work. His later life was consumed by conflict with the Catholic Church over his support of the Copernican system. In 1632, Galileo published his best-selling book, *Dialogo sopra i due massimi sistemi del mondo* ("Dialogue Concerning the Two Chief World Systems"). The *Dialogo* presents a brilliant philosopher named Salviati as the champion of the Copernican heliocentric view of the universe. The defender of an Earth-centered universe, Simplicio—who uses arguments made by the classical Greek philosophers and the pope—sounds silly and ignorant.

Galileo, a religious man with two daughters in a convent, thought he had the Catholic Church's tacit approval for his book. But when he placed several of the pope's geocentric arguments in the unflattering mouth of Simplicio, the perceived attack on the pope got the church's attention. Galileo was put on trial for heresy, sentenced to prison, and eventually placed under house arrest. To escape a harsher sentence, Galileo was forced to publicly recant his belief in the Copernican theory that Earth moves around the Sun. According to one story, as he left the courtroom, Galileo stamped his foot on the ground and muttered, "And yet it moves!"

The *Dialogo* was placed on the pope's Index of Prohibited Books, along with Copernicus's *De Revolutionibus*. Nevertheless, Galileo's work traveled across Europe, was translated into other languages, and was read by other scientists. Galileo spent his final years compiling his research on inertia and other ideas into the book *Discourses and Mathematical Demonstrations Relating to Two New Sciences*, which was published in 1638 in Holland, outside the Catholic Church's jurisdiction. (In 1992, Pope John Paul II apologized for the "Galileo Case.")

CHECK YOUR UNDERSTANDING **3.3**

> Which of Galileo's astronomical observations did Copernicus's model explain better than Ptolemy's? (a) sunspots; (b) craters on the Moon; (c) the moons of Jupiter; (d) the apparent size and phases of Venus

3.4   Newton's Three Laws Help Explain How Celestial Bodies Move

Empirical laws, such as Kepler's laws, describe *what* happens, but they do not explain *why*. Kepler described the orbits of planets as ellipses, but he did not explain why they should be so. To take that next step in the scientific process, scientists use basic physical principles and the tools of mathematics to derive the empirically determined laws. Alternatively, a scientist might start with physical laws and predict relationships, which are then verified or falsified through experiment and observation. If those predictions are verified, the scientist may have determined something fundamental about how the universe works.

Sir Isaac Newton (1642–1727—**Figure 3.20**) took that next step in explaining the nature of motion. Newton was a student of mathematics at Cambridge University when it closed because of the Great Plague and students were sent home to the safer countryside. Over the next 2 years, he studied on his own, and at the age of 23 he invented calculus, which would become crucial to his development of the physics of motion. (The German mathematician Gottfried Leibniz independently developed calculus around the same time.)

▶‖ **AstroTour:** Velocity, Acceleration, Inertia

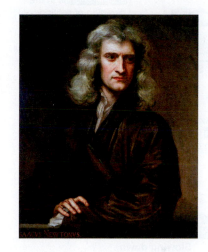

Figure 3.20 Sir Isaac Newton formulated three laws of motion.

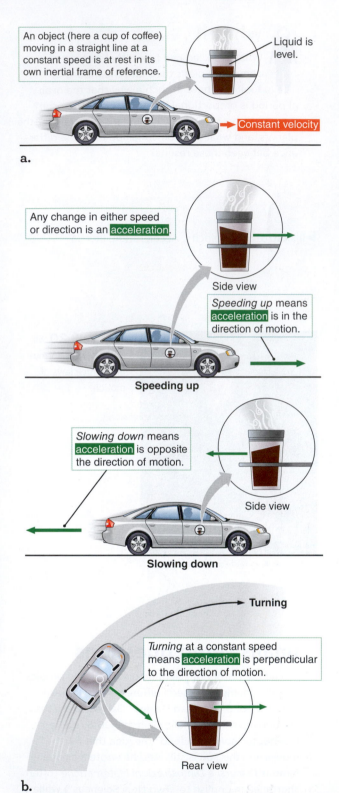

An object (here a cup of coffee) moving in a straight line at a constant speed is at rest in its own inertial frame of reference.

Liquid is level.

Constant velocity

a.

Any change in either speed or direction is an acceleration.

Side view

Speeding up means acceleration is in the direction of motion.

Speeding up

Slowing down means acceleration is opposite the direction of motion.

Side view

Slowing down

Turning

Turning at a constant speed means acceleration is perpendicular to the direction of motion.

Rear view

b.

Figure 3.21 a. An object moving in a straight line at a constant speed is at rest in its own inertial frame of reference. **b.** Any change in an object's velocity is an acceleration. When you are driving, for example, any time your speed changes or you follow a curve in the road, you are accelerating. (Throughout the text, velocity arrows are red and acceleration arrows are green.)

Building on the work of Kepler, Galileo, and others, Newton proposed three physical laws that govern the motions of all objects in the sky and on Earth. To understand how the planets and all other celestial bodies move, you must understand these three laws.

Newton's First Law: Objects at Rest Stay at Rest; Objects in Motion Stay in Motion

A **force** (***F***) is a push or a pull on an object. Two or more forces can oppose one another such that they are perfectly balanced and cancel out. For example, gravity pulls down on you as you sit in your chair, but the chair pushes up on you with an exactly equal and opposite force. As a result, you stay motionless. Forces that cancel out do not affect an object's motion. When forces combine to produce an effect, we often use the term *net force*, or sometimes just *force*.

Imagine that you are driving a car, and your book is on the seat next to you. A rabbit runs across the road in front of you, so you hit the brakes hard. You feel the seat belt tighten to restrain you. At the same time, your book flies off the seat and hits the dashboard. You have just experienced what Newton describes in his first law of motion. **Inertia** is an object's tendency to maintain its state—either of uniform motion or of rest—until a net force pushes or pulls it. In the stopping car, you did not hit the dashboard because the force of the seat belt slowed you down. The book, however, hit the dashboard because no such force acted upon it.

Newton's first law of motion describes inertia and states that an object in motion tends to stay in motion, in the same direction, until a net force acts upon it; and an object at rest tends to stay at rest until a net force acts upon it. Galileo's law of inertia became the cornerstone of physics as Newton's first law.

Recall from Section 2.1 the concept of a frame of reference. Within a frame of reference, only the relative motions between objects have any meaning. Without external clues, you cannot tell the difference between sitting still and traveling at constant speed in a straight line. For example, if you close your eyes while riding in the passenger seat of a quiet car on a smooth road, you feel as though you are sitting still. For you, the book on the seat beside you was "at rest," whereas a person standing by the side of the road would see the book moving past at the same speed as you and the car. People in a car approaching you would see the book moving even faster than the person by the side of the road—namely, at the speed they are traveling plus the speed you are traveling! All those perspectives are equally valid, and all those speeds of the book are correct when measured in the appropriate reference frame.

A reference frame moving in a straight line at a constant speed is called an **inertial frame of reference**. Any inertial frame of reference is as good as another. In the inertial frame of reference of a cup of coffee, for example, **Figure 3.21a** shows the cup is at rest in its own frame even if the car is moving quickly down the road.

Newton's Second Law: Motion Is Changed by Forces

What happens if a net force does act? In the earlier example, you were traveling in the car, and your motion slowed when the force of the seat belt acted upon you. Forces change an object's motion—by changing either the speed or the direction. That effect reflects **Newton's second law of motion**: if a net force acts on an object, the object's motion changes.

In the driver's seat of a car, you have several controls, including an accelerator and a brake pedal, which you use to make the car speed up or slow down. A *change in speed* is one way the car's motion can change. But you also have the steering wheel in your hands. When you are moving down the road and you turn the wheel, your speed does not necessarily change, but the direction of your motion does. A *change in direction* also is a kind of change in motion.

Together, an object's speed and direction are called **velocity (*v*)**. "Traveling at 50 kilometers per hour (km/h)" indicates speed, whereas "traveling north at 50 km/h" indicates velocity. **Acceleration (*a*)** describes the changes in an object's velocity. For example, if you go from 0 to 100 km/h in 4 seconds, you feel a strong push from the seat back as it shoves your body forward, causing you to accelerate along with the car. However, if you take 2 minutes to go from 0 to 100 km/h, the acceleration is so slight that you hardly notice it.

Partly because a car's gas pedal is often called the accelerator, some people think *acceleration* always means that an object is speeding up. In physics, however, *any change in speed or direction is an acceleration.* **Figure 3.21b** illustrates that point by showing what happens to the coffee in a cup as the car speeds up, slows down, or turns. Slamming on your brakes and going from 100 to 0 km/h in 4 seconds is just as much an acceleration as going from 0 to 100 km/h in 4 seconds. Similarly, the acceleration you experience as you go through a tight turn at a constant speed is every bit as real as the acceleration you feel when you slam your foot on the accelerator or the brake. Whether speeding up, slowing down, turning left, or turning right—if you are not moving in a straight line at a constant speed, you are accelerating.

Newton's second law of motion says that a net force causes acceleration. An object's acceleration depends on two things. First, as shown in **Figure 3.22**, the acceleration depends on the strength of the net force acting on the object to change its motion. If the forces acting on the object do *not* add up to zero, a net force is present and the object accelerates (Figure 3.22a). The stronger the net force, the greater the acceleration (Figure 3.22b). If you push on something twice as hard, it experiences twice as much acceleration. Push on something three times as hard and its acceleration will be three times as great. The acceleration occurs in the direction the net force points. Push an object away from you, and it will accelerate away from you.

An object's acceleration also depends on its inertia. You can push some objects easily, such as an empty box from a new refrigerator. But you can't easily shove an actual refrigerator around, even though it is about the same size as the box. Figure 3.22c shows that the greater the mass, the greater the inertia, and the *less* acceleration that will occur in response to the same net force. That relationship among acceleration (*a*), force (*F*), and mass (*m*) is expressed mathematically in **Working It Out 3.3**.

Newton's Third Law: Whatever Gets Pushed Pushes Back

Imagine that you are standing on a skateboard and pushing yourself along with your foot. Each shove of your foot against the ground sends you faster along your way. But why? You accelerate because as you push on the ground, the ground pushes back on you.

Part of Newton's genius was his ability to see patterns in such everyday events. Newton realized that *every* time one object exerts a force on another, the second object exerts a matching force on the first. That second force is as strong as the first force but is in the opposite direction. When you accelerate yourself on the skateboard,

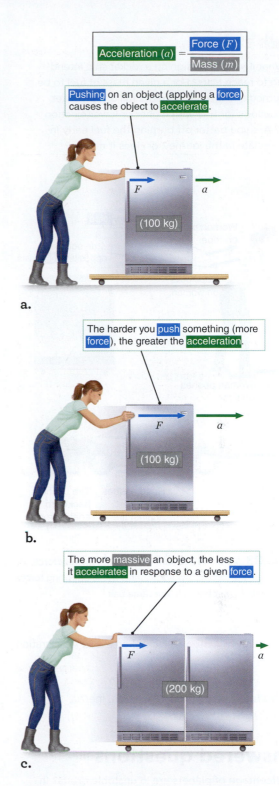

$$\text{Acceleration } (a) = \frac{\text{Force } (F)}{\text{Mass } (m)}$$

Pushing on an object (applying a force) causes the object to accelerate.

a.

The harder you push something (more force), the greater the acceleration.

(100 kg)

b.

The more massive an object, the less it accelerates in response to a given force.

(200 kg)

c.

Figure 3.22 According to Newton's second law of motion, an object's acceleration is the force acting on the object divided by the object's mass. (Throughout the text, force arrows are blue.)

what if . . .

What if you are designing a rocket ship intending to reach Mars? For a given mass of fuel to be burned and ejected from the tail of the rocket, should you look for fuel with higher or lower ejection velocity? Are you better off burning the fuel early in the journey, late in the journey, or does it matter?

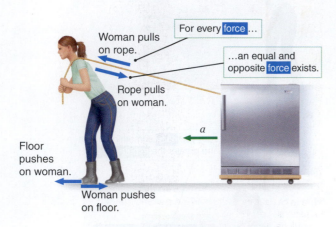

Figure 3.23 Newton's third law states that for every force, an equal and opposite force is always present. Those opposing forces always act on the two objects in the same pair.

🎥 **Astronomy in Action:** Velocity, Force, and Acceleration

▶❙❙ **AstroTour:** Velocity, Acceleration, Inertia

unanswered questions

What percentage of planets are in unstable orbits? In younger planetary systems, planets might migrate in their orbits because of the presence of other massive planets nearby. We explain in Chapter 7 that Uranus and Neptune might have migrated in that way. Some planets have been discovered moving through the galaxy without any obvious orbit around a star. These planets did not originate in the stable orbits we see in our own Solar System.

you push backward on Earth, and Earth pushes you forward. As shown in **Figure 3.23**, when a woman moves a load on a cart by pulling a rope, the rope pulls back, and when a car tire pushes back on the road, the road pushes forward on the tire. Similarly, when Earth pulls on the Moon, the Moon pulls on Earth, and when a rocket engine pushes hot gases out of its nozzle, those hot gases push back on the rocket, propelling it into space.

All those force pairs exemplify **Newton's third law of motion**, which says forces always come in pairs, with those two forces always equal in strength but opposite in direction. The forces in those pairs always act on two objects. Your weight pushes down on the floor, and the floor pushes back up on you with the same amount of force. For every force, an equal force in the opposite direction is *always* there.

CHECK YOUR UNDERSTANDING 3.4

If a planet moves in a perfectly circular orbit around the Sun, is that planet accelerating? (a) Yes, because it is constantly changing its speed. (b) Yes, because it is constantly changing its direction. (c) No, because its speed is not constantly changing. (d) No, because planets do not accelerate.

Origins: Planets and Orbits

In addition to the planets in our own Solar System, researchers have detected thousands of planets orbiting stars other than our own Sun. Those planets' orbits can be calculated and understood by applying the same three Kepler's laws that we have discussed here. Astrobiologists think that the distance of a planet's orbit around its star affects the chances of life developing.

A planet close to its star receives more energy than a planet farther out. If Earth were closer to the Sun, it would be hotter throughout the year—probably so hot that water would evaporate and no longer exist as a liquid. If Earth were farther from the Sun, it would be colder and all surface water would probably freeze. Liquid water was crucial for life to form on Earth, so some astronomers look for planets at a distance from their star such that liquid water can exist (that distance varies depending on the star's temperature and size).

Next, recall from Figure 3.14a that Earth's elliptical orbit differs from a circle by less than 2 percent. Thus, Earth's distance from the Sun varies by only about 5 million km throughout the year; and, as we saw in Chapter 2, seasons change on Earth because of the tilt of Earth's axis, not because of slight changes in distance from the Sun. Mars has about the same axial tilt as Earth, but the orbit of Mars is more eccentric, so Mars has greater seasonal variation. The distance between Mars and the Sun varies from 1.38 AU (207 million km) to 1.67 AU (249 million km)—an eccentricity of 9 percent (or ~40 million km). As a result, the seasons on Mars are unequal. They are shorter when Mars is closer to the Sun and moving faster, and longer when Mars is farther from the Sun and moving slower. The inequality of the martian seasons affects the overall stability of the planet's temperature and climate. When we look at planets orbiting other stars, we see that many have orbital eccentricities even higher than that of Mars and therefore large variations in temperature.

Earth is at the right distance from the Sun to have temperatures that permit water to be liquid, and its orbital eccentricity is low enough that the average planetary temperature does not change much during its orbit. Those orbital characteristics have contributed to making the conditions on Earth suitable for life to develop.

working it out 3.3

Using Newton's Laws

Your acceleration is calculated by dividing how much your velocity changes by how long that change takes to happen:

$$\text{Acceleration} = \frac{\text{How much velocity changes}}{\text{How long the change takes to happen}}$$

For example, if an object's speed goes from 5 to 15 meters per second (m/s), the change in velocity is 10 m/s. If that change happens in 2 seconds, the acceleration is given by

$$a = \frac{15\,\text{m/s} - 5\,\text{m/s}}{2\,\text{s}} = 5\,\text{m/s}^2$$

To determine how an object's motion is changing, we need to know two things: what net force is acting on the object and the object's resistance to that force. We can put that idea into equation form as follows:

$$(\text{An object's acceleration}) = \frac{(\textit{The force acting to change the object's motion})}{(\textit{The object's resistance to that change})}$$
$$= \frac{Force}{Mass}$$

Newton's second law above is often written as Force = mass × acceleration, or $F = ma$. The units of force are called **newtons (N)**, so $1\,\text{N} = 1\,\text{kg m/s}^2$.

Suppose you are holding two blocks of the same size, but the block in your right hand has twice the mass of the block in your left hand. When you drop the blocks, they both fall with the same acceleration, as Galileo showed, and they hit your two feet at the same time. Which will hit with more force: the block falling onto your right foot or the one falling onto your left foot? The block in your right hand, with twice the mass, will hit your right foot with twice the force that the other block hits your left foot.

To see how Newton's three laws of motion work together, study **Figure 3.24**. An astronaut is adrift in space, motionless with respect to the nearby space station. With no tether to pull on, how can the astronaut get back to the station? Suppose the 100-kg astronaut throws a 1-kg wrench directly away from the station at a speed of 10 m/s. Newton's second law says that to change the wrench's motion, the astronaut must apply a force to it in the direction away from the station. Newton's third law says that the wrench must therefore push back on the astronaut with as much force but in the opposite direction. The force of the wrench on the astronaut causes the astronaut to begin drifting toward the station. How fast will the astronaut move? Turn to Newton's second law again. Because the astronaut has more mass, he or she will accelerate less than the wrench will. A force that causes the 1-kg wrench to accelerate

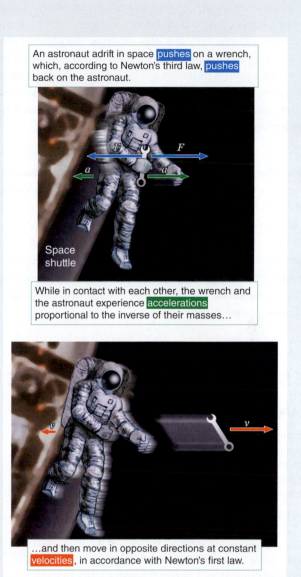

An astronaut adrift in space pushes on a wrench, which, according to Newton's third law, pushes back on the astronaut.

Space shuttle

While in contact with each other, the wrench and the astronaut experience accelerations proportional to the inverse of their masses…

…and then move in opposite directions at constant velocities, in accordance with Newton's first law.

Figure 3.24 According to Newton's laws, if an astronaut adrift in space throws a wrench, the two will move in opposite directions. Their speeds will depend on their masses: the same force will produce a smaller acceleration of a more massive object than of a less massive object. (Acceleration and velocity arrows not drawn to scale.)

to 10 m/s will have much less effect on the 100-kg astronaut. Because acceleration equals force divided by mass, the 100-kg astronaut will experience only 1/100 as much acceleration as the 1-kg wrench. The astronaut will drift toward the station, but only at the leisurely rate of 1/100 × 10 m/s, or 0.1 m/s.

SUMMARY

Early astronomers hypothesized that Earth was stationary at the center of the Solar System. Later astronomers realized that a Sun-centered Solar System was much simpler and could explain what people observed in the sky. Planets, such as Jupiter, orbit the Sun, not Earth. Kepler's laws describe the planets' elliptical orbits around the Sun, including details about how fast a planet travels at various points in its orbit. Those laws helped Newton develop his laws of motion, which govern how all objects move (not just orbiting ones). Orbital semimajor axis, eccentricity, and stability may affect a planet's suitability to foster life.

(1) **Compare the geocentric and heliocentric models of the Solar System.** Earth's motion is hard to detect, so before the Copernican Revolution, most people accepted a geocentric model of the Solar System, in which all objects orbit Earth. But in that model, the planets' apparent retrograde motion was hard to explain. Copernicus created the first comprehensive mathematical model of the Solar System with the Sun at the center, called a heliocentric model. His model explained apparent retrograde motion as a phenomenon caused when an inner planet passes an outer planet in their orbits.

(2) **Use Kepler's laws to describe how objects in the Solar System move.** Using Tycho's observational data, Kepler developed empirical rules to describe the motions of the planets. Kepler's three laws state that (1) planets move in elliptical orbits around the Sun; (2) planets move fastest when closest to the Sun and slowest when farthest from the Sun, so that the planets sweep out equal areas in equal times; and (3) the orbital period (P) of a planet squared equals the semimajor axis (A) of its orbit cubed: $P^2 = A^3$.

(3) **Explain how Galileo's astronomical discoveries convinced him that the heliocentric model was correct.** Galileo used telescopes to observe sunspots, craters on the Moon, and moons orbiting Jupiter. He also saw Venus going through phases, like the Moon, but changing its apparent size in each phase. The observations of Venus were hard to explain with Ptolemy's geocentric model but were consistent with a heliocentric model.

(4) **Describe the physical laws of motion discovered by Galileo and Newton.** Galileo studied the physics of falling objects and discovered the principle of inertia. Newton's laws state that (1) objects do not change their motion unless they experience a net force, (2) force = mass × acceleration, and (3) every force has an equal and opposite force. Net forces cause accelerations (changes in motion). Inertia resists changes in motion.

QUESTIONS AND PROBLEMS

TEST YOUR UNDERSTANDING

1. An *empirical science* is one based on
   a. hypothesis.
   b. calculus.
   c. computer models.
   d. observed data.

2. When Earth catches up to a slower-moving outer planet and passes it in its orbit, the outer planet
   a. exhibits retrograde motion.
   b. slows down because it feels Earth's gravitational pull.
   c. becomes dimmer as it passes through Earth's shadow.
   d. moves into a more elliptical orbit.

3. Copernicus's model of the Solar System was superior to Ptolemy's because
   a. it had a mathematical basis and computed the spacing of the planets.
   b. it was much more accurate.
   c. it did not require epicycles.
   d. it fit the telescopic data better.

4. An inferior planet is one that is
   a. smaller than Earth.
   b. larger than Earth.
   c. closer to the Sun than Earth.
   d. farther from the Sun than Earth.

5. The time a planet takes to come back to the same position in space in relation to the Sun is called its _____ period.
   a. synodic
   b. sidereal
   c. heliocentric
   d. geocentric

6. Suppose a planet is discovered orbiting a star in a highly elliptical orbit. While the planet is close to the star, it moves _____, but while it is far away, it moves _____.
   a. faster; slower
   b. slower; faster
   c. retrograde; prograde
   d. prograde; retrograde

7. If a superior planet is observed from Earth to have a synodic period of 1.2 years, what is its sidereal period?
   a. 0.54 years
   b. 1.8 years
   c. 4.0 years
   d. 6.0 years

8. A net force must be acting when an object
   a. accelerates.
   b. changes direction but not speed.
   c. changes speed but not direction.
   d. all of the above

9. For Earth, $P^2/A^3 = 1.0$ (in appropriate units). Suppose a new dwarf planet is discovered that is 14 times farther from the Sun than Earth is. For that planet,
   a. $P^2/A^3 = 1.0$.
   b. $P^2/A^3 > 1.0$.
   c. $P^2/A^3 < 1.0$.
   d. we can't know the value of P^2/A^3 without more information.

10. Galileo observed that Venus had phases that correlated with the size of its image in his telescope. From that information, you may conclude that Venus
    a. is the center of the Solar System.
    b. orbits the Sun.
    c. orbits Earth.
    d. orbits the Moon.

11. Kepler's second law says that
    a. planetary orbits are ellipses with the Sun at one focus.
    b. the square of a planet's orbital period equals the cube of its semimajor axis.
    c. net forces cause changes in motion.
    d. planets move fastest when closest to the Sun.

12. The average speed of a new planet in orbit has been reported to be 33 km/s. When it is closest to its star it moves at 31 km/s, and when it is farthest from its star it moves at 35 km/s. Those data must be in error because
    a. the average speed is far too fast.
    b. Kepler's third law says the planet has to sweep out equal areas in equal times, so the speed of the planet cannot change.
    c. Kepler's second law says the planet must move fastest when closest to its star, not when farthest away.
    d. with those numbers, the square of the orbital period will not be equal to the cube of the semimajor axis.

13. Galileo observed that Jupiter has moons. From that information, you may conclude that
    a. Jupiter is the center of the Solar System.
    b. Jupiter orbits the Sun.
    c. Jupiter orbits Earth.
    d. some things do not orbit Earth.

14. If you start from rest and accelerate at 10 mph/s and end up traveling at 60 mph, how long did it take?
    a. 1 second
    b. 6 seconds
    c. 60 seconds
    d. 0.6 seconds

15. Planets with high eccentricity may be unlikely candidates for life because
    a. the speed varies too much.
    b. the period varies too much.
    c. the temperature varies too much.
    d. the orbit varies too much.

THINKING ABOUT THE CONCEPTS

16. ★ **WHAT AN ASTRONOMER SEES** In Figure 3.2, Mars appears larger in the part of the loop that is retrograde, when Mars is moving from left to right. Why does Mars appear larger at this time? What conclusion can you make about the cause of the changing size of Venus in Figure 3.19?

17. Study Figure 3.2. During normal motion, does Mars move toward the east or west? Which direction does it travel when in apparent retrograde motion? For how many days did Mars move in apparent retrograde motion? If one of the martian missions were photographing *Earth* in the sky during those days, what would it have observed?

18. Copernicus and Kepler engaged in what is called empirical science. What do we mean by *empirical*?

19. Explain why Saturn's synodic period is very close to 1 Earth year (a sketch may help).

20. Experiment with falling objects as Galileo did. When you drop pairs of objects with different masses, do they reach the ground at the same time? Do they hit the ground with the same force? Does that approach work with a sheet of paper or a tissue—why or why not?

21. A planet's orbital speed around the Sun varies. When is the planet moving fastest in its orbit? When is it moving slowest?

22. The Moon's orbit around Earth is elliptical, too, with an eccentricity of 0.05. How does that value compare with the eccentricity of Earth's orbit? How do those elliptical orbits explain the types of solar eclipses discussed in Chapter 2?

23. Galileo came up with the concept of inertia. What do we mean by *inertia*?

24. If Kepler had lived on Mars, would he have deduced the same empirical laws for the motion of the planets? Explain.

25. What is the difference between speed and velocity? Between velocity and acceleration?

26. When involved in an automobile collision, a person not wearing a seat belt will move through the car and often strike the windshield directly. Which of Newton's laws explains why the person continues forward, even though the car stopped?

27. When riding in a car, we can sense changes in speed or direction through the forces that the car applies on us. Do we wear seat belts in cars and airplanes to protect us from speed or from acceleration? Explain your answer.

28. An astronaut standing on Earth can easily lift a wrench having a mass of 1 kg, but not a scientific instrument with a mass of 100 kg. On the International Space Station, an astronaut can manipulate both, although the scientific instrument responds much more slowly than the wrench. Explain why.

29. The Process of Science Figure illustrates that scientific ideas are always open to challenge. Construct an argument that the constant process of challenging and falsifying ideas is a strength of science, rather than a weakness. ★

30. How might you expect conditions on Earth to be different if the eccentricity of its orbit were 0.17 instead of 0.017?

APPLYING THE CONCEPTS

31. Dwarf planet Ceres is 2.77 AU from the Sun. Its synodic period is 1.278 years. What is its sidereal period? ●—●—●
    a. Make a prediction: There is enough information in this question to calculate the sidereal (orbital) period in two different ways! How do you expect those answers to compare?
    b. Calculate: Use Working It Out 3.1 to find the sidereal period in years.
    c. Calculate: Use Kepler's law to find the sidereal period in years.
    d. Check your work: Compare your results for (a) and (b).

32. Suppose you are riding an electric scooter with a top speed of 32 km/hr. A typical electric scooter takes about ten seconds to accelerate to this speed. What is the acceleration of an electric scooter? ●—●—●
    a. Make a prediction: Do you expect this acceleration to be larger or smaller than the acceleration due to gravity (9.8 m/s²)? You may think you have no idea about this, but if you compare riding an electric scooter to falling off of a building, you might be able to guess.
    b. Calculate: Follow Working It Out 3.3 to calculate the acceleration of the scooter. Be careful about the units you are using!
    c. Check your work: Compare your answer to the acceleration due to gravity on Earth (9.8 m/s²).

33. Venus has a synodic period of 1.6 years. What is the sidereal period of Venus?
    a. Make a prediction: Is Venus closer or farther from the Sun than Earth? Therefore, do you expect its sidereal (orbital) period to be longer or shorter than Earth's? Will it be much longer or shorter?
    b. Follow Working It Out 3.1 to find the sidereal period, P, of Venus. Choose carefully which equation to use for Venus.
    c. Check your work by comparing your answer to your prediction.
    d. Repeat these steps and this calculation for Mars, which has a synodic period of 2.1 years. Again, think carefully before choosing which equation to use from Working It Out 3.1.

34. Suppose a dwarf planet is discovered orbiting the Sun with a semimajor axis of 50 AU. What would be the planet's orbital period?
    a. Make a prediction: Table 3.2 lists the period and semimajor axis of many objects in the Solar System. Examine this table to predict whether this dwarf planet will have an orbital period in the tens, hundreds, or thousands of years. ★
    b. Calculate: Use Kepler's third law to find the orbital period in years.
    c. Check your work: Compare your answer to the period of some of the objects in Table 3.2. ★

35. Suppose you are pushing a small refrigerator of mass 50 kg on wheels. You push with a force of 100 N. Assume the refrigerator starts at rest. How long will the refrigerator accelerate before it is moving faster than you can run (about 10 m/s)?
    a. Make a prediction: Do you expect this time to be small (a few seconds) or large (tens of minutes)? Study Working It Out 3.3. This is a two-step problem; you'll first have to find the acceleration, and then find the time. What units do you expect for each of these answers? Do you expect the acceleration due to your push to be larger or smaller than the acceleration due to Earth's gravity (9.8 m/s₂)?
    b. Calculate: What is the refrigerator's acceleration?
    c. Check your work: Does your acceleration have the correct units (m/s₂), and is this a reasonable acceleration?
    d. Calculate: Assume the refrigerator starts at rest. How long will the refrigerator accelerate before it is traveling at 10 m/s?
    e. Check your work: Does your answer have the correct units of seconds, and is it a reasonable length of time?

36. Is the graph in Figure 3.16 linear or logarithmic? From the data on the graph, find the approximate semimajor axis and period of Saturn. Show your work. ★

37. Study Figure 3.19, which shows that Venus's apparent size changes as it goes through phases. Approximately how many times larger is Venus in the sky at the thinnest crescent than at the gibbous phase shown? Approximately how many times closer is Venus to us at the phase of that thinnest crescent than at the gibbous phase? ★

38. The orbital period of Uranus is 84 years. Compute the semimajor axis of its orbit. How much time passes between oppositions of Uranus?

39. Suppose you read online that "experts have discovered a new planet with a distance from the Sun of 4 AU and a period of 3 years." Use Kepler's third law to argue that this is impossible.

40. Show, as Galileo did, that Kepler's third law applies to the four Galilean moons of Jupiter by calculating P^2 divided by A^3 for each moon. (Data on the moons can be found in Appendix 4.)

41. In a period of 3 months, a planet travels 30,000 km with an average speed of 3.8 m/s. Some time later, the same planet travels 65,000 km in 3 months. How fast is the planet traveling at that later time? During which period is the planet closer to the Sun?

42. If you were on Mars, how often would you see retrograde motion of *Earth* in the martian night sky? (You can view a picture at https://mars.jpl.nasa.gov/allaboutmars/nightsky /retrograde.)

43. The elliptical orbit of a comet that a spacecraft recently visited is 1.24 AU from the Sun at its closest approach and 5.68 AU from the Sun at its farthest.
    a. Sketch the comet's orbit. When is it moving fastest? Slowest?
    b. What is the semimajor axis of its orbit? How long does the comet take to go around the Sun?
    c. What is the distance from the Sun to the "center" of the ellipse? What is the eccentricity of the comet's orbit?

44. You are driving down a straight road at a speed of 90 km/h, and you see another car approaching you at a speed of 110 km/h along the road.
    a. With respect to your own frame of reference, how fast is the other car approaching you?
    b. With respect to the other driver's frame of reference, how fast are you approaching the other driver's car?

45. Sketch the orbit of Mars by using the information provided in the "Origins: Planets and Orbits" section of the chapter for the closest and farthest distances of Mars from the Sun.
    a. What is the orbit's major axis? Its semimajor axis?
    b. What is the distance from the "center" of the orbit to the Sun? Compute the orbit's eccentricity. Compare that value with the eccentricity of Earth's orbit.

EXPLORATION Kepler's Laws

digital.wwnorton.com/astro7

In this Exploration, we examine how Kepler's laws apply to the orbit of Mercury. Visit the Digital Resources Page and on the Student Site open the "Planetary Orbits" Interactive Simulation in Chapter 3. That simulator animates the orbits of the planets, enabling you to control the simulation speed, as well as several other parameters. Here we focus on exploring Mercury's orbit, but you may wish to spend some time examining the orbits of other planets as well.

Kepler's First Law

In the top left panel, use the drop-down menu to select "Mercury" and then click "OK." Click the "Kepler's First Law" tab at the bottom of the control panel. Use the radio buttons to select "Show Empty Focus" and "Show Center."

1. How would you describe the shape of Mercury's orbit?

_____

_____

Deselect "Show Empty Focus" and "Show Center," and select "Show Semiminor Axis" and "Show Semimajor Axis." Under "Visualization Options," select "Show Grid."

2. Use the grid markings to estimate the ratio of the semiminor axis to the semimajor axis.

_____

_____

3. Calculate the eccentricity of Mercury's orbit from that ratio by using $e = [1 - (\text{Ratio})^2]^{1/2}$.

_____

_____

Kepler's Second Law

Click on "Reset" in the bottom left panel, set parameters for Mercury, and click "OK." Then click on the "Kepler's Second Law" tab at the bottom of the control panel. Click "Play."

4. Observe the speed of the planet in its orbit. Where does the planet travel fastest? Where does the planet travel slowest?

_____

_____

Kepler's Third Law

Click on "Reset" in the bottom left panel, set parameters for Mercury, and then click on the "Kepler's Third Law" tab at the bottom of the control panel. Select "Show Solar System Orbits" in the "Visualization Options" panel. Study the graph. Use the eccentricity slider to change the simulated planet's eccentricity. Make the eccentricity first smaller and then larger.

5. Did anything in the graph change?

_____

_____

6. What do your observations of the graph tell you about the dependence of the period on the eccentricity?

_____

_____

Click on "Reset," and set the parameters back to those for Mercury. Now use the semimajor axis slider to change the semimajor axis of the simulated planet.

7. What happens to the period when you make the semimajor axis smaller?

_____

_____

8. What happens when you make it larger?

_____

_____

9. What do those results tell you about the dependence of the period on the semimajor axis?

_____

_____

Gravity

In order for an object to move in a circle, a force must act upon it. This is as true for objects held by a string as it is for objects held by gravity. Tie a piece of string around a water bottle or an object of similar weight. Before you whirl this object over your head, predict whether whirling faster or whirling slower will require more force to hold onto the string. Go outside and whirl the bottle around over your head. First, whirl the bottle just fast enough so that the circle is horizontal. Then, whirl it as fast as you can. Compare how difficult it is to hold onto the string in each case. Afterward, make a sketch that shows the relative sizes of the force and velocity arrows for each of these two cases.

EXPERIMENT SETUP

SLOWER

FASTER

First, spin the bottle just fast enough so that the circle is horizontal.

Then, spin it as fast as you can.

PREDICTION

I predict that the string will be ☐ **easier** ☐ **harder** to hold when the water bottle is faster.

SKETCH OF RESULTS

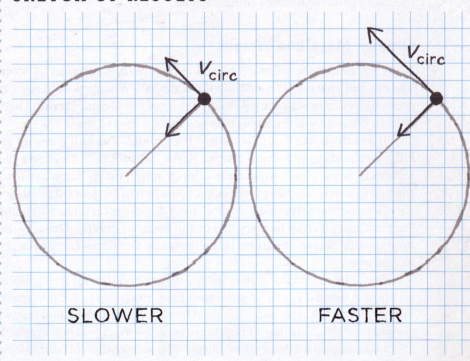

V_{circ}

V_{circ}

SLOWER

FASTER

4

Here in Chapter 4, we explore the physical laws that explain the regular patterns in the motions of the planets. Because the Sun is far more massive than all the other parts of the Solar System combined, its gravity shapes the motions of every object in its vicinity, from the almost circular orbits of some planets to the extremely elongated orbits of comets.

LEARNING GOALS

By the end of this chapter, you should be able to:

(1) Identify the elements of Newton's universal law of gravitation.

(2) Use the laws of motion and gravitation to relate planetary orbits, trajectories of rocket flights, and the escape velocity from a system.

(3) Explain how gravitational forces from the Sun and Moon create Earth's tides.

(4) Describe how tidal forces affect solid bodies.

4.1 Gravity Is an Attractive Force between Objects

In Chapter 3, you learned about Kepler's work on the movement of the planets around the Sun and Newton's laws of motion. Here in Chapter 4, you will build on those concepts to understand Newton's universal law of gravitation. Although some properties of gravity were observed before Newton, his work connected the everyday phenomenon of falling objects to the motion of the planets around the Sun. Newton's law of gravity, when combined with Newton's own laws of motion, inevitably produce Kepler's empirical laws.

Gravity, Mass, and Weight

Many forces that we see in everyday life involve direct contact between objects. The cue ball on a pool table strikes the eight ball. When a person runs, their shoe presses directly against the pavement. When physical contact occurs between two objects, the source of the forces between them is easy to see.

If you drop a ball, it falls toward the ground. The ball picks up speed as it falls, accelerating downward toward Earth. According to Newton's second law, an accelerating object must have a force acting on it. But what force causes the ball to accelerate? The force causing the ball to fall toward Earth is an example of a force that acts at a distance without direct contact. The ball falling toward Earth is accelerating in response to the force of gravity. **Gravity**, one of the fundamental forces of nature, is the mutually attractive force between objects having mass.

Recall from Chapter 3 that Galileo discovered that all freely falling objects accelerate toward Earth at the same rate, regardless of their mass. Drop a marble and a book at the same time and from the same height, and they will hit the ground together. Note that air resistance becomes a factor at higher speeds, but the effect is negligible for dense, slow objects. The acceleration of falling objects due to gravity near the surface of Earth, which Galileo also measured experimentally, is usually written as g (lowercase) and has an average value on the surface of Earth of 9.8 meters per second squared (m/s²).

Newton's second law (Section 3.4) states that acceleration (a) equals force (F) divided by mass (m), or $a = F/m$. The acceleration due to gravity can be the same for all objects only if the value of the force divided by the mass is the same for all objects. In other words, an object twice as massive has double the gravitational force acting on it, an object three times as massive has triple the gravitational force acting on it, and so on. If all objects, regardless of mass, fall with the same acceleration, the gravitational *force* on an object must be determined by the object's *mass*.

The gravitational force acting on an object attracted by a planet is called the object's **weight**. In common language, we often use the words *weight* and *mass* interchangeably. To be more scientifically precise, scientists use *mass* to refer to the amount of matter in an object and *weight* to refer to the force that the planet's gravitational pull exerts on that object. On the surface of Earth, weight equals mass multiplied by the acceleration of gravity at Earth's surface, g:

$$F_{weight} = m \times g$$

where F_{weight} is an object's weight in newtons (N), the metric unit of force; m is the object's mass in kilograms (kg); and g is Earth's constant for acceleration due to gravity, 9.8 m/s². On Earth, then, an object with a *mass* of 1 kg has a *weight* of 9.8 N (**Figure 4.1**).

Your mass is the same no matter what planet or moon you are on, but your weight will depend on the local acceleration due to gravity. On the Moon, for example, the acceleration due to gravity is 1.6 m/s², which is about one-sixth of its value on Earth, so a 1-kg mass has a weight of 1.6 N on the Moon. Similarly, your weight on the Moon would be about one-sixth of your weight on Earth.

The value of g varies slightly across Earth's surface, ranging from 9.78 m/s² at the equator to 9.83 m/s² at the poles. That variation exists because Earth is not a perfect sphere: its rotation makes it flatter at the poles, so the radius of Earth is smaller there. A similar effect operates in locations at high altitude, causing g to be slightly smaller there.

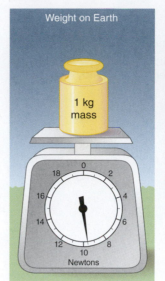

Figure 4.1 On the Moon, a mass of 1 kg has ⅙ the weight (displayed in newtons) that it has on Earth.

Newton's Law of Gravity

As Newton told the story, he saw an apple fall from a tree to the ground. He reasoned that if gravity is a force that depends on mass, a gravitational force should exist between *any* two masses, including between a falling apple and Earth. That great insight came from applying his third law of motion to gravity. According to Newton's third law (Chapter 3), every force is paired with another force that is equal and opposite to it. Therefore, if Earth exerts a force of 9.8 N on a 1-kg mass sitting on its surface, that 1-kg mass must also exert a force of 9.8 N on Earth. Drop a 7-kg bowling ball and it falls toward Earth, but at the same time Earth falls toward the 7-kg bowling ball. We don't notice the planet's motion because Earth is very massive, so it has a lot of inertia. In the time a 7-kg bowling ball takes to fall to the ground from a height of 1 kilometer (km), Earth has "fallen" toward the bowling ball by only a tiny fraction of the size of an atom.

Newton reasoned that if doubling the mass of any object doubles the gravitational force between the object and Earth, doubling the mass of Earth ought to do the same thing. In short, the gravitational force between Earth and an object equals the product of the two masses multiplied by *something*:

Gravitational force = Something × Mass of Earth × Mass of object

unanswered questions

What range of gravities will support human life? Humans have evolved to live on Earth's surface, in Earth's "surface gravity," but what happens when humans go elsewhere? What are the limits for our hearts, lungs, eyes, and bones? At the higher end of human tolerance, fighter pilots have been trained to experience about 10 times the normal surface gravity on Earth for very short periods (too long and they black out). Astronauts who spend several months in near-weightless conditions experience medical problems such as bone loss. On the Moon or Mars, humans will weigh much less than on Earth. Many science fiction tales have been written about what happens to children born on a space station or on another planet or moon with low surface gravity: would their hearts and bodies ever be able to adjust to the higher surface gravity of Earth, or must they stay in space forever?

If the mass of the object is 2 times greater, the force of gravity will be 2 times greater. Likewise, if the mass of Earth happened to be 3 times what it is, the force of gravity also would have to be 3 times greater. If the mass of both Earth and the object were greater by those amounts, the gravitational force would increase by a factor of 2×3, or 6 times. Because objects fall toward the center of Earth, we know that gravity is an attractive force acting along a line between the two masses.

If gravity is a force that depends on mass, a gravitational force should exist between *any* two masses. Suppose we have two masses—call them mass 1 and mass 2 (or m_1 and m_2 for short). The gravitational force between them is *something* multiplied by the product of the masses:

$$\text{Gravitational force between two objects} = \text{Something} \times m_1 \times m_2$$

We have gotten this far just by combining Galileo's observations of falling objects with (1) Newton's laws of motion and (2) Newton's belief that Earth is a mass just like any other mass. But what's that "something" in the previous expression?

Kepler had already thought about that question. He reasoned that because the Sun is the focal point for planetary orbits, the Sun must exert an influence over the motions of the planets. Kepler speculated that the influence must grow weaker with distance from the Sun because the planets closer to the Sun moved much faster than the ones farther away. Kepler did not know about forces or inertia or gravity as the cause of celestial motion, but he thought that geometry alone suggested how that solar "influence" might change for planets progressively farther from the Sun.

To see why the influence must become weaker, imagine you have a certain amount of paint to spread in a 2-mm-thick layer over the surface of a small sphere. The surface area of a sphere depends on the square of the sphere's radius: double the radius of a sphere, and the sphere's surface area becomes 4 times what it was. Suppose you wish to paint a larger sphere, with twice the radius of the small sphere. The paint must cover 4 times as much area, and the thickness of the paint will be only a fourth of what it was on the smaller sphere. If you triple the radius of the sphere, the sphere's surface will be 9 times larger and the coat of paint will be only one-ninth as thick.

Kepler reasoned that as the influence of the Sun extended farther and farther into space, that influence would have to spread out to cover the surface of a larger and larger imaginary sphere centered on the Sun. The Sun's influence should diminish with the square of the distance from the Sun—this is an example of a relationship known as an **inverse square law**.

Kepler had an interesting idea, but not a scientific hypothesis with testable predictions. He couldn't explain how the Sun influences the planets, and he lacked the mathematical tools to calculate how an object would move under such an influence. Newton, however, had both. If gravity is a force between *any* two objects, a gravitational force should then exist between the Sun and each of the planets. If that gravitational force were the same as Kepler's "influence," gravity might behave according to an inverse square law.

Newton's expression for gravity came to look like this:

$$\text{Gravitational force between two objects} = \text{Something} \times \frac{m_1 \times m_2}{(\text{Distance between objects})^2}$$

The expression still has a "something," and that something is a constant of proportionality. That constant determines the strength of gravity between objects, and it is the same for all pairs of objects. Newton named it the **universal gravitational constant**. He estimated the gravitational constant, written as G (uppercase), by using Galileo's measurement of g, estimates of Earth's radius, and a guess at the mass of Earth by assuming it had about the same density as typical rocks. Not until about a century later was the actual value of G first measured. Today the value of G is accepted as 6.67×10^{-11} N m²/kg² or its equivalents: 6.67×10^{-11} m³/(kg s²) or 6.67×10^{-20} km³/(kg s²).

A Universal Law of Gravitation

Newton's **universal law of gravitation**, illustrated in **Figure 4.2**, states that gravity is a force between any two objects having mass. The force due to gravity has the following properties:

1. It is an attractive force acting along a straight line between the two objects.

2. It is proportional to the mass of one object (m_1) multiplied by the mass of the other object (m_2). If we double m_1, the force (F) increases by a factor of 2. Likewise, if we double m_2, F increases by a factor of 2.

3. It is inversely proportional to the square of the distance r between the centers of the two objects. The graph in **Figure 4.3** shows that if we double r, F decreases by a factor of 4. If we triple r, F falls by a factor of 9.

Written as a mathematical formula, the universal law of gravitation is

$$F_{\text{grav}} = G \times \frac{m_1 \times m_2}{r^2}$$

where F_{grav} is the force of gravity between two objects; m_1 and m_2 are the masses of objects 1 and 2, respectively; r is the distance between the centers of mass of the two objects; and G is the universal gravitational constant. The relationship between the force of gravity and the masses of and distance between two objects is further explored in **Working It Out 4.1**.

Gravity pulls you toward the center of Earth. Gravity holds the planets and stars together and keeps the thin blanket of air we breathe close to Earth's surface. Gravity holds the planets, including Earth, in orbit around the Sun. Gravity caused a vast interstellar cloud of gas and dust to collapse 4.5 billion years ago to form the Sun, Earth, and the rest of the Solar System. Gravity binds colossal groups of stars into galaxies. Gravity shapes space and time, and it can affect the ultimate fate of the universe. We return often to the concept of gravity because it is central to an understanding of the universe.

CHECK YOUR UNDERSTANDING 4.1a

If the distance between Earth and the Sun were cut in half, the gravitational force between them would: (a) decrease by a factor of 4; (b) decrease by a factor of 2; (c) increase by a factor of 2; (d) increase by a factor of 4.

Answers to Check Your Understanding questions are in the back of the book.

Gravity Differs from Place to Place within an Object

As you sit reading this book, you are exerting a gravitational attraction on every other fragment of Earth, just as every other fragment of Earth is exerting a gravitational attraction on you. Your

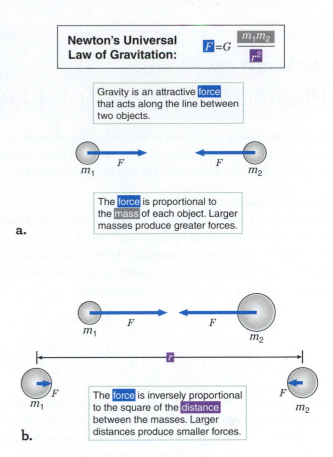

Newton's Universal Law of Gravitation: $F = G \dfrac{m_1 m_2}{r^2}$

Gravity is an attractive force that acts along the line between two objects.

a.

The force is proportional to the mass of each object. Larger masses produce greater forces.

The force is inversely proportional to the square of the distance between the masses. Larger distances produce smaller forces.

b.

Figure 4.2 Gravity is an attractive force between two objects. The force of gravity depends on the masses of the objects, m_1 and m_2, and the distance, r, between them.

Figure 4.3 As two objects move apart, the gravitational force between them decreases by the inverse square of the distance between them.

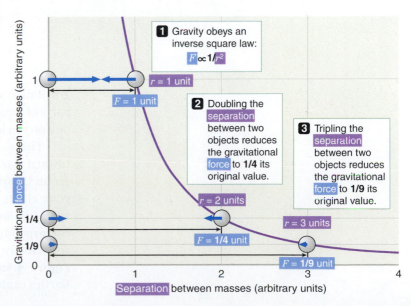

1 Gravity obeys an inverse square law: $F \propto 1/r^2$

$r = 1$ unit
$F = 1$ unit

2 Doubling the separation between two objects reduces the gravitational force to **1/4** its original value.

3 Tripling the separation between two objects reduces the gravitational force to **1/9** its original value.

$r = 2$ units
$r = 3$ units
$F = 1/4$ unit
$F = 1/9$ unit

Gravitational force between masses (arbitrary units)

Separation between masses (arbitrary units)

working it out 4.1

Playing with Newton's Laws of Motion and Gravitation

For any two objects, the force of gravity is directly proportional to the masses and inversely proportional to the *square* of the distance between them. Let's look at a few examples of how to use this equation:

Changing the Distance

How would the gravitational force between Earth and the Moon change if the distance between them were doubled? In this example, the masses of Earth and the Moon stay the same, but r becomes $2r$. We can calculate how the force changes by writing the equation for distance r and again for distance $2r$, and then taking a ratio to compare them:

$$\frac{F_{\text{grav at distance } 2r}}{F_{\text{grav at distance } r}} = \frac{G \times \dfrac{M_{\text{Earth}} M_{\text{Moon}}}{(2r)^2}}{G \times \dfrac{M_{\text{Earth}} M_{\text{Moon}}}{r^2}}$$

We can cancel out the constant G and the masses of Earth and the Moon, which do not change. Then we need to multiply both the numerator and denominator of the fraction by r^2, remembering that both the 2 and the r get squared in $(2r)^2 = 4r^2$. The equation becomes

$$\frac{F_{\text{grav at distance } 2r}}{F_{\text{grav at distance } r}} = \frac{\cancel{G} \times \cancel{M_{\text{Earth}}} \cancel{M_{\text{Moon}}}}{\cancel{G} \times \cancel{M_{\text{Earth}}} \cancel{M_{\text{Moon}}}} \times \frac{r^2}{(2r)^2}$$

$$\frac{F_{\text{grav at distance } 2r}}{F_{\text{grav at distance } r}} = \frac{r^2}{4r^2} = \frac{1}{4}$$

Doubling the distance reduced the force by a factor of 4; that is, the force is ¼ as strong.

Gravitational Acceleration

We have two ways to think about the gravitational force that Earth exerts on an object with mass m located on the surface of Earth. Recall Newton's second law of motion: $F = m \times a$. Here we are specifically considering the gravitational force and the acceleration due to gravity, so we might write $F_{\text{grav}} = m \times g$. The other way to think about the force is from the perspective of the universal law of gravitation:

$$F_{\text{grav}} = G \times \frac{M_{\text{Earth}} \times m}{R_{\text{Earth}}^2}$$

Here, M_{Earth} is the mass of Earth, and R_{Earth} is the radius of Earth. The two expressions describing that force must be equal. Therefore,

$$m \times g = G \times \frac{M_{\text{Earth}} \times m}{R_{\text{Earth}}^2}$$

The mass m is on both sides of the equation, so we can cancel it out. The equation then becomes

$$g = G \times \frac{M_{\text{Earth}}}{R_{\text{Earth}}^2}$$

This expression shows that Earth's mass and radius determine the gravitational acceleration (g) experienced by an object of mass m on the surface of Earth. The mass of the object itself (m) appears nowhere in the expression, so changing m does not affect the gravitational *acceleration* of an object on Earth.

gravitational interaction is strongest with the parts of Earth closest to you. The parts of Earth on the other side of our planet are much farther from you, so their pull on you is correspondingly less.

The net effect of all those forces is to pull you (or any other object) toward Earth's center. If you drop a hammer, it falls directly toward the ground. Earth is nearly spherical, so for every piece of Earth pulling you toward your right, a corresponding piece of Earth is pulling you toward your left with just as much force. For every piece of Earth pulling you forward, a corresponding piece of Earth is pulling you backward. Because Earth is almost spherically symmetric, all those "sideways" forces (see the blue arrows in **Figure 4.4a**) cancel out, leaving behind an overall force that points toward Earth's center (**Figure 4.4b**).

Some parts of Earth are closer to you and others are farther away, but an average distance exists between you and all the small fragments of Earth pulling on you. That average distance is the distance between you and the center of Earth. As illustrated in Figure 4.4b, the overall pull that you feel is the same as it would be if the entire mass of Earth were concentrated at a single point located at the very center of the planet.

That relationship is true for any spherically symmetric object. Outside the object, its gravity behaves as though all the mass of that object were concentrated at a point at its center. That relationship is important in many applications. For example, when you estimate your weight on another planet, you are calculating the force of gravity between you and the planet. The "distance" in the gravitational equation will be the distance between you and the center of the planet, which is just the radius of the planet plus your altitude.

You can think of Earth as a collection of small masses, each of which feels a gravitational attraction toward every other small part of Earth. The mutual gravitational attraction that occurs among all parts of the same object is called *self-gravity*. Self-gravity of a spherically symmetric object, like a planet or a roughly spherical cloud, is calculated much the same way as in the preceding discussion. For any bit of mass within the object, all the parts that are closer to the center pull that bit of mass toward the center, and the gravitational force acts as though they actually were at the center. All the parts that are farther away from the center than the bit of mass pull in different directions, and their forces cancel out.

CHECK YOUR UNDERSTANDING 4.1b

If Earth shrank to a smaller radius but kept the same mass, would the gravitational force between Earth and the Moon: (a) decrease; (b) increase; or (c) stay the same? Would everyone's weight at Earth's surface: (a) increase; (b) decrease; or (c) stay the same?

4.2 An Orbit Is One Body "Falling Around" Another

Newton used his laws of motion and his proposed law of gravity to predict the paths of planetary orbits. His calculations showed that those orbits should be ellipses with the Sun at one focus, that planets should move faster when closer to the Sun, and that the square of the period of a planet's orbit should vary as the cube of the semimajor axis of that elliptical orbit. Thus, Newton's universal law of gravitation indicated that planets should orbit the Sun in just the way that Kepler's empirical laws described. By explaining Kepler's laws, Newton found important corroboration for his law of gravitation.

Gravity and Orbits

Newton's laws tell us how forces change an object's motion and how objects interact with one another through gravity. To know where an object will be at any given time, we have to "add up" changes in the object's motion over time. Newton invented calculus to do just that, but here we aim for a conceptual understanding of how gravity causes orbits.

In Newton's time, the closest thing to making a heavy object fly was shooting cannonballs out of a cannon, so he used cannonballs in "thought experiments"

what if . . .

What if you jump from a very high tower toward a very deep body of water at an amusement park? As you fall, how would your experience compare to that of an astronaut in orbit?

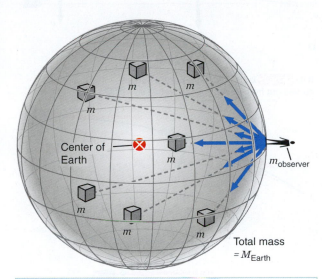

An object on the surface of a spherical mass (such as Earth) feels a gravitational attraction toward each small part of the sphere.

a.

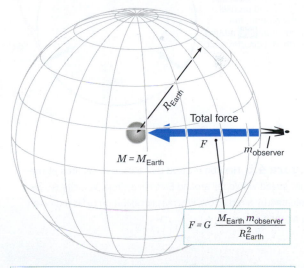

$$F = G \frac{M_{Earth} \, m_{observer}}{R_{Earth}^2}$$

The net force is the same as if we scooped up the mass of the entire sphere and concentrated it at a point at the center.

b.

Figure 4.4 Outside a sphere, the net gravitational force due to a spherical mass is the same as the gravitational force from the same mass concentrated at a point at the center of the sphere.

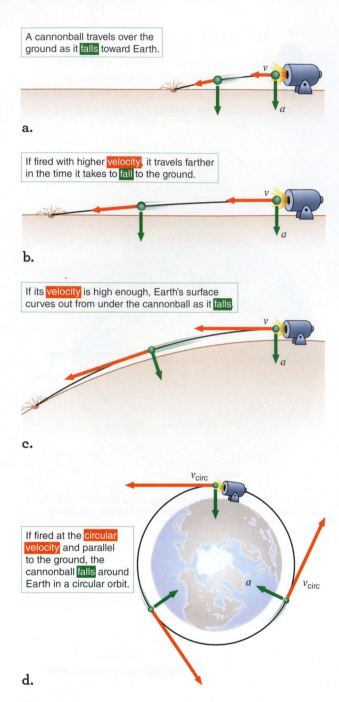

A cannonball travels over the ground as it **falls** toward Earth.

a.

If fired with higher **velocity**, it travels farther in the time it takes to **fall** to the ground.

b.

If its **velocity** is high enough, Earth's surface curves out from under the cannonball as it **falls**.

c.

If fired at the **circular velocity** and parallel to the ground, the cannonball **falls** around Earth in a circular orbit.

d.

Figure 4.5 Newton realized that a cannonball fired at the right speed would fall around Earth in a circle. Velocity (*v*) is indicated by a red arrow, whereas acceleration (*a*) is indicated by a green arrow.

about planetary motions. A dropped cannonball falls directly to the ground, like any other mass does. A cannonball fired out of a cannon that is level with the ground behaves differently, however, as shown in **Figure 4.5a**. The cannonball still falls to the ground in the same amount of time as it does when it is dropped, but while falling it also travels a horizontal distance over the ground, following a curved path. The faster the cannonball moves when it is fired from the cannon, the farther it will go before it hits the ground (**Figure 4.5b**).

To travel through air, the cannonball must push air out of its way—an effect normally referred to as *air resistance*—which slows it down. But we can ignore such real-world complications in this thought experiment. Instead imagine that, having inertia, the cannonball continues along its course until it runs into something. The faster the cannonball moves when it is fired, the farther it goes before hitting the ground. If the cannonball flies far enough, Earth's surface curves out from under it, as shown in **Figure 4.5c**. Eventually a speed is reached (**Figure 4.5d**) at which the cannonball is flying so fast that the surface of Earth curves away from the cannonball at the same rate at which the cannonball is falling toward Earth. When that occurs, the cannonball, which always falls toward the center of Earth, is orbiting. An **orbit** is the path of one object that freely falls around another.

Why do astronauts appear to float freely about the cabin of a spacecraft? It is not because they have escaped Earth's gravity; Earth's gravity is what holds them in their orbit. Instead, the answer lies in Galileo's early observation that all objects fall with the same acceleration, regardless of their mass. The astronauts and the spacecraft are both in orbit around Earth, moving in the same direction, at the same speed, and experiencing the same gravitational acceleration, so they fall around Earth together. **Figure 4.6** demonstrates that point. The astronaut is orbiting Earth just as the spacecraft is orbiting Earth. On the surface of Earth, your body tries to fall toward the center of Earth, but the ground gets in the way. You feel your weight when you are standing on Earth because the ground pushes on you to oppose the force of gravity, which pulls you downward. In the spacecraft, however, nothing interrupts the astronaut's fall because the spacecraft is falling around Earth in the same orbit as the astronaut. The astronaut is in **free fall**, falling freely in Earth's gravity. The **Process of Science Figure** illustrates the universality of Newton's law of gravitation.

When two objects are closer to having the same mass—such as dwarf planet Pluto and its moon Charon, or a large planet and a star, or two stars—both objects experience significant accelerations in response to their mutual gravitational attraction. The two objects are both orbiting about a common point located between them, called the **center of mass**, so we now must think of them as falling around each other. Each mass is moving on its own elliptical orbit around the two objects' mutual center of mass. From measuring the size and period of any orbit, we can calculate the *sum* of the masses of the orbiting objects. Almost all knowledge about the masses of astronomical objects comes directly from applying Newton's version of Kepler's third law.

What Velocity Is Needed to Reach Orbit?

How fast must Newton's cannonball be fired for it to fall around the world? The cannonball would be in **uniform circular motion**, meaning it would move along a circular path at constant speed. That type of motion is discussed in more depth in Appendix 8. Another example of uniform circular motion is a ball whirling around your head on a string. If you let go of the string, the ball will fly off in a

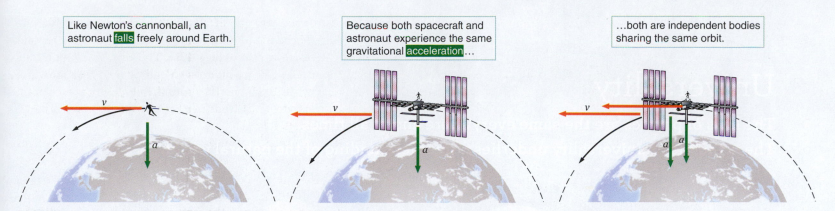

Like Newton's cannonball, an astronaut **falls** freely around Earth.

Because both spacecraft and astronaut experience the same gravitational **acceleration**...

...both are independent bodies sharing the same orbit.

Figure 4.6 A "weightless" astronaut has not escaped Earth's gravity. Rather, an astronaut and a spacecraft share the same orbit as they fall around Earth together.

straight line in whatever direction it is traveling at the time, just as Newton's first law predicts for an object in motion, as shown in **Figure 4.7a**. The string prevents the ball from flying off by constantly changing the direction the ball is traveling. The central force of the string on the ball is called a **centripetal force**: a force toward the center of the circle. Using a more massive ball, speeding up its motion, and making the string shorter so that the turn is tighter all are ways to increase the force needed to keep a ball moving in a circle.

For Newton's cannonball (or a satellite), no string holds the ball in its circular motion. Instead, gravity supplies the force, as illustrated in **Figure 4.7b**. The force of gravity must be just enough to keep the satellite moving on its circular path. Because that force has a specific strength, the satellite must therefore be moving at a particular speed around the circle, which we call its **circular velocity** (v_{circ}). If the satellite were moving at any other velocity, it would not be moving in a circular orbit. Remember the cannonball: if it moves too slowly, it will drop below the circular path and hit the ground. Similarly, if the cannonball moves too fast, its motion will carry it above the circular orbit. Only a cannonball moving at just the right velocity—the circular velocity—will fall around Earth on a circular path (see Figure 4.5d).

Newton's thought experiment became a reality in 1957, when the Soviet Union (the USSR) launched the first artificial object to orbit Earth. The USSR used a rocket to lift Sputnik 1, an object about the size of a basketball, high enough

▶❙❙ **AstroTour:** Newton's Laws and Universal Gravitation

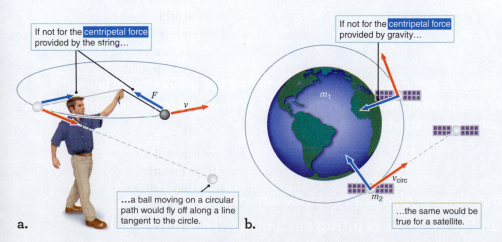

If not for the centripetal force provided by the string...

...a ball moving on a circular path would fly off along a line tangent to the circle.

If not for the centripetal force provided by gravity...

...the same would be true for a satellite.

a.   **b.**

Figure 4.7 a. A string provides the centripetal force that keeps a ball moving in a circle. (We are ignoring the smaller force of gravity that also acts on the ball.) **b.** Gravity provides the centripetal force that holds a satellite in a circular orbit about Earth.

Universality

The laws of physics are the same everywhere and at all times.
The principle of universality underlies our understanding of the natural world.

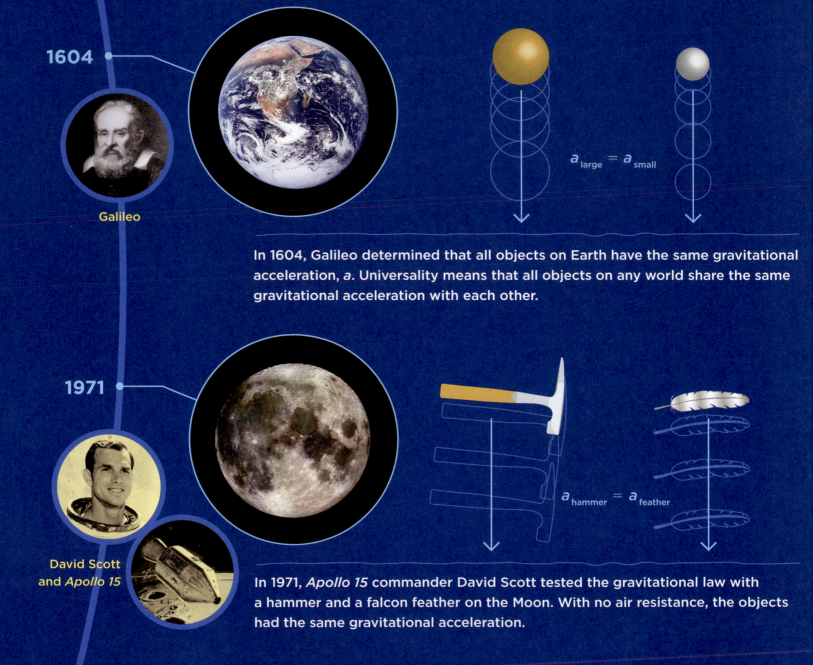

1604

Galileo

$$a_{large} = a_{small}$$

In 1604, Galileo determined that all objects on Earth have the same gravitational acceleration, a. Universality means that all objects on any world share the same gravitational acceleration with each other.

1971

David Scott and *Apollo 15*

$$a_{hammer} = a_{feather}$$

In 1971, *Apollo 15* commander David Scott tested the gravitational law with a hammer and a falcon feather on the Moon. With no air resistance, the objects had the same gravitational acceleration.

The physical laws that apply to falling objects also apply to planets orbiting the Sun, to stars orbiting within the galaxy, and to galaxies orbiting each other. This universality allows us to draw conclusions about very distant objects.

above Earth's upper atmosphere that air resistance wasn't an issue. Sputnik 1 achieved a high enough speed that it fell around Earth, just like Newton's imaginary cannonball.

When one object is falling around a much more massive object, we say that the less massive object is a **satellite** of the more massive object. Planets are satellites of the Sun, and moons are natural satellites of planets. Newton's imaginary cannonball and Sputnik 1 were satellites (*sputnik* means "satellite" in Russian). An Earth-orbiting spacecraft and the astronauts inside it are independent satellites of Earth that share the same orbit.

The Shape of Orbits

Some Earth satellites travel a circular path at constant speed. Just like the ball on the string, satellites traveling at the circular velocity always stay the same distance from Earth, neither speeding up nor slowing down in orbit. But what would happen if the satellite then fired its rockets and started traveling *faster* than the circular velocity? Earth's pull is as strong as ever, but because the satellite has greater speed, Earth's gravity does not bend the satellite's path sharply enough to hold it in a circle. When that happens, the satellite begins to climb above a circular orbit.

As the distance between the satellite and Earth increases, the satellite slows down. Think about a ball thrown upward into the air, as shown in **Figure 4.8a**. As the ball climbs higher, the pull of Earth's gravity opposes its motion, slowing the ball down. The ball climbs more and more slowly until its vertical motion stops for an instant and then is reversed; the ball then begins to fall back toward Earth, speeding up along the way. A satellite does the same thing. As the satellite climbs above a circular orbit and begins to move away from Earth, Earth's gravity opposes the satellite's outward motion, slowing the satellite down. The farther the satellite is from Earth, the more slowly the satellite moves—just like the ball thrown into the air. Also just like the ball, the satellite reaches a maximum height on its curving path and then begins falling back toward Earth. When it does, Earth's gravity speeds the satellite up as it gets closer and closer to Earth. The satellite's orbit has changed from circular to elliptical.

Any object in an elliptical orbit, including a planet orbiting the Sun (**Figure 4.8b**), will therefore lose speed as it pulls away from what it is orbiting and then gain that speed back again as gravity causes it to fall inward toward what it is orbiting. Gravity, therefore, explains Kepler's second law from Section 3.2—namely, that a planet moves fastest when closest to the Sun and slowest when farthest from the Sun.

Newton's laws do more than explain Kepler's laws: they predict different types of orbits beyond Kepler's empirical experience. **Figure 4.9a** shows a series of satellite orbits, each with the same point of closest approach to Earth but with different velocities at that point, as indicated in **Figure 4.9b**. The more speed a satellite has at its closest approach to Earth, the farther the satellite can pull away from Earth, and the more eccentric its orbit becomes. As long as it remains elliptical, no matter how eccentric, the orbit is called a **bound orbit** because the satellite is gravitationally bound to the object it is orbiting.

In that sequence of faster and faster satellites, a point comes at which the satellite is moving so fast that gravity cannot reverse its outward motion, so the

▶❚❚ **AstroTour:** Elliptical Orbit

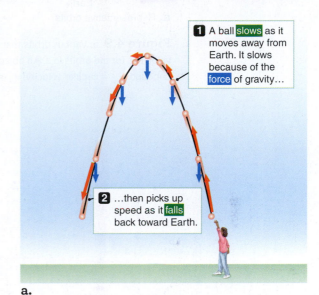

1 A ball slows as it moves away from Earth. It slows because of the force of gravity...

2 ...then picks up speed as it falls back toward Earth.

a.

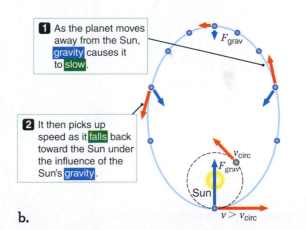

1 As the planet moves away from the Sun, gravity causes it to slow.

2 It then picks up speed as it falls back toward the Sun under the influence of the Sun's gravity.

F_{grav}

v_{circ}

F_{grav}

Sun

$v > v_{circ}$

b.

Figure 4.8 a. A ball thrown into the air slows as it climbs away from Earth and then speeds up as it heads back toward Earth. **b.** A planet in an elliptical orbit around the Sun does the same thing. (Although no planet has an orbit as eccentric as the one shown here, the orbits of comets can be far more eccentric.) A planet in an elliptical orbit will move fastest at its closest approach to the Sun. That speed is faster than the speed of a planet in a circular orbit of that radius.

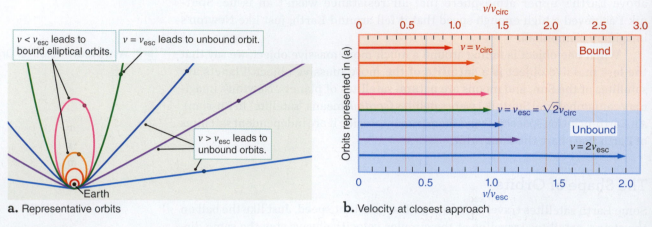

a. Representative orbits

b. Velocity at closest approach

Figure 4.9 **a.** Various orbits that share the same point of closest approach but differ in velocity at that point.   **b.** Closest-approach velocities for the orbits in a. An object's velocity at closest approach determines the orbit's shape and whether the orbit is bound or unbound. v_{circ} = circular velocity; v_{esc} = escape velocity.

object travels away from Earth, never to return. The lowest speed at which an object can permanently leave the gravitational grasp of another mass is called the **escape velocity**, v_{esc}. Once a satellite's velocity at closest approach equals or exceeds v_{esc}, and it is no longer gravitationally bound to the object it was orbiting, we say it travels on an **unbound orbit**.

As Figure 4.9a shows, an object with a velocity *less* than the escape velocity (v_{esc}) has an elliptically shaped orbit and follows the same path over and over again, whereas unbound orbits do not close like an ellipse. An object such as a comet on an unbound orbit makes only a single pass around the Sun and then continues away from the Sun into deep space, never to return. Circular velocity and escape velocity are further explored in **Working It Out 4.2**.

Estimating Mass by Using Newton's Version of Kepler's Laws

Newton's calculations opened up a new way to investigate the universe. He showed that the same physical laws that describe the flight of a cannonball on Earth—or the legendary apple falling from a tree—also describe the motions of the planets through the heavens. His laws of motion and gravitation predict all three of Kepler's empirical laws of planetary motion. Newton's version of Kepler's laws can be used to estimate the mass of the Sun from the orbit of Earth, as demonstrated in **Working It Out 4.3**.

When Earth, a much less massive object, is orbiting the Sun, a much more massive object, the Sun's gravity has a strong influence on Earth, but Earth's gravity has little effect on the Sun. Therefore, the Sun's motion due to Earth's orbit around it is tiny.

Astronomy in Action: Center of Mass

CHECK YOUR UNDERSTANDING **4.2**

Rank the circular velocities of the following objects from smallest to largest: (a) a 5-kg object orbiting Earth halfway to the Moon; (b) a 10-kg object orbiting Earth just above Earth's surface; (c) a 15-kg object orbiting Earth at the same distance as the Moon.

working it out 4.2

Circular Velocity and Escape Velocity

Circular Velocity

In Appendix 8, we show that the circular velocity (v_{circ}) is given by

$$v_{circ} = \sqrt{\frac{GM}{r}}$$

where M is the mass of the orbited object, r is the radius of the circular orbit, and G is the universal gravitational constant. A cannonball moving at just the right velocity—the circular velocity—will fall around Earth on a circular path.

We can use that equation to show how fast Newton's cannonball would have to travel to stay in its circular orbit. The average radius of Earth is 6370 km, the mass of Earth is 5.97×10^{24} kg, and the gravitational constant is 6.67×10^{-20} km³/(kg s²). Inserting those values into the expression for v_{circ}, we get

$$v_{circ} = \sqrt{\frac{[6.67 \times 10^{-20}\,\text{km}^3/(\text{kg s}^2)] \times (5.97 \times 10^{24}\,\text{kg})}{6370\,\text{km}}} = 7.91\,\text{km/s}$$

To stay in its circular orbit, Newton's cannonball would have to be traveling about 8 kilometers per second (km/s)—more than 28,000 kilometers per hour (km/h). That's well beyond the range of a typical cannon, but rockets routinely reach those speeds.

Now let's compare that speed with the speed needed to launch a satellite from the Moon into orbit just above the lunar surface. The radius of the Moon is 1740 km, and its mass is 7.35×10^{22} kg. Those values give a considerably lower circular velocity than the one for Earth:

$$v_{circ} = \sqrt{\frac{(6.67 \times 10^{-20}\,\text{km}^3/(\text{kg s}^2) \times (7.35 \times 10^{22}\,\text{kg})}{1740\,\text{km}}} = 1.68\,\text{km/s}$$

Escape Velocity

Sending a spacecraft to another planet requires launching it with a velocity greater than Earth's escape velocity. The escape velocity is a factor of $\sqrt{2}$, or approximately 1.41, times the circular velocity. That relation can be expressed as

$$v_{esc} = \sqrt{2} \times v_{circ} = \sqrt{\frac{2GM}{R}}$$

Using the numbers in the above example, we can calculate the escape velocity from the surface of Earth:

$$v_{esc} = \sqrt{2} \times v_{circ} = 1.41 \times 7.91\,\text{km/s} = 11.2\,\text{km/s}$$

To leave Earth, a rocket must have a speed of at least 11.2 km/s, or 40,100 km/h.

As with weight, the escape velocity from other astronomical objects is different from the escape velocity from Earth. Ida is a small asteroid orbiting the Sun between the orbits of Mars and Jupiter. Ida has an average radius of 15.7 km and a mass of 4.2×10^{16} kg. Therefore,

$$v_{esc} = \sqrt{\frac{2 \times [6.67 \times 10^{-20}\,\text{km}^3/(\text{kg s}^2)] \times (4.2 \times 10^{16}\,\text{kg})}{15.7\,\text{km}}}$$

$$v_{esc} = 0.019\,\text{km/s} = 68\,\text{km/h}$$

A baseball thrown at about 130 km/h (81 mph) would easily escape from Ida's surface and fly off into interplanetary space.

4.3 Tidal Forces Are Caused by Gravity

The rise and the fall of the oceans are called Earth's **tides**. Coastal dwellers long ago noted that the strength of the tides varies with the phase of the Moon. Tides are strongest during a new or a full Moon and are weakest during first quarter or third quarter Moon. Here in Section 4.3, we see how tides result from differences between the strength of the gravitational pull of the Moon and Sun on one part of Earth in comparison with their pull on other parts of Earth.

 **Astronomy in Action:** Tides

Tides and the Moon

Recall from Figure 4.4 that each small part of an object feels a gravitational attraction toward every other small part of the object, and that self-gravity differs from place to place. In addition, each small part of an object feels a gravitational attraction

working it out 4.3

Calculating Mass from Orbital Periods

Newton's Version of Kepler's Third Law

The time a planet takes to complete one orbit around the Sun (the orbital period, P) equals the distance traveled divided by the planet's speed. For simplicity, let's assume the orbit is circular. Thus, the time an object takes to make one trip around the Sun is the circumference of the circle ($2\pi r$) divided by the object's speed (v). The speed of the planet must be equal to the circular velocity discussed in Working It Out 4.2. Combining these two relationships, we have

$$\text{Orbital period } (P) = \frac{\text{Circumference of orbit}}{\text{Circular velocity}} = \frac{2\pi r}{\sqrt{\dfrac{GM_{\text{Sun}}}{r}}}$$

Squaring both sides of the equation gives

$$P^2 = \frac{4\pi^2 r^2}{\dfrac{GM_{\text{Sun}}}{r}} = \frac{4\pi^2}{GM_{\text{Sun}}} \times r^3$$

The square of the period of an orbit is equal to a constant ($4\pi^2/GM_{\text{Sun}}$) multiplied by the cube of the radius of the orbit. That is Kepler's third law ($P^2 = \text{constants} \times A^3$) applied to circular orbits. When Kepler used Earth units of years and astronomical units for the planets, he was taking a ratio of their periods and orbital radii with those for Earth, so the constants canceled out. Using calculus, Newton similarly could derive Kepler's third law for elliptical orbits with semimajor axis A instead of radius r:

$$P^2 = \frac{4\pi^2}{GM_{\text{Sun}}} \times A^3$$

Mass of the Sun

If we can measure the size and period of any orbit, we can use Newton's universal law of gravitation to calculate the mass of the object being orbited. To do so, we rearrange Newton's form of Kepler's third law above to read

$$M = \frac{4\pi^2}{G} \times \frac{A^3}{P^2}$$

Everything on the right side of that equation is either a constant (4, π, and G) or a measurable quantity (the semimajor axis A and period P of an orbit). The left side of the equation is the mass of the object at the focus of the ellipse. For example, we can find the mass of the Sun by noting the period and semimajor axis of the orbit of a planet around the Sun. Let's use the numbers for Earth. Whenever we have an equation with G, putting everything else into the same units as G (km, kg, s) tends to simplify the math. So first we must compute the number of seconds in 1 year: $P = 1$ yr $= 365.24$ day/yr $\times$ 24 h/day $\times$ 60 min/h $\times$ 60 s/min $= 3.16 \times 10^7$ s. The semimajor axis $A = 1$ AU $= 1.5 \times 10^8$ km. Then the mass of the Sun can be computed:

$$M_{\text{Sun}} = \frac{4\pi^2}{G} \times \frac{A^3}{P^2} = \frac{4\pi^2}{6.67 \times 10^{-20} \text{ km}^3/(\text{kg s}^2)} \times \frac{(1.5 \times 10^8 \text{ km})^3}{(3.16 \times 10^7 \text{ s})^2}$$

$$M_{\text{Sun}} = 2.0 \times 10^{30} \text{ kg}$$

If we had used the period and semimajor axis of any other planet in the Solar System, we would have gotten the same result for the mass of the Sun.

▶▶ **Interactive Simulation:** Tides on Earth

▶❚❚ **AstroTour:** Tides and the Moon

toward every other mass in the universe, and those external forces also differ from place to place within the object.

The Moon pulls more strongly on the part of Earth that is closest (the near side) than it does on the part of Earth that's farthest (the far side). The pull of the Moon on the near side of Earth is about 7 percent greater than its pull on the far side of Earth. This is because the side of Earth that faces the Moon is closer to the Moon than the rest of Earth is, so it feels a stronger-than-average gravitational attraction toward the Moon. In contrast, the side of Earth facing away from the Moon is farther than average from the Moon, so it feels a weaker-than-average attraction toward the Moon.

To understand the consequence of that variation in the pull of the Moon, imagine three rocks being pulled by gravity toward the Moon. A rock closer to the Moon feels a stronger force than a rock farther from the Moon. Now suppose the three rocks are connected by springs (**Figure 4.10a**). As the

Reading Astronomy News

Earth Has Acquired a Brand-New Moon That's about the Size of a Car

Leah Crane

In the Solar System, whether an orbit is stable depends on interactions between objects that orbit the Sun. Sometimes even Earth gets a temporary extra moon!

Earth might have a tiny new moon. On 19 February, astronomers at the Catalina Sky Survey in Arizona spotted a dim object moving quickly across the sky. Over the next few days, researchers at six more observatories around the world watched the object, designated 2020 CD3, and calculated its orbit, confirming that it has been gravitationally bound to Earth for about three years.

An announcement posted by the Minor Planet Center, which monitors small bodies in space, states that "no link to a known artificial object has been found," implying that

it is most likely an asteroid caught by Earth's gravity as it passed by.

This is just the second asteroid known to have been captured by our planet as a mini-moon—the first, 2006 RH120, hung around between September 2006 and June 2007 before escaping.

Our new moon is probably between 1.9 and 3.5 meters across, or roughly the size of a car, making it no match for Earth's primary moon. It circles our planet about once every 47 days on a wide, oval-shaped orbit that mostly swoops far outside the larger moon's path.

The orbit isn't stable, so eventually 2020 CD3 will be flung away from Earth. "It is heading away from the Earth-moon system as we speak," says Grigori Fedorets at Queen's University Belfast in the UK, and it looks likely it will escape in April.

However, there are several different simulations of its trajectory and they don't all agree—we will need more observations to

accurately predict the fate of our mini-moon and even to confirm that it is definitely a temporary moon and not a piece of artificial space debris. "Our international team is continuously working to constrain a better solution," says Fedorets.

QUESTIONS

1. About how large (in meters) is the mini-moon? How large is this in feet?
2. How much time elapsed between when the mini-moon began orbiting Earth and its discovery in 2020? Is that disturbing? Why or why not?
3. What is the predicted ultimate fate of this mini-moon?
4. Look up this mini-moon on the Internet, by searching for it by name: "2020 CD3." What has happened to the mini-moon since this article was written?

Source: https://www.newscientist.com/article/2235427-earth-has-acquired-a-brand-new-moon-thats-about-the-size-of-a-car/.

rocks are pulled toward the Moon, the purple rock pulls away from the red rock, and the red rock pulls away from the blue rock. Therefore, the differences in the gravitational forces they feel will stretch *both* springs (**Figure 4.10b**). Now, instead of springs, imagine that the rocks are at different places on Earth (**Figure 4.10c**). On the side of Earth away from the Moon, the force is smaller (as indicated by the shorter arrow), so that part gets left behind (**Figure 4.10d**). Those differences in the Moon's gravitational attraction on different parts of Earth are called **tidal forces**. Tidal forces cause an object to stretch.

Figure 4.11 shows how Earth is stretched, causing a tidal bulge. The Moon is not pulling on the far side of Earth as hard as it is pulling on the planet as a whole. The far side of Earth is "left behind" as the rest of the planet is pulled more strongly toward the Moon. Figure 4.11 shows that a net force also is squeezing inward on Earth in the direction perpendicular to the line between Earth and the

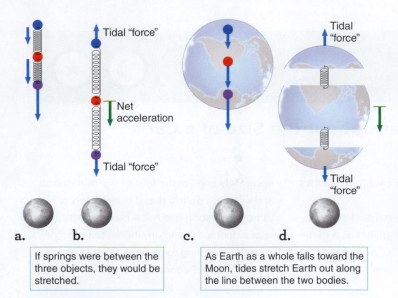

a. **b.** **c.** **d.**

If springs were between the three objects, they would be stretched.

As Earth as a whole falls toward the Moon, tides stretch Earth out along the line between the two bodies.

Figure 4.10 **a.** Imagine three objects connected by springs. **b.** The springs are stretched as though forces were pulling outward on each end of the chain. **c.** Similarly, three locations on Earth experience different gravitational attractions toward the Moon. **d.** The difference in the Moon's gravitational attraction across Earth causes Earth's tides.

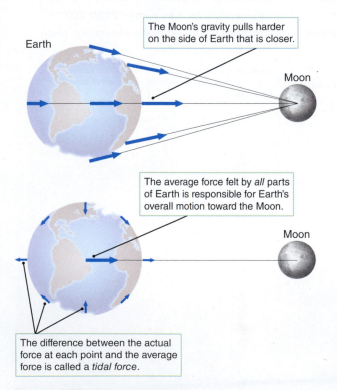

Earth

The Moon's gravity pulls harder on the side of Earth that is closer.

Moon

The average force felt by *all* parts of Earth is responsible for Earth's overall motion toward the Moon.

Moon

The difference between the actual force at each point and the average force is called a *tidal force.*

Figure 4.11 Tidal forces stretch Earth along the line between Earth and the Moon but compress Earth perpendicular to that line.

Moon. Together, the stretching by tidal forces along the line between Earth and the Moon and the squeezing by tidal forces perpendicular to that line distort Earth's shape, like a rubber ball caught in the middle of a tug-of-war.

If the surface of Earth were perfectly smooth and covered with an ocean of uniform depth and Earth did not rotate, the Moon would pull our oceans into an elongated **tidal bulge** like that shown in **Figure 4.12a**. The water would be at its deepest on the side toward the Moon and on the side away from the Moon, and the water would be at its shallowest midway between. But Earth is *not* covered with perfectly uniform oceans, and Earth *does* rotate. As any point on Earth rotates through the ocean's tidal bulges, that point experiences the ebb and flow of the tides. In addition, friction between the spinning Earth and its tidal bulge drags the oceanic tidal bulge around in the direction of Earth's rotation, as illustrated in **Figure 4.12b**.

Figure 4.12c shows the tides you would experience throughout a day. Recall that Earth's rotation period of one day is much shorter than the Moon's orbital period of about a month. Over one day, the Moon moves only a little way in its orbit, but you travel around Earth once. Each day, the Moon rises 50 minutes later than it did on the previous day, so the time between the Moon's return to the same position in the sky from the previous day is 24 h 50 m. Begin as the rotating Earth carries you through the tidal bulge on the Moonward side of the planet. Because Earth's rotation drags the tidal bulge, the Moon is not exactly overhead but is instead high in the western sky. When you are at the high point in the tidal bulge, the ocean around you has risen higher than average—called a *high tide*. The rotation of Earth then carries you around to a point where the ocean is lower than average—called a *low tide*. Later, you pass through the region where ocean water is "left behind" (with respect to Earth as a whole) in the tidal bulge on the side of Earth that is away from the Moon. At this high tide, the Moon at that time is hidden from view on the far side of Earth. After the Moon has risen above the eastern horizon, you pass through a second low tide. After 24 h 50 m—the amount of time the Moon takes to return to the same point in the sky from which it started—you again pass through the tidal bulge on the near side of the planet. Each of these tides occurs about $6\frac{1}{4}$ hours after the previous one, because there are four tides, and $\frac{1}{4}$ of 24 h 50 m is about $6\frac{1}{4}$ hours. That is the age-old pattern by which mariners have lived their lives for millennia: the twice-daily coming and going of high tide, shifting through the day in lockstep with the passing of the Moon.

Ocean currents and local geography, such as the shapes of Earth's shorelines and ocean basins, complicates the simple picture of tides. In addition, ocean-wide oscillations occur, similar to water sloshing around in a basin. As the oceans respond to the tidal forces from the Moon, they flow around the various landmasses that break up the water covering our planet. Some places, such as the Mediterranean Sea and the Baltic Sea, are protected from tides by their relatively small sizes and the narrow passages connecting those bodies of water to the larger ocean. In other places, the shape of the land funnels the tidal surge from a large region of ocean into a relatively small area, concentrating its effect, as at eastern Canada's Bay of Fundy (**Figure 4.13**).

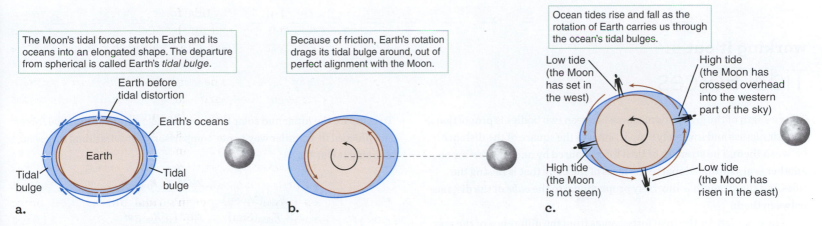

Figure 4.12 **a.** Tidal forces pull Earth and its oceans into a tidal bulge **b.** Earth's rotation pulls its tidal bulge slightly out of alignment with the Moon. **c.** As Earth's rotation carries us through those bulges, we experience the ocean tides. The magnitude of the tides has been exaggerated here for clarity. In these figures, the observer is looking down from above Earth's North Pole. Sizes and distances are not to scale.

Solar Tides

Tides resulting from the pull of the Moon are called **lunar tides**, but the Sun also influences Earth's tides. The gravitational pull of the Sun causes Earth to stretch along a line pointing approximately in the direction of the Sun. The side of Earth closer to the Sun is pulled toward the Sun more strongly than the side of Earth away from the Sun, just as the side of Earth closest to the Moon is pulled more strongly toward the Moon. Tides on Earth resulting from differences in the gravitational pull of the Sun are called **solar tides**. The absolute strength of the Sun's pull on Earth is nearly 200 times greater than the strength of the Moon's pull on Earth. However, the Sun's gravitational attraction does not change by much from one side of Earth to the other because the Sun is much farther away than the Moon. As a result, solar tides are only about half as strong as lunar tides (**Working It Out 4.4**).

Figure 4.13 The world's most extreme tides are found in the Bay of Fundy in eastern Canada. Water rocks back and forth in the bay with a period of about 13 hours, close to the 12.5-hour period of the tides. The shape of the basin amplifies the tides so that the difference in water depth between **a.** low tide and **b.** high tide is extreme—typically about 14.5 meters and as much as 16.6 meters.

working it out 4.4

Tidal Forces

The strength of the gravitational force between two bodies is proportional to their masses and inversely proportional to the square of the distance between them. The strength of tidal forces caused by one body acting on another is also proportional to the mass of the body that is raising the tides, but the strength is inversely proportional to the *cube* of the distance between them.

The equation for the tidal force comes from the difference of the gravitational force on one side of a body compared with the force on the other side. We can approximate the tidal force of the Moon acting on a mass m on the surface of Earth with

$$F_{tidal}(\text{Moon}) = \frac{2GM_{\text{Moon}}R_{\text{Earth}}m}{d^3_{\text{Earth-Moon}}}$$

where R_{Earth} is Earth's radius and $d_{\text{Earth-Moon}}$ is the distance between Earth and the Moon.

Let's compare the Moon's tidal force acting on Earth (given by the preceding equation) with the Sun's tidal force acting on Earth, which is given by the following equation:

$$F_{tidal}(\text{Sun}) = \frac{2GM_{\text{Sun}}R_{\text{Earth}}m}{d^3_{\text{Earth-Sun}}}$$

To compare the lunar and solar tides, we can take a ratio of the tidal forces and proceed in a similar way to our comparison of gravitational forces in Working It Out 4.1:

$$\frac{F_{tidal}(\text{Moon})}{F_{tidal}(\text{Sun})} = \frac{\dfrac{2GM_{\text{Moon}}R_{\text{Earth}}m}{d^3_{\text{Earth-Moon}}}}{\dfrac{2GM_{\text{Sun}}R_{\text{Earth}}m}{d^3_{\text{Earth-Sun}}}}$$

Canceling out the constant G and the terms common in both equations (m and R_{Earth}) gives

$$\frac{F_{tidal}(\text{Moon})}{F_{tidal}(\text{Sun})} = \frac{M_{\text{Moon}}}{M_{\text{Sun}}} \times \frac{d^3_{\text{Earth-Sun}}}{d^3_{\text{Earth-Moon}}} = \frac{M_{\text{Moon}}}{M_{\text{Sun}}} \times \left(\frac{d_{\text{Earth-Sun}}}{d_{\text{Earth-Moon}}}\right)^3$$

Using the values from Appendixes 2 and 4, $M_{\text{Moon}} = 7.35 \times 10^{22}$ kg, $M_{\text{Sun}} = 2 \times 10^{30}$ kg, $d_{\text{Earth-Moon}} = 384,400$ km, and $d_{\text{Earth-Sun}} = 1.5 \times 10^8$ km, gives

$$\frac{F_{tidal}(\text{Moon})}{F_{tidal}(\text{Sun})} = \frac{7.35 \times 10^{22} \text{ kg}}{2 \times 10^{30} \text{ kg}} \times \left(\frac{1.5 \times 10^8 \text{ km}}{384,400 \text{ km}}\right)^3 = 2.2$$

Thus, the tidal force between Earth and the Moon is 2.2 times stronger than the tidal force between Earth and the Sun. The Sun is still an important factor, though, and that's why the tides change depending on the alignment of the Moon and the Sun with Earth.

what if . . .

What if the Moon's orbit was gradually shrinking? As the Moon's orbital radius shrank to half, or even less, of its current size, how would the tides change, and what effects might that have on life on Earth?

Solar and lunar tides interact. When the Moon and the Sun are lined up with Earth (**Figure 4.14a**), at either new or full Moon, the lunar and solar tides on Earth overlap. That alignment creates more extreme tides; both higher high tides and lower low tides. The extreme tides near the new or full Moon are called **spring tides**—not because of the season but because the water appears to spring out of the sea. Conversely, when the Moon, Earth, and Sun make a right angle (**Figure 4.14b**), at the Moon's first and third quarters, the lunar and solar tidal forces stretch Earth in different directions, creating less extreme tides known as **neap tides**. The word *neap* is derived from the Saxon word *neafte*, which means "scarcity": at those times of the month, shellfish and other food gathered in the tidal region are less accessible because the low tide is higher than at other times. Neap tides are only about half as strong as average tides and only a third as strong as spring tides.

CHECK YOUR UNDERSTANDING 4.3

Rank the following in order of weakest to strongest total tides: (a) new Moon in July; (b) first quarter Moon in July; (c) full Moon in January; (d) third quarter Moon in January.

4.4 Tidal Forces Affect Solid Bodies

In Section 4.3, you learned how the liquid of Earth's oceans moves in response to the tidal forces from the Moon and Sun. But those tidal forces also affect the solid body of Earth. As Earth rotates through its tidal bulge, the solid body of the planet is constantly being deformed by tidal forces. Earth is somewhat elastic (like a rubber ball), and tidal stresses cause a vertical displacement of about 30 centimeters (cm) between high tide and low tide, or roughly a third of the displacement of the oceans.

Deforming the shape of a solid object requires energy. For a practical demonstration of that fact, hold a rubber ball in your hand and squeeze and release it a few dozen times; it will gradually become warmer. The energy from the deformation is converted into thermal energy because of friction. Similarly, for Earth, friction from the tidal deformation opposes and takes energy from the rotation of Earth, causing Earth to gradually slow. Earth's internal friction adds to the slowing caused by friction between Earth and its oceans as the planet rotates through the tidal bulge of the oceans. As a result, Earth's days are currently getting about 1.7 milliseconds (ms) longer every century. That sounds insignificant, but it adds up: when dinosaurs ruled 200 million years ago, the day was closer to 23 hours long, and 200 million years into the future, the day will be close to 25 hours.

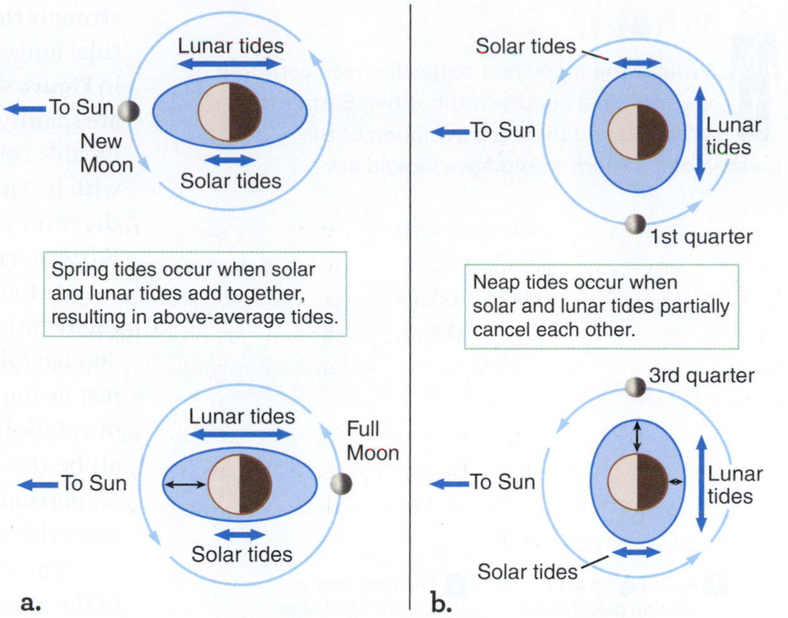

Figure 4.14 Solar tides are about half as strong as lunar tides. The interactions of solar and lunar tides result in either **a.** spring tides when they are added together or **b.** neap tides when they partially cancel each other.

Spring tides occur when solar and lunar tides add together, resulting in above-average tides.

Neap tides occur when solar and lunar tides partially cancel each other.

Tidal Locking

Other solid bodies besides Earth experience tidal forces. For example, the Moon has no bodies of liquid to make tidal forces obvious, but its shape is distorted in the same manner as Earth's. Because of Earth's much greater mass, Earth's tidal effects on the Moon are stronger than the Moon's tidal effects on Earth. Given that the average tidal deformation of Earth is about 30 cm, the average tidal deformation of the Moon should be about 6 meters. However, what we actually observe on the Moon is a tidal bulge of about 20 meters. That unexpectedly large displacement exists because the Moon's tidal bulge was "frozen" into its relatively rigid crust at an earlier time, when the Moon was closer to Earth and tidal forces were much stronger than they are today. Planetary scientists sometimes call that deformation the Moon's *fossil tidal bulge.*

Recall that the Moon's rotation period equals its orbital period. That synchronous rotation of the Moon is a result of **tidal locking**. Early in the Moon's history, the period of its rotation was almost certainly different from its orbital period. As the Moon rotated through its extreme tidal bulge, however, friction within the Moon's crust was tremendous, rapidly slowing the Moon's rotation. After a fairly short time, the period of the Moon's rotation equaled the period of its orbit. When its orbital and rotation periods became equalized, the Moon no longer rotated with respect to its tidal bulge. Instead, the Moon and its tidal bulge rotated *together*, in lockstep with the Moon's orbit around Earth. As illustrated in **Figure 4.15**, that scenario continues today as the tidally distorted Moon orbits Earth, always keeping the same face and the long axis of its tidal bulge toward Earth.

Tidal forces affect not only the rotations of the Moon and Earth but also their orbits. Because of its tidal bulge, Earth is not perfectly spherical. Therefore, the material in Earth's tidal bulge on the side nearer the Moon pulls on the Moon more

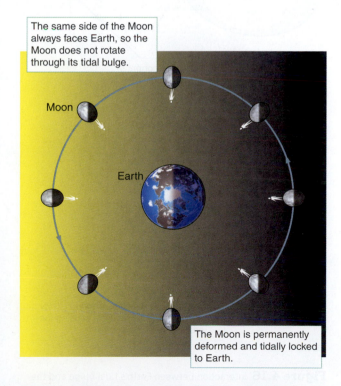

The same side of the Moon always faces Earth, so the Moon does not rotate through its tidal bulge.

The Moon is permanently deformed and tidally locked to Earth.

Figure 4.15 Tidal forces due to Earth's gravity lock the Moon's rotation to its orbital period.

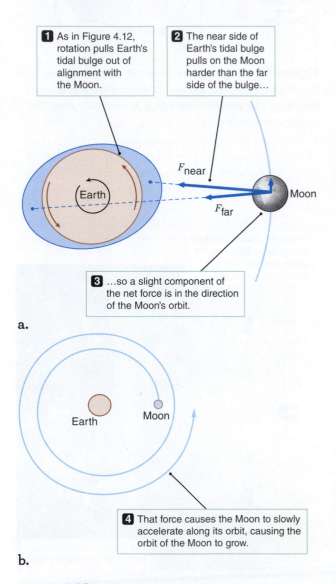

1 As in Figure 4.12, rotation pulls Earth's tidal bulge out of alignment with the Moon.

2 The near side of Earth's tidal bulge pulls on the Moon harder than the far side of the bulge…

F_{near}

Earth

Moon

F_{far}

3 …so a slight component of the net force is in the direction of the Moon's orbit.

a.

Earth   Moon

4 That force causes the Moon to slowly accelerate along its orbit, causing the orbit of the Moon to grow.

b.

Figure 4.16 Interaction between Earth's tidal bulge and the Moon causes the Moon to accelerate in its orbit and the Moon's orbit to grow.

strongly than does material in the tidal bulge on the back side of Earth. Because the tidal bulge on the Moonward side of Earth "leads" the Moon somewhat, as shown in **Figure 4.16**, the gravitational attraction of the bulge causes the Moon to accelerate slightly along the direction of its orbit around Earth and simultaneously causes Earth's rotation to slow down. The rotation of Earth is dragging the Moon along with it. The acceleration of the Moon in the direction of its orbital motion causes the orbit of the Moon to grow larger. At present, the Moon is drifting away from Earth at a rate of 3.83 cm per year.

As the Moon grows more distant, the lunar month gets about 0.014 second longer each century. If that increase in the radius of the Moon's orbit were to continue long enough (about 50 billion years), Earth would become tidally locked to the Moon, just as the Moon is now tidally locked to Earth. At that point, the Earth's period of rotation, the Moon's period of rotation, and the Moon's orbital period would all be the same—about 47 of our current days—and the Moon would be about 43 percent farther from Earth than it is today. That situation will never actually occur, however, because the Sun will have burned itself out long before then.

The effects of tidal forces are apparent throughout the Solar System. Most of the moons in the Solar System are tidally locked to their parent planets, and for dwarf planet Pluto and its largest moon, Charon, each is tidally locked to the other.

Tidal locking is only one way that orbits and rotations can be coupled. Tidal forces have coupled the planet Mercury's rotation to its very elliptical orbit around the Sun. Unlike the Moon's synchronous rotation, however, Mercury spins on its axis three times for every two trips around the Sun. The period of Mercury's orbit—87.97 Earth days—is exactly $1\frac{1}{2}$ times the 58.64 days that Mercury takes to spin once on its axis. When Mercury comes to the point in its orbit that is closest to the Sun, one hemisphere faces the Sun, and then in the next orbit, the other hemisphere faces the Sun.

Tidal Forces on Many Scales

We normally think of the effects of tidal forces as small in comparison with the force of gravity holding an object together, yet tidal effects can be extremely destructive. Consider for a moment the fate of a small moon, asteroid, or comet that wanders too close to a massive planet such as Jupiter or Saturn. All objects in the Solar System larger than about a kilometer in diameter are held together by their self-gravity. However, the self-gravity of a small object such as an asteroid, a comet, or a small moon is feeble. In contrast, the tidal forces close to a massive object such as Jupiter can be very strong. If the tidal forces trying to tear an object apart become greater than the self-gravity trying to hold the object together, the object will break into pieces.

The **Roche limit** is the distance at which a planet's tidal forces become greater than the self-gravity of a smaller object—such as a moon, asteroid, or comet—causing the object to break apart. For a smaller object having the same density as the planet, the Roche limit is about 2.45 times the planet's radius. Such an object bound together solely by its own gravity can remain intact when it is outside a planet's Roche limit, but not when it is inside the limit. Objects such as the International Space Station and other Earth satellites are not torn apart, even though they orbit well within Earth's Roche limit, because chemical and physical bonds hold them together, not just self-gravity.

Tidal forces exist throughout the Solar System and the universe. Any time two objects of significant size or two collections of objects interact gravitationally, the gravitational forces will differ from one place to another within the objects, giving rise to tidal effects. Tidal disruption of small bodies is the source of the particles that make up the rings of the giant planets. Tidal interactions can cause material from one star in a binary pair to be pulled onto the other star. Tidal effects can strip stars from clusters consisting of thousands of stars. Galaxies can pass close enough together to strongly interact gravitationally. When that happens, as in **Figure 4.17**, tidal effects can grossly distort both galaxies. Tidal forces even play a role in shaping huge collections of galaxies—the largest known structures in the universe.

CHECK YOUR UNDERSTANDING 4.4

The Moon always keeps the same face toward Earth because of: (a) tidal locking; (b) tidal forces from the Sun; (c) tidal forces from Earth and the Sun.

Origins: Tidal Forces and Life

Earth's rotation is slowing down as the Moon slowly moves away into a larger orbit, so in the distant past the Moon was closer and Earth rotated faster. Tides at that time would have been stronger and the interval between high tides would have been shorter. Tides are also affected by the configuration of the continents and oceans on Earth, so precisely how much faster Earth rotated billions of years ago is unknown. What is known, though, is that the stronger and more frequent tides would have provided additional energy to the oceans of the young Earth.

Scientists debate whether life on Earth originated deep in the ocean, on the surface of the ocean, or in hot springs on land (see Chapter 24). The tides shaped the regions in the margins between land and ocean, such as tide pools and coastal flats. Some researchers think that those border regions, which alternate between wet and dry with the tides, could have been places where concentrations of biochemicals periodically became high enough for more complex reactions to take place. Those complex reactions were important to the earliest life. Later, those border regions may have been important as advanced life moved from the sea to the land.

Elsewhere in the Solar System, the giant planets Jupiter and Saturn are far from the Sun and thus very cold. Both Jupiter and Saturn have many moons, and the closest moons would experience strong tidal forces from their respective planets. Several of those moons are thought to have a liquid ocean underneath an icy surface. Tidal forces from Jupiter or Saturn supply the heat to keep the water in a liquid state. Astrobiologists think that those subsurface liquid oceans, such as the ones on Saturn's moon Enceladus, are perhaps the most probable location for life elsewhere in the Solar System.

On Earth, solar tides are about half as strong as lunar tides. A planet with a closer orbit would experience much stronger tidal forces from its star. Many planets detected outside the Solar System have orbits very close to their stars (see Chapter 7), so those exoplanets experience strong tidal forces. The planets might be tidally locked, too, so they have a synchronous rotation like the Moon does around Earth, with one side of the planet always facing the star and one side facing away. How might life on Earth have evolved differently if half the planet was in perpetual night and half in perpetual day?

Figure 4.17 ★ WHAT AN ASTRONOMER SEES
An astronomer looking at this image will instantly identify that there are two galaxies here, indicated by the bright white blobs. She will see that these galaxies are in motion, because the tidal "tails" that stream roughly diagonally across the image are characteristic of tidal interactions between galaxies. These are made of stars, gas, and dust from each galaxy that have been left behind as the galaxies come together. To an astronomer, this picture is dynamic, providing a snapshot of a time during which these galaxies are evolving from two distinct galaxies into one larger one.

Credit: NASA, Holland Ford (JHU), the ACS Science Team and ESA. https://esahubble.org/images/heic0206b/. https://creativecommons.org/licenses/by/4.0/.

SUMMARY

Objects stay in orbit because of gravity. Newton's laws of motion and his proposed law of gravity predict the paths of planetary orbits and explain Kepler's laws. Newton's calculations showed that those orbits should be ellipses with the Sun at one focus, that planets should move faster when closer to the Sun, and that the square of the period of a planet's orbit should vary as the cube of the semimajor axis of that elliptical orbit. Newton also mathematically confirmed Galileo's conclusion that falling objects have an accelerated motion independent of their mass. Tidal forces provide energy to Earth's oceans. Tide pools on Earth may have been a site of early biochemical reactions. Some moons in the outer Solar System might have liquid water because of tidal heating from their respective planets.

① **Identify the elements of Newton's universal law of gravitation.** Gravity is a force between any two objects with mass. Gravity is one of the fundamental forces of nature—it binds the universe together. The force of gravity is proportional to the mass of each object and inversely proportional to the square of the distance between them.

② **Use the laws of motion and gravitation to relate planetary orbits, trajectories of rocket flights, and the escape velocity from a system.**

An orbit is one body "falling around" another. Planets orbit the Sun in elliptical orbits. All objects affected by gravity move either in bound elliptical orbits or unbound paths. If an object acquires escape velocity, it is moving fast enough to be on an unbound path. Orbits and trajectories are ultimately given their shape by the gravitational attraction of the objects involved, which in turn reflects the masses of those objects.

③ **Explain how gravitational forces from the Sun and Moon create Earth's tides.** Tides on Earth are the result of differences between how hard the Moon and Sun pull on one part of Earth in comparison with their pull on other parts of Earth. The primary cause of tides is the Moon, which stretches out Earth. The tides are the strongest when the Sun, Moon, and Earth are aligned. As Earth rotates, tides rise and fall twice each day.

④ **Describe how tidal forces affect solid bodies.** Tidal forces lock the Moon's rotation to its orbit around Earth. Tidal forces can break up an object if it gets too close to a more massive object. Tidal forces are observed throughout the universe—in planets and moons, pairs of stars, and interacting galaxies.

QUESTIONS AND PROBLEMS

TEST YOUR UNDERSTANDING

1. In Newton's universal law of gravitation, the force is
   a. proportional to both masses.
   b. proportional to the radius.
   c. proportional to the radius squared.
   d. inversely proportional to the orbiting mass.

2. If we move a satellite to a new, higher altitude above Earth, the satellite would now
   a. have a faster speed.
   b. have a higher mass.
   c. have a lower mass.
   d. have a slower speed.

3. An object in a(n) _____ orbit in the Solar System will remain in its orbit forever. An object in a(n) _____ orbit will escape from the Solar System.
   a. unbound; bound
   b. circular; elliptical
   c. bound; unbound
   d. elliptical; circular

4. Compared with your mass on Earth, your mass on the Moon would be
   a. lower because the Moon is smaller than Earth.
   b. lower because the Moon has less mass than Earth.
   c. higher because of the combination of the Moon's mass and size.
   d. the same—mass doesn't change.

5. If you went to Mars, your weight would be
   a. higher because you are closer to the center of the planet.
   b. lower because Mars has two small moons instead of one big moon, so less tidal force is present.
   c. lower because the lower mass and smaller radius of Mars combine to a lower surface gravity.
   d. the same as on Earth.

6. Venus has about 80 percent of Earth's mass and about 95 percent of Earth's radius. Your weight on Venus would be
   a. 20 percent more than on Earth.
   b. 20 percent less than on Earth.
   c. 10 percent more than on Earth.
   d. 10 percent less than on Earth.

7. The connection between gravity and orbits enables astronomers to measure the _____ of stars and planets.
   a. distances
   b. sizes
   c. masses
   d. compositions

8. If the Moon were twice as massive, how would the strength of lunar tides change?
   a. The highs would be higher, and the lows would be lower.
   b. Both the highs and the lows would be higher.
   c. The highs would be lower, and the lows would be higher.
   d. Nothing would change.

9. If Earth had half its current radius, how would the strength of lunar tides change?
   a. The highs would be higher, and the lows would be lower.
   b. Both the highs and the lows would be higher.
   c. The highs would be lower, and the lows would be higher.
   d. Both the highs and the lows would be lower.

10. If the Moon were 2 times closer to Earth than it is now, the gravitational force between Earth and the Moon would be
    a. 2 times stronger.
    b. 4 times stronger.
    c. 8 times stronger.
    d. 16 times stronger.

11. If the Moon were 2 times closer to Earth than it is now, the tides would be
    a. 2 times stronger.
    b. 4 times stronger.
    c. 8 times stronger.
    d. 16 times stronger.

12. If two objects are tidally locked to each other,
    a. the tides always stay on the same place on each object.
    b. the objects always remain in the same place in each other's sky.
    c. the objects are falling together.
    d. both a and b would be true.

13. Spring tides occur only when
    a. the Sun is near the vernal equinox.
    b. the Moon's phase is new or full.
    c. the Moon's phase is first quarter or third quarter.
    d. it is either spring or fall.

14. If an object crosses from farther to closer than the Roche limit, it
    a. can no longer be seen.
    b. begins to accelerate very quickly.
    c. slows down.
    d. may be torn apart.

15. Self-gravity is
    a. the gravitational pull of a person.
    b. the force that holds objects such as people and lamps together.
    c. the gravitational interaction of all the parts of a body.
    d. the force that holds objects on Earth.

THINKING ABOUT THE CONCEPTS

16. Both Kepler's laws and Newton's laws tell us something about the motion of the planets, but they have a fundamental difference. What is that difference?

17. Explain the difference between circular velocity and escape velocity. Which of those must be larger? Why?

18. Explain the difference between weight and mass.

19. Weight on Earth is proportional to mass. Weight is proportional to mass on the Moon, too, but the multiplier is different on the Moon from what it is on Earth. Why? Explain why that difference does not violate the universality of physical law, as described in the Process of Science Figure. ☽

20. Suppose a satellite has its orbit disrupted so that it begins traveling at a speed of twice the circular velocity. Study Figure 4.9 to determine if the orbit of the satellite remains bound, or if it is unbound. ☽

21. What is the advantage of launching satellites from spaceports located near the equator? Would you expect satellites to be launched to the east or to the west? Why?

22. Explain how to use celestial orbits to estimate an object's mass. What observational quantities do you need to make that estimation?

23. What determines the strength of gravity at various radii between Earth's center and its surface?

24. The best time to dig for clams along the seashore is when the ocean tide is at its lowest. What phases of the Moon would be best for clam digging? What would be the best times of day during those phases?

25. The Moon is on the meridian at your seaside home, but your tide calendar does not show that it is high tide. Use Figure 4.12 to explain that apparent discrepancy. ☽

26. We may have an intuitive feeling for why lunar tides raise sea level on the side of Earth facing the Moon, but why is sea level also raised on the side facing away from the Moon?

27. Tides raise and lower the level of Earth's oceans. Can they do the same for Earth's landmasses? Explain your answer.

28. Lunar tides raise the ocean surface by less than 1 meter. How can tides as large as 5–10 meters occur?

29. Most commercial satellites are well inside the Roche limit as they orbit Earth. Why are they not torn apart?

30. ★ WHAT AN ASTRONOMER SEES   In Figure 4.17, two galaxies are merging. The tidal tail of the galaxy on the right shows its trajectory and indicates that these galaxies are about to pass each other. Why, then, would an astronomer say these galaxies are going to merge? ☽

APPLYING THE CONCEPTS

31. Mars has about one-tenth the mass of Earth and about half of Earth's radius. What is the value of gravitational acceleration on the surface of Mars in comparison with that on Earth? Compare your estimated mass and weight on Mars with your mass and weight on Earth. Do Hollywood movies showing people on Mars accurately portray that difference in weight? ●–●–●
    a. Make a prediction: Mars is less massive, but also smaller than Earth. Will these factors work together, or will they somewhat cancel out? Do you expect the gravitational acceleration on Mars to be very much bigger, very much smaller, or about the same as the gravitational acceleration on Earth?
    b. Set up a ratio. Because the first part of the question asks for a number "in comparison," the question is asking for a ratio. Use the formula for gravitational acceleration on Earth, rewrite it for Mars, and then set up to find the ratio of g_{Mars}/g_{Earth}.
    c. Solve the right-hand-side of the ratio.
    d. Check your work by comparing your answer to your prediction.
    e. Answer the other questions in the problem; be sure to think about the difference between mass and weight!

32. Earth speeds along at 29.8 km/s in its orbit. Neptune's nearly circular orbit has a radius of 4.5×10^9 km, and Neptune takes 164.8 Earth years to make one trip around the Sun. Calculate how fast Neptune moves along in its orbit. ●—●—●
 a. Make a prediction: Neptune is much farther from the Sun than Earth. Do you expect Neptune to move faster or slower in its orbit?
 b. Calculate: Follow Working It Out 4.2 to calculate Neptune's circular velocity. (Ponder: Why can you use the "circular velocity" formula here?)
 c. Check your work: Check the units of your final answer, and compare your answer to your prediction to check your work.

33. Venus's circular velocity is 35.03 km/s, and its orbital radius is 1.082×10^8 km. Use that information to calculate the mass of the Sun. ●—●—●
 a. Make a prediction: Do you expect the mass of the Sun to be very large or very small? What units do you expect to have for "mass"?
 b. Calculate: Follow Working It Out 4.3 to calculate the mass of the Sun.
 c. Check your work: Check your units and compare to your prediction to check your work. In this instance, the answer is well known, so you can also compare your answer to the actual mass of the Sun to see if your answer is approximately in agreement. (Ponder: You have made an assumption in this problem that the orbit of Venus is perfectly circular. Given that assumption, might it be reasonable for your answer to be slightly "off" from the accepted value?)

34. Some astrologers claim that your destiny is determined by the "influence" of the planets that are rising above the horizon at the moment of your birth. Compare the tidal force of Jupiter (mass = 1.9×10^{27} kg; distance = 7.8×10^8 km) with that of the doctor in attendance at your birth (mass = 80 kg; distance = 1 meter = 0.001 km). ●—●—●
 a. Make a prediction: Consider that in our discussion of the tides, we only worried about the Sun and the Moon. Given that, what might you conclude about the tidal force of Jupiter at the distance of Earth—is it likely to be larger or smaller?
 b. Calculate: Follow Working It Out 4.4 to calculate the ratio of the tidal force from Jupiter to the tidal force from the doctor.
 c. Check your work: Compare your answer to your prediction.

35. At the surface of Earth, the escape velocity is 11.2 km/s. What would be the escape velocity at the surface of a very small asteroid having a radius 10^{-4} that of Earth and a mass 10^{-12} that of Earth?
 a. Make a prediction: This asteroid is less massive, but also much smaller than Earth. Will these factors work together, or will they somewhat cancel out? Do you expect the escape velocity from an asteroid to be very much bigger, very much smaller, or about the same as the escape velocity from Earth?
 b. Set up a ratio: All the information in the problem is given in terms of those values for Earth. Follow Working It Out 4.2 to set up a ratio of the escape velocity for the asteroid divided by the escape velocity from Earth. On the right-hand side, the mass of Earth and the radius of Earth will divide out.
 c. Calculate: Solve for the escape velocity of the asteroid.
 d. Check your work: Compare your answer to your prediction to check your work.

36. When a spacecraft is sent to Mars, it is first launched into an Earth orbit with circular velocity.
 a. Describe the shape of that orbit.
 b. What minimum velocity must we give the spacecraft to send it on its way to Mars?

37. Earth's average radius is 6370 km and its mass is 5.97×10^{24} kg. Show that the acceleration of gravity at the surface of Earth is 9.81 m/s².

38. Using 6370 km for Earth's radius, compare the gravitational force acting on a NASA rocket when it is sitting on its launchpad with the gravitational force acting on it when it is orbiting 350 km above Earth's surface.

39. The International Space Station travels on a nearly circular orbit 350 km above Earth's surface. What is its orbital speed?

40. Using the last equation in Working It Out 4.1, calculate the value of g on Earth. Compare that result with the measured value.

41. How long does Newton's cannonball, moving at 7.9 km/s just above Earth's surface, take to complete one orbit around Earth?

42. The asteroid Ida (mass = 4.2×10^{16} kg) is attended by a tiny asteroidal moon, Dactyl, which orbits Ida at an average distance of 90 km. If you ignore the mass of the tiny moon, what is Dactyl's orbital period in hours?

43. Suppose you go skydiving.
 a. Just as you fall out of the airplane, what is your gravitational acceleration?
 b. Would that acceleration be bigger, smaller, or the same if you were strapped to a flight instructor and so had twice the mass?
 c. Just as you fall out of the airplane, what is the gravitational force on you? (Assume that your mass is 70 kg.)
 d. Would that force be bigger, smaller, or the same if you were strapped to a flight instructor and so had twice the mass?

44. Assume that a planet just like Earth is orbiting the bright star Vega at a distance of 1 astronomical unit (AU). Vega has twice the mass of the Sun.
 a. How long in Earth years will that planet take to complete one orbit around Vega?
 b. How fast is the Earth-like planet traveling in its orbit around Vega?

45. Suppose that in the past the Moon was 80 percent of the distance from Earth that it is now. Calculate how much stronger the lunar tides would have been. How would the neap and spring tides be different from now?

EXPLORATION   Newton's Laws

digital.wwnorton.com/astro7

In the Exploration of Chapter 3, we used the "Planetary Orbits" Interactive Simulation to explore Kepler's laws for Mercury. Now that we know how Newton's laws explain why Kepler's laws describe orbits, we will revisit the simulator to explore the Newtonian features of Mercury's orbit. Visit the Digital Resources Page and on the Student Site open the "Planetary Orbits" Interactive Simulation in Chapter 3.

Acceleration

To begin exploring the simulation, set parameters for "Mercury" in the top left panel and then click "OK." Click the "Newtonian Features" tab at the bottom of the control panel. Select "Show Solar System Orbits" and "Show Grid" under "Visualization Options." Slow down the animation rate, and select the "Play" button.

Examine the graph at the bottom of the panel.

1. Where is Mercury in its orbit when the acceleration is smallest?

_____

_____

2. Where is Mercury in its orbit when the acceleration is largest?

_____

_____

3. What are the values of the largest and smallest accelerations?

_____

_____

In the "Newtonian Features" graph, mark the boxes for "Vector" and "Line" that correspond to the acceleration. Checking these boxes will insert an arrow that shows the direction of the acceleration and a line that extends the arrow.

4. To what Solar System object does the arrow point?

_____

_____

5. In what direction is the force on the planet?

_____

Velocity

Examine the graph at the bottom of the panel again.

6. Where is Mercury in its orbit when the velocity is smallest?

_____

_____

7. Where is Mercury in its orbit when the velocity is largest?

_____

_____

8. What are the values of the largest and smallest velocities?

_____

_____

Add the velocity vector and line to the simulation by clicking on the boxes below the graph window. Study the resulting arrows carefully.

9. Are the velocity and the acceleration always perpendicular (is the angle between the arrows always 90°)?

_____

_____

10. If the orbit were a perfect circle, what would be the angle between the velocity and the acceleration?

_____

_____

Hypothetical Planet

In the top left panel, change the semimajor axis to 0.8 AU.

11. How does that imaginary planet's orbital period now compare with Mercury's? (Check on the Kepler's Third Law tab.)

_____

_____

Now change the semimajor axis to 0.1 AU.

12. How does the planet's orbital period now compare with Mercury's?

_____

_____

13. Summarize your observations of the relationship between the orbital period of an orbiting object and its semimajor axis.

_____

_____

_____

Light bends as it passes through a gap or past a sharp edge. To see this, wrap some tape around the top of a pencil, and then hold another pencil tightly against it to make a small slit. Hold the slit between the pencils up to your eye, and look through it at an LED light, such as the power light on your television or from some other electronic source. You will see a bright line perpendicular to the slit, caused by the light bending as it passes each side of the gap. (If it's difficult to see, you may need to darken the rest of the room.) Rotate the slit, and you will see that the line rotates, too. If you look carefully, you will see a series of bumps in the line. The slit is spreading the light out into a spectrum, and in some places the light overlaps to add together, whereas in other places it overlaps to cancel out. Because the spectrum of an LED has only one color in it, the bumps are very clear and of only one color. Now try this again with a small light source that has multiple colors (like the flashlight of your phone, which appears white). It's more difficult to see, but now the bumps are rainbow colored, with red at one end and blue at the other. You can try this out with all kinds of light sources, to see how they are the same or different.

EXPERIMENT SETUP

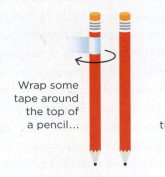

Wrap some tape around the top of a pencil…

SLIT →

…and then hold another pencil upside down and tightly against it to make a small slit.

LED power light on your television or other electronics.

Hold the slit between the pencils up to your eye, and look through it at the LED light. You will see a bright line perpendicular to the slit, caused by the light bending as it passes each side of the gap.

PREDICTION

I predict that the bumps on each side of the slit between the pencils will move

☐ **farther apart**   ☐ **closer together**

if I make the slit smaller by squeezing the pencils closer together.

SKETCH OF RESULTS

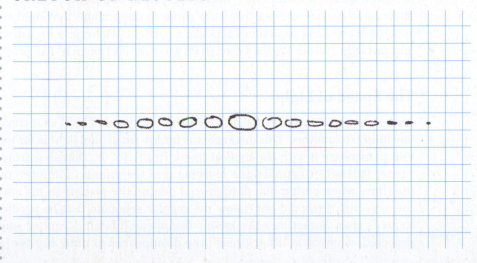

5

Our knowledge of the universe beyond Earth comes from light emitted, absorbed, or reflected by astronomical objects. Light carries information about the temperature, composition, and speed of the objects. Light also tells us about the nature of the material that the light passed through on its way to Earth. Light, however, plays a far larger role in astronomy than as a messenger. Light is one of the primary means by which energy is transported throughout the universe. Stars, planets, and vast clouds of gas and dust filling the space between the stars heat up as they absorb light and cool off as they emit light. Light carries energy generated in the heart of a star outward through the star and off into space. Light transports energy from the Sun outward through the Solar System, heating the planets; and light carries energy away from each planet, allowing each one to cool. The balance between those two processes establishes each planet's temperature and therefore a planet's possible suitability for life.

LEARNING GOALS

An astronomer must try to understand the universe by studying the light and other particles that reach Earth from distant objects. By the end of this chapter, you should be able to:

1. Compare the wave and particle properties of light, and describe the electromagnetic spectrum.

2. Describe how to measure the chemical composition of distant objects by using the unique spectral lines of different types of atoms.

3. Apply the Doppler effect and use it to measure the motion of distant objects.

4. Explain how the spectrum of light that an object emits depends on its temperature.

5. Differentiate luminosity from brightness, and illustrate how distance affects each.

5.1 Light Brings Us the News of the Universe

Is light a **wave**—a disturbance that travels from one point to another—or is it made up of particles? Scientists have come to understand that light sometimes acts like a wave and sometimes acts like a particle. We begin with a discussion of how fast light travels. We then discuss its wavelike and particle-like properties.

The Speed of Light

In the early 1600s, Galileo tried to measure the speed of light as it traveled from one hilltop to another. This distance was far too small, and light is far too fast, for him to measure it with the tools at hand. In the 1670s, Danish astronomer Ole Rømer (1644–1710) studied the movement of the moons of Jupiter, measuring the times when each moon disappeared behind the planet. To his amazement, the observed times did not follow the regular schedule that he predicted from Kepler's laws. Sometimes the moons disappeared behind Jupiter sooner than expected, whereas at other times they disappeared behind Jupiter later than

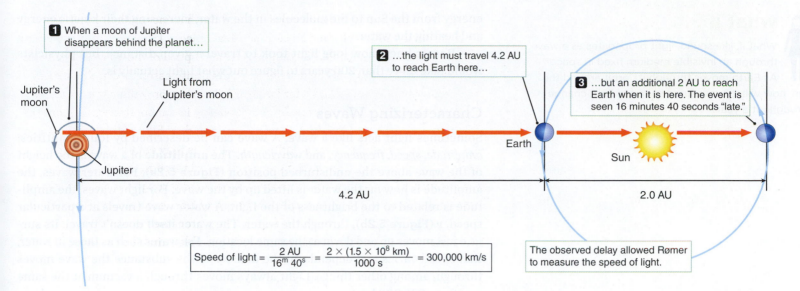

1. When a moon of Jupiter disappears behind the planet…

2. …the light must travel 4.2 AU to reach Earth here…

3. …but an additional 2 AU to reach Earth when it is here. The event is seen 16 minutes 40 seconds "late."

Jupiter's moon

Light from Jupiter's moon

Jupiter

Earth

Sun

4.2 AU

2.0 AU

$$\text{Speed of light} = \frac{2\ \text{AU}}{16^m\ 40^s} = \frac{2 \times (1.5 \times 10^8\ \text{km})}{1000\ \text{s}} = 300{,}000\ \text{km/s}$$

The observed delay allowed Rømer to measure the speed of light.

Figure 5.1 Danish astronomer Ole Rømer realized that apparent differences between the predicted and observed orbital motions of Jupiter's moons depend on the distance between Earth and Jupiter. He used those observations to measure the speed of light. (The superscript letters in "$16^m\ 40^s$" stand for minutes and seconds of time, respectively.)

expected. Rømer realized that the difference depended on where Earth was in its orbit. If he began tracking the moons when Earth was closest to Jupiter, the moons were almost 17 minutes "late" by the time Earth was farthest from Jupiter. When Earth was once again closest to Jupiter, the moons again passed behind Jupiter at the predicted times.

Rømer correctly concluded that his observations did not represent a failure of Kepler's laws. Instead, he was seeing the first clear evidence that light travels at a finite speed. As shown in **Figure 5.1**, the moons appeared "late" when Earth was farther from Jupiter because of the time needed for light to travel the extra distance between the two planets. Over the course of Earth's yearly trip around the Sun, the distance between Earth and Jupiter changes by 2 astronomical units (AU). The speed of light equals that distance divided by Rømer's 16.7-minute delay, or about 3×10^5 kilometers per second (km/s). The value that Rømer actually announced in 1676 was a bit on the low side—2.25×10^5 km/s—because the size of Earth's orbit was not well known. Modern measurements of the speed of light give a value of 2.99792458×10^5 km/s in a **vacuum** (a region of space devoid of matter). The speed of light in a vacuum is one of nature's fundamental constants, usually written as c (lowercase). The speed of light through any medium, such as air or glass, is always less than c. The International Space Station moves around Earth at a speed of about 28,000 kilometers per hour (km/h), taking 91 minutes to complete one orbit. Light travels almost 40,000 times faster than that and can circle Earth in only $1/7$ of a second.

Because light is so fast, and its speed is constant, it is convenient to express cosmic distances in terms of the time it takes light to travel that far, in units such as light-seconds, light-hours, or light-years. Light takes $1\frac{1}{4}$ seconds to travel between Earth and the Moon, so the Moon is $1\frac{1}{4}$ *light-seconds* from Earth. The Sun is $8\frac{1}{3}$ *light-minutes* away, and the next-nearest star is $4\frac{1}{3}$ *light-years* distant. A *light-year*—how far light travels in 1 year—is about 9.5 trillion km.

While traveling at that high speed, light carries energy from place to place. **Energy** is the ability to do work, and it comes in many forms. **Kinetic energy (E_k)** is the energy of moving objects. **Thermal energy** is closely related to kinetic energy and is the sum of all the random motion of atoms, molecules, and particles in some region, which we characterize by its temperature. For example, when light from the Sun strikes a body of water, the water heats up. Light carried that

unanswered questions

Has the speed of light always been 300,000 km/s? Some theoretical physicists have questioned whether light traveled much faster earlier in the history of our universe. The observational evidence that may test that idea comes from studying the spectra of the most distant objects—whose light has been traveling for billions of years—and determining whether billions of years ago chemical elements absorbed light differently from how they do today. So far, no evidence exists that the speed of light has changed.

what if . . .

What if, like sound, light propagates as a wave through an invisible medium fixed in space? As Earth moves through its orbit around the Sun, how will the measured speed of light change throughout a year?

energy from the Sun to the molecules in the water, increasing their kinetic energy and heating the water.

Rømer knew how long light took to travel a given distance, but physicists would take more than 200 years to figure out what light actually is.

Characterizing Waves

Sometimes light acts like a wave. A wave can be described by four quantities: *amplitude*, *speed*, *frequency*, and *wavelength*. The **amplitude** of a wave is the height of the wave above the undisturbed position (**Figure 5.2a**). For water waves, the amplitude is how far the water is lifted up by the wave. For light waves, the amplitude is related to the brightness of the light. A water wave travels at a particular speed, v (**Figure 5.2b**), through the water. The water itself doesn't travel; its surface just moves up and down at the same location. For waves such as those in water, that speed varies and depends on the density of the substance the wave moves through, among other things. Light always moves through a vacuum at the same speed, $c \approx 300{,}000$ km/s.

The distance from one crest of a wave to the next is the **wavelength**, usually denoted by the Greek letter lambda, λ (**Figures 5.2c** and **5.2d**). The number of wave crests passing a point in space each second is the wave's **frequency**, f. The unit of frequency is cycles per second, which is called **hertz (Hz)** after the 19th-century physicist Heinrich Hertz (1857–1894), who was the first to experimentally confirm theoretical predictions about electromagnetic radiation.

Figures 5.2c and 5.2d show that waves with longer (larger) wavelengths have lower (smaller) frequencies, whereas waves with shorter (smaller) wavelengths have higher (larger) frequencies. Higher-frequency waves carry more energy. Imagine standing on an ocean beach with the waves lapping up against you: the amount of energy you feel from the waves increases as the frequency of the ocean waves increases.

Waves travel a distance of one wavelength each cycle, so the speed of a wave can be found by multiplying the wavelength and the frequency; $v = \lambda f$. The speed of light in a vacuum is always c, so once the wavelength or frequency is known, its frequency or wavelength can be found from this equation. Because light travels at constant speed, its wavelength and frequency are inversely proportional to each other. Thus, if the wavelength increases, the frequency decreases,

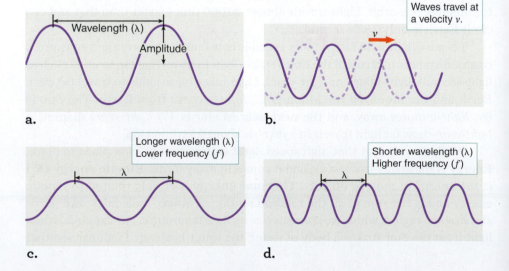

Figure 5.2 A wave is characterized by **a.** the distance from one peak to the next (wavelength, λ), the maximum deviations from the medium's undisturbed state (amplitude), and **b.** the velocity (v) at which the wave pattern travels from one place to another. The wavelength and the frequency (f) of the wave are inversely proportional: when one is large, the other is small, as shown in **c.** and **d.** In an electromagnetic wave, the amplitude is the maximum strength of the electric field, and the speed of light is written as c.

and vice versa. We revisit the idea of a wave in multiple contexts throughout this text—from seismic waves inside Earth to light to pressure waves in stars.

Light as an Electromagnetic Wave

In the late 19th century, the Scottish physicist James Clerk Maxwell (1831–1879) introduced the concept that electricity and magnetism are two components of the same physical phenomenon. An **electric force** is the push or pull between electrically charged particles that make up atoms, such as protons and electrons, arising from their electric charges. Particles with opposite charges attract, whereas those with like charges repel. A **magnetic force** is a force between electrically charged particles arising from their motion.

To describe those electric and magnetic forces, which act at a distance from the charges that create them, Maxwell used the concept of a "field." He thought of electric charges as creating a field around themselves that interacts with other charges, which then experience a force. The **electric field** is caused by a stationary charge and determines the electric force on a charge at any point in space. The **magnetic field** is caused by a moving charge and determines the magnetic force acting on a moving charge at any point in space.

Maxwell summarized the behavior of electric fields and magnetic fields in four elegant equations. Among other things, those equations indicate that a changing electric field causes a magnetic field and that a changing magnetic field causes an electric field. An acceleration (that is, a change) in the motion of a charged particle causes a changing electric field, which causes a changing magnetic field, which causes a changing electric field, and so on. This interaction is illustrated in **Figure 5.3**. Once the process starts, a self-sustaining procession of oscillating electric and magnetic fields moves out in all directions through space. In other words, an accelerating charged particle gives rise to an **electromagnetic wave**. Maxwell's equations also predict the speed at which an electromagnetic wave should travel, which agrees with the measured speed of light (c).

Maxwell's wave description of light also suggests how light originates and how it interacts with matter. When a drop of water falls from the faucet into a sink full of water, it causes a disturbance, or wave, like the one shown in **Figure 5.4a**. The wave moves outward as a ripple on the surface of the water. Similarly, an oscillating particle carrying a charge causes electromagnetic waves that move out through space away from their source in much the same way (**Figure 5.4b**). The ripples in the sink, however, are distortions of the water's surface, and they require a **medium**—a substance to travel through. Light waves do not require a medium. They move through empty space—what we call a vacuum.

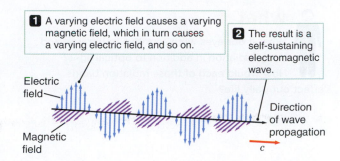

1 A varying electric field causes a varying magnetic field, which in turn causes a varying electric field, and so on.

2 The result is a self-sustaining electromagnetic wave.

Electric field

Magnetic field

Direction of wave propagation

c

Figure 5.3 An electromagnetic wave consists of oscillating electric and magnetic fields that are perpendicular both to each other and to the direction in which the wave travels.

Figure 5.4 a. A drop falling into water generates waves that move outward across the water's surface. b. Similarly, an oscillating (accelerated) electric charge generates electromagnetic waves that move away at the speed of light.

a.

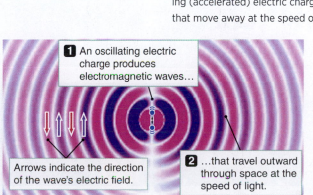

1 An oscillating electric charge produces electromagnetic waves…

Arrows indicate the direction of the wave's electric field.

2 …that travel outward through space at the speed of light.

b.

what if . . .

What if our eyes were sensitive to infrared and X-ray radiation in addition to optical light? How would each of those radiation bands affect our daily life?

Now imagine that a cork is floating in the sink (**Figure 5.5a**). The cork remains stationary until the ripple from the dripping faucet reaches it. As the ripple passes by, the rising and falling water causes the cork to rise and fall. That can happen only if the wave is carrying energy—a conserved quantity that can give objects and particles the ability to change their state. Light waves similarly carry energy through space and cause other electrically charged particles to vibrate, as in **Figure 5.5b**.

The Electromagnetic Spectrum

For visible light, different wavelengths correspond to different colors. Most light signals are made up of many wavelengths. For example, a rainbow is created when white light interacts with water droplets and is spread out by wavelength into its component colors. Red light has a long wavelength, and therefore a low frequency (and a low energy), compared to violet light, which has a short wavelength and a high frequency (and a high energy). Light spread out by wavelength, as in a rainbow, is called a **spectrum**. A commonly used unit for the wavelength of light is the **nanometer (nm)**. A nanometer is one-billionth (10^{-9}) of a meter.

The light-sensitive cells in our eyes respond to visible light, but light can have wavelengths much shorter or much longer than what our eyes can perceive. The whole range of wavelengths of light—collectively called the **electromagnetic spectrum**—is shown in **Figure 5.6**. Most of the electromagnetic spectrum—and therefore most of the information in the universe—is invisible to the human eye. Specialized detectors of various kinds are required to detect light outside the visible range, as we explain in Chapter 6.

Refer to Figure 5.6 as we tour the electromagnetic spectrum, beginning with the shortest wavelengths and working our way to the longest ones. The very shortest wavelengths of light are called **gamma rays**, or sometimes gamma radiation. Because such radiation has the shortest wavelengths, it has the highest frequency and the highest energy, so it penetrates matter easily. Wavelengths between 0.1 and 40 nm are called **X-rays**. You have probably encountered X-rays at the dentist's office or in a hospital's emergency room—X-ray light has enough energy to penetrate skin and muscle but is stopped by denser bone. **Ultraviolet (UV) radiation** has wavelengths between 40 and about 380 nm—longer than X-rays but shorter than visible light. You are familiar with that type of light if you have ever tanned or been sunburned. Thus, UV light has enough energy to penetrate your skin, but not much deeper.

Figure 5.5 a. When waves moving across the surface of water reach a cork, they cause the cork to bob up and down. **b.** Similarly, a passing electromagnetic wave causes an electric charge to oscillate in response to the wave.

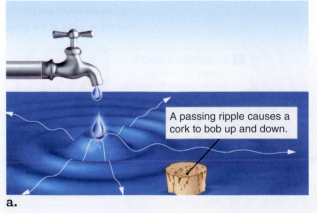

A passing ripple causes a cork to bob up and down.

a.

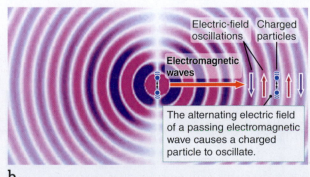

Electric-field oscillations   Charged particles

Electromagnetic waves

The alternating electric field of a passing electromagnetic wave causes a charged particle to oscillate.

b.

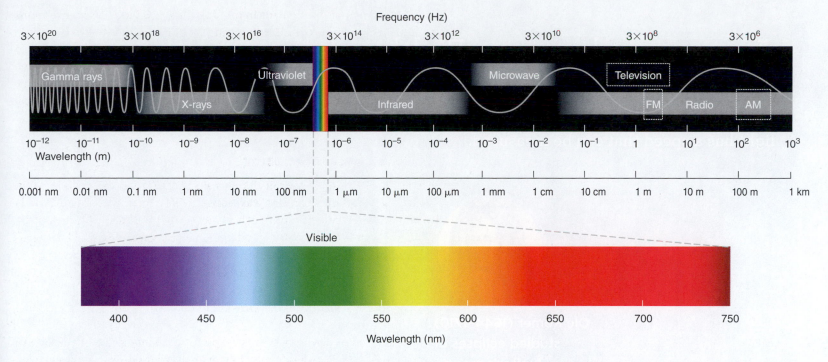

Figure 5.6 By convention, the electromagnetic spectrum is broken into loosely defined regions ranging from gamma rays to radio waves. Throughout the rest of this book, a labeled icon appears below individual astronomical images to identify what part of the spectrum was used to take the image: gamma rays (G), X-rays (X), ultraviolet (U), visible (V), infrared (I), or radio (R). If more than one region of the spectrum was used, multiple labels are highlighted in the icon.

Visible light, the type that the human eye is sensitive to, lies between violet (about 380 nm) and red (750 nm). Stretched out between violet and red are the other colors of the rainbow.

Infrared (IR) radiation (or infrared light) has longer wavelengths (and less energy) than the reddest wavelengths in the visible range. You often feel infrared radiation as heat. When you hold your hand next to a hot stove, for example, some of the heat you feel is carried to your hand by infrared radiation emitted from the stove. In that sense, you could think of your skin as being a giant infrared eyeball—it is sensitive to infrared wavelengths. Infrared radiation is also used in television remote controls, and night vision goggles detect infrared radiation from warm objects such as animals. A useful unit for infrared light is the **micron (μm**, where μ is the Greek letter mu). One micron is 1000 nm, or one-millionth (10^{-6}) of a meter. Infrared wavelengths are longer than about 0.75 microns (the red end of the visible range), and shorter than 500 microns.

Microwave radiation has even longer wavelengths (and less energy) than those of infrared radiation. The microwave in your kitchen heats the water in food by using the light of those wavelengths. The longest-wavelength light, which has wavelengths longer than a few centimeters, is called **radio waves**. The light of those wavelengths in the form of FM, AM, television, and cell phone signals is used to transmit information.

★ **WHAT AN ASTRONOMER SEES** An astronomer looking at this figure will notice that there are two axes shown: frequency is on top and wavelength is on the bottom. She will pause to notice that wavelength is measured in meters and frequency is measured in hertz (Hz). Both scales are logarithmic, so that one tick mark corresponds to a factor of 10 change in the frequency or wavelength. The visible portion of the spectrum is tiny, compared to the rest of the spectrum, and has been "zoomed" below the main graph. Yellow light marks the middle of this visible part of the spectrum, at about 550 nm, just a bit larger than 3×10^{14} Hz.

CHECK YOUR UNDERSTANDING **5.1a**

Rank the following in order from longest wavelength to shortest wavelength: (a) gamma rays; (b) visible light; (c) infrared light; (d) ultraviolet light; (e) radio waves

Answers to Check Your Understanding questions are in the back of the book.

Light as a Particle

Up to now, we've discussed light as a wave but sometimes it acts like a particle. A particle might have mass and sometimes has charge. An electron, for example, is a subatomic

Agreement between Fields

Scientists working on very different problems in different fields all found the same results: light has a speed that can be measured.

Ole Rømer (1644–1710) studied eclipses of Jupiter's moons.

Rømer calculated the speed of light from the eclipse delays of Jupiter's moons.

In 1727, James Bradley (1693–1762) studied the apparent motions of stars, which appeared to make small circles because of the relative motion of Earth.

A half century after Rømer's measurement, Bradley's motion studies led to a more accurate measurement of the speed of light.

$$c = 1/(\sqrt{\mu_0 \varepsilon_0})$$

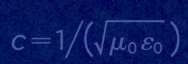

James Clerk Maxwell (1831–1879) studied electricity and magnetism.

Maxwell determined that the speed of electromagnetic waves must be a constant, and is equal to the speed of light.

$$t' = \frac{t}{\sqrt{1 - \frac{v^2}{c^2}}}$$

Einstein (1879–1955) explored the consequences of the fact that the speed of light is the same for all observers.

Predictions from this theory have been tested repeatedly since 1905.

Astronomers and physicists converged on an understanding that photons travel at the "speed of light"—a fundamental constant of the universe that is the same for all observers.

particle that carries a negative charge. If it is moving, a particle has velocity, which means that it also has momentum and kinetic energy. Particles are often visualized as tiny baseballs. This is not entirely correct, but it's useful for distinguishing particle-like behaviors from wavelike behaviors. Particles are *quantized*; like baseballs, they come in individual units. **Quantum mechanics** is a branch of physics that deals with particles and the quantization of energy and of other properties of matter.

The work of Albert Einstein and other scientists showed that light not only acts like a wave, but also acts like a particle. In 1905, Einstein explained the *photoelectric effect*. This effect occurs when light (photo-) strikes a surface and liberates electrons (-electric). The photoelectric effect has two important properties. First, electrons are only liberated when the light has a frequency (and therefore energy) above a certain threshold. The value of the threshold frequency depends on the type of material in the surface. The more the incoming light's frequency is above the threshold, the faster the liberated electrons move. Second, as long as the frequency of the light is above the threshold, the rate at which electrons are emitted depends only on the brightness of incoming light, not the frequency. In order to explain these observations, Einstein proposed that light is made up of massless particles called **photons** (*phot-* means "light," as in *photograph*; and *-on* signifies a particle). Photons always travel at the speed of light (**Process of Science Figure**), and they carry energy. This concept explains the photoelectric effect because it means that each electron interacts with only one photon. If a photon has enough energy to lift the electron out of the atom, any extra energy is carried away in the kinetic energy of the electron. At a fixed frequency above the threshold, brighter light has more photons, and liberates more electrons, but all those electrons have the same "extra" energy, so they have the same speed. This work earned Einstein the Nobel Prize in Physics in 1921.

Several times, we've noted that higher-frequency light has more energy. In fact, energy and frequency are directly proportional, so if a photon has twice as much energy, it has twice the frequency. The constant of proportionality between the energy (E) and the frequency (f) is called Planck's constant (h), which is equal to 6.63×10^{-34} joule-seconds (a joule is a unit of energy):

$$E = hf$$

The wavelength (λ) and frequency (f) of electromagnetic waves are inversely proportional (remember: $c = \lambda f$), so the photon energy is inversely proportional to the wavelength: $E = hc/\lambda$. High-energy visible light is blue, whereas low-energy visible light is red. **Working It Out 5.1** explores the relationships among the wavelength, frequency, and energy of light.

In the particle description of light, the electromagnetic spectrum is a spectrum of photon energies. The higher the frequency of the electromagnetic wave, the greater the energy carried by each photon. Photons of shorter wavelength (higher frequency) carry more energy than that carried by photons of longer wavelength (lower frequency). For example, photons of blue light carry more energy than photons of longer-wavelength red light. Ultraviolet photons carry more energy than photons of visible light, and X-ray photons carry more energy than ultraviolet photons. The lowest-energy photons are radio wave photons. For a beam of red light to carry just as much energy as a beam of blue light, the red beam must have more photons than the blue beam. This concept is illustrated in **Figure 5.7**, which also shows that the relationship between pennies and quarters is similar. The number of pennies required to pay for an item is larger than the number of quarters, because each quarter carries more value.

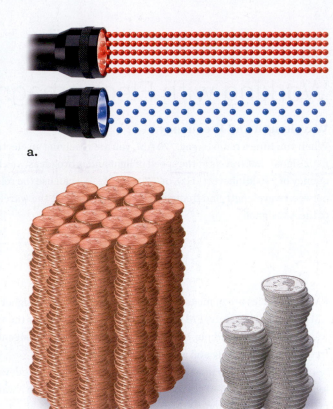

a.

b.

Figure 5.7 a. Photons of red light carry less energy than photons of blue light, so it takes more red photons than blue photons to make a beam of a particular intensity. **b.** Similarly, pennies are worth less than quarters, so it takes more pennies than quarters to add up to $10.

▶❚❚ **AstroTour:** Light as a Wave, Light as a Photon

working it out 5.1

Working with Electromagnetic Radiation

Wavelength and Frequency

When you tune a radio to, say, 770 AM, you are receiving an electromagnetic signal that travels at the speed of light and is broadcast at a frequency of 770 kilohertz (kHz), or 7.7×10^5 Hz. We can use the relationship between wavelength and frequency, $c = \lambda f$, to calculate the wavelength of the AM signal:

$$\lambda = \frac{c}{f} = \left(\frac{3 \times 10^8 \, \text{m/s}}{7.7 \times 10^5 / \text{s}} \right) = 390 \, \text{m}$$

FM frequencies are in megahertz (MHz), a factor of 1000 higher than AM stations. Therefore, FM wavelengths are much shorter than AM wavelengths. For example, a station at 89.5 FM broadcasts signals with a frequency of 89.5 MHz and a wavelength of 3.4 m.

The human eye is most sensitive to light in green and yellow wavelengths, about 500–590 nm. If we examine green light with a wavelength of 530 nm, we can compute its frequency:

$$f = \frac{c}{\lambda} = \left(\frac{3 \times 10^8 \, \text{m/s}}{530 \times 10^{-9} \, \text{m}} \right) = 5.66 \times 10^{14} / \text{s} = 5.66 \times 10^{14} \, \text{Hz}$$

That frequency corresponds to 566 *trillion* wave crests passing by each second.

Photon Energy

How does the energy of an X-ray photon with a wavelength of 1 nm compare with the energy of a visible-light photon with a wavelength of 530 nm (the same green light used in the previous calculation)? The equation for the energy of a photon is $E = hf$. Because $f = c/\lambda$, substituting c/λ for f yields the inverse relationship, $E = hc/\lambda$. Because we are making a *comparison*, we can take a ratio, and then the constants h and c cancel out:

$$\frac{E_{\text{X-ray photon}}}{E_{\text{visible photon}}} = \frac{hc/\lambda_{1 \, \text{nm}}}{hc/\lambda_{530 \, \text{nm}}} = \frac{hc}{hc} \times \frac{\lambda_{500 \, \text{nm}}}{\lambda_{1 \, \text{nm}}} = \frac{530 \, \text{nm}}{1 \, \text{nm}} = 530$$

The X-ray photon has 530 times the energy of the visible-light photon.

CHECK YOUR UNDERSTANDING 5.1b

As wavelength increases, the energy of a photon _____ and its frequency _____. (a) increases; decreases (b) increases; increases (c) decreases; decreases (d) decreases; increases

5.2 The Quantum View of Matter Explains Spectral Lines

Matter is anything that occupies space and has mass. Light and matter interact, and that interaction allows us to detect matter even at great distances in space. To understand that interaction, we must first understand the building blocks of matter. Here in Section 5.2, we review atomic structure and the process by which astronomers identify the chemical elements in astronomical objects.

Atomic Structure

The matter around you is composed of atoms. An atom is composed of a central massive **nucleus**, which contains **protons** with a positive charge and **neutrons**, which have no charge. A cloud of negatively charged **electrons** surrounds the nucleus. Atoms with the same number of protons are all of the same type of

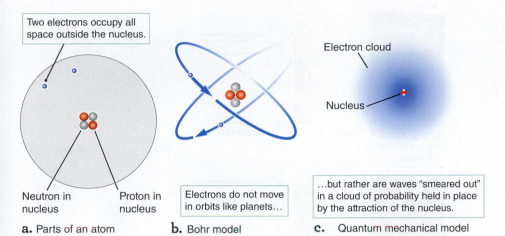

Two electrons occupy all space outside the nucleus.

Neutron in nucleus

Proton in nucleus

a. Parts of an atom

Electrons do not move in orbits like planets…

b. Bohr model

Electron cloud

Nucleus

…but rather are waves "smeared out" in a cloud of probability held in place by the attraction of the nucleus.

c.   Quantum mechanical model

Figure 5.8 a. An atom (here, helium) is made up of a nucleus consisting of positively charged protons and electrically neutral neutrons and is surrounded by much less massive negatively charged electrons. **b.** Atoms are often drawn as miniature "solar systems," but this Bohr model is incorrect. **c.** Electrons are actually smeared out around the nucleus in quantum mechanical clouds of probability.

chemical **element**. For example, an atom with two protons, shown in **Figure 5.8a**, is the element helium. An atom with six protons is the element carbon, one with eight protons is the element oxygen, and so forth. An element may have many **isotopes**: atoms with the same number of protons but different numbers of neutrons. **Molecules** are groups of atoms bound together by shared electrons. A single teaspoon of water contains about 10^{23} molecules—about as many molecules as the number of stars in the observable universe.

Electrons have much less mass than protons or neutrons, so almost all the mass of an atom is found in its nucleus. The Danish physicist Niels Bohr (1885–1962) proposed a model of the atom in 1913 in which a massive nucleus sits in the center and the smaller electrons orbit around it, much as planets orbit around the Sun (**Figure 5.8b**). For an atom to be electrically neutral, it must have the same number of electrons as protons.

The **Bohr model**, however, is an incomplete description of the atom. Just as waves of light have particle-like properties, particles of matter also have wavelike properties. Once that was understood, the Bohr model was modified so that the positively charged nucleus is surrounded by electron "clouds" or "waves," as shown in **Figure 5.8c**. That model conveys the concept that it is not possible to know precisely where an electron is in its orbit. The wave characteristics of particles make it impossible to pin down simultaneously both their exact location and their exact velocity; some uncertainty will always exist. That is why a featureless cloud is used to represent electrons in orbit around an atomic nucleus.

▶❙❙ **AstroTour:** Atomic Energy Levels and the Bohr Model

Atomic Energy Levels

Each atom has a series of energy states, like the shelves of the bookcase depicted in **Figure 5.9a**. The energy of an atom might correspond to the energy of one state or to the energy of the next state, but the energy of the atom is never found between the two states, just as a book can be on only one shelf at a time and cannot be partly on one shelf and partly on another. A given atom may have many energy states available to it, but those states are *discrete*. When electrons in an atom gain or lose energy, the atom shifts from one energy state to another. Because the atom's energy states are discrete, the electrons must gain or lose energy in particular amounts, corresponding to the difference between energy levels in the atom. An electron cannot gain an amount of energy that puts the atom between states.

Figure 5.9 **a.** Energy states of an atom are analogous to shelves in a bookcase. You can move a book from one shelf to another, but books can never be placed between shelves. **b.** Atoms exist in one allowed energy state or another but never in between. The ground state is the lowest possible level.

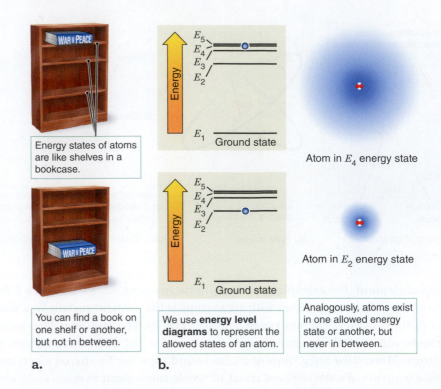

Energy states of atoms are like shelves in a bookcase.

Atom in E_4 energy state

Atom in E_2 energy state

You can find a book on one shelf or another, but not in between.

We use **energy level diagrams** to represent the allowed states of an atom.

Analogously, atoms exist in one allowed energy state or another, but never in between.

a.   **b.**

Astronomers keep track of the allowed states of an atom by using energy level diagrams, as shown in **Figure 5.9b**. Both the bookcase and the energy level diagram are simplifications of the possible energies of a three-dimensional system. The lowest possible energy state for an atom is called the **ground state**. When the atom is in the ground state, the electron has its minimum energy. It can't give up any more energy to move to a lower state because a lower state does not exist. An atom will remain in its ground state forever unless it gets energy from outside itself. In the bookcase analogy, a book sitting on the bottom shelf at the floor is in its ground state. It has nowhere left to fall, and it cannot jump to one of the higher shelves on its own.

Energy levels above the ground state are called **excited states**. Just as a book on an upper shelf might fall to a lower shelf, an atom in an excited state might **decay** to a lower state by getting rid of some of its extra energy. A common way for an atom to do that is to emit a photon. The photon carries away exactly the amount of energy that the atom loses as it goes from the higher energy state to the lower energy state. Similarly, atoms moving from a lower energy state to a higher energy state can absorb only certain specific energies. Unlike in the analogy of the bookcase, where you might happen to see the book between shelves as it falls, the transition of an atom between energy states is instantaneous: When the atom drops from a higher state to a lower one, the difference in energy between the two states is carried off all at once. We commonly refer to this change as a transition of the electron between energy states, and we may say that "an electron dropped from state 3 to state 2, emitting a photon."

To make an analogy with money, suppose you have a penny (1 cent), a nickel (5 cents), and a dime (10 cents), totaling 16 cents. Now imagine that you give away the nickel and are left with 11 cents. You never had exactly 13 cents or 13.6 cents. You had 16 cents and then 11 cents. Atoms don't accept and give away money to change energy states, but they do accept and give away photons with well-defined energies.

Emission Spectra   Imagine a hypothetical atom that has only two available energy states. The energy of the lower energy state (the ground state) is E_1, and the energy of the higher energy state (the excited state) is E_2. The energy levels of that atom

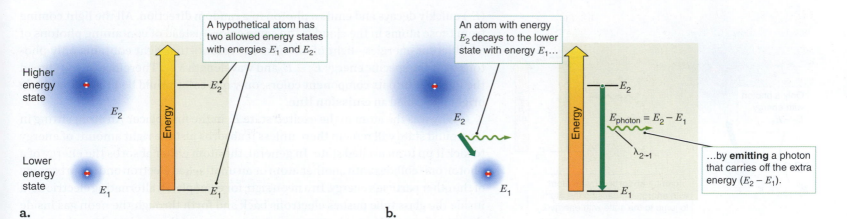

A hypothetical atom has two allowed energy states with energies E_1 and E_2.

Higher energy state E_2

Lower energy state E_1

a.

An atom with energy E_2 decays to the lower state with energy E_1...

$E_{photon} = E_2 - E_1$

$\lambda_{2\to1}$

...by **emitting** a photon that carries off the extra energy ($E_2 - E_1$).

b.

can be represented in an energy level diagram such as the one in **Figure 5.10a**. An atom in the excited state moves to the ground state by getting rid of the "extra" energy all at once. In that **emission** process, a photon is released.

In **Figure 5.10b**, the green arrow pointing straight down indicates that the atom went from the higher state (E_2) to the lower state (E_1). The atom lost an amount of energy equal to $E_2 - E_1$, the difference between the two states. Because energy is never truly lost or created, the energy the atom lost has to show up somewhere. Here, the energy shows up in the form of a photon emitted by the atom. The energy of the emitted photon, E_{photon}, matches the energy lost by the atom; that is, $E_{photon} = E_2 - E_1$.

An atom can emit photons with energies corresponding *only* to the difference between two of its allowed energy states. Because the energy of a photon is related to the frequency or wavelength of electromagnetic radiation, a photon of energy $E_{photon} = E_2 - E_1$ has a specific frequency $f_{2\to1}$ and a specific wavelength $\lambda_{2\to1}$ for the transition from level 2 to level 1. Therefore, those emitted photons have a specific color, and every photon emitted in any transition from E_2 to E_1 will have that color. The energy level structure of an atom determines the wavelengths of the photons it emits—the color of the light that the atom gives off.

Figure 5.11 illustrates the light coming from a cloud of gas consisting of the hypothetical two-state atoms in Figure 5.10. Any atom in the higher energy state

Figure 5.10 **a.** The energy levels of a hypothetical two-level atom. **b.** A photon with energy $E_{photon} = E_2 - E_1$ is emitted when an atom in the higher energy state (E_2) decays to the lower energy state (E_1).

Figure 5.11 A cloud of gas containing atoms with two energy states, E_1 and E_2, emits photons with an energy $E_{photon} = E_2 - E_1$, which appear in the spectrum (right) as a single bright *emission line*.

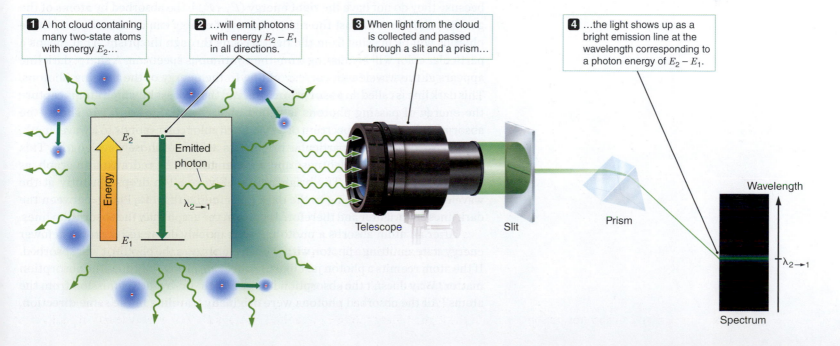

1 A hot cloud containing many two-state atoms with energy E_2...

2 ...will emit photons with energy $E_2 - E_1$ in all directions.

3 When light from the cloud is collected and passed through a slit and a prism...

4 ...the light shows up as a bright emission line at the wavelength corresponding to a photon energy of $E_2 - E_1$.

E_2

Emitted photon

$\lambda_{2\to1}$

E_1

Telescope

Slit

Prism

Wavelength

$\lambda_{2\to1}$

Spectrum

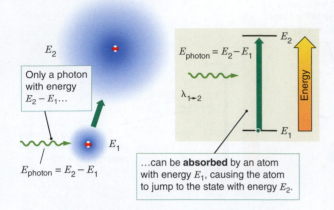

Figure 5.12 An atom in a lower energy state (E_1) may absorb a photon of energy $E_{photon} = E_2 - E_1$, leaving the atom in a higher energy state (E_2).

(E_2) quickly decays and emits a photon in a random direction. All the light coming from those atoms in the cloud is the same color. Instead of containing photons of all different energies—light of all different colors—that light contains only photons with the specific energy $E_2 - E_1$ and wavelength $\lambda_{2\rightarrow1}$. Therefore, if you spread the light out into its component colors, only one color would be present—a single bright line called an **emission line**.

Why was the atom in the excited state E_2 in the first place? An atom sitting in its ground state will remain there unless it absorbs just the right amount of energy to kick it up to an excited state. In general, the atom either absorbs the energy of a photon or it collides with another atom or an unattached electron and absorbs some of the other particle's energy. In a neon sign, for example, an alternating electric field inside the glass tube pushes electrons back and forth through the neon gas inside the tube. Some of those electrons crash into atoms of the gas, knocking them into excited states. The atoms then drop back down to their ground states by emitting photons, causing the gas inside the tube to glow a red-orange color characteristic of the element neon.

Absorption Spectra An atom in a low energy state can absorb the energy of a passing photon and move to a higher energy state, as shown in **Figure 5.12**. Once again, the energy required to go from E_1 to E_2 is the difference in energy between the two states, $E_2 - E_1$. The only photons that can excite atoms from E_1 to E_2 are photons with exactly $E_{photon} = E_2 - E_1$. Those absorbed photons have precisely the same energy as the photons emitted by the atoms when they decay from E_2 to E_1. That is not a coincidence. The energy difference between the two levels is the same whether the atom is emitting a photon or absorbing one, so the energy of the photon involved is the same in either case. In both cases, the photons have a corresponding frequency ($f_{1\rightarrow2} = E_{photon}/h$) and wavelength ($\lambda_{1\rightarrow2} = hc/E_{photon}$), so they have the same color.

In **Figure 5.13a**, white light (which has all wavelengths of photons in it) passes directly through a glass prism and spreads out into a rainbow of colors, called a **continuous spectrum**. If the white light passes through a cool cloud composed of the hypothetical gas of two-state atoms (**Figure 5.13b**), however, some photons will be absorbed. Almost all the photons will pass through the cloud of gas unaffected because they do not have the right energy ($E_2 - E_1$) to be absorbed by atoms of the gas. But photons with just the right amount of energy can be absorbed, so those photons will be missing from the light passing through the prism. That means a particular color will be missing from the continuous spectrum. A sharp, dark line appears at the wavelength corresponding to the energy of the missing photons. This dark line is called an **absorption line**, and the process by which atoms capture the energy of passing photons is called **absorption**. **Figure 5.14a** shows the absorption lines in the spectrum of a star. Each color corresponds to a wavelength, and the spectrum has a particular brightness at each of those wavelengths. This relationship between brightness and wavelength can be redrawn as a graph, as shown in **Figure 5.14b**. The brightness in Figure 5.14b drops abruptly at the wavelengths corresponding to the dark lines in Figure 5.14a. Places between the dark lines are brighter and therefore higher on the graph than the absorption lines.

After an atom absorbs a photon, it may quickly decay to its previous lower energy state, emitting a photon with the same energy as the photon it just absorbed. If the atom reemits a photon just like the one it absorbed, why does the absorption matter? Why doesn't the absorption line get "filled in" by the reemission from the atoms? All the absorbed photons were originally traveling in the same direction,

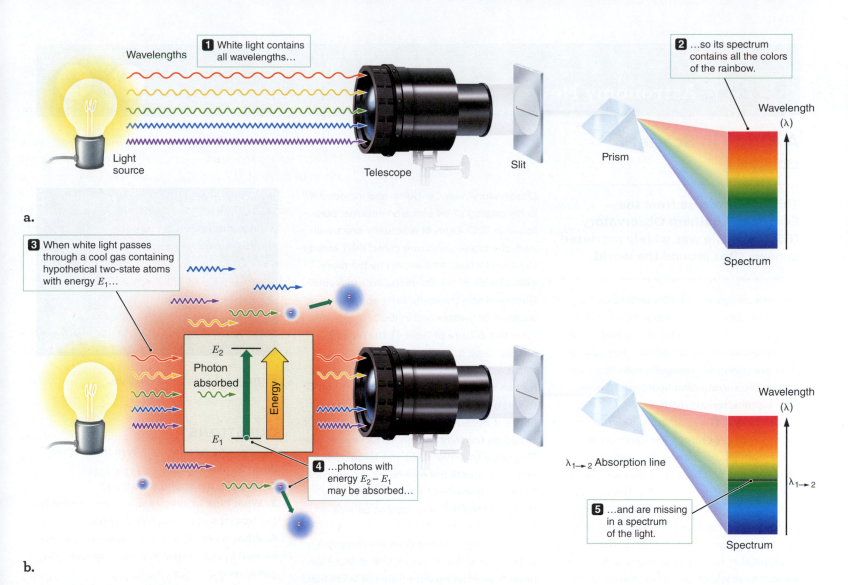

a.

1 White light contains all wavelengths...

Wavelengths

Light source

Telescope

Slit

Prism

2 ...so its spectrum contains all the colors of the rainbow.

Wavelength (λ)

Spectrum

b.

3 When white light passes through a cool gas containing hypothetical two-state atoms with energy E_1...

E_2

Photon absorbed

Energy

E_1

4 ...photons with energy $E_2 - E_1$ may be absorbed...

Telescope

Wavelength (λ)

$\lambda_{1 \rightarrow 2}$ Absorption line

$\lambda_{1 \rightarrow 2}$

5 ...and are missing in a spectrum of the light.

Spectrum

Figure 5.13 **a.** When passed through a prism, white light produces a spectrum containing all colors. **b.** When light of all colors passes through a cloud of hypothetical two-state atoms, photons with energy $E_{photon} = E_2 - E_1$ may be absorbed, leading to the dark absorption line in the spectrum.

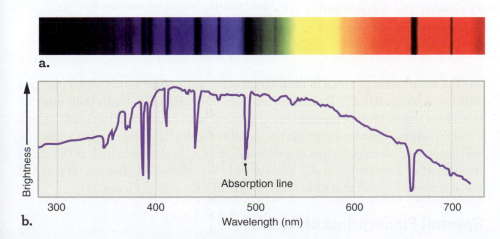

a.

Brightness

Absorption line

300   400   500   600   700

Wavelength (nm)

b.

Figure 5.14 Absorption lines in the spectrum of a star as an image **a.** and a graph **b.**

Reading Astronomy News

A Study in Scarlet

ESO

This press release from the European Southern Observatory (ESO) in Chile was widely reprinted in news sites around the world.

This new image from ESO's La Silla Observatory in Chile reveals a cloud of hydrogen called Gum 41 (**Figure 5.15**). In the middle of this little-known nebula, brilliant hot young stars are giving off energetic radiation that causes the surrounding hydrogen to glow with a characteristic red hue.

This area of the southern sky, in the constellation of Centaurus (The Centaur), is home to many bright nebulae, each associated with hot newborn stars that formed out of the clouds of hydrogen gas. The intense radiation from the stellar newborns excites the remaining hydrogen around them, making the gas glow in the distinctive shade of red typical of star-forming regions. Another famous example of this phenomenon is the Lagoon Nebula, a vast cloud that glows in similar bright shades of scarlet.

The nebula in this picture is located some 7300 light-years from Earth. Australian astronomer Colin Gum discovered it on photographs taken at the Mount Stromlo

Observatory near Canberra, and included it in his catalog of 84 emission nebulae, published in 1955. Gum 41 is actually one small part of a bigger structure called the Lambda Centauri Nebula, also known by the more exotic name of the Running Chicken Nebula. Gum died at a tragically early age in a skiing accident in Switzerland in 1960.

In this picture of Gum 41, the clouds appear to be quite thick and bright, but this is actually misleading. If a hypothetical human space traveler could pass through this nebula, it is likely that they would not notice it because—even at close quarters—it would be too faint for the human eye to see. This helps to explain why this large object had to wait until the mid-twentieth century to be discovered—its light is spread very thinly and the red glow cannot be well seen visually.

This new portrait of Gum 41—likely one of the best so far of this elusive object—has been created using data from the Wide Field Imager (WFI) on the MPG/ESO 2.2-meter telescope at the La Silla Observatory in Chile. It is a combination of images taken through blue, green, and red filters, along with an image using a special filter designed to pick out the red glow from hydrogen.

Figure 5.15 The Gum 41 Nebula.

QUESTIONS

1. How long has the light from that nebula taken to reach us?

2. Why are the young stars blue?

3. What type of spectra would you expect to get from the stars and from the gas?

4. What is happening to the electrons in the excited hydrogen gas in Gum 41 to make the gas glow red?

5. Would you be able to see Gum 41 from your location? Why or why not?

Source: "Photo Release: A Study in Scarlet," ESO.org, April 16, 2014. CC by 4.0.
Photo credit: ESO, https://www.eso.org/public /images/eso1413a/. https://creativecommons.org /licenses/by/4.0/.

▶❚❚ **AstroTour:** Atomic Energy Levels and Light Emission and Absorption

but the atom emits photons in random directions. In other words, some photons with energies equal to $E_2 - E_1$ are diverted from their original paths by their interactions with atoms. Most of those diverted photons will come out of the cloud in some other direction. If you look through the cloud at a white light, as in Figure 5.13b, you will observe an absorption line at a wavelength of $\lambda_{2\to1}$, but if you look at the cloud from another direction, you will observe only the redirected photons, so you will see an emission line at that same wavelength, like the one in Figure 5.11.

Spectral Fingerprints of Atoms

In the mid-19th century, Gustav Kirchoff (1824–1887) first observed emission, continuous, and absorption spectra from the three types of sources shown in

Figures 5.11 and 5.13. He did not know about energy levels in atoms, so he could not create a theory about emission. He summarized his findings as three empirical laws. Kirchoff, together with Robert Bunsen (1811–1899), concluded that the dark-line (absorption) spectrum of the Sun was the "reverse" of the bright-line (emission) spectrum that would be produced by the Sun's atmosphere alone and identified some of the elements on the Sun. (A few years later the element helium was discovered in the solar spectrum.) Others observed the stars. Noteworthy among them were William (1824–1910) and Margaret Lindsay (1848–1915) Huggins, who published an atlas of stellar spectra in 1899 and who showed that the types of atoms seen in the stars are the same as those found on Earth. Spectra are how we know what makes up the stars and planets, and that we on Earth are composed of the same elements.

An atom's allowed energy states are determined by the interactions among the electrons and the nucleus. Real atoms can occupy many more than just two possible energy states; therefore, any given type of atom can emit and absorb photons at many wavelengths. An atom with three energy states, for example, might jump from state 3 to state 2, or from state 3 to state 1, or from state 2 to state 1. The three distinct emission lines in the spectrum from a gas made up of those atoms would have wavelengths of $hc/(E_3 - E_2)$, $hc/(E_3 - E_1)$, and $hc/(E_2 - E_1)$, respectively.

For example, every neutral hydrogen atom consists of a nucleus containing one proton, plus a single electron in a cloud surrounding the nucleus. Therefore, every hydrogen atom has the same energy states available to it, and all hydrogen atoms have the same emission and absorption lines. **Figure 5.16a** shows the energy

Figure 5.16 a. The energy states of the hydrogen atom. Decays to level E_2 emit photons in the visible part of the spectrum. **b.** This visible emission spectrum is what you might see if you looked at the light from a hydrogen lamp projected through a prism onto a screen. **c.** The brightness at every wavelength can be measured to produce a graph of the brightness of spectral lines versus their wavelength. **d.** Emission spectra from gaseous sodium, helium, neon, and mercury.

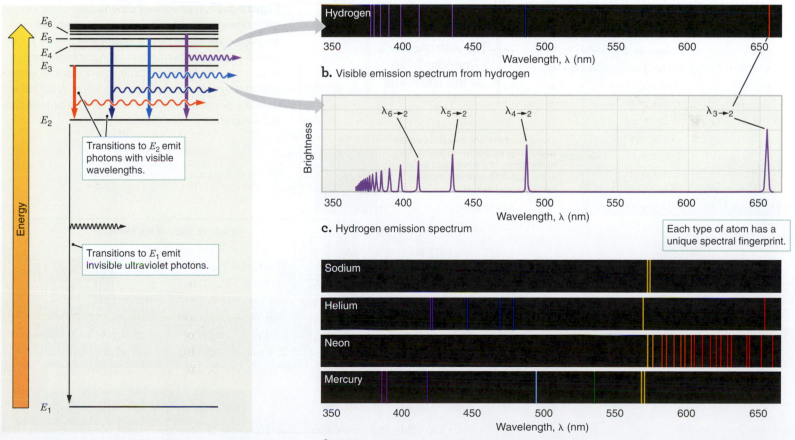

a. Energy states of the hydrogen atom

b. Visible emission spectrum from hydrogen

c. Hydrogen emission spectrum

d. Emission spectra for sodium, helium, neon, and mercury

Transitions to E_2 emit photons with visible wavelengths.

Transitions to E_1 emit invisible ultraviolet photons.

Each type of atom has a unique spectral fingerprint.

what if . . .

What if the abundance of elements on Earth were similar to what we see in the Sun? How might this have affected the evolution of life on Earth?

level diagram of hydrogen. **Figure 5.16b** shows the visible emission spectrum from hydrogen, and **Figure 5.16c** displays that same information as a graph.

Each type of atom—that is, each chemical element—has a unique set of available energy states and therefore a unique set of wavelengths at which it can emit or absorb radiation. **Figure 5.16d** shows the emission spectra of the elements sodium, helium, neon, and mercury. Those unique sets of wavelengths serve as unmistakable spectral "fingerprints" for each chemical element.

Cecilia Payne-Gaposchkin applied this analysis to the stars in 1925. She figured out the types of atoms (or molecules) in distant objects by looking at the spectra of light from those objects. If the spectral lines of hydrogen, helium, carbon, oxygen, or any other element are visible in the light from a distant object, then that element is present in that object. Payne-Gaposchkin took the analysis further, by studying the strengths of spectral lines. The "strength" of a line describes how deep and wide it is, if it's an absorption line, or how high and wide it is if it's an emission line. The strength of a line is determined in part by how many atoms of that type are present in the source. The strengths of the lines from different types of atoms in the spectrum of a distant object can be used to infer the relative amounts of elements that make up the object. This analysis led Payne-Gaposchkin to the then-controversial conclusion that the universe is primarily made of hydrogen and helium. Astronomers use the relative abundance of the elements in the Sun (called the **solar abundance**) as a standard reference. Hydrogen (H) is the most abundant element in the Sun (**Figure 5.17**), followed by helium (He) and 13 others. Those 15 elements make up 99.99 percent of the mass of the Sun. The other elements on the regular periodic table (in the lower right) make up

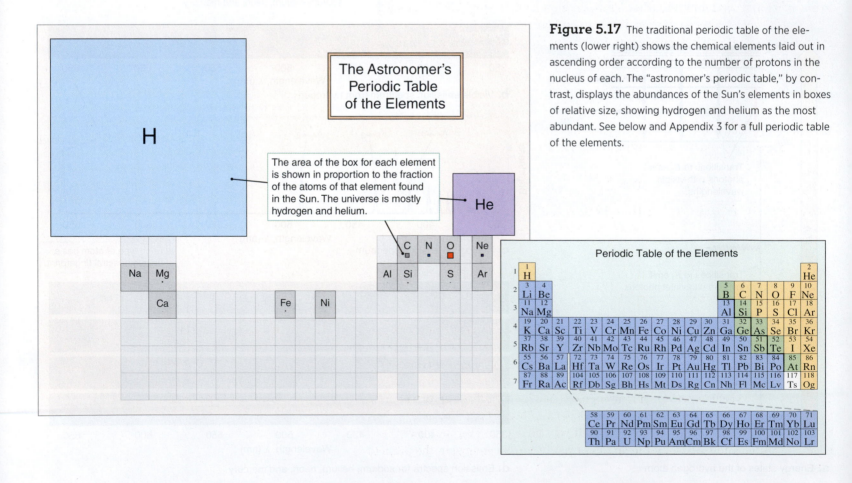

Figure 5.17 The traditional periodic table of the elements (lower right) shows the chemical elements laid out in ascending order according to the number of protons in the nucleus of each. The "astronomer's periodic table," by contrast, displays the abundances of the Sun's elements in boxes of relative size, showing hydrogen and helium as the most abundant. See below and Appendix 3 for a full periodic table of the elements.

less than 0.01 percent of the mass of the Sun. In addition, by looking at the relative strengths of lines from the same element, astronomers often can determine the temperature, density, and pressure of the material.

Excitement and Decay

If a book on a level shelf is not disturbed, it will sit there forever; something must *cause* the book to fall off the shelf. However, while sometimes an atom in a higher energy state can be "stimulated" into emitting a photon, usually nothing causes the atom to jump to the lower energy state. Instead, the atom decays *spontaneously*. Although scientists can determine on average how long a specific atom is likely to remain in the excited state, they cannot predict exactly when a particular atom will decay. We can observe the moment when an atom decays, but beforehand, we can only use the average lifetime to roughly estimate when the atom might decay. Rather than use the average lifetime, scientists typically use the "half-life" of an atomic state; this describes the time until half of all the atoms will have decayed.

Toys that glow in the dark are an example of spontaneous decay. Photons in sunlight or from a lightbulb are absorbed by phosphorescent atoms in the toy, exciting those atoms to upper energy states. The excited states of the atoms in the toy last for many seconds. If the half-life of an excited state is a minute, then on average half of the atoms will have decayed within one minute. It is impossible to say exactly which atoms will decay in which minute, but about half of the trillions and trillions of atoms in the toy will decay within 1 minute, and the brightness of the glow from the toy will have dropped to half of what it was. After each minute, half of the remaining excited atoms decay, and the glow from the toy drops to half of what it was 1 minute earlier. Thus, the glow from the toy slowly fades away.

In deep space, where atoms can remain undisturbed for long periods, certain excited states of atoms last, on average, for tens of millions of years or even longer. An atom may have been in such an excited energy state for a few seconds, a few hours, or 50 million years when, in an instant, it decays to the lower energy state without anything having caused it to do so. Space is extremely large, and has a lot of atoms in it, so even rare atomic decays are regularly observed. Physicists can calculate the *probability* that a decay will take place somewhere along a line of sight through space, but cannot predict when any individual atom will decay.

CHECK YOUR UNDERSTANDING 5.2

How can spectra tell us the chemical composition of a distant star?

5.3 The Doppler Shift Indicates Motion Toward or Away from Us

You have already seen that light can reveal a wealth of information about the physical state of material located tremendous distances away. Here in Section 5.3, we explain how light can be used to determine whether a distant astronomical object is moving away from us or toward us, and at what speed.

Have you ever listened to an ambulance speed by with its siren blaring? As the ambulance approaches, the siren has a certain high pitch, but as it passes by, the pitch of the siren drops noticeably. If you close your eyes and listen, you have no trouble knowing when the ambulance passed; the change in the pitch of

what if . . .

We've discussed the changing pitch you hear as an ambulance speeds by. What if the ambulance accelerated and reached, or even exceeded, the speed of sound? What might you hear?

Astronomy in Action: Doppler Shift

AstroTour: The Doppler Effect

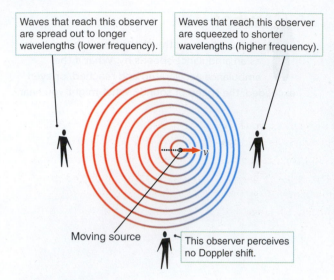

Waves that reach this observer are spread out to longer wavelengths (lower frequency).

Waves that reach this observer are squeezed to shorter wavelengths (higher frequency).

Moving source

This observer perceives no Doppler shift.

Figure 5.18 Motion of a light or sound source relative to an observer may cause waves to be spread out (redshifted, for light, or lower in pitch, for sound) or squeezed together (blueshifted, for light, or higher in pitch, for sound). A change in the wavelength of light or the frequency of sound is called a *Doppler shift*.

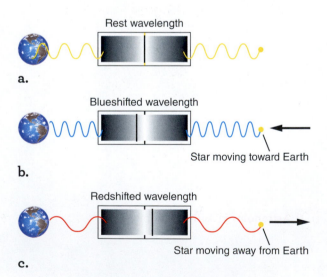

Rest wavelength

a.

Blueshifted wavelength

Star moving toward Earth

b.

Redshifted wavelength

Star moving away from Earth

c.

Figure 5.19 a. Spectral lines from a star at rest will be observed at their rest wavelength. **b.** Spectral lines from a star moving towards Earth will be blueshifted, while **c.** spectral lines from a star moving away from Earth will be redshifted.

its siren indicates that it has passed you by. You do not even need an ambulance to hear that effect. The sound of normal traffic behaves in the same way. As a car drives past, the pitch of the sound that it makes suddenly drops. (The same effect would happen if you were the one driving past a stationary siren.)

Like light, sound is a wave. The pitch of a sound is like the color of light: it is determined by the wavelength or, equivalently, the frequency of the sound wave. Sound waves with higher frequencies and shorter wavelengths have higher pitch. Sound waves with lower frequencies and longer wavelengths have lower pitch. When an object is moving toward you, the waves that it emits "crowd together" in front of the object. You can see how that process works in **Figure 5.18**, which shows the locations of successive wave crests emitted by a moving object. If you are standing in front of the object, the waves that reach you have a shorter wavelength and therefore a higher frequency than the waves given off by the object when it is not moving. Conversely, if an object is moving away from you, the waves reaching you from the object are spread out (longer λ, lower f). That change in frequency as a result of motion is known as the **Doppler effect** and is named after physicist Christian Doppler (1803–1853).

The Doppler effect applies to light waves as well as sound waves. If a star is at rest relative to you, then it emits light with the **rest wavelength** (λ_{rest}), as shown in **Figure 5.19a**. If a star is moving *toward* you, the light reaching you from the star has a shorter wavelength than its rest wavelength. The light is "bluer" than the rest wavelength, and the light is **blueshifted**, as shown by the blue waves in **Figure 5.19b**. In contrast, light from a star moving *away* from you is shifted to longer, redder wavelengths. The light is **redshifted**, as shown by the red waves in **Figure 5.19c**. The amount by which the wavelength of light is shifted by the Doppler effect is called the light's **Doppler shift**, which depends on the speed of the object emitting the light; faster objects have larger shifts.

The Doppler shift provides information only about the **radial velocity** (v_r) of the object, which is the part of the motion that is toward you or away from you. The radial velocity is the rate at which the distance between you and the object is changing: if v_r is positive, the object is getting farther away from you; if v_r is negative, the object is getting closer. At the moment an ambulance is passing you, it is getting neither closer nor farther away, so the pitch you hear is the same as the pitch heard by the crew riding on the truck (see the observer at the bottom of Figure 5.18). Similarly, a distant object moving slowly across the sky does not move toward or away from you, and so its light will not be Doppler shifted from your point of view.

Doppler shifts are easiest to measure when an object has prominent spectral lines. These lines are shifted, just like the rest of the light. Because the spectral lines from each element have a particular pattern, astronomers can find that pattern no matter how far it has been shifted from its rest position. The amount of the shift indicates how rapidly the object is moving toward or away from Earth. To determine that velocity, astronomers first identify a spectral line from a certain chemical element. The rest wavelengths (λ_{rest}) of most spectral lines have been measured in a lab on Earth. An astronomer measures the observed wavelength (λ_{obs}) in the spectrum of the distant object. The difference between the rest wavelength and the observed wavelength is used to find the object's radial velocity, as shown in detail in **Working It Out 5.2**.

CHECK YOUR UNDERSTANDING 5.3

Which of the following Doppler shifts indicates the fastest-approaching object?
(a) 0.04 nm; (b) 0.06 nm; (c) –0.04 nm; (d) –0.06 nm

working it out 5.2

Making Use of the Doppler Effect

The Doppler formula for objects moving at a radial velocity (v_r) much less than the speed of light is given by

$$v_r = \frac{\lambda_{obs} - \lambda_{rest}}{\lambda_{rest}} \times c$$

A prominent spectral line of hydrogen atoms has a rest wavelength, λ_{rest}, of 656.3 nm (see Figure 5.16b). Suppose that you measure the wavelength of that line in the spectrum of a distant object and find that instead of seeing the line at 656.3 nm, you see the line at a wavelength, λ_{obs}, of 659.0 nm. What is its radial velocity? Using the above equation,

$$v_r = \frac{659.0 \text{ nm} - 656.3 \text{ nm}}{656.3 \text{ nm}} \times (3 \times 10^5 \text{ km/s})$$

$$v_r = 1200 \text{ km/s}$$

Thus, the object is moving away from you with a radial velocity of 1200 km/s.

Suppose instead that you know the velocity and want to compute the wavelength at which you would observe the spectral line. Earth's nearest stellar neighbor, Proxima Centauri, is moving toward us at a radial velocity of −21.6 km/s. What is the observed wavelength, λ_{obs}, of a magnesium line in Proxima Centauri's spectrum that has a rest wavelength, λ_{rest}, of 517.27 nm? We can rearrange the Doppler formula to solve for λ_{obs}:

$$\lambda_{obs} = \left(1 + \frac{v_r}{c}\right) \times \lambda_{rest}$$

$$\lambda_{obs} = \left(1 + \frac{-21.6 \text{ km/s}}{3 \times 10^5 \text{ km/s}}\right) \times 517.27 \text{ nm} = 517.23 \text{ nm}$$

Although the observed Doppler blueshift ($\lambda_{obs} - \lambda_{rest} = 517.23 - 517.27$) is only −0.04 nm, it is easily measured with modern instrumentation.

5.4 Temperature Affects the Spectrum of Light That an Object Emits

The balance between heating and cooling in an object determines its temperature. If an object's temperature is constant, the heating and cooling must be in balance. Here in Section 5.4, we examine that balance and see how we can use it to predict the temperatures of planets and stars.

Astronomy in Action: Changing Equilibrium

Equilibrium and Balance

In a tug-of-war contest between two perfectly matched teams, each team pulls on the rope, but the force of one team's pull is only enough to match, not overcome, the force exerted by the other team. A picture taken now and another taken 5 minutes from now would not differ in any significant way. In that static equilibrium, opposing forces balance each other exactly. Static equilibrium can be stable, unstable, or neutral. A marble in a bowl is in a stable equilibrium: if it moves, it will return to its original position at the bottom of the bowl. A book standing on its edge, unsupported on either side, is in an unstable equilibrium: if you nudge it, it will fall over, not settle back into its original position. When an unstable equilibrium is disturbed, it moves further away from equilibrium rather than back toward it.

Equilibrium can be dynamic, so that one source of change is exactly balanced by another source of change, and the configuration of the system remains the same. In **Figure 5.20**, a can with a hole near the bottom has been placed under an

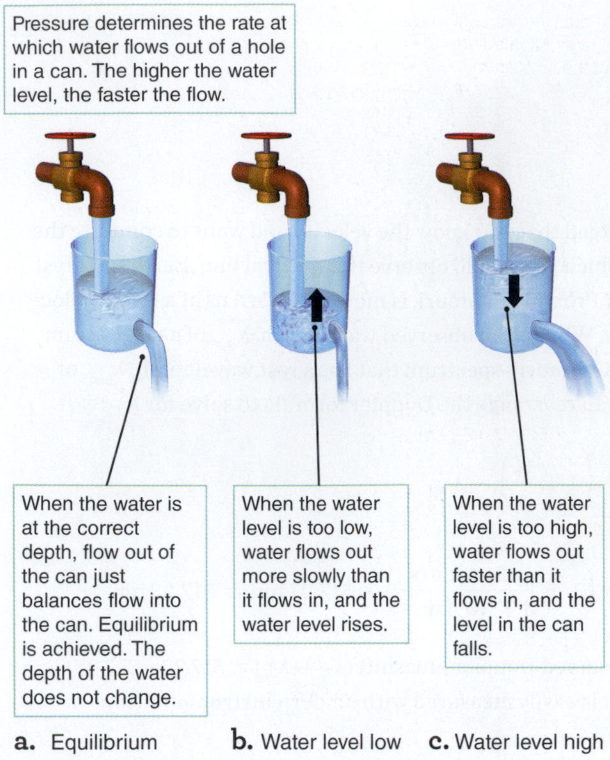

Pressure determines the rate at which water flows out of a hole in a can. The higher the water level, the faster the flow.

When the water is at the correct depth, flow out of the can just balances flow into the can. Equilibrium is achieved. The depth of the water does not change.

When the water level is too low, water flows out more slowly than it flows in, and the water level rises.

When the water level is too high, water flows out faster than it flows in, and the level in the can falls.

a. Equilibrium   **b.** Water level low   **c.** Water level high

Figure 5.20 The relative rates at which water flows into and out of a can determine the water level in the can. This is an example of dynamic equilibrium. The pressure in the can changes with the depth of the water, increasing or decreasing the flow through the hole at the bottom. **a.** When the water flows out at the same rate it flows in, the water level is in equilibrium. **b.** If the water flows out more slowly than it flows in, the water level will rise, increasing the pressure until the rates are equal. **c.** Conversely, if the water flows out more quickly than it flows in, the water level will fall, decreasing the pressure until the rates are equal.

open water faucet. This system provides an example of dynamic equilibrium, because the depth of the water in the can determines how fast water pours out through the hole near the bottom. When the water reaches just the right depth (Figure 5.20a), water pours out of the hole in the bottom of the can at the same rate it pours into the top of the can from the faucet. The water leaving the can balances the water entering, and the system is in equilibrium. If you took a picture now and another picture in a few minutes, little of the water in the can would be the same, but the pictures would be indistinguishable.

Suppose the level of the water in the can is too low (Figure 5.20b). Since the depth determines the rate at which water flows through the hole, water will not flow out of the bottom of the can fast enough to balance the water flowing in. The water level will begin to rise until the depth increases so much that the amount going out equals the amount coming in. Conversely, if the water level in the can is too high (Figure 5.20c), water will flow out of the can faster than it flows in. The water level will begin to fall until the amount going out equals the amount coming in. This is a dynamic equilibrium. One source of change is balanced by another source of change, so that the system finds a new state of equilibrium.

Your body is heated by the release of chemical energy inside it. For your body temperature to remain stable, the heating must be balanced by cooling. Typically, your body cools by radiating energy from your skin. If a given day is particularly hot, you may also perspire, so that your skin cools by evaporation of water on its surface.

A system is in **thermal equilibrium** if its heating (energy in) is balanced by its cooling (energy out). Planets have a dynamic but stable thermal equilibrium, and electromagnetic radiation plays a crucial role in maintaining that balance. Energy from sunlight heats the surface of a planet, driving its temperature up, whereas the planet emits thermal radiation into space, cooling it down. For a planet to remain at the same average temperature over time, the energy it radiates into space must exactly balance the energy it absorbs from the Sun. **Figure 5.21** shows how the equilibrium temperature of a planet is analogous to the water level in Figure 5.20. We return to planetary equilibrium later in the chapter. Many kinds of equilibrium exist besides thermal equilibrium, some of which we encounter later in the book.

Temperature

In everyday life, we define hot and cold subjectively: something is hot or cold when it feels hot or cold. When we measure *temperature*, we specify degrees on a thermometer, but the way we define a degree is arbitrary.

The air around you is composed of vast numbers of atoms and molecules. Those particles are moving about every which way. Some particles move slowly, and some move more rapidly, but all of them are constantly in motion. The kinetic energy (E_K) of a particle is given by $E_K = \frac{1}{2}mv^2$, where m is the mass of the particle and v is its velocity. **Temperature** is a measurement of the average kinetic energy of all the atoms and molecules in an object. The more energetically the atoms or molecules are bouncing about, the higher the object's temperature. In fact, the random motions of atoms and molecules are often called their **thermal motions** to emphasize the connection between those motions and temperature. **Figure 5.22** shows that when the temperature of a gas is increased, the kinetic energy is increased, in which case the atoms move faster.

The atoms and molecules in a solid body (like you) cannot move about freely like a gas, but they still move back and forth around their average location, and temperature measures the amount of that movement. If an object is hotter than you are, thermal energy flows from that object into you. At the atomic level, that means the object's atoms are bouncing more energetically than the atoms in your body are, so if you touch the object, its atoms collide with your atoms, causing the atoms in your body to move faster. Your body gets hotter as thermal energy flows from the object to you. At the same time, those collisions rob the particles in the object of some of their energy. Their motions slow down, and the hotter object cools. Heating processes increase the average thermal energy of an object's particles, whereas cooling processes decrease the average thermal energy of those particles.

There are three commonly used scales for measuring temperature. If you grew up in the United States, you probably think of temperatures in degrees Fahrenheit (°F), whereas if you grew up almost anywhere else in the world, you think of temperatures in degrees Celsius (°C). On the Fahrenheit scale, water at sea level has a freezing point of 32°F and a boiling point of 212°F. On the Celsius scale, water freezes at 0°C and boils at 100°C. Because the number of degrees between freezing and boiling on those two scales is different (180°F vs. 100°C), a 1-degree change measured in °F is not the same as a 1-degree change measured in °C.

As the motions of the particles in an object slow down, the temperature decreases more and more. The lowest possible temperature, at which all thermal motions would stop, is called **absolute zero**. Absolute zero corresponds to −273.15°C and −459.57°F. It also is the zero point of the **Kelvin temperature scale**. The size of one unit on the Kelvin scale, called a **kelvin (K)**, is the same as the Celsius degree. To convert between °C and K, just add 273.15 to the temperature in °C. Thus, water freezes at 273.15 K and water boils at 373.15 K.

Scientists use the Kelvin temperature scale because when temperatures are measured in kelvins, the average thermal energy of particles is proportional to the measured temperature. Thus, the average thermal energy of the atoms in an object with a temperature of 200 K is twice the average thermal energy of the atoms in an object with a temperature of 100 K. Just as the thermal energy cannot be negative, the Kelvin scale has no negative temperatures. Note that increments on the Kelvin temperature scale are called "kelvins" and not "degrees Kelvin."

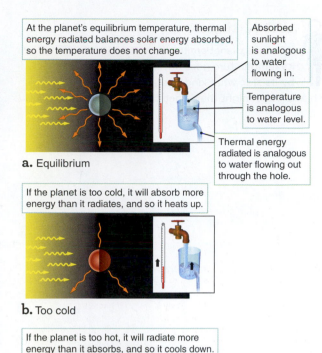

At the planet's equilibrium temperature, thermal energy radiated balances solar energy absorbed, so the temperature does not change.

Absorbed sunlight is analogous to water flowing in.

Temperature is analogous to water level.

Thermal energy radiated is analogous to water flowing out through the hole.

a. Equilibrium

If the planet is too cold, it will absorb more energy than it radiates, and so it heats up.

b. Too cold

If the planet is too hot, it will radiate more energy than it absorbs, and so it cools down.

c. Too hot

Figure 5.21 Planets are heated by absorbing sunlight (and sometimes by internal heat sources) and cooled by emitting thermal radiation into space. In the absence of other sources of heating or means of cooling, the equilibrium between those processes determines the temperature of the planet. Compare parts **a**, **b**, and **c** of this figure with those of Figure 5.20. Just as water level adjusts in a can until the rate at which water flows out is equal to the rate at which it flows in, the temperature of a planet will adjust until the rate at which energy flows out is equal to the rate at which it flows in.

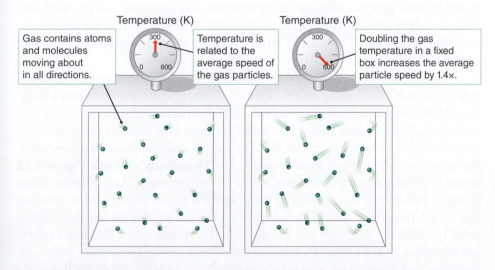

Gas contains atoms and molecules moving about in all directions.

Temperature (K)

Temperature is related to the average speed of the gas particles.

Temperature (K)

Doubling the gas temperature in a fixed box increases the average particle speed by 1.4×.

Figure 5.22 Higher gas temperatures correspond to faster-moving atoms.

Temperature, Luminosity, and Color

We have seen the way discrete atoms emit and absorb radiation, which leads to a useful understanding of emission lines and absorption lines that tell us about the physical state and motion of distant objects. But not all objects have spectra dominated by discrete spectral lines. As you saw in Figure 5.13a, if you pass the light from a lightbulb through a prism, instead of discrete bright and dark bands you will see light spread out smoothly from the blue end of the spectrum to the red. Similarly, if you look closely at the spectrum of the Sun, you will see absorption lines, and you will see light smoothly spread out across all colors of the spectrum—the continuous spectrum noted earlier. How is this continuous spectrum produced?

A dense material is a collection of charged particles that are constantly jostled as their thermal motions cause them to run into their neighbors. The hotter the material is, the more violently its particles are jostled; they speed up, slow down, and change direction. In short, they are accelerated. Each particle will be accelerated differently, across an entire range of changes of speed and direction. Recall from Figure 5.4 that a charged particle radiates anytime it accelerates. The energy of the radiation (and therefore the frequency and the wavelength) depends on how much the particle is accelerated, so the jostling of particles that results from their thermal motions causes them to emit an entire range of electromagnetic radiation. Any material dense enough for its particles to be jostled by their neighbors emits light simply because of its temperature. That kind of continuous radiation is called **thermal radiation**.

Luminosity is the total amount of light emitted each second (energy per second, measured in watts, W) from a source. The hotter the object, the more energetically the charged particles within it move, and the more energy they emit in the form of electromagnetic radiation. In other words, *an object is more luminous when it is hotter*.

As an object gets hotter, the thermal motions of its particles become more energetic, producing not only more, but also more energetic photons. As the average energy of the photons that it emits increases, the average wavelength of the emitted photons becomes shorter, and the light from the object gets bluer. Thus, *hotter objects are bluer*. If you heat a piece of metal, the metal will glow—first a dull red, then orange, and then yellow. The hotter the metal becomes, the more the highly energetic blue photons become mixed with the less energetic red photons, and the color of the light shifts from red toward blue. The light becomes more intense and bluer as the metal becomes hotter. Changing the temperature of an object causes both its luminosity and its color to change in its spectrum.

Imagine an idealized object that emits light only because of its temperature, independent of its composition. This kind of idealized object is called a **blackbody**. Blackbodies emit exactly as much thermal radiation as they absorb from their surroundings. Physicist Max Planck (1858–1947) graphed the intensity of the emitted radiation from a blackbody across all wavelengths and obtained the characteristic curves that we now call **Planck spectra** or **blackbody spectra**. Figure 5.23 shows blackbody spectra for objects at several temperatures. As the temperature changes, two effects can be seen in these spectra: hotter objects are both more luminous—the entire spectrum is higher on the graph, and bluer—the highest point (peak) of the spectrum moves left to shorter wavelengths. These two effects are described by the Stefan-Boltzmann law, which relates luminosity with temperature, and Wien's law, which relates temperature with color. We'll explain both of these in the next subsection.

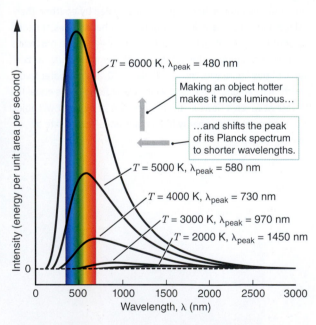

Figure 5.23 This graph shows blackbody spectra emitted by sources with temperatures of 2000, 3000, 4000, 5000, and 6000 K. At higher temperatures, the peak of the spectrum shifts toward shorter wavelengths, and the amount of energy radiated per second from each square meter of the source increases.

Blackbody Laws

Spectra from stars such as the Sun and the thermal radiation from planets often closely resemble blackbody spectra. The spectra of both types of objects can be described by the Stefan-Boltzmann law and Wien's law.

Stefan-Boltzmann Law   It would be difficult to directly measure all the photons emitted by Earth in all possible directions. But it is relatively easy to measure the number of photons leaving a small area. That number can then be multiplied by the area of the entire Earth to find Earth's luminosity. The **flux** ($\mathcal{F}$) is the amount of energy radiated by each square meter of the surface of an object each second. Multiplying the flux by the total surface area gives the luminosity. According to the **Stefan-Boltzmann law**, the flux is given by ($\mathcal{F} = \sigma T^4$, where σ (the Greek letter sigma) is the **Stefan-Boltzmann constant**. It equals 5.67×10^{-8} W/(m² K⁴), where 1 watt (W) = 1 joule per second (J/s).

The Stefan-Boltzmann law was discovered in the laboratory by physicist Josef Stefan (1835–1893) and derived mathematically by his student Ludwig Boltzmann (1844–1906). The Stefan-Boltzmann law says that an object rapidly becomes more luminous as its temperature increases. If the temperature of an object doubles, the amount of energy being radiated each second increases by a factor of $2^4 = 16$. If the temperature of an object increases by a factor of 3, the energy being radiated by the object each second increases by a factor of $3^4 = 81$. A lightbulb with a filament temperature of 3000 K radiates 16 times as much light as it would if the filament temperature were 1500 K. Even modest changes in temperature can result in large changes in the luminosity of an object.

Wien's Law   Look again at Figure 5.23. The wavelength where the blackbody spectrum is at its peak, λ_{peak}, indicates the wavelength where the electromagnetic radiation from an object is greatest. As the temperature, T, increases, the peak of the spectrum shifts toward shorter, bluer, wavelengths. For example, $\lambda_{peak} = 970$ nm for a 3000 K object, but only 480 nm for a 6000 K object. The physicist Wilhelm Wien (1864–1928) found that the peak wavelength in the spectrum is inversely proportional to the temperature of the object ($\lambda_{peak} \propto 1/T$). **Wien's law** says that if you double the temperature, the peak wavelength becomes half of what it was. If you increase the temperature by a factor of 3, the peak wavelength becomes a third of what it was. The Stefan-Boltzmann law and Wien's law are further explored in **Working It Out 5.3**. We use both laws later in the chapter to estimate the temperatures of the planets.

Astronomy in Action: Wien's Law

CHECK YOUR UNDERSTANDING 5.4

When you look at the sky on a dark night you see stars of different colors. Rank them from coolest to hottest. (a) orange; (b) red-orange; (c) yellow; (d) red; (e) blue

5.5 The Brightness of Light Depends on the Luminosity and Distance of the Source

Whereas luminosity is the amount of light *leaving* a source, the **brightness** is the amount of light *arriving* at a particular location. The brightness depends on the luminosity and the distance of the light source. For example, if you needed more light to read this book, you could replace the bulb in your lamp with a more

working it out 5.3

Working with the Stefan-Boltzmann Law and Wien's Law

The Stefan-Boltzmann law can be used to estimate the flux and luminosity of Earth. Earth's average temperature is 288 K, so the flux from its surface is

$$\mathcal{F} = \sigma T^4$$
$$\mathcal{F} = (5.67 \times 10^{-8}\,\text{W/m}^2\,\text{K}^4) \times (288\,\text{K})^4$$
$$\mathcal{F} = 390\,\text{W/m}^2$$

The luminosity is the flux multiplied by the surface area (A) of Earth. The surface area of a spherical object is given by $4\pi R^2$, where R is the radius of the object. The radius of Earth is 6378 km, or 6.378×10^6 meters, so the luminosity is

$$L = \mathcal{F} \times A = \mathcal{F} \times 4\pi R^2$$
$$L = (390\,\text{W/m}^2) \times [4\pi (6.378 \times 10^6\,\text{m})^2]$$
$$L \approx 2 \times 10^{17}\,\text{W}$$

Earth emits the equivalent of the energy used by 2,000,000,000,000,000 (2 million billion) hundred-watt lightbulbs. That value is still not anywhere close to the amount emitted by the Sun.

If astronomers measure the spectrum of an object emitting thermal radiation and find where the peak in the spectrum is, Wien's law can be used to calculate the temperature of the object. Wien's law can be written as

$$T = \frac{2{,}900{,}000\,\text{nm K}}{\lambda_{\text{peak}}}$$

What, for example, is the surface temperature of the Sun? The spectrum of the light coming from the Sun peaks at a wavelength of $\lambda_{\text{peak}} = 500$ nm, so

$$T = \frac{2{,}900{,}000\,\text{nm K}}{500\,\text{nm}} = 5800\,\text{K}$$

What is the peak wavelength at which Earth radiates? Using Earth's average temperature of 288 K in Wien's law gives

$$\lambda_{\text{peak}} = \frac{2{,}900{,}000\,\text{nm K}}{288\,\text{K}} = 10{,}100\,\text{nm} = 10.1\,\mu\text{m}$$

Thus, Earth's radiation peaks in the infrared region of the spectrum.

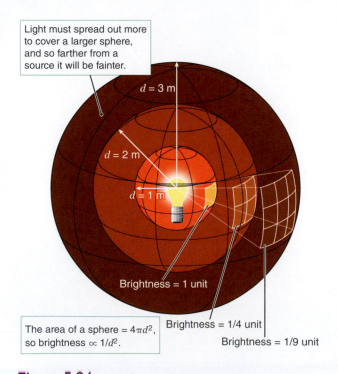

Light must spread out more to cover a larger sphere, and so farther from a source it will be fainter.

$d = 3$ m

$d = 2$ m

$d = 1$ m

Brightness = 1 unit

The area of a sphere = $4\pi d^2$, so brightness $\propto 1/d^2$.

Brightness = 1/4 unit

Brightness = 1/9 unit

Figure 5.24 Light obeys an inverse square law as it spreads away from a source. Twice as far means one-fourth as bright, three times as far means one-ninth as bright, and so on.

luminous bulb or you could move the book closer to the light. Conversely, if a light were too bright, you could move away from it, or replace the bulb with a less luminous one.

Suppose you had a piece of cardboard that measured 1 meter by 1 meter. To make the light falling on the cardboard twice as bright, you would need to double the number of photons that hit the cardboard each second. Tripling the brightness of the light would mean increasing the number of photons hitting the cardboard each second by a factor of 3, and so on. Brightness depends on the number of photons falling on each square meter of a surface each second.

Now imagine a lightbulb sitting at the center of a spherical shell (**Figure 5.24**). Photons from the bulb travel in all directions and strike the inside of the shell. To find the number of photons landing on each square meter of the shell during each second—that is, to determine the brightness of the light—take the *total* number of photons given off by the lightbulb each second and divide by the area over which those photons have to be spread. The surface area of a spherical shell is given by the formula $A = 4\pi d^2$, where d is the distance between the bulb and the surface of the sphere. The number of photons striking one square meter each second is equal to the total number of photons emitted each second divided by the surface area $4\pi d^2$.

Imagine that you increase the size of the spherical shell. As the shell becomes larger, the photons from the lightbulb must spread out to cover a larger surface area. Each square meter of the shell receives fewer photons each second, so the brightness of the light decreases. If the shell's surface is moved twice as far from the light, d is twice as large, so the area over which the light must spread increases by a factor of $2^2 = 2 \times 2 = 4$. The photons from the bulb spread out over 4 times as much area, so the number of photons falling on each square meter each second becomes $\frac{1}{4}$ of what it was. If the surface of the sphere is 3 times as far from the light, the area over which the light must spread increases by a factor of $3^2 = 3 \times 3 = 9$, and the number of photons per second falling on each square meter becomes $\frac{1}{9}$ of what it was originally. The brightness of the light from an object is inversely proportional to the square of the distance from the object. Twice as far means one-fourth as bright. (You have seen this "inverse square" relationship before, when you learned about gravity in Chapter 4.)

The idea of photons streaming and spreading onto a surface from a light explains why brightness follows an inverse square law. In practice, however, talking about the average *energy* falling on a surface each second, rather than the number of photons, is usually more useful to an astronomer.

The luminosity of an object is the total number of photons given off by the object multiplied by the energy of each photon. So, instead of thinking about how the number of photons must spread out to cover the surface of a sphere, we can think about how the energy carried by the photons must spread out to cover the surface of a sphere. The brightness of the light is the amount of energy falling on a square meter in a second, and it equals the luminosity L divided by the area of the sphere, which depends on the radius squared. That tells us, for example, that the brightness of the Sun on a given planet depends on the inverse square of the planet's distance from the Sun. That relationship factors in when we estimate the equilibrium temperatures of the planets in **Working It Out 5.4.**

CHECK YOUR UNDERSTANDING 5.5

The average distance of Mars from the Sun is 1.4 AU. How bright is the Sun on Mars compared with its brightness on Earth? (a) 1.4 times brighter; (b) about 2 times brighter; (c) about 2 times fainter; (d) 1.4 times fainter

Origins: Temperatures of Planets

In the previous chapters, we discussed how a planet's axial tilt and the shape of its orbit affect its temperature and thus its prospects for life. Now let's get more specific about the temperatures of planets, using what you learned in this chapter about thermal radiation. For a planet at an equilibrium temperature, the energy radiated by a planet exactly balances the energy absorbed by the planet. If the planet is hotter than that equilibrium temperature, it will radiate energy faster than it absorbs sunlight, and its temperature will decrease. If the planet is cooler than that temperature, it will radiate energy slower than it absorbs sunlight, and its temperature will increase.

Planets at different distances from the Sun will have different temperatures, and the temperature should be inversely proportional to the square root of the distance, as you saw in Working It Out 5.4. **Figure 5.25** plots the actual and predicted temperatures of nine solar system objects. Each vertical orange bar shows the range of temperatures found on the surface of the object or, for the giant planets, at the top of the planet's clouds. The black dots show the predictions made using the equation in Working It Out 5.4. For most objects, the predictions are not too far off, indicating that our basic understanding of *why* planets and dwarf planets have the

what if . . .

What if the Sun's luminosity suddenly increased by a factor of two? We know the new equilibrium temperature of Earth would be higher and would be independent of Earth's radius. Do you think the length of time to achieve that new equilibrium will depend on Earth's radius?

Astronomy in Action: Inverse Square Law

working it out 5.4

Using Radiation Laws to Calculate Equilibrium Temperatures of Planets

The temperature of a planet is determined by a balance between the amount of sunlight being absorbed and the amount of energy being radiated back into space. We begin with the amount of sunlight being absorbed. When viewed from the Sun, a planet looks like a circular disk with a radius equal to the radius of the planet, R_{planet}. The area of the planet that is lit by the Sun is

$$\text{Absorbing area of planet} = \pi R_{planet}^2$$

The amount of energy striking a planet also depends on the brightness of sunlight at the distance at which the planet orbits. The brightness of sunlight at a distance d from the Sun is equal to the luminosity of the Sun (L_{Sun}, in watts) divided by $4\pi d^2$:

$$\text{Brightness of sunlight} = \frac{L_{Sun}}{4\pi d^2}$$

A planet does not absorb *all* the sunlight that falls on it. **Albedo**, a, is the fraction of the sunlight that reflects from a planet. The corresponding fraction of the sunlight absorbed by the planet is 1 minus the albedo $(1 - a)$. A planet covered in snow would have a high albedo (close to 1), whereas a planet covered by black rocks would have a low albedo, close to 0:

$$\text{Fraction of sunlight absorbed} = 1 - a$$

We can now calculate the energy absorbed by the planet each second. Writing that relationship as an equation, we say that

$$\begin{pmatrix} \text{Energy absorbed} \\ \text{by the planet} \\ \text{each second} \end{pmatrix} = \begin{pmatrix} \text{Absorbing} \\ \text{area of} \\ \text{the planet} \end{pmatrix} \times \begin{pmatrix} \text{Brightness} \\ \text{of sunlight} \end{pmatrix} \times \begin{pmatrix} \text{Fraction} \\ \text{of sunlight} \\ \text{absorbed} \end{pmatrix}$$

$$= \pi R_{planet}^2 \times \frac{L_{Sun}}{4\pi d^2} \times (1-a)$$

Now let's turn to the other piece of the equilibrium: the amount of energy that the planet radiates away into space each second. We can calculate that amount by multiplying the number of square meters of the planet's total surface area by the energy radiated by each square meter each second. The surface area for the planet is given by $4\pi R_{planet}^2$. According to the Stefan-Boltzmann law, the energy radiated by each square meter each second is given by σT^4. Thus,

$$\begin{pmatrix} \text{Energy radiated} \\ \text{by the planet} \\ \text{each second} \end{pmatrix} = \begin{pmatrix} \text{Surface} \\ \text{area of} \\ \text{the planet} \end{pmatrix} \times \begin{pmatrix} \text{Energy radiated} \\ \text{per square meter} \\ \text{per second} \end{pmatrix}$$

$$= 4\pi R_{planet}^2 \times \sigma T^4$$

If the planet's temperature is to remain stable—not heating up or cooli down—then each second the "energy radiated" must be equal to "energ absorbed":

$$\begin{pmatrix} \text{Energy radiated} \\ \text{by the planet} \\ \text{each second} \end{pmatrix} = \begin{pmatrix} \text{Energy absorbed} \\ \text{by the planet} \\ \text{each second} \end{pmatrix}$$

When we set those two quantities equal to each other, we arrive at the following expression:

$$4\pi R_{planet}^2 \sigma T^4 = \pi R_{planet}^2 \frac{L_{Sun}}{4\pi d^2}(1 - a)$$

Canceling out πR_{planet}^2 on both sides, and rearranging the equation to pu on one side and everything else on the other, gives

$$T^4 = \frac{L_{Sun}(1 - a)}{16\sigma\pi d^2}$$

If we take the fourth root of each side, we get

$$T = \left(\frac{L_{Sun}(1 - a)}{16\sigma\pi d^2}\right)^{1/4}$$

Putting in the appropriate numbers for the known luminosity of the Su L_{Sun}, and the constants π and σ yields a simpler equation:

$$T = 279\,\text{K} \times \left(\frac{1 - a}{d_{AU}^2}\right)^{1/4}$$

where d_{AU} is the distance of the planet from the Sun in astronomical un

To use that equation, we need to know a planet's distance from the S and its average albedo. For a blackbody ($a = 0$) at 1 AU from the Sun, th temperature is 279 K. For Earth, with an albedo of 0.3 and a distance fr the Sun of 1 AU, the temperature is

$$T = 279\,\text{K} \times \left(\frac{1 - 0.3}{1^2}\right)^{1/4} = 255\,\text{K}$$

(Calculator hint: To take a fourth root, you can take the square root twi or use the x^y button with $y = 0.25$.)

Earth is cooler than a blackbody at 1 AU from the Sun because its av age albedo is greater than zero. If Earth's albedo changed or the Sun's luminosity changed, that would affect the result. When we examine pla around other stars, we must use the luminosity of the particular star in the equation, instead of the Sun's luminosity, so the temperature at 1 A will be different from what it is for Earth.

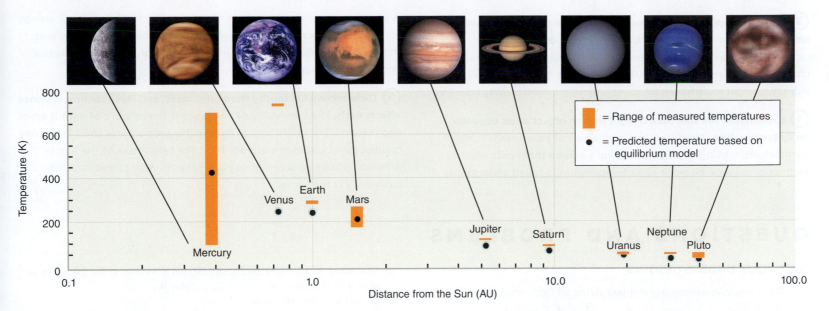

Figure 5.25 Predicted temperatures for the planets and dwarf planet Pluto are based on the equilibrium between absorbed sunlight and thermal radiation into space. Those temperatures are compared with ranges of observed surface temperatures.

temperatures they have is probably pretty good. The data for Mercury, Mars, and Pluto agree particularly well.

Sometimes, however, the predictions are wrong. For Earth, the actual measured temperature is a bit higher than the predicted temperature, and for Venus the actual surface temperature is much higher than the prediction.

The predicted values assume that the temperature of the planet is the same everywhere, but planets are likely to be hotter on the day side than on the night side. The predictions also assume that a planet's only source of energy is sunlight and that the fraction of sunlight reflected is constant over the surface of each planet. The values also incorporate the assumption that the planets absorb and radiate energy into space as blackbodies.

The discrepancies between the calculated and the measured temperatures of some of the planets indicate that for those planets, some or all of those assumptions are incorrect. For example, the planet may have its own source of energy besides sunlight, or it may have an atmosphere that traps more solar energy and increases the temperature of the planet. Understanding the temperatures of planets makes it possible to hypothesize which planets are more suitable for life.

SUMMARY

Light carries both information and energy throughout the universe. The speed of light in a vacuum is 300,000 km/s; nothing can travel faster. Visible light is only a tiny portion of the entire electromagnetic spectrum. Atoms absorb and emit radiation at unique wavelengths, giving them spectral fingerprints. A planet's temperature depends on its distance from its star, its albedo, and the luminosity of its star.

(1) **Compare the wave and particle properties of light, and describe the electromagnetic spectrum.** Light is simultaneously a stream of particles called photons and an electromagnetic wave. Different types of electromagnetic radiation, from gamma rays to visible light to radio waves, are electromagnetic waves that differ in frequency and wavelength.

(2) **Describe how to measure the chemical composition of distant objects by using the unique spectral lines of different types of atoms.** Light can reveal the identity of the chemical elements present in matter. The electron energy levels of each element have different (unique) spacings, and the wavelengths of the photons that each element emits correspond to the differences in those levels. As a result, we can identify different chemical elements and molecules in distant objects.

③ **Apply the Doppler effect and use it to measure the motion of distant objects.** Because of the Doppler effect, light from receding objects is redshifted to longer wavelengths, and light from approaching objects is blueshifted to shorter wavelengths. The wavelength shifts of the spectral lines indicate how fast an astronomical object is moving toward or away from Earth.

④ **Explain how the spectrum of light that an object emits depends on its temperature.** Temperature is a measure of how energetically particles are moving in an object. A light source that emits electromagnetic radiation because of its temperature is called a blackbody.

A blackbody emits a continuous spectrum. The total amount of energy emitted is proportional to the temperature to the fourth power, and the peak wavelength, which determines its color, is inversely proportional to the temperature.

⑤ **Differentiate luminosity from brightness, and illustrate how distance affects each.** The luminosity of an object is the amount of light it emits each second. The brightness of an object is proportional to its luminosity divided by its distance squared. Thus, the brightness of the Sun is different when measured from each Solar System planet, but the luminosity is the same.

QUESTIONS AND PROBLEMS

TEST YOUR UNDERSTANDING

1. If the Sun instantaneously stopped giving off light, what would happen on Earth?
   a. Earth would immediately get dark.
   b. Earth would get dark 8 minutes later.
   c. Earth would get dark 27 minutes later.
   d. Earth would get dark 1 hour later.

2. Why is an iron atom a different element from a sodium atom?
   a. A sodium atom has fewer neutrons in its nucleus than an iron atom has.
   b. An iron atom has more protons in its nucleus than a sodium atom has.
   c. A sodium atom is bigger than an iron atom.
   d. A sodium atom has more electrons.

3. Suppose an atom has three energy levels, specified in arbitrary units as 10, 7, and 5. In those units, which of the following energies might an emitted photon have? (Select all that apply.)
   a. 3
   b. 2
   c. 5
   d. 4

4. When a boat moves through the water, the waves in front of the boat bunch up, whereas the waves behind the boat spread out. That is an example of
   a. the Bohr model.
   b. the wave nature of light.
   c. emission and absorption.
   d. the Doppler effect.

5. As a blackbody becomes hotter, it also becomes _____ and _____.
   a. more luminous; redder
   b. more luminous; bluer
   c. less luminous; redder
   d. less luminous; bluer

6. Which of the following factors directly influences the temperature of a planet? (Choose all that apply.)
   a. the luminosity of the Sun
   b. the distance from the planet to the Sun
   c. the albedo of the planet
   d. the size of the planet

7. Two stars are of equal luminosity. Star A is 3 times as far from you as star B. Star A appears _____ star B.
   a. 9 times brighter than
   b. 3 times brighter than
   c. the same brightness as
   d. $\frac{1}{3}$ as bright as
   e. $\frac{1}{9}$ as bright as

8. When less energy radiates from a planet, its _____ increases until a new _____ is achieved.
   a. temperature; equilibrium
   b. size; temperature
   c. equilibrium; size
   d. temperature; size

9. How does the speed of light in a medium compare with the speed in a vacuum?
   a. The speed is the same in both a medium and a vacuum because the speed of light is a constant.
   b. The speed in the medium is always faster than the speed in a vacuum.
   c. The speed in the medium is always slower than the speed in a vacuum.
   d. The speed in the medium may be faster or slower, depending on the medium.

10. When an electron moves from a higher energy level in an atom to a lower energy level,
    a. a continuous spectrum is emitted.
    b. a photon is emitted.
    c. a photon is absorbed.
    d. a redshifted spectrum is emitted.

11. In Figure 5.16, the red photons come from the transition from E_3 to E_2. Those photons will have the _____ wavelengths because they have the _____ energy compared with that of the other photons. ⭐
    a. shortest; least
    b. shortest; most
    c. longest; least
    d. longest; most

12. Star A and star B appear equally bright in the sky. Star A is twice as far away from Earth as star B. How do the luminosities of stars A and B compare?
    a. Star A is 4 times as luminous as star B.
    b. Star A is 2 times as luminous as star B.
    c. Star B is 2 times as luminous as star A.
    d. Star B is 4 times as luminous as star A.

13. What is the surface temperature of a star that has a peak wavelength of 290 nm?
    a. 1000 K
    b. 2000 K
    c. 5000 K
    d. 10,000 K
    e. 100,000 K

14. If a planet is in thermal equilibrium,
    a. no energy is leaving the planet.
    b. no energy is arriving on the planet.
    c. the amount of energy leaving equals the amount of energy arriving.
    d. the temperature is very low.

15. The temperature of an object has a specific meaning as it relates to the object's atoms. A high temperature means that the atoms
    a. are very large.
    b. are moving very fast.
    c. are all moving together.
    d. have a lot of energy.

THINKING ABOUT THE CONCEPTS

16. ★ WHAT AN ASTRONOMER SEES   Suppose that you read about a new observation of an object that combines data from the ultraviolet and the radio parts of the spectrum. Use Figure 5.6 to determine the approximate wavelength and frequency ranges included in this observation. Why would astronomers want to combine multiple parts of the spectrum when observing an object? 👁

17. The speed of light in a vacuum is 3×10^5 km/s. Can light travel at a lower speed? Explain your answer.

18. Is light a wave or a particle or both? Explain your answer.

19. If any of the experiments mentioned in the Process of Science Figure had *not* agreed with the others, what would that mean for the conclusion that light has a finite, constant speed? 👁

20. If photons of blue light have more energy than photons of red light, how can a beam of red light carry as much energy as a beam of blue light?

21. Patterns of emission or absorption lines in spectra can uniquely identify individual atomic elements. How can the positive identification of atomic elements be used to test the validity of the cosmological principle discussed in Chapter 1?

22. An atom in an excited state can drop to a lower energy state by emitting a photon. Can we predict exactly how long the atom will remain in the higher energy state? Explain your answer.

23. Spectra of astronomical objects show both bright and dark lines. Describe what those lines indicate about the atoms responsible for the spectral lines.

24. Astronomers describe certain celestial objects as being *redshifted* or *blueshifted*. What do those terms indicate about the objects?

25. An object somewhere near you is emitting a pure tone at middle C on the octave scale (262 Hz). You, having perfect pitch, hear the tone as A above middle C (440 Hz). Describe the motion of that object relative to where you are standing.

26. During a popular art exhibition, the museum staff finds it necessary to protect the artwork by limiting the total number of viewers in the museum at any particular time. New viewers are admitted at the same rate that others leave. Is that system an example of static equilibrium or of dynamic equilibrium? Explain.

27. A favorite object for amateur astronomers is the double star Albireo, with one of its components a golden yellow and the other a bright blue. What do those colors tell you about the relative temperatures of the two stars?

28. The stars you see in the night sky cover a large range of brightness. What does that range tell you about the distances of the various stars? Explain your answer.

29. Why is it not surprising that sunlight peaks in the "visible"?

30. In Figure 5.25, why is the range of temperatures so much greater for Mercury than for the other planets? 👁

APPLYING THE CONCEPTS

31. Your microwave oven cooks by vibrating water molecules at a frequency of 2.45 gigahertz (GHz), or 2.45×10^9 Hz. What is the wavelength of the microwave's electromagnetic radiation? 🟢—🟢—🟢
    a. Make a prediction: Do you expect this wavelength to be longer or shorter than visible wavelengths (which are a few hundred nanometers)?
    b. Calculate: Follow Working It Out 5.1 to convert the given frequency into a wavelength.
    c. Check your work: Does your answer agree with your prediction? Does it lie in the microwave region of the spectrum?

32. Assume that an object emitting a pure tone of 440 Hz is on a vehicle approaching you at a speed of 25 m/s. If the speed of sound at this particular atmospheric temperature and pressure is 340 m/s, what will be the frequency of the sound that you hear? 🟢—🟢—🟢
    a. Make a prediction: Do you expect the wavelength of sound from an approaching vehicle to be longer or shorter? So then, do you expect the frequency to be higher or lower?
    b. Calculate: Follow Working It Out 5.2 to calculate the observed sound frequency.
    c. Check your answer: Check your units, and compare your answer to your prediction.

33. The Sun has a radius of 6.96×10^5 km and a blackbody temperature of 5780 K. Calculate the Sun's luminosity. 🟢—🟢—🟢
    a. Make a prediction: Do you expect the Sun's luminosity to be small, large, very large, or extremely large? Suppose that you calculate that the Sun's luminosity is 100 W; would you believe this answer?
    b. Calculate: Follow Working It Out 5.3 to calculate the Sun's luminosity.
    c. Check your work: Verify that you have the correct units for luminosity in your answer, and compare the luminosity you calculate to your prediction and to the known luminosity of the Sun.

34. Suppose our Sun had 10 times its current luminosity. What would the average blackbody surface temperature of Earth be if Earth had the same albedo? ●–●–●
    a. Make a prediction: Do you expect Earth to be hotter or colder than it currently is, if the Sun were 10 times as luminous?
    b. Calculate: Follow Working It Out 5.4 to calculate the surface temperature of Earth if the Sun were 10 times as luminous.
    c. Check your work: Verify that your answer has the correct units for temperature, and compare it to your prediction.

35. Some of the hottest stars known have a blackbody temperature of 100,000 K. What is the peak wavelength of their radiation? What type of radiation is it?
    a. Make a prediction: Given that these are the hottest stars, do you expect the peak wavelength to be longer than red wavelengths, or shorter than blue wavelengths?
    b. Calculate: Follow Working It Out 5.3 to calculate the peak wavelength of the hottest stars.
    c. Check your work: Verify that you have the correct units for wavelength in your answer, and compare your numerical value to your prediction.

36. You observe a spectral line of hydrogen at a wavelength of 502.3 nm in a distant galaxy. The rest wavelength of that line is 486.1 nm. What is the radial velocity of that galaxy? Is it moving toward you or away from you?

37. If half of the phosphorescent atoms in a glow-in-the-dark toy give up a photon every 30 minutes, how bright (relative to its original brightness) will the toy be after 2 hours?

38. How bright would the Sun appear to be from Neptune, 30 AU from the Sun, compared with its brightness as seen from Earth? The spacecraft *Voyager 1* is now about 140 AU from the Sun and heading out of the Solar System. Compare the brightness of the Sun as seen from *Voyager 1* with that seen from Earth.

39. You are tuned to 790 on AM radio. That station is broadcasting at a frequency of 790 kHz (7.90×10^5 Hz). You switch to 98.3 on FM radio. That station is broadcasting at a frequency of 98.3 MHz (9.83×10^7 Hz).
    a. What are the respective wavelengths of the AM and FM radio signals?
    b. Which broadcasts at higher frequencies, AM or FM?
    c. What are the respective photon energies of the two broadcasts?

40. On a dark night you notice that a distant lightbulb happens to have the same brightness as a firefly 5 meters away from you. If the lightbulb is a million times more luminous than the firefly, how far away is the lightbulb?

41. Two stars appear to have the same brightness, but one star is 3 times more distant than the other. How much more luminous is the more distant star?

42. A panel with an area of 1 square meter (m^2) is heated to a temperature of 500 K. How many watts is the panel radiating into its surroundings?

43. Your body emits radiation at a temperature of about 37°C.
    a. What is that temperature in kelvins? What is the peak wavelength, in microns, of your emitted radiation? In what region of the spectrum is this?
    b. If you assume an exposed body surface area of 0.25 m^2, how many watts of power do you radiate?

44. A planet with no atmosphere at 1 AU from the Sun would have an average blackbody surface temperature of 279 K if it absorbed all the Sun's electromagnetic energy falling on it (albedo = 0).
    a. What would the average temperature on that planet be if its albedo were 0.1, typical of a rock-covered surface?
    b. What would the average temperature be if its albedo were 0.9, typical of a snow-covered surface?

45. The orbit of Eris, a dwarf planet, carries it out to a maximum distance of 97.7 AU from the Sun. If you assume an albedo of 0.8, what is the average temperature of Eris when it is farthest from the Sun?

EXPLORATION Light as a Wave

digital.wwnorton.com/astro7

Visit the Digital Resources Page, and on the Student Site open the "Light as a Wave, Light as a Photon" AstroTour in Chapter 5. Watch the first section and then click through, using the "Play" button, until you reach "Section 2 of 3."

Here we explore the following questions: How many properties does a wave have? Are any of those properties related to each other?

Work your way to the experimental section, where you can adjust the properties of the wave. Watch the simulation for a moment to see how fast the frequency counter increases.

1 Increase the wavelength by pressing the arrow key. What happens to the rate of the frequency counter?

_____

_____

2 Reset the simulation and then decrease the wavelength. What happens to the rate of the frequency counter?

_____

_____

3 How are the wavelength and frequency related to each other?

_____

_____

4 Imagine that you increase the frequency instead of the wavelength. How should the wavelength change when you increase the frequency?

_____

_____

5 Reset the simulation, and increase the frequency. Did the wavelength change in the way you expected?

_____

_____

6 Reset the simulation, and increase the amplitude. What happens to the wavelength and the frequency counter?

_____

_____

7 Decrease the amplitude. What happens to the wavelength and the frequency counter?

_____

_____

8 Is the amplitude related to the wavelength or frequency?

_____

_____

9 Why can't you change the speed of this wave?

_____

_____

The Tools of the Astronomer

Some telescopes use lenses to change the path of light. The properties of the lens determine the appearance of the image. A glass full of water may be used as a lens. Fold a piece of paper so that it stands up. Draw a horizontal arrow (pointing to the side) on the vertical part of the paper. Place the arrow about 3 inches behind a clear empty glass and observe the arrow through the glass. Make a prediction about what you will see if you fill the glass with water. As you fill the glass with water, watch the arrow through the glass. Write down your observations. Make a sketch, drawn from above, of the path of the light rays as they leave the two ends of the arrow and pass through the empty glass on their way to your eye. Repeat the sketch for the water-filled glass. Move the glass closer to and farther away from the arrow. Write down your observations of any changes that occur at the different distances.

EXPERIMENT SETUP

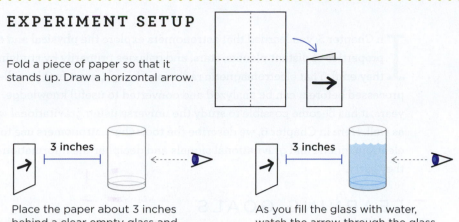

Fold a piece of paper so that it stands up. Draw a horizontal arrow.

Place the paper about 3 inches behind a clear empty glass and observe the arrow through the glass.

As you fill the glass with water, watch the arrow through the glass. Write down your observations.

Make a sketch, drawn from above, of the path of the light rays as they leave the two ends of the arrow and pass through the empty glass on their way to your eye. Repeat the sketch for the water-filled glass.

CLOSER   **FARTHER**

Move the glass closer to and farther away from the arrow. Write down your observations of any changes that occur at the different distances.

PREDICTION

I predict that when I view the arrow through the glass, that it will be:

_____

SKETCH OF RESULTS (in progress)

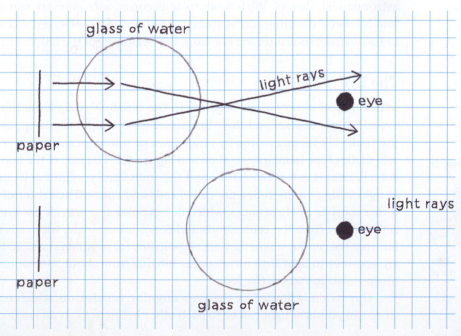

6

In Chapter 5, you learned that astronomers explore the physical and chemical properties of distant planets, stars, and galaxies primarily by studying the light they emit. That electromagnetic radiation, though, must first be collected and processed before it can be analyzed and converted to useful knowledge. In recent years, it has become possible to study the universe using gravitational waves as well. Here in Chapter 6, we describe the tools that astronomers use to collect electromagnetic and gravitational signals and decipher the information that they carry.

LEARNING GOALS

By the end of this chapter, you should be able to:

(1) Compare how the two main types of optical telescopes gather and focus light.

(2) Summarize the main types of detectors that are used on telescopes.

(3) Explain why some wavelengths of radiation must be observed from high, dry, and remote observatories on Earth, or from space.

(4) Explain the benefits of sending spacecraft to study the planets and moons of our Solar System.

(5) Describe other astronomical tools, such as gravitational wave detectors, that contribute to the study of the universe.

6.1 The Optical Telescope Revolutionized Astronomy

The development of **telescopes**—devices for collecting and focusing light—in the 17th century greatly increased the amount of light that can be collected from astronomical objects. With modern telescopes, astronomers can detect light that has been traveling across space for billions of years—even electromagnetic radiation from soon after the Big Bang, the beginning of the universe itself.

The Eye

Astronomical observations began with the human eye. Information about the overall colors of stars and their brightness in the night sky is apparent even to the "naked" eye, unassisted by binoculars or telescopes or filters. A simplified schematic of the human eye is shown in **Figure 6.1**. The part of the human eye that detects light is called the retina, and the individual receptor cells that respond to light falling on the retina are called rods and cones. The center of the human retina consists solely of cones, which detect color and provide the greatest visual acuity. Away from the center, rods and cones intermingle, with rods dominating far from the center, where they are responsible for peripheral vision. Human eyes are sensitive to light with wavelengths ranging from about 380 nanometers (nm) (deep violet) to 750 nm (far red).

Our vision is limited by the eye's **angular resolution**, which refers to how close two points of light can be to each other before we can no longer distinguish

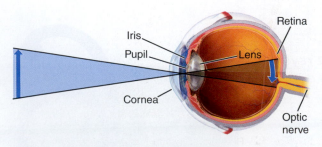

Figure 6.1 A schematic view of the human eye creating an image of an object (the blue arrow).

them. Unaided, the best human eyes can resolve objects separated by 1 **arcminute** (1/60 of a degree), an angular distance of about 1/30 the diameter of the full Moon. (A more in-depth description of angular units—radians, degrees, arcminutes, and arcseconds—can be found in Appendix 1.) That may seem small, but thousands of stars and galaxies may reside within a patch of sky with that diameter.

Refracting Telescopes

Optical telescopes come in two primary types: **refracting telescopes**, which use lenses, and **reflecting telescopes**, which use mirrors. For all telescopes, the "size" of the telescope refers to the diameter of the largest mirror (or lens), known as the **primary mirror** (or lens) which determines the light-collecting area. That diameter is called the **aperture**. The light-gathering power of a telescope is proportional to the area of its opening—that is, to the square of its diameter. The larger the aperture, the more light the telescope can collect. A "1-meter telescope" has a primary mirror (or lens) that is 1 meter in diameter. The aperture of the human eye is 6–7 millimeters (mm).

In the late 13th century, craftsmen in Venice were making small lentil-shaped disks of glass that could be mounted in frames and worn over the eyes to improve vision. More than 300 years later, Hans Lippershey (1570–1619), a spectacle maker living in the Netherlands, put two of his lenses together in a tube. With that new instrument, he saw distant objects magnified and could see farther. Galileo Galilei heard news of that invention and constructed one of his own. Recall from Chapter 3 that by the early 1600s, Galileo had become the first to see the phases of Venus and the moons of Jupiter and was among the first to see craters on the Moon. He also was the first to realize that the Milky Way is made up of many individual stars. The refracting telescope—one that uses lenses—quickly revolutionized the science of astronomy.

Recall from Chapter 5 that the speed of light is constant in a vacuum, but through a medium such as air or glass, the speed of light is always lower. As light enters a new medium, its speed changes. As shown in the diagram in **Figure 6.2a**, if the light traveling in the direction of the green line strikes a surface at an angle, some of the crest of the wave (red lines) arrives at the surface earlier and some arrives later. **Figure 6.2b** shows an actual light ray passing into and out of a medium (here, glass). The ray bends each time the medium changes. When the medium through which it travels changes, that bending of light is called **refraction**. Refraction is the basis for the refracting telescope.

The amount of refraction depends on the properties of the medium—what it's made of, its temperature, even the pressure that it's under. A medium's **index of refraction** (n) is equal to the ratio of the speed of light in a vacuum (c) to the speed of light in the medium (v). That relationship can be expressed by the equation $n = c/v$. For example, most glass has an index of refraction of approximately 1.5, so the speed of light in glass is 300,000 kilometers per second (km/s) divided by 1.5, or 200,000 km/s. The amount of refraction depends on the index of refraction of the materials involved and the angle at which the light strikes.

The primary lens in a refracting telescope is a simple convex lens, called the **objective lens (Figure 6.3)**, whose curved surfaces refract the light from a distant object. That refracted light forms an image on the telescope's **focal plane**, which is perpendicular to the *optical axis*—the path that light takes through the center of the lens. Because the telescope's glass lens is curved, light at the outer edges of the lens strikes the surface at a different angle than light near the center. As a result,

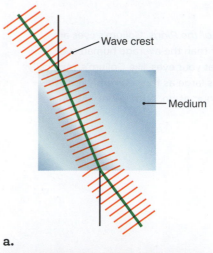

a.

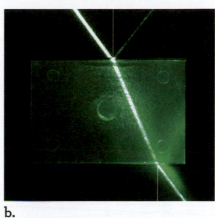

b.

Figure 6.2 a. When wavefronts enter a new medium, they bend in a new direction relative to a line perpendicular to the surface (black lines). **b.** An actual light ray entering and leaving a medium.

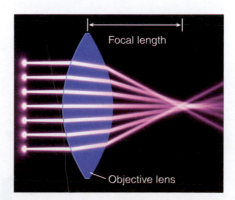

Figure 6.3 For a curved lens such as the one shown, refraction causes the light to focus to a point. That point is in a slightly different location for different wavelengths (colors) of light.

light at the outer edges of the lens is refracted more than light near its center. The lens concentrates the light rays entering the telescope, bringing them to a sharp focus at a distance called the **focal length**. Sometimes focal length is specified on a telescope as *focal ratio* (or "f-number"), which equals focal length divided by aperture; that term may be familiar to you from the lenses used in photography. Aperture and focal length are the two most important parameters of a telescope.

Figure 6.4 illustrates how focal length affects the image created by a refracting telescope. Figure 6.4a shows the light from two stars passing through a lens and converging at the focal plane of the lens. Figure 6.4b shows the same situation for a lens with a longer focal length. Longer focal lengths increase the size and separation of objects in the focal plane. In some telescopes, a second lens makes it possible for the user to view the image directly. The focal length of this interchangeable **eyepiece** determines the magnification (**Working It Out 6.1**). In modern research telescopes, however, the images are sent directly to a camera or other detector.

Refracting telescopes have two major shortcomings. First, physical limits constrain the size of refracting telescopes. The larger the area of the objective lens, the more light-gathering power it has and the fainter the stars we can observe. Larger objective lenses are heavier, however, and the weight of a massive lens can bend the tube that holds it. Refracting telescopes grew in size until the 1897 completion of the Yerkes 1-meter refractor (**Figure 6.5**), the world's largest refracting telescope. Located in Williams Bay, Wisconsin, the Yerkes telescope consists of a 450-kilogram (kg) objective lens mounted at the end of a 19.2-meter tube.

The second major shortcoming of refracting telescopes is **chromatic aberration (Figure 6.6)**. Starlight is made up of all the colors of the rainbow, and each color refracts at a slightly different angle because the index of refraction depends on the wavelength of the light. Shorter (bluer) wavelengths are refracted more than longer (redder) wavelengths (Figure 6.6a). That wavelength-dependent difference in refraction, which spreads the white light out into its spectral colors, is called **dispersion**. Dispersion causes bluer light to come to a shorter focus than that of the longer visible wavelengths, creating chromatic aberration. In a refracting telescope with a simple convex lens, chromatic aberration produces haloed images around the star. That issue was addressed in early telescopes by increasing the focal length, but then the telescope requires a longer tube. Now manufacturers of quality cameras and telescopes use a **compound lens** composed of two types of glass to substantially correct for chromatic aberration (Figure 6.6b).

Figure 6.4 a. A refracting telescope uses a lens to collect and focus light from two stars, forming images of the stars on its focal plane. **b.** A telescope with a longer focal length produces larger images.

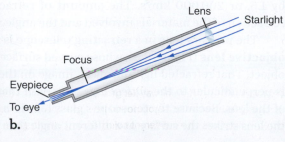

Figure 6.5 a. The Yerkes 1-m refractor is the world's largest refracting telescope. **b.** The main parts of a refractor.

Telescope Aperture and Magnification

If you are shopping for a telescope, you need to consider its aperture and magnification.

Aperture

The light-gathering power of a telescope is proportional to the area of its lens or mirror, $\pi \times (D/2)^2$, and thus to the square of the aperture (D). The amount of light a telescope collects increases as the aperture increases. How does the light-gathering power of a telescope with a diameter of 200 mm, or 8 inches, compare with that of the pupil of your eye, which is about 6 mm in the dark?

$$\text{Light-gathering power of telescope} = \frac{\pi}{4} \times (200\,\text{mm})^2$$

and

$$\text{Light-gathering power of eye} = \frac{\pi}{4} \times (6\,\text{mm})^2$$

So, to compare:

$$\frac{\text{Light-gathering power of telescope}}{\text{Light-gathering power of eye}} = \frac{\frac{\pi}{4}(200\,\text{mm})^2}{\frac{\pi}{4}(6\,\text{mm})^2} = \left(\frac{200}{6}\right)^2 = 1111$$

Thus, an 8-inch telescope has more than 1000 times the light-gathering power of your eye.

Comparing that 8-inch telescope with the Keck 10-meter telescope shows why bigger is better: 200 mm = 0.2 meter, and we cancel out the $\pi/4$ again to obtain

$$\frac{\text{Light-gathering power of Keck}}{\text{Light-gathering power of 8-inch telescope}} = \left(\frac{10\,\text{m}}{0.2\,\text{m}}\right)^2 = 2500$$

Even larger telescopes, 25–40 meters in diameter, are under construction.

Magnification

Most telescopes have a set focal length and come with a collection of eyepieces. The magnification of the image in the telescope is given by

$$\text{Magnification} = \frac{\text{Telescope focal length}}{\text{Eyepiece focal length}}$$

Suppose the focal length of the 200-mm telescope in the preceding example is 2000 mm. Combined with the focal length of a standard eyepiece, 25 mm, that telescope will give the following magnification:

$$\text{Magnification} = \frac{2000\,\text{mm}}{25\,\text{mm}} = 80$$

That telescope and eyepiece combination has a magnifying power of 80, meaning that a crater on the Moon will appear 80 times larger in the telescope's eyepiece than it does when viewed by the naked eye. An eyepiece that has a focal length of 8 mm will have about 3 times more magnifying power, a magnification of 250.

A higher magnification will not necessarily let you see an object better. A faint and fuzzy image will not look clearer when magnified. Although a larger aperture collects more light so that you can observe fainter objects, magnification spreads that light out again, making faint objects more difficult to see; there is an inherent trade-off between aperture and magnification. If you are buying a telescope, opt for the larger aperture, and a collection of eyepieces that let you change the magnification to suit your observation target.

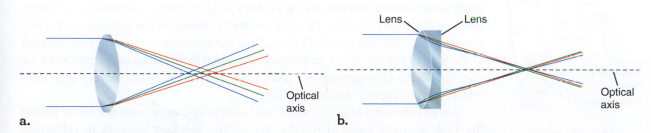

Figure 6.6 **a.** Different wavelengths of light come to a focus at different places along the optical axis of a simple lens, causing chromatic aberration. **b.** A compound lens using two types of glass with different indices of refraction can compensate for much of the chromatic aberration, so different colors of light all come to a focus at the same point.

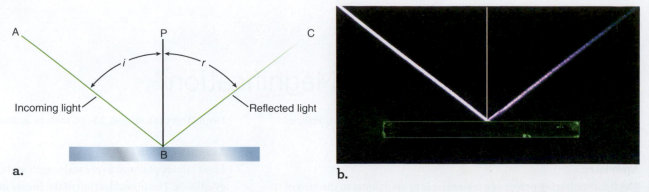

Figure 6.7 **a.** When a ray of incoming light (AB) shines on a flat surface, it reflects from the surface, becoming the reflected ray BC. The angle between AB and PB, the perpendicular to the surface, is the angle of incidence (*i*). The angle between BC and PB is the angle of reflection (*r*). The angles of incidence and reflection are always equal. **b.** Light from a laser beam is reflected from a flat glass surface.

Reflecting Telescopes

When light encounters a different medium, some of the light will bounce off of the surface, and remain in the original medium. This effect, called **reflection**, is the basis for reflecting telescopes. As shown in **Figure 6.7a**, the angle of the incoming light and the angle of outgoing light are always equal. You have experienced this when viewing an ordinary flat mirror (**Figure 6.7b**). A reflected image in a mirror looks like the original image, and is not stretched or distorted. Reflecting telescopes solve the two problems of refracting telescopes. Because mirrors can be supported from the back, they can be much heavier without affecting the telescope's performance. Because the angle of reflection does not depend on the wavelength of light, chromatic aberration is not a problem in reflecting telescopes.

In 1668, Isaac Newton designed a reflecting telescope, which uses mirrors instead of lenses (**Figure 6.8a**). The important parts of Newton's reflecting telescope are shown schematically in **Figure 6.8b**. To make that reflecting telescope, Newton cast a 2-inch primary mirror made of copper and tin and polished it to a special curvature. He then placed the primary mirror at the bottom of a tube with a secondary flat mirror mounted above it at a 45° angle. The second mirror directed the focused light to an eyepiece on the outside of the tube. The curve of the primary mirror of any reflecting telescope is shaped so that parallel rays of light will strike the mirror at different angles, and reflect them all so that they cross at the focal length of the mirror, as shown by the blue rays in Figure 6.8b.

The light path from the primary mirror to the focal plane can be "folded" by using a **secondary mirror**, which enables a significant reduction in the length and weight of the telescope. In many modern telescopes, the primary mirror has a hole so that light can pass back through it; the eyepiece or camera is on the back of the tube of the telescope, as shown in **Figure 6.9**. For this configuration, or for the Newtonian arrangement in Figure 6.8b, the tube is much shorter than for a refracting telescope of similar aperture.

Large reflecting telescopes did not become common until the latter half of the 18th century. Since then, the size of the primary mirrors in reflecting telescopes has grown larger every decade. Primary mirrors can be supported from the back, and they can be made thinner and therefore less massive than the objective lenses found in refracting telescopes. The limitation on the size of reflecting

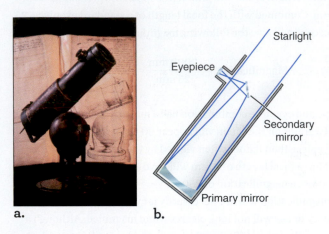

Figure 6.8 **a.** Newton's reflecting telescope. **b.** A Newtonian focus telescope has a curved primary and a flat secondary mirror, as shown in the sketch. The eyepiece is on the side of the tube.

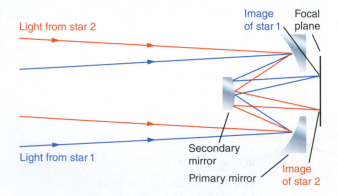

Figure 6.9 Large reflecting telescopes often use Cassegrain focus, in which a secondary mirror directs the light back through a hole in the concave parabolic primary mirror to an accessible focal plane behind the primary mirror.

telescopes is the cost of their fabrication and support structure. **Table 6.1** lists the world's largest optical telescopes. All are reflecting telescopes. The largest single mirrors constructed today are 8 meters in diameter. Larger reflecting telescopes use an array of smaller segments. For example, the primary mirror of each of the 10-meter twin Keck telescopes has 36 hexagon-shaped segments that are each 1.8 meters in diameter (**Figure 6.10**). Located on 4205-meter-high Mauna Kea in Hawai'i, the Keck telescopes are among the world's largest reflecting telescopes. Each one can gather 4 million times more light than the human eye.

▶❚❚ **AstroTour:** Geometric Optics and Lenses

| Table 6.1 | The World's Largest Optical Telescopes |

Mirror Diameter (meters)	Telescope	Sponsor(s)	Location	Operational Date
39.3	Extremely Large Telescope (ELT)	European Southern Observatory (Europe, Chile, Brazil)	Cerro Armazones, Chile	Under construction
30.0	Thirty Meter Telescope (TMT)	International collaboration led by Caltech, U. of California, U. of Hawai'i, China, Japan, India, and Canada	Mauna Kea, Hawai'i, or Canary Islands, Spain	TBD
24.5	Giant Magellan Telescope (GMT)	Carnegie Institution, Arizona State U., Harvard U., Smithsonian Institution, U. of Arizona, U. of Texas, Texas A&M U., U. of Chicago, São Paulo Research Foundation (FAPESP), Australian National U., Astronomy Australia Ltd., and Korea Astronomy and Space Science Institute	Cerro Las Campanas, Chile	Under construction
10.4	Gran Telescopio Canarias (GTC)	Spain, Mexico, U. of Florida	Canary Islands	2007
10	Keck I	Caltech, U. of California, NASA	Mauna Kea, Hawai'i	1993
10	Keck II	Caltech, U. of California, NASA	Mauna Kea, Hawai'i	1996
∼10	South African Large Telescope (SALT)	South Africa, USA, UK, Germany, Poland, New Zealand, India	Sutherland, South Africa	2005
10	Hobby-Eberly Telescope (HET)	U. of Texas, Penn State U., Stanford U., Germany	Mount Fowlkes, Texas	1999
8.4 × 2	Large Binocular Telescope (LBT)	U. of Arizona, Ohio State U., Italy, Germany, Arizona State, and others	Mount Graham, Arizona	2008
8.4	Large Synoptic Survey Telescope (LSST)	National Science Foundation, Dept. of Energy, and other partners through the private LSST corporation	Cerro Pachón, Chile	Under construction
8.2	Subaru Telescope	Japan	Mauna Kea, Hawai'i	1999
8.2 × 4	Very Large Telescope (VLT)	European Southern Observatory	Cerro Paranal, Chile	2000
8.1	Gemini North	USA, UK, Canada, Chile, Brazil, Argentina, Australia	Mauna Kea, Hawai'i	1999
8.1	Gemini South	USA, UK, Canada, Chile, Brazil, Argentina, Australia	Cerro Pachón, Chile	2000
6.5	MMT	Smithsonian Institution, U. of Arizona	Tucson, Arizona	2000
6.5	Magellan I	Carnegie Institution, U. of Arizona, Harvard U., U. of Michigan, MIT	Cerro Las Campanas, Chile	2000
6.5	Magellan II	Carnegie Institution, U. of Arizona, Harvard U., U. of Michigan, MIT	Cerro Las Campanas, Chile	2002

Figure 6.10 Each of the Keck 10-m reflectors uses an aligned group of 36 hexagonal mirrors to collect light.

a.　　　　　　b.

Figure 6.11 Angular resolution is the ability to separate two images that appear close together. When angular resolution is lower **a.**, the two images blend together. When angular resolution is higher **b.**, individual images can be distinguished.

CHECK YOUR UNDERSTANDING 6.1a

Why are all large astronomical telescopes reflectors (choose all reasons that apply)? (a) chromatic aberration is minimized; (b) they are not as heavy; (c) they can be shorter; (d) only one surface of a mirror needs polishing.

Answers to Check Your Understanding questions are in the back of the book.

Optical and Atmospheric Limitations

The **angular resolution** (sometimes just "resolution") of a telescope determines how close two points of light can be to each other before they are indistinguishable (**Figure 6.11**). Review Figure 6.4a to see the path followed by rays of light from two distant stars as they pass through the lens of a refracting telescope. Figure 6.4b illustrated that increasing the focal length increases the size of and separation between the images that a telescope produces. That is one reason why telescopes provide a much clearer view of the stars than that obtained with the naked eye. The focal length of a human eye is typically about 20 mm, whereas telescopes used by professional astronomers often have focal lengths of tens or even hundreds of meters. Such telescopes make images that are far larger than those formed by the human eye, so they contain far more detail.

Focal length explains only one difference between the angular resolution of telescopes and that of the unaided eye. The other difference results from the wave nature of light. **Figure 6.12** shows what happens when light waves spread out from the edges of the lens or mirror as they pass through the aperture of a telescope. The distortion that occurs as light passes the edge of an opaque object is called **diffraction**. Diffraction diverts some of the light from its path, slightly blurring the image made by the telescope. Some of these light rays will overlap, and the waves will **interfere**, producing bright and dark "fringes." The degree of blurring depends on the wavelength of the light and the telescope's aperture. The larger the aperture, the smaller the problem diffraction poses. The best angular resolution that a given telescope can theoretically achieve is known as the **diffraction limit** (Working It Out 6.2).

Larger telescopes have better angular resolution and can distinguish objects that appear closer together. Theoretically, the 10-meter Keck telescopes have a

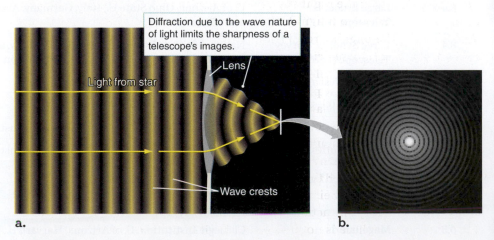

a.　　　　　　　　　　　　　　　　　　　　　　b.

Figure 6.12 **a.** Light waves from a star are diffracted by the edges of a telescope's lens or mirror. **b.** That diffraction causes the stellar image to be blurred, limiting a telescope's ability to resolve objects.

working it out 6.2

Diffraction Limit

The best possible angular resolution (θ) that can be obtained with a telescope is called the diffraction limit. That limit is determined by the ratio of the wavelength of light passing through the telescope (λ) to the diameter of the aperture (D):

$$\theta = 2.06 \times 10^5 \left(\frac{\lambda}{D}\right) \text{arcsec}$$

The constant, 2.06×10^5, has units of arcseconds (arcsec). An **arcsecond** is a tiny angular measure found by first dividing the sky into 360 degrees, then dividing each degree by 60 to get arcminutes, and then dividing each arcminute by 60 to get arcseconds. An arcsecond is 1/1800 of the size of the Moon in the sky, or about the size of a tennis ball if you could see it from 13 km (8 miles) away.

Both λ and D must be expressed in the same units—usually meters (m). The smaller the ratio of λ/D, the better the angular resolution. For example, the size of the human pupil (see Figure 6.1) ranges from about 2 mm in bright light to 8 mm (0.008 m) in the dark. Visible (green) light has a wavelength (λ) of 550 nm—that is, 550×10^{-9} m, or 5.5×10^{-7} m. Using those values for the aperture and the wavelength gives

$$\theta = 2.06 \times 10^5 \left(\frac{5.5 \times 10^{-7} \text{ m}}{0.008 \text{ m}}\right) \text{arcsec} = 14 \text{ arcsec}$$

or about $\frac{1}{4}$ arcmin for the theoretical best angular resolution. The typical angular resolution of the human eye is 1 arcmin. We do not achieve the best possible angular resolution with our eyes because the physical properties of our eyes are not perfect.

How does the angular resolution of the human eye compare with that of the Hubble Space Telescope in the visible part of the spectrum? The aperture of the Hubble Space Telescope is 2.4 m. Substituting that value for D and again using visible (green) light gives

$$\theta = 2.06 \times 10^5 \left(\frac{5.5 \times 10^{-7} \text{ m}}{2.4 \text{ m}}\right) \text{arcsec} = 0.047 \text{ arcsec}$$

which is about 300 times better than the theoretical best angular resolution of the human eye.

diffraction-limited angular resolution of 0.0113 arcsec in visible light. If your eyes had this angular resolution, you could read this text 15 km away. But for telescopes with apertures larger than about a meter, Earth's atmosphere stands in the way of this ideal angular resolution. If you have ever looked out across a large asphalt parking lot on a summer day, you have probably seen the distant horizon shimmer as light is bent this way and that by turbulent bubbles of warm air rising off the hot pavement. The problem of the shimmering atmosphere is less pronounced when we look overhead, but the same phenomenon causes the twinkling of stars in the night sky. As telescopes magnify the angular diameter of an object, they also magnify the shimmering effects of the atmosphere. The limit on the angular resolution of a telescope on the surface of Earth caused by that atmospheric distortion is called **astronomical seeing**. One advantage of launching telescopes such as the Hubble Space Telescope into orbit around Earth is that placing them above Earth's atmosphere means they are no longer hindered by astronomical seeing.

To understand the effect Earth's atmosphere has on angular resolution, consider light from a distant star that arrives at the top of Earth's atmosphere with flat, parallel wave crests. If Earth's atmosphere were perfectly uniform, the crests would remain flat as they reached the objective lens or primary mirror of a ground-based telescope. After making their way through the telescope's optical system, the crests would produce a tiny diffraction disk in the focal plane, as shown in Figure 6.12b. But Earth's atmosphere is not uniform. It is filled with bubbles of air that have slightly different temperatures from those of their surroundings. Different temperatures mean different densities, and different densities mean different refractive properties, so each bubble bends light differently. Those air bubbles act as

unanswered questions

Will telescopes be placed on the Moon? The Moon has no atmosphere to make stars twinkle, cause weather, or block certain wavelengths of light from reaching its surface. All parts of the Moon have days and nights that last for 2 Earth weeks each. China had a small (15 cm) ultraviolet telescope operated from Earth on their *Chang'e 3* lander. One proposal is for placing a small array of radio telescopes on the side of the Moon facing Earth. These would study the Sun and the solar wind. Eventually, telescopes would go on the far side of the Moon, which faces away from the light and radio radiation of Earth. On the far side one might put an array of hundreds of radio telescopes that would be deployed to study the earliest formation of stars and galaxies. Another proposal is for the Lunar Liquid Mirror Telescope (LLMT), with a diameter of 20–100 m, to be located at one of the Moon's poles. Gravity would settle the rotating liquid into the necessary parabolic shape, and those liquid mirror telescopes are much simpler than are arrays of telescopes with large glass mirrors. Astronomers debate whether telescopes on the Moon would be easier to service and repair than those in space and whether problems caused by lunar dust would outweigh any advantages.

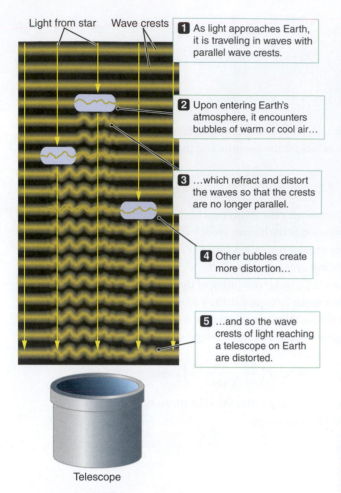

Light from star   Wave crests

1 As light approaches Earth, it is traveling in waves with parallel wave crests.

2 Upon entering Earth's atmosphere, it encounters bubbles of warm or cool air…

3 …which refract and distort the waves so that the crests are no longer parallel.

4 Other bubbles create more distortion…

5 …and so the wave crests of light reaching a telescope on Earth are distorted.

Telescope

Figure 6.13 Bubbles of warmer or cooler air in Earth's atmosphere distort the wavefront of light from a distant object.

weak lenses, and by the time the waves reach the telescope they are far from flat, as shown in **Figure 6.13**. Instead of a tiny diffraction disk, the image in the telescope's focal plane is distorted and swollen, degrading the angular resolution.

Modern technology has improved ground-based telescopes with computer-controlled **adaptive optics** that compensate for much of the atmosphere's distortion. First, an optical device within the telescope constantly measures the wave crests. To calibrate those systems, a laser beam is shone into the atmosphere to create an artificial "star" in the same direction as the telescope is pointing. Then, before reaching the telescope's focal plane, the light from a target star is reflected off yet another mirror, which has a flexible surface. A computer analyzes the light and bends the flexible mirror so that it accurately corrects for the distortion of the artificial star caused by the air bubbles. **Figure 6.14** shows an example of an image corrected by adaptive optics. The widespread use of adaptive optics has made the image quality of ground-based telescopes competitive with the quality of Hubble images from space at some wavelengths. Telescopes also can have active optics, in which actuators on the rear of the mirror correct for factors such as wind, temperature, and sagging from gravity.

Observatory Locations

What makes a good location for a telescope on Earth? As you could deduce from Table 6.1, astronomers look for sites that are high, dry, and dark. The best sites are far away from the lights of cities; in locations with little moisture, humidity,

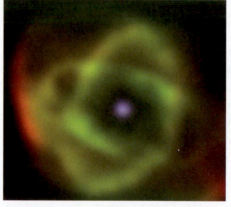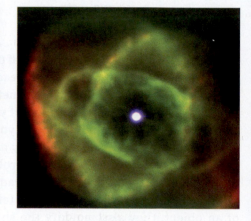

Figure 6.14 These images of the Cat's Eye Nebula from the Palomar Observatory telescope without (left) and with (right) adaptive optics show the benefit of the technique.

★ **WHAT AN ASTRONOMER SEES** An astronomer, looking at this image, would instantly identify it as a nebula (a "cloud"), because it is extended rather than pointlike, and even in the sharper image, is a little bit "fuzzy." This indicates that the light she is seeing comes from an extended cloud of dust and gas. She will also take particular note of the white dot, which indicates a central star. Because of the color, she will know that this is a relatively hot star (recall Wien's law from Chapter 5). An astronomer will notice the green and red colors of the nebula, and take note of them. An astronomer will be pleased that the image on the right allows her to see fine detail regarding the distribution of material in the nebula, including where it is red and where it is green. Without further information, she will not know for certain whether these colors are "true" colors that indicate composition (that there is line emission from particular elements in those particular colors), or whether the color represents some other aspect, like the spectral region of the observations or the velocity of the dust and gas. While it is likely that, for a nebula of this type, the colors indicate composition, even a professional astronomer cannot be certain of it without more information.

or rain; and where the atmosphere is relatively still. Telescopes are located as high as possible, above a significant part of Earth's atmosphere, which distorts images and blocks infrared and microwave light. Many telescopes are situated on remote, high mountaintops surrounded by desert or ocean. Recall from Chapter 2 that the stars that can be seen throughout the year depend on latitude, and only at the equator would a telescope have access to all the stars in the sky. But because equatorial latitudes have tropical weather—wet, humid, and stormy—they are poor locations for telescopes. To cover the entire sky, astronomers have built telescopes in both northern and southern locations. In the United States, large telescopes are located in California, Arizona, New Mexico, Texas, and Hawai'i. The largest southern-sky observatories are found in Chile, South Africa, and Australia. The twin Gemini telescopes, designed to be a matched pair, are located in Hawai'i in the Northern Hemisphere and in Chile in the Southern Hemisphere.

Newer and larger telescopes are planned for many of the same locations listed in Table 6.1. The 8-m Vera C. Rubin Observatory (formerly LSST) is under construction in Cerro Pachón in Chile, current site of the Gemini South telescope. The Giant Magellan Telescope (GMT), consisting of seven 8.4-m mirrors in a pattern equivalent to a 24.5-m mirror, is being constructed at Cerro Las Campanas in Chile. The Thirty Meter Telescope (TMT) is planned for Mauna Kea in Hawai'i or the Canary Islands, and the European Southern Observatory (ESO) is building the 39-m Extremely Large Telescope (ELT) at Cerro Armazones in Chile (**Figure 6.15**). As telescopes get larger—and more expensive—international collaboration becomes increasingly important.

Today's professional astronomers rarely look through the eyepiece of a telescope because they learn much more and make better use of observing time by permanently recording an object's image at a variety of wavelengths or seeing its light spread out into a revealing spectrum. Some astronomers no longer travel to telescopes at all, instead observing remotely from the base of the mountain or far away at their own institutions.

Professional and amateur astronomers alike are concerned about loss of the dark sky. As cities and suburbs around the world grow and expand, the use of outdoor artificial light becomes more widespread. Pictures from space show how bright many areas of Earth are at night (**Figure 6.16**). In the United States, two-thirds of the population resides in an area too bright to see the Milky Way in the sky at night, and it has been estimated that by 2025 the continental United States will have almost no dark skies. Increased air pollution also dims the view of the night sky in many locations. The U.S. National Park Service now advertises evening astronomy programs in natural, unpolluted dark skies as one of the reasons to visit some parks. Several international astronomy associations are working with UNESCO (the United Nations Educational, Scientific and Cultural Organization) to promote the "right to starlight," arguing that for historical, cultural, and scientific reasons, it would be a huge loss if humanity could no longer view the stars. Those organizations are encouraging countries to create starlight reserves and starlight parks where people can experience increasingly rare dark skies and a natural nocturnal environment.

Satellites are becoming an increasing hazard to dark skies. Many are bright enough to be seen with the naked eye. As more and more of them are launched, astronomers have to work hard to figure out how to avoid having their images of astronomical objects ruined by satellite tracks across the sky. Famously, SpaceX's Starlink satellites have already impacted astronomical observations

Figure 6.15 An artist's rendering of the European Extremely Large Telescope, a 39-m reflecting telescope under construction in Chile. The telescope will become operational in 2025. Credit: ESO/L. Calçada, https://www.eso.org/public/images /eso1440e/?lang=no. https://creativecommons.org/licenses /by/4.0/.

Figure 6.16 This satellite image of Earth at night shows that few populated areas are free from light pollution.

Figure 6.17 This image of the sky taken from the Lowell Observatory shows some stars and distant galaxies. The most noticeable features, however, are the bright streaks of more than two dozen of SpaceX's Starlink satellites passing overhead, a few weeks after they launched. Satellites such as these increasingly cause problems for astronomical observations.

(see **Figure 6.17**). By the end of 2020, SpaceX had launched more than 800 of these satellites into orbit around Earth, but it eventually plans to launch more than 10,000. Astronomers are working with SpaceX on plans to mitigate the effect by painting the satellites black, for example.

CHECK YOUR UNDERSTANDING **6.1b**

In practice, the smallest angular size that you can resolve with a 10-inch telescope is governed by the: (a) blurring caused by Earth's atmosphere; (b) diffraction limit of the telescope; (c) size of the primary mirror; (d) magnification of the telescope.

6.2 Optical Detectors and Instruments Used with Telescopes

Beginning in the 1800s, the development of film photography, and later digital photography, revolutionized astronomy, allowing astronomers to detect fainter and more-distant objects than it's possible to detect with the eye alone. Here in Section 6.2, we examine some of the more common types of detectors.

Integration Time and Quantum Efficiency

Originally, the retina of the human eye was the only astronomical detector. The limit of the faintest stars we can see with our unaided eyes is determined in part by two factors that are characteristic of all detectors: *integration time* and *quantum efficiency*.

Integration time is the limited time interval during which the eye can add up photons. It is analogous to the time the shutter is left open on a camera when taking a picture. The brain "reads out" the information gathered by the eye about every 100 milliseconds (ms). When anything happens faster than that, the human eye cannot detect the faster speed. If two images on a computer screen appear 30 ms apart, you will see them as a single image because your eyes will add up (or integrate) whatever they see over an interval of 100 ms or less. If the images occur 200 ms apart, however, you will see them as separate images. That relatively brief integration time is the most important factor limiting our nighttime vision. Stars too faint to be seen with the unaided eye are those from which you receive too few photons for your eyes to process in 100 ms.

Quantum efficiency measures how likely it is that any photon will produce a signal. For the human eye, 10 photons must strike a cone within 100 ms to activate a single response. So, the quantum efficiency of our eyes is about 10 percent. For every 10 events, then, the eye sends one signal to the brain. Together, integration time and quantum efficiency determine the rate at which photons must arrive at the retina before the brain says, "I see something." Astronomers seek to use detectors with longer integration times and higher quantum efficiency than our eyes can achieve.

From Photographic Plates to Charge-Coupled Devices

For more than two centuries after the invention of the telescope, astronomers struggled with the problem of **surface brightness**. Only *point sources* such as stars appear brighter in a telescope. Extended astronomical objects, such as the

Moon, appear bigger in the eyepiece, but the light is spread over that larger image, so their surfaces are no brighter than they appear to the unaided eye. Even when astronomers built larger telescopes, nebulae and galaxies appeared larger, but the details of these faint objects remained elusive. The problem was not with the telescopes but with the limitations of optics and the human eye. Only with the longer exposure times made possible by the invention of photography and the later development of electronic cameras could astronomers finally discern intricate details in faint objects.

In 1840, John W. Draper (1811–1882), a New York chemistry professor, created the earliest known astronomical photograph (**Figure 6.18**). By the late 1800s, astronomers had filled thousands of photographic plates with permanent images of planets, nebulae, and galaxies. The quantum efficiency of most photographic systems used in astronomy was poorer than that of the human eye—typically 1–3 percent. But unlike the eye, photography can overcome poor quantum efficiency by leaving the shutter open on the camera, increasing the integration time to many hours of exposure. Photography enabled astronomers to record and study objects invisible to the human eye. However, the response of photography to light is not linear, especially at long exposures, so if you doubled the exposure time, you did not get an image twice as bright. By the middle of the 20th century, the search was on for electronic detectors that would overcome photography's problems in sensitivity, spectral range, and nonlinearity.

In 1969, scientists at Bell Laboratories invented a detector called a **charge-coupled device (CCD)**. By the late 1970s, the CCD had become the detector of choice in almost all astronomical-imaging applications. CCDs are linear, so doubling the exposure means you record twice as much light. As a result, they are good for measuring objects that vary in brightness, as well as for faint objects that require long exposures. CCDs have a quantum efficiency far superior to that of either photography or the eye—up to 80 percent at some wavelengths. That improvement dramatically increases our ability to view faint objects with short exposure times.

A CCD is an ultrathin wafer of silicon—less than the thickness of a human hair—that is divided into a two-dimensional array of picture elements, or **pixels** (**Figure 6.19a**). When a photon strikes a pixel, it liberates an electron from an atom and creates a small electric charge within the silicon. At very low light levels, this response can be dwarfed by thermal "noise"—electrons that are moving freely because the camera is warmer than absolute zero. Liquid nitrogen or helium is used to cool some CCDs down to very low temperatures to reduce this thermal noise. Conversely, at very high light levels, it's possible to basically run out of electrons to liberate with another photon. Any further photons that arrive will not register. Just as a sponge can become saturated with water, so that it can absorb no more water droplets, a CCD camera can become saturated with light so that it can register no more photons. Between these two limits, the digital signal that flows to the computer is proportional to the accumulated charge, so the CCD is a linear device.

The output from a CCD is a digital signal that can either be sent directly from the telescope to image-processing software or stored electronically for later analysis. Nearly every spectacular astronomical image in ultraviolet (UV), visible, or infrared (IR) wavelength that you find online was recorded by a CCD in a telescope either on the ground or in space. The first astronomical CCDs were small arrays containing a few hundred thousand pixels. The larger CCDs used in astronomy today may contain arrays totaling more than 100 million pixels (**Figure 6.19b**). Still larger arrays are under development as ever-faster computing

Figure 6.18 A photograph of the Moon taken by John W. Draper in 1840. This is one of the first known photographs of an astronomical object.

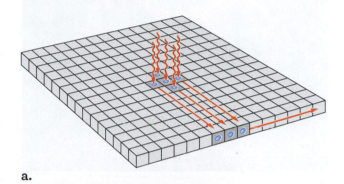

a.

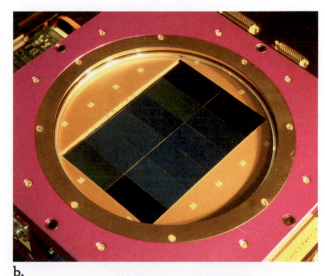

b.

Figure 6.19 **a.** In this simplified diagram of a charge-coupled device (CCD), photons from a star land on pixels (represented by gray squares) and produce free electrons within the silicon. The electron charges are electronically moved sequentially to the collecting register at the bottom. Each row is then moved out to the right to an electronic amplifier, which converts the electric charge of each pixel into a digital signal. **b.** This large CCD (about 6 inches across) contains 12,288 × 8192 pixels.

a.

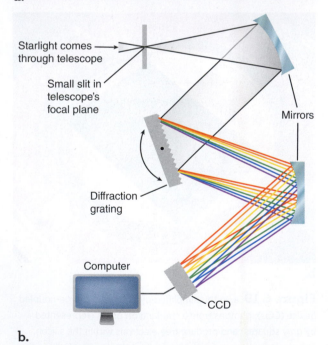

Starlight comes through telescope

Small slit in telescope's focal plane

Mirrors

Diffraction grating

Computer

CCD

b.

Figure 6.20 **a.** A spectrum is created by the reflection of light from the closely spaced tracks of a CD. **b.** In a grating spectrograph, light goes through the telescope and then a slit, where it is reflected to the diffraction grating and split into components. The spectrum is recorded on the CCD.

power keeps up with image-processing demands. CCDs are found in many everyday devices, too, such as digital cameras, digital video cameras, and camera phones.

Your cell phone takes color pictures by using a grid of CCD pixels arranged in groups of three. Each pixel in a group is constructed to respond to only a particular range of colors—to only red light, for example. That also is true for digital image displays. You can see for yourself if you place a small drop of water on the screen of your smartphone or tablet and turn it on. The water droplet magnifies the grid of pixels so that you can see them individually. In a camera, this grid degrades the angular resolution because each spot in the final image requires three pixels of information. Astronomers are motivated to maximize angular resolution, so instead of making a camera that takes color pictures, they use a camera that takes grayscale images at higher angular resolution. They put filters in front of the camera to allow light of only particular colors to pass through. Color pictures are constructed by taking multiple pictures in different filters, using software to color each one, and then carefully aligning and overlapping them to produce beautiful and informative images. Sometimes the colors are "true"; that is, they are close to the colors you would see if you were actually looking at the object with your eyes. Other times "false" colors represent different portions of the electromagnetic spectrum and tell you the temperature or composition of different parts of the object. Using changeable filters instead of designated color pixels gives astronomers greater flexibility and better angular resolution.

Spectrographs

Spectroscopy is the study of an object's *spectrum* (plural: *spectra*)—its electromagnetic radiation split into component wavelengths. **Spectrographs** (sometimes called **spectrometers**) are instruments that take the spectrum of an object and then record it. The first spectrographs used prisms to disperse the light. Modern spectrographs use a **diffraction grating**, which is made by engraving closely spaced lines on glass to disperse incoming light into a spectrum. **Figure 6.20a** shows the light reflected from a CD or DVD; the closely spaced tracks act as a grating and create a spectrum. **Figure 6.20b** shows a schematic of a grating spectrograph: Light from an astronomical object enters a telescope and passes through a slit. The light is reflected onto the diffraction grating, which creates a spectrum like the one shown in Figure 5.14. A CCD captures the spectrum for later analysis. Some modern spectrographs use bundles of optical fibers, or masks with multiple slits, to obtain spectra simultaneously from multiple objects in the field of view of the telescope.

CHECK YOUR UNDERSTANDING 6.2

CCD cameras have much higher quantum efficiency than other detectors. Therefore, CCD cameras: (a) can collect photons for longer times; (b) can collect photons of different energies; (c) can generate a signal from fewer photons; (d) can split light into different colors.

6.3 Astronomers Observe in Wavelengths Beyond the Visible

Astronomers use telescopes that observe at all the wavelengths of the electromagnetic spectrum. Recall from Section 5.4 that Wien's law states that an object's temperature can be found from the peak wavelength of its continuous spectrum.

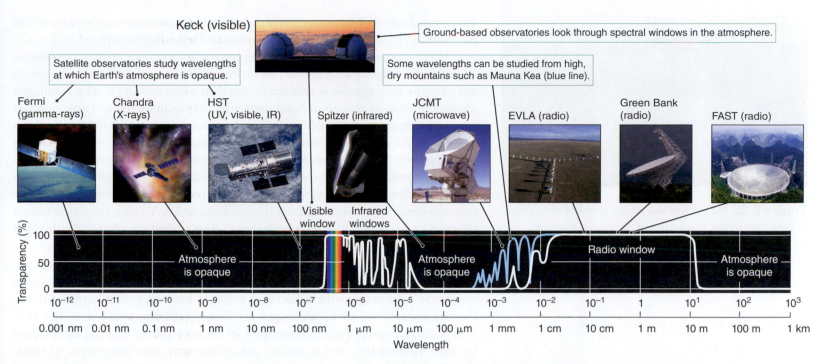

Figure 6.21 Earth's atmosphere blocks most electromagnetic radiation. Fermi = Fermi Gamma-ray Space Telescope (orbiting); Chandra = Chandra X-ray Observatory (orbiting); HST = Hubble Space Telescope (orbiting); Keck = Keck Observatory (Hawai'i); Spitzer = Spitzer Space Telescope (orbiting); JCMT = James Clerk Maxwell Telescope (Hawai'i); EVLA = Expanded Very Large Array (New Mexico); Green Bank = Robert C. Byrd Green Bank Telescope (West Virginia); FAST = Five-hundred-meter Aperture Spherical Telescope (Guizhou, China).

Hot objects emit the most light in the X-ray or gamma-ray regions of the spectrum, whereas cool objects emit the most light in the radio or infrared regions of the spectrum. Radio or infrared telescopes are used to study cool objects, such as clouds of dust, whereas X-ray or gamma-ray telescopes are used to study violently hot gas. Not all of these spectral regions are accessible from the ground. **Figure 6.21** shows Earth's **atmospheric windows** and which parts of the spectrum they let in. The largest window is in radio wavelengths, including microwaves at the short-wavelength end of the radio window. Radio telescopes can be built on the ground. However, gamma-ray, X-ray, ultraviolet, and most infrared light from astronomical objects fails to reach the ground because it is partially or completely absorbed by ozone, water vapor, carbon dioxide, and other molecules in Earth's atmosphere. Light at those wavelengths has to be observed from space.

Radio Telescopes

Karl Jansky (1905–1950), a young physicist working for Bell Laboratories in the early 1930s, identified a radio source in the Milky Way in the direction of the galactic center, in the constellation Sagittarius. Jansky's discovery marked the birth of radio astronomy, and in his honor, the basic unit for the strength of a radio source is called the **jansky** (Jy). A few years later, Grote Reber (1911–2002), a radio engineer and ham radio operator, built his own radio telescope and conducted the first survey of the sky at radio frequencies. Reber was largely responsible for the rapid advancement of radio astronomy in the post–World War II era.

Most radio telescopes are large, steerable dishes, typically tens of meters in diameter (**Figure 6.22a**). The world's largest single-dish radio telescope is the FAST (Five-hundred-meter Aperture Spherical Telescope) dish in China (**Figure 6.22b**).

Figure 6.22 **a.** The Green Bank telescope in West Virginia is the world's largest steerable radio telescope. **b.** The FAST radio telescope in China is the world's largest single-dish telescope.

Figure 6.23 The EVLA in New Mexico combines signals from 27 telescopes so that they act as one "very large" telescope.

This large single-dish telescope is not steerable, so it can observe only sources that pass within 20° of the zenith as Earth's rotation carries them overhead.

As large as radio telescopes are, they have relatively poor angular resolution. Recall that a telescope's angular resolution (θ) is determined by the ratio λ/D and that the best angular resolution is obtained when θ is small. As a result, the angular resolution gets poorer and poorer as that ratio increases (which occurs when λ increases and/or D decreases). Radio telescopes have diameters (D) much larger than the apertures of most optical telescopes. However, the wavelengths (λ) of radio waves range from about 1 centimeter (cm) to 10 meters, or up to several hundred thousand times greater than the wavelengths of visible light, which makes the ratio larger. Radio telescopes are thus limited by the very long wavelengths they are designed to receive. Even though China's FAST radio telescope has an effective aperture of 300 m, its angular resolution is similar to that of the unaided human eye—about 1 arcmin.

Radio astronomers have developed a clever way to improve angular resolution. Mathematically combining the signals from two radio telescopes turns them into a telescope with a diameter equal to the separation between them. For example, if two 10-m telescopes are located 1000 m apart, then the D in λ/D is 1000, not 10. That combination of two (or more) telescopes is called an **interferometer**, and it makes use of the wavelike properties of light. Usually, several telescopes are used in an arrangement called an **interferometric array**. Through the use of very large arrays, radio astronomers can better observe bright sources and exceed the angular resolution possible with optical telescopes.

The Expanded Very Large Array (EVLA) in New Mexico (**Figure 6.23**) is an interferometric array made up of 27 movable dishes spread out in a Y-shaped configuration up to 36 km across. At a wavelength of 10 cm, that array reaches angular resolutions of less than 1 arcsec. The Very Long Baseline Array (VLBA) uses 10 radio telescopes spread out over more than 8000 km from the Virgin Islands in the Caribbean to Hawai'i in the Pacific. At a wavelength of 10 cm, that array can attain angular resolutions of better than 0.003 arcsec. A radio telescope put into near-Earth orbit as part of a Space Very Long Baseline Interferometer (SVLBI) overcomes even that limit. The Event Horizon Telescope combines many of the most advanced existing radio telescopes, from Greenland to the South Pole, to make an *Earth-sized* interferometer. For a few nights each year, all the telescopes observe the same object, with a combined angular resolution that may be good enough to image objects near the center of the Milky Way.

Some radio telescopes use many small dishes. The Atacama Large Millimeter/submillimeter Array (ALMA; **Figure 6.24**), located at an elevation of 5000 m in the Atacama Desert in Chile, was completed in 2013. That project, an international collaboration of astronomers from Europe, North America, East Asia, and Chile, consists of 66 dishes with diameters of 12 m and 7 m for observations in the 0.3- to 9.6-mm-wavelength range. The Square Kilometre Array (SKA) is designed to have *thousands* of small radio dishes, which together will act as one dish with a collecting area of 1 square kilometer (km²). Twenty countries are supporting that telescope, which will be located in Australia and South Africa.

Optical telescopes also can be combined in an array to yield angular resolutions greater than those of single telescopes, although for technical reasons the individual units cannot be spread as far apart as radio telescopes. The Very Large Telescope Interferometer (VLTI) in Chile, operated by ESO, combines four

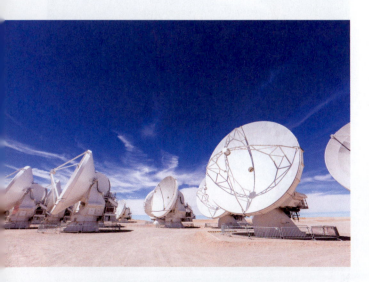

Figure 6.24 The new Atacama Large Millimeter/submillimeter Array (ALMA) telescope in the Atacama Desert in northern Chile has many international partners.

8-m telescopes with four movable 1.8-m auxiliary telescopes. It has a baseline of up to 200 m, yielding angular resolution of about 0.001 arcsec. The six-telescope Center for High Angular Resolution Astronomy (CHARA) array in California works in the visible and near-infrared regions. It has a baseline of 330 m with angular resolution of 0.0003 arcsec.

Infrared Telescopes

Molecules such as water vapor in Earth's atmosphere block infrared photons from reaching astronomical telescopes on the ground, so telescopes that observe in the infrared [0.75–30 microns (μm)] are at the highest locations. Mauna Kea, a dormant volcano and home of the Mauna Kea Observatories (MKO), rises 4205 m above the Pacific Ocean. At that altitude, the MKO telescopes sit above 40 percent of Earth's atmosphere; but more important, 90 percent of Earth's atmospheric water vapor lies below. Still, for the infrared astronomer, the remaining 10 percent is troublesome.

Airborne observatories overcome atmospheric absorption of infrared light by placing telescopes above most of the water vapor in the atmosphere. NASA's Stratospheric Observatory for Infrared Astronomy (SOFIA) (**Figure 6.25**), a joint project with the German Aerospace Center (DLR), is a modified 747 airplane that carries a 2.5-m telescope and works in the far-infrared region of the spectrum, from 1 to 650 μm. It flies in the stratosphere at an altitude of about 12 km, above 99 percent of the water vapor in Earth's lower atmosphere. Because airplanes are highly mobile, SOFIA can observe in both the Northern and Southern Hemispheres. Other infrared wavelengths must be observed from space.

Orbiting Observatories

Gaining full access to the complete electromagnetic spectrum requires getting completely above Earth's atmosphere. The first astronomical satellite was the British Ariel 1, launched in 1962 to study solar UV and X-ray radiation. Today, many orbiting astronomical telescopes cover the electromagnetic spectrum from gamma rays to microwaves, with more in the planning stage (**Table 6.2**). Optical telescopes, such as the 2.4-m Hubble Space Telescope (HST; see Figure 6.21), operate successfully at low Earth orbit, 600 km above Earth's surface. Launched in 1990, HST has been the workhorse for UV, visible, and IR space astronomy for more than 30 years. Low Earth orbit is also the region where the International Space Station and many scientific satellites orbit.

For certain other satellites and space telescopes, however, 600 km is not high enough. The Chandra X-ray Observatory (see Figure 6.21), NASA's X-ray telescope, cannot see through even the tiniest traces of atmosphere and therefore orbits more than 16,000 km above Earth's surface. NASA's Spitzer Space Telescope, an infrared telescope, is so sensitive that it needs to be completely free from Earth's own infrared radiation. The solution was to put it into a *solar* orbit, trailing tens of millions of kilometers behind Earth. The James Webb Space Telescope (JWST), scheduled to succeed the HST, will observe primarily in infrared wavelengths. It will be located 1.5 million miles away from Earth, orbiting the Sun at a fixed distance from the Sun and Earth.

Orbiting telescopes located above the atmosphere are unaffected by atmospheric image distortions, weather, or brightening night skies. But space observatories are much more expensive than ground-based observatories and can be difficult or

what if . . .

What if human life had developed on a planet containing almost no atmosphere? How might astronomy have developed differently?

a.

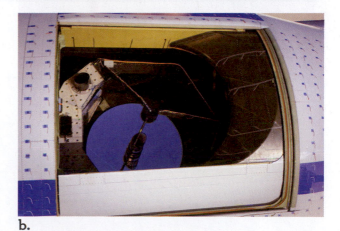

b.

Figure 6.25 SOFIA is a 2.5-m infrared telescope mounted in a Boeing 747 aircraft.

Table 6.2	Selected Space Observatories		
Telescope	**Sponsor(s)**	**Description**	**Year**
Hubble Space Telescope (HST)	NASA, ESA	Optical, infrared, ultraviolet observations	1990–
Chandra X-ray Observatory	NASA	X-ray imaging and spectroscopy	1999–
X-ray Multi-Mirror Mission (XMM-Newton)	ESA	X-ray spectroscopy	1999–
Galaxy Evolution Explorer (GALEX)	NASA	Ultraviolet observations	2003–2012
Spitzer Space Telescope	NASA	Infrared observations	2004–2020
Neil Gehrels Swift Observatory	NASA	Gamma-ray bursts	2004–
Convection Rotation and Planetary Transits (COROT) space telescope	CNES (France)	Planet finder	2006–2013
Fermi Gamma-ray Space Telescope	NASA, European partners	Gamma-ray imaging and gamma-ray bursts	2008–
Planck telescope	ESA	Cosmic microwave background radiation	2009–2013
Herschel Space Observatory	ESA	Far-infrared and submillimeter observations	2009–2013
Kepler telescope	NASA	Planet finder	2009–2018
Solar Dynamics Observatory (SDO)	NASA	Sun, solar weather	2010–
RadioAstron	Russia, international collaborators	Very-long-baseline interferometry in space	2011–
Nuclear Spectroscopic Telescopic Array (NuSTAR)	NASA	High-energy X-ray	2012–
Gaia	ESA	Optical, digital 3D space camera	2013–
Transiting Exoplanet Survey Satellite (TESS)	NASA, MIT, SAO	Survey entire sky for transiting planets around nearby and bright stars	2018–
CHEOPS (CHaracterising ExOPlanet Satellite)	ESA	High-precision brightness measurements of stars known to have planets	2019–
James Webb Space Telescope (JWST)	NASA, ESA, Canadian Space Agency	Primarily infrared	2021 (scheduled)

impossible to repair. The HST required several servicing missions, but such missions are impossible for the observatories in more distant Earth orbits. Ground-based telescopes at even the most remote mountaintop locations can receive shipments of replacement parts in a few days; space telescopes cannot. Some wavelengths can be observed from space only, but issues of cost and repair are why ground-based telescopes remain more prevalent for wavelengths that can be observed from the ground.

CHECK YOUR UNDERSTANDING 6.3

Which of the following is the biggest disadvantage of putting a telescope in space? (a) Astronomers don't have as much control in choosing what to observe. (b) Astronomers have to wait until the telescopes come back to Earth to get their images. (c) Space telescopes can observe only in certain parts of the electromagnetic spectrum. (d) Space telescopes are much more expensive and riskier than similar ground-based telescopes.

6.4 Planetary Spacecraft Explore the Solar System

Recall from Chapter 2 that everyone always sees the same face of the Moon from Earth because the Moon's orbital and rotational periods are equal. The first view of the "far" side was in 1959, when the Soviet flyby mission *Luna 3* sent back pictures showing that the far side of the Moon was very different from its Earth-facing half. No matter how powerful our ground-based or Earth-orbiting telescopes, sometimes we need to send a spacecraft for a different view.

Spacecraft have now visited all of our Solar System's planets and some of their moons, as well as some comets and asteroids, providing the first close-up views of those distant worlds. The study of the Solar System from space is an international collaboration involving NASA, the European Space Agency (ESA), the Russian Federal Space Agency (Roscosmos), the Japan Aerospace Exploration Agency (JAXA), the China National Space Administration (CNSA), and the Indian Space Research Organisation (ISRO). Other countries may soon join the endeavor. In this section, we look at the types of spacecraft used to explore our Solar System.

Flybys and Orbiters

Exploration of the Solar System began with a reconnaissance phase, using spacecraft to fly by or orbit a planet or other body. A **flyby** is a spacecraft that first approaches and then continues flying past the target. As those spacecraft speed by, instruments aboard them briefly probe the physical and chemical properties of the target and its environment.

Flyby missions are the most common first phase of exploration. They cost less than orbiters or landers and are easier to design and execute. Flyby spacecraft, such as *Voyager 1* and *Voyager 2*, can sometimes visit several worlds during their travels (**Figure 6.26**). The downside of flyby missions is that the spacecraft must

Figure 6.26 The two *Voyager* spacecraft **a.** flew past the outer planets and **b.** have recently passed the boundary of our Solar System.

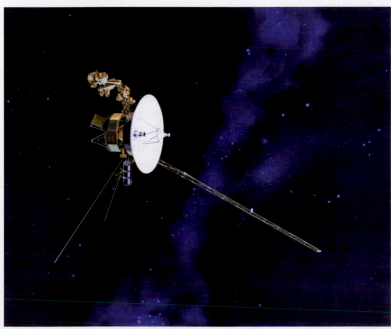

a.

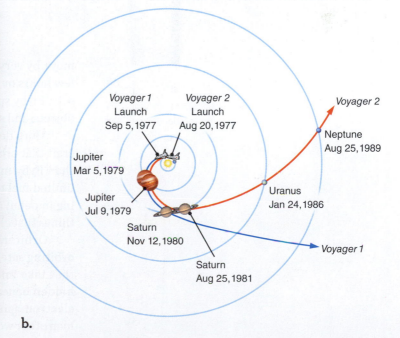

b.

Reading Astronomy News

NASA's *Perseverance* Mars Rover Extracts First Oxygen from Red Planet

NASA, *April 21, 2021*

In the 21st Century, missions to other planets do more than observe and take data; they are beginning to explore resource extraction and utilization.

The growing list of "firsts" for *Perseverance*, NASA's newest six-wheeled robot on the Martian surface, includes converting some of the Red Planet's thin, carbon dioxide–rich atmosphere into oxygen. A toaster-size, experimental instrument aboard *Perseverance* called the Mars Oxygen In-Situ Resource Utilization Experiment (MOXIE) accomplished the task. The test took place April 20, the 60th Martian day, or sol, since the mission landed Feb. 18.

While the technology demonstration is just getting started, it could pave the way for science fiction to become science fact—isolating and storing oxygen on Mars to help power rockets that could lift astronauts off the planet's surface. Such devices also might one day provide breathable air for astronauts themselves. MOXIE is an exploration technology investigation—as is the Mars Environmental Dynamics Analyzer (MEDA) weather station—and is sponsored by NASA's Space Technology Mission Directorate (STMD) and Human Exploration and Operations Mission Directorate.

"This is a critical first step at converting carbon dioxide to oxygen on Mars," said Jim Reuter, associate administrator for STMD. "MOXIE has more work to do, but the results from this technology demonstration are full of promise as we move toward our goal of one day seeing humans on Mars. Oxygen isn't just the stuff we breathe. Rocket propellant depends on oxygen, and future explorers will depend on producing propellant on Mars to make the trip home."

For rockets or astronauts, oxygen is key, said MOXIE's principal investigator, Michael Hecht of the Massachusetts Institute of Technology's Haystack Observatory.

To burn its fuel, a rocket must have more oxygen by weight. Getting four astronauts off the Martian surface on a future mission would require approximately 15,000 pounds (7 metric tons) of rocket fuel and 55,000 pounds (25 metric tons) of oxygen. In contrast, astronauts living and working on Mars would require far less oxygen to breathe. "The astronauts who spend a year on the surface will maybe use one metric ton between them," Hecht said.

Hauling 25 metric tons of oxygen from Earth to Mars would be an arduous task. Transporting a one-ton oxygen converter—a larger, more powerful descendant of MOXIE that could produce those 25 tons—would be far more economical and practical.

Mars' atmosphere is 96 percent carbon dioxide. MOXIE works by separating oxygen atoms from carbon dioxide molecules, which are made up of one carbon atom and two oxygen atoms. A waste product,

move by very swiftly due to the physics of orbits. Thus, they are limited to just a few hours or at most a few days in which to conduct close-up studies of their targets. Flyby spacecraft give astronomers their first close-up views of Solar System objects, and sometimes the data obtained are then used to plan follow-up studies.

More detailed reconnaissance work is done by **orbiters**, which are spacecraft that orbit around their target. Such missions are intrinsically more difficult than flyby missions because they have to make risky maneuvers and use their limited fuel to change their speed to enter an orbit. But orbiters can linger, looking in detail at more of the surfaces of the objects they are orbiting and studying things that change with time, such as planetary weather.

Orbiters use remote-sensing instrumentation like that used by Earth-orbiting satellites to study our own planet. Those instruments include cameras that take images at different wavelength ranges, radar that can map surfaces hidden beneath obscuring layers of clouds, and spectrographs that analyze the electromagnetic spectrum. Those instruments enable planetary scientists to map other worlds, measure the heights of mountains, identify geological features

carbon monoxide, is emitted into the Martian atmosphere.

The conversion process requires high levels of heat to reach a temperature of approximately 1470 degrees Fahrenheit (800 Celsius). To accommodate this, the MOXIE unit is made with heat-tolerant materials. These include 3D-printed nickel alloy parts, which heat and cool the gases flowing through it, and a lightweight aerogel that helps hold in the heat. A thin gold coating on the outside of MOXIE reflects infrared heat, keeping it from radiating outward and potentially damaging other parts of *Perseverance*.

In this first operation, MOXIE's oxygen production was quite modest—about 5 grams, equivalent to about 10 minutes worth of breathable oxygen for an astronaut. MOXIE is designed to generate up to 10 grams of oxygen per hour.

This technology demonstration was designed to ensure the instrument survived the launch from Earth, a nearly seven-month journey through deep space, and touchdown with *Perseverance* on Feb. 18. MOXIE is expected to extract oxygen at least nine more times over the course of a Martian year (nearly two years on Earth).

These oxygen-production runs will come in three phases. The first phase will check out and characterize the instrument's function, while the second phase will run the instrument in varying atmospheric conditions, such as different times of day and seasons. In the third phase, Hecht said, "we'll push the envelope"—trying new operating modes, or introducing "new wrinkles, such as a run where we compare operations at three or more different temperatures."

"MOXIE isn't just the first instrument to produce oxygen on another world," said Trudy Kortes, director of technology demonstrations within STMD. It's the first technology of its kind that will help future missions "live off the land," using elements of another world's environment, also known as in-situ resource utilization.

"It's taking regolith, the substance you find on the ground, and putting it through a processing plant, making it into a large structure, or taking carbon dioxide—the bulk of the atmosphere—and converting it into oxygen," she said. "This process allows us to convert these abundant materials into useable things: propellant, breathable air, or, combined with hydrogen, water."

QUESTIONS

1. What does MOXIE do?

2. In this test, MOXIE produced 5.4 grams of oxygen in two hours; enough to sustain an astronaut for ten minutes. How much oxygen is needed to sustain an astronaut for one hour?

3. In future tests, MOXIE will produce 10 grams of oxygen per hour. Will MOXIE produce oxygen fast enough to sustain an astronaut?

4. In order for humans to travel to Mars, what other activities (besides sustaining breathing astronauts) require oxygen?

5. NASA plans to have MOXIE run nine other tests on Mars. Why is it important that MOXIE conduct these other tests?

Source: https://www.nasa.gov/press-release/nasa-s-perseverance-mars-rover-extracts-first-oxygen-from-red-planet.

and rock types, watch weather patterns develop, measure the composition of atmospheres, and get a general sense of the place. Additional instruments make measurements of the extended atmospheres and space environment through which they travel.

Landers, Rovers, and Atmospheric Probes

Reconnaissance spacecraft provide a wealth of information about a planet, but no better way exists to explore a planet than doing so within a planet's atmosphere or on solid ground. Spacecraft have landed on the Moon, Mars, Venus, Saturn's large moon Titan, and several asteroids and comets. Those spacecraft have taken pictures of planetary surfaces, measured surface chemistry, and conducted experiments to determine the physical properties of the surface rocks and soils.

Using **landers**—spacecraft that touch down and remain on the surface—has several disadvantages. Because of the expense, only a few landings in limited areas are practical. With that limitation, the results may apply only to the small area

Figure 6.27 The robotic rover *Curiosity* took this selfie on the surface of Mars.

Figure 6.28 This photograph of "Earthrise" over the limb of the Moon was taken by the astronauts on *Apollo 8*.

unanswered questions

Will further human exploration of the Solar System occur within your lifetime? Since *Apollo 17*, in 1972, humans have not returned to the Moon or traveled to other planets or moons in the Solar System. Sending humans to the worlds of the Solar System is much more complicated, risky, and expensive than sending robotic spacecraft. Humans need life support such as air, water, and food. Radiation in space can be dangerous. Furthermore, human explorers would expect to return to Earth, whereas most spacecraft do not come back. Astronomers and space scientists have heated debates about human spaceflight versus robotic exploration. Some argue that true exploration requires that human eyes and brains actually go there; others argue that the costs and risks are too high for the potential additional scientific knowledge. Beyond basic exploration, we also do not know whether humans will ever permanently colonize space.

around the landing site. Imagine, for example, what a different picture of Earth you might get from a spacecraft that landed in Antarctica, as opposed to a spacecraft that landed in a volcano or on the floor of a dry riverbed. Sites to be explored with landed spacecraft must be carefully chosen on the basis of reconnaissance data. Some landers have wheels and can explore the vicinity of the landing site. Such remote-controlled vehicles, called **rovers**, were used first by the Soviet Union on the Moon four decades ago and more recently by the United States and China on Mars. **Figure 6.27** shows a self-portrait of the *Curiosity* rover on Mars.

Atmospheric probes descend into the atmospheres of planets and continually measure and send back data on temperature, pressure, and wind speed, along with other properties, such as chemical composition. Atmospheric probes have survived all the way to the solid surfaces of Venus and of Saturn's moon Titan, sending back streams of data during their descent. An atmospheric probe sent into Jupiter's atmosphere never reached that planet's surface because, as you will learn later in the book, Jupiter does not have a solid surface in the same sense that terrestrial planets and moons do. After sending back its data, the Jupiter probe eventually melted and vaporized as it descended into the hotter layers of the planet's atmosphere.

Sample Returns

If you pick up a rock from the side of a road, you might learn a lot from the rock by using tools that you could carry in your pocket or in your car. Much better, though, would be to pick up a few samples and carry them back to a laboratory equipped with a full range of state-of-the-art instruments that can measure chemical composition, mineral type, age, and other information needed to reconstruct the story of your rock sample's origin and evolution. The same is true of Solar System exploration. One of the most powerful methods for investigating remote objects is to collect samples of the objects and bring them back to Earth for study. So far, only samples of the Moon, a comet, and streams of charged particles from the Sun have been collected and returned to Earth. Scientists have found meteorites on Earth that are pieces of Mars that were blasted loose when objects crashed into that world. In the next 10 years, robotic "sample and return" missions may go to Mars.

The missions discussed so far in this section have all been conducted with robotic spacecraft. The only spacecraft that took people to another world were the *Apollo* missions to the Moon. That program ran from 1961 to 1972 and included several missions before the first Moon landing on July 20, 1969. The *Apollo 8* astronauts took the magnificent picture of Earth viewed over the surface of the Moon (**Figure 6.28**). Each mission from *Apollo 11* through *Apollo 17* had three astronauts—two to land on the Moon and one to remain in orbit. *Apollo 13* did not reach the Moon but returned to Earth safely. Twelve American astronauts walked on the Moon between 1969 and 1972 and brought back a total of 382 kg of rocks and other material.

The return of extraterrestrial samples to Earth is governed by international treaties and standards to ensure that the samples do not contaminate our world. For example, before the lunar samples that the *Apollo* missions brought back could be studied, they (and the astronauts) had to be quarantined and tested for the presence of alien life-forms. The same international standards apply to spacecraft landing elsewhere. The goal of those standards is to avoid transporting life-forms from Earth to another planet; we do not want to "discover" life that we, in fact, introduced. In addition, we do not want to potentially harm life that may exist on other planets.

With many missions under way and others on the horizon, robotic exploration of the Solar System is an ongoing, dynamic activity. Appendix 5 summarizes some recent and current missions. We discuss some of those missions in the relevant chapter. Information on the latest discoveries can be found on mission websites and in science news sources.

CHECK YOUR UNDERSTANDING 6.4

Spacecraft are the most effective way to study planets in our Solar System because: (a) planets move too fast across the sky for us to image them well from Earth; (b) planets cannot be imaged from Earth; (c) spacecraft can collect more information than is available just from images from Earth; (d) space missions are easier than long observing campaigns.

6.5 Other Tools Contribute to the Study of the Universe

High-profile space missions have sent back stunning images and data from across the electromagnetic spectrum. But astronomers use other tools as well, including particle accelerators and colliders, neutrino and gravitational-wave detectors, and super-computers.

Particle Accelerators

Ever since the early years of the 20th century, physicists have been peering into the structure of the atom by observing what happens when small particles collide. By the 1930s, physicists had developed the technology to accelerate charged subatomic particles such as protons to very high speeds and then observe what happens when they slam into a target. From such experiments, physicists have discovered many kinds of subatomic particles and learned about their physical properties. High-energy particle colliders have proved to be an essential tool for physicists studying the basic building blocks of matter.

Astronomers have realized that to understand the very largest structures seen in the universe, it is important to understand the physics that took place during the earliest moments in the universe, when everything was extremely hot and dense. High-energy particle colliders that physicists use today are designed to approach the energies of the early universe. The effectiveness of particle accelerators is determined by the energy they can achieve and the number of particles they can accelerate. Modern particle colliders such as the Large Hadron Collider near Geneva, Switzerland (**Figure 6.29**), reach very high energies. Particles also can be studied from space. The Alpha Magnetic Spectrometer, installed on the International Space Station in 2011, searches for some of the most exotic forms of matter, such as dark matter, antimatter, and high-energy particles called cosmic rays.

Neutrinos and Gravitational Waves

The **neutrino** is an elusive elementary particle that plays a major role in the physics of the interiors of stars. Neutrinos are extremely difficult to detect. In less time than you take to read this sentence, a thousand trillion (10^{15}) solar neutrinos from

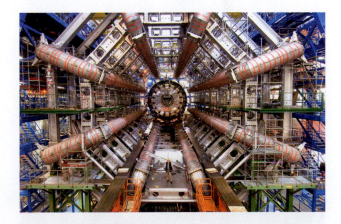

Figure 6.29 The ATLAS particle detector at CERN's Large Hadron Collider near Geneva, Switzerland. The instrument's enormous size is evident from the person standing near the bottom center of the picture.
Credit: Photograph: Maximilien Brice, © 2005-2021 CERN, http://cds.cern.ch/record/910381. https://creativecommons.org/licenses/by/4.0/.

what if . . .

What if, instead of detecting several black hole or neutron star merger events per year, LIGO had seen only one such event during its three years of observing: What might you reasonably conclude from that result?

the Sun are passing through your body, even during the night. Neutrinos are so nonreactive with matter that they can pass right through Earth (and you) as though it (or you) weren't there at all. To be observed, a neutrino has to interact with a detector. Neutrino detectors typically record only one of every 10^{22} (10 billion trillion) neutrinos passing through them, but that's enough to reveal processes deep within the Sun or the violent death of a star 160,000 light-years away.

Experiments designed to look for neutrinos originating outside Earth are buried deep underground in mines or caverns or under the ocean or ice to ensure that only neutrinos are detected. For example, the ANTARES experiment uses the Mediterranean Sea as a neutrino telescope. Detectors located 2.5 km under the sea, off the coast of France, observe neutrinos that originated in objects visible in southern skies and passed through Earth. In the IceCube neutrino observatory located at the South Pole in Antarctica, the neutrino detectors are 1.5–2.5 km under the ice, and they observe neutrinos that originated in objects visible in northern skies (**Figure 6.30**).

Another elusive phenomenon is the **gravitational wave**. Gravitational waves are disturbances in a gravitational field, similar to the waves that spread out from the disturbance you create when you toss a pebble onto the quiet surface of a pond. Several facilities, including the Laser Interferometer Gravitational-Wave Observatory (LIGO) and the European VIRGO interferometer, have been constructed to detect gravitational waves (**Process of Science Figure**). Recently, LIGO has detected several pairs of coalescing binary black holes and neutron stars, as we explore in Chapter 18.

Computers

Astronomers use powerful computers to gather, analyze, and interpret data. A single CCD image may contain millions of pixels, with each pixel displaying roughly 30,000 levels of brightness. That adds up to several trillion pieces of information in each image. To analyze their data, astronomers typically do calculations for *every single pixel* of an image to remove unwanted contributions from Earth's atmosphere or to correct for instrumental effects. Astronomers conduct many types of sky surveys—in which one or more telescopes survey a specific part of the sky—yielding thousands of images that need to be analyzed.

High-performance computers also play an essential role in generating and testing theoretical models of astronomical objects. Even when we completely understand the underlying physical laws that govern the behavior of a particular object, often the object is so complex that calculating its properties and behavior would be impossible without the assistance of high-performance computers. As discussed in Chapter 4, for example, you can use Newton's laws to compute the orbits of two stars that are gravitationally bound to each other because their orbits take the form of simple ellipses. However, understanding the orbits of the several hundred billion stars that make up the Milky Way Galaxy is not so easy, even though the underlying physical laws are the same.

Computer modeling is used to determine the interior properties of stars and planets, including Earth. Although astronomers cannot see beneath the surfaces of those bodies, they have a surprisingly good understanding of their interiors, which we describe in later chapters. Astronomers start a model by assigning well-understood physical properties to tiny volumes within a planet or star. The computer assembles an enormous number of those individual elements into an overall representation. The result is a rather good picture of what the interior of the star or planet is like.

Figure 6.30 The IceCube neutrino telescope at the South Pole, Antarctica.

Technology and Science Are Symbiotic

Scientists have been searching for waves that carry gravitational information for nearly 100 years, but the accuracy of their measurements has been limited by the available technology.

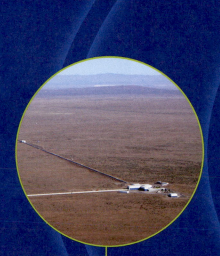

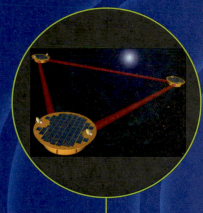

Einstein predicts that changing gravitational fields should produce gravitational waves. He calculates the size of the effect, and concludes that it will be impossible to detect them.

1916

c. 1970

Joseph Weber constructs precision-machined bars of metal that should "ring" as a gravitational wave passes by. His results were never confirmed.

1990s

Construction begins on LIGO, an observatory which uses lasers to precisely measure the change in the length of long tubes as gravitational waves pass by.

2015

LIGO detects gravitational waves for the first time, announcing the discovery in early 2016, 100 years after Einstein's prediction. This first detection has been followed by many more, so many that detections are now nearly routine.

Future (2034?)

LISA, a gravitational wave observatory in space, will be more sensitive; it will be able to detect much weaker events than LIGO.

Technology and science develop together. New technologies enable humans to ask new scientific questions. Asking new scientific questions pushes the development of better instrumentation and new technologies.

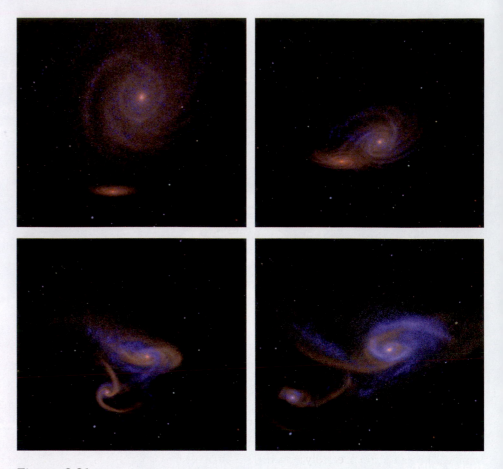

Figure 6.31 These images show supercomputer simulations of the collision of two galaxies. Astronomers compare simulations such as these with telescopic observations.

Astronomers also use high-performance computers to study how astronomical objects, systems of objects, and the universe as a whole evolve. For example, astronomers create models of galaxies and then run computer simulations to study how those galaxies might change over billions of years. **Figure 6.31** shows a simulation of the collision of two galaxies. The results of the computer simulations are then compared with telescopic observations. If the simulations do not match the observations, the model is adjusted and the simulations are run again until general agreement exists between them.

CHECK YOUR UNDERSTANDING **6.5**

High-performance computers have become one of an astronomer's most important tools. Which of the following require the use of that type of computer? (Choose all that apply.) (a) analyzing images taken with very large CCDs; (b) generating and testing theoretical models; (c) pointing a telescope from object to object; (d) studying how astronomical objects or systems evolve

Origins: Microwave Telescopes Detect Radiation from the Big Bang

In this chapter, we explored the tools of the astronomer, from basic optical telescopes to instruments that observe in different wavelengths. Now let's examine in more detail one type of telescope that has aided the study of the history of the universe. Recall from Chapter 1 that astronomers think the universe originated with a hot Big Bang. The multiple strands of evidence for that conclusion are discussed in Chapter 21. Here, we look at one piece: the observation of faint microwave radiation left over from the early hot universe. Two Bell Laboratories physicists, Arno Penzias (1933–) and Robert Wilson (1936–), were working on satellite communications when they first detected that radiation in 1964 with a microwave antenna in New Jersey. Today, we routinely use cell phones and handheld GPS devices that communicate directly with satellites, but at the time, that capability was at the limit of technology.

Penzias and Wilson needed a very sensitive microwave telescope for the work they were doing for Bell Labs because any spurious signals coming from the telescope itself might wash out the faint signals bounced off a satellite. To that end, they were working hard to eliminate all possible sources of microwave radiation originating from within their instrument, including keeping the telescope free of bird droppings. No matter how carefully they tried to eliminate sources of extraneous noise, they always still detected a faint signal at microwave wavelengths. That faint signal was the same in every direction and turned out to be from the Big Bang. Penzias and Wilson shared the 1978 Nobel Prize in Physics for discovering the **cosmic microwave background radiation (CMB)** left over from the Big Bang itself.

Since 1964, astronomers from around the world have designed increasingly precise instruments to measure that radiation from the ground, from high-altitude balloons, from rockets, and from satellites. The Russian experiment RELIKT-1, launched in 1983, found some limits on the variation of the CMB. The COBE (Cosmic Background Explorer) satellite, launched in 1989, showed that the spectrum of that radiation precisely matched that of a blackbody with a temperature of 2.73 K—exactly what was predicted for the radiation left over from the Big Bang. (Compare **Figure 6.32** with the curves in Figure 5.23.) The data also showed some slight differences in temperature—small fractions of a degree—over the map of the sky. Those slight variations tell us about how the universe evolved from one dominated by radiation to one that contains structures such as galaxies, stars, planets, and us. John Mather and George Smoot shared the 2006 Nobel Prize in Physics for that work.

In 1998 and 2003, a high-altitude balloon experiment called BOOMERANG (short for "balloon observations of millimetric extragalactic radiation and geophysics") flew over Antarctica at an altitude of 42 km to study CMB variations and estimate the overall geometry of the universe. The *WMAP* (Wilkinson Microwave Anisotropy Probe) satellite, launched in 2001, created an even more detailed map of the temperature variations in that radiation, yielding more precise values for the age and shape of the universe and the presence of dark matter and dark energy. The Planck space telescope, operated from 2009 to 2013 by the European Space Agency, was much more sensitive than *WMAP* and studied those CMB variations in even more precise detail. The Atacama Cosmology Telescope and Simons Array (Chile) and the South Pole Telescope (Antarctica) study that radiation to look for evidence of when galaxy clusters formed. Those experiments and observations have opened up the current era of precision cosmology, in which astronomers can make detailed models of how the universe was born, eventually leading to stars, planets, and us.

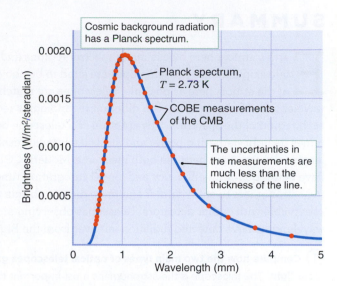

Figure 6.32 This graph shows the spectrum of the cosmic microwave background radiation (CMB) as measured by the COBE satellite (red dots). A steradian is a unit of solid angle. The uncertainty in the measurement at each wavelength is much less than the size of a dot. The line running through the data is a Planck blackbody spectrum with a temperature of 2.73 K.

SUMMARY

Earth's atmosphere blocks many spectral regions and distorts telescopic images, so telescopes are sited to be above as much of the atmosphere as possible. Telescopes are matched to the wavelengths of observation, with different technologies required for each region of the spectrum. The aperture of a telescope both determines its light-gathering power and limits its angular resolution; larger telescopes are better in both measures. Modern CCD cameras have better quantum efficiency and longer integration times, allowing astronomers to study fainter and more distant objects than were observable with earlier detectors. Telescopes observing at microwave wavelengths have detected radiation left over from the Big Bang.

(1) **Compare how the two main types of optical telescopes gather and focus light.** The telescope is the astronomer's most important tool. Ground-based telescopes that observe in visible wavelengths are either refractors (lenses) or reflectors (mirrors). All large astronomical telescopes are reflectors. Large telescopes collect more light and have better angular resolution. The diffraction limit is the limiting angular resolution of a telescope.

(2) **Summarize the main types of detectors that are used on telescopes.** Photography improved the ability of astronomers to record details of faint objects seen in telescopes. CCDs are today's astronomical detector of choice because they are much more linear, have a broader spectral response, and can send electronic images directly to a computer. Spectrographs are specialized instruments that take the spectrum of an object to reveal what the object is made of and many other physical properties.

(3) **Explain why some wavelengths of radiation must be observed from high, dry, and remote observatories on Earth, or from space.** Radio, near-infrared, and optical telescopes can see through our atmosphere. Those types of telescopes can be arrayed to greatly increase angular resolution. Putting telescopes in space solves problems created by Earth's atmosphere.

(4) **Explain the benefits of sending spacecraft to study the planets and moons of our Solar System.** Most of what is known about the planets and their moons comes from observations by spacecraft. Flyby and orbiting missions obtain data from space, and landers and rovers collect data from the ground.

(5) **Describe other astronomical tools, such as gravitational wave detectors, that contribute to the study of the universe.** Astronomers also use particle accelerators, neutrino detectors, and gravitational-wave detectors to study the universe. High-performance computers are essential to acquiring, analyzing, and interpreting astronomical data.

QUESTIONS AND PROBLEMS

TEST YOUR UNDERSTANDING

1. If one telescope has an aperture of 20 cm, and another has an aperture of 30 cm, and if aperture size is the only difference, then which should you choose, and why?
   a. The 20 cm, because the light-gathering power will be better.
   b. The 20 cm, because the image size will be larger.
   c. The 30 cm, because the light-gathering power will be better.
   d. The 30 cm, because the image size will be larger.

2. Study Figure 6.21. Which of the following can be observed from Earth's surface? (Choose all that apply.) ⊙★
   a. radio waves
   b. gamma radiation
   c. far UV light
   d. X-ray light
   e. visible light

3. Match the following properties of telescopes (lettered) with their corresponding definitions (numbered).
   a. aperture
   b. angular resolution
   c. focal length
   d. chromatic aberration
   e. diffraction
   f. interferometer
   g. adaptive optics

   (1) two or more telescopes connected to act as one
   (2) distance from lens to focal plane
   (3) diameter
   (4) ability to distinguish close objects
   (5) computer-controlled correction for atmospheric distortion
   (6) color-separating effect
   (7) smearing effect due to sharp edge

4. Two 10-m telescopes, separated by 85 m, can operate as an interferometer. What is its angular resolution when it observes in the infrared at a wavelength of 2 microns?
   a. 0.01 arcsec
   b. 0.005 arcsec
   c. 0.2 arcsec
   d. 0.05 arcsec

5. Arrays of radio telescopes can produce much better angular resolution than single-dish telescopes can because they work based on the principle of
   a. reflection.
   b. refraction.
   c. diffraction.
   d. interference.

6. Refraction is caused by
   a. light bouncing off a surface.
   b. light changing colors as it enters a new medium.
   c. light changing speed as it enters a new medium.
   d. two light beams interfering.

7. The light-gathering power of a 4-m telescope is _____ than that of a 2-m telescope.
   a. 4 times larger
   b. 8 times larger
   c. 16 times smaller
   d. 2 times smaller

8. Improved angular resolution is helpful to astronomers because
   a. they often want to look in detail at small features of an object.
   b. they often want to look at very distant objects.
   c. they often want to look at many objects close together.
   d. all of the above

9. The part of the human eye that acts as the detector is the
   a. retina.
   b. pupil.
   c. lens.
   d. iris.

10. Cameras that use adaptive optics provide images with better angular resolution primarily because
    a. they operate above Earth's atmosphere.
    b. deformable mirrors are used to correct the blurring due to Earth's atmosphere.
    c. composite lenses correct for chromatic aberration.
    d. they simulate a much larger telescope.

11. The advantage of an interferometer is that
    a. the angular resolution is dramatically improved.
    b. the focal length is dramatically increased.
    c. the light-gathering power is dramatically increased.
    d. diffraction effects are dramatically decreased.
    e. chromatic aberration is dramatically decreased.

12. The angular resolution of a ground-based telescope is usually determined by
    a. diffraction.
    b. the focal length.
    c. refraction.
    d. atmospheric seeing.

13. A grating can spread white light out into a spectrum of colors because of the property of
    a. reflection.
    b. interference.
    c. dispersion.
    d. diffraction.

14. Why would astronomers put telescopes in airplanes?
    a. to get the telescopes closer to the stars
    b. to get the telescopes above most of the water vapor in Earth's atmosphere
    c. to be able to observe one object for more than 24 hours without stopping
    d. to allow the telescopes to observe the full spectrum of light

15. If we could increase the quantum efficiency of the human eye, doing so would
    a. allow humans to see a larger range of wavelengths.
    b. allow humans to see better at night or in other low-light conditions.
    c. increase the angular resolution of the human eye.
    d. decrease the angular resolution of the human eye.

THINKING ABOUT THE CONCEPTS

16. Galileo's telescope used simple lenses. What is the primary disadvantage of using a simple lens in a refracting telescope?

17. The largest astronomical refractor has an aperture of 1 m. List several reasons why building a larger refractor with twice that aperture would be impractical.

18. Your camera may have a zoom lens, ranging between wide angle (short focal length) and telephoto (long focal length). How does the size of an object in the camera's focal plane differ between wide angle and telephoto?

19. Optical telescopes reveal much about the nature of astronomical objects. Why do astronomers also need information provided by gamma-ray, X-ray, infrared, and radio telescopes?

20. For light reflecting from a flat surface, the angles of incidence and reflection are the same. That is also true for light reflecting from the curved surface of a reflecting telescope's primary mirror. Sketch a curved mirror and several of those reflecting rays.

21. Consider two optically perfect telescopes having different diameters but the same focal length. Is the image of a star larger or smaller in the focal plane of the larger telescope? Explain your answer.

22. Study the Process of Science Figure. Make a flowchart for the symbiosis between technology and science that led to the detection of gravitational waves. ⊛

23. ★ WHAT AN ASTRONOMER SEES   Use Figure 6.14 as an example to explain adaptive optics and describe how they improve a telescope's image quality. ⊛

24. Explain integration time and quantum efficiency and how each contributes to the detection of faint astronomical objects.

25. Some people believe that we put astronomical telescopes on high mountaintops or in orbit because doing so gets them closer to the objects they are observing. Explain what is wrong with that popular misconception, and give the actual reason telescopes are located in those places.

26. Humans have sent various kinds of spacecraft—including flybys, orbiters, and landers—to all the planets in our Solar System. Explain the advantages and disadvantages of each of those types of spacecraft.

27. If Earth has meteorites that are pieces of Mars, why is going to Mars and bringing back samples of the martian surface so important?

28. Humans had a first look at the far side of the Moon as recently as 1959. Why had we not seen it earlier—when Galileo first observed the Moon with his telescope in 1610?

29. Where are neutrino detectors located? Why are neutrinos so difficult to detect?

30. Why do telescopes in space give a better picture of the leftover radiation from the Big Bang?

APPLYING THE CONCEPTS

31. Many amateur astronomers start out with a 4-inch (aperture) telescope and then graduate to a 16-inch telescope. By what factor does the light-gathering power of the telescope increase with that upgrade? ●—●—●

   a. Make a prediction: Do you expect the 16-inch telescope to have a larger or smaller light-gathering power than the 4-inch telescope? Do you expect the "factor" that you are solving for to be more or less than one? Study Working It Out 6.1. Will you need to change the units of inches to meters in order to find this factor?

   b. Calculate: Find the ratio of the light-gathering power of the 16-inch telescope to the 4-inch telescope.

   c. Check your work: Verify that the factor you have calculated is dimensionless (the units have canceled out), and compare it to your prediction.

32. Assume that you have a telescope with an aperture of 1 m. What is the telescope's theoretical angular resolution when you are observing in the near-infrared region of the spectrum ($\lambda = 1000$ nm)? ●—●—●

   a. Make a prediction: Study Working It Out 6.2. Is this wavelength in the near-infrared region longer or shorter than the wavelength of green light used there? Is the diameter of this telescope more or less than the diameter of Hubble Space Telescope? Which term do you expect to dominate in this calculation; that is, do you expect the angular resolution you calculate to be smaller or larger than that of Hubble Space Telescope?

   b. Calculate: Use the angular resolution equation to calculate the angular resolution of this telescope at this wavelength.

   c. Check Your Work: Verify that the units of your answer are correct, and compare your result to your prediction.

33. Compare the light-gathering power of the Thirty Meter Telescope with that of the dark-adapted human eye (aperture, 8 mm).

34. The diameter of the full Moon in the focal plane of an average amateur's telescope (focal length, 1.5 m) is 13.8 mm. How big would the Moon be in the focal plane of a very large astronomical telescope (focal length, 250 m)?

   a. Make a prediction: Study the magnification equation in Working It Out 6.1. Do telescopes with longer focal lengths magnify more or less than telescopes with shorter focal lengths? Do you expect the Moon to be larger or smaller in the focal plane of the very large astronomical telescope?

   b. Calculate: Set up a ratio of the magnification of the Moon in the larger telescope to the magnification in the smaller telescope (assume the eyepiece used is the same in both cases). Then use that ratio of magnifications to find the size of the Moon in the larger telescope.

   c. Check your work: Verify that the unit of your answer is a unit of length, and compare your answer to your prediction.

35. The angular resolution of the human eye is about 1.5 arcmin. What would the aperture of a radio telescope (observing at 21 cm) have to be to have that angular resolution? Even though the atmosphere is transparent at radio wavelengths, humans do not see light in the radio range. Using your calculations and logic, explain why.

36. Assume that the maximum aperture of the human eye, D, is approximately 8 mm and the average wavelength of visible light, λ, is 5.5×10^{-4} mm.

   a. Calculate the diffraction limit of the human eye in visible light.

   b. How does the diffraction limit compare with the actual angular resolution of 1–2 arcmin (60–120 arcsec)?

   c. To what do you attribute the difference?

37. Study the photograph of light entering and leaving a block of refractive material in Figure 6.2b. Use a protractor to measure the angles of the green light as it enters the block and as it leaves the block. How are those angles related? ◉

38. One of the earliest astronomical CCDs had 160,000 pixels, each recording 8 bits (256 levels of brightness). A new generation of astronomical CCDs may contain a billion pixels, each recording 15 bits (32,768 levels of brightness). Compare the number of bits of data that each of those two CCD types produces in a single image.

39. Consider a CCD with a quantum efficiency of 80 percent and a photographic plate with a quantum efficiency of 1 percent. If an exposure time of 1 hour is required to photograph a celestial object with a given telescope, how much observing time would be saved by substituting a CCD for the photographic plate?

40. The VLBA uses an array of radio telescopes ranging across 8000 km of Earth's surface from the Virgin Islands to Hawai'i.

   a. Calculate the angular resolution of the array when radio astronomers are observing interstellar water molecules at a microwave wavelength of 1.35 cm.

   b. How does that angular resolution compare with the angular resolution of two large optical telescopes separated by 100 m and operating as an interferometer at a visible wavelength of 550 nm?

41. When operational, the Space VLBI may have a baseline of 100,000 km. What will be the angular resolution when studying interstellar molecules emitting at a wavelength of 17 mm from a distant galaxy?

42. The *Mars Reconnaissance Orbiter* (*MRO*) flies at an average altitude of 280 km above the martian surface. If its cameras have an angular resolution of 0.2 arcsec, what is the size of the smallest objects that the *MRO* can detect on the martian surface?

43. At this writing, *Voyager 1* is about 140 astronomical units (AU) from Earth, continuing to record its environment as it departed the boundary of our Solar System.

   a. How far away (in kilometers) is *Voyager 1*?

   b. How long do observational data take to come back to us from *Voyager 1*?

   c. How does *Voyager 1*'s distance from Earth compare with that of the nearest star (other than the Sun)?

44. Gravitational waves travel at the speed of light. Their speed, wavelength, and frequency are related as $c = \lambda \times f$. If we were to observe a gravitational wave from a distant cosmic event with a frequency of 10 hertz (Hz), what would be the wavelength of the gravitational wave?

45. Compute the peak of the blackbody spectrum with a temperature of 2.73 K. What region of the spectrum is this?

EXPLORATION Geometric Optics and Lenses

digital.wwnorton.com/astro7

Visit the Digital Resources Page and on the Student Site open the "Geometric Optics and Lenses" AstroTour in Chapter 6. Read through the animation until you reach the optics simulation, pictured in **Figure 6.33**. The simulator shows a converging lens and a pencil. Rays come from the pencil on the left of the converging lens, pass through the lens, and make an image to the right of the lens. The view that would be seen by an observer at the position of the eye is shown in the circle at upper right. Initially, when the pencil is at position 2.3 and the eye is at position 2.0, the pencil is out of focus and blurry.

1 Is the eraser at the top or the bottom of the actual pencil? (That answer becomes important later.)

Using the red slider in the upper left of the window, try moving the pencil to the right. Pause when the observer's eye sees a recognizable pencil (even if it's still blurry).

2 Does the eye see the pencil right side up or upside down?

3 This is somewhat analogous to the view through a telescope. The objects are very far from the lenses, and the observer sees things upside down in the telescope. If an object in your field of view is at the top of the field and you want it in the center, should you move the telescope up or down?

Now return the pencil to position 2.3. Use the red slider in the lower right of the window to move the eye closer to the lens (to the left).

4 At what distance does the image of the pencil first become crisp and clear?

5 Is the pencil right side up or upside down?

6 In practice at the telescope, we do not move the observer back (away from the eyepiece) to bring the image into focus. Why not?

7 Instead of moving the observer, we use a focusing knob to move the lens in the eyepiece, which brings the image into focus. Imagine that you are looking through the eyepiece of a telescope and the image is blurry. You turn the focusing knob and things get blurrier! What should you try next?

8 Now imagine that you get the image focused just right, so it is crisp and sharp. The next person to use the telescope wears glasses and insists that the image is blurry. When you look through the telescope again, though, the image is still crisp. Explain why your experiences differ.

Step through the animation to the next picture. Carefully study the two telescopes shown and the path the light takes through them.

9 Which telescope has a longer focal length: the top one or the bottom one?

10 Which telescope produces an image with the red and the blue stars more separated: the top one or the bottom one?

11 A longer focal length is an advantage in one sense, but it's not the entire story. What are some disadvantages of a telescope with a very long focal length?

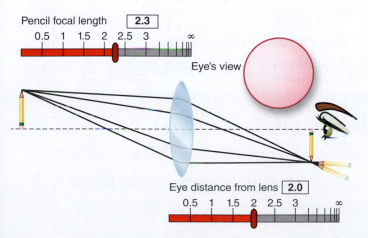

Pencil focal length 2.3

0.5 1 1.5 2 2.5 3 ∞

Eye's view

Eye distance from lens 2.0

0.5 1 1.5 2 2.5 3 ∞

Figure 6.33 Use this simulation to change the position of the object of the eye to explore what will be seen for various configurations.

The Formation of Planetary Systems

Hydrostatic equilibrium is a balance between forces. The stability of clouds, planets, and stars depends on it. So does the stability of an inflated balloon. Blow up a balloon and tie it off. Wrap a string around the widest part of the balloon, and use a marker to trace the string on the balloon. Mark on the string where it overlaps; this indicates the balloon's circumference at room temperature. Use a ruler to measure the circumference. Predict how the circumference of the balloon will change if it is warmed above room temperature and if it is cooled below room temperature. Next, place the balloon in a sink, pot, or bucket and cover it completely with hot water (but not boiling!). After 15 minutes, measure the circumference again. What was the circumference this time? How did it change? Now, place the balloon someplace cold, such as a freezer. After 15 minutes, wrap the string around the balloon again, following the line you made previously. How did the circumference change this time? Use your data to sketch a graph with the circumference on the *y*-axis and the relative temperature on the *x*-axis.

EXPERIMENT SETUP

Blow up a balloon and tie it off. Wrap a string around the widest part of the balloon.

Use a marker to trace the string on the balloon.

Mark on the string where it overlaps.

Submerge the balloon in hot water (but not boiling!) for 15 minutes. Measure the circumference of the balloon.

 00:15

Place the balloon someplace cold for 15 minutes. Measure the circumference of the balloon.

00:15

PREDICTION

The circumference will be
☐ larger ☐ smaller ☐ the same
when the balloon is cooled. It will be
☐ larger ☐ smaller ☐ the same
when the balloon is heated.

SKETCH OF RESULTS (in progress)

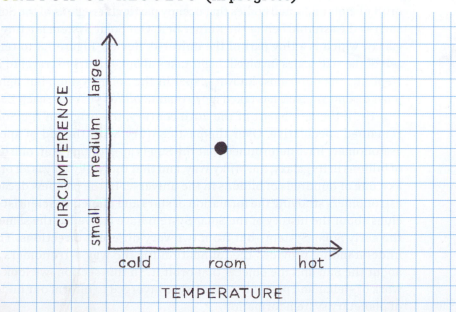

CIRCUMFERENCE — small · medium · large

TEMPERATURE — cold · room · hot

The planetary system containing Earth—our Solar System—is a by-product of the birth of the Sun, but the physical processes that shaped the formation of the Solar System are not unique to it. The same processes have formed many other multiplanet systems. Here in Chapter 7, we examine how planetary systems are born and evolve.

LEARNING GOALS

By the end of this chapter, you should be able to:

(1) Describe how our understanding of planetary system formation developed from the work of both planetary and stellar scientists.

(2) Discuss the role of gravity and angular momentum in explaining why planets orbit the Sun in a plane and why they revolve in the same direction that the Sun rotates.

(3) Explain how temperature at different locations in the protoplanetary disk affects the composition of planets, moons, and other bodies.

(4) Discuss the processes that resulted in the formation of planets and other objects in our Solar System.

(5) Describe how astronomers both find planets around other stars and determine exoplanet properties.

▶❚❚ **AstroTour:** Solar System Formation

7.1 Planetary Systems Form around a Star

Earth is part of a collection of **planets**—large, round isolated bodies that orbit a star. Astronomers call a system of planets surrounding a star a **planetary system**. The Solar System, shown in **Figure 7.1**, is the planetary system that includes Earth, seven other planets, and the Sun. The system also includes moons that orbit planets and small bodies that occupy particular regions of the Solar System, such as the asteroid belt or the Kuiper Belt. Our Solar System is a tiny part of our galaxy, which is a tiny part of the universe. Review Figure 1.3 to remind yourself of the size scales involved. Light takes about 4 hours to travel to Earth from Neptune, the outermost known planet in the Solar System, but light from the most distant galaxies has taken more than 13 *billion* years to reach Earth.

Until the latter part of the 20th century, the origin of the Solar System remained speculative. Over the past century, with the aid of spectroscopy, astronomers have determined that the Sun is an ordinary star, one of hundreds of billions in its galaxy, the Milky Way, and that the Milky Way is an ordinary galaxy, one of hundreds of billions in the universe. In the past few decades, stellar astronomers studying the formation of stars and planetary scientists analyzing clues about the history of the Solar System have arrived at the same picture of the early Solar System—but from two very different directions. That unified understanding provides the foundation for the way astronomers now think about the Sun and the objects that orbit it. In this section, we look at how the work of stellar and planetary scientists converged to inform our understanding of planetary system formation.

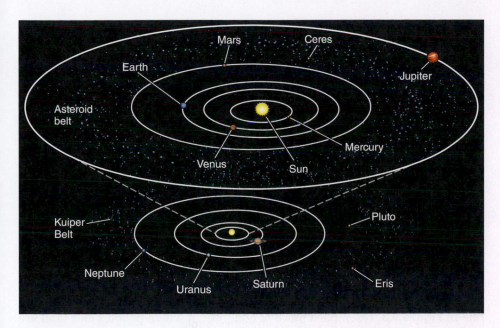

Figure 7.1 Our Solar System includes planets, moons, and other small bodies. Sizes and distances shown are not to scale.

The Nebular Hypothesis

The first plausible theory for the formation of the Solar System, the **nebular hypothesis**, was proposed in 1755 by the German philosopher Immanuel Kant (1724–1804) and conceived independently in 1796 by the French astronomer Pierre-Simon Laplace (1749–1827). Kant and Laplace argued that a rotating cloud of interstellar gas, or **nebula** (Latin for "cloud"), gradually collapsed and flattened to form a disk with the Sun at its center. Surrounding the Sun were rings of material from which the planets formed. That configuration would explain why the planets orbit the Sun in the same direction in the same plane. The nebular hypothesis remained popular throughout the 19th century, and those basic principles of the hypothesis are still retained today.

Our modern theory of planetary system formation calculates the conditions required for a cloud of interstellar gas to collapse under the force of its own self-gravity to form stars. Recall from Chapter 4 that self-gravity is the gravitational attraction between the parts of an object, such as a planet or star, that pulls all the parts toward the object's center. That inward force is opposed by either structural strength (for rocks that make up terrestrial planets) or the outward force resulting from gas pressure and radiation pressure within a star. If the outward force is less than self-gravity, the object contracts; if it is greater, the object expands. In a stable object, the inward and outward forces are balanced.

In support of the nebular hypothesis, disks of gas and dust have been observed surrounding young stellar objects (**Figure 7.2**). From that observational evidence, stellar astronomers have shown that, much as a spinning ball of pizza dough spreads out to form a flat crust, the cloud that produces a star—the Sun, for

Figure 7.2 **a.** Hubble Space Telescope image of a disk around newly formed stars. The dark band is the silhouette of the disk seen edge on. Bright regions are dust illuminated by the star's light. The jets are shown in green. **b.** The Atacama Large Millimeter /submillimeter Array (ALMA) obtained this image of a protoplanetary disk around the star HL Tau. This image shows substructures and possibly planets in the system's dark patches.

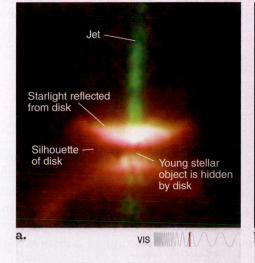

Jet

Starlight reflected from disk

Silhouette of disk

Young stellar object is hidden by disk

a.   VIS

b.   RADIO

what if . . .

What if astronomers find another planetary system where the planets orbit the central star but align in two separate planes which are oblique to each other? What would such an observation tell you about the nebular hypothesis?

Figure 7.3 Meteorites are the surviving pieces of Solar System fragments that land on planets. This meteorite formed from many smaller components that stuck together.

unanswered questions

How typical is the Solar System? Only within the past decade have astronomers found other systems containing four or more planets, and so far, the observed distributions of large and small planets in those multiplanet systems have looked different from those of the Solar System. Computer simulations of planetary system formation suggest that a system with stable orbits and a planetary distribution like those of the Solar System may develop only rarely. Improved supercomputers can run more complex simulations, which can be compared with the observations to better understand how solar systems are configured.

example—collapses first into a rotating disk. Material in the disk eventually suffers one of three fates: it travels inward onto the forming star at its center, it remains in the disk itself to form planets and other objects, or it is ejected back into interstellar space.

Planetary Scientists and the Convergence of Evidence

While astronomers were working to understand star formation, other groups of scientists with very different backgrounds were piecing together the history of the Solar System. Planetary scientists, geochemists, and geologists looking at the current structure of the Solar System inferred what some of its early characteristics must have been. The orbits of all the planets lie very close to a single plane, so the early Solar System must have been flat. In addition, all the planets orbit the Sun in the same direction, so the material from which the planets formed must have been orbiting the Sun in the same direction as well.

To find out more, scientists study samples of the very early Solar System. Rocks that fall to Earth from space, known as **meteorites**, include pieces of material left over from the Solar System's youth. Many meteorites, such as the one in **Figure 7.3**, resemble a piece of concrete in which pebbles and sand are mixed with a much finer filler, suggesting that the larger bodies in the Solar System must have grown from the aggregation of smaller bodies. That finding suggests an early Solar System in which the young Sun was surrounded by a flattened disk of both gaseous and solid material. Our Solar System formed from that swirling disk of gas and dust.

As astronomers and planetary scientists compared notes, they realized they had arrived at the same picture of the early Solar System from two completely different directions. The rotating disk from which the planets formed was the remains of the disk that had accompanied the formation of the Sun. Earth, along with all the other orbiting bodies that make up the Solar System, formed from the remnants of an *interstellar cloud* that collapsed to form the local star, the Sun. The connection between the formation of stars and the origin and later evolution of the Solar System is one of the cornerstones of both astronomy and planetary science—a central theme of our understanding of our Solar System (see the **Process of Science Figure**).

CHECK YOUR UNDERSTANDING 7.1

Which of the following pieces of evidence support the nebular hypothesis? (Choose all that apply.) (a) Planets orbit the Sun in the same direction. (b) The Solar System is relatively flat. (c) Earth has a large Moon. (d) We observe disks of gas and dust around other stars.

Answers to Check Your Understanding questions are in the back of the book.

7.2 The Solar System Began with a Disk

Planets form in a disk around young stars, but what are some of the specifics of the process? **Figure 7.4** illustrates the young Solar System as it appeared roughly 5 billion years ago. At that time, the Sun was still a **protostar**—a large ball of gas but not yet hot enough in its center to be a star. As the cloud of interstellar gas collapsed to form the protostar, its gravitational energy was converted into heat energy and radiation. Surrounding the protostellar Sun was a flat, orbiting disk

Converging Lines of Inquiry

Why is the Solar System a disk, with all planets orbiting in the same direction? Scientists from different disciplines often contribute to the solution to a problem. These scientists approach the problem from different directions and perspectives, so when they arrive at the same conclusion, that is compelling evidence that they are all on the right track.

Mathematicians:

Suggested the nebular hypothesis—that a collapsing rotating cloud formed the Solar System.

Stellar Astronomers:

Tested the nebular hypothesis, seeking evidence for or against.

Found dust and gas around young stars.

Observed this gas and dust to be in the shape of disks.

Planetary Scientists:

Tested the nebular hypothesis, seeking evidence for or against.

Studied meteorites, which showed that the planets formed from many smaller bodies.

Beginning from the same fundamental observations about the shape of the Solar System, theorists (the mathematicians), stellar astronomers, and planetary scientists **converged on the nebular theory** that stars and planets form together from a collapsing cloud of gas and dust.

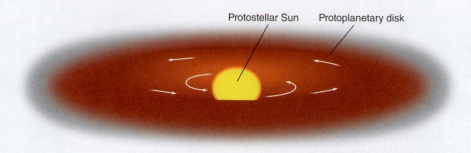

Protostellar Sun   Protoplanetary disk

Figure 7.4 Think of the young Sun as being surrounded by a flat, rotating disk of gas and dust that was flared at its outer edge.

Astronomy in Action: Angular Momentum

Figure 7.5 A figure-skater relies on the principle of conservation of angular momentum to change the speed at which she spins.

of gas and dust. Each bit of the material in that thin disk orbited the Sun in accordance with the same laws of motion and gravitation that govern the orbits of the planets. The disk around the Sun—like the disks that astronomers see today surrounding protostars elsewhere in our galaxy—is called a **protoplanetary disk**. The disk is so "fluffy" and low-density compared to the protostar that it probably contained less than 1 percent of the mass of the star forming at its center, but that amount was more than enough to account for the bodies that make up the Solar System today.

The Collapsing Cloud and Angular Momentum

Conservation of *angular momentum* causes protoplanetary disks to form. **Angular momentum** is a conserved quantity associated with a revolving or rotating system and depends on both the velocity and distribution of the system's mass. The angular momentum of an isolated system is always conserved; that is, it remains unchanged unless acted on by an external force. A figure-skater spinning on the ice (**Figure 7.5**), like any other rotating object, has some amount of angular momentum. Unless frictional forces act to reduce her angular momentum, she will always have the same amount of angular momentum.

The amount of angular momentum depends on three factors:

1. How fast the object is rotating. The faster an object is rotating, the more angular momentum it has.

2. The mass of the object. If a similarly sized bowling ball and a basketball are spinning at the same speed, the bowling ball has more angular momentum because it has more mass.

3. How the mass of the object is distributed relative to the spin axis—that is, how spread out the object is. For an object of a given mass and rate of rotation, the more spread out it is, the more angular momentum it has. A spread-out object rotating slowly might have the same angular momentum as a compact object rotating rapidly.

Both an ice-skater and a collapsing interstellar cloud are subject to the **conservation of angular momentum**: the angular momentum must remain the same in the absence of an external force. For angular momentum to be conserved, a change in one of the three quantities (the rate of spin, mass, or distribution of mass) must be accompanied by a compensating change in another quantity. Because an ice-skater's mass doesn't change, for example, she can control how rapidly she spins by pulling in or extending her arms or legs. As she pulls in her arms to become more compact, she changes her distribution of mass and must spin faster to maintain the same angular momentum. When her arms are held tightly in front of her and one leg is wrapped around the other, the skater's spin becomes a blur. She finishes with a flourish by throwing her arms and leg out—an action that abruptly slows her spin by spreading out her mass. The skater's angular momentum remains constant throughout the maneuver. Similarly (see **Figure 7.6**), the cloud that formed our Sun rotated faster and faster as it collapsed, just as the ice-skater speeds up when she pulls in her arms.

That description, however, presents a puzzle. Suppose the Sun formed from a typical cloud—one about a light-year across and rotating so slowly that completing one rotation took a million years. By the time the cloud collapsed to the size of

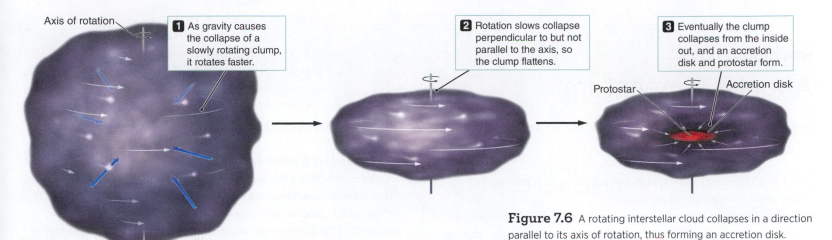

Axis of rotation

1 As gravity causes the collapse of a slowly rotating clump, it rotates faster.

2 Rotation slows collapse perpendicular to but not parallel to the axis, so the clump flattens.

3 Eventually the clump collapses from the inside out, and an accretion disk and protostar form.

Protostar   Accretion disk

Figure 7.6 A rotating interstellar cloud collapses in a direction parallel to its axis of rotation, thus forming an accretion disk.

the Sun today, it would have been spinning so fast that one rotation would occur every 0.6 second. That rate is more than 3 million times faster than our Sun actually spins. At that rate of rotation, the Sun would tear itself apart. It *appears* that angular momentum was *not* conserved in the actual formation of the Sun— but that can't be right because angular momentum *must* be conserved. We must be missing something. Where did the angular momentum go?

The Formation of an Accretion Disk

To understand how angular momentum is conserved in disk formation, we must think in three dimensions. Imagine that the ice-skater bends her knees, compressing herself downward instead of bringing her arms toward her body. As she does so, she again makes herself less spread out, but her rate of spin does not change because no part of her body has become any closer to the axis of spin. Similarly, as shown in Figure 7.6, a clump of a molecular cloud can flatten out without speeding up by collapsing parallel to its axis of rotation. Instead of collapsing into a ball, the interstellar cloud flattens into a disk. As the cloud collapses, its self-gravity increases and the inner parts begin to fall freely inward, raining down on the growing disk at the center. The outer portions of the cloud lose the support of the collapsed inner portion, and they start falling inward, too. As that material makes its final inward plunge, it lands on a thin, rotating disk—called an **accretion disk**—that forms from the accretion of material around a massive object.

The formation of accretion disks, shown in **Figure 7.7**, is common in the universe. As material falls onto the disk at an angle, it impacts material coming up to the disk from below. Over time, as more collisions occur, these perpendicular motions gradually cancel out. But the part of the motion parallel to the disk remains unchanged. This is much like two football players approaching from opposite sides and colliding as they jump for the ball. Their motion across the field cancels out, but their motion down the field remains, so they fall to the ground closer to the goal. In the case of accretion disks, the material flattens out and continues to rotate, which conserves angular momentum.

Thus, the angular momentum of the infalling material is transferred to the accretion disk. The rotating accretion disk has a radius of hundreds of astronomical units, and that is *thousands* of times greater than the radius of the star that will

▶❚❚ **AstroTour:** Traffic Circle Analogy

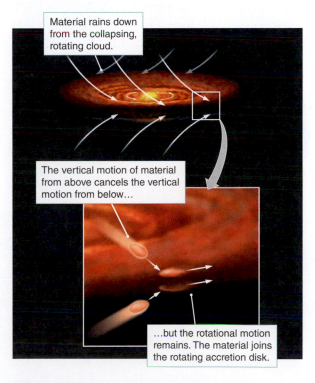

Material rains down from the collapsing, rotating cloud.

The vertical motion of material from above cancels the vertical motion from below…

…but the rotational motion remains. The material joins the rotating accretion disk.

Figure 7.7 Gas from a rotating cloud falls inward from opposite sides, piling up onto a rotating disk.

working it out 7.1

Angular Momentum

In its simplest form, the angular momentum (L) of a system is given by

$$L = m \times v \times r$$

where m is the mass, v is the speed at which the mass is moving, and r represents how spread out the mass is.

Let's apply this relationship to the angular momentum of Jupiter in its orbit about the Sun. The angular momentum from one body orbiting another is called *orbital* angular momentum, $L_{orbital}$. The mass (m) of Jupiter is 1.90×10^{27} kilograms (kg), the speed of Jupiter in orbit (v) is 1.31×10^4 meters per second (m/s), and the radius of Jupiter's orbit (r) is 7.79×10^{11} meters. Putting all that together gives

$$L_{orbital} = (1.90 \times 10^{27}\,\text{kg}) \times (1.31 \times 10^4\,\text{m/s}) \times (7.79 \times 10^{11}\,\text{m})$$

$$L_{orbital} = 1.94 \times 10^{43}\,\text{kg m}^2/\text{s}$$

Calculating the *spin* angular momentum of a spinning object, such as a skater, a planet, a star, or an interstellar cloud, is more complicated. Here, we must add up the individual angular momenta of *every tiny mass element* within the object. For a uniform sphere, the spin angular momentum is

$$L_{spin} = \frac{4\pi m R^2}{5P}$$

where R is the radius of the sphere and P is the rotation period of its spin.

Let's compare Jupiter's orbital angular momentum with the Sun's spin angular momentum to investigate the distribution of angular momentum in the Solar System. The Sun's radius is 6.96×10^8 meters, its mass is 1.99×10^{30} kg, and its rotation period is 24.5 days = 2.12×10^6 seconds. If we assume that the Sun is a uniform sphere, the spin angular momentum of the Sun is

$$L_{spin} = \frac{4 \times \pi \times (1.99 \times 10^{30}\,\text{kg}) \times (6.96 \times 10^8\,\text{m})^2}{5 \times (2.12 \times 10^6\,\text{s})}$$

$$L_{spin} = 1.14 \times 10^{42}\,\text{kg m}^2/\text{s}$$

$L_{orbital}$ of Jupiter is about 17 times greater than L_{spin} of the Sun. Thus, most of the angular momentum of the Solar System now resides in the orbits of its major planets.

For a collapsing sphere to conserve L_{spin}, its rotation period P must be proportional to R^2. As with the skater, when a sphere decreases in radius, its rotation period decreases; that is, it spins faster.

what if . . .

What if you observe a close pair of stars forming a binary system? How would you expect the disk around such a pair of stars to differ from the disk around a single star?

eventually form at its center. Therefore, most of the angular momentum in the original interstellar cloud ends up in the accretion disk rather than in the central protostar (see **Working It Out 7.1** for an example of the relevant calculation).

Most of the matter that lands on the accretion disk either becomes part of the star or is ejected back into interstellar space in the form of jets or other outflows (**Figure 7.8**). Those jets are bipolar—they come in pairs aligned along an axis. Material swirling in the bipolar jets carries angular momentum away from the accretion disk in the general direction of the poles of the rotation axis. However, a small amount of material is left behind in the disk. The objects in that leftover disk—the dregs of the process of star formation—form planets and other objects that orbit the star.

Formation of Large Objects

Random motions of the gas within the protoplanetary disk eventually push the smaller grains of solid material toward larger grains. As that happens, the smaller grains stick to the larger grains. The "sticking" process among smaller grains is due to the same static electricity that causes dust bunnies to grow under your bed. Starting out at only a few microns (μm) across—about the size of particles

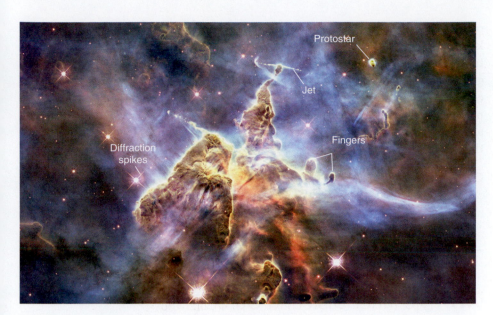

Protostar

Jet

Diffraction
spikes

Fingers

Figure 7.8 ★ WHAT AN ASTRONOMER SEES This gorgeous image from the Hubble Space Telescope shows a portion of the Carina Nebula. An astronomer looking at this image will immediately notice the colors, because color often indicates where different atoms or molecules are present. She will not necessarily know which atoms or molecules are represented by these colors because different astronomers will use a different palette. But even without reading any background on the image, she will know that there is something different about the hazy blue areas and the hazy pink or green ones. An astronomer will recognize and then mostly ignore the diffraction spikes (mentioned in Chapter 6) that form an X around each bright star. She will notice the brown clumps of material that are too dense to see through. These high-density regions indicate that star formation might be happening in this nebula, and this will be confirmed by the small oval protostar in the upper right corner and by the dense "fingers" that stick out in various places. These fingers point the way toward a source of interstellar wind outside the image to the upper right. That wind has eroded away the less dense material around these denser regions and may have triggered star formation by compressing the material at the top of each finger. Each finger is a dense blob of material that creates a wind "shadow" behind it. A new star has just formed at the top of the finger near the middle of the image. An astronomer will identify this new star because of the thin jets of material that are being ejected in opposite directions; new stars sometimes create such jets.

in smoke—the slightly larger bits of dust grow to the size of pebbles and then to clumps the size of boulders, which are not as easily pushed around by gas (**Figure 7.9**). When clumps grow to about 100 meters across, the objects are so far apart that they collide less often, and their growth rate slows down but does not stop. Within a protoplanetary disk, the larger dust grains become larger at the expense of the smaller grains.

For two large clumps to stick together rather than explode into many small pieces, they must bump into each other very gently: collision speeds must be about 0.1 m/s or less. If your stride is about a meter long, a collision speed of 0.1 m/s would correspond to only one step every 10 seconds. The process does not yield larger and larger bodies with every collision. When violent collisions occur in an accretion disk, larger clumps break back into smaller pieces. But over a long time, large bodies do form.

Objects continue to grow by "sweeping up" smaller objects that get in their way. Those objects can eventually measure up to several hundred meters across. As the clumps reach the size of about a kilometer, a different process becomes important. Those kilometer-sized objects are massive enough that their gravity pulls

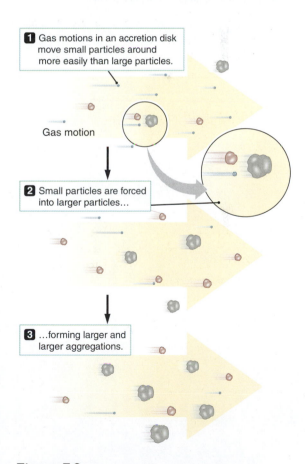

1 Gas motions in an accretion disk move small particles around more easily than large particles.

Gas motion

2 Small particles are forced into larger particles…

3 …forming larger and larger aggregations.

Figure 7.9 Motions of gas in a protoplanetary disk blow smaller particles of dust into larger particles, making the larger particles larger still. That process continues, eventually creating objects many meters in size.

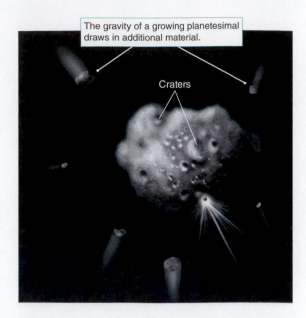

The gravity of a growing planetesimal draws in additional material.

Craters

Figure 7.10 The gravity of a planetesimal is strong enough to attract surrounding material, which causes the planetesimal to grow more rapidly. Some of this additional material impacts with enough energy to make craters on the planetesimal.

on nearby bodies, as shown in **Figure 7.10**. Those bodies of rock and ice are known as **planetesimals** ("tiny planets") and eventually combine with one another to form planets. Besides chance collisions with other objects, a planetesimal's gravity can now pull in and capture small objects outside its direct path. That process speeds up the growth of planetesimals, so larger planetesimals quickly consume most of the remaining bodies near their orbits. The final survivors of that process are large enough to be called planets. As with the major bodies in orbit around the Sun, some of the planets may be relatively small and others quite large.

CHECK YOUR UNDERSTANDING 7.2

Where does most of the angular momentum of the original cloud go? It (a) goes into the orbital angular momentum of planets; (b) goes into the star; (c) goes into the spin of the planets; (d) is lost along the jets from the star.

7.3 The Inner Disk and Outer Disk Formed at Different Temperatures

In the Solar System, the inner planets are small and mostly rocky, whereas the outer planets are very large and mostly gaseous. That distinct difference between the inner and outer Solar System can be explained by how the local disk environment affects the formation process. In this section, we examine these differences.

Energy in the Disk

The accretion disks surrounding young stars form from interstellar material that may have a temperature of only a few kelvins, but the disks themselves reach temperatures of hundreds of kelvins or more. Astronomers want to understand what heats up the disk around a forming star so that we can calculate how hot those disks get.

According to the law of **conservation of energy**, the total amount of energy in a system must remain constant unless energy is added to or taken away from the system from the outside. The form the energy takes, however, can change. With that concept in mind, why does gas falling on the disk make the disk hot? Imagine you are working against gravity by lifting a heavy object, such as a bowling ball. Lifting the bowling ball takes energy, and the law of conservation of energy states that energy is never lost. Where does that energy go? The energy is stored and changed into a form called **gravitational potential energy**. If you drop the bowling ball, it falls, and as it falls, it speeds up. The gravitational potential energy that was stored is converted to energy of motion, which is called **kinetic energy**. When the bowling ball hits the floor, it stops suddenly. The bowling ball loses its energy of motion, so what form does that energy take now? Some of the energy is converted into the sound the bowling ball makes when it hits the floor, and some goes into heating and distorting the floor. But much of the energy is converted into **thermal energy**. The atoms and molecules that make up the bowling ball are moving around within the bowling ball a bit faster than they were before the bowling ball hit, so the bowling ball and its surroundings, including the floor, grow a tiny bit warmer. Similarly, as gas falls toward the disk surrounding a protostar, gravitational potential energy is converted first to kinetic energy, causing the gas to pick up speed. When the gas hits the disk and stops suddenly, that kinetic energy turns into thermal energy.

Material falling onto the accretion disk around a forming star causes the disk to heat up, too (**Figure 7.11**). The amount of heating depends on *where* the material hits the disk. Material hitting the inner part of the disk (the *inner disk*) has fallen farther and picked up greater speed within the gravitational field of the forming star than material hitting the disk farther out. Like a brick dropped from a tall building, material striking the inner disk is moving quite rapidly when it hits, so it heats the inner disk to high temperatures. In contrast, material falling onto the outer part of the disk (the *outer disk*) is moving much more slowly, like a brick dropped from just a foot or so above the ground. As a result, the temperature at the outermost parts of the disk is not much higher than that of the original interstellar cloud. Stated another way: material falling onto the inner disk has more gravitational potential energy to convert into thermal energy than does material falling onto the outer disk.

The energy released as material falls onto the disk is not the only source of thermal energy in the disk. Even before the nuclear reactions that will one day power the new star have ignited, the conversion of gravitational energy into thermal energy drives the temperature at the surface of the protostar to several thousand kelvins, and it drives the luminosity of the huge ball of glowing gas to many times the luminosity of the present-day Sun. For the same reasons why Mercury is hot whereas Pluto is not (see Chapter 5), the radiation streaming outward from the protostar at the center of the disk drives the temperature in the inner parts of the disk even higher, increasing the difference in temperature between the inner and outer parts of the disk.

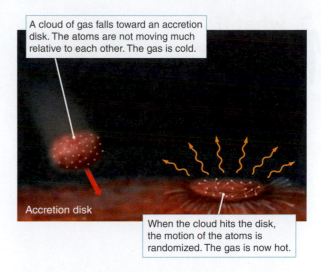

A cloud of gas falls toward an accretion disk. The atoms are not moving much relative to each other. The gas is cold.

Accretion disk

When the cloud hits the disk, the motion of the atoms is randomized. The gas is now hot.

Figure 7.11 Atoms in a gas fall together until they hit the accretion disk, at which point their motions become randomized, raising the temperature of the gas.

The Compositions of Planets

Temperature affects whether a material exists as a solid, a liquid, or a gas. On a hot summer day, ice melts and water quickly evaporates; on a cold winter night, water in your breath freezes into tiny ice crystals. Metals and rocky materials, such as iron, **silicates** (minerals containing silicon and oxygen), and carbon, remain solid even at high temperatures. Substances that can withstand high temperatures without melting or being vaporized are called **refractory materials**. Other materials, such as water, ammonia, and methane, remain in a solid form only if their temperature is very low. Those materials, which become gases at moderate temperatures, are called **volatile materials** (or *volatiles* for short). Astronomers generally call the solid form of any volatile material an **ice**.

Differences in temperature from place to place within the protoplanetary disk significantly affect the makeup of the dust grains in the disk. **Figure 7.12** illustrates that only refractory substances exist in the hottest parts of the disk—the area closest to the protostar. In the inner disk, dust grains are composed

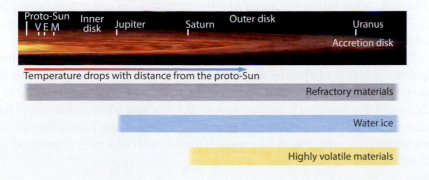

Proto-Sun
V E M
Inner disk
Jupiter
Saturn
Outer disk
Uranus
Accretion disk

Temperature drops with distance from the proto-Sun

Refractory materials

Water ice

Highly volatile materials

Figure 7.12 Differences in temperature within a protoplanetary disk determine the composition of dust grains that then evolve into planetesimals and planets. The colored bars show that refractory materials are found throughout the disk, whereas water ice is found only outside Jupiter's orbit, and highly volatile materials are found only outside Saturn's orbit. Shown here are the proto-Sun and the orbits of Venus (V), Earth (E), Mars (M), Jupiter, Saturn, and Uranus.

what if . . .

What if you observed an Earth-mass planet that is far from its central star, but has almost no atmosphere at all? What could you conclude about this planet's formation history?

almost entirely of refractory materials. Some substances can survive in solid form somewhat farther out, including some hardier volatiles, such as water ice and certain chemical compounds that are **organic** (meaning that they contain molecules with a carbon–hydrogen bond). Those solids add to the materials that make up dust grains. In the coldest, outermost parts of the accretion disk, far from the central protostar, highly volatile components such as methane, ammonia, and carbon monoxide ices and other organic molecules survive only in solid form. The differences in composition of dust grains within the disk are reflected in the composition of the planets formed from that dust. Planets that form closer to the central star tend to be made up mostly of refractory materials, such as rock and metals, but are deficient in volatiles. Planets that form farther from the central star contain not only refractory materials but also large quantities of ices and organic materials.

In the Solar System, the inner planets are composed of rocky material surrounding metallic cores of iron and nickel. Objects in the outer Solar System, including moons, giant planets, and comets, are composed largely of ices of various types. But not all planetary systems are so neatly organized as our Solar System. When planets around other stars were first discovered, they appeared to be very different, with large planets close to their respective stars. Astronomers now think that **chaotic** encounters—in which a small change in the initial state of a system can lead to a large change in the final state of the system—may change the organization of planetary compositions. In a process called **planet migration**, the force of gravity from all the nearby objects can move some planets so that they end up far from the place of their birth. In our Solar System, for example, Uranus and Neptune originally may have formed nearer to the orbits of Jupiter and Saturn but were then driven outward to their current locations by gravitational encounters with Jupiter and Saturn. A planet also can migrate when it gives up some of its orbital angular momentum to the disk material that surrounds it. Such a loss of angular momentum causes the planet to slowly spiral inward toward the central star. Thus, the order of planets in a system can change.

Formation of an Atmosphere

Once a solid planet has formed, it may continue to grow by capturing gas from the protoplanetary disk. To do so, it must act quickly. Young stars and protostars emit fast-moving particles and intense radiation that can quickly disperse the gaseous remains of the accretion disk. Gaseous planets such as Jupiter probably have only about 10 million years to form and to grab whatever gas they can. Because of their strong gravitational fields, more massive young planets can capture more of the hydrogen and helium gases that makes up the bulk of the disk. What follows is much like the formation of a star and protoplanetary disk, but on a smaller scale—namely, gas from a mini accretion disk moves inward and falls onto the planet.

The gas that a planet captures when it forms—primarily hydrogen and helium—is called the planet's **primary atmosphere**. The primary atmosphere of a large planet can be more massive than the solid body, as with Jupiter. Some of the solid material in the mini accretion disk might stay behind to coalesce into larger bodies in much the same way that particles of dust in the protoplanetary disk came together to form planets. The result is a mini "solar system"—a group of moons that orbit about the planet.

A less massive planet also may capture some gas from the protoplanetary disk, only to lose it later. The gravity of small planets may be too weak to hold low-mass gases such as hydrogen or helium. Even if a small planet can gather some hydrogen and helium from its surroundings, that primary atmosphere will not last long. In the inner solar system, the temperatures are higher, so the hydrogen and helium atoms are moving faster than in the outer solar system and will escape from a small planet. The atmosphere that remains around a small planet such as Earth is a **secondary atmosphere**, which forms later in the life of a planet. Volcanism is one important source of a secondary atmosphere because it releases heavier and thus slower-moving gases, such as carbon dioxide, water vapor, and other gases trapped in the planet's interior. In addition, volatile-rich comets that formed in the outer parts of the disk fall inward toward the new star long after its planets have formed, and they sometimes collide with planets. **Comets** are icy planetesimals that survive planetary accretion. They may serve as a significant source of water, organic compounds, and other volatile materials on planets close to the central star.

CHECK YOUR UNDERSTANDING 7.3

In our Solar System, the inner planets are rocky because: (a) the original cloud had more rocky material near the center; (b) warm temperatures in the inner disk caused the inner planetesimals to be formed of only rocky material; (c) the inner disk filled a smaller volume, and so it was denser; (d) the hydrogen and helium atoms were too low mass to remain in the inner disk.

7.4 The Formation of Our Solar System

Nearly 5 billion years ago, the Sun was still a protostar surrounded by a protoplanetary disk of gas and dust. During the next few hundred thousand years, much of the dust in the disk had collected into planetesimals—clumps of rock and metal near the emerging Sun, and aggregates of rock, metal, ice, and organic materials farther from the Sun. In this section, we look at the formation of the types of planets in our own Solar System.

The Terrestrial Planets

Within the inner 5 astronomical units (AU) of the disk, several rock and metal planetesimals quickly grew larger to become the dominant masses in their orbits. With their ever-strengthening gravitational fields, they either captured most of the remaining planetesimals or ejected them from the inner part of the disk. **Figure 7.13** shows some results from a computer simulation of how that might have happened. The dominant planetesimals became planet-sized bodies with masses ranging between 5 percent and 100 percent of Earth's mass. Those dominant planetesimals evolved into the **terrestrial planets**, which are rocky, Earth-like planets. Today, the surviving terrestrial planets are Mercury, Venus, Earth, and Mars. Earth's Moon is often grouped with those terrestrial planets because of its similar physical and geological properties, even though it is not a planet itself and formed differently. It is possible that one or two other planets or large moons formed in the young Solar System but were later destroyed.

For several hundred million years after the four surviving terrestrial planets formed, leftover pieces of debris still in orbit around the Sun continued to rain

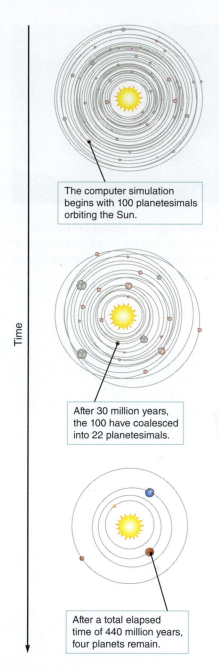

Time

The computer simulation begins with 100 planetesimals orbiting the Sun.

After 30 million years, the 100 have coalesced into 22 planetesimals.

After a total elapsed time of 440 million years, four planets remain.

Figure 7.13 Computer models simulate how material in the protoplanetary disk became clumped into the planets over time. Only a few planets remain at the end.

Figure 7.14 Large impact craters on Mercury (and on other solid bodies throughout the Solar System) record the final days of the Solar System's youth, when planets and planetesimals grew as smaller planetesimals rained down on their surfaces.

down on the surfaces of those planets. Today, we can still see the scars of those early impacts on the cratered surfaces of all the terrestrial planets (**Figure 7.14**). That rain of debris continues even today, but at a much lower rate.

Before the proto-Sun became a true star, gas in the inner part of the protoplanetary disk was still plentiful. During that early period the two larger terrestrial planets, Earth and Venus, may have held on to weak primary atmospheres of hydrogen and helium, but those thin atmospheres were soon lost to space. The terrestrial planets did not develop thick atmospheres until the formation of the secondary atmospheres that now surround Venus, Earth, and Mars. Mercury's size and proximity to the Sun and the Moon's small mass prevented those bodies from retaining significant secondary atmospheres.

The Giant Planets

Beyond 5 AU from the Sun, in a much colder part of the accretion disk, planetesimals combined to form several bodies with masses about 5–20 times that of Earth (5–20 M_{Earth}). Those planet-sized objects formed from planetesimals containing volatile ices and organic compounds in addition to rock and metal. In a process astronomers call **core accretion–gas capture**, mini accretion disks formed around those planetary cores, capturing massive amounts of hydrogen and helium and funneling that material onto the planets. Four such massive bodies became the cores of the **giant planets**—Jupiter, Saturn, Uranus, and Neptune. Those giant planets are many times the mass of any terrestrial planet.

Jupiter's massive solid core captured and retained the most gas—roughly 300 M_{Earth}. The other outer planetary cores captured less hydrogen and helium, perhaps because their cores were less massive or because less gas was available to them. Saturn ended up with less than 100 M_{Earth} of gas, and Uranus and Neptune grabbed less than 20 M_{Earth} of gas.

The core accretion model indicates that a Jupiter-like planet could take up to 10 million years to accumulate. Some planetary scientists think that our protoplanetary disk could not have survived long enough to form gas giants such as Jupiter through the general process of core accretion. All the gas may have dispersed in roughly half that time, cutting off Jupiter's supply of hydrogen and helium. An alternative explanation is a process called *disk instability*, in which the protoplanetary disk suddenly and quickly fragments into massive clumps equivalent to those of a large planet. Both core accretion and disk instability may have played a role in the formation of our own and other planetary systems.

During the formation of the planets, gravitational energy was converted into thermal energy as individual atoms and molecules moved faster. That conversion warmed the gas surrounding the cores of the giant planets. Proto-Jupiter and proto-Saturn probably became so hot that they glowed a deep red color, similar to the heating element on an electric stove. Their internal temperatures may have been even higher.

In the mini accretion disks surrounding the giant planets, some of the remaining material combined into small bodies, which became moons. A **moon** is any natural satellite in orbit about a planet or asteroid. The composition of the moons that formed around the giant planets followed the same trend as that of the planets that formed around the Sun: the innermost moons formed under the hottest conditions and therefore contained the smallest amounts of volatile material. For example, the closest of Jupiter's many moons may have experienced high temperatures from nearby Jupiter's glowing so intensely that it would have evaporated most of the volatile substances in the inner part of its mini accretion disk.

Remaining Planetesimals

Not all planetesimals in the disk became planets. For example, dwarf planets orbit the Sun but have not cleared other, smaller bodies from their orbits. Ceres and Pluto (Figure 7.1) are dwarf planets. More dwarf planets, along with many smaller bodies, are found in the Kuiper Belt, beyond Pluto's orbit. Asteroids are small bodies found inside Jupiter's orbit around the Sun; most are located in the main asteroid belt between the orbits of Mars and Jupiter. Jupiter's gravity kept the region between Jupiter and Mars so stirred up that most planetesimals there never formed a large planet.

Planetesimals persist to this day in the outermost part of the Solar System as well. Formed in a deep freeze, those objects have retained most of the highly volatile materials found in the grains present when the accretion disk formed. Unlike the crowded inner part of the disk, the outermost parts had planetesimals too sparsely distributed for large planets to grow. Icy planetesimals in the outer Solar System that survived planetary accretion remain today as **comet nuclei**. The frozen, distant dwarf planets Pluto and Eris are especially large examples of those residents of the outer Solar System.

Many Solar System objects show evidence of cataclysmic impacts that reshaped worlds, suggesting that the early Solar System must have been a remarkably violent and chaotic place. The dramatic difference in the terrain of the northern and southern hemispheres on Mars, for example, has been interpreted as the result of one or more colossal collisions. The leading theory for the origin of our Moon is that it resulted from the collision of an object with Earth. Mercury has a crater on its surface from an impact so devastating that it caused the crust to buckle on the opposite side of the planet. In the outer Solar System, one of Saturn's moons, Mimas, has a crater roughly one-third the diameter of the moon itself. Uranus suffered one or more collisions violent enough to knock it on its side. As a result, its equatorial plane is tilted at almost a right angle to its orbital plane. Other examples are discussed in later chapters.

CHECK YOUR UNDERSTANDING 7.4

Suppose that astronomers found a rocky, terrestrial planet beyond the orbit of Neptune. What is the most likely explanation for its origin? (a) It formed close to the Sun and migrated outward. (b) It formed in that location and was not disturbed by migration. (c) It formed later in the Sun's history than other planets. (d) It is a captured planet that formed around another star.

7.5 Planetary Systems Are Common

When astronomers turn their telescopes to young nearby stars, they see disks of the same type from which the Solar System formed. When the light from the central star is blocked (**Figure 7.15**), evidence of the planetary disk is observed. The physical processes that led to the formation of the Solar System should be commonplace wherever new stars are being born. Compared with stars, however, planets are small and dim objects. They shine primarily by reflection and therefore are millions to billions of times fainter than their host stars. Thus, they were difficult to find until advances in telescope detector technology in the 1990s enabled astronomers to discover them through indirect methods. In 1995, astronomers announced the first confirmed **exoplanet**—a planet orbiting around a star other than the Sun.

what if . . .

What if astronomers had observed that all other systems were similar to our own, with rocky planets closer to the star and gas giants farther away? What would that tell us about star formation in general?

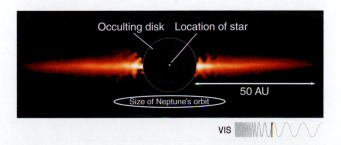

Figure 7.15 An edge-on dust disk around a star is seen extending outward to 60 AU from the young (12-million-year-old) star AU Microscopii. The star itself, whose brilliance would otherwise overpower the dust disk, is hidden behind an opaque mask (called an "occulting disk") placed in the telescope's focal plane. The star's position is represented by the dot.

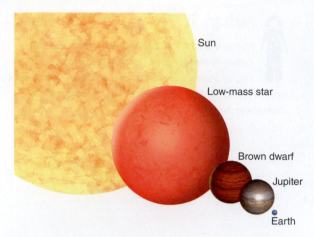

Figure 7.16 A comparison of the diameters of the Sun, a low-mass star, a brown dwarf, Jupiter, and Earth.

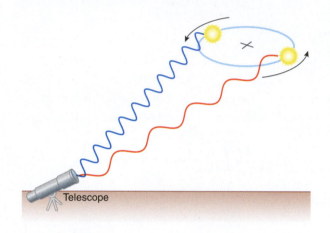

Figure 7.17 Doppler shifts observed in the spectrum of a star as it moves around its common center of gravity with its planet. The spectral lines are blueshifted as the star moves toward us and redshifted as it moves away from us.

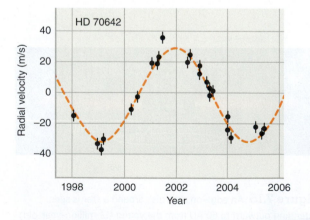

Figure 7.18 Radial velocity data for a star with a planet. A positive velocity is motion away from the observer, whereas a negative velocity is motion toward the observer. The plot repeats every 5.7 years as the planet completes another orbit around the star.

(These are sometimes called "extrasolar planets," although "exoplanet" is more accepted today.) Today, the number of known exoplanets has grown to the thousands, and new discoveries occur almost daily.

The International Astronomical Union (IAU) currently defines an exoplanet as an object that orbits a star other than the Sun and has a mass less than 10–13 Jupiter masses (10–13 M_{Jup}). Objects more massive than 10–13 M_{Jup} but less massive than 0.08 solar masses (0.08 M_{Sun}; about 80 M_{Jup}) are **brown dwarfs**. Objects more massive than 0.08 M_{Sun} are defined as stars. **Figure 7.16** compares the diameters of typical objects of each class.

The Search for Exoplanets

The first planets were discovered indirectly, by observing their gravitational tug on the central star. As technology has improved, other methods have become more productive. Astronomers now have direct images of planets orbiting stars and have taken spectra of some exoplanets to observe the composition of their atmospheres. Almost certainly, between the time we write this and the time you read it, new discoveries will have been made. The field is advancing extremely quickly: more than 100 projects are searching for exoplanets from the ground and from space. We will now look at each discovery method.

The Radial Velocity Method   As a planet orbits a star, the planet's gravity tugs the star around ever so slightly. If the star has a radial velocity (Chapter 5), an observable Doppler shift may appear in the spectrum of the star. **Figure 7.17** illustrates that motion. When the star is moving toward us (negative radial velocity), the light is blueshifted; when the star is moving away from us (positive radial velocity), the light is redshifted. That pattern of radial velocity repeats. After detecting those changes in the radial velocity (**Figure 7.18**), astronomers can infer the existence of the planet and find the planet's mass and distance from the star.

The smallest planet that can be found depends on the precision of the observations. For example, in our Solar System, Jupiter's mass is greater than the mass of all the other planets, asteroids, and comets combined, so Jupiter is the planet with the largest effect on the Sun. Both the Sun and Jupiter orbit a common center of gravity (sometimes called center of mass—the location where the effect of one mass balances the other) that lies just outside the surface of the Sun, as shown in **Figure 7.19**. Jupiter tugs the Sun around in a circle with a speed of 12 m/s. Alien astronomers would find that the Sun's radial velocity varies by ±12 m/s, with a period equal to Jupiter's orbital period of 11.86 years. From that information, the astronomers would rightly conclude that the Sun has at least one planet with a mass comparable to Jupiter's. Without greater precision, the observers would be unaware of the other, less massive planets. To detect Saturn, they would need to improve the precision of their measurements to 2.7 m/s. Earth would not be detectable unless the aliens could detect motions as small as 0.09 m/s.

The precision of radial velocity instruments has been about 0.3 m/s. The technique enabled astronomers to detect giant planets around solar-type stars, but not yet to find planets with masses similar to Earth's. Finding the signal of the Doppler shift in the noise of the observation requires the star to be quite bright in our sky. A new instrument at the Very Large Telescope is expected to improve precision and be able to detect smaller planets. **Working It Out 7.2** explains more about the spectroscopic radial velocity method.

working it out 7.2

Estimating the Size of a Planet's Orbit

In the spectroscopic radial velocity method, the star is moving about its center of mass, and its spectral lines are Doppler-shifted accordingly. Recall from Figure 7.19 that an alien astronomer looking toward the Solar System would observe a shift in the wavelengths of the Sun's spectral lines—caused by the presence of Jupiter—of ~12 m/s.

Figure 7.18 showed the radial velocity data for a star with a planet discovered with that method. How do astronomers use that method to estimate the distance (A) of the planet from the star? Recall from Chapter 4 that Newton generalized Kepler's law relating the period of an object's orbit to the orbital semimajor axis:

$$P^2 = \frac{4\pi^2}{G} \times \frac{A^3}{M}$$

where A is the semimajor axis of the orbit, P is its period, and M is the combined mass of the two objects. To find A, we rearrange the equation as follows:

$$A^3 = \frac{G}{4\pi^2} \times M \times P^2$$

According to the graph of radial velocity observations in Figure 7.18, the period of the orbit is 5.7 years. A year has 3.16×10^7 seconds, so

$$P = 5.7 \text{ yr} \times (3.16 \times 10^7 \text{ s/yr}) = 1.8 \times 10^8 \text{ s}$$

The mass of the star is much greater than the mass of the planet, so the combined masses of the star and the planet can be approximated as the mass of the star, which here is about equal to the mass of the Sun, 2×10^{30} kg. (Stellar masses can be estimated from their spectra.) The gravitational constant G is 6.67×10^{-20} km³/(kg s²). Plugging in the numbers gives

$$A^3 = \frac{6.67 \times 10^{-20} \dfrac{\text{km}^3}{\text{kg s}^2}}{4\pi^2} \times (2 \times 10^{30} \text{ kg}) \times (1.8 \times 10^8 \text{ s})^2$$

$$A^3 = 1.1 \times 10^{26} \text{ km}^3$$

Taking the cube root,

$$A = 4.8 \times 10^8 \text{ km}$$

If we convert that number into astronomical units (where 1 AU = 1.5 $\times 10^8$ km), the semimajor axis of the orbit of this planet is

$$A = \frac{4.8 \times 10^8 \text{ km}}{1.5 \times 10^8 \text{ km/AU}} = 3.2 \text{ AU}$$

The planet is 3.2 times farther from its star than Earth is from the Sun.

The Transit Method   From Earth it is sometimes possible to see the inner planets Mercury and Venus transit, or pass in front of, the Sun. An alien located somewhere in the plane of Earth's orbit would see Earth pass in front of the Sun and could infer the existence of Earth by detecting the 0.009 percent drop in the Sun's brightness during the transit. Similarly, for astronomers on Earth to observe a planet passing in front of a star, Earth must lie nearly in the orbital

Astronomy in Action: Doppler Shift

Interactive Simulation: Radial Velocity

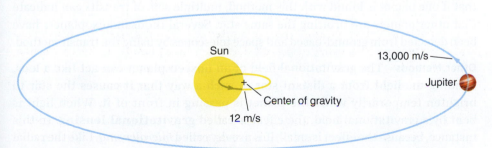

Figure 7.19 The Sun and Jupiter orbit around a common center of gravity (+), which lies just outside the Sun's surface. Spectroscopic measurements made by an extrasolar astronomer would reveal the Sun's radial velocity varying by ±12 m/s over 11.86 years, which is Jupiter's orbital period. Jupiter travels around its orbit at a speed of 13,000 m/s.

working it out 7.3

Estimating the Radius of an Exoplanet

The masses of exoplanets can often be estimated using Kepler's laws and the conservation of angular momentum. When planets are detected with the transit method, astronomers can estimate the radius of an exoplanet. In that method, astronomers look for planets that eclipse their stars and then observe how much the star's light decreases during that eclipse (see Figure 7.20). When Venus or Mercury transits the Sun, a black circular disk is visible on the face of the circular Sun. During the transit, the amount of light from the transited star is reduced by the area of the circular disk of the planet divided by the area of the circular disk of the star:

$$\text{fractional reduction in light} = \frac{\text{Area of disk of planet}}{\text{Area of disk of star}} = \frac{\pi R_{\text{planet}}^2}{\pi R_{\text{star}}^2} = \frac{R_{\text{planet}}^2}{R_{\text{star}}^2}$$

Then, to solve for the radius of the planet, astronomers need an estimate of the radius of the star and a measurement of the fractional reduction in

light during the transit. The radius of a star is estimated from the surface temperature and the luminosity of the star.

Kepler-11, for example, is a system of at least six planets that transit a star. The radius of the star, R_{star}, is estimated to be 1.1 times the radius of the Sun, or $1.1 \times (7.0 \times 10^5 \text{ km}) = 7.7 \times 10^5 \text{ km}$. The light from the Kepler-11 star is observed to decrease by 0.077 percent, or 0.00077 (see Figure 7.20), from planet Kepler-11c. What is the radius of Kepler-11c?

$$0.00077 = \frac{R_{\text{Kepler-11c}}^2}{R_{\text{star}}^2} = \frac{R_{\text{Kepler-11c}}^2}{(7.7 \times 10^5 \text{ km})^2}$$

$$R_{\text{Kepler-11c}}^2 = 4.6 \times 10^8 \text{ km}^2 \quad \text{thus} \quad R_{\text{Kepler-11c}} = 2.1 \times 10^4 \text{ km}$$

Dividing $R_{\text{Kepler-11c}}$ by R_{Earth} (6400 km) shows that $R_{\text{Kepler-11c}} = 3.3\, R_{\text{Earth}}$.

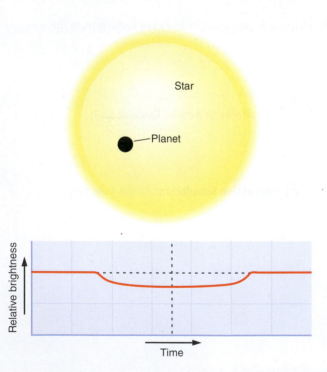

Figure 7.20 A planet that passes in front of a star blocks some of the light coming from the star's surface, causing the brightness of the star to decrease slightly. (The decrease in brightness is exaggerated here.)

plane of that planet. When an exoplanet passes in front of its parent star, the light from the star diminishes by a tiny amount (**Figure 7.20**). This is the **transit method** of detecting exoplanets, in which we observe the dimming of starlight as a planet passes in front of its parent star. Whereas the radial velocity method gives us the mass of the planet and its orbital distance from a star, the transit method provides the radius of a planet. **Working It Out 7.3** shows how the radii are estimated.

Current ground-based technology limits the sensitivity of the transit method to about 0.1 percent of a star's brightness. Telescopes in space improve the sensitivity because smaller dips in brightness can be measured. The small French COROT telescope (27 cm) discovered 32 planets during its 6 years of operation (2007–2013). NASA's 0.95-meter Kepler telescope has discovered many planets and has found thousands more candidates that are being investigated further. **Figure 7.21** shows that if one planet is found with this method, multiple sets of transits can indicate that other planets are orbiting the same star. Several thousand exoplanets have been detected from ground-based and space telescopes by using the transit method.

Other Methods   The gravitational field of an unseen planet can act like a lens, bending the light from a distant star in such a way that it causes the star to brighten temporarily while the planet is passing in front of it. When light is bent by a gravitational field, the effect is called **gravitational lensing**. In this instance, because the effect is small, it is usually called *microlensing*. Like the radial velocity method, microlensing provides an estimate of the mass of the planet. To date, about 60 exoplanets have been found with this technique.

Planets also may be detected by *astrometry*—precisely measuring the position of a star in the sky. If the system is viewed from "above," the star moves in a mini-orbit as the planet pulls it around. That motion is generally tiny and therefore very

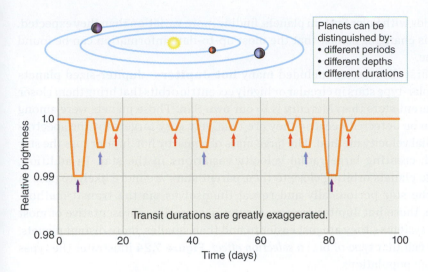

Figure 7.21 Multiple planets can be detected through multiple transits with different changes in brightness. The colored arrows point to the changes in the total light as the three planets (red, blue, and purple) transit the star.

difficult to measure. For systems viewed from above the plane of the planet's orbit, however, none of the prior methods will work because the planet neither passes in front of the star nor causes a shift in its speed along the line of sight. Space missions such as the Gaia observatory, launched in 2013 by the European Space Agency, conduct such observations.

Direct imaging involves taking a picture of the planet directly. The technique is conceptually straightforward but is technically difficult because it involves searching for a relatively faint planet in the overpowering glare of a bright star—a challenge far more difficult than looking for a star in a clear, bright daytime sky. Even when an object is detected by direct imaging, an astronomer must still determine whether the observed object is actually a planet. Suppose we detect a faint object near a bright star. Could it be a more distant star that just happens to be in the line of sight? Future observations could tell whether the object shares the bright star's motion through space, but it also could be a brown dwarf rather than a true planet. An astronomer would need to make further observations to determine the object's mass.

Some planets have been discovered through that method with large ground-based telescopes operating in the infrared region of the spectrum, using adaptive optics. **Figure 7.22** is an infrared image of Beta Pictoris b. Hubble Space Telescope observations of Fomalhaut, a bright naked-eye star only 25 light-years away, revealed a 3 Jupiter–mass planet in the dusty debris ring about 17 billion km from the central star (**Figure 7.23**). A related form of direct observation involves separating the spectrum of a planet from the spectrum of its star to obtain information about the planet directly. Large ground-based telescopes have obtained spectra of the atmospheres of some exoplanets and have found, for example, carbon monoxide and water in those atmospheres.

Types of Exoplanets

Searches for exoplanets have been remarkably successful. Between the discovery of the first (in 1995) and this writing, more than 4000 more have been confirmed, and thousands more candidates are under investigation. As the number of observed systems with single and multiple planets increases, astronomers can compare

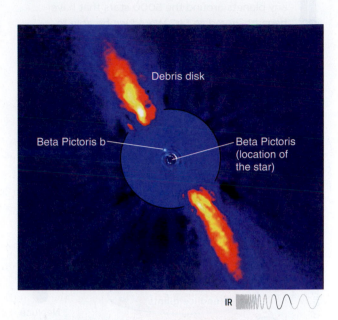

Figure 7.22 The planet Beta Pictoris b is seen orbiting within a dusty debris disk that surrounds the bright naked-eye star Beta Pictoris. The planet's estimated mass is 8 times that of Jupiter. The star is hidden behind an opaque mask, and the planet appears through a semitransparent mask used to subdue the brightness of the dusty disk.
Credit: ESO/A.-M. Lagrange et al., https://www.eso.org/public/images /eso0842b/. https://creativecommons.org/licenses/by/4.0/.

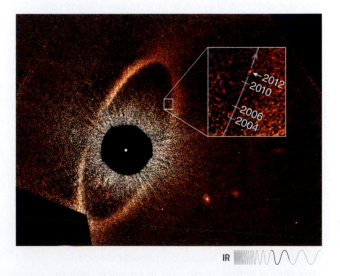

Figure 7.23 A Hubble Space Telescope image of Fomalhaut b, seen here in its 2012 position in the orbit. Prior positions are indicated by the tic marks on the white line overlaid on the orbit. The parent star, hidden by an obscuring mask, is about a billion times brighter than the planet.

what if . . .

What if astronomers had not yet discovered any planets around the 5000 stars that have been observed so far? Would we be able to conclude (with current technology) that there are no planets around the 100 billion stars in the Milky Way? Why or why not?

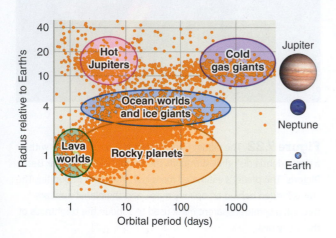

Figure 7.24 Different populations of exoplanets. (Although Kepler uses the transit method, its discoveries are shown separately, in orange.) Jupiter-sized planets can be close to their star and hot, or they can be far from their star and cold. Neptune-sized planets are probably composed of water and ice. Earth-sized planets have higher density, indicating that they are made of rock.

unanswered questions

How Earth-like must a planet be before scientists declare it to be "another Earth"? An editorial in the science journal *Nature* cautioned that scientists should define "Earth-like" in advance—before multiple discoveries of planets "similar" to Earth are announced and a media frenzy ensues. Must a planet be of similar size and mass, be located in the habitable zone, and have spectroscopic evidence of liquid water before we call it "Earth 2.0"?

those worlds with Solar System planets, finding more variation than they expected. The field is changing so fast that the most up-to-date information can be found only online.

The first discoveries included many **hot Jupiters**—Jupiter-sized planets orbiting solar-type stars in circular or highly eccentric orbits that bring them closer to their parent stars than Mercury is to our own Sun. Those planets were among the first to be detected because they are relatively easy targets for the spectroscopic radial velocity method. The large mass of a nearby hot Jupiter tugs the star very hard, creating large radial velocity variations in the star. In addition, those large planets orbiting close to their parent stars are more likely to pass in front of the star periodically and reveal themselves via the transit method. Therefore, those hot Jupiter systems are not necessarily representative of most planetary systems—they are just easier to find than smaller, more distant planets. Scientists call that type of bias a *selection effect*. **Figure 7.24** illustrates the types of exoplanet populations.

Astronomers were surprised by the hot Jupiters because, according to the theory of planet formation described earlier in the chapter, those giant, volatile-rich planets should not have been able to form so close to their parent stars. From theories based on the Solar System, astronomers expected that Jupiter-type planets should form in the more distant, cooler regions of the protoplanetary disk, where the volatiles that make up much of their composition can survive. Hot Jupiters may form much farther from their parent stars and later migrate inward to a closer orbit. That migration may be caused by an interaction with gas or planetesimals in which orbital angular momentum is transferred from the planet to its surroundings, allowing it to spiral inward.

The radial velocity method yields an estimate of the mass of a planet (see Working It Out 7.2), and the transit method yields an estimate of the size of a planet (see Working It Out 7.3). With both pieces of information, the density (mass divided by the volume) of the planet can be computed to determine whether the planet is mostly gaseous (low density) or mostly rock (high density). Many of the new planets discovered by Kepler are mini-Neptunes (gaseous planets with masses of 2–10 M_{Earth}) or **super-Earths** (rocky planets more massive than Earth). **Figure 7.25** illustrates the formation of those two types.

Planets with longer orbital periods, and therefore larger orbits, can be discovered only when the observations have gone on long enough to observe more than one complete orbit. Some of the exoplanets have highly elliptical orbits compared with those in the Solar System. Planets have been found with orbits that are highly tilted with respect to the plane of the rotation of their star, and some planets move in orbits whose direction is opposite that of their star's rotation. Multiple-planet systems have been observed in which the larger mini-Neptunes alternate with smaller super-Earths. The multiple-planet systems that have been found with the transit method reside in flat systems like our own, offering further evidence that the planets formed in a flat protoplanetary disk around a young star. But the current hypothesis to explain the Solar System's inner, small rocky planets and outer, large gaseous planets may not apply to all other planetary systems.

In addition, some planets, discovered through microlensing, don't have a star at all. Those planets may have been ejected from their solar systems after they formed and are no longer in gravitationally bound orbits around their stars. Others are "circumbinary"; they orbit two stars that form a binary star system. A scroll through NASA's exoplanet archive reveals that new planet discoveries are confirmed

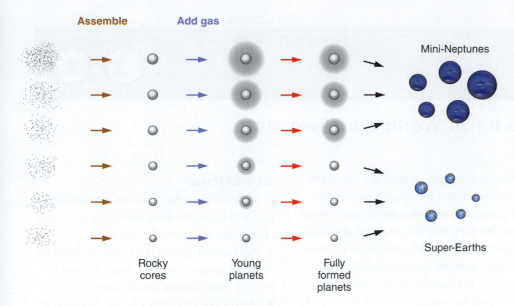

Figure 7.25 Many exoplanets are between the sizes of Earth and Neptune ($\sim 4R_E$). "Super-Earths" began with smaller rocky cores and less material, whereas "mini-Neptunes" had larger cores and collected more gas while forming.

nearly every week. The frequent new discoveries requiring revisions of existing theories make exoplanets one of the most exciting topics in astronomy today.

We return to exoplanets at several points in this book—namely, when we review the planets in our own Solar System, when we consider the types of stars, and when we discuss the search for Earth-like planets in Chapter 24.

CHECK YOUR UNDERSTANDING 7.5

What is the most common method for discovering an Earth-mass planet around a star? (a) Doppler spectroscopy; (b) direct imaging; (c) transit; (d) astrometric

Origins: The Search for Earth-Sized Planets

The discovery of planetary systems, many different from the Solar System, shows us that the formation of planets often, and perhaps always, accompanies the formation of stars. The implications of that conclusion are profound. Planets are a common by-product of star formation. In a galaxy of 200 billion stars and a universe of hundreds of billions of galaxies, how many planets (and moons) might exist? And with all those planets in the universe, how many might have conditions suitable for the particular category of chemical reactions that we refer to as "life"? (We return to that point in Chapter 24.)

The Kepler Mission was developed by NASA to find Earth-sized and larger planets in orbit about a variety of stars. Kepler was a 0.95-meter telescope with 42 CCD detectors that observed approximately 150,000 stars in 100 square degrees of sky to look for planetary transits. To confirm a planetary detection, transits needed to be observed three times with repeatable changes in brightness, duration of transit times, and computed orbital period. Kepler could detect a dip of 0.01 percent in the brightness of a star—sensitive enough to detect an

Reading Astronomy News

NASA's TESS Mission Uncovers Its 1st World With Two Stars

NASA

Discovering more planets means that we discover more unusual ones; like this planet that orbits two stars at once

The TOI 1338 system lies 1,300 light-years away in the constellation Pictor. The two stars orbit each other every 15 days. One is about 10% more massive than our Sun, while the other is cooler, dimmer and only one-third the Sun's mass.

Newly discovered TOI 1338 b is the only known planet in this binary star system. It's around 6.9 times larger than Earth, or between the sizes of Neptune and Saturn. The planet orbits in almost exactly the same plane as the stars, so it experiences regular stellar eclipses.

Scientists use the observations from TESS to generate graphs of how the brightness of stars change over time. When a planet transits in front of its star from our perspective, its passage causes a dip in the star's brightness.

Planets orbiting two stars are more difficult to detect than those orbiting one. TOI 1338 b's transits are irregular, between every 93 and 95 days, and vary in depth and duration thanks to the orbital motion of its stars. TESS only sees the transits crossing the larger star; the transits of the smaller star are too faint to detect.

"These are the types of signals that algorithms really struggle with," said lead author Veselin Kostov, a research scientist at the SETI Institute and Goddard. "The human eye is extremely good at finding patterns in data, especially non-periodic patterns like those we see in transits from these systems."

After identifying TOI 1338 b, the research team used a software package called eleanor, named after Eleanor Arroway, the central character in Carl Sagan's novel "Contact," to confirm the transits were real.

TOI 1338 had already been studied from the ground by radial velocity surveys. Kostov's team used this archival data to analyze the system and confirm the planet.

QUESTIONS

1. The period of the two stars in the binary pair is 15 days. Are these two stars very close to each other, or very far apart?

2. The planet's period is 95 days. Is the planet's orbit larger or smaller than the orbit of the two stars around one another?

3. Make a sketch of this system, including the two stars and the planet. Draw circles to show how the orbits are related, and whether or not they cross.

4. The article states that the discovery was confirmed using archival data from ground-based telescopes. Speculate: Why was the planet not discovered prior to this, using that ground-based data?

5. This system provides a window into the naming conventions for stars and planets. Study the names of the stars and planet. How are the two stars of a binary star system designated? How is the first planet in a planetary system designated?

Source: https://www.nasa.gov/feature/goddard /2020/nasa-s-tess-mission-uncovers-its-1st-world -with-two-stars.

Earth-sized planet. Kepler identified the first Earth-sized planets in 2011. Stars with transiting planets detected by Kepler also are observed spectroscopically to obtain radial velocity measurements that can lead to an estimate of a planet's mass. If a planet's radius and mass are known, the planet's density (mass per volume) can be estimated, too. From the density, astronomers can get a sense of whether the planet is composed primarily of gas, rock, ice, water, or some mixture of those.

On Earth, liquid water was essential for life to form and evolve. Because life on Earth is the only example of life for which we have evidence, we do not know whether liquid water is a cosmic requirement, but it is a place to start. The Kepler Mission searched for rocky planets at the right distance from

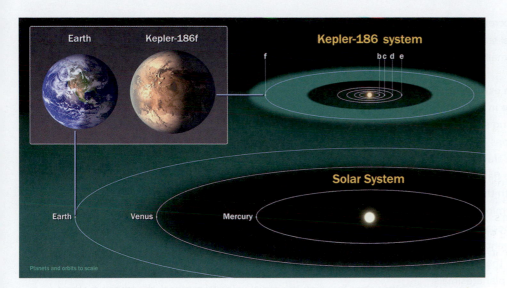

Figure 7.26 Artist's conception of the Kepler-186 system. Located about 500 light-years from Earth, this system has a planet in its habitable zone. The Solar System is shown for comparison. The Kepler-186 star is cooler than the Sun, so its habitable zone is closer to the star.

their stars to permit the existence of liquid water, a distance known as the **habitable zone**. The location and width of the zone varies depending on the temperature of the star. If a planet is too close to its star, water will exist only as a vapor; if too far, water will be frozen as ice. In the Solar System, Earth is the only planet currently in the habitable zone. Although announcements of new planets often state whether the planet is in the habitable zone, just being in the zone doesn't guarantee that the planet actually has liquid water—or that the planet is inhabited! An example of an Earth-sized planet in a habitable zone is shown in **Figure 7.26**.

Kepler identified thousands of planet candidates, some in the habitable zones of their respective stars. The candidates are confirmed by follow-up observations of more transits or of radial velocities before they are officially announced as planet detections. Citizen scientists contribute to that search through the Exoplanet Explorers project.

The Transiting Exoplanet Surveying Satellite (TESS) mission, launched in 2018, monitors 200,000 nearby stars for dimming caused by planetary transits. The mission is beginning to yield thousands more planetary candidates, with subsequent confirmations taking substantially longer. The James Webb Space Telescope (JWST), scheduled for launch in 2021, will observe in the infrared and should be able to detect gases in the atmospheres of exoplanets.

SUMMARY

Stars and their planetary systems form from collapsing interstellar clouds of gas and dust, following the laws of gravity and conservation of angular momentum. Conservation of angular momentum produces an accretion disk around a protostar that often fragments to form multiple planets, as well as smaller objects such as asteroids and dwarf planets, through the gradual accumulation of material into larger and larger objects. Multiple methods exist for finding planets around other stars, and such planets are now thought to be very common. This field of study is evolving very quickly due to advances in technology.

① **Describe how our understanding of planetary system formation developed from the work of both planetary and stellar scientists.**
Planetary scientists observed the motions of planets in our Solar System, as well as compositions of planets and meteorites. Stellar scientists observed the life cycle of stars and the nebula that precede and follow the time when a star fuses hydrogen. From these observations, the two groups developed the story of planetary system formation. Planets are a common by-product of star formation, and many stars are surrounded by planetary systems. Gravity pulls clumps of gas and dust together, causing them to shrink and heat up. Angular momentum must be

conserved, leading to both a spinning central star and an accretion disk that rotates and revolves in the same direction as the central star. Solar System meteorites show that larger objects build up from smaller objects.

(2) Discuss the role of gravity and angular momentum in explaining why planets orbit the Sun in a plane and why they revolve in the same direction that the Sun rotates. As particles orbit the forming star, gravity pulls the dust and gas inward toward the center. Collisions between particles that are traveling upward in their orbit with those traveling downward in their orbit cause the cloud of dust and gas to flatten into a plane. Conservation of angular momentum determines both the speed and the direction of the revolution of the objects in the forming system. Dust grains in the protoplanetary disk first stick together because of collisions and static electricity. As those objects grow, they eventually have enough mass to attract other objects gravitationally. Once that occurs, they begin emptying the space around them. Collisions of planetesimals lead to the formation of planets.

(3) Explain how temperature at different locations in the protoplanetary disk affects the composition of planets, moons, and other bodies. The temperature is higher near the central protostar, forcing volatile elements, such as water, to evaporate and leave the inner part of the disk. Planets in the inner part of the disk will have fewer volatiles than those in the outer part of the disk. The gas that a planet captures when it forms is the planet's primary atmosphere. Less massive planets lose their primary atmospheres and then form secondary atmospheres.

(4) Discuss the processes that resulted in the formation of planets and other objects in our Solar System. In the current model of the formation of the Solar System, solid terrestrial planets formed in the inner disk, where temperatures were high, whereas giant gaseous planets formed in the outer disk, where temperatures were low. Dwarf planets such as Pluto formed in the asteroid belt and in the region beyond the orbit of Neptune. Asteroids and comet nuclei remain today as leftover debris.

(5) Describe how astronomers both find planets around other stars and determine exoplanet properties. Astronomers find planets around other stars by using the radial velocity method, the transit method, microlensing, astrometry, and direct imaging. As technology has improved, the number and variety of known exoplanets has increased dramatically, with thousands of planets and planet candidates discovered orbiting other stars near the Sun within the Milky Way Galaxy in just the past few years.

QUESTIONS AND PROBLEMS

TEST YOUR UNDERSTANDING

1. Place the following events in the order that corresponds to the formation of a planetary system.
   a. Gravity collapses a cloud of interstellar gas.
   b. A rotating disk forms.
   c. Small bodies collide to form larger bodies.
   d. A stellar wind "turns on" and sweeps away gas and dust.
   e. Primary atmospheres form.
   f. Primary atmospheres are lost.
   g. Secondary atmospheres form.
   h. Dust grains stick together by static electricity.

2. If the radius of an object's orbit is halved, and angular momentum is conserved, what must happen to the object's speed?
   a. It must be halved.   c. It must be doubled.
   b. It must stay the same.   d. It must be squared.

3. Unlike the giant planets, the terrestrial planets formed when
   a. the inner Solar System was richer in heavy elements than the outer Solar System.
   b. the inner Solar System was hotter than the outer Solar System.
   c. the outer Solar System took up more volume than the inner Solar System, so more material was available to form planets.
   d. the inner Solar System was moving faster than the outer Solar System.

4. The terrestrial planets and the giant planets have different compositions because
   a. the giant planets are much larger.
   b. the terrestrial planets formed closer to the Sun.
   c. the giant planets are made mostly of solids.
   d. the terrestrial planets have few moons.

5. The spectroscopic radial velocity method preferentially detects
   a. large planets close to the central star.
   b. small planets close to the central star.
   c. large planets far from the central star.
   d. small planets far from the central star.
   e. The method detects all those planets equally well.

6. The concept of disk instability was developed to solve the problem that
   a. Jupiter-like planets migrate after formation.
   b. not enough gas was in the Solar System to form Jupiter.
   c. the early solar nebula probably dispersed too soon to form Jupiter.
   d. Jupiter consists mostly of volatiles.

7. Because angular momentum is conserved, an ice-skater who throws her arms out will
   a. rotate more slowly.   c. rotate at the same rate.
   b. rotate more quickly.   d. stop rotating.

8. Clumps grow into planetesimals by
   a. gravitationally pulling in other clumps.
   b. colliding with other clumps.
   c. attracting other clumps with opposite charge.
   d. conserving angular momentum.

9. The transit method preferentially detects
   a. large planets close to the central star.
   b. small planets close to the central star.
   c. large planets far from the central star.
   d. small planets far from the central star.
   e. The method detects all those planets equally well.

10. If the radius of a spherical object is halved, what must happen to the period so that the spin angular momentum is conserved?
    a. It must be divided by 4.
    d. It must double.
    b. It must be halved.
    e. It must be multiplied by 4.
    c. It must stay the same.

11. Which of the following affects the angular momentum of a spherical object? (Choose all that apply.)
    a. radius
    c. rotation speed
    b. mass
    d. temperature

12. The planets in the inner part of the Solar System are made primarily of refractory materials; the planets in the outer Solar System are made primarily of volatiles. That difference occurs because
    a. refractory materials are heavier than volatiles, so they sank farther into the nebula.
    b. no volatiles were in the inner part of the accretion disk.
    c. the volatiles on the inner planets were lost soon after the planets formed.
    d. the outer Solar System has gained more volatiles from space since formation.

13. If scientists want to find out about the composition of the early Solar System, the best objects to study are
    a. the terrestrial planets.
    b. the giant planets.
    c. the Sun.
    d. asteroids and comets.

14. The direction of revolution in the plane of the Solar System was determined by
    a. the plane of the galaxy in which the Solar System sits.
    b. the direction of the gravitational force within the original cloud.
    c. the direction of rotation of the original cloud.
    d. the amount of material in the original cloud.

15. A planet in the "habitable zone"
    a. is close to the central star.
    b. is far from the central star.
    c. is the same distance from its star as Earth is from the Sun.
    d. is at a distance where liquid water can exist on the surface.

THINKING ABOUT THE CONCEPTS

16. ★ WHAT AN ASTRONOMER SEES   In Figure 7.8, identify an unlabeled finger, jet, and diffraction spike. Make a sketch of the image, and label the locations of the features you have identified. 👁

17. What is the source of the material that now makes up the Sun and the rest of the Solar System?

18. Describe the different ways by which stellar astronomers and planetary scientists each came to the same conclusion about how planetary systems form.

19. What is a protoplanetary disk? What are two reasons why the inner part of the disk is hotter than the outer part?

20. Physicists describe certain properties, such as angular momentum and energy, as being *conserved*. Does this property mean conservation laws imply that an individual object can never lose or gain angular momentum or energy? Explain your reasoning.

21. The Process of Science Figure in this chapter makes the point that different areas of science must agree with one another. Suppose that a few new exoplanets are discovered that appear not to have formed from the collapse of a stellar nebula (for example, the planetary orbits might be in random orientations). What will scientists do with that new information? 👁

22. How does the law of conservation of angular momentum control a figure-skater's rate of spin?

23. What is an accretion disk?

24. Look under your bed, the refrigerator, or any similar place for dust bunnies. Once you find them, blow one toward another. Watch carefully and describe what happens as they meet. What happens if you repeat that action with additional dust bunnies? Will those dust bunnies ever have enough gravity to begin pulling themselves together? If they were in space instead of on the floor, might that happen? What force prevents their mutual gravity from drawing them together into a "bunny-tesimal" under your bed?

25. Why do we find rocky material everywhere in the Solar System but find large amounts of volatile material only in the outer regions?

26. Why could the four giant planets collect massive gaseous atmospheres, whereas the terrestrial planets could not? Explain the source of the secondary atmospheres surrounding the terrestrial planets.

27. Describe four methods that astronomers use to search for exoplanets. What are the limitations of each method; that is, what circumstances are necessary to detect a planet by each method?

28. Why is it difficult to obtain an image of an exoplanet?

29. Many of the first exoplanets that astronomers found orbiting other stars were giant planets with Jupiter-like masses and with orbits located very close to their parent stars. Explain why those characteristics are a selection effect of the discovery method.

30. How has the Kepler telescope found Earth-like planets, and what do astronomers mean by "Earth-like"?

APPLYING THE CONCEPTS

31. Compare Earth's orbital angular momentum with its spin angular momentum by using the following values: $m = 5.97 \times 10^{24}$ kg, $v = 29.8$ kilometers per second (km/s), $r = 1$ AU, $R = 6378$ km, and $P = 1$ day. Assume that Earth is a uniform body. What fraction does each component (orbital and spin) contribute to Earth's total angular momentum? Refer to Working It Out 7.1. 🟡–🟢–🟢
    a. Make a prediction: Study Working It Out 7.1. Do you expect Earth to have more angular momentum in its spin, or in its orbit?
    b. Calculate: Follow Working It Out 7.1 to calculate Earth's orbital angular momentum and its spin angular momentum. Add these together to find the total angular momentum. Then divide each component by the total to find the fraction of the angular momentum that is accounted for by that component.
    c. Check your work: Verify in both cases that you obtain the correct units for angular momentum (kg m²/s), and compare your results to your prediction.

32. A planet has been found to orbit a 1-M_{Sun} star in 200 days. What is the semimajor axis of this exoplanet's orbit? Compare the semimajor axis of this exoplanet with that of the planets around our own Sun. What temperatures must that planet experience? 🟢—⚫—🟢
  a. Make a prediction: Is 200 days longer or shorter than Earth's orbital period? Therefore, do you expect the semimajor axis to be larger or smaller than 1 AU?
  b. Calculate: Follow Working It Out 7.2 to find the semimajor axis of this exoplanet.
  c. Check your work: Verify that your semimajor axis has correct units, and compare your result to your prediction.
  d. Evaluate: Given the semimajor axis of this planet's orbit, what do you expect the temperature to be like on the surface?

33. The star Kepler-11 has an orbital radius of 1.1 R_{Sun}. If the planet has a radius of 4.5 R_{Earth}, by what percentage does the brightness of Kepler-11 decrease when that planet transits the star? 🟢—⚫—🟢
  a. Make a prediction: Study Working It Out 7.3. Do you expect a large or a small drop in brightness of the star when the planet transits?
  b. Calculate: Follow Working It Out 7.3 to calculate the percent reduction in brightness of Kepler-11.
  c. Check your work: Verify that all the units cancel out in your percentage, and compare your result to your prediction.

34. Jupiter has a mass equal to 318 times Earth's mass, an orbital radius of 5.2 AU, and an orbital velocity of 13.1 km/s. Earth's orbital velocity is 29.8 km/s. What is the ratio of Jupiter's orbital angular momentum to that of Earth?
  a. Make a prediction: Consider the equation for orbital angular momentum. Do you expect Jupiter to have more or less orbital angular momentum than Earth? Do you expect the ratio of Jupiter's orbital angular momentum to Earth's orbital angular momentum to be greater or smaller than one?
  b. Calculate: Set up the ratio of orbital angular momenta, using letters first. Divide out terms that appear in top and bottom. (Hint: Given the problem statement, does Earth's mass appear in both the top and bottom?) Then use your calculator to solve for the value of the ratio.
  c. Check your work: Verify that the final ratio is dimensionless (all the units should have divided out). Compare your result with your prediction.

35. The asteroid Vesta has a diameter of 530 km and a mass of 2.7×10^{20} kg. Calculate the density (mass/volume) of Vesta.
  a. Make a prediction: The density of water is 1000 kg/m3, whereas that of rock is about 2500 kg/m3. Do you expect the density of Vesta to be similar to the density of rock, or similar to the density of water?
  b. Calculate: Use the formula for density (mass/volume) to find the density of Vesta. You may need to recall that the volume of a sphere is (4/3)ϖR3.
  c. Check your work: Verify that your answer has correct units of density, and compare it to the densities of water and rock.

36. In Figure 7.18, what is the maximum radial velocity of HD 70642 in meters per second? Convert that number to miles per hour (mph). How does that value compare with the speed at which Earth orbits the Sun (67,000 mph)? ⊛

37. Use Appendix 4 to answer the following:
  a. What is the total mass of all the planets in the Solar System, expressed in Earth masses (M_{Earth})?
  b. What fraction of that total planetary mass is Jupiter?
  c. What fraction does Earth represent?

38. Venus has a radius 0.950 times that of Earth and a mass 0.815 times that of Earth. Venus's rotation period is 243 days. What is the ratio of Venus's spin angular momentum to that of Earth? Assume that Venus and Earth are uniform spheres.

39. Suppose you can measure radial velocities of about 0.3 m/s. Suppose, too, you are observing a spectral line with a wavelength of 575 nanometers (nm). How large a shift in wavelength would a radial velocity of 0.3 m/s produce?

40. Using data in Appendix 4, compute the densities of Venus, Jupiter, and Neptune. Compare your answers with the densities of rock, water, and gas.

41. Recalling Kepler's laws, put the three planets in Figure 7.21 in order from fastest to slowest. Compare the duration of the transits in Figure 7.21. Why does the outermost planet have the longest duration? ⊛

42. Earth tugs the Sun around as it orbits, but that effect (only 0.09 m/s) is much smaller than that of any known exoplanet. How large a shift in wavelength does that effect cause in the Sun's spectrum at 500 nm?

43. If an alien astronomer observed a plot of the light curve as Jupiter passed in front of the Sun, by how much would the Sun's brightness drop during the transit?

44. Kepler detected a planet with a diameter of 1.7 Earth (D_{Earth}).
  a. How much larger is the volume of that planet than Earth's?
  b. Assume that the density of the planet is the same as Earth's. How much more massive is it than Earth?

45. The planet COROT-11b was discovered using the transit method, and astronomers have followed up with radial velocity measurements, so both its radius (1.43 R_{Jup}) and its mass (2.33 M_{Jup}) are known. The density provides a clue about whether the object is gaseous or rocky.
  a. What is the mass of the planet in kilograms?
  b. What is the planet's radius in meters?
  c. What is the planet's volume?
  d. What is the planet's density? How does that density compare with the density of water (1000 kg/m³)? Is the planet likely to be rocky or gaseous?

EXPLORATION Exploring Exoplanets

digital.wwnorton.com/astro7

Visit the Student Site at the Digital Resources page and open the Radial Velocity Simulation in Chapter 7. This applet has a number of different panels that allow you to experiment with the variables that are important for measuring radial velocities. Compare the views shown in the various panels with the colored arrows in the first panel to see where an observer would stand to see the view shown. Start the animation (press "Play") and allow it to run while you watch the planet orbit its star from each of the views shown. Stop the animation, and in the "Sample Systems" panel select "Option A."

1. Is Earth's view of this system most nearly like the "side view" or most nearly like the "orbit view"?

2. Is the orbit of this planet circular or elongated?

3. Study the radial velocity graph in the upper right panel. The blue curve shows the radial velocity of the star over a full period. What is the maximum radial velocity of the star?

4. The horizontal axis of the graph shows the "phase," or fraction of the period. A phase of 0.5 is halfway through a period. The vertical red line indicates the phase shown in views in the upper left panel. Start the animation to see how the red line sweeps across the graph as the planet orbits the star. The period of this planet is 365 days. How many days pass between the minimum radial velocity and the maximum radial velocity?

5. When the planet moves away from Earth, the star moves toward Earth. The sign of the radial velocity tells the direction of the motion (toward or away). Is the radial velocity of the star positive or negative at this time in the orbit? If you could graph the radial velocity of the planet at this point in the orbit, would it be positive or negative?

In the "Sample Systems" window, select "Option B:"

6. What has changed about the orbit of the planet as shown in the views in the upper left panel?

7. When is the planet moving fastest—when it is close to the star or when it is far from the star?

8. When is the star moving fastest—when the planet is close to it or when it is far away?

9. Explain how an astronomer would determine, from a radial velocity graph of the star's motion, whether the orbit of the planet was in a circular or elongated orbit.

10. Study the "Earth View" panel. Would this planet be a good candidate for a transit observation? Why or why not?

Taking the Measure of Stars

The stars in the night sky have different brightnesses for reasons that you will learn about here in Chapter 13. Study the star chart for the current season, found in Appendix 7. About how many more faint stars (magnitude 4) than bright stars (magnitude 1) are there in this chart? Notice the scale bar that shows how the size of the dot relates to the star's magnitude. Next, count the number of stars of each magnitude and make a table of your results. Does this more careful analysis agree with your estimate? Take this star chart outside on a clear night, and find all the stars with magnitude 1 that are above the horizon. This will orient you to the sky. Now find some stars with magnitude 2, 3, 4, and so on. If you can find no stars fainter than magnitude 3, then the "limiting magnitude" of your observing site is 3. What is the limiting magnitude of your observing site on this date? What sources of light are making it difficult to see stars fainter than this?

EXPERIMENT SETUP

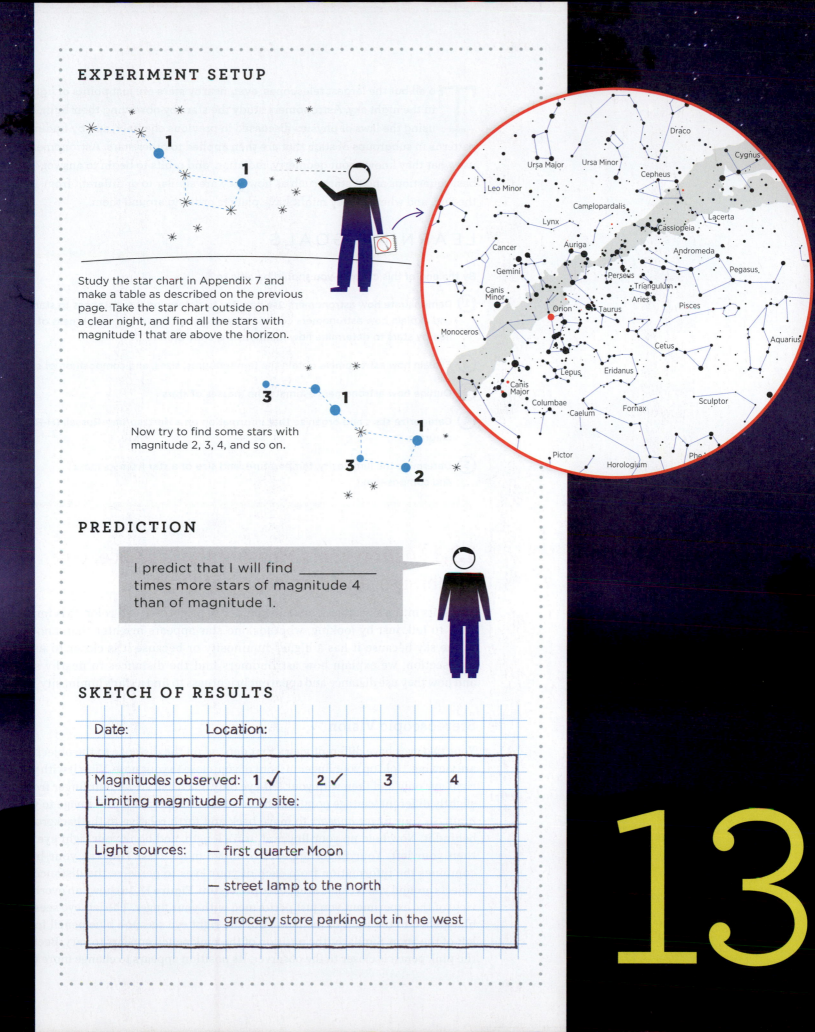

Study the star chart in Appendix 7 and make a table as described on the previous page. Take the star chart outside on a clear night, and find all the stars with magnitude 1 that are above the horizon.

Now try to find some stars with magnitude 2, 3, 4, and so on.

PREDICTION

I predict that I will find _____ times more stars of magnitude 4 than of magnitude 1.

SKETCH OF RESULTS

Date: _____ Location: _____

Magnitudes observed: 1 ✓   2 ✓   3   4
Limiting magnitude of my site: _____

Light sources: — first quarter Moon

— street lamp to the north

— grocery store parking lot in the west

13

To all but the largest telescopes, even nearby stars are just points of light in the night sky. Astronomers study the stars by observing their light, by using the laws of physics discussed in previous chapters, and by finding patterns in subgroups of stars that are then applied to other stars. Astronomers use what they know about geometry, radiation, and orbits to begin to answer basic questions about stars, such as how they are similar to or different from the Sun, and whether they might have planets orbiting around them.

LEARNING GOALS

By the end of this chapter, you should be able to:

(1) Demonstrate how astronomers use parallax to determine the distances to stars, and explain how astronomers combine these distances with the brightness of nearby stars to determine how luminous the stars are.

(2) Explain how astronomers obtain the temperatures, sizes, and composition of stars.

(3) Outline how astronomers estimate the masses of stars.

(4) Categorize stars and organize that information on a Hertzsprung-Russell (H-R) diagram.

(5) Determine the luminosity, temperature, and size of a star from its mass and composition.

13.1 Astronomers Measure the Distance, Brightness, and Luminosity of Stars

The stars in the sky differ from one another in brightness and color. It is impossible to tell, just by looking, whether one star appears brighter than another in the sky because it has a higher luminosity or because it is closer to us. In this section, we explain how astronomers find the distances to nearby stars and how they use distance and apparent brightness to find a star's luminosity.

Stereoscopic Vision

Your two eyes have different views that depend on the distance to the object you are viewing. Hold up your finger in front of you, close to your nose. View it with your right eye only and then with your left eye only. Each eye views your finger from a slightly different vantage point. Each eye sends a slightly different image to your brain, so your finger *appears* to move back and forth relative to the background behind it. Now hold up your finger at arm's length, and blink your right eye and then your left. Your finger appears to move much less. The way your brain combines the information from each of your eyes to perceive the distances to objects around you is called **stereoscopic vision**. **Figure 13.1a** shows an overhead view of the experiment you just performed with your finger. The left eye sees the blue pencil to the right of the lamp. But the right eye sees the blue pencil to the left of the lamp. Similarly, the position of the pink pencil appears to vary. Because the pink pencil is closer to the observer, its position appears to change more than

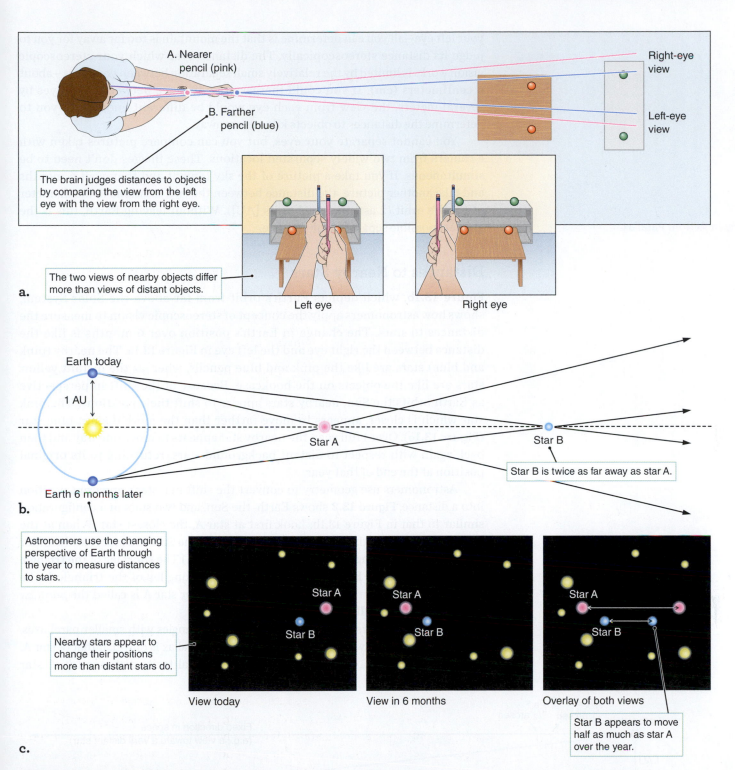

A. Nearer pencil (pink)

B. Farther pencil (blue)

Right-eye view

Left-eye view

The brain judges distances to objects by comparing the view from the left eye with the view from the right eye.

The two views of nearby objects differ more than views of distant objects.

a.

Left eye

Right eye

Earth today

1 AU

Earth 6 months later

b.

Star A

Star B

Star B is twice as far away as star A.

Astronomers use the changing perspective of Earth through the year to measure distances to stars.

Nearby stars appear to change their positions more than distant stars do.

Star A

Star B

View today

Star A

Star B

View in 6 months

Star A

Star B

Overlay of both views

Star B appears to move half as much as star A over the year.

c.

Figure 13.1 **a.** Stereoscopic vision enables you to determine the distance to an object by comparing the view from each eye. **b.** Similarly, comparing views from different places in Earth's orbit enables astronomers to determine the distances to stars. **c.** As Earth moves around the Sun, the apparent positions of nearby stars change more than the apparent positions of more distant stars. (Diagram not to scale.)

the position of the blue pencil—seeming to move from the right of the blue pencil to the left of the blue pencil.

Stereoscopic vision enables you to judge the distances of objects as far away as a few hundred meters, but beyond that it is of little use. Your right eye's view of a mountain several kilometers away is indistinguishable from the view seen by your left eye—all you can determine is that the mountain is too far away for you to judge its distance stereoscopically. The distance over which your stereoscopic vision works is limited by the relatively small separation between your eyes—about 6 centimeters (cm). If you could increase the distance between your eyes by several meters, the view from each eye would be different enough for you to determine the distances to objects kilometers away.

You cannot separate your eyes, but you can compare pictures taken with a camera from two widely separated locations. These images don't need to be simultaneous. If you take a picture of the sky tonight and then wait 6 months and take another picture, the distance between the two locations is the diameter of Earth's orbit (2 astronomical units [AU]). Without leaving Earth, this is the greatest possible separation obtainable.

Astronomy in Action: Parallax

Distances to Nearby Stars

Figure 13.1b, which depicts Earth's orbit from far above the Solar System, shows how astronomers apply the concept of stereoscopic vision to measure the distances to stars. The change in Earth's position over 6 months is like the distance between the right eye and the left eye in Figure 13.1a. The nearby (pink and blue) stars are like the pink and blue pencils, whereas the distant yellow stars are like the objects on the bookcase. Because of the shift in perspective as Earth orbits the Sun, nearby stars appear to shift their positions. The pink star which is closer, appears to move farther than the more distant blue star (**Figure 13.1c**). Over 1 full year, the nearby star appears to move one way and then back again with respect to distant background stars, returning to its original position at the end of that year.

Astronomers use geometry to convert the shift in a star's apparent position into a distance. **Figure 13.2** shows Earth, the Sun, and two stars in a configuration similar to that in Figure 13.1b. Look first at star A, the closest star. When at the top of the figure, Earth forms a right triangle with the Sun and star A at the other corners. (A right triangle is one with a 90° angle in it.) The short leg of the triangle is the distance from Earth to the Sun, 1 AU. The long leg of the triangle is the distance from the Sun to star A. The small angle near star A is called the *parallax angle*, or simply **parallax**, of the star.

More distant stars make longer and skinnier triangles with smaller parallaxes. Star B is twice as far away as star A, and so star B's parallax is only half that of star A. The parallax (p) of a star is inversely proportional to its distance (d). A star

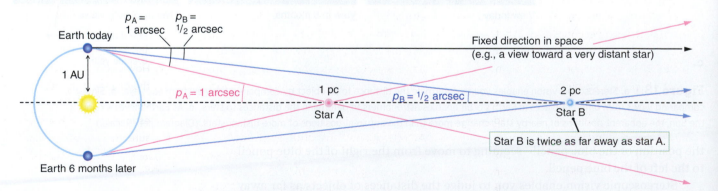

Figure 13.2 The parallax (p) of a star is inversely proportional to its distance. More distant stars have smaller parallaxes. (Diagram not to scale.)

Reading Astronomy News

NASA's *New Horizons* Conducts the First Interstellar Parallax Experiment

June 11, 2020

To make the most precise parallax measurements, scientists need the longest possible baseline. In 2020, they took advantage of a rare opportunity to use an unusually long baseline.

For the first time, a spacecraft has sent back pictures of the sky from so far away that some stars appear to be in different positions than we see from Earth.

More than four billion miles from home and speeding toward interstellar space, NASA's *New Horizons* has traveled so far that it now has a unique view of the nearest stars.

On April 22–23, the spacecraft turned its long-range telescopic camera to a pair of the closest stars, Proxima Centauri and Wolf 359, showing just how they appear in different places than we see from Earth. Scientists have long used this "parallax effect"—how a star appears to shift against its background when seen from different locations—to measure distances to stars.

"No human eye can detect these shifts," Stern said.

But when *New Horizons* images are paired with pictures of the same stars taken on the same dates by telescopes on Earth, the parallax shift is instantly visible. The combination yields a 3-D view of the stars "floating" in front of their background star fields.

"The *New Horizons* experiment provides the largest parallax baseline ever made—over 4 billion miles—and is the first demonstration of an easily observable stellar parallax," said Tod Lauer, New Horizons science team member from the National Science Foundation's

National Optical-Infrared Astronomy Research Laboratory who coordinated the parallax demonstration.

Working in Stereo

Lauer, *New Horizons* Deputy Project Scientist John Spencer, of SwRI, and science team collaborator, astrophysicist, Queen guitarist and stereo imaging enthusiast Brian May created the images that clearly show the effect of the vast distance between Earth and the two nearby stars.

"It could be argued that in astro-stereoscopy—3-D images of astronomical objects—NASA's *New Horizons* team already leads the field, having delivered astounding stereoscopic images of both Pluto and the remote Kuiper Belt object Arrokoth," May said. "But the latest *New Horizons* stereoscopic experiment breaks all records. These photographs of Proxima Centauri and Wolf 359—stars that are well-known to amateur astronomers and science fiction aficionados alike—employ the largest distance between viewpoints ever achieved in 180 years of stereoscopy!"

The companion Earth-based images of Proxima Centauri and Wolf 359 were provided by Edward Gomez of the Las Cumbres Observatory, operating a remote telescope at Siding Spring Observatory in Australia, and astronomers John Kielkopf, University of Louisville, and Karen Collins, Harvard and Smithsonian Center for Astrophysics, operating a remote telescope at Mt. Lemmon Observatory in Arizona.

An Interstellar Navigation First

Throughout history, navigators have used measurements of the stars to establish their position on Earth. Interstellar navigators can

do the same to establish their position in the galaxy, using a technique that *New Horizons* has demonstrated for the first time.

At the time of the observations, New Horizons was more than 4.3 billion miles (about 7 billion kilometers) from Earth, where a radio signal, traveling at the speed of light, needed just under 6 hours and 30 minutes to reach home.

Launched in 2006, *New Horizons* is the first mission to Pluto and the Kuiper Belt. It explored Pluto and its moons in July 2015—completing the space-age reconnaissance of the planets that started 50 years earlier—and continued on its unparalleled voyage of exploration with the close flyby of Kuiper Belt object Arrokoth in January 2019. *New Horizons* will eventually leave the Solar System, joining the *Voyagers* and *Pioneers* on their paths to the stars.

QUESTIONS

1. How many times longer was this baseline than the usual baseline, which is about the diameter of Earth's orbit?

2. Why did the team choose to measure the parallax of nearby stars, whose distances are already well known?

3. Go online to the source of the article to find the stereoscopic images of Proxima Centauri, which have been made into a "video." Where, in the image, is the shifting star?

4. How does improving the accuracy of the distances to nearby stars from trigonometric parallax affect astronomers' estimates of farther stars' distances, measured using spectroscopic parallax?

Source: http://pluto.jhuapl.edu/News-Center/News-Article .php?.

What if you measured the distances to all the visible stars in the constellation Sagittarius? Would you expect the distances to be similar or very different for these stars, and what would that imply?

3 times farther away than star A has a parallax 1/3 of star A's parallax. A star 10 times farther away has a parallax 1/10 of star A's parallax.

The parallaxes of real stars are tiny. Recall from Chapter 2 that the full circle of the sky can be divided into 360 degrees. The apparent diameter of the full Moon in the sky averages about half a degree. Just as an hour on the clock is divided into minutes and seconds, a degree of sky can be divided into arcminutes and arcseconds. An **arcminute** (abbreviated **arcmin**) is 1/60 of a degree. An **arcsecond** (abbreviated **arcsec**) is 1/60 of an arcminute and 1/3600 of a degree. To give you a better sense of these scales, an arcsecond is approximately the angle formed by the diameter of a golf ball viewed 9 km away from you.

We often use units of light-years to indicate distances to stars (Chapter 1). One light-year is the distance that light travels in 1 year—about 9.5×10^{12} (9.5 trillion) kilometers (km). Another common unit for discussing distances to stars and galaxies is the **parsec (pc)**, which is equal to 3.26 light-years (or 206,265 AU). The term is short for *parallax second*—a star at a distance of 1 parsec has a parallax of 1 arcsecond.

When astronomers began to measure the parallax angles of stars, they discovered that even the closest stars are very far away (**Working It Out 13.1**). F. W. Bessel (1784–1846) made the first successful measurement of stellar parallax in 1838, reporting a parallax of 0.314 arcsec for the star 61 Cygni. That finding indicated that 61 Cygni was 3.2 pc away, or 660,000 times as far away as the Sun. With that one measurement, Bessel increased the known volume of the universe by a factor of 10,000. Today, astronomers know of at least 60 stars, some in multiple-star systems, within 5 pc (16.3 light-years) of the Sun. In the neighborhood of the Sun, each star or star system has on average a volume of about 300 cubic light-years of space to itself.

Most stars are so far away that the parallax angle is too small to measure using ground-based telescopes, which are limited by Earth's atmosphere. In the 1990s, the European Space Agency's Hipparcos satellite measured the positions and parallaxes of over 100,000 stars with an uncertainty of about ±0.001 arcsec. Then, in 2013, the European Space Agency launched *Gaia*, a space mission that has been studying stellar parallaxes. *Gaia*'s second data release in 2018 included parallaxes of more than 1 billion stars, with uncertainties of ±0.00004 to ±0.0007 arcsec, depending on the star's brightness.

An uncertainty in the parallax will mean a corresponding uncertainty in the distance. For example, a star measured by Hipparcos to have a parallax of 0.004 ± 0.001 arcsec has a parallax between 0.003 and 0.005 arcsec. That range of parallaxes gives a corresponding distance range of 200–333 pc from Earth; a range of 133 pc. If the parallax measured by *Gaia* is 0.004 ± 0.0001 arcsec, then the distance range is 244–256 parsecs; a range of only 12 pc. Astronomers are very excited about these recent data from *Gaia*, which yield very small uncertainties compared to previously measured distances. Nevertheless, nearly all stars are too far away to find their distances by parallax. Other methods of measuring distance to stars too remote for parallax are discussed later in this chapter.

Luminosity, Brightness, and Distance

The stars in Earth's sky have noticeably different brightnesses, where brightness corresponds to the amount of energy falling on a square meter of area each second in the form of electromagnetic radiation (Chapter 5). Although a star's brightness can be measured directly, it does not immediately give much information about

working it out 13.1

Parallax and Distance

If star B in Figure 13.2 is twice as far away from us as star A, then star B will have half the parallax of star A. That is, the parallax of a star (p) is inversely proportional to its distance (d):

$$p \propto \frac{1}{d} \quad \text{or} \quad d \propto \frac{1}{p}$$

A star with a parallax of 1 arcsec is at a distance of 1 pc (Figure 13.2). The inverse proportionality between distance and parallax becomes

$$\left(\begin{array}{c} \text{Distance measured} \\ \text{in parsecs} \end{array} \right) = \cfrac{1}{\left(\begin{array}{c} \text{Parallax measured} \\ \text{in arcseconds} \end{array} \right)}$$

or

$$d(\text{pc}) = \frac{1}{p(\text{arcsec})}$$

Suppose that the parallax of a star is measured to be 0.5 arcsec. The distance can be found by

$$d(\text{pc}) = \frac{1}{0.5} = 2 \, \text{parsecs}$$

Similarly, a star with a measured parallax of 0.01 arcsec is located at a distance of 1/0.01 = 100 pc.

Astronomers use the unit parsecs because doing so makes the relationship between distance and parallax easier to manage than when the distance is measured in light-years. One parsec equals 206,265 AU (see Appendix 1), which is 3.09×10^{13} km or 3.26 light-years.

After the Sun, the next-closest star to Earth is Proxima Centauri. Located at a distance of 4.24 light-years, Proxima Centauri is a faint member of a system of three stars called Alpha Centauri. What is that star's parallax? First, we convert the distance to parsecs:

$$d = 4.24 \, \text{light-years} \times \frac{1 \, \text{parsec}}{3.26 \, \text{light-years}} = 1.30 \, \text{parsecs}$$

Then,

$$p(\text{arcsec}) = \frac{1}{1.30 \, \text{pc}} = 0.77 \, \text{arcsec}$$

Even the closest star to the Sun has a parallax of only about 3/4 arcsec.

---

the star itself. As illustrated in **Figure 13.3**, a star may appear bright only because it is nearby. Conversely, a faint star may actually be very luminous but appear faint because it is far away.

Astronomers measure the brightness of stars by comparing them with one another. The system they use dates back 2100 years, when the Greek astronomer Hipparchus classified stars according to their brightness: the brightest stars he could see were magnitude 1, and the faintest were magnitude 6. The details of his system, still in use today, are discussed in **Working It Out 13.2** and Appendix 7.

To learn about the actual properties of a star, astronomers need to know the total energy radiated by a star each second—the star's luminosity. Recall from Chapter 5 that the brightness of an object that has a known luminosity L and is located at a distance d is given by

$$\text{Brightness} = \frac{\text{Total light emitted per second}}{\text{Area of sphere of radius } d} = \frac{\text{Luminosity}}{4\pi d^2}$$

You can rearrange that equation, moving the quantities you can measure (distance and brightness) to the right-hand side and the quantity you would like to know (luminosity) to the left, to get

$$\text{Luminosity} = 4\pi d^2 \times \text{Brightness}$$

That equation is used to find how luminous a star must be to appear as bright as it does when seen from Earth.

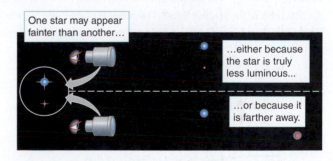

Figure 13.3 The brightness of a star visible in our sky depends on both its luminosity—how much light it emits—and its distance.

working it out 13.2

The Magnitude System

The **magnitude** system of brightness for celestial objects has been adjusted since Hipparchus. Astronomers have defined 1st magnitude stars to be exactly 100 times brighter than 6th magnitude stars. The system has been extended to stars brighter than 1st magnitude by using zero and negative numbers. Traditionally, Vega, a bright star in the constellation Lyra, is considered to have a magnitude of 0. A negative magnitude signifies that an object is *brighter* than Vega. For example, Sirius, the brightest star in the night sky, has a magnitude of −1.46. Venus can be as bright as magnitude −4.4, or about 15 times brighter than Sirius and bright enough to cast a shadow. The full Moon's magnitude is −12.6, and the Sun's is −26.7 (**Figure 13.4**). Hipparchus must have had typical eyesight because an average person under dark skies can see stars only as faint as 6th magnitude. Today, telescopes extend our vision to very faint objects. The Hubble Space Telescope (HST) can take long exposures and detect stars as faint as 30th magnitude.

With five steps between the 1st and 6th magnitudes, each step is equal to the fifth root of 100, or $100^{1/5}$, approximately 2.512. The magnitude system is logarithmic, but instead of the usual base 10, it is base 2.512. Fifth magnitude stars are 2.512 times brighter than 6th magnitude stars, and 4th magnitude stars are $2.512 \times 2.512 = 6.310$ times brighter than 6th magnitude stars. The brightness ratio between any two stars is equal to $(2.512)^N$, where N is the magnitude difference between them.

The limit of the Hubble Space Telescope is magnitude 30, whereas the limit of the human eye is magnitude 6. The difference is 24 magnitudes, so the HST can detect stars that are $(2.512)^{24} = 4 \times 10^9$, or 4 billion, times fainter than the magnitude 6 that the naked eye can see. Or, if we compare the Sun and the Moon, the Sun is 14 magnitudes brighter, or $(2.512)^{14} = 4 \times 10^5$; the Sun is 400,000 times brighter than the full Moon. (For more detailed calculations and a table of magnitudes and brightness differences, see Appendix 7.)

The magnitude of a star, as we have discussed it so far, is called the star's **apparent magnitude** because it is the star's brightness as it *appears* in Earth's sky. If all stars were located exactly 10 pc (32.6 light-years) away from Earth, the brightness of each star would be proportional to its luminosity. If the distance from Earth to a star is known, astronomers compute how bright the star would appear if it were located at 10 pc. A star's **absolute magnitude**— its apparent magnitude at a distance of 10 pc—specifies the star's luminosity.

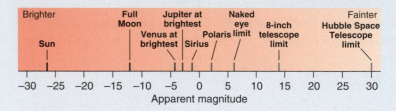

Figure 13.4 Apparent magnitude indicates the apparent brightness of an object in our sky. The brightest objects have a negative apparent magnitude, while telescopes have extended the observable range to fainter objects with higher magnitudes.

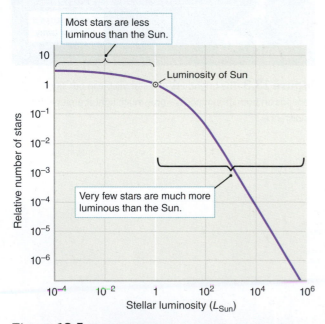

Figure 13.5 The distribution of the luminosities of stars is plotted logarithmically, with increments in powers of 10.

Stars have a wide range of luminosities—the most luminous are more than 10 billion (10^{10}) times more luminous than the least luminous. The Sun provides a convenient unit when measuring the properties of other stars, including their luminosity. The measured luminosity of the Sun is $L_{Sun} = 3.8 \times 10^{26}$ watts (W). The most luminous stars are a million times more luminous than the Sun ($10^6 L_{Sun}$). Very few stars are this luminous. The least luminous stars have luminosities less than 1/10,000 that of the Sun ($10^{-4} L_{Sun}$). Most stars are less luminous than the Sun. The graph in **Figure 13.5** shows how many stars of a particular luminosity exist for every star like the Sun. For every star like the Sun, there are about 3 stars that are 1/100 as luminous as the Sun. For every star like the Sun, there is only roughly 0.1 star that is 90 times as luminous. It's awkward to think about 0.1 star, so we often turn this around, to say that for every star that is 90 times as luminous as the Sun, there are 10 stars like the Sun. (Distances for the nearest stars are obtained from their parallaxes; other methods—discussed later in this chapter—are used for the more distant stars.) Two everyday concepts—stereoscopic vision and the fact that objects appear brighter when closer—have provided the tools needed to measure the distance and luminosity of the closest stars. In Section 13.2, we explain how the laws of radiation described in Chapter 5 reveal still more about stars.

CHECK YOUR UNDERSTANDING **13.1**

Stars A and B appear equally bright, but star A is twice as far away from us as star B. Which of the following is true? (a) Star A is twice as luminous as star B. (b) Star A is 4 times as luminous as star B. (c) Star B is twice as luminous as star A. (d) Star B is 4 times as luminous as star A. (e) Star A and star B have the same luminosity because they have the same brightness.

Answers to Check Your Understanding questions are in the back of the book.

13.2 Astronomers Can Determine the Temperature, Size, and Composition of Stars

The gas in the outer layers of stars is dense enough that the radiation from a star comes close to obeying the same laws as the radiation from solid objects, such as the heating element on an electric stove. We can therefore use what you already learned about blackbody radiation (Chapter 5) to understand the radiation from stars. According to the Stefan-Boltzmann law, if two objects are the same size, then the hotter object is more luminous. And, according to Wien's law, the hotter object is also bluer. In this section, we explain how astronomers use those two laws to measure the temperatures and sizes of stars. We also provide a more detailed discussion of the absorption and emission of spectral lines mentioned in Chapter 5 to learn about the composition of stars.

Wien's Law Revisited: The Color and Surface Temperature of Stars

Wien's law relates an object's temperature to the peak wavelength of its spectrum (see Working It Out 5.3 for details); as the object's surface temperature increases, the light that it emits gets bluer. Stars with especially hot surfaces (tens of thousands of kelvins) are blue, stars with especially cool surfaces (a few thousand kelvins) are red, and the stars in between, such as the Sun (about 6000 K), are yellow-white. A more precise measurement of a star's spectrum can be used to find the wavelength at which the spectrum peaks, and then Wien's law can tell you the temperature of the star's surface. A star's color tells you about the temperature only at the surface because that layer is giving off most of the radiation that you see. (Stellar interiors are far hotter, as we discuss in Chapter 14.)

    Astronomers can determine the temperature of a star without a spectrum. They often measure the colors of stars by comparing the brightness at two specific wavelengths. The brightness is usually measured through an optical **filter**—sometimes just a piece of colored glass—that lets through only a small range of wavelengths. Two of the most common are a blue filter, which allows light with wavelengths of about 440 nanometers (nm) to pass through, and a "visual" (yellow-green) filter, which allows light with wavelengths of about 550 nm to pass through. The ratio of brightness between the blue and visual filters is called the **color index** of the star (see Appendix 7 for more details). From a pair of pictures of a group of stars, each taken through a different filter, astronomers can find an approximate value of the surface temperature of every star in the picture—perhaps hundreds or even thousands—all at once. Based on this type of analysis, most stars have surface temperatures lower than that of the Sun.

▶❚❚ **AstroTour:** Stellar Spectrum

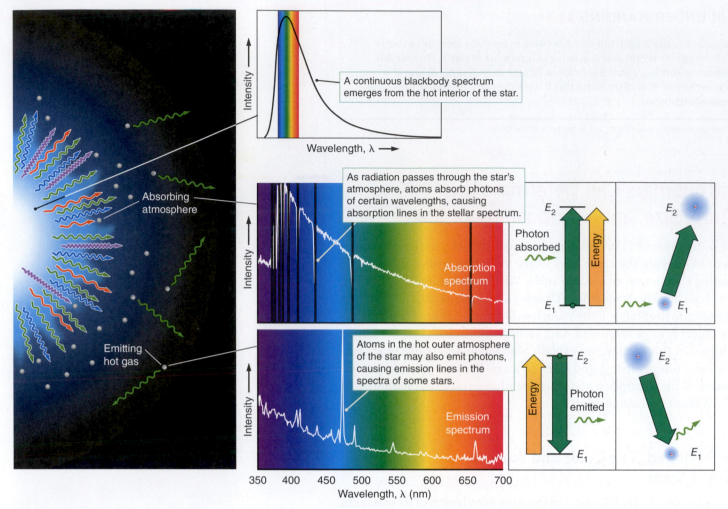

A continuous blackbody spectrum emerges from the hot interior of the star.

Wavelength, λ →

As radiation passes through the star's atmosphere, atoms absorb photons of certain wavelengths, causing absorption lines in the stellar spectrum.

Absorption spectrum

E_2 E_1 Photon absorbed Energy E_2 E_1

Atoms in the hot outer atmosphere of the star may also emit photons, causing emission lines in the spectra of some stars.

Emission spectrum

Energy E_2 E_1 Photon emitted E_2 E_1

Absorbing atmosphere

Emitting hot gas

350   400   450   500   550   600   650   700
Wavelength, λ (nm)

Figure 13.6 Absorption and emission lines both appear in the spectra of stars. The Planck blackbody spectrum is the light emitted from a hot object just because it is hot. As that light passes through a gas, some of it is absorbed, producing an absorption spectrum. Hot gas also emits light and produces emission lines in the spectra of some stars.

Classifying Stars by Surface Temperature

Although the hot "surface" of a star emits radiation with a spectrum very close to a smooth Planck blackbody curve, that light must then escape through the outer layers of the star's atmosphere. The atoms and molecules in the cooler layers of the star's atmosphere leave their absorption line fingerprints in the escaping light (**Figure 13.6**). Under some circumstances, the atoms and molecules in the star's atmosphere, along with any gas that might be found near the star, can also produce emission lines in stellar spectra. Absorption and emission lines complicate how astronomers use the laws of Planck blackbody radiation to interpret light from stars, but spectral lines provide a wealth of information about the state and composition of the gas in a star's atmosphere.

The spectra of stars were first classified during the late 1800s, long before stars, atoms, and radiation were well understood. Stars were classified by the appearance of the dark bands (now known to be absorption lines) seen in their spectra. The original ordering of that classification was arbitrarily based on the prominence of particular absorption lines known to be associated with the element hydrogen. Stars with the strongest hydrogen lines were denoted *A stars*, stars with weaker hydrogen lines were denoted *B stars*, and so on.

Annie Jump Cannon (1863–1941) led an effort at the Harvard College Observatory to examine and classify the spectra of hundreds of thousands of stars systematically. She dropped many of the earlier spectral types, keeping only seven that were later reordered on the basis of surface temperatures. Spectra of stars of several spectral types are shown in **Figure 13.7**. The hottest stars, with surface temperatures above 30,000 K, are denoted *O stars*. O stars have only weak absorption lines from hydrogen and helium. The coolest stars—*M stars*—have temperatures as low as about 2800 K. M stars have absorption lines from many types of atoms and molecules. The sequence of **spectral types** of stars, from hottest to coolest, is O, B, A, F, G, K, M. That sequence has undergone several modifications, most recently to add cooler objects known as brown dwarfs with spectral types L, T, and Y.

Astronomers divide the main spectral types into a finer sequence of subclasses by adding numbers to the letter designations. For example, the hottest B stars are B0 stars, slightly cooler B stars are B1 stars, and so on. The coolest B stars are B9 stars, which are only slightly hotter than A0 stars. The boundaries between spectral types are not always easy to determine. A hotter-than-average G star is very similar to a cooler-than-average F star. The Sun is a G2 star.

In Figure 13.7, notice that hot stars emit more blue light than cool stars, as described by Wien's law. Notice also that the absorption lines in their spectra are different. The temperature of the gas in the atmosphere of a star affects the state of the atoms in that gas, affecting in turn the energy level transitions available to absorb radiation (see Section 5.2 to review atomic energy levels). In O stars, the temperature is so high that most atoms have had one or more electrons stripped from them by energetic collisions within the hot gas. As a result, few transitions are available in the visible part of the electromagnetic spectrum, making the visible spectrum of an O star relatively featureless. At lower temperatures, more atoms are in energy states that can absorb light in the visible part of the spectrum, so the visible spectra of cooler stars are far more complex than the spectra of O stars.

All absorption lines have a temperature at which they are strongest. For example, absorption lines from hydrogen are most prominent at temperatures of about 10,000 K, the surface temperature of an A star. At the very lowest stellar temperatures, atoms in a star's atmosphere react with one another, forming molecules. Molecules such as titanium oxide (TiO) are responsible for much of the absorption in the atmospheres of cool M stars.

Because different spectral lines are formed at different temperatures, astronomers can use those absorption lines to measure a star's temperature. The surface temperatures of stars measured in that way agree extremely well with the surface temperatures of stars measured by using Wien's law, confirming once again that the physical laws that apply on Earth apply to stars as well.

The Composition of Stars

Most variations in the lines of a particular chemical element seen in stellar spectra are due to temperature, but the details of the absorption and emission lines found in starlight also reveal a star's composition.

Each type of atom has different energy levels and thus different spectral lines (Chapter 5). The spectral line patterns are measured in laboratories on Earth and then used to identify the atoms (or molecules) present in stars. For example, if a star

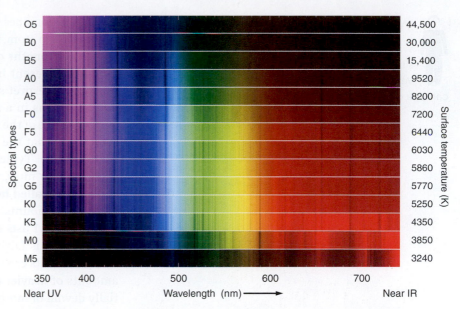

Figure 13.7 Visible light spectra of stars with different spectral types, ranging from hot, blue O stars to cool, red M stars. Hotter stars are more luminous at shorter wavelengths. The dark lines are absorption lines. These spectra have been adjusted to show only the brightest part of each spectrum. Comparing the O stars to the M stars makes clear that as the temperature decreases, the brightest part of the spectrum moves from the left (blue end of the spectrum) to the right (red end of the spectrum).

What if you observe two stars that have very similar peak wavelengths for their spectra but very different spectral line strengths? What can you conclude from your observations?

has absorption lines that correspond to the energy difference between two levels in the calcium atom, we know that calcium is present in the star's atmosphere.

The strengths of various absorption lines tell us not only what kinds of atoms are present in the gas but also the amount of each. However, we must carefully interpret spectra to account properly for the temperature and density of the gas in a star's atmosphere. Recall Figure 5.17—the "astronomer's periodic table of the elements." In 1925, Cecilia Payne-Gaposchkin (1900–1979) discovered that the composition of stars is very different from the composition of Earth. Typically, stars are composed of 90 percent hydrogen. Helium accounts for most of what remains, and all the other chemical elements, collectively called **heavy elements** or **massive elements**, are present in only very small amounts.

Table 13.1 lists the elements that make up the Sun's atmosphere. The Sun's composition is fairly typical for stars nearby, but the percentages of specific heavy elements can vary tremendously from star to star. Some stars have smaller amounts of heavier elements than the Sun. The existence of such stars, essentially devoid of more massive elements, provides important clues about the origin of chemical elements and the chemical evolution of the universe. Many of the atoms that make up Earth and its atmosphere (such as iron, silicon, nitrogen, oxygen, and carbon) exist as only a small percentage of the Sun.

By applying the physics of atoms and molecules to the study of stellar absorption lines, astronomers can accurately determine not only surface temperatures of stars but also pressures, chemical compositions, magnetic-field strengths, and other physical properties. In addition, by using the Doppler shift of emission and absorption lines, astronomers can measure rotation rates, atmospheric motions, expansion and contraction, "winds" driven away from stars, and other dynamic properties of stars. Stellar spectroscopy is one of the most powerful tools astronomers have for exploring the universe.

| Table 13.1 | Relative Amounts of Chemical Elements in the Atmosphere of the Sun |

Element	Percentage of Atoms in the Sun	Percentage of Sun's Mass
Hydrogen	92.5	74.5
Helium	7.4	23.7
Oxygen	0.064	0.82
Carbon	0.039	0.37
Neon	0.012	0.19
Nitrogen	0.008	0.09
Silicon	0.004	0.09
Magnesium	0.003	0.06
Iron	0.003	0.16
Sulfur	0.001	0.04
Total of others	0.001	0.03

The Stefan-Boltzmann Law and Finding the Sizes of Stars

Stars are so far away that most cannot be imaged as more than point sources. Finding a star's radius, then, involves putting together other observable measurements. The Stefan-Boltzmann law relates an object's temperature to the amount of light that it emits from every square meter of its surface. The "from every square meter" part is important: if two stars have the same temperature and are the same distance from us, the larger one will be brighter—namely, it emits more light because it has more surface area. Unfortunately, stars are rarely at the same distance from us, or at the same temperature, so typically a few more steps are necessary to figure out which star is larger.

First, astronomers find a star's temperature directly, either from its color through Wien's law (**Figure 13.8a**) or from the strength of its spectral lines. The temperature of a star's surface is one factor that influences its luminosity. Second, astronomers find the distance to the star, perhaps from parallax. Third, astronomers find the luminosity from its brightness (easily measured) and its distance. Once an astronomer has determined both the temperature and the luminosity of the star, she is able to calculate the star's radius using the *luminosity-temperature-radius relationship*, which connects the three quantities, as shown in **Working It Out 13.3**.

As the Stefan-Boltzmann law states, if a large star and a small star are the same temperature, they will emit the same energy from every patch of surface. However,

Wien's law:
Blue stars are hot;
red stars are cool.

a.

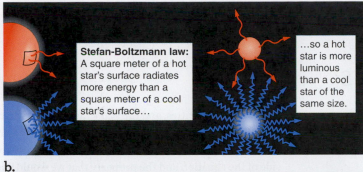

Stefan-Boltzmann law:
A square meter of a hot
star's surface radiates
more energy than a
square meter of a cool
star's surface…

…so a hot
star is more
luminous
than a cool
star of the
same size.

b.

Measure luminosity
from distance
and brightness.

Calculate
how big
the star
must be.

λ peak

Intensity

Measure temperature
from color.

c.   Wavelength

Figure 13.8 **a.** The temperature of a star can be found from its color using Wien's law. **b.** The luminosity depends on both the temperature and the size of the star. Blue stars are hotter; red stars are cooler. **c.** Once the temperature and the luminosity are known, the size of the star can be calculated.

the large star has more patches, so it will be more luminous altogether. Conversely, if two stars are the same size, but one is hotter than the other, the hot star will emit more light (**Figure 13.8b**). A small, hot star might even be more luminous than a larger cool star. Astronomers combine the temperature and the luminosity to find the star's radius (**Figure 13.8c**). A low-luminosity hot star must be small, whereas a high-luminosity cool star must be large.

Astronomers have used the luminosity-temperature-radius relationship to estimate the radii of many thousands of stars. The range of stellar sizes is smaller than the range of stellar luminosities: the radius of the largest star is 100,000 times larger than the radius of the smallest star. As with stellar luminosities, it's convenient to use the Sun as a unit of measure for size. The Sun's radius, designated R_{Sun}, is 696,000 km. One of the smallest types of stars, called white dwarfs, have radii only about 1 percent of the Sun's radius ($0.01\,R_{Sun}$)—about the size of Earth. The largest stars, called red supergiants, can have radii more than 1000 times that of the Sun ($1000\,R_{Sun}$). Many more stars are toward the small end of that range—smaller than the Sun—than giant stars.

CHECK YOUR UNDERSTANDING **13.2**

If star A has twice the surface temperature of the Sun but has the same luminosity, the radius of star A must be _____ the radius of the Sun. (a) 16 times; (b) 4 times; (c) 1/2; (d) 1/4

working it out 13.3

Estimating the Sizes of Stars

According to the Stefan-Boltzmann law (Chapter 5), the amount of energy radiated each second by each square meter of a star's surface is equal to the constant σ multiplied by the star's surface temperature raised to the fourth power:

$$\left(\begin{array}{c} \text{Energy radiated each} \\ \text{second by } 1\,\text{m}^2 \text{ of surface} \end{array} \right) = \sigma T^4$$

To find the total amount of light radiated each second by a star, we need to multiply the radiation emitted per second from each square meter by the number of square meters of the star's surface:

$$\left(\begin{array}{c} \text{Energy radiated} \\ \text{each second} \end{array} \right) = \left(\begin{array}{c} \text{Energy radiated each} \\ \text{second by } 1\,\text{m}^2 \text{of surface} \end{array} \right) \times \left(\begin{array}{c} \text{Surface} \\ \text{area} \end{array} \right)$$

The left-hand term in that equation—the total energy emitted by the star per second (in units of joules per second [J/s] = watts [W])—is the star's luminosity, L. The middle term—the energy radiated by each square meter of the star per second (in units of joules per square meter per second [J/m²/s])—can be replaced with the σT^4 factor from the Stefan-Boltzmann law. The remaining term—the number of square meters covering the star's surface—is the surface area of a sphere, $A_{\text{sphere}} = 4\pi R^2$ (in units of square meters [m²]), where R is the star's radius.

If we replace the words in the equation with the appropriate mathematical expressions for the Stefan-Boltzmann law and the area of a sphere, the equation for the luminosity becomes

$$L = \sigma T^4 \times 4\pi R^2$$

Combining gives

$$L = 4\pi R^2 \sigma T^4 \text{ J/s (W)}$$

That last equation is called the **luminosity-temperature-radius relationship** for stars. Because the constants (4, π, and σ) do not change, the star's luminosity is proportional only to $R^2 T^4$. Make a star 3 times as large, and its surface area becomes $3^2 = 9$ times as large. Because 9 times as much area is present to radiate, 9 times as much radiation is emitted. Make a star twice as hot, and each square meter of the star's surface radiates $2^4 = 16$ times as much energy. Thus, larger, hotter stars are more luminous than smaller, cooler stars.

Now, how large must a star of a given temperature be to have a total luminosity of L? The star's luminosity (L) and temperature (T) are measurable quantities, and the star's radius (R) is what we want to know. We can rearrange the previous equation, moving the properties that we know how to measure (temperature and luminosity) to the right-hand side of the equation and the property that we would like to know (the star's radius) to the left-hand side. After doing some algebra, we find that

$$R = \sqrt{\frac{L}{4\pi\sigma T^4}} = \frac{1}{T^2} \times \sqrt{\frac{L}{4\pi\sigma}}$$

Again, the right-hand side of the equation contains only things that we know or can measure. The constants 4, π, and σ are always the same. We can find L, the star's luminosity, from the measurements of the star's brightness and parallax (although only for nearby stars with known parallax). The star's surface temperature is T, which can be measured from its color. From the relationship of those measurements, we now know something new: the size of the star.

If we compare two stars, the constants all cancel out, leaving L, T, and R:

$$\frac{L_{\text{star}\,1}}{L_{\text{star}\,2}} = \frac{R^2_{\text{star}\,1}}{R^2_{\text{star}\,2}} \times \frac{T^4_{\text{star}\,1}}{T^4_{\text{star}\,2}}$$

Rearranging the terms, and taking the square root on both sides:

$$\frac{R_{\text{star}\,1}}{R_{\text{star}\,2}} = \sqrt{\frac{L_{\text{star}\,1}}{L_{\text{star}\,2}}} \times \frac{T^2_{\text{star}\,2}}{T^2_{\text{star}\,1}}$$

How does the Sun compare with the second-brightest star in the constellation Orion, a red star called Betelgeuse? From its spectrum, we know that Betelgeuse's surface temperature T is about 3500 K. Its distance is about 200 pc, and from that and its brightness, its luminosity is estimated to be 140,000 times that of the Sun. What can we say about the size of Betelgeuse? Using the preceding equation, we can determine the following:

$$\frac{R_{\text{Betelgeuse}}}{R_{\text{Sun}}} = \sqrt{\frac{L_{\text{Betelgeuse}}}{L_{\text{Sun}}}} \times \frac{T^2_{\text{Sun}}}{T^2_{\text{Betelgeuse}}}$$

$$\frac{R_{\text{Betelgeuse}}}{R_{\text{Sun}}} = \sqrt{\frac{140,000}{1}} \times \frac{5,800^2}{3,500^2} = \approx 370 \times 2.7 \approx 1,000$$

Thus, Betelgeuse has a radius about 1000 times larger than that of the Sun, making it a *supergiant*.

13.3 Measuring the Masses of Stars in Binary Systems

Determining a star's mass is difficult. Astronomers cannot use the amount of light from a star or the star's size as a measure of its mass, because stars can swell, contract, heat up, or cool down as they age. However, more massive stars *always* have stronger gravity. When astronomers are trying to determine the masses of astronomical objects, they almost always wind up looking for the effects of gravity. In Chapter 4, you learned that Kepler's laws of planetary motion are the result of gravity and that the properties of a planet's orbit can be used to measure the mass of the Sun. Similarly, astronomers can study two stars that orbit each other to determine their masses.

Many stars in our galaxy are in systems consisting of several stars moving about under the influence of their mutual gravity. Most of those systems are **binary stars**, in which two stars orbit each other in elliptical orbits with many of the same properties as the elliptical orbits described by Kepler's laws. In fact, Newton's version of Kepler's third law can be used to find the masses of the stars in a binary system, as we will demonstrate. However, most low-mass stars are single, and most stars are low-mass stars, so most stars are single and their mass cannot be calculated. We next explain how astronomers find the masses of stars in binary systems.

Binary Star Orbits

The **center of mass** is the balance point of a system. If two objects were sitting on a seesaw in a gravitational field, the support of the seesaw would have to be directly under the center of mass for the objects to balance (**Figure 13.9**). When Newton applied his laws of motion to the problem of orbits, he found that two objects must move in elliptical orbits around each other and that their common center of mass lies at one focus shared by both ellipses (**Figure 13.10**). The center of mass, which lies along the line between the two objects, remains stationary. The two objects will always be found on exactly opposite sides of the center of mass.

Imagine a binary star, as shown in **Figure 13.11a**. As seen from above, two stars orbit the common center of mass. Star 1, which is less massive, must complete its orbit in the same time as star 2, which is more massive. Because the less massive star has farther to go around the center of mass, it must be moving *faster* than the more massive star. If an astronomer can find the stars' velocities in their orbits, she can find each star's mass. However, in this face-on view, the Doppler shift (Chapter 5) cannot be measured because all the motion is in the plane of the sky, and none is toward or away from the observer.

When a system is edge-on to the observer, however, the astronomer can take advantage of the Doppler shift to find out about the motion. **Figure 13.11b** shows observations of the combined system's spectrum for each position in Figure 13.11a, if it were edge-on instead of face-on. Because the two stars are always on exactly opposite sides of their center of mass, they always move in opposite directions. Thus, when star 2 approaches, star 1 recedes. Because of the Doppler effect, the light coming from star 2 will be shifted to *shorter* wavelengths as it approaches, so the light will be *blue*shifted, and the light coming from star 1 will be shifted to *longer* wavelengths as it recedes, so the light will be *red*shifted. Half an orbital period later, the situation is reversed: lines from star 1 are blueshifted, and lines from star 2 are redshifted.

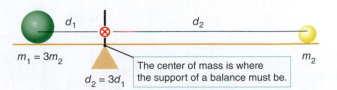

Figure 13.9 The center of mass of two objects is the "balance" point on a line joining the centers of two masses.

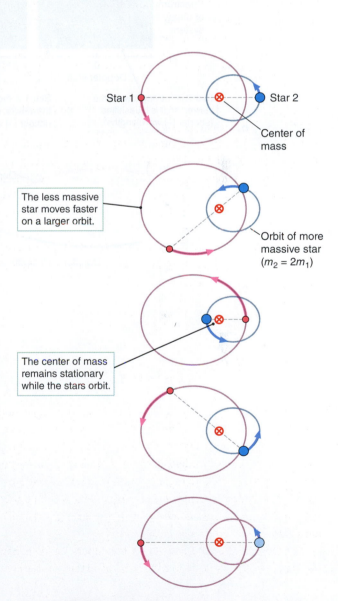

Figure 13.10 In a binary star system, the two stars orbit on elliptical paths about their common center of mass. Here, the blue star has twice the mass of the red one. The eccentricity of the orbits is 0.5. The time steps between the frames are equal.

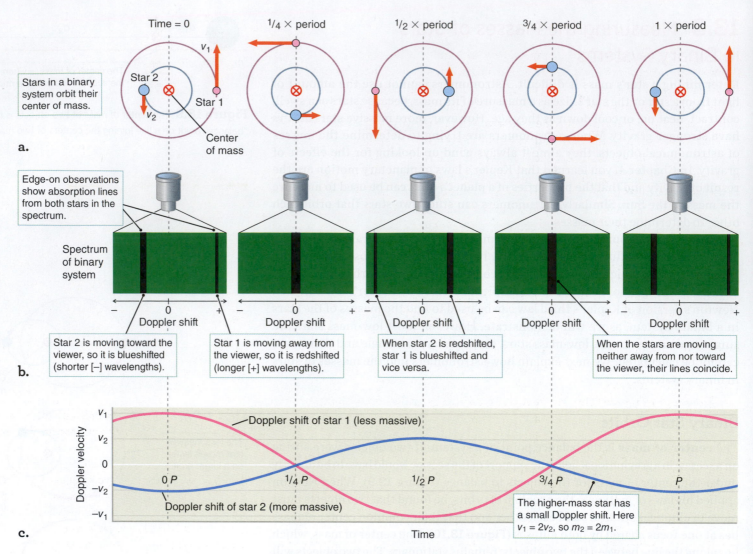

Figure 13.11 a. The view from "above" the binary system shows that both stars are on circular orbits around a common center of mass. **b.** The spectrum of the combined system (seen edge-on) shows that the spectral lines of each star shift back and forth. **c.** Graphing the Doppler shift of star 1 with star 2 versus time reveals that star 2 has half the maximum Doppler shift, so star 2 is twice as massive as star 1. *P* is the period of the orbit.

The less massive star, star 1, has a larger orbit, but both stars must orbit in the same amount of time in order to stay on opposite sides of the center of mass. Therefore, star 1 must move more quickly in order to travel around this larger orbit in the same amount of time that star 2 takes to complete a smaller orbit: there is an inverse proportion between velocity and mass. The velocities can be determined by finding the maximum value of the Doppler shift in a series of spectra taken throughout the orbit. **Figure 13.11c** compares the velocity obtained for star 1 with the velocity obtained for star 2. Because the velocity is inversely proportional to the mass, that comparison gives the ratio of the masses of the two stars:

$$\frac{v_1}{v_2} = \frac{m_2}{m_1}$$

By observing the system's spectrum, we can find from that equation the relative masses of the two stars—star 2 is 2 times as massive as star 1—but we can't find the actual mass of either star from those observations alone.

The Masses of Binary Stars

In Chapter 4, we ignored the complexity of the motion of two objects around their common center of mass, because one object was so much more massive than the other. Now, however, that very complexity enables us to measure the masses of the two stars in a binary system. If we can measure the binary system's period and the average separation between the two stars, Newton's version of Kepler's third law gives us the total mass in the system: the sum of the two masses. Because the analysis in the previous subsection gives us the ratio of the two masses, we now have two relationships between two unknowns. We have all we need to determine the mass of each star separately. In other words, if we know that star 2 is 2 times as massive as star 1, and we know that star 1 and star 2 together are 3 times as massive as the Sun, we can calculate separate values for the masses of star 1 and star 2.

Depending on the type of system, there are several ways to measure the average separation and the orbital period. In a **visual binary** system, because the system is close enough to Earth, and the stars are far enough from each other, we can take pictures that show the two stars separately (**Figure 13.12**). Then, astronomers can directly measure the shapes and periods of the two stars' orbits just by watching them as they orbit each other. Those values can be used with Doppler measurements of the stars' radial (line-of-sight) velocities to solve for the ratio of the two masses.

In most binary systems, however, the two stars are so close together and so far away from us that we cannot see the stars separately. The identification of those stars as binary systems is indirect and comes from observing periodic variations in the *light* from the star or from observing periodic changes in the star's *spectrum*. If we view a binary system nearly edge-on, so that one star passes in front of the other, it is called an **eclipsing binary**. An observer will see a repeating dip in brightness as one star passes in front of (eclipses) the other (**Figure 13.13**). If the stars have different temperatures, a repeating pattern will occur, showing a smaller dip in brightness when the hotter star eclipses the cooler one, followed by a larger dip in brightness when the cooler star eclipses the hotter one. The pattern of those dips also gives an estimate of the two stars' relative sizes (radii). That procedure for identifying binary systems—similar to the transit method for finding exoplanets discussed in Chapter 7—works only when the system is viewed nearly edge-on. The Kepler space telescope has discovered thousands of eclipsing binaries in addition to finding new exoplanets.

In a **spectroscopic binary** system, the two stars' spectral lines show periodic changes as they are Doppler-shifted away from each other, first in one direction and then in the other (Figure 13.11). The orbit's period is determined from the time it takes for a set of spectral lines to go from approaching to receding and back again. The stars' orbital velocities and the period of the orbit give the orbit's size because distance equals velocity multiplied by time. Consequently, astronomers can estimate the combined masses of the two stars. To calculate the individual masses, an estimate of the orbit's tilt is needed. Thus, spectroscopic binary masses are more approximate than those in eclipsing binary systems.

A binary system can fall into more than one of those three categories, regardless of how it was originally discovered. If a spectroscopic binary system is also a visual or eclipsing binary, the orbit and masses of the stars can be completely

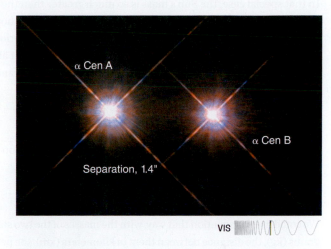

VIS

Figure 13.12 The two stars of this visual binary are resolved. These stars are two components of Alpha Centauri, the nearest star system to the Sun.

★ **WHAT AN ASTRONOMER SEES** When looking at images of stars, astronomers know that the brighter stars appear larger on the image. In this image, then, α Cen A is significantly brighter than α Cen B. Knowing that these two stars are part of a system of stars, an astronomer will further conclude that they are the same distance away. She will then know that α Cen A is not only brighter but also more luminous than α Cen B.

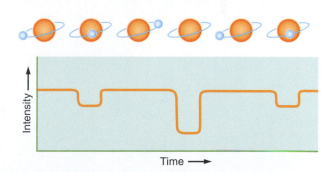

Figure 13.13 In an eclipsing binary system, the system is viewed nearly edge-on, so that the stars repeatedly pass behind each other, blocking some of the light. When the blue star passes in front of the larger, cooler star, less light is blocked than when the smaller, hotter blue star passes behind the red star. The shape of the dips in the light curve of an eclipsing binary can reveal information about the relative size and surface brightness of the two stars.

working it out 13.4

Measuring the Mass of an Eclipsing Binary Pair

In Working It Out 4.3, we used Newton's version of Kepler's third law to calculate the Sun's mass by observing the orbital period of one of its planets. In that special case, the Sun's mass is so much greater than the mass of the planet that the planet's mass is negligible. For two stars, however, both masses are significant, and we need to keep both in the equations. Newton showed that if two objects with masses m_1 and m_2 are in orbit about each other, the period of the orbit, P, is related to the average distance between the two masses, the semimajor axis A, by the equation

$$P^2 = \frac{4\pi^2 A^3}{G(m_1 + m_2)}$$

This equation can be rearranged into the following expression for the sum of the two objects' masses:

$$m_1 + m_2 = \frac{4\pi^2}{G} \times \frac{A^3}{P^2}$$

We could use the equation that way, with the masses of the two stars in kilograms (kg), the distance between them in kilometers (km), the period of their orbit in seconds (s), and the gravitational constant G in km³/(kg s²). However, astronomers often think about stellar masses in units of the Sun's mass. If we divide that equation by the "mass of the Sun" equation in Working It Out 4.3,

$$M_{\text{Sun}} = \frac{4\pi^2}{G} \times \frac{A^3}{P^2}$$

(where M_{Sun} = mass of the Sun, $A = 1$ AU, and $P = 1$ year), the constants cancel out and the equation simplifies to

$$\frac{m_1}{M_{\text{Sun}}} + \frac{m_2}{M_{\text{Sun}}} = \frac{A_{\text{AU}}^3}{P_{\text{years}}^2}$$

Therefore, if we know both m_1/m_2 from measuring velocities by Doppler shifts and $m_1 + m_2$ from the observed orbital properties, we can solve for the separate values of m_1 and m_2.

Suppose you are an astronomer studying a binary star system. After observing the star for several years, you accumulate the following information about the system:

1. The star is an eclipsing binary.
2. The period of the orbit is 2.63 years.
3. Star 1 has a Doppler velocity that varies between $+20.4$ and -20.4 km/s.
4. Star 2 has a Doppler velocity that varies between $+6.8$ and -6.8 km/s.
5. The stars are in circular orbits. You know that because the Doppler velocities about the star are symmetric—that is, for each star the approach and recession speeds are equal.

Those data are summarized in **Figure 13.14**. The star is an eclipsing binary, so the star's orbit is edge-on to your line of sight. The Doppler velocities tell you the total orbital velocity of each star, and you determine the size of the orbits by using the relationship

$$\text{Distance} = \text{Speed} \times \text{Time}$$

In one orbital period, star 1 travels around a circle—a distance of

$$d = (20.4 \text{ km/s}) \times (2.63 \text{ yr}) = 53.7 \text{ km} \times \text{yr/s}$$

solved (**Working It Out 13.4**). Historically, most stellar masses were measured for stars in eclipsing binary systems rather than for those in visual or spectroscopic binaries. New observational capabilities, however, have increased the number of known visual binaries by greatly improving the ability to directly see the stars in a binary. Accurate mass measurements have been obtained for several hundred binary stars, about half of which are eclipsing binaries. The range of stellar masses found in that way is not nearly as great as the range of stellar luminosities. The least massive stars have masses of about $0.08\ M_{\text{Sun}}$, whereas the most massive stars appear to have masses up to about $200\ M_{\text{Sun}}$.

CHECK YOUR UNDERSTANDING 13.3

Which of the following properties must be measured to determine the masses of stars in a typical binary system? (Choose all that apply.) (a) the period of the two stars' orbits; (b) the average separation between the two stars; (c) the two stars' radii; (d) the two stars' velocities.

Multiply by the number of seconds in a year:

$$d = 53.7 \frac{\text{km} \times \text{yr}}{\text{s}} \times \frac{3.16 \times 10^7 \text{s}}{\text{yr}} = 1.70 \times 10^9 \text{km}$$

That distance is the circumference of the star's orbit, or 2π times the radius of the star's orbit, A_1. Thus, star 1 is following an orbit with a radius of

$$A_1 = \frac{d}{2\pi} = \frac{1.70 \times 10^9 \text{km}}{2\pi} = 2.7 \times 10^8 \text{km}$$

To convert that result to astronomical units, use the relation $1\,\text{AU} = 1.50 \times 10^8 \text{km}$:

$$A_1 = 2.7 \times 10^8 \text{km} \times \frac{1\,\text{AU}}{1.50 \times 10^8 \text{km}} = 1.8\,\text{AU}$$

A similar analysis of star 2 shows that its orbit has a radius of $A_2 = 0.6\,\text{AU}$.

Next, apply Newton's version of Kepler's third law. Because the stars are always on opposite sides of the center of mass, the semimajor axis $A_{\text{AU}} = 1.8\,\text{AU} + 0.6\,\text{AU} = 2.4\,\text{AU}$. Because you know A and the period P (measured as 2.63 years), you can calculate the total mass of the two stars:

$$\frac{m_1}{M_{\text{Sun}}} + \frac{m_2}{M_{\text{Sun}}} = \frac{(A_{\text{AU}})^3}{(P_{\text{years}})^2} = \frac{(2.4)^3}{(2.63)^2} = 2.0$$

Thus, the combined mass of the two stars is twice the mass of the Sun. To sort out the stars' individual masses, use the measured velocities and the fact that the mass and velocity are inversely proportional:

$$\frac{m_2}{m_1} = \frac{v_1}{v_2} = \frac{20.4\,\text{km/s}}{6.8\,\text{km/s}} = 3.0$$

Star 2 is 3 times as massive as star 1. In mathematical terms, $m_2 = 3 \times m_1$. Substituting into the equation

$$m_1 + m_2 = 2.0\,M_{\text{Sun}}$$

gives

$$m_1 + 3m_1 = 2.0\,M_{\text{Sun}}$$

or $4m_1 = 2.0\,M_{\text{Sun}}$, so $m_1 = 0.5\,M_{\text{Sun}}$. Because $m_2 = 3 \times m_1$, $m_2 = 1.5\,M_{\text{Sun}}$. Star 1 has a mass of $0.5\,M_{\text{Sun}}$, and star 2 has a mass of $1.5\,M_{\text{Sun}}$. You have just found the masses of two distant stars.

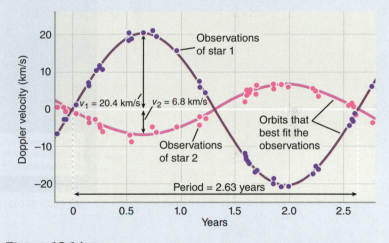

Figure 13.14 Doppler velocities of the stars in an eclipsing binary are used to measure the masses of the stars.

13.4 The Hertzsprung-Russell Diagram Is the Key to Understanding Stars

Measuring basic properties of stars is only the first step in understanding them; the next step is to look for patterns in their properties. In the early 20th century, Ejnar Hertzsprung (1873–1967) and Henry Norris Russell (1877–1957) independently studied the properties of stars. Both astronomers plotted stars' luminosities versus their surface temperatures—a diagram that came to be known as the *Hertzsprung-Russell diagram*, or simply the **H-R diagram**. We use H-R diagrams often to study stars. In this section, we provide a first look at that important diagram and how stars are organized within it.

The H-R Diagram

The H-R diagram (**Figure 13.15**) is a graph of luminosity versus temperature that can be used to study how stars change over time. The spectral type is plotted on

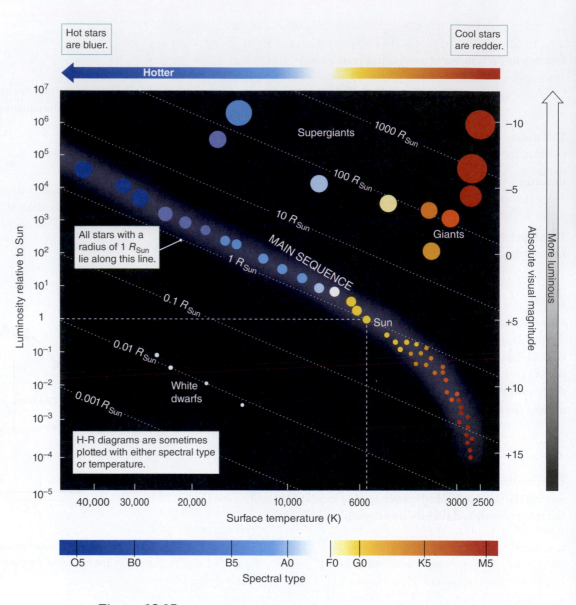

Figure 13.15 The Hertzsprung-Russell (H-R) diagram is used to plot the properties of stars. More luminous stars are at the top of the diagram. Hotter stars are on the left. Stars of the same radius (*R*) lie along the dotted lines moving from upper left to lower right. Absolute magnitudes are discussed in Working It Out 13.2 and Appendix 7.

▶‖ **AstroTour:** H-R Diagram

▶▶ **Interactive Simulation:** H-R Diagram

the horizontal axis (the *x*-axis), along with the surface temperature plotted backward: temperature is higher on the left and lower on the right. Hot, blue stars are on the left side of the H-R diagram, whereas cool, red stars are on the right. Temperature is plotted logarithmically. For example, consider two points on the H-R diagram. One represents a star with a surface temperature of 40,000 K, and the other a star with a surface temperature of 20,000 K—a temperature change of a factor of 2. The interval along the axis between those points is the same size as the interval between points representing stars with surface temperatures of 10,000 and 5000 K—also a temperature change of a factor of 2. The horizontal axis is sometimes labeled with another characteristic that corresponds to temperature, such as the color.

Stars' luminosities are plotted along the vertical axis (the *y*-axis). More luminous stars are toward the top of the diagram; less luminous stars are toward the bottom. Sometimes the luminosity axis is labeled with the absolute visual magnitude instead of luminosity, as shown on the right-hand y-axis. As with the temperature axis, luminosities are plotted logarithmically. Here, each step along the left-hand y-axis corresponds to a multiplicative factor of 10 in the luminosity. To understand why the plotting is done that way, recall that the most luminous stars are 10 billion times more luminous than the least luminous stars, yet all those stars must fit on the same plot.

Each point on the H-R diagram is specified by a surface temperature and luminosity. Therefore, we can use the luminosity-temperature-radius relationship described earlier in the chapter to find a star's radius at that point as well. A star in the upper right corner of the H-R diagram is very cool, so each square meter of its surface radiates only a small amount of energy. That star is extremely luminous, too. It must be huge, then, to account for its high luminosity, despite the feeble radiation coming from each square meter of its surface. Conversely, a star in the lower left corner of the H-R diagram is very hot, which means that a large amount of energy is coming from each square meter of its surface. That star has a very low overall luminosity, however, so it must be very small. Moving up and to the right takes you to larger and larger stars; moving down and to the left takes you to smaller and smaller stars. All stars of the same radius lie along slanted lines across the H-R diagram. Astronomers can note a star's properties—its temperature, color, size, and luminosity—from a glance at its position on the H-R diagram. The discovery and study of those patterns led to an understanding of the astrophysics of stars (see the **Process of Science Figure**).

The Main Sequence

Figure 13.16 shows 4 million stars plotted on an H-R diagram. The data are based on observations of stars within 5000 light-years of the Sun, which makes them near enough for the *Gaia* satellite to obtain parallax measurements. A quick look at the diagram immediately shows what was first discovered in the original diagrams of Hertzsprung and Russell. About 90 percent of the stars in the sky lie in a well-defined region running across the H-R diagram from lower right to upper left, known as the **main sequence**. On the left end of the main sequence are the O stars: hotter, larger, and more luminous than the Sun. On the right end of the main sequence are the M stars: cooler, smaller, and fainter than the Sun. If you know where a star lies on the main sequence, you know its approximate luminosity, surface temperature, and size.

The H-R diagram offers a useful method for finding the distance to main-sequence stars. Astronomers can determine whether a star is on the main sequence by looking at the absorption lines in its spectrum. The spectral type, too, is determined from the spectral lines, and that spectral type indicates the star's temperature. Once that value on the x-axis is known, we can then read up to the main sequence and then across to the y-axis to find the star's luminosity. Recall that the luminosity, brightness, and distance are all connected, so we can find the star's distance by comparing its luminosity, obtained from the H-R diagram, with its apparent brightness. That method of determining distances to main-sequence stars from the spectra, luminosity, and brightness of stars is called **spectroscopic parallax**. Details of that method are discussed in Appendix 7. Despite the similarity between the names, the method is very different from the parallax method discussed in Section 13.1. Spectroscopic parallax is useful to much larger distances than the geometric method, although it is less precise.

what if . . .

What if Hertzsprung and Russell had made their diagram a plot of stellar radius versus temperature? How would the main sequence appear in such a diagram?

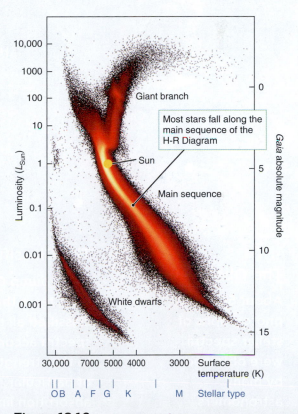

Figure 13.16 An H-R diagram for 4 million stars within 5000 light-years of the Sun plotted from data obtained by the *Gaia* satellite clearly shows the main sequence. Most of the stars lie along this band running from the lower right of the diagram toward the upper left. Credit: ESA/Gaia/DPAC, CC BY-SA 3.0 IGO, https://www.esa.int/ESA _Multimedia/Images/2018/04/Gaia_s_Hertzsprung-Russell_diagram. https://creativecommons.org/licenses/by-sa/3.0/igo/.

Science Is Collaborative

It took decades, and the contributions of dozens of people, all working toward a common goal, to understand the meaning behind stellar data.

Henry Norris Russell

Cecilia Payne-Gaposchkin

Ejnar Hertzsprung

Meghnad Saha

Annie Jump Cannon

1800s
The Observations

About 500,000 photographs of stellar spectra were obtained by many astronomers at many telescopes.

1900s
The Classification

Annie Jump Cannon led a team that classified all the spectra according to the strengths of particular absorption lines at particular wavelengths.

1910s
The Graph

Hertzsprung and Russell independently developed what will later be called the H-R diagram. They did not understand why the *x*-axis, ordered O-B-A-F-G-K-M, revealed a nice band across the middle of the diagram. Russell hypothesized this result must be caused by a single stellar characteristic.

1920s
The Understanding

Meghnad Saha showed that the stellar characteristic in question was temperature. Cecilia Payne-Gaposchkin showed that stars were mostly composed of hydrogen and helium. Modern astrophysics was born; others went on to develop the understanding of stellar atmospheres.

Scientific discoveries sometimes seem to occur suddenly. Instead, new scientific knowledge usually results from the effort of many people working for many years to solve a problem.

From a combination of observations of binary star masses, parallax, luminosity measurements, and mathematical models, astronomers have determined that stars of different masses lie on different parts of the main sequence. Stellar mass increases smoothly from the lower right to the upper left along the main sequence (**Figure 13.17**). If a main-sequence star is *less* massive than the Sun, it is also smaller, cooler, redder, and less luminous than the Sun and is located to the lower right of the Sun on the main sequence. Conversely, if a main-sequence star is *more* massive than the Sun, it is also larger, hotter, bluer, and more luminous than the Sun and is located to the upper left of the Sun on the main sequence. The mass of a star determines where on the main sequence the star will lie.

Table 13.2 summarizes the properties of the spectral classes of main-sequence stars. *All* main-sequence stars with a mass of 1 M_{Sun} are G2 stars like the Sun and have the *same* surface temperature, size, and luminosity as the Sun. Similarly, if a main-sequence star is classified as B0, it has a surface temperature of about 30,000 K, a luminosity about 32,500 times that of the Sun, a mass of about 17.5 M_{Sun}, and a radius of about 6.7 R_{Sun}. If a different main-sequence star is classified as M5, then it has a surface temperature of 3170 K, a luminosity of about 0.008 L_{Sun}, a mass of about 0.21 M_{Sun}, and a radius of about 0.29 R_{Sun}.

The relationship between the mass and the luminosity of stars is very sensitive. Relatively small differences in the masses of stars result in large differences in their main-sequence luminosities. From determining the luminosities of binary stars with measured mass, a relationship between the mass and luminosity emerged. That **mass-luminosity relationship**, usually expressed as

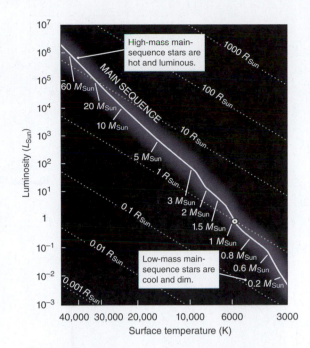

Figure 13.17 Mass determines the location of a star along the main sequence.

Table 13.2	Properties of Main-Sequence Stars			
Spectral Type	Temperature (K)	Mass (M_{Sun})	Radius (R_{Sun})	Luminosity (L_{Sun})
O5	42,000	60	13	500,000
B0	30,000	17.5	6.7	32,500
B5	15,200	5.9	3.2	480
A0	9800	2.9	2.0	39
A5	8200	2.0	1.8	12.3
F0	7300	1.6	1.4	5.2
F5	6650	1.4	1.2	2.6
G0	5940	1.05	1.06	1.25
G2 (Sun)	5780	1.00	1.00	1.0
G5	5560	0.92	0.93	0.8
K0	5150	0.79	0.93	0.55
K5	4410	0.67	0.80	0.32
M0	3840	0.51	0.63	0.08
M5	3170	0.21	0.29	0.008

unanswered questions

What is the *upper* limit for stellar mass? Both theory and observation have shown that the *lower* limit for stellar mass is approximately 0.08 M_{Sun} with a temperature of about 2000 K. However, neither theory nor observation has yielded a definitive value for the upper limit. Many astronomers believe that the upper limit lies somewhere around 150–200 M_{Sun}. The very first stars that formed in the universe might have been even larger.

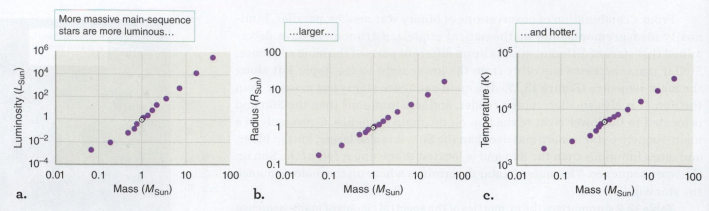

Figure 13.18 These graphs plot **a.** luminosity, **b.** radius, and **c.** temperature versus mass for stars along the main sequence. The mass of a main-sequence star determines its other properties.

$L \propto M^{3.5}$, is shown in **Figure 13.18a**. The exact exponent varies from about 3 to 4 for different ranges of stellar masses, but the method is useful for estimating masses of single stars. The mass also correlates to the size of a star (**Figure 13.18b**) and to its temperature (**Figure 13.18c**).

The mass and chemical composition of a main-sequence star determine how large it is, what its surface temperature is, how luminous it is, what its internal structure is, how long it will live, how it will evolve, and what its final fate will be. A star must have a balance between gravity trying to hold the star together and the energy released by nuclear reactions within the star trying to blow it apart. A star's mass determines the strength of its gravity, which in turn determines how much energy must be generated in its interior to prevent it from collapsing under its own weight. The mass of a star determines where that balance is struck.

Stars *Not* on the Main Sequence

Although 90 percent of stars are main-sequence stars, some stars are found in the upper right portion of the H-R diagram, well above the main sequence (see Figure 13.15). Those stars, which must be luminous, cool, and large, with radii hundreds or thousands of times the radius of the Sun, are called giants or supergiants. At the other extreme are stars found in the far lower left corner of the H-R diagram. Those stars are the tiny white dwarfs, comparable to the size of Earth. Their small surface areas explain why they have such low luminosities, despite having high temperatures.

Stars that lie off the main sequence on the H-R diagram can be identified by their luminosities (determined by their distance) or by slight differences in their spectral lines. The width of a star's spectral lines indicates the density and surface pressure of gas in the star's atmosphere. In general, denser stars have broader lines. Puffed-up stars above the main sequence have lower densities and lower surface pressure and narrower absorption lines than main-sequence stars.

When using the H-R diagram to estimate the distance to a star by the spectroscopic parallax method, astronomers must know whether the star is on, above, or below the main sequence to find the star's luminosity. The spectral

unanswered questions

Are planets with life likely to be orbiting around main-sequence stars of type M? Those low-luminosity stars are the most common type in the Milky Way, and the Kepler telescope has detected many planets orbiting such stars. However, the habitable zone of an M star is very close to the star, so that the planet may be strongly affected by streams of charged particles blowing off the star. Unless the planet has a strong protective magnetic field, that radiation may decrease or completely strip a planet of its atmosphere. Another complication is that a close-in planet is likely to be tidally locked to the star, so that one hemisphere of the planet receives light and the other hemisphere is permanently dark. That imbalance of light and heat might make the planet uninhabitable.

Table 13.3	**Taking the Measure of Stars**

Property	Methods
Luminosity	• For a star with a known distance, measure the brightness and then apply the inverse square law of radiation: $$\text{Luminosity} = 4\pi \times \text{Distance}^2 \times \text{Brightness}$$ • For a star *without* a known distance, take a spectrum of the star to determine its spectral and luminosity classes, plot them on an H-R diagram, and read the luminosity from the diagram.
Temperature	• Measure the star's color index by using blue and visual filters. Use Wien's law to relate the color to a temperature. • Take a spectrum of the star, and estimate the temperature from its spectral class by noting which spectral lines are present.
Distance	• For a relatively nearby star (within a few hundred parsecs), measure the star's parallax shift over the year. • For a more distant star, use the spectroscopic parallax method to find the luminosity from the H-R diagram. Then determine the distance from the luminosity and brightness.
Size	• For a few of the largest and closest stars, measure the size directly or by the length of the eclipse in eclipsing binary stars. • From the width of the star's spectral lines, estimate the luminosity class (supergiant, giant, or main sequence) for a star of given temperature. • For a star with known luminosity and temperature, use the Stefan-Boltzmann law to calculate the star's radius (the luminosity-temperature-radius relationship).
Mass	• Measure the motions of the stars in a binary system, use those to determine the stars' orbits, and then apply Newton's form of Kepler's third law. • For a non-binary star, use the mass-luminosity relationship to estimate the mass from the luminosity.
Composition	• Analyze the lines in the star's spectrum to measure chemical composition.

line widths of stars both on and off the main sequence indicate **luminosity class**, which tells us the star's relative *size* within each spectral class. Supergiant stars, the largest stars that we see, are luminosity class I, bright giants are class II, giants are class III, subgiants are class IV, main-sequence stars are class V, and white dwarfs are class WD. Luminosity classes I–IV lie above the main sequence, whereas class WD falls below and to the left of the main sequence (**Figure 13.19**). Thus, the complete spectral classification of a star includes both its spectral type (indicating temperature and color) and its luminosity class (indicating relative size).

The existence of the main sequence, together with the fact that the mass of a main-sequence star determines where on the sequence it will lie, creates a grand pattern that makes it possible to understand stars in fundamental ways. The existence of stars that do *not* follow that pattern raises yet more questions. In the coming chapters, we show that the main sequence tells us what stars are and how they work, and that stars off the main sequence reveal how stars form, evolve, and die. **Table 13.3** summarizes the techniques that astronomers use to determine some of the basic properties of stars. Of the properties listed in the table, only temperature, distance, and composition can be *measured*. Luminosity must be *inferred* from the H-R diagram or calculated from distance and brightness, and

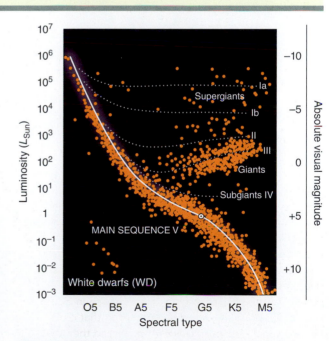

Figure 13.19 Stellar luminosity classes indicate the size (radius) of a star at each spectral type.

size and mass must be *calculated*. Other measurable properties include brightness, color, spectral type, and parallax shift.

CHECK YOUR UNDERSTANDING 13.4

Choose the two qualities that describe a star located in the lower right of the H-R diagram: (a) hot; (b) cool; (c) high luminosity; (d) low luminosity.

Origins: Habitable Zones

How might a star's basic property, such as its luminosity, color, mass, or surface temperature, affect the chance of a planet with life being in orbit around that star? The only known life is that on planet Earth, where liquid water was essential for life to form and evolve. Whether liquid water is an absolute requirement for life elsewhere is not known, but the presence of water is a good starting point for determining where to look. So astronomers look for planets at the right distance from their stars to have a planetary temperature that permits water to exist in a liquid state on their surfaces—a range of distances known as the **habitable zone**. On planets whose orbits are closer to their star than the habitable zone, water would exist only as a vapor—if at all. On planets that have orbits beyond the habitable zone, water would be permanently frozen as ice.

An important factor for estimating the temperature of a planet is the brightness of the sunlight that falls on that planet (see the Chapter 5 "Origins" section and Working It Out 5.4). That factor depends on the star's luminosity and the planet's distance from the star. In the Solar System, the habitable zone ranges from ~0.9 to ~1.4 AU, which includes Earth but just misses Venus and Mars. Main-sequence stars less luminous than the Sun are cooler and have narrower habitable zones, minimizing the chance that a planet will form within that slender zone. Main-sequence stars more massive than the Sun are hotter and have larger habitable zones. **Figure 13.20** illustrates those zones around Sun-like, hotter, and cooler stars.

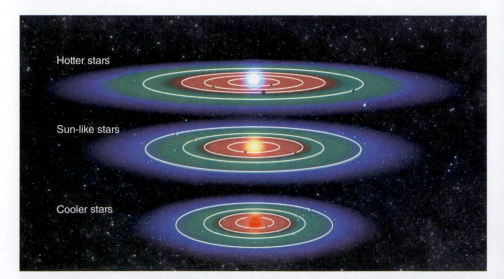

Figure 13.20 The distance and extent of a habitable zone (green) surrounding a star depends on the star's temperature. Regions too close to the star are too hot (red) and those too far away are too cold (blue) to permit the existence of liquid water. The orbits of Mercury, Venus, Earth, and Mars have been drawn around these stars for scale.

Astronomers are now finding planets in the habitable zones of their respective stars. Methods of planet detection, as discussed in Chapter 7, work best when the planet is close to its star. As of this writing, the Kepler Mission has identified and confirmed a few dozen planets in habitable zones by using the transit method and has found more candidate planets that need to be confirmed.

The distance from a star at a certain temperature is not the only consideration for whether a planet has water. The presence of a planetary atmosphere also is a factor. More massive planets can retain their atmospheres, which can trap heat and raise the planet's temperature, as we saw in Chapter 5 for Venus and Earth. Smaller planets have a lower gravitational pull and may not be able to retain an atmosphere. In addition, some habitable zones may be heated by planets, not stars. Some of the giant planets in the cold, outer part of our own Solar System have moons with liquid water (Chapters 11 and 24). The heat keeping the water liquid is due to the gravity of the nearby planet, not from the Sun.

Finally, *habitable* does not mean *inhabited*; it means only that the planet is at the right distance from its star that it could have liquid water. Identifying planets in their habitable zone is a first step to selecting which planets are most interesting for further study.

SUMMARY

Finding the distances to stars is a difficult but important task for astronomers. Parallax and spectroscopic parallax are two methods that astronomers use to determine distances to stars. Brightness and distance can be used to obtain the luminosity. Careful study of the light from a star, including its spectral lines, gives the temperature, size, and composition of the star. The study of binary systems gives the masses of stars of various spectral types, which we can extend to all stars of the same spectral type. The H-R diagram shows the relationship among the various physical properties of stars. The mass of a star is the major determining factor in its evolution. The habitable zone is the distance from a star in which a planet could have the right temperature for liquid water to exist on its surface. Stars of different luminosities and temperatures have habitable zones of different widths at different distances from the star.

① **Demonstrate how astronomers use parallax to determine the distances to stars, and explain how astronomers combine these distances with the brightness of nearby stars to determine how luminous the stars are.** The distance to a nearby star is measured by finding its parallax—by measuring how the star's apparent position changes in the sky over a year. The nearest star (other than the Sun) is about 4.2 light-years (1.3 parsecs) away. The brightness of a star in the sky can be measured directly, and brightness and distance can be used to obtain the star's luminosity—how much light the star emits.

② **Explain how astronomers obtain the temperatures, sizes, and composition of stars.** The color of a star depends on its temperature; blue stars are hotter, whereas red stars are cooler. The radius can be computed from the temperature and luminosity of the star. Small, cool stars greatly outnumber large, hot stars. Spectral lines carry a great deal of information about a star, including what chemical elements and molecules are present in the star.

③ **Outline how astronomers estimate the masses of stars.** The motions of stars are observed in a binary system. Newton's universal law of gravitation and Kepler's laws connect the stars' motion to the forces they experience. Comparing the accelerations of the stars to the forces that cause them yields the masses of the stars in the system.

④ **Categorize stars and organize that information on a Hertzsprung-Russell (H-R) diagram.** The H-R diagram shows the relationship among the various physical properties of stars. Temperature increases to the left, so that hotter stars lie on the left side of the diagram, whereas cooler stars lie on the right. Luminosity increases vertically, so that the most luminous stars lie near the top of the diagram. A star's luminosity class and temperature indicate its size. The mass and composition of a main-sequence star determine its luminosity, temperature, and size. Ninety percent of stars lie along the main sequence.

⑤ **Determine the luminosity, temperature, and size of a star from its mass and composition.** The mass and composition of a main-sequence star determine its position on the H-R diagram. The main sequence on the H-R diagram is actually a sequence of masses. That position connects its other properties such as its luminosity, temperature, and size.

QUESTIONS AND PROBLEMS

TEST YOUR UNDERSTANDING

1. Star A and star B are nearly the same distance from Earth. If star A is half as bright as star B, which of the following statements must be true?
   a. Star B is farther away than star A.
   b. Star B is twice as luminous as star A.
   c. Star B is hotter than star A.
   d. Star B is larger than star A.

2. Star A and star B are two nearby stars. If star A is blue and star B is red, which of the following statements must be true?
   a. Star A is hotter than star B.
   b. Star A is cooler than star B.
   c. Star A is farther away than star B.
   d. Star A is more luminous than star B.

3. Star A and star B are two stars nearly the same distance from Earth. If star A is blue and star B is red, but they have equal brightness, which of the following statements is true?
   a. Star A is more luminous than star B.
   b. Star A is larger than star B.
   c. Star A is smaller than star B.
   d. Star A is less luminous than star B.

4. What does it most likely mean when a star has very weak hydrogen lines and is blue?
   a. The star is too hot for hydrogen lines to form.
   b. The star has no hydrogen.
   c. The star is too cold for hydrogen lines to form.
   d. The star is moving too fast to measure the lines.

5. Star A and star B are a binary system. If the Doppler shift of star A's absorption lines is 3 times the Doppler shift of star B's absorption lines, which of the following statements is true?
   a. Star A is 3 times as massive as star B.
   b. Star A is one-third as massive as star B.
   c. Star A is closer than star B.
   d. The binary pair is moving toward Earth, but star A is farther away.

6. Star A and star B are two red stars at nearly the same distance from Earth. If star A is many times brighter than star B, which of the following statements is true?
   a. Star A is a main-sequence star, whereas star B is a red giant.
   b. Star A is a red giant, whereas star B is a main-sequence star.
   c. Star A is hotter than star B.
   d. Star A is a white dwarf, whereas star B is a red giant.

7. Star A and star B are two blue stars at nearly the same distance from Earth. If star A is many times brighter than star B, which of the following statements is true?
   a. Star A is a main-sequence star, whereas star B is a red giant.
   b. Star A is a main-sequence star, whereas star B is a blue giant.
   c. Star A is a white dwarf, whereas star B is a blue giant.
   d. Star A is a blue giant, whereas star B is a white dwarf.

8. In which region of an H-R diagram would you find the main-sequence stars with the widest habitable zones?
   a. upper left
   b. upper right
   c. center
   d. lower left
   e. lower right

9. If star A is more massive than star B, and both are main-sequence stars, star A is _____ than star B. (Choose all that apply.)
   a. more luminous
   b. less luminous
   c. hotter
   d. colder
   e. larger
   f. smaller

10. A telescope on Mars could measure the distances to more stars than can be measured from Earth because
    a. the resolution of the telescope would be better.
    b. Mars has a thin atmosphere.
    c. it would be closer to the stars.
    d. the parallax "baseline" would be longer.

11. Star A and star B are two nearby stars. If star A has a parallax angle 4 times as large as star B's, which of the following statements is true?
    a. Star A is one-quarter as far away as star B.
    b. Star A is 4 times as far away as star B.
    c. Star A has moved through space one-quarter as far as star B.
    d. Star A has moved through space 4 times as far as star B.

12. If star A appears twice as bright as star B but is also twice as far away, star A is _____ as luminous as star B.
    a. 8 times   c. twice
    b. 4 times   d. half

13. In Table 13.1, the percentage of hydrogen in the Sun decreases when changing from percentage by number of atoms to percentage by mass, but the percentage of helium increases. Why?
    a. Hydrogen is more massive than helium.
    b. Helium is more massive than hydrogen.
    c. Hydrogen is located in a different part of the Sun.
    d. The mass of hydrogen is hard to measure.

14. Capella (in the constellation Auriga) is the sixth-brightest star in the sky. When viewed with a high-powered telescope, it turns out that Capella is actually two pairs of binary stars: the first pair are G-type giants, whereas the second pair are M-type main-sequence stars. What color does Capella appear to be?
    a. red
    b. yellow
    c. blue
    d. The color cannot be determined from the given information.

15. An eclipsing binary system has a primary eclipse (star A is eclipsed by star B) that is deeper (more light is removed from the light curve) than the secondary eclipse (star B is eclipsed by star A). What does that information tell you about stars A and B?
   a. Star A is hotter than star B.
   b. Star B is hotter than star A.
   c. Star B is larger than star A.
   d. Star B is moving faster than star A.

THINKING ABOUT THE CONCEPTS

16. The distances of nearby stars are determined by their parallaxes. Why are the distances of stars farther from Earth more uncertain?

17. To know certain properties of a star, you must first determine the star's distance. For other properties, knowing its distance is unnecessary. Explain why an astronomer does or does not need to know a star's distance to determine each of the following properties: size, mass, temperature, color, spectral type, and chemical composition.

18. Albireo, in the constellation Cygnus, is a visual binary system whose two components can easily be seen with even a small, amateur telescope. Viewers describe the brighter star as "golden" and the fainter one as "sapphire blue."
   a. What does that description tell you about the relative temperatures of the two stars?
   b. What does that description tell you about their respective sizes?

19. Very cool stars have temperatures around 2500 K and emit Planck spectra with peak wavelengths in the red part of the spectrum. Do those stars emit any blue light? Explain your answer.

20. The stars Betelgeuse and Rigel are both in the constellation Orion. Betelgeuse appears red, whereas Rigel is bluish white. To the eye, the two stars seem equally bright. If you can compare the temperature, luminosity, or size from just that information, do so. If not, explain why.

21. Explain why the stellar spectral types (O, B, A, F, G, K, M) are not in alphabetical order. What sequence of temperatures is defined by those spectral types?

22. Other than the Sun, the only stars whose mass astronomers can measure *directly* are those in eclipsing or visual binary systems. Why? How do astronomers estimate the masses of stars that are not in eclipsing or visual binary systems?

23. Once the mass of a certain spectral type of star located in a binary system has been determined, it can be assumed that all other stars of the same spectral type and luminosity class have the same mass. Why is that a reasonable assumption?

24. ★ WHAT AN ASTRONOMER SEES   Suppose that Figure 13.12 showed a third star in the system of stars, with the third star being much smaller than the other two in the image. What could you conclude about that third component's brightness, distance, and luminosity? 👁★

25. As discussed in the Process of Science Figure, scientific advances often require the participation of scientists from all over the world, working on the same problem over many decades, even centuries. Compare that mode of "collaboration" with collaborations in your courses (perhaps on final projects or papers). What mechanisms must be in place to allow scientists to collaborate across space and time in that way? ★

26. How would our ability to measure stellar parallax change if we were on Mars? What if we were on Venus or Jupiter?

27. Figure 13.7 has an absorption line at about 410 nm that is weak for O stars and weak for G stars but very strong in A stars. That particular line is due to the transition from the second excited state of hydrogen up to the sixth excited state. Why is that line weak in O stars? Why is it weak in G stars? Why is it strongest in the middle of the range of spectral types? 👁

28. Which kinds of binary systems are detected edge-on? Which kinds are detected face-on?

29. In Figure 13.10, two stars orbit a common center of mass. 👁
   a. Explain why star 2 has a smaller orbit than star 1.
   b. Re-sketch that picture for the case in which star 1 has a very low mass—perhaps close to that of a planet.
   c. Re-sketch that picture for the case in which star 1 and star 2 have the same mass.

30. If our Sun were a blue main-sequence star, and Earth was still 1 AU from the Sun, would you expect Earth to be in the habitable zone? What about if our Sun were a red main-sequence star?

APPLYING THE CONCEPTS

31. Betelgeuse (in Orion) has a parallax of 0.00451 ± 0.00080 arcsec, as measured by the Hipparcos satellite. What is the distance to Betelgeuse, and what is the uncertainty in that measurement? 🟢🟢🟢
   a. Make a prediction: Working It Out 13.1 gives a few examples of the relationship between the parallax and the distance. Study those numbers for a moment, then compare the parallax of Betelgeuse to the numbers in Working It Out 13.1. Do you expect the distance to Betelgeuse to be more or less than 100 pc?
   b. Calculate: Find the distance to Betelgeuse, and then use the uncertainty of 0.00080 to find the range in distances that result from this uncertainty. (Hint: Adding 0.00080 to 0.00451 gives a slightly larger parallax, which corresponds to a slightly smaller distance and gives one end of the range.)
   c. Check your work: Compare your answer for the distance to Betelgeuse to your prediction, and check that the distance you calculate lies within the range of distances that you calculated from the uncertainty.

32. Wolf 359 has an apparent magnitude of 13.44, whereas Barnard's Star has an apparent magnitude of 9.53. Which star is brighter, and how many times brighter is it? ●—●—●

   a. Make a prediction: Roughly how many apparent magnitudes difference is there in the brightness of these two stars? Each magnitude corresponds to a factor of between 2 and 3 in brightness. What is the range of reasonable answers?

   b. Calculate: Follow Working It Out 13.2 to find the ratio of brightnesses from the difference in magnitudes.

   c. Check your work: Compare your answer to the range of reasonable answers that you found in part a. Does your answer lie in this range?

33. Suppose that you observe a star to have a luminosity 12,000 L_{Sun} and a temperature of 4500 K. What is the radius of this star in units of R_{Sun}? ●—●—●

   a. Make a prediction: Identify the approximate location of this star on the H-R diagram in Figure 13.17. Roughly what radius do you expect to calculate?

   b. Calculate: Follow Working It Out 13.3 to calculate the radius of this star.

   c. Check your work: Compare your answer to your prediction.

34. In an unusual twist of events, you discover a system that is almost exactly like the binary star system described in Working It Out 13.4! The only difference is that the speed of the second star (whose mass and velocity have a subscript "2") is $v_2 = 10.2$ km/s. What are the masses of the two stars in the system? ●—●—●

   a. Make a prediction: This star moves *faster* than the original star 2 from the system in Working It Out 13.4, but the size of the orbit is the same. Do you expect to find that star 1 in this system is more or less massive than the original star 1?

   b. Calculate: Follow the last few steps of Working It Out 13.4 to find the mass of the two stars in this binary system.

   c. Check your work: Compare your answer to your prediction.

35. Barnard's Star has the highest known motion across the sky for any star; it moves 10.3 arcseconds per year as it orbits the center of the Milky Way. This star is located 1.8 pc away. What is its parallax? Is this larger or smaller than the angle of its movement each year across the sky?

   a. Make a prediction: Working It Out 13.1 gives a few examples of the relationship between the parallax and the distance. Study those numbers for a moment, then compare the parallax of Barnard's Star to the numbers in Working It Out 13.1. Do you expect the parallax angle for this star to be "large" (something like 1), or "small" (something like 0.0001)?

   b. Calculate: Find the parallax of Barnard's Star. Is this larger or smaller than the 10.3 arcseconds that Barnard's Star moves every year?

   c. Check your work: Compare your answer to your prediction to check your work.

36. Suppose that Figure 13.1b included a third star, located 4 times as far away as star A. How much less than star A would the third star appear to move each year? How much less than star B? ⊛

37. Suppose you see an object jump from side to side by half a degree as you blink back and forth between your eyes. How much farther away is an object that moves only one-third of a degree?

38. Figure 13.5 is plotted logarithmically on both axes. The luminosities are in units of solar luminosities. ⊛

   a. How much more luminous than the Sun is a star on the far right side of the plot?

   b. How much less luminous than the Sun is a star on the far left side of the plot?

39. Compared with the Sun, how luminous, large, and hot is a star that has 10 times the mass of the Sun? Use Figure 13.17 to answer that question. ⊛

40. Sirius, the brightest star in the sky, has a parallax of 0.379 arcsec. What is its distance in parsecs? In light-years? How long does the light take to reach Earth?

41. Sirius is actually a binary pair of two A-type stars. The brighter star is called the "Dog Star" and the fainter is called the "Pup Star" because Sirius is in the constellation Canis Major (meaning "big dog"). The Dog Star appears about 6800 times brighter than the Pup Star, even though both stars are at the same distance from Earth. Compare the temperatures, luminosities, and sizes of those two stars.

42. Sirius and its companion orbit around a common center of mass with a period of 50 years. The mass of Sirius is 2 times the mass of the Sun.

   a. If the orbital velocity of the companion is 2.35 times greater than that of Sirius, what is the mass of the companion?

   b. What is the semimajor axis of the orbit?

43. The star Achernar has a Hipparcos parallax of 0.02339 arcsec and appears about as bright as Betelgeuese (Problem 31) in the sky. Which star is actually more luminous? Betelgeuse appears reddish, whereas Achernar appears bluish, so which star is hotter? Why is it hotter?

44. The Sun is about 16 trillion (1.6×10^{13}) times brighter than the faintest stars visible to the naked eye.

   a. How far away (in astronomical units) would an identical solar-type star be if it were just barely visible to the naked eye?

   b. What would be its distance in light-years?

45. Find the peak wavelength of blackbody emission for a star with a temperature of about 10,000 K. In what region of the spectrum does that wavelength fall? What color is that star?

EXPLORATION H-R Diagram

digital.wwnorton.com/astro7

Visit the Digital Resources Page and on the Student Site open the "H-R Diagram" Interactive Simulation for Chapter 13. This simulation enables you to compare stars on the H-R diagram in two ways. You can compare an individual star (marked by a red *X*) to the Sun by varying its properties in the box in the left half of the window. Or you can compare groups of the nearest and brightest stars. Play around with the controls for a few minutes to familiarize yourself with the simulation.

Begin by exploring how changes to the properties of the individual star change its location on the H-R diagram. First, press the "Reset" button.

Decrease the temperature of the star by dragging the temperature slider to the left. Notice that the luminosity remains the same. Because the temperature has decreased, each square meter of star surface must be emitting less light. What other property of the star changes to keep the total luminosity of the star constant?

Predict what will happen when you move the temperature slider all the way to the right. Now do it. Did the star behave as you expected?

1 As you move to the left across the H-R diagram, what happens to the radius?

_____

_____

_____

2 What happens to the radius as you move to the right?

_____

_____

_____

Press "Reset" and experiment with the luminosity slider.

3 As you move up on the H-R diagram, what happens to the radius?

_____

_____

4 What happens to the radius as you move down?

_____

_____

_____

Press "Reset" again and predict how you would have to adjust the slider bars to move your star into the red giant portion of the H-R diagram (upper right). Adjust the slider bars until the star is in that area. Were you correct?

5 How would you adjust the slider bars to move the star into the white dwarf area of the H-R diagram (lower left)?

_____

_____

Press the "Reset" button and explore the right-hand side of the window. Add the nearest stars to the graph by clicking their radio button. Using what you have learned so far, compare the temperatures and luminosities of those stars with the Sun (marked by the yellow circle on the graph).

6 Are the nearest stars generally hotter or cooler than the Sun?

_____

7 Are the nearest stars generally more or less luminous than the Sun?

_____

Press the radio button for the brightest stars. That action will add the brightest stars in the sky to the plot. Compare those stars with the Sun.

8 Are the brightest stars generally hotter or cooler than the Sun?

_____

9 Are the brightest stars generally more or less luminous than the Sun?

_____

10 How do the temperatures and luminosities of the brightest stars in the sky compare with the temperatures and luminosities of the nearest stars? Does that information support the claim in the chapter that there are more low-luminosity stars than high-luminosity stars? Explain.

_____

_____

_____

_____

Our Star—The Sun

The Sun is more massive *and* significantly larger than everything else in the Solar System. You can measure the diameter of the Sun with a heavy sheet of paper or poster board, a long piece of string, a ruler, and a little help from a friend. On a sunny day, poke a tiny hole in the paper. Standing with her back to the Sun, your friend should hold up the paper so that the Sun shines through the hole. An image of the Sun will be projected onto a second piece of paper on the ground within the shadow of the paper. If the paper is close to the ground, the image will be small. If the paper is far from the ground, the image will be larger, but it will not be as bright. Move the paper until you can accurately measure the diameter (d) of the image on the ground with your ruler, in millimeters. Convert this number to meters. Use the long piece of string to measure how far (h) the paper is from the ground, by first marking the distance on the piece of string, and then using the ruler to measure the marked distance in meters. (You can use a tape measure, instead, if you have one.) To calculate the diameter of the Sun, all you need are these two measurements and the distance from Earth to the Sun (in meters), which is available in your textbook or on the Internet.

EXPERIMENT SETUP

On a sunny day, poke a tiny hole in the paper. Standing with her back to the Sun, your friend should hold up the paper so that the Sun shines through the hole. An image of the Sun will be projected onto the second piece of paper on the ground within the shadow of the paper.

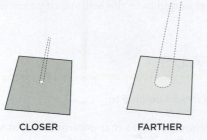

CLOSER **FARTHER**

If the paper is close to the ground, the image will be small. If the paper is far from the ground, the image will be larger, but it will not be as bright. Move the paper until you can accurately measure the diameter (*d*) of the image on the ground with your ruler, in millimeters. Convert this number to meters.

Use a long piece of string to measure how far (*h*) the paper is from the ground, by first marking the distance on the piece of string, and then using the ruler to measure the marked distance in meters.

Sun

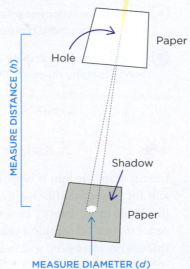

Paper

Hole

MEASURE DISTANCE (*h*)

Shadow

Paper

MEASURE DIAMETER (*d*)

SKETCH OF RESULTS

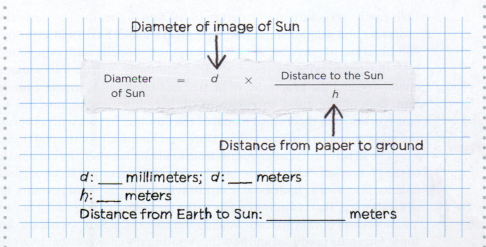

Diameter of image of Sun

↓

$$\text{Diameter of Sun} = d \times \frac{\text{Distance to the Sun}}{h}$$

↑

Distance from paper to ground

d: ___ millimeters; *d*: ___ meters
h: ___ meters
Distance from Earth to Sun: _____ meters

14

Because the Sun is the only star close to Earth, much of our detailed knowledge about stars has come from studying the Sun. In Chapter 13, we looked at the physical properties of distant stars, including their mass, luminosity, size, temperature, and chemical composition. Here in Chapter 14, we ask fundamental questions about Earth's local star: Where does the Sun get its energy? How does energy move through the Sun? What is its atmosphere like? How has its luminosity changed over the billions of years since the Solar System formed?

LEARNING GOALS

By the end of this chapter, you should be able to:

(1) Describe the balance between the forces that determine the structure of the Sun.

(2) Diagram how mass is converted to energy in the Sun's core and estimate how long the Sun will take to use up its fuel.

(3) Sketch a physical model of the Sun's interior, and list the ways that energy moves outward from the Sun's core toward its surface.

(4) Relate observations of solar neutrinos and seismic vibrations on the surface of the Sun to astronomers' models of the Sun.

(5) Describe the solar activity cycles of 11 and 22 years, and relate those cycles to the Sun's changing magnetic field and solar phenomena such as sunspots and flares.

14.1 The Sun Is Powered by Nuclear Fusion

Energy from the Sun is responsible for daylight, for Earth's weather and seasons, and for terrestrial life itself. At a luminosity of 3.85×10^{26} watts (W), the Sun produces more energy in a second than all the power plants on Earth could generate in a half-million years. In this section, we explain how the Sun releases this energy from deep within its core.

Hydrostatic Equilibrium

Like Earth's structure, the structure of the Sun is governed by several physical processes and relationships. Geologists learn about Earth's interior by using a combination of physics, detailed computer models, and experiments that test the predictions of those models. Similarly, astronomers use physics, chemistry, and the properties of matter and radiation to create a model of the Sun. One of the great successes of 20th century astronomy has been the construction of a theoretical model of the Sun that agrees with observations of the mass, composition, size, temperature, and luminosity of the real thing.

The structure of the Sun results from a balance between the outward force due to the pressure of radiation produced inside and the inward force of gravity. That balance is known as **hydrostatic equilibrium**. The outward pressure is a result of local energy that is finding its way to the Sun's surface from deep in its interior. To understand how hydrostatic equilibrium affects the Sun, we need to know how those forces are produced and how they continually change to balance each other.

The balance between the forces due to radiation pressure and gravity is illustrated in **Figure 14.1**. The Sun is a huge ball of hot gas. Deep in the Sun's interior, the outer layers press downward because of gravity, producing a large inward force. To maintain balance, the outward force due to pressure must be equally large. If the forces due to gravity exceeded the forces due to radiation pressure, the Sun would collapse. Conversely, if radiation pressure were greater than gravity, the Sun would blow itself apart. At every point within the Sun's interior, the radiation pressure must be just enough to hold up the weight of all the layers above that point. If the Sun were not in a stable hydrostatic equilibrium, forces within it would not be in balance, and the size of the Sun would change accordingly.

Hydrostatic equilibrium becomes an even more powerful concept when combined with the way gases behave. Deeper in the Sun's interior, the weight of the material above becomes greater, and hence the radiation pressure increases. In a gas, higher pressure means higher density, higher temperature, or both. **Figure 14.2** shows how conditions vary inside the Sun. The graphs in Figure 14.2b, which are based on calculations, show that as the radiation pressure increases toward the Sun's center, the density and temperature of the gas increase as well.

CHECK YOUR UNDERSTANDING 14.1a

Hydrostatic equilibrium in the Sun means that: (a) the Sun does not change; (b) the Sun absorbs and emits equal amounts of energy; (c) the outward force from radiation pressure balances the weight of overlying layers; (d) energy produced in the core per unit time equals energy emitted at the surface per unit time.

Answers to Check Your Understanding questions are in the back of the book.

Nuclear Fusion

A second fundamental balance within the Sun is the balance of energy (see Figure 14.1). Stars like the Sun are remarkably stable objects. To remain in balance, the Sun must produce just enough energy in its interior to replace the energy radiated away from its surface. That energy balance tells us how much energy must be produced in the Sun's interior and how that energy finds its way from the interior to the Sun's surface, where it is radiated away. Theoretical models of stellar evolution indicate that the Sun's luminosity is increasing, but very, very slowly. From geological records and these models, astronomers estimate that the Sun's luminosity 4.5 billion years ago was about 70 percent of its current luminosity.

One of the most basic questions facing the pioneers of stellar astrophysics was how the Sun and other stars get their energy. In the 19th century, physicists proposed that the Sun was slowly shrinking and that the core was heating up as a result of that gravitational contraction. Calculations soon showed, however, that gravitational contraction could power the Sun for only millions of years. Geological and biological evidence available at the time suggested that Earth was tens of millions or hundreds of millions of years old. In the early 20th century, radiometric dating suggested that Earth was more than a billion years old, and therefore gravitational contraction could not be the source of the Sun's energy. In the 1930s, using theoretical and laboratory physics, nuclear physicists concluded that the Sun's energy

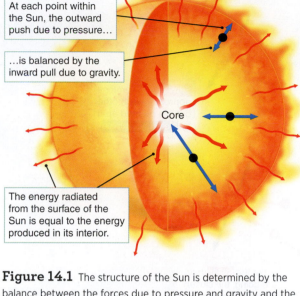

At each point within the Sun, the outward push due to pressure…

…is balanced by the inward pull due to gravity.

Core

The energy radiated from the surface of the Sun is equal to the energy produced in its interior.

Figure 14.1 The structure of the Sun is determined by the balance between the forces due to pressure and gravity and the balance between the energy generated in its core and the energy radiated from its surface.

Figure 14.2 a. This cutaway figure shows how the fraction of radius given in the *x*-axis of the graphs in b. is measured. The energy produced by the Sun is generated in the Sun's core. **b.** Pressure, density, and temperature all increase toward the center of the Sun.

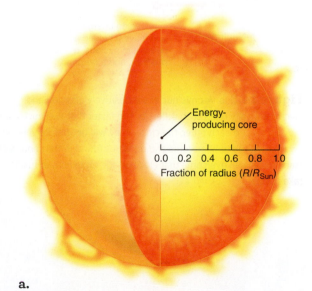

Energy-producing core

0.0 0.2 0.4 0.6 0.8 1.0
Fraction of radius (R/R_Sun)

a.

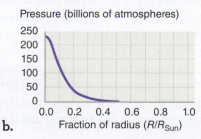

Pressure (billions of atmospheres)

0.0 0.2 0.4 0.6 0.8 1.0
Fraction of radius (R/R_Sun)

Density (thousands of kg/m³)

0.0 0.2 0.4 0.6 0.8 1.0
Fraction of radius (R/R_Sun)

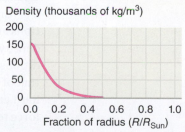

Temperature (millions of K)

0.0 0.2 0.4 0.6 0.8 1.0
Fraction of radius (R/R_Sun)

b.

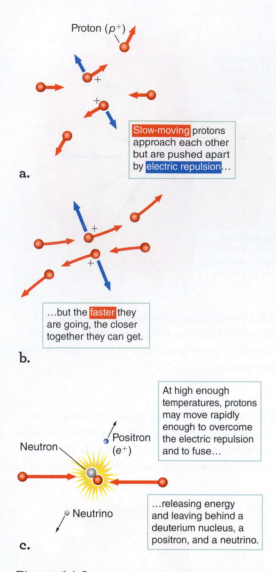

Proton (p⁺)

a.

Slow-moving protons approach each other but are pushed apart by electric repulsion...

b.

...but the faster they are going, the closer together they can get.

At high enough temperatures, protons may move rapidly enough to overcome the electric repulsion and to fuse...

Neutron

Positron (e⁺)

Neutrino

...releasing energy and leaving behind a deuterium nucleus, a positron, and a neutrino.

c.

Figure 14.3 **a.** Atomic nuclei are positively charged and electrostatically repel each other. **b.** The faster that two nuclei are moving toward each other, the closer they will get before veering away. **c.** At the temperatures and densities found in the centers of stars, nuclei can overcome that electrostatic repulsion, so fusion takes place.

comes from nuclear reactions at its core, capable of powering the star for billions of years.

The nucleus of a hydrogen atom consists of a single proton and, in most cases, no neutrons (Chapter 5). The nuclei of all other atoms are built from a mixture of protons and neutrons. Helium nuclei, for example, consist of two protons and usually two neutrons. Protons have a positive electric charge, whereas neutrons have no electric charge. Because like charges repel, and the closer they are the stronger the force, therefore all of the protons in an atomic nucleus continually repel each other with a tremendous force. The nuclei of atoms should fly apart because of electric repulsion. They don't, though, because they are held together by the **strong nuclear force**, which overcomes that repulsion. However, the strong nuclear force acts only over very short distances, on the order of 10^{-15} meters—about the size of the atomic nucleus, or about a hundred-thousandth the size of an atom.

Compared with the energy required to free an electron from an atom, the amount of energy required to tear a nucleus apart is enormous. Conversely, when an atomic nucleus (with a mass up to the nucleus of iron) is formed from component parts, energy is released. **Nuclear fusion**—the process of combining two less massive atomic nuclei into a single, more massive atomic nucleus—occurs when atomic nuclei are brought close enough together for the strong nuclear force to overcome the force of electric repulsion (**Figure 14.3**). Many kinds of nuclear fusion can occur in stars. In main-sequence stars like the Sun, the primary energy generation process is the fusion of hydrogen into helium. That process is sometimes called "hydrogen burning," even though it has nothing to do with fire or other chemical combustion. The fusion of hydrogen into helium takes several steps, but the net result is that four hydrogen nuclei become one helium nucleus plus energy.

The energy produced in nuclear reactions comes from converting mass to energy. The exchange rate between mass and energy is given by Einstein's equation, $E = mc^2$, in which E is energy, m is mass, and c^2 is the speed of light squared. For any nuclear reaction, we can determine the mass turned into energy by calculating the mass lost. To find that lost mass, we subtract the mass of the outputs from the mass of the inputs. In **hydrogen fusion**, the inputs are four hydrogen nuclei, and the output is a helium nucleus plus energy. The mass of four separate hydrogen nuclei is 1.007 times greater than the mass of a single helium nucleus (with two protons and two neutrons); so, when hydrogen fuses to make helium, 0.7 percent of the mass of the hydrogen is converted to energy.

Although each fusion reaction produces a small amount of energy, the Sun's total mass is very large, so a great deal of hydrogen is available to fuse. When the amount of energy produced by nuclear fusion is compared with the luminosity of the Sun, we see that those reactions can power the Sun for 10 billion years—a time frame longer than the 4.6-billion-year age of the Solar System (measured from radioactive dating of meteorites from space and the Moon). Details of that calculation are provided in **Working It Out 14.1**.

Energy is produced in the Sun's innermost region, the **core**, where the conditions are the most extreme. The density of matter in the core is about 150 times the density of water, and the temperature is about 15 million kelvin (K). Under those conditions, the atomic nuclei have tens of thousands of times more kinetic energy than that of atoms at room temperature, so that many of them can slam into each other hard enough to overcome the electrostatic repulsion, allowing the strong nuclear force to act (Figure 14.3c). In hotter and denser gases, these collisions happen more

working it out 14.1

The Source of the Sun's Energy

Like all stars, the Sun has a lifetime limited by the amount of fuel available. We can calculate how long the Sun will live by comparing the mass involved in nuclear fusion with the amount of mass available. Converting four hydrogen nuclei (protons) into a single helium nucleus results in a loss of mass. The mass of a single hydrogen nucleus is 1.6726×10^{-27} kilogram (kg). So, four hydrogen nuclei have a mass of 4 times that, or 6.6904×10^{-27} kg. The mass of a helium nucleus is 6.6447×10^{-27} kg, which is less than the mass of the four hydrogen nuclei. The amount of mass lost, m, is

$$m = 6.6904 \times 10^{-27} \text{ kg} - 6.6447 \times 10^{-27} \text{ kg} = 0.0457 \times 10^{-27} \text{ kg}$$

We can write that result as 4.57×10^{-29} kg—a mass loss of about 0.7 percent. Converting 0.7 percent of the mass of the hydrogen into energy might not seem very efficient—until we compare it with other sources of energy and discover that it is millions of times more efficient than even the most efficient chemical reactions.

Using Einstein's equation $E = mc^2$, where c is the speed of light $(3 \times 10^8 \text{ m/s})$, along with the definition of a joule $(1 \text{ J} = 1 \text{ kg m}^2/\text{s}^2)$, we can calculate the energy released by that mass-to-energy conversion:

$$E = mc^2 = (4.57 \times 10^{-29} \text{ kg}) \times (3.00 \times 10^8 \text{ m/s})^2 = 4.11 \times 10^{-12} \text{ J}$$

Each reaction that takes four hydrogen nuclei and turns them into a helium nucleus releases 4.11×10^{-12} J of energy, which doesn't seem like very much. Atoms are very small, however, so fusing a single kilogram of hydrogen into helium releases about 6.3×10^{14} J of energy—about the equivalent of the chemical energy released in burning 100,000 barrels of oil. To see how much the Sun must be fusing per second to produce

its current luminosity, we divide the Sun's luminosity by that amount of energy per kilogram:

$$\frac{\text{Luminosity of Sun}}{\text{Energy per kilogram}} = \frac{3.9 \times 10^{26} \text{ J/s}}{6.3 \times 10^{14} \text{ J/kg}} = 6.2 \times 10^{11} \text{ kg/s}$$

For the Sun to produce as much energy as it does, it must convert roughly 620 billion kg of hydrogen into helium every second (and about 4 billion kg of matter—0.7 percent—is converted to energy in the process). The Sun has been fusing hydrogen at that rate for at least the age of Earth and the Solar System—4.6 billion years. How much longer will the Sun last?

Astronomers estimate that only 10 percent of the Sun's total mass will ever be involved in fusion because the other 90 percent will never get hot enough or dense enough for the strong nuclear force to make fusion happen. Ten percent of the mass of the Sun is $(0.1) \times (2 \times 10^{30})$ kg, or 2×10^{29} kg. That is the amount of "fuel" the Sun has available. The Sun consumes hydrogen at a rate of 620 billion kg/s, so each year the Sun consumes about:

$$M_{\text{year}} = (6.2 \times 10^{11} \text{ kg/s}) \times (3.16 \times 10^7 \text{ s/yr}) \approx 2 \times 10^{19} \text{ kg/yr}$$

If we know how much fuel the Sun has $(2 \times 10^{29}$ kg), and we know how much the Sun fuses each year $(2 \times 10^{19}$ kg/yr), we can divide the amount by the rate to find the lifetime of the Sun:

$$\text{Lifetime} = \frac{M_{\text{fuel}}}{M_{\text{year}}} = \frac{2 \times 10^{29} \text{ kg}}{2 \times 10^{19} \text{ kg/yr}} = 10^{10} \text{ yr}$$

When the Sun was formed, it had enough fuel to power it for about 10 billion years. The Sun is nearly halfway through its lifetime of hydrogen fusion.

---

often. For that reason, the rate of nuclear fusion reactions is extremely sensitive to the temperature and the density of the gas, which is why those energy-producing collisions are concentrated in the Sun's core. Half the energy produced by the Sun is generated within the inner part of the core: the inner 9 percent of the Sun's radius, or less than 0.1 percent of the Sun's volume.

The conversion of four hydrogen nuclei to one helium nucleus is the most significant source of energy in main-sequence stars. Hydrogen is the most abundant element in the universe, so it is the most abundant source of nuclear fuel at the beginning of a star's lifetime. Hydrogen also is the easiest atom to fuse. Hydrogen nuclei—protons—have an electric charge of +1. The electric barrier that must be overcome to fuse protons is the repulsion of one proton against another. To fuse two carbon nuclei, for example, requires overcoming the repulsion of the six protons in one carbon nucleus that are pushing against the six protons in the other carbon nucleus. As a result, the repulsion between two carbon nuclei is 36 times

▶❚❚ **AstroTour:** The Solar Core

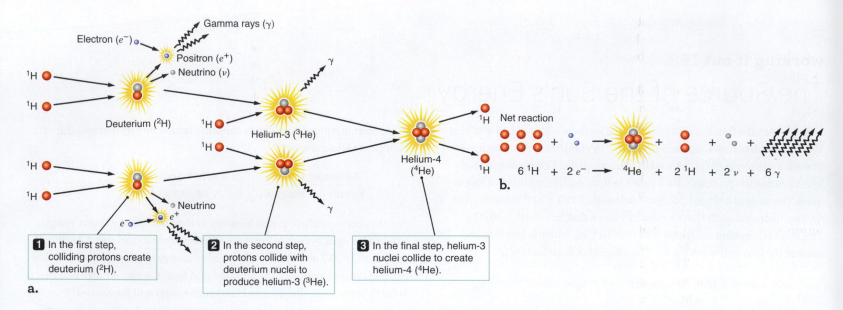

Figure 14.4 a. In the Sun, about 85 percent of the energy produced comes from the branch of the proton-proton chain illustrated here. **b.** The net reaction, in which four hydrogen atoms fuse to make a helium atom, is the primary source of energy for the Sun and all other main-sequence stars. The energy is released in the form of gamma rays, neutrinos, and kinetic energy.

▶▶ **Interactive Simulation:** Proton-Proton Chain

unanswered questions

Will nuclear fusion become a major source of energy production on Earth? Scientists have been working on controlled nuclear fusion for more than 60 years, since the first hydrogen bombs were developed. So far, however, replicating the conditions inside the Sun has posed too many difficulties. Nuclear fusion requires that we have hydrogen isotopes at very high temperature, density, and pressure, just as when a hydrogen bomb explodes. However, controlled nuclear fusion requires that we confine that material long enough to get more energy out than we put in. Several major experiments have attempted to fuse isotopes of hydrogen. So far, hydrogen fusion reactions have not been sustained long enough to be a commercially viable energy source; South Korea's KSTAR project set a record by sustaining the reaction for 20 seconds. An alternative approach is to fuse an isotope of helium, ³He, which has only three particles in the nucleus (two protons and one neutron). On Earth, ³He is found in very limited supply. It is in much greater abundance on the Moon, however, so some people propose setting up mining colonies on the Moon to extract ³He for use in fusion reactions on Earth or possibly even on the Moon.

stronger than that between two hydrogen nuclei. Therefore, hydrogen fusion occurs at a much lower temperature than any other type of nuclear fusion.

The Proton-Proton Chain

To test the theory that the Sun shines because of nuclear fusion, astronomers can analyze the predicted by-products of the nuclear reactions. In the Sun's core and in other low-mass stars, hydrogen fusion takes place in a series of nuclear reactions called the **proton-proton chain**, which has three branches. The most important branch, responsible for about 85 percent of the energy generated in the Sun, consists of the three steps illustrated in **Figure 14.4**. Each step produces particles and/or energy in the form of light. We begin by following the creation of the helium nucleus and then go back to find out what happens to the other products of the reaction.

Follow along in Figure 14.4a as we step through the proton-proton chain. The hydrogen nucleus consists of just one proton and is written as ¹H (where H is the element symbol for hydrogen and 1 is the *atomic mass number*—the total number of protons and neutrons in the nucleus). In the first step, two protons fuse. During that process, one proton is transformed into a neutron. To conserve spin and charge, two particles are emitted: a positively charged particle called a **positron** (e^+) and a neutral particle called a **neutrino** (ν). The positron is then annihilated when it encounters an electron, thereby producing energy in the form of gamma-ray photons (γ). The new atomic nucleus formed by the first step in the chain consists of a proton and a neutron. Recall from Chapter 5 that an *isotope* of an element has the same number of protons but a different number of neutrons. Thus, the new atomic nucleus, called **deuterium,** is still hydrogen because it has only one proton, but it is written as ²H because its atomic mass number is 2 (1 proton + 1 neutron = 2).

In the second step of the proton-proton chain, another proton slams into the deuterium nucleus, forming the nucleus of an isotope of helium, ³He, consisting of two protons and one neutron. The energy released in that step is carried away as a highly energetic gamma-ray photon. Those first two steps are shown twice in Figure 14.4a because two ³He nuclei are needed to produce a single ⁴He nucleus.

In the third and final step of the proton-proton chain, two ³He nuclei collide and fuse, producing an ordinary ⁴He nucleus (consisting of 2 protons and

2 neutrons) and ejecting two protons (^{1}H) in the process. The energy released in that step is the kinetic energy of the helium nucleus and two ejected protons. Overall, four hydrogen nuclei have combined to form one helium nucleus, as summarized in Figure 14.4b.

Now let's go back and look at what happens to the other products of the reaction. In step 1, a positron—a particle of antimatter—is produced. **Antimatter** particles have the same mass as a corresponding matter particle but have opposite values of other properties, such as charge. The positron (e^+) is the antimatter counterpart of an electron (e^-). When matter (electrons) and antimatter (positrons) meet, they annihilate each other, and their total mass is converted to energy in the form of gamma-ray photons (γ). That's what happens to the emitted positrons inside the Sun, and the emitted photons from the annihilation carry away part of the energy released when the two protons fuse. Those photons heat the surrounding gas. The gamma rays emitted in step 2 similarly heat the gas. The thermal energy produced in the core of the Sun typically takes one hundred thousand years or more to find its way to the Sun's surface, and so the light we see from the Sun indicates what the Sun's core was doing a very long time ago.

The neutrino emitted in step 1 has a very different fate. Neutrinos are particles that have no charge, have very little mass, and travel at nearly the speed of light. They interact so weakly with ordinary matter that the neutrino escapes from the Sun without further interactions with any other particles. The core of the Sun is buried beneath 700,000 kilometers of dense, hot matter, yet the Sun is transparent to neutrinos—essentially all of them travel into space as though the outer layers of the Sun did not exist. Because they travel at nearly the speed of light, neutrinos from the center of the Sun arrive at Earth after 8.3 minutes. Therefore, we can use them to probe what the Sun is doing today.

The dominant branch of the proton-proton chain can be written symbolically as follows:

Step 1: ^{1}H + ^{1}H → ^{2}H + e^+ + ν and then e^+ + e^- → γ + γ

Step 2: ^{2}H + ^{1}H → ^{3}He + γ

Step 3: ^{3}He + ^{3}He → ^{4}He + ^{1}H + ^{1}H

The rate of the proton-proton chain reaction depends on both temperature and density. At the temperature and pressure that exist within the Sun's core, the reaction rate is relatively slow, which is fortunate for life on Earth. If the hydrogen fused quickly, the Sun would have exhausted its supply long ago, and life might not have had time to evolve.

While the reactions in Figure 14.4 generate about 85 percent of the Sun's energy, variations of the proton-proton chain account for the other 15 percent. The most common variation happens in step 3, in which ^{3}He fuses with an existing ^{4}He to create beryllium (^{7}Be), which decays to lithium (^{7}Li) and energy, and then the ^{7}Li plus one ^{1}H become two ^{4}He. In a less common variation, the beryllium combines with hydrogen to become boron (^{8}B), which then decays to beryllium and then to two ^{4}He. In both variations, ultimately four hydrogen nuclei become one helium nucleus.

CHECK YOUR UNDERSTANDING 14.1b

When hydrogen fuses into helium, energy is released from: (a) gravitational collapse; (b) the conversion of mass to energy; (c) the increase in pressure; (d) the decrease in the gravitational field.

what if . . .

What if you had a neutrino detector in your classroom? How would you expect the signal of solar neutrinos to vary between night and day?

Astronomy in Action: Random Walk

14.2 Energy Is Transferred from the Interior of the Sun

Although geologists cannot travel deep inside Earth to find out how it is structured, they can build a model of its interior by using data on how seismic waves travel during earthquakes. Similarly, astronomers can create a model of the Sun's interior by using their knowledge of the balance of forces and energy within the Sun and an understanding of how energy moves from one place to another. Those models can be tested by observations of waves traveling through the Sun and by studying neutrinos from the Sun.

Energy Transport

Some of the energy released by hydrogen fusion in the Sun's core escapes directly into space in the form of neutrinos. However, most of the energy heats the solar interior and then moves outward through the Sun to the surface, a process known as **energy transport**. Energy transport, a key determinant of the Sun's structure, can occur by *conduction*, *convection*, or *radiation*.

Conduction is important primarily in solids. When you pick up a hot object, for example, your fingers are heated by conduction. That happens because energetic thermal vibrations of atoms and molecules cause neighboring atoms and molecules to vibrate more rapidly as well. Conduction is typically ineffective in a gas because the atoms and molecules are too far apart to transmit vibrations to one another efficiently. Conduction does not play a key role in transporting energy from the Sun's core to its surface, but conduction is relevant when we discuss dying stars in Chapters 16 and 17.

In the Sun, energy is transported by convection and radiation through zones, as shown in **Figure 14.5**. The mechanism of energy transport from the center of the Sun outward depends on the decreasing temperature and density as the radius increases. First, energy moves outward through the inner layers of the Sun as radiation in the form of photons. Next, energy moves by convection in parcels of gas. Finally, energy radiates from the Sun's surface as light. We look at each process in turn.

Near the core, **radiation** transfers energy from hotter to cooler regions via photons (**Figure 14.6**), which carry the energy with them. Recall from your study of radiation in Chapter 5 that the hotter region contains more (and more energetic) photons than the cooler region. More photons move from the hotter, very crowded region to the cooler, less crowded region than in the reverse direction. A net transfer of photons and photon energy occurs from the hotter region to the cooler region, and radiation carries energy outward from the Sun's core.

The transfer of energy from one point to another by radiation also depends on how freely radiation can move from one point to another within a star. The degree to which matter blocks the flow of photons through it is called **opacity**. The opacity of a material depends on many things, including the density of the material, its composition, its temperature, and the wavelength of the photons moving through it.

Energy transfer by radiation is most efficient in regions with low opacity. The **radiative zone** (see Figure 14.5) is the region in the inner part of the Sun in which the opacity is relatively low, and radiation carries the energy produced in the core outward through the star. That radiative zone extends about 70 percent of the way out toward the surface of the Sun. Even though the region's opacity is low enough

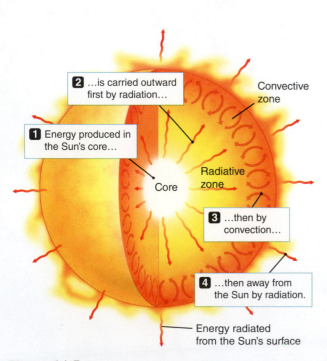

2 ...is carried outward first by radiation...

1 Energy produced in the Sun's core...

Convective zone

Core

Radiative zone

3 ...then by convection...

4 ...then away from the Sun by radiation.

Energy radiated from the Sun's surface

Figure 14.5 The interior of the Sun is divided into zones on the basis of where energy is produced and how it is transported outward.

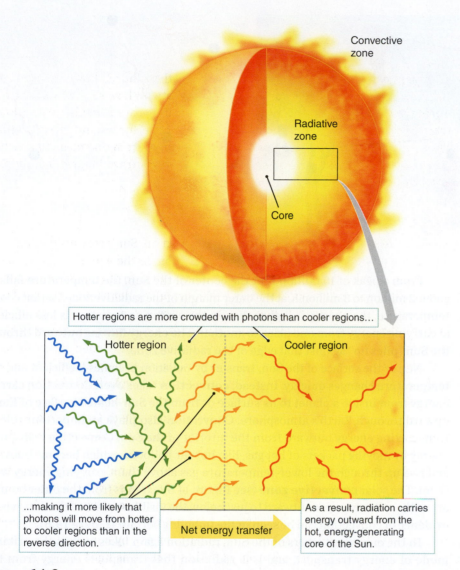

Figure 14.6 Higher-temperature regions deep within the Sun contain more radiation than do lower-temperature regions farther out. Although radiation flows in both directions, more radiation flows from the hotter regions to the cooler regions than from the cooler regions to the hotter regions. Therefore, radiation carries energy outward from the inner parts of the Sun. For simplicity, this illustration includes only a few common photons. Photons of wavelengths representing all colors are present in all regions, with more of all kinds in the hotter regions and fewer of all kinds in the cooler regions.

for radiation to dominate convection as an energy transport mechanism, photons still travel only a short distance within the region before being absorbed and then reemitted, or scattered by matter, much like a beach ball being batted about by a crowd of people (**Figure 14.7**). Each interaction sends the photon in an unpredictable direction—not necessarily toward the surface of the star. The distances between interactions are so short that, on average, the energy of a gamma-ray photon produced in the interior of the Sun takes about one hundred thousand or more years to find its way to the outer layers of the Sun. Opacity holds energy inside the Sun and lets it seep away only slowly. As it travels, the gamma-ray photon is gradually converted to lower-energy photons, emerging mostly as optical and infrared radiation from the surface.

Figure 14.7 **a.** When a crowd of people plays with a beach ball, the ball never travels very far before someone hits it, turning it in another direction. A ball often takes a long time to make its way from one edge of the crowd to the other. **b.** Similarly, a photon traveling through the Sun takes a long time to make its way out of the Sun.

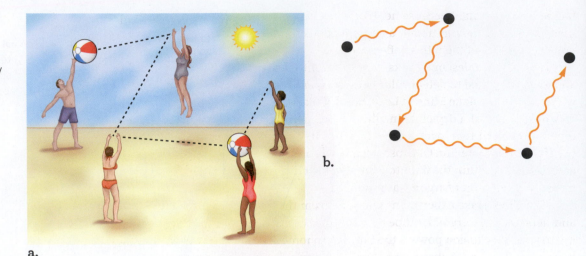

a.

From a peak of 15 million K at the center of the Sun, the temperature falls to about 2 million to 3 million K at the outer margin of the radiative zone. At that cooler temperature, the opacity is higher than in the center, so radiation is less efficient at carrying energy from one place to another. The energy flowing outward through the Sun "piles up" against that edge of the radiative zone.

Nearer the surface of the Sun, transfer by radiation becomes inefficient and the temperature changes quickly. Instead, **convection** takes over. Convection carries energy from inside a planet to its surface or from the Sun-heated surface of Earth upward through Earth's atmosphere. Convection also plays an important role in transporting energy outward from the interior of the Sun. Convection transports energy by moving packets of hot gas, like hot-air balloons, which become buoyant and rise up through the lower-temperature gas above them, carrying energy with them. The solar **convective zone** (see Figure 14.5) extends from the outer boundary of the radiative zone outward to just below the visible surface of the Sun, where evidence of convection can be seen in the bubbling surface (**Figure 14.8**).

In the outermost layers of the Sun, radiation again takes over as the primary mode of energy transport, and it is radiation that transports energy from the Sun's outermost layers off into space.

Observing Neutrinos from the Core of the Sun

The model of energy production and energy transport in the Sun, as just presented, correctly matches observed global properties of the Sun such as its size, temperature, and luminosity. The nuclear fusion model of the Sun predicts exactly which nuclear reactions should be occurring in the Sun's core and at what rate. The nuclear reactions that make up the proton-proton chain produce a vast number of neutrinos. Because neutrinos barely interact with other ordinary matter, almost all the neutrinos produced in the heart of the Sun travel freely through the outer parts of the Sun and on into space as though the outer layers of the Sun were not there. Solar neutrinos produced in the core of the Sun, traveling at nearly the speed of light, take only 8.3 minutes to reach Earth—much quicker than the 100,000-year journey of photons leaving the core.

Neutrinos interact so weakly with matter that they are extremely difficult to observe. Nevertheless, an extremely large number of nuclear reactions take place in the Sun, so the Sun produces an enormous number of neutrinos. As you read this sentence, about 400 trillion solar neutrinos are passing through your body. That

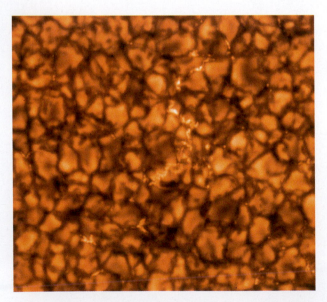

Figure 14.8 The top of the Sun's convective zone shows the bubbling of the surface caused by rising and falling packets of gas.

happens even at night because neutrinos easily pass through Earth, too. With that many neutrinos moving about, a neutrino detector does not have to detect a very large percentage of them to be effective.

A neutrino telescope looks very different from other telescopes. The first apparatus designed to detect solar neutrinos was built 1500 meters underground, within the Homestake Mine in Lead, South Dakota. Astronomers filled a tank with 100,000 gallons of a dry-cleaning fluid called perchloroethylene (C_2Cl_4). Astronomers predicted that, over 2 days, roughly 10^{22} solar neutrinos passed through the Homestake detector. Of those, on average only one neutrino interacted with a chlorine atom within the fluid to form a radioactive isotope of argon. Over time, a measurable amount of argon was produced.

The Homestake experiment operated from the late 1960s to the early 1990s and detected that argon isotope—evidence of neutrinos from the Sun, confirming that nuclear fusion powers the Sun. Astronomers noticed, however, that they were measuring only about one-third as many solar neutrinos as predicted by solar models. The difference between the predicted and measured number of solar neutrinos was called the **solar neutrino problem**.

One possible explanation of the solar neutrino problem was that the working model of the structure of the Sun was somehow wrong. That possibility seemed unlikely, however, because of the many other successful predictions of the solar model. A second possibility was that an understanding of the neutrino itself was incomplete. The neutrino was long thought to have zero mass, like photons, and to travel at the speed of light. But if neutrinos actually do have a tiny amount of mass, particle physics suggests that solar neutrinos should oscillate—alternate back and forth—among three kinds of neutrinos: the electron, muon, and tau neutrinos. According to that explanation, early neutrino experiments could detect only the electron neutrino and, consequently, observed only about a third of the expected number of neutrinos. Since then, many other neutrino detectors have been built, each using different reactions to detect neutrinos of different energies or different types. Experiments at high-energy physics labs, nuclear reactors, and neutrino telescopes around the world have shown that neutrinos do have a nonzero mass and do oscillate among neutrino types.

Solving the solar neutrino problem is a good example of how science works—how a better model of the neutrino showed that the solar neutrino problem was real and not merely an experimental mistake, and how a single set of anomalous observations was later confirmed by other, more sophisticated experiments. All that effort led to a better understanding of basic physics (see the **Process of Science Figure**).

Probing the Sun's Interior

Models of Earth's interior predict how density and temperature change from place to place. Those differences affect the seismic waves traveling through Earth, bending the paths that they travel. Geologists test models of Earth's interior by comparing measurements of seismic waves from earthquakes with model predictions of how seismic waves should travel through the planet.

Just as geologists use seismic waves from earthquakes to probe the interior of Earth, solar physicists use the surface oscillations of the Sun to probe the solar interior. The science that uses solar oscillations to study the Sun is called **helioseismology**. Detailed observations of the Doppler shifts caused by the motions of material from place to place across the Sun's surface show that the Sun vibrates

what if . . .

What if a typical photon produced in the Sun's center took 1000 rather than 100,000 years to escape from the Sun's surface? What could be different about this hypothetical Sun?

Learning from Failure

The first detection of solar neutrinos raised more questions than it answered.

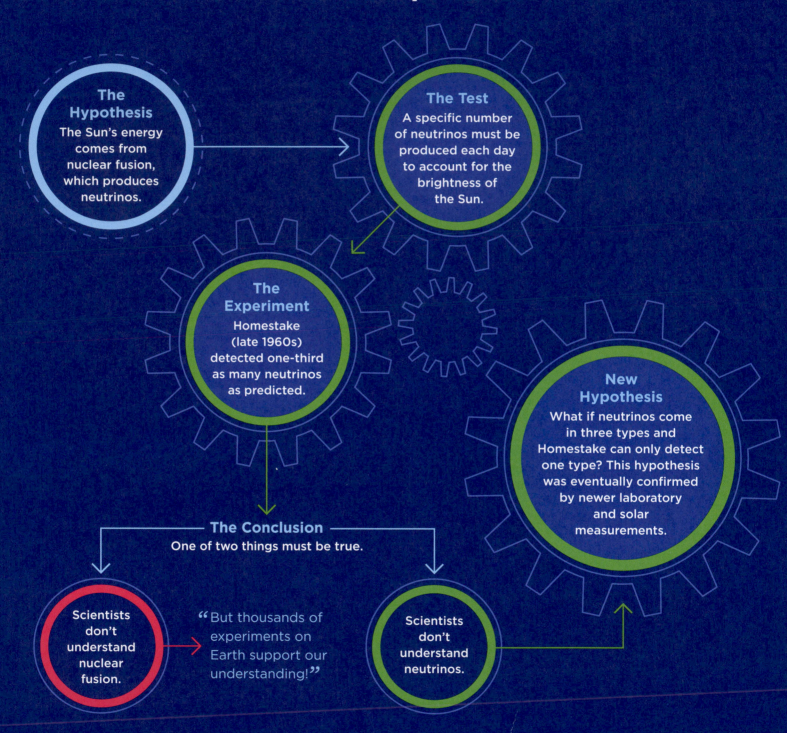

The Hypothesis
The Sun's energy comes from nuclear fusion, which produces neutrinos.

The Test
A specific number of neutrinos must be produced each day to account for the brightness of the Sun.

The Experiment
Homestake (late 1960s) detected one-third as many neutrinos as predicted.

New Hypothesis
What if neutrinos come in three types and Homestake can only detect one type? This hypothesis was eventually confirmed by newer laboratory and solar measurements.

The Conclusion
One of two things must be true.

Scientists don't understand nuclear fusion.

"But thousands of experiments on Earth support our understanding!"

Scientists don't understand neutrinos.

Part of the "scientific attitude" is to find failure exciting. When experiments do not turn out as expected, good scientists get excited because there is something new to understand!

or rings, something like a bell that has been struck. Unlike a well-tuned bell—which vibrates primarily at one frequency—the vibrations of the Sun are very complex. In the Sun, many frequencies of vibrations occur simultaneously, causing some parts of the Sun to bulge outward and some to draw inward. Those motions help researchers probe what lies below. **Figure 14.9** illustrates the motions of the different parts of the Sun, with red and blue areas moving in opposite directions. Some waves are amplified and some are suppressed, depending on how they overlap as they travel through the Sun. Astronomers study those waves by using the Doppler effect (see Chapter 5), which distinguishes between parts of the Sun that move toward the observer and those that move away.

To detect the disturbances of helioseismic waves on the surface of the Sun, astronomers must measure Doppler shifts of less than 0.1 m/s while detecting changes in brightness of only a few parts per million at any given location on the Sun. Tens of millions of wave motions are possible within the Sun. Some waves travel around the circumference of the Sun, yielding information about the density of the upper convection zone. Other waves travel through the interior of the Sun, revealing the Sun's density structure close to its core. Still others travel inward toward the center of the Sun, until they are bent by the changing solar density and return to the surface.

All those wave motions are going on at the same time, so sorting them out requires computer analysis of long, unbroken strings of solar observations from several sources. The Global Oscillation Network Group (GONG) is a network of six solar observation stations spread around the world that enables astronomers to observe the Sun's surface approximately 90 percent of the time.

To interpret helioseismology data, scientists compare the measurements of the strength, frequency, and wavelengths of the waves against predicted vibrations calculated from models of the solar interior. That technique serves as a powerful test of solar interior models and has led both to some surprises and to improvements in the models. For example, some scientists proposed that the solar neutrino problem might be solved if the models had overestimated the amount of helium in the Sun. That explanation was ruled out by analyzing the waves that penetrate to the Sun's core. Helioseismology showed that the value for opacity used in early solar models was too low. That realization led astronomers to recalculate the location of the bottom of the convective zone. Both theory and observation now put the base of the convective zone at 70 percent of the way out from the Sun's center, with an uncertainty in that number of less than 0.5 percent.

Working back and forth between observation and theory has enabled astronomers to probe the Sun's otherwise inaccessible interior. We now know that the energy is produced by nuclear fusion deep in the core and that it moves outward by radiation to a point about 70 percent of the Sun's radius. Then it travels outward by convection to the surface. We also know how the temperature, density, and pressure change with radius and how those factors change the opacity at different distances from the center. Even though researchers usually cannot sample directly or set up controlled experiments, collaboration between theory and observation is essential to observational sciences such as astronomy.

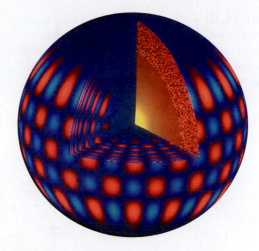

Figure 14.9 The interior of the Sun rings like a bell as helioseismic waves move through it. This figure shows one particular mode of the Sun's vibration, in false color. Red indicates regions where gas is traveling inward, whereas blue indicates regions where gas is traveling outward. Astronomers observe those motions via Doppler shifts.

CHECK YOUR UNDERSTANDING 14.2

How do neutrinos help us understand what is going on in the Sun's core? (a) Neutrinos from distant objects pass through the Sun, probing the interior. (b) Neutrinos from the Sun pass easily through Earth. (c) Neutrinos created in fusion reactions at the Sun's core easily escape.

Figure 14.10 a. The components of the Sun's atmosphere are located above the convective zone. **b.** The density and temperature of the Sun's atmosphere change abruptly at the boundary between the chromosphere and corona. The *y*-axes are logarithmic.

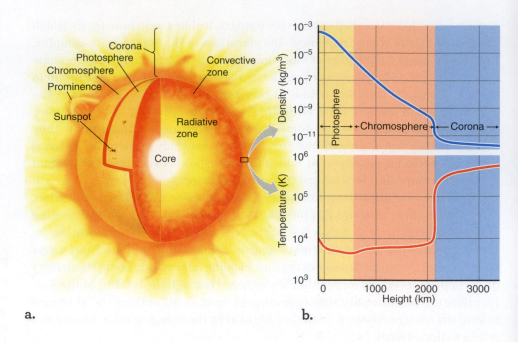

a.   b.

The Sun is "limb darkened." It is dimmer near its edge because near its edge we see the Sun at a steep angle and so do not see deeply into its atmosphere.

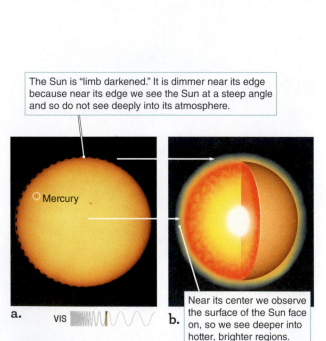

Near its center we observe the surface of the Sun face on, so we see deeper into hotter, brighter regions.

a.   VIS ╫╫╫╫╫〜〜〜   b.

Figure 14.11 a. When viewed in visible light, the Sun appears to have a sharp outline, even though it has no true surface. The center of the Sun appears brighter, whereas the limb of the Sun is darker—an effect known as limb darkening. The small black dot indicated by the white circle is Mercury, transiting in front of the Sun. **b.** Looking at the center of the Sun allows us to see deeper into the Sun's interior than we do when looking at the edge of the Sun. Because higher temperature means more luminous radiation, the center of the Sun appears brighter than its limb.

14.3 The Atmosphere of the Sun

Beyond the convective zone lie the outer layers of the Sun, collectively known as the Sun's atmosphere. Those layers, shown in **Figure 14.10a**, include the *photosphere*, the *chromosphere*, and the *corona*. We can observe those layers of the Sun directly by using telescopes and satellites. Observations of the Sun's atmosphere are important because activity in the Sun's atmosphere has consequences for human infrastructure such as power grids and satellites in orbit around Earth.

Unlike Earth, the Sun has no solid surface. Its apparent surface is like a fog bank on Earth. After people walk into a fog bank and disappear from view, you would say they were inside the fog bank, even though they never passed through a definite boundary. The apparent surface of the Sun is similar. Light from the Sun's surface can escape directly into space, so we can see it. Light from below the Sun's surface cannot escape directly into space, so we cannot see it.

At the base of the atmosphere is the **photosphere**: the apparent surface of the Sun or its visible boundary. That is where features such as sunspots can be seen. Above the photosphere is the **chromosphere**, a region of strong emission lines. The top layer is the **corona**, which can be viewed during a solar eclipse as a halo around the Sun. In the Sun's atmosphere, the density of the gas drops very rapidly with increasing altitude. The graphs in **Figure 14.10b** show how density and temperature change across the Sun's atmosphere. The temperature increases sharply across the boundary between the chromosphere and the corona, whereas the density falls sharply across the same boundary. In this section, we explore each layer, beginning at the bottom with the photosphere.

The Photosphere

The **effective temperature** of a star is the temperature found from the spectrum of the light emitted from the photosphere. While underlying layers are much hotter, the effective temperature is an observable quantity that is useful for comparing stars with one another; when astronomers mention the temperature of a star, they are typically referring to the effective temperature. The effective temperature of the Sun is calculated from the Sun's luminosity and radius by using the Stefan-Boltzmann law (see Chapter 5). The photosphere has an effective temperature of 5780 K, ranging from 6600 K to 4500 K over a 500-km-thick zone. The Sun appears to have a well-defined surface and a sharp outline when viewed from Earth because 500 km does not look very thick when viewed from a distance of 150 million km.

In **Figure 14.11a**, the Sun appears fainter near its edges than near its center, an effect called **limb darkening**. That effect is an artifact of the structure of the Sun's photosphere. When looking near the edge of the Sun, you are looking through the photosphere at a steep angle. As a result, you do not see as deeply into the Sun as when you are looking directly down through the photosphere near the center of the Sun's disk. The light from the **limb** of the Sun comes from a shallower layer that is cooler and fainter, as shown in **Figure 14.11b**.

In the Sun's atmosphere, the gas density drops very rapidly with increasing altitude. All visible solar phenomena take place in the Sun's atmosphere. Most of the radiation from below the Sun's photosphere is absorbed by matter and reemitted at the photosphere as a blackbody spectrum.

As we examine the Sun's structure in more detail, however, that simple description of the spectra of stars turns out to be incomplete. Light from the solar photosphere must escape through the upper layers of the Sun's atmosphere, which affects the spectrum we observe. In Chapter 13, we discussed the presence of absorption lines in the spectra of stars. Now we can take a closer look at how those absorption lines form. As photospheric light travels upward, atoms in the solar atmosphere absorb the light at discrete wavelengths, forming absorption lines. Because the Sun appears so much brighter than any other star, its spectrum can be studied in far more detail, so specially designed telescopes and high-resolution spectrometers have been built specifically to study the Sun's light. The solar spectrum is shown in **Figure 14.12**. Absorption lines from more than 70 elements have been identified. Analysis of those lines forms the basis for much of astronomers' knowledge of the solar atmosphere, including the Sun's composition. That is also the starting point for understanding the atmospheres and spectra of other stars.

The Chromosphere and Corona

The temperature falls from 6600 K at the photosphere's bottom to 4400 K at its top. At that point, the trend reverses and the temperature slowly begins to climb, rising to about 6000 K at a height of 1500 km above the top of the photosphere (see Figure 14.10b). This region of increasing temperature is called the chromosphere (**Figure 14.13a**). Why the chromosphere's temperature increases from the bottom to the top is not well understood, but it may be caused by magnetic waves propagating through the region and depositing their energy at the top of the chromosphere.

The chromosphere was discovered in the 19th century during observations of total solar eclipses (**Figure 14.13b**). The chromosphere is seen most strongly at the solar limb as a source of emission lines, especially the Hα line (the "hydrogen alpha line"), which is produced when an electron falls from the third energy state of hydrogen to the second energy state. The deep red color of the Hα line is what gives the chromosphere its name ("chromosphere" means "the place where color

what if . . .

What if the temperature at the base of the corona were 10 million K rather than 1 million K? How would the light from this region be different in this hypothetical Sun?

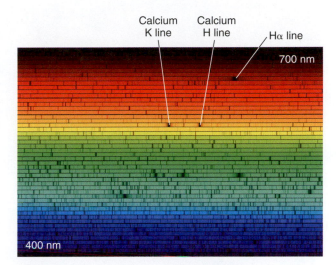

Figure 14.12 This high-resolution spectrum of the Sun stretches from 400 nanometers (nm) in the lower left corner to 700 nm in the upper right corner and shows black absorption lines.

★ **WHAT AN ASTRONOMER SEES** An astronomer will know that this spectrum was produced by passing the Sun's light through a prism-like device and then cutting and folding the single long spectrum (from blue to red) into rows so that it will fit in a single image taken by a camera. She will notice the particularly strong absorption lines, which show up as dark blotches. The one near the top in the red part of the spectrum is particularly noticeable, and an astronomer who looks at a lot of spectra will recognize this line by its color: It is the hydrogen alpha (Hα) line, marking the transition of electrons from the third down to the second energy state of hydrogen. She might also recognize the strong "calcium H" and "calcium K" lines in the orange part of the spectrum. These are recognizable from the combination of their color, their relative strength, and their nearness to each other. The Sun's spectrum is crowded with absorption lines, and an astronomer will immediately know that the outer layers of the Sun are cooler than the layers deep down, because atoms in those outer layers are absorbing energy as it makes its way out from the Sun. Credit: Nigel Sharp, NOAO/NSO/Kitt Peak FTS/AURA/NSF, https://noirlab .edu/public/images/noao-sun/. https://creativecommons.org/licenses/by/4.0.

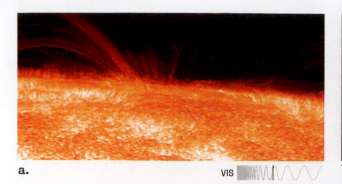

a.   VIS

b.   VIS

c.   VIS

Figure 14.13 **a.** This spacecraft image of the Sun shows fine structure in the chromosphere extending outward from the photosphere. **b.** The chromosphere (seen in pink here) is visible during a total eclipse. **c.** This eclipse image shows the Sun's corona, consisting of million-kelvin gas that extends for millions of kilometers beyond the surface of the Sun.

comes from"). The element helium was discovered in 1868 from a spectrum of the chromosphere of the Sun nearly 30 years before it was found on Earth. Helium is named after *helios*, the Greek word for "Sun."

At the top of the chromosphere, across a transition region only about 100 km thick, the temperature suddenly soars, whereas the density abruptly drops (see Figure 14.10b). Above that transition lies the outermost region of the Sun's atmosphere, the corona, where temperatures reach 1 million to 2 million K. The corona is thought to be heated by magnetic fields and micro solar flares. NASA's *Parker Solar Probe* is currently in orbit around the Sun, investigating the flow of energy that heats and accelerates the solar corona and solar wind. As a result, scientists hope to be better able to predict changes in the solar wind known as "space weather," which affects satellites, spacecraft, and astronauts, even here in Earth orbit.

The Sun's corona has been known since ancient times: It is visible with the naked eye during total solar eclipses as an eerie glow stretching several solar radii beyond the Sun's surface (**Figure 14.13c**). Because it is so hot, the solar corona also is a strong source of X-rays. Those X-ray photons, which are invisible to the human eye, have so much energy that many electrons are stripped away from nuclei, leaving atoms in the corona highly ionized.

CHECK YOUR UNDERSTANDING 14.3

The Sun's surface appears sharp in visible light because: (a) the photosphere is cooler than the layers below it; (b) the photosphere is thinner than the other layers in the Sun; (c) the photosphere is less dense than the convection zone; (d) the Sun has a distinct surface.

14.4 The Atmosphere of the Sun Is Very Active

The Sun's atmosphere is a very turbulent place. The best-known features on the Sun's surface are relatively dark blemishes in the solar photosphere, called **sunspots**. Sunspots come and go over time, though they remain long enough for us to determine the rotation rate of the Sun. Those spots are associated with **active regions**: loops of material and explosions that fling particles far out into the Solar System. Long-term patterns have been observed in the variations of sunspots and active regions, revealing that the Sun's magnetic field is constantly changing.

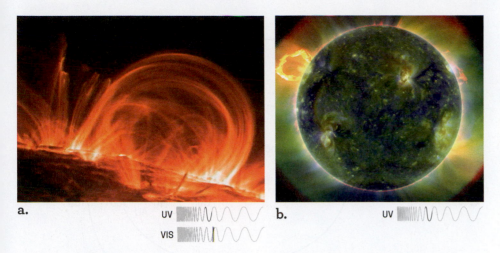

Figure 14.14 **a.** This close-up image of the Sun shows the tangled structure of coronal loops. **b.** This image is a combination of several extreme ultraviolet images of the Sun from the Solar Dynamics Observatory (SDO). Coronal holes are dark in these images, which indicates that they are cooler and less dense than their surroundings. Several solar prominences are also visible in this image, as well as the wispy corona.

Solar Activity Is Caused by Magnetic Effects

The Sun's magnetic field (see Chapter 5) causes virtually all the structure seen in the Sun's atmosphere. High-resolution images of the Sun show *coronal loops* that make up much of the Sun's lower corona (**Figure 14.14a**). That texture is the result of magnetic structures called flux tubes. Magnetic fields are responsible for much of the corona's structure as well. The corona is far too hot to be held in by the Sun's gravity, but over most of the Sun's surface, coronal gas is confined by magnetic loops with both ends firmly anchored deep within the Sun. The magnetic field in the corona acts almost like a network of rubber bands that coronal gas is free to slide along but cannot cross. In contrast, about 20 percent of the Sun's surface is covered by an ever-shifting pattern of **coronal holes**, large regions where the magnetic field points outward, away from the Sun, and where coronal material is free to stream away into interplanetary space as the solar wind. In extreme ultraviolet images of the Sun, we see coronal holes as dark regions, which indicates that they are cooler and less dense than their surroundings (**Figure 14.14b**).

The relatively steady part of the solar wind consists of lower-speed flows with velocities of about 350 km/s and higher-speed flows with velocities up to about 700 km/s. The higher-speed flows originate in coronal holes. Depending on their speed, particles in the solar wind take 2–5 days to reach Earth. Often, 2–5 days after a coronal hole passes across the center of the face of the Sun, the speed and density of the solar wind reaching Earth increases. The solar wind drags the Sun's magnetic field along with it (**Figure 14.15**), so the magnetic field in the solar wind gets "wound up" by the Sun's rotation. Consequently, the magnetic field has a spiral structure resembling the stream of water from a rotating lawn sprinkler.

The effects of the solar wind are felt throughout the Solar System. The solar wind blows the tails of comets away from the Sun, shapes the magnetospheres of the planets, and supplies the energetic particles that power Earth's spectacular auroral displays. Using space probes, astronomers have observed the solar wind extending out to 100 astronomical units (AU) from the Sun. But the solar wind does not go on forever. The farther it gets from the Sun, the more it spreads out.

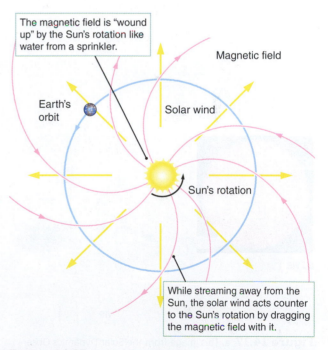

The magnetic field is "wound up" by the Sun's rotation like water from a sprinkler.

Magnetic field

Earth's orbit

Solar wind

Sun's rotation

While streaming away from the Sun, the solar wind acts counter to the Sun's rotation by dragging the magnetic field with it.

Figure 14.15 The solar wind (yellow arrows) streams away from active areas and coronal holes on the Sun. As the Sun rotates, the solar wind takes on a spiral structure, much like the spiral of water that streams away from a rotating lawn sprinkler.

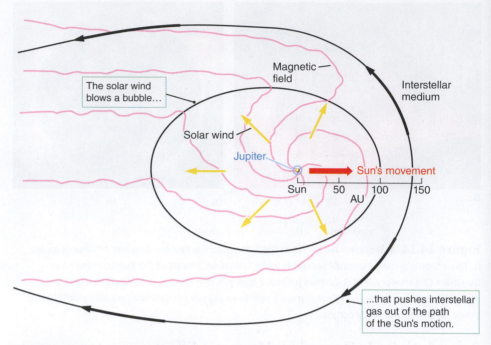

Figure 14.16 The solar wind (yellow arrows) streams away from the Sun for about 100 AU until it finally piles up against the pressure of the interstellar medium through which the Sun is traveling. The *Voyager 1* spacecraft crossed that boundary in 2012.

Just like radiation, the density of the solar wind follows an inverse square law. At a distance of about 100 AU from the Sun, the solar wind stops abruptly. Here it piles up against the pressure of the **interstellar medium**, the gas and dust that lie between stars in a galaxy. **Figure 14.16** shows the region of space over which the solar wind is measured. The *Voyager 1* and *Voyager 2* spacecraft have crossed the outer edge of that boundary and sent back the first direct measurements of true interstellar space. The *Interstellar Boundary Explorer* spacecraft, launched in 2008, is exploring the region, too.

Sunspots and Changes in the Sun

Sunspots have been noted since antiquity. Telescopic observations of sunspots date back almost 400 years, and records exist of naked-eye observations by Chinese, Greek, and medieval astronomers centuries before that. *But remember: Never look directly at the Sun!* Direct viewing through a commercial solar filter is safe, as is projecting the image through a telescope or binoculars onto a surface such as paper and looking only at the projection. Many websites have live images of the Sun viewed through ground and space telescopes. Sunspots are places where material is trapped at the surface of the Sun by magnetic-field lines. When that material cools, convection cannot carry it downward, so it makes a cooler (and therefore darker) spot on the surface of the Sun. **Figure 14.17** shows a large sunspot group. Sunspots appear dark, but only in contrast to the brighter surface of the Sun (**Working It Out 14.2**).

Early telescopic observations of sunspots made during the 17th century led to the discovery of the Sun's rotation, which has an average period of about 27 days as seen from Earth and 25 days relative to the stars. Because Earth orbits the Sun in the same direction that the Sun rotates, observers on Earth see a

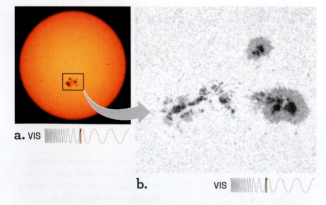

Figure 14.17 **a.** This image from the Solar Dynamics Observatory, taken in 2010, shows a large sunspot group. Sunspots are magnetically active regions cooler than the surrounding surface of the Sun. **b.** This high-resolution view shows the sunspots in this group.

= =

working it out 14.2

Sunspots and Temperature

Sunspots are about 1500 K cooler than their surroundings. What does that lower temperature tell us about their luminosity? Think back to the Stefan-Boltzmann law in Chapter 5. The flux, $\mathcal{F}$, from a blackbody is proportional to the fourth power of the temperature, T:

$$\mathcal{F} = \sigma T^4$$

The constant of proportionality is the Stefan-Boltzmann constant, σ, which has a value of $5.67 \times 10^{-8}\ \mathrm{W/(m^2\ K^4)}$.

The flux is the amount of energy coming from a square meter of surface every second. How much less energy comes out of a sunspot than out of the rest of the Sun? Using 4500 and 6000 K for the temperatures of a typical sunspot and the surrounding photosphere, respectively, we can set up two equations:

$$\mathcal{F}_{spot} = \sigma T^4_{spot} \quad \text{and} \quad \mathcal{F}_{surface} = \sigma T^4_{surface}$$

We could solve each of those separately and then divide the value of $\mathcal{F}_{spot}$ by $\mathcal{F}_{surface}$ to find out how much fainter the sunspot is, but solving for the ratio of the fluxes eliminates σ:

$$\frac{\mathcal{F}_{spot}}{\mathcal{F}_{surface}} = \frac{\sigma T^4_{spot}}{\sigma T^4_{surface}} = \frac{T^4_{spot}}{T^4_{surface}} = \left(\frac{T_{spot}}{T_{surface}}\right)^4$$

Plugging in our values for T_{spot} and $T_{surface}$ gives

$$\frac{\mathcal{F}_{spot}}{\mathcal{F}_{surface}} = \left(\frac{4500\ \mathrm{K}}{6000\ \mathrm{K}}\right)^4 = 0.32$$

Finally, multiplying both sides by $\mathcal{F}_{surface}$ gives

$$\mathcal{F}_{spot} = 0.32\mathcal{F}_{surface}$$

So, the amount of energy coming from a square meter of sunspot every second is about one-third as much as the amount of energy coming from a square meter of surrounding surface every second. In other words, the sunspot is about one-third as bright as the surrounding photosphere. If you could cut out the sunspot and place it elsewhere in the sky, it would be brighter than the full Moon.

---

slightly longer rotation period. Observations of sunspots also show that the Sun's rotation period is shorter at its equator than at higher latitudes, which means that the equator rotates faster. This effect is called **differential rotation**. Differential rotation is possible only because the Sun is not a solid object.

Figure 14.18 shows the structure of a sunspot on the surface of the Sun. A sunspot consists of an inner dark core called the **umbra**, surrounded by a less dark region called the **penumbra**. These words are the same as those used when discussing brighter and darker areas within shadows (see Figure 2.24), but sunspots are not shadows. The words are simply used to refer to the brighter and darker regions of the sunspot. The penumbra shows an intricate radial pattern, reminiscent of a flower's petals. Sunspots are caused by magnetic fields thousands of times greater than the magnetic field at Earth's surface. They occur in pairs connected by loops in the magnetic field. Sunspots range in size from a few tens of kilometers across up to complex groups that may contain several dozen individual spots and span as much as 150,000 km. The largest sunspot groups can be seen without a telescope.

Although sunspots occasionally last 100 days or longer, half of all sunspots come and go in about 2 days, and 90 percent are gone within 11 days. The number and distribution of sunspots change in a pattern averaging 11 years called the **sunspot cycle**. **Figure 14.19a** shows data for several recent cycles. At the beginning of a cycle, sunspots appear at solar latitudes of about 30° north and south of the solar equator. Over the following years, sunspots are found closer to the equator as their number increases to a maximum and then declines. As the last few sunspots approach the equator, new sunspots again begin appearing at middle

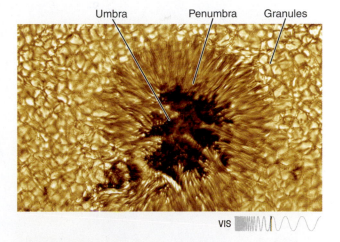

Umbra   Penumbra   Granules

VIS

Figure 14.18 This very high-resolution view of a sunspot shows the dark umbra surrounded by the lighter penumbra. The solar surface around the sunspot bubbles with separate cells of hot gas called *granules*. The smallest features are about 100 km across, whereas each granule is about the size of the state of Texas. Credit: NOIRLab, https://noirlab.edu/public/images/noao9808a/, https://creativecommons.org/licenses/by/4.0/.

Figure 14.19 a. The number of sunspots varies, as shown in this graph of the past few solar cycles. **b.** The "solar butterfly" diagram shows the fraction of the Sun covered by sunspots at each latitude. The data are color coded to show the percentage of the strip at that latitude that is covered in sunspots at that time: black, 0 to less than 0.1 percent; red, 0.1–1.0 percent; yellow, greater than 1.0 percent. **c.** The Sun's magnetic poles flip every 11 years. Yellow indicates magnetic north, whereas blue indicates magnetic south.

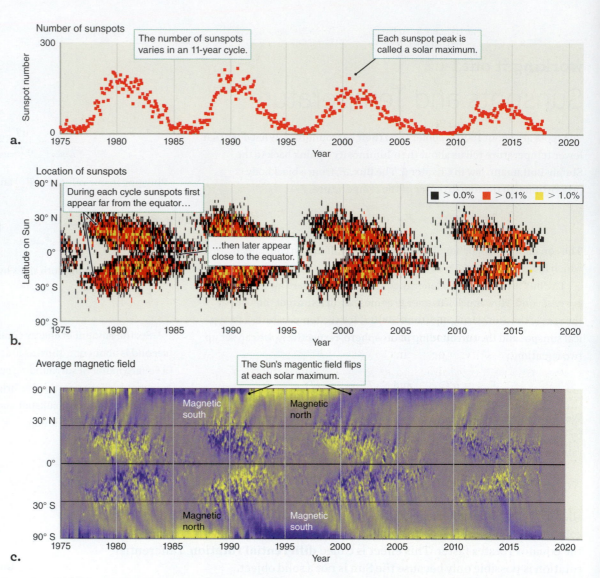

latitudes, and the next cycle begins. **Figure 14.19b** shows the number of sunspots at a given latitude plotted against time. The resulting diagram of opposing diagonal bands is called the sunspot "butterfly diagram."

In the early 20th century, solar astronomer George Ellery Hale (1868–1938) was the first to show that the 11-year sunspot cycle is actually half of a 22-year magnetic cycle during which the direction of the Sun's magnetic field reverses after each 11-year sunspot cycle. **Figure 14.19c** shows how the average strength of the magnetic field at every latitude has changed over more than 45 years. The direction of the Sun's magnetic field flips at the maximum of each sunspot cycle. Sunspots come in pairs, with one spot (the leading sunspot) in front of the other with respect to the Sun's rotation. In one 11-year sunspot cycle, the leading sunspot in each pair tends to be a north magnetic pole, whereas the trailing sunspot tends to be a south magnetic pole. In the next 11-year sunspot cycle, that polarity is reversed, so the leading sunspot in each pair is a south magnetic pole, whereas the trailing sunspot tends to be a north magnetic pole. The transition between those two magnetic polarities occurs near the peak of each sunspot cycle. Magnetic activity on the Sun affects the photosphere, chromosphere, and corona.

The graph in **Figure 14.20** shows 400 years of sunspot observations. The 11-year cycle is neither perfectly periodic nor especially reliable. The time between peaks in the number of sunspots actually varies between about 9.7 and 11.8 years. The

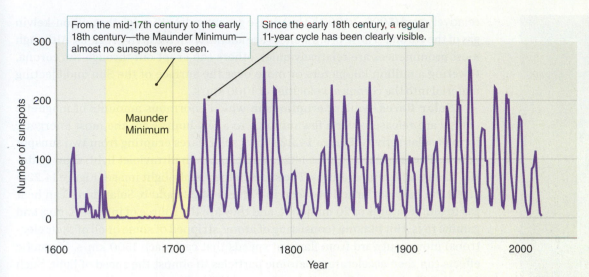

From the mid-17th century to the early 18th century—the Maunder Minimum—almost no sunspots were seen.

Since the early 18th century, a regular 11-year cycle has been clearly visible.

Figure 14.20 Sunspots have been observed for hundreds of years. In this plot, the 11-year cycle in the number of sunspots (half of the 22-year solar magnetic cycle) is clearly visible. Sunspot activity varies greatly. The period from the middle of the 17th century to the early 18th century, when almost no sunspots were seen, is called the *Maunder minimum*.

number of spots seen during a given cycle fluctuates as well, and in some periods sunspot activity has disappeared almost entirely. An extended lull in solar activity, called the **Maunder minimum**, lasted from 1645 to 1715. Typically, about six peaks of solar activity occur in 70 years, but virtually no sunspots were seen during the Maunder minimum, and auroral displays were less frequent than usual.

Sunspots are only one of several phenomena that follow the Sun's 22-year cycle of magnetic activity. The peaks of the cycle, called **solar maxima**, are times of intense activity. Sunspots are often accompanied by a brightening of the solar chromosphere that is seen most clearly in emission lines such as Hα. Those bright regions are known as solar active regions. The magnificent loops arching through the solar corona, shown in **Figure 14.21**, are solar **prominences**, magnetic flux tubes of

what if . . .

What if the Sun had only a very small magnetic field at the surface? How might that change the appearance of the Sun?

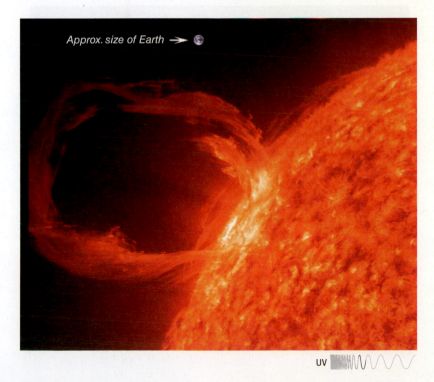

Approx. size of Earth ➡

UV

Figure 14.21 Solar prominences are magnetically supported arches of hot gas that rise high above active regions on the Sun. Here, you can see a close-up view at the base of a large prominence. An image of Earth is included for scale (it is not actually that close to the Sun).

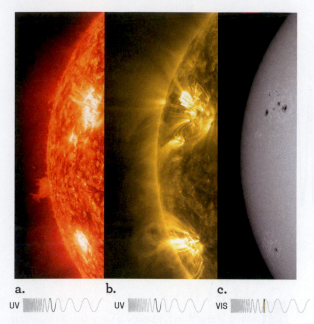

Figure 14.22 The Solar Dynamics Observatory (SDO) observed these active regions of the Sun that produced solar flares in August 2011. **a.** Activity near the surface at 60,000 K is visible in extreme ultraviolet light (along with a prominence rising up from the Sun's edge). **b.** Viewed at other ultraviolet wavelengths, many looping arcs and plasma heated to about 1 million K become visible. **c.** The dark spots in this image are the magnetically intense sunspots that are the sources of all the activity.

relatively cool (5000–10,000 K) but dense gas extending through the million-kelvin gas of the corona. Those prominences are anchored in the active regions. Although most prominences are relatively quiet, others can erupt out through the corona, towering a million kilometers or more over the surface of the Sun and ejecting material into the corona at velocities of 1000 km/s.

Solar flares—violent eruptions in which enormous amounts of magnetic energy are released over a few minutes to a few hours—are the most energetic form of solar activity. **Figure 14.22** shows solar flares erupting from two sunspot groups. The images in Figures 14.22a and b, taken in ultraviolet light, show material at very high temperatures. The spots in the visible-light image (Figure 14.22c) are at the base of the activity seen in Figures 14.22a and b. Solar flares can heat gas to temperatures of 20 million K, and they are the source of intense X-rays and gamma rays. Hot **plasma** (consisting of atoms stripped of some or all of their electrons) moves outward from flares at speeds that can reach 1500 km/s. Magnetic effects can then accelerate subatomic particles to almost the speed of light. Such events, called **coronal mass ejections (CMEs)** (**Figure 14.23**), send powerful bursts of energetic particles outward through the Solar System. CMEs occur about once per week during the minimum of the sunspot cycle and as often as several times per day near the maximum of the cycle.

Solar Activity Affects Earth

The amount of solar radiation received at the distance of Earth from the Sun has been measured to be 1361 watts per square meter (W/m²) on average. Satellite

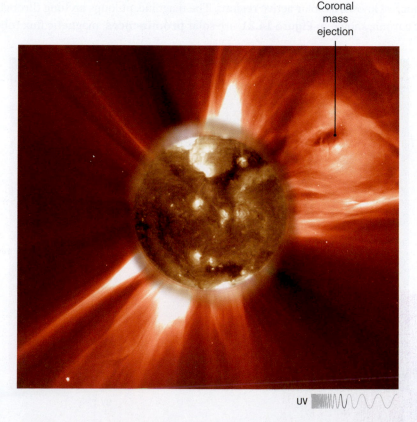

Coronal mass ejection

Figure 14.23 This Solar and Heliospheric Observatory (SOHO) image shows a coronal mass ejection (upper right); a simultaneously recorded ultraviolet image of the solar disk is superimposed.

Reading Astronomy News

Carrington-Class CME Narrowly Misses Earth

Dr. Tony Phillips, Science@NASA

Last month (April 8–11, 2014), scientists, government officials, emergency planners, and others converged on Boulder, Colorado, for NOAA's Space Weather Workshop—an annual gathering to discuss the perils and probabilities of solar storms.

The current solar cycle is weaker than usual, so you might expect a correspondingly low-key meeting. On the contrary, the halls and meeting rooms were abuzz with excitement about an intense solar storm that narrowly missed Earth.

"If it had hit, we would still be picking up the pieces," says Daniel Baker of the University of Colorado, who presented a talk entitled *The Major Solar Eruptive Event in July 2012: Defining Extreme Space Weather Scenarios*.

The close shave happened almost two years ago. On July 23, 2012, a plasma cloud or "CME" rocketed away from the Sun as fast as 3000 km/s, more than 4 times faster than a typical eruption. The storm tore through the Earth's orbit, but fortunately Earth wasn't there. Instead it hit the *STEREO-A* spacecraft. Researchers have been analyzing the data ever since, and they have concluded that the storm was one of the strongest in recorded history. "It might have been stronger than the Carrington Event itself," says Baker.

The Carrington Event of September 1859 was a series of powerful CMEs that hit Earth head-on, sparking Northern Lights as far south as Tahiti. Intense geomagnetic storms caused global telegraph lines to spark, setting fire to some telegraph offices and

disabling the "Victorian Internet." A similar storm today could have a catastrophic effect on modern power grids and telecommunication networks. According to a study by the National Academy of Sciences, the total economic impact could exceed $2 trillion or 20 times greater than the costs of a Hurricane Katrina. Multi-ton transformers fried by such a storm could take years to repair and impact national security.

A recent paper in *Nature Communications* authored by UC Berkeley space physicist Janet G. Luhmann and former postdoc Ying D. Liu describes what gave the July 2012 storm Carrington-like potency. For one thing, the CME was actually *two* CMEs separated by only 10 to 15 minutes. This double storm cloud traveled through a region of space that had been cleared out by another CME only four days earlier. As a result, the CMEs were not decelerated as much as usual by their transit through the interplanetary medium.

Had the eruption occurred just one week earlier, the blast site would have been facing Earth, rather than off to the side, so it was a relatively narrow escape.

When the Carrington Event enveloped Earth in the 19th century, technologies of the day were hardly sensitive to electromagnetic disturbances. Modern society, on the other hand, is deeply dependent on Sun-sensitive technologies such as GPS, satellite communications, and the Internet.

"The effect of such a storm on our modern technologies would be tremendous," says Luhmann.

During informal discussions at the workshop, Nat Gopalswamy of the Goddard Space Flight Center noted that "without NASA's *STEREO* probes, we might never have known the severity of the 2012 superstorm. This shows the value of having 'space weather buoys' located all around the Sun."

It also highlights the potency of the Sun even during so-called "quiet times." Many observers have noted that the current solar cycle is weak, perhaps the weakest in 100 years. Clearly, even a weak solar cycle can produce a very strong storm. Says Baker, "We need to be prepared."

QUESTIONS

1. What is a coronal mass ejection (CME)?

2. Why would a CME cause disruptions on Earth?

3. Explain how the 2012 storm missed Earth by 1 week.

4. More sensationalistic headlines for this story claimed that Earth almost "was sent back to the Dark Ages." What did they mean by that exaggeration?

5. Go to the NASA press release for this event (https://science.nasa.gov/science-news/science-at-nasa/2014/23jul_superstorm/), and click to watch the 4-minute "Science-Cast" video. What happened during the Carrington CME in 1859? Is that video effective at communicating the science information to the nonspecialist?

Source: https://science.nasa.gov/science-news/science-at-nasa/2014/02may_superstorm/.

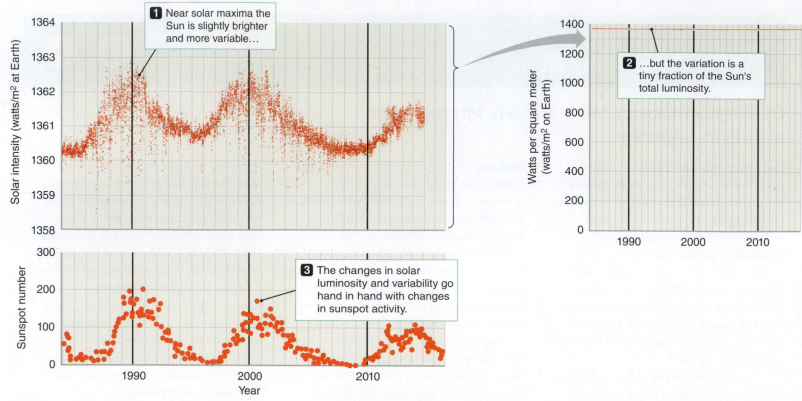

Figure 14.24 Measurements taken by satellites show that the amount of light from the Sun changes slightly.

measurements of the amount of radiation coming from the Sun (**Figure 14.24**) show that this value varies by as much as 0.2 percent over periods of a few weeks, as dark sunspots in the photosphere and bright spots in the chromosphere move across the disk. Overall, however, the increased radiation from active regions on the Sun more than makes up for the reduction in radiation from sunspots. On average, the Sun seems to be about 0.1 percent brighter during the peak of a solar cycle than it is at its minimum.

Solar activity affects Earth in many ways. Solar active regions are the source of most of the Sun's extreme ultraviolet and X-ray emissions—energetic radiation that heats Earth's upper atmosphere and, during periods of increased solar activity, causes Earth's upper atmosphere to expand. When that happens, the swollen upper atmosphere can significantly increase the atmospheric drag on spacecraft orbiting at relatively low altitudes, such as that of the Hubble Space Telescope, causing their orbits to decay. As a result, periodic boosts have been necessary to keep the Hubble Space Telescope in its orbit.

Earth's magnetosphere is the result of the interaction between Earth's magnetic field and the solar wind. Increases in the solar wind accompanying solar activity, especially CMEs directed at Earth, can disrupt Earth's magnetosphere. Spectacular auroras can accompany such events, as can magnetic storms that have disrupted electric power grids and caused blackouts across large regions. CMEs emitted in the direction of Earth also hinder radio communication and navigation, and they can damage sensitive satellite electronics, including communication satellites. In addition, energetic particles accelerated in solar flares pose one of the greatest dangers to human exploration of space.

Detailed observations from the ground and from space help astronomers understand the complex nature of the solar atmosphere. The Solar and Heliospheric Observatory (*SOHO*) spacecraft is a joint mission between NASA and the European

unanswered questions

Are large timescale variations in Earth's climate—ice ages—related to solar activity? Solar activity affects Earth's upper atmosphere and may affect weather patterns as well. Variations in the amount of radiation from the Sun might be responsible for past variations in Earth's climate. Current models indicate that observed variations in the Sun's luminosity could account for only about 0.1-K differences in Earth's average temperature—much less than the effects due to the ongoing buildup of carbon dioxide in Earth's atmosphere. Triggering the onset of an ice age may require a sustained drop in global temperatures of only 0.2–0.5 K, so astronomers are continuing to investigate a possible link between solar variability and long-timescale changes in Earth's climate.

Space Agency (ESA). Because of the combined gravity of Earth and the Sun, *SOHO* moves in lockstep with Earth at a location approximately 1,500,000 km from Earth that is almost directly in line between Earth and the Sun. *SOHO* carries 12 scientific instruments that monitor the Sun and measure the solar wind upstream of Earth. In addition, NASA's Solar Dynamics Observatory (SDO) studies the solar magnetic field to predict when major solar events will occur, instead of simply responding after they happen.

CHECK YOUR UNDERSTANDING 14.4

Sunspots appear dark because: (a) they have very low density; (b) magnetic fields absorb most of the light that falls on them; (c) they are regions of very high pressure; (d) they are cooler than their surroundings.

Origins: The Solar Wind and Life

Energetic particles accelerated in solar flares need to be considered when astronauts are orbiting Earth in a space station or traveling to the Moon or farther. Earth's magnetic field protects life on the surface from those energetic particles, which travel along the magnetic-field lines to Earth's poles, creating the auroras. But the Moon does not have that protection because its magnetic field is very weak. Astronauts on the lunar surface would be exposed to as much radiation as astronauts traveling in space. The strength of the solar wind varies with the solar cycle, as noted in Section 14.4, so the exposure danger varies as well.

The Solar System is surrounded by the **heliosphere (Figure 14.25)**, in which the solar wind blows against the interstellar medium and clears out an area like the inside of a bubble. As the Sun and Solar System move through the Milky Way Galaxy, passing in and out of interstellar clouds, that heliosphere protects the entire Solar System from galactic high-energy particles known as cosmic rays that originate primarily in high-energy explosions of massive dying stars. When the Sun is in its lower-activity state, the heliosphere is weaker, so more galactic cosmic rays enter the Solar System. In addition, the intensity of those cosmic rays depends on where the Sun and Solar System are located in their orbit about the center of the Milky Way Galaxy.

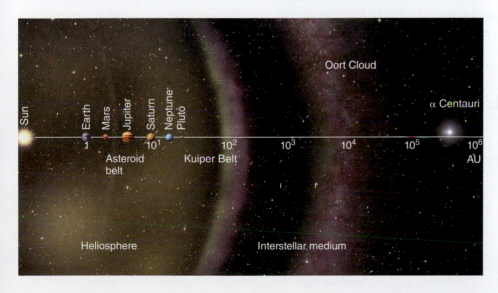

Figure 14.25 The heliosphere of the Sun is a bubble of charged particles covering the Solar System. The heliosphere is formed by the solar wind blowing against the interstellar medium. Both of the *Voyager* spacecraft are now past 100 AU. The scale, indicating how far objects are from the Sun, is logarithmic. Recall that a logarithmic graph has steps that are powers of ten: 10, 100, 1000, 10,000, etc.

Some scientists have theorized that at times when the Sun was quiet and the heliosphere was weaker than average, and the Solar System was passing through a particular part of the galaxy, the cosmic-ray flux in the Solar System—and on Earth—increased. That increased flux possibly led to a disruption in Earth's ozone layer and possibly contributed to a mass extinction in which many species died out on Earth.

Thus, in addition to heat and light on Earth, the extension of the Sun through the solar wind may have affected the evolution of life on Earth—and may affect the ability of humans to live and work in space.

SUMMARY

The forces due to pressure and gravity balance each other in hydrostatic equilibrium, maintaining the Sun's structure. Nuclear reactions converting hydrogen to helium are the source of the Sun's energy. Energy created in the Sun's core moves outward to the surface, first by radiation and then by convection. The solar wind may adversely affect astronauts located in space or on planets that lack a protective magnetic field, but it also has protected the Solar System from galactic high-energy cosmic-ray particles.

(1) **Describe the balance between the forces that determine the structure of the Sun.** The outward radiation pressure of the hot gas inside the Sun balances the inward pull of gravity at every point. That balance is dynamically maintained. An energy balance is also maintained, with the energy produced in the Sun's core balancing the energy lost from the surface.

(2) **Diagram how mass is converted to energy in the Sun's core and estimate how long the Sun will take to use up its fuel.** In the Sun's core, mass is converted to energy via the proton-proton chain. When four hydrogen atoms fuse to form one helium atom, some mass is lost. That mass is released as energy, nearly all of which leaves the Sun either as photons or as neutrinos. Neutrinos are elusive, almost massless particles that interact only very weakly with other matter. Observations of neutrinos confirm that nuclear fusion is the Sun's primary energy source. The innermost 10 percent of the Sun will participate in the fusion process, limiting the Sun's main-sequence lifetime to 10 billion years.

(3) **Sketch a physical model of the Sun's interior, and list the ways that energy moves outward from the Sun's core toward its surface.** The Sun's interior is divided into zones defined by how energy is transported in that region. Energy moves outward through the Sun by radiation and by convection.

(4) **Relate observations of solar neutrinos and seismic vibrations on the surface of the Sun to astronomers' models of the Sun.** The Sun has multiple layers, each with a characteristic pressure, density, and temperature. Neutrinos directly probe the interior of the Sun. That model of the interior of the Sun has been tested by helioseismology, in much the same way that the model of Earth's interior has been tested by seismology.

(5) **Describe the solar activity cycles of 11 and 22 years, and relate those cycles to the Sun's changing magnetic field and solar phenomena such as sunspots and flares.** Activity on the Sun follows a cycle that peaks every 11 years but takes 22 full years for the magnetic field to reverse. Sunspots are photospheric regions cooler than their surroundings, and they reveal the cycles in solar activity. Material streaming away from the Sun's corona creates the solar wind, which moves outward through the Solar System until it meets the interstellar medium. Solar storms, including flares and other ejections of mass from the corona, produce auroras and can disrupt power grids and damage satellites.

QUESTIONS AND PROBLEMS

TEST YOUR UNDERSTANDING

1. The physical model of the Sun's interior has been confirmed by observations of
   a. neutrinos and seismic vibrations.
   b. sunspots and solar flares.
   c. neutrinos and positrons.
   d. sample returns from spacecraft.
   e. sunspots and seismic vibrations.

2. Place in order the following steps in the fusion of hydrogen into helium. If two or more steps happen simultaneously, use an equals sign (=).
   a. A positron is emitted.
   b. One gamma ray is emitted.
   c. Two hydrogen nuclei are emitted.
   d. Two ^{3}He collide and become ^{4}He.
   e. Two hydrogen nuclei collide and become ^{2}H.
   f. Two gamma rays are emitted.
   g. A neutrino is emitted.
   h. One deuterium nucleus and one hydrogen nucleus collide and become ^{3}He.

3. Sunspots, flares, prominences, and coronal mass ejections are all caused by
   a. magnetic activity on the Sun.
   b. electrical activity on the Sun.
   c. the interaction of the Sun's magnetic field and the interstellar medium.
   d. the interaction of the solar wind and Earth's magnetic field.
   e. the interaction of the solar wind and the Sun's magnetic field.

4. The structure of the Sun is determined by not only the balance between the forces due to _____ and gravity but also the balance between energy generation and energy _____.
   a. radiation pressure; production
   b. radiation pressure; loss
   c. ions; loss
   d. solar wind; production

5. In the proton-proton chain, four hydrogen nuclei are converted to a helium nucleus. That does not happen spontaneously on Earth because the process requires
   a. vast amounts of hydrogen.
   b. very high temperatures and densities.
   c. hydrostatic equilibrium.
   d. very strong magnetic fields.

6. The solar neutrino problem pointed to a fundamental gap in our knowledge of
   a. nuclear fusion.
   b. neutrinos.
   c. hydrostatic equilibrium.
   d. magnetic fields.

7. Sunspots change in number and location during the solar cycle. That phenomenon is connected to
   a. the rotation rate of the Sun.
   b. the temperature of the Sun.
   c. the magnetic field of the Sun.
   d. the tilt of the axis of the Sun.

8. Suppose an abnormally large amount of hydrogen suddenly fused in the core of the Sun. Which of the following would be observed first?
   a. The Sun would become brighter.
   b. The Sun would swell and become larger.
   c. The Sun would become bluer.
   d. The Sun would emit more neutrinos.

9. The solar corona has a temperature of 1 million to 2 million K, whereas the photosphere has a temperature of only about 6000 K. Why isn't the corona much, much brighter than the photosphere?
   a. The magnetic field traps the light.
   b. The corona emits only X-rays.
   c. The photosphere is closer to us.
   d. The corona has a much lower density.

10. The Sun rotates once every 25 days relative to the stars. The Sun rotates once every 27 days as seen from Earth. Why are those two numbers different?
    a. The stars are farther away.
    b. Earth is smaller.
    c. Earth moves in its orbit as the Sun rotates.
    d. The Sun moves relative to the stars.

11. Place the following regions of the Sun in order of increasing radius.
    a. corona
    b. core
    c. radiative zone
    d. convective zone
    e. chromosphere
    f. photosphere
    g. a sunspot

12. Coronal mass ejections
    a. carry away 1 percent of the mass of the Sun each year.
    b. are caused by breaking magnetic fields.
    c. are always emitted in the direction of Earth.
    d. are unimportant to life on Earth.

13. As energy moves out from the Sun's core toward its surface, it first travels by _____, then by _____, and then by _____.
    a. radiation; conduction; radiation
    b. conduction; radiation; convection
    c. radiation; convection; radiation
    d. radiation; convection; conduction

14. Energy is produced primarily in the center of the Sun because
    a. the strong nuclear force is too weak elsewhere.
    b. that's where neutrinos are created.
    c. that's where most of the helium is.
    d. the temperature and density are high enough in the core.

15. The solar wind pushes on the magnetosphere of Earth, changing its shape, because
    a. the solar wind is so dense.
    b. the magnetosphere is so weak.
    c. the solar wind contains charged particles.
    d. the solar wind is so fast.

THINKING ABOUT THE CONCEPTS

16. Explain how hydrostatic equilibrium acts to keep the Sun at its constant size, temperature, and luminosity.

17. Two of the three atoms in a molecule of water (H_2O) are hydrogen. Why are Earth's oceans not fusing hydrogen into helium and setting Earth ablaze?

18. Why are neutrinos so difficult to detect?

19. Explain the proton-proton chain through which the Sun generates energy by converting hydrogen to helium.

20. On Earth, nuclear power plants use *fission* to generate electricity. In fission, a heavy element such as uranium is broken into many atoms, where the total mass of the fragments is less than that of the original atom. Explain why fission could not be powering the Sun today.

21. If an abnormally large amount of hydrogen suddenly fused in the core of the Sun, what would happen to the rest of the Sun? Would the Sun change as seen from Earth?

22. Study the Process of Science Figure. If the follow-up experiments did not detect the other types of neutrinos, what would have been the next step for scientists at that point? ★

23. What is the solar neutrino problem, and how was it solved?

24. How are orbiting satellites and telescopes affected by the Sun?

25. Describe the solar corona. Under what circumstances can it be seen without special instruments?

26. ★ **WHAT AN ASTRONOMER SEES** In Figure 14.12, an astronomer would identify hydrogen and calcium lines immediately. Explain how these lines may have been identified the first time, and how experience might lead an astronomer to be able to instantly recognize them. ★

27. In the proton-proton chain, the mass of four protons is slightly greater than the mass of a helium nucleus. Explain what happens to that "lost" mass.

28. What have sunspots revealed about the Sun's rotation?

29. Why are different parts of the Sun best studied at different wavelengths? Which parts are best studied from space?

30. Why is studying the interaction of the solar wind with the interstellar medium important?

APPLYING THE CONCEPTS

31. The Sun shines by converting mass into energy according to $E = mc^2$. Show that if the Sun produces 3.85×10^{26} J of energy per second, it must convert 4.3 billion kg (4.3×10^9 kg) of mass per second into energy. ●—●—●
  a. Make a prediction: Is your final answer a rate or a mass? Therefore, what will the final units of your answer be?
  b. Calculate: Use the equation $E = mc^2$ to find the mass lost every second.
  c. Check your work: Verify that your units are correct and that you found the numerical answer stated in the problem statement.

32. If a sunspot has a temperature of 4800 K, while the surrounding photosphere has a temperature of 5780 K, what is the ratio of fluxes between the sunspot and the photosphere? ●—●—●
  a. Make a prediction: In Working It Out 14.2, you learned that the flux is proportional to T^4. Do you expect that the ratio of fluxes in this case will be larger or smaller than 1?
  b. Calculate: Follow Working It Out 14.2 to find the ratio of fluxes.
  c. Check your work: Verify that your answer is dimensionless (the units cancel out), and compare your answer to your prediction and to the answer in Working It Out 14.2.

33. Assume that the Sun has been producing energy at a constant rate over its lifetime of 4.6 billion years (1.4×10^{17} seconds). The current mass of the Sun is 2×10^{30} kg, and in Problem 31, you may have verified that the Sun loses mass at 4.3×10^{30} kg/s. What fraction of its current mass has been converted into energy over the lifetime of the Sun?
  a. Make a prediction: If the Sun has lost a large fraction of its mass over its lifetime, would that have had a large or small effect on planetary orbits? Therefore, do you expect this fraction of current mass to be large or small?
  b. Calculate: First, find the amount of mass lost by the Sun over its lifetime, and then express that as a fraction of the current mass.
  c. Check your work: Verify that your answer is dimensionless (all the units have canceled out), and compare your answer to your prediction.

34. The hydrogen bomb represents an effort to create a process similar to what takes place in the core of the Sun. The energy released by a 5-megaton hydrogen bomb is 2×10^{16} J. How much mass did Earth lose each time a 5-megaton hydrogen bomb was exploded?
  a. Make a prediction: The Sun loses about 4 billion kg of mass every second, to produce the amount of energy it emits. Do you expect your answer to be on that same scale (billions of kg) or much smaller?
  b. Calculate: Use $E = mc^2$ to find the mass lost in a 5-megaton hydrogen bomb explosion.
  c. Check your work: Verify that your answer has units of kg, and compare your answer to your prediction.

35. If a sunspot appears only 70 percent as bright as the surrounding photosphere, and the photosphere has a temperature of approximately 5780 K, what is the temperature of the sunspot?
  a. Make a prediction: What is a reasonable number for the temperature of a sunspot? A few kelvins? Hundreds of kelvins? Thousands

of kelvins? Millions of kelvins? Should the temperature that you calculate for a sunspot be larger or smaller than the temperature of the surrounding photosphere?
  b. Calculate: Follow Working It Out 14.2 in reverse to solve for the temperature of a sunspot from the ratio of brightness.
  c. Check your work: Verify that your answer has units of kelvins and compare your answer to your prediction.

36. In Figure 14.10, density and temperature are both graphed versus height. ✦
  a. Is the height axis linear or logarithmic? How do you know?
  b. Is the density axis linear or logarithmic? How do you know?
  c. Is the temperature axis linear or logarithmic? How do you know?

37. Use the data in Figures 14.19b and c to present an argument that sunspots occur in regions of strong magnetic field. ★

38. Study Figure 14.17a and Figure 14.20. ◉
  a. Estimate the fraction of the Sun's surface covered by the large sunspot group in Figure 14.17a. (Remember that you are seeing only one hemisphere of the Sun.)
  b. From the graph in Figure 14.20, estimate the average number of sunspots that occurs at solar maximum.
  c. On average, what fraction of the Sun could be covered by sunspots at solar maximum? Is that a large fraction?
  d. Compare your conclusion with the graph of intensity in Figure 14.24. Does that graph make sense to you?

39. Assume that the Sun's mass is about 300,000 Earth masses and that its radius is about 100 times that of Earth. The density of Earth is about 5500 kg/m³.
  a. What is the average density of the Sun?
  b. How does that value compare with the density of Earth? With the density of water?

40. Suppose our Sun was an A5 main-sequence star, with twice the mass and 12 times the luminosity of the Sun, a G2 star. How long would that A5 star fuse hydrogen to helium? What would that mean for Earth?

41. Imagine that the source of energy inside the Sun changed abruptly.
  a. How long would it take before a neutrino telescope detected the event?
  b. When would a visible-light telescope see evidence of the change?

42. On average, how long do particles in the solar wind take to reach Earth from the Sun if they are traveling at an average speed of 400 km/s?

43. Verify the claim made at the start of this chapter that the Sun produces more energy per second than all the electric power plants on Earth could generate in a half-million years. Estimate or look up how many power plants are on the planet and how much energy an average power plant produces. Be sure to account for different kinds of power, such as coal, nuclear, and wind.

44. Let's examine the reason why the Sun cannot power itself by chemical reactions. Using Working It Out 14.1 and the fact that an average chemical reaction between two atoms releases 1.6×10^{-19} J, estimate how long the Sun could emit energy at its current luminosity. Compare that estimate with the known age of Earth.

45. The Sun could get energy from gravitational contraction for a time period of ($GM_{Sun}^2/R_{Sun}L_{Sun}$). How long would the Sun last at its current luminosity? (Pay careful attention to units!)

EXPLORATION The Proton-Proton Chain

digital.wwnorton.com/astro7

The proton-proton chain powers the Sun by fusing hydrogen into helium. That fusion process produces several types of particles as by-products, as well as energy. In this Exploration, we study the steps of the proton-proton chain in detail, with the intent of helping you keep them straight.

Visit the Digital Resources Page on the Student Site and open the "Proton-Proton Chain" Interactive Simulation in Chapter 14.

Watch the animation all the way through once.

Play the animation again, pausing after the first collision. Two hydrogen nuclei (both positively charged) have collided to produce a new nucleus with only one positive charge.

1 Which particle carried away the other positive charge?

_____

_____

2 What is a neutrino? Did the neutrino enter the reaction, or was the neutrino produced in the reaction?

_____

_____

_____

Compare the interaction on the top with the interaction on the bottom.

3 Did the same reaction occur in each instance?

_____

_____

Resume playing the animation, pausing it after the second collision.

4 What two types of nuclei entered the collision? What type of nucleus resulted?

_____

_____

5 Was charge conserved in that reaction, or did a particle have to carry charge away?

_____

_____

6 What is a gamma ray? Did the gamma ray enter the reaction, or was it produced by the reaction?

_____

_____

_____

Resume the animation again, and allow it to run to the end.

7 What nuclei enter the final collision? What nuclei are produced?

_____

_____

_____

8 In chemistry, a catalyst is a species that facilitates a reaction but is not used up in the process. Do any nuclei act like catalysts in the proton-proton chain?

_____

_____

Make a table of inputs and outputs. Which particles in the final frame of the animation were inputs to the reaction? Which were outputs? Fill in your table with those inputs and outputs.

9 Which outputs are converted into energy that leaves the Sun as light?

_____

_____

10 Which outputs could become involved in another reaction immediately?

_____

_____

11 Which output is likely to stay in that form for a very long time?

_____

_____

The Interstellar Medium and Star Formation

I saw these clouds.

teapot

On a clear night, take a star chart to a dark site and carefully compare your chart to the sky. Notice that there is a shaded section on the star chart. This is the plane of the Milky Way. Find this plane, and carefully sketch it onto your observation log. Include and label a constellation or two to clarify the directions of your sketch. Take the time to carefully identify and note dark "lanes" in the brighter wash of the Milky Way. These are places where there is more stuff, not less. In fact, you are directly observing the effects of the interstellar medium, blocking your view of more distant stars!

EXPERIMENT SETUP

Take a star chart to a dark site and carefully compare your chart to the sky. Notice that there is a shaded section on the star chart. Find this plane and carefully sketch it onto your observation log.

Date May 15, 2021
Time 11:00 PM
Location

Include and label a constellation or two to clarify the directions of your sketch. Take the time to carefully identify and note dark "lanes" in the brighter wash of the Milky Way.

PREDICTION

I expect to see a ☐ **bright** ☐ **faint** wash of light oriented nearly ☐ **parallel** ☐ **perpendicular** to the ecliptic.

SKETCH OF RESULTS

Observation Log

Date May 15, 2021   Time 11:00 PM
Object Milky Way
Location Springville City Park
Sky conditions Some clouds to the East, then clear

Comments
Hard to see... Saw the Milky Way best out of the corner of my eye. Had to be patient. Compared my sketch to a photo at home, and could see some of the same dark dust lanes!

15

To understand star formation, astronomers have observed many stars at various stages of development. Stars form from clouds of dust and gas, which collapse until nuclear fusion begins. The most massive stars take tens of thousands of years to form, whereas the least massive stars take hundreds of millions of years. In this chapter, we discuss the interstellar environment from which stars of all masses form. We then focus on the forming star—the *protostar*—and explain how it becomes a star.

LEARNING GOALS

By the end of the chapter, you should be able to:

(1) Categorize the types and states of material that exist in the space between the stars and describe how that material is detected.

(2) Explain the conditions under which a cloud of gas can contract into a stellar system and contrast the roles that gravity and angular momentum play in forming stars and planets.

(3) List the steps in the evolution of a protostar, and summarize how the mass of a protostar affects its evolution.

(4) Reconstruct the track of a protostar as it evolves to a main-sequence star on the Hertzsprung-Russell (H-R) diagram.

15.1 The Interstellar Medium Fills the Space between the Stars

The space between the stars contains thin gas, interrupted by giant clouds of cool gas and dust. Those clouds can be observed in the visible part of the spectrum in three ways: they absorb light from objects behind them, they emit light from excited atoms, or they reflect starlight from nearby stars. The thin gas is more difficult to observe directly, and most often is observed in the far-infrared or radio wavelengths of the electromagnetic spectrum. Together, the clouds and the thin gas are known as the **interstellar medium**.

Stars interact with the interstellar medium. Stars form from the interstellar medium and return much of their material back to it when they die. Outflows from dying stars pile up material in their path like snow in front of a snowplow, clearing out vast hot bubbles. Those hot bubbles of high-pressure gas compress clouds, driving up their densities and triggering the formation of new stars. Interstellar clouds are destroyed by those violent events. In turn, new clouds are formed from the swept-up gas. Those clouds go on to produce new stars. In addition, energy from stars heats and stirs the interstellar medium. For example, ultraviolet radiation from massive, hot stars warms nearby gas that then pushes outward into its surroundings. In this section, we survey the varied states of the interstellar medium.

The Composition and Density of the Interstellar Medium

The chemical composition of the interstellar medium near the Sun is similar to the Sun's chemical composition (see Table 13.1)—that is, hydrogen accounts for about 90 percent of the number of atomic nuclei, and the remaining 10 percent is almost all helium. The more massive elements account for only 0.1 percent of the atomic nuclei, which corresponds to about 2 percent of the *mass* in the interstellar medium (because those atoms are more massive than hydrogen or helium, they account for a larger fraction of the mass in the interstellar medium than they do of the number of individual atoms). Roughly 99 percent of interstellar matter is a gas, consisting of individual atoms or molecules moving about freely, as the molecules in the air do.

The air that you breathe has a density 100 billion billion times greater than the average density of the interstellar medium. Each cubic centimeter (cm^3) of the air around you contains about 2.7×10^{19} molecules. The local interstellar medium has an average density of about 0.1 atom/cm^3. Stated another way, imagine a cylinder with the diameter of your fist that stretches from the Solar System to the center of our galaxy, 26,000 light-years away (**Figure 15.1**). Now imagine a cylinder with the same diameter on Earth that stretches between your eye and the ground you are standing on. The amount of material in both cylinders is about the same.

Interstellar Dust and Its Effects on Light

In the interstellar medium, about half of the atoms that are more massive than helium (about 1 percent of the total mass) are found in solid grains, called **interstellar dust**. Interstellar dust begins to form when larger atoms such as iron, silicon, and carbon stick together to form grains in dense, relatively cool environments such as the outer atmospheres of cool, red giant stars. Once those grains are in the interstellar medium, other atoms and molecules stick to them, forming solid grains that are more like the particles of soot from a candle flame than like the dust that collects on a windowsill. Ranging in size from little more than large molecules up to particles about 1 micron (μm; 10^{-6} meters) across, it would take several hundred average interstellar grains to span the thickness of a single human hair.

Interstellar dust is extremely effective at absorbing and diverting light, so the view of distant objects is affected even by the low-density interstellar medium. That effect is called **interstellar extinction**. Go out on a dark summer night in the Northern Hemisphere (or on a dark winter night in the Southern Hemisphere) and look closely at the Milky Way, visible as a band of diffuse light running through the constellation Sagittarius. A dark "lane" is running roughly down the middle of that bright band, splitting it in two, as shown in **Figure 15.2a**. That dark band is caused by interstellar extinction, dust that dims the light from distant stars. Recall the comparison in Figure 15.1. If the interstellar medium were compressed to the same density as air, it would be so full of dust (as opposed to gas) that you could not easily see your hand 10 centimeters (cm) in front of your face. When spread out between Earth and the distant stars, the interstellar dust dims light just as effectively. Although interstellar extinction is most noticeable in the dark lanes in the Milky Way, it has a lesser but still important effect on starlight coming from all directions in the galaxy.

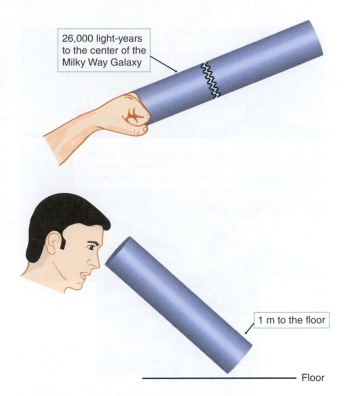

26,000 light-years to the center of the Milky Way Galaxy

1 m to the floor

Floor

Figure 15.1 The interstellar medium is so diffuse that it would take a cylinder the diameter of your fist, stretching 26,000 light-years, all the way to the center of the Milky Way Galaxy, to encompass as many particles as are in the air contained in a similar cylinder that stretches only between your eye and the ground.

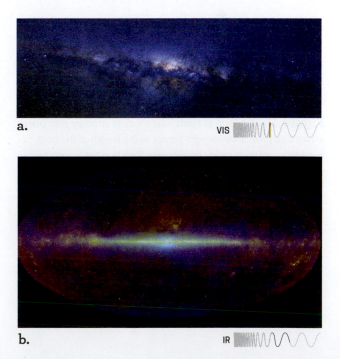

a.   VIS

b.   IR

Figure 15.2 a. This all-sky picture of the Milky Way was taken in visible light. The dark splotches blocking the view are dusty interstellar clouds. The center of this and all the other all-sky images in this chapter is the center of the Milky Way. **b.** This all-sky picture was taken in the near infrared. Infrared radiation penetrates the interstellar dust, providing a clearer view of the stars in the disk of the Milky Way.

Figure 15.3 a. Just as boats interact most strongly with ocean waves that have a wavelength similar to the length of the boat, **b.** particles interact most strongly with light that has wavelengths near the same size as the particle.

Figure 15.2 shows two images of the Milky Way Galaxy: one taken in visible light, the other taken in the infrared (IR). Those two images are *all-sky images*, because they portray the entire sky surrounding Earth. They have been oriented so that the disk of the Milky Way, in which the Solar System is embedded, runs horizontally across the center of the image. The dark clouds that block the shorter-wavelength visible light (Figure 15.2a) seem to have vanished in the longer-wavelength infrared image (Figure 15.2b). That observation shows that interstellar extinction affects different regions of the electromagnetic spectrum differently. When astronomers observe in long-wavelength regions of the spectrum, such as infrared or radio wavelengths, they can make observations through the clouds to the center of the galaxy and beyond.

To understand why dust obscures short-wavelength radiation but not long-wavelength radiation, think about waves on the ocean, as shown in **Figure 15.3a**. Imagine you are on the ocean in a boat. If the ocean waves have a wavelength much longer than your boat, the waves cause you to move slowly up and down. But that is all that happens: No energy is transferred from the wave to the boat, so the wave travels on, unaffected by the interaction. The situation is different if the wavelength of the waves is close in size to your boat's length. If the waves are roughly half the size of your boat, for example, then the front of the boat may be on a wave crest while the back of the boat is in a trough or vice versa. The boat will tip wildly back and forth as the waves pass by, and energy will be transferred from the wave to the boat; the wave loses energy in the interaction. If the boat's size and the wavelength of the waves are the right match, even low waves will rock the boat. Comparing those two situations shows that the wave loses very little energy when it is much bigger than the boat. But the wave loses a lot of energy when it is of similar size or smaller than the boat, and the wave can be blocked by the boat if the boat is large enough in comparison.

The interaction of light with matter is more involved than that of a boat rocking on the ocean, but the same basic idea applies, as shown in **Figure 15.3b**. Tiny interstellar dust grains effectively block the transmission of short-wavelength ultraviolet light and blue light because they have wavelengths comparable to or smaller than the typical size of dust grains. Our view of the Milky Way in those wavelengths is limited. In contrast, longer-wavelength infrared and radio radiation interact less strongly with the tiny interstellar dust grains, so in those portions of the spectrum we get a more complete view of the Milky Way.

The presence of dust significantly affects an object's spectrum, as shown in **Figure 15.4**. Extinction occurs at all wavelengths (Figure 15.4a), causing an object viewed through dust to be fainter at all wavelengths than it would be otherwise. Extinction also affects short-wavelength blue light more than it affects long-wavelength red light, so the object appears less blue than it really is (Figures 15.4b and c). Removing the blue light causes an object to appear more red, so that effect is called **reddening**. Correcting for the reddening effect of dust can be one of the most difficult parts of interpreting astronomical observations, often adding to uncertainty in the measurement of an object's properties.

At longer infrared wavelengths, interstellar extinction from small gas particles is less effective, but larger dust particles still play an important role. In Chapter 5, we discussed how Wien's law can be used to relate an object's temperature to the wavelength at which it shines most brightly, and how the equilibrium between absorbed sunlight and emitted thermal radiation determines the temperatures of the terrestrial planets (see Figure 5.25).

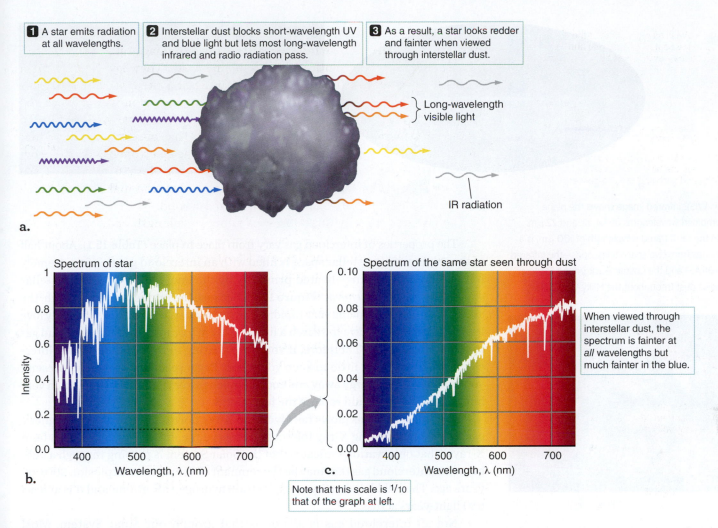

1 A star emits radiation at all wavelengths.

2 Interstellar dust blocks short-wavelength UV and blue light but lets most long-wavelength infrared and radio radiation pass.

3 As a result, a star looks redder and fainter when viewed through interstellar dust.

Long-wavelength visible light

IR radiation

a.

Spectrum of star

Spectrum of the same star seen through dust

When viewed through interstellar dust, the spectrum is fainter at *all* wavelengths but much fainter in the blue.

b.

c.

Note that this scale is 1/10 that of the graph at left.

Figure 15.4 **a.** The wavelengths of ultraviolet and blue light are close to the size of interstellar grains, so the grains effectively block that light. Grains are less effective at blocking longer-wavelength light. Therefore, **b.** the spectrum of a star **c.** appears fainter and redder when seen through an interstellar cloud.

As **Figure 15.5** illustrates, a similar equilibrium is at work in interstellar space, where dust is heated by starlight and surrounding gas to temperatures of tens to hundreds of kelvins. According to Wien's law, dust with a temperature of 100 K shines most strongly at a wavelength of 29 μm, whereas cooler dust —at 10 K—shines most strongly at a longer wavelength (**Working It Out 15.1**). Such warm dust produces much of the diffuse light present in infrared observations. **Figure 15.6a**, from NASA's Wide-field Infrared Survey Explorer (WISE), shows the sky at combined wavelengths of 3.4, 12, and 22 μm. That light is produced by dust hotter than 100 K. In **Figure 15.6b**, a far-infrared image of the sky shows the Milky Way's dark, dusty, cooler clouds shining at a wavelength of 100 μm.

Temperatures and Densities of Interstellar Gas

The gas and dust in interstellar space are roughly evenly divided between dense regions called **interstellar clouds** and the **intercloud gas** spread between them. The clouds fill about 2 percent of the volume of interstellar space, whereas the remaining 98 percent is filled with thin gas.

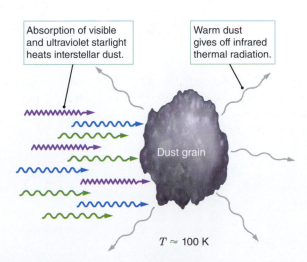

Absorption of visible and ultraviolet starlight heats interstellar dust.

Warm dust gives off infrared thermal radiation.

Dust grain

$T \approx 100$ K

Figure 15.5 The temperature of interstellar grains depends on the equilibrium between absorbed and emitted radiation.

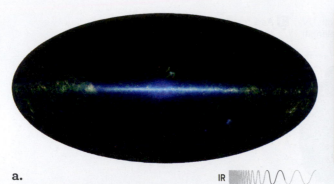

a.

Far infrared

b.

IR 〜〜〜

Figure 15.6 **a.** This WISE infrared image shows the plane of the Milky Way at combined wavelengths of 3.4, 12, and 22 μm. **b.** This all-sky image, in the far-infrared wavelength of 100 μm, is a combination of two images from two space telescopes—the Infrared Astronomical Satellite (IRAS) and the Cosmic Background Explorer (COBE). The image shows dust throughout the Milky Way Galaxy.

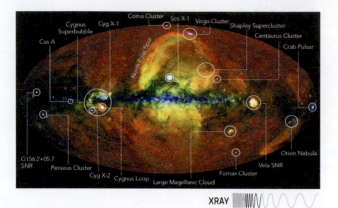

XRAY 〜〜〜

Figure 15.7 Many of the bright spots in this image are distant X-ray sources, including objects such as bubbles of very hot, high-pressure gas surrounding the sites of recent explosions of supernovae. Supernovae, which mark the death of a high-mass star, are more common in the disk of the Milky Way, where most stars are located.

The properties of intercloud gas vary from place to place (**Table 15.1**). About half of the volume of interstellar space is filled with an intercloud gas that is extremely hot—millions of kelvins—heated primarily by the energy of tremendous stellar explosions called *supernovae* (**Figure 15.7**). Because the temperature is so high, the atoms in the gas are moving very rapidly. The gas density, however, is extremely low; typically, you would have to search a liter (1000 cm^3) or more of hot intercloud gas to find a single atom. Therefore, if you were floating in that million-kelvin intercloud gas, it would do little to keep you warm. Because so few atoms are present, you would radiate energy away and cool off much faster than the rate at which very hot gas around you could replace the lost energy.

Hot intercloud gas glows faintly in the X-ray portion of the electromagnetic spectrum, and orbiting X-ray telescopes observe the entire sky aglow with faint X-rays. That observation indicates that the Solar System is passing through a bubble of hot intercloud gas that may be the remnant of a supernova explosion 300,000 years ago. The bubble's density is about 0.005 hydrogen atom/cm^3 and it is at least 650 light-years across.

Not all intercloud gas is as hot as that around our Solar System. Most other intercloud gas is "warm," with a temperature of about 8000 K and a density

Table 15.1   Typical Properties of Components of the Interstellar Medium

Component	Temperature (K)	Number Density (atoms/cm³)	Size of Cube per Gram* (km)	State of Hydrogen
Hot intercloud gas	~1 million	~0.005	~8000	Ionized
Warm intercloud gas	~8000	0.01–1	~800	Ionized or neutral
Cold intercloud gas	~100	1–100	~80	Neutral
Interstellar clouds	~10	100–1000	~8	Molecular or neutral

*This is the length of one side of the cube of space you would need to search to find 1 gram of the material. It is another way of thinking about density.

working it out 15.1

Dust Glows in the Infrared

The temperature of interstellar dust can be found from its spectrum. Wien's law, discussed in Chapter 5, relates an object's temperature to the peak wavelength (λ_{peak}) of its emitted radiation. For warm dust at a temperature of 100 K (recall that 1 μm = 10^{-6} meter = 1000 nanometers [nm]):

$$\lambda_{peak} = \frac{2900 \text{ μm K}}{T} = \frac{2900 \text{ μm K}}{100 \text{ K}} = 29 \text{ μm}$$

The constant in the numerator (2900 μm K) is the same as in Chapter 5, but written here in more convenient units.

For cooler dust, at 10 K:

$$\lambda_{peak} = \frac{2900 \text{ μm K}}{T} = \frac{2900 \text{ μm K}}{10 \text{ K}} = 290 \text{ μm}$$

The temperature and the peak wavelength are inversely proportional, so if the temperature decreases, the peak wavelength increases. For the temperatures common for dust in the interstellar medium, the peak wavelength is in the far-infrared region of the electromagnetic spectrum.

---

ranging from about 0.01 to 1 atom/cm³. Ultraviolet starlight with wavelengths shorter than 91.2 nm has enough energy to **ionize** hydrogen—that is, enough to strip the electron away from the atom, leaving only the positively charged nucleus (the proton). About half of the warm intercloud gas is kept ionized by ultraviolet starlight. That ionized gas "uses up" the ultraviolet light from stars, so the remaining half of warm intercloud gas is protected and remains in an un-ionized state. In a similar fashion, the ozone layer in Earth's upper atmosphere shields the planet's surface from ultraviolet light from the Sun.

Interstellar gas is found between the stars. One way to look for both warm and hot interstellar gas is to study the spectra of distant stars. Most commonly, that gas produces absorption lines in the spectra of distant stars when atoms in the gas absorb starlight at particular wavelengths. Those absorption lines can be used to find the temperature, density, and chemical composition of the gas. Regions of warm, ionized intercloud gas, as in supernova remnants, can also produce emission lines when protons and electrons constantly recombine into hydrogen atoms. When a proton and an electron combine to form a neutral hydrogen atom, energy is emitted in the form of electromagnetic radiation. Typically, the resulting hydrogen atom is left in an excited state (see Chapter 5). The atom then drops down to lower and lower energy states, emitting a photon at each step. Therefore, warm, ionized interstellar gas glows in emission lines characteristic of hydrogen, with the Hα (hydrogen alpha) line often the strongest. That line occurs in the red part of the spectrum at a wavelength of 656.3 nm. Other elements undergo a similar process.

The faint, diffuse emission in **Figure 15.8** comes mostly from warm (about 8000 K), ionized intercloud gas glowing in Hα. The bright spots are called **H II regions** ("H two regions") because the hydrogen atoms are ionized (that is, H II atoms are ionized, whereas H I atoms are neutral). Hot luminous O and B stars produce enough ultraviolet radiation to ionize even relatively dense interstellar clouds around them, forming the H II regions, which are roughly spherical in shape. H II regions indicate areas of active star formation because O stars do not live long enough to move very far from where they formed. H II regions around O stars are the very clouds from which those stars were born.

what if . . .

What if an O star is born and fully ionizes the gas cloud around itself, forming an H II region. What would be the fate of dust particles that were present in the gas cloud?

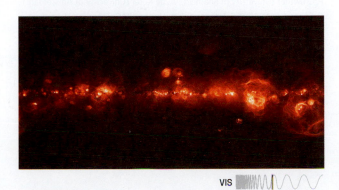

VIS

Figure 15.8 Warm interstellar gas (about 8000 K) glows in the Hα line of hydrogen. This image of the Hα emission from much of the northern sky reveals the complex structure of the interstellar medium.

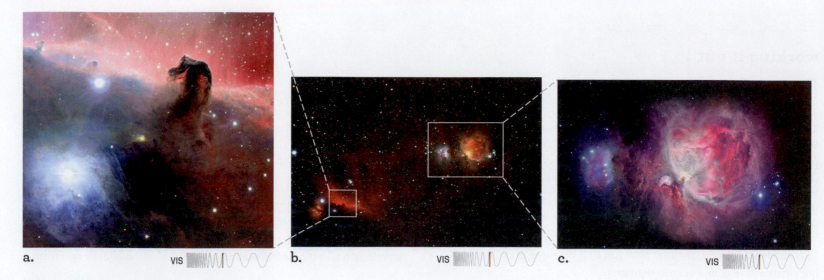

a.   VIS   b.   VIS   c.   VIS

Figure 15.9 The Orion Nebula is only a small part of the larger Orion star-forming region. The dark Horsehead Nebula is seen at the lower left of image **b.** and in **a.** The circular halos around the bright stars in a. are a photographic artifact. **c.** The Orion Nebula is seen as a glowing region of interstellar gas surrounding a cluster of young, hot stars. New stars are still forming in the dense clouds surrounding the nebula.

Credit (part c.): Stefan Seip – photomeeting.de.

One of the closest H II regions to the Sun is the Orion Nebula, located 1340 light-years from the Sun in the constellation Orion (**Figure 15.9**). Almost all the ultraviolet light that powers that nebula comes from a single hot star, and only a few hundred stars are forming in its immediate vicinity. In contrast, a pair of dense star clusters containing thousands of hot, luminous stars power a giant H II region called 30 Doradus, located in the Large Magellanic Cloud, a small companion galaxy to the Milky Way located 160,000 light-years away (**Figure 15.10**). If 30 Doradus were as close as the Orion Nebula, it would be bright enough in the nighttime sky to cast shadows.

Warm, *neutral* hydrogen gas also produces an emission line, although it gives off radiation differently compared to warm, *ionized* hydrogen. Many subatomic

Figure 15.10 The 30 Doradus nebula is found in a nearby galaxy to the Milky Way. This nebula contains a pair of star clusters of different ages, one of which is elongated. This implies that these two star clusters may be in the early stages of colliding within the nebula.

Credit: NASA, ESA, D. Lennon and E. Sabbi (ESA/STScI), J. Anderson, S.E. de Mink, R. van der Marel, T. Sohn, and N. Walborn (STScI), N. Bastian (Excellence Cluster, Munich), L. Bedin (INAF, Padua), E. Bressert (ESO), P. Crowther (University of Sheffield), A. de Koter (University of Amsterdam), C. Evans (UKATC/STFC, Edinburgh), A. Herrero (IAC, Tenerife), N. Langer (AifA, Bonn), I. Platais (JHU), and H. Sana (University of Amsterdam) Upper Right Image Credit: NASA, ESA, R. O'Connell (University of Virginia), and the WFC3 Science Oversight Committee, https://esahubble.org/images /opo1235d/. https://creativecommons.org/licenses/by/4.0/.

VIS

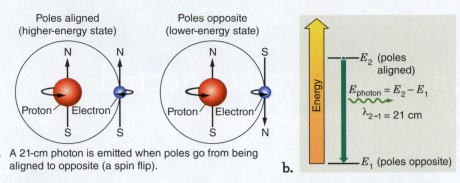

a. A 21-cm photon is emitted when poles go from being aligned to opposite (a spin flip).

b.

Figure 15.11 **a.** A slight difference in energy occurs when the poles of the proton and electron are aligned compared with when they are opposite. **b.** This energy difference corresponds to a photon with a wavelength of 21 cm.

particles, including protons and electrons, have a property called *spin* that causes them to behave as though each particle has a bar magnet, with a north and a south pole, built into it. As demonstrated in **Figure 15.11**, a hydrogen atom can exist in one of two configurations: either the magnetic "poles" of the proton and electron point in opposite directions or they are aligned. Those configurations have different energies. When the two "magnets" point in the same direction, the atom has slightly more energy than when they point in the opposite direction. If left undisturbed long enough, a hydrogen atom in the higher-energy aligned state will spontaneously jump to the lower-energy unaligned state, emitting a photon in the process. The energy difference between the two magnetic spin states of a hydrogen atom is extremely small, so the emitted photon has a 21-cm wavelength, which falls in the radio region of the spectrum. Interactions between atoms in the gas will later bump the hydrogen atoms back to the higher-energy state, refreshing the supply of atoms that can produce the 21-cm line.

This transition is very rare. On average, you would have to wait about 11 million years for an individual hydrogen atom in the higher-energy state to jump spontaneously to the lower-energy state and give off a photon. But the universe has a lot of hydrogen, so at any given time, many atoms are making that transition. In **Figure 15.12**, the sky is aglow with 21-cm radiation from neutral hydrogen. Because of its long wavelength, 21-cm radiation freely penetrates dust in the interstellar medium, enabling astronomers to observe neutral hydrogen throughout the galaxy. Measurements of the line's Doppler shift indicate how fast the emitting gas is moving toward us or away from us. Those two attributes make the 21-cm line of neutral hydrogen important for understanding the Milky Way's structure (see the **Process of Science Figure**).

Regions of Cool, Dense Gas

Most interstellar clouds are composed primarily of isolated neutral hydrogen atoms and are much cooler and denser than the warm intercloud gas. Interstellar clouds have temperatures of about 100 K and densities of about 1–100 atoms/cm³. On average, atoms in interstellar clouds are moving at about 20 kilometers per second (km/s). In contrast, hot gas in the interstellar medium has atoms moving at thousands of kilometers per second.

On Earth, finding atoms in isolation is uncommon, because most atoms are bound up in molecules. In most of interstellar space, however, including most interstellar clouds, molecules do not survive long. If interstellar gas is too hot, molecules soon collide with other molecules or atoms and are torn apart. The temperature in

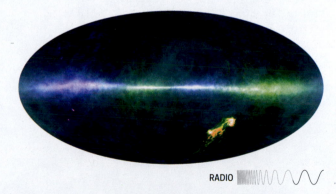

RADIO

Figure 15.12 This radio image of the sky shows the distribution of neutral hydrogen gas throughout the galaxy. Brightness indicates the density of hydrogen, while color gives the direction the gas is moving relative to the Sun. Radio waves penetrate interstellar dust, allowing astronomers to probe the structure of the galaxy. The prominent orange structures at lower right are the Large and the Small Magellanic Clouds—satellite galaxies of the Milky Way.

Credit: © ESO, 2016, https://www.aanda.org/articles/aa/full_html/2016/10/aa29178-16/F3.html. https://creativecommons.org/licenses/by/4.0/.

All Branches of Science Are Interconnected

Studies of the natural world on the smallest scales and the largest scales inform one another.

Atomic physics

Scientists studying atoms and quantum mechanics—the underlying principles that govern the behavior of atoms—found that the electron in the hydrogen atom makes a rare transition, releasing radiation with a wavelength of 21 cm. This is so rarely observed on Earth that it is known as a *forbidden transition.*

Radio astronomy

Because space is so large, there are vast numbers of hydrogen atoms along the line of sight in every direction. There are so many that even exceptionally rare events happen often enough to be detectable. The forbidden transition is commonly observed by radio astronomers and maps out the neutral hydrogen in the Milky Way.

Radio astronomers use the physics of tiny atoms to understand the behavior of an enormous galaxy, which is 10^{30} times larger than an atom.

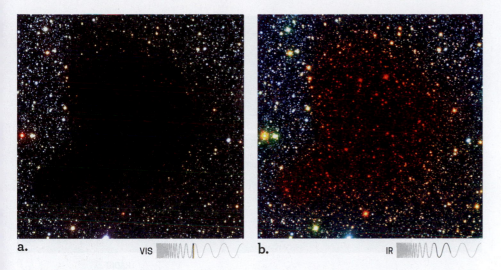

Figure 15.13 In visible light, interstellar molecular clouds are seen in silhouette against a background of stars and glowing gas.   **a.** Light from background stars is blocked by dust and gas in nearby Barnard 68, a dense, dark molecular cloud.   **b.** Infrared wavelengths can penetrate much of that gas and dust, as seen in this false-color image of Barnard 68.
Credit (part a.): ESO, https://www.eso.org/public/images/eso0102a/. https://creativecommons.org/licenses/by/4.0/; (part b.): ESO, https://www.eso.org/public/images/eso0102b/. https://creativecommons.org/licenses/by/4.0/.

a neutral hydrogen cloud may be low enough for some molecules to survive, but ultraviolet photons with enough energy to break molecules apart can penetrate neutral hydrogen clouds. The inner regions of the densest interstellar clouds are known as **molecular clouds** because in those regions, dust blocks almost all of the ionizing radiation, and so molecules persist.

Molecular clouds range from a few solar masses to 10 million solar masses. The smallest molecular clouds may be less than half a light-year across; the largest may be more than a thousand light-years. **Giant molecular clouds** are typically about 100–200 light-years across and have masses a few hundred thousand times that of the Sun. The Milky Way Galaxy contains several thousand giant molecular clouds and many more smaller ones. Despite that large number of clouds, molecular clouds fill only about 0.1 percent of interstellar space. Such clouds may be rare, but they are extremely important because stars form in molecular clouds.

In images such as those of **Figure 15.13**, the dust in molecular clouds causes a silhouette in visible light against a background of stars. Long-wavelength infrared radiation passes through that dust, revealing sources inside and behind the cloud. Inside such clouds it is dark and usually very cold, typically only about 10 K. Most of those clouds have densities of about 100–1000 molecules/cm^3, but densities as high as 10^{10} molecules/cm^3 have been observed. Even at 10^{10} molecules/cm^3, that gas is still less than a billionth as dense as the air around you. In that cold, relatively dense environment, atoms combine to form a wide variety of molecules.

Like atoms, molecules also make transitions between energy states. Molecular energy states are determined by the way the molecules rotate or vibrate. Each type of molecule is unique in its properties, and thus unique in its energy states, and so molecular emission lines are useful in the same way that atomic emission lines are useful. The wavelengths of emission lines from molecules are an unmistakable fingerprint of the kinds of molecules responsible for them. Because some of the transitions are in the radio or infrared portion of the spectrum, those molecules can be detected even deep inside a molecular cloud. Observations of molecular lines reveal the innermost workings of the densest and most opaque interstellar clouds. By far the most common component of molecular clouds is molecular hydrogen (H_2), which consists of two hydrogen atoms and is the smallest possible molecule. Cold molecular hydrogen is very difficult to observe directly because it has no emission lines in the long-wavelength regions of the spectrum.

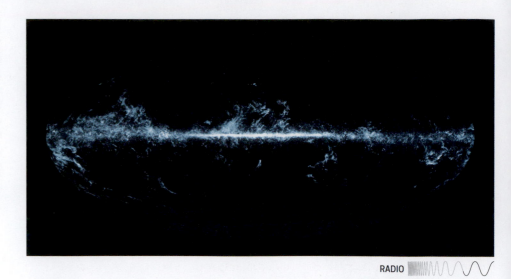

RADIO

Figure 15.14 This all-sky image from the Planck observatory, a spacecraft that observed in the microwave region of the spectrum, shows the distribution of carbon monoxide (CO). Because CO is linked to the presence of molecular hydrogen, locations that are dense in CO are also dense in molecular hydrogen. These maps of CO therefore identify molecular clouds where stars are born.

In addition to molecular hydrogen, approximately 150 other molecules have been observed in interstellar space. They range from very simple structures such as carbon monoxide (CO), to complex organic compounds such as methanol (CH_3OH), to molecules with large carbon chains. Very large carbon molecules, made of hundreds of individual atoms, have sizes between those of large interstellar molecules and small interstellar grains. Among the more important molecules is CO. The ratio of CO to H_2 is relatively constant, as far as has been tested, and interstellar CO (**Figure 15.14**) is often used to estimate the amounts and distribution of interstellar H_2, which is more difficult to observe directly.

CHECK YOUR UNDERSTANDING 15.1

When visible light from an object passes through the interstellar medium (choose all that apply): (a) the object appears dimmer; (b) the object appears bluer; (c) the object appears brighter; (d) the object appears redder.

Answers to Check Your Understanding questions are in the back of the book.

15.2 Stars Form in Molecular Clouds

Stars and planets form from large clouds of dust and gas in the interstellar medium. In this section, we describe the first steps of that process as a cloud begins to contract and fragment to form stars.

Self-Gravity in the Molecular Cloud

As shown in **Figure 15.15**, each gas parcel in an interstellar cloud experiences a gravitational attraction from every other part of the cloud. Adding together all those forces, accounting for the direction of each, results in the net force on the gas parcel. For each gas parcel, the net force always points toward the cloud's center of mass, so the gas parcel will begin to move toward the center of mass. The gravitational attraction between all the parts of a cloud is called *self-gravity*,

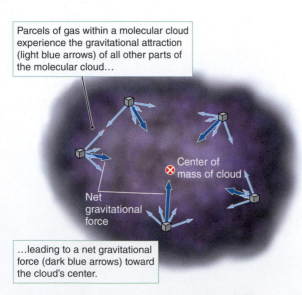

Parcels of gas within a molecular cloud experience the gravitational attraction (light blue arrows) of all other parts of the molecular cloud…

Center of mass of cloud

Net gravitational force

…leading to a net gravitational force (dark blue arrows) toward the cloud's center.

Figure 15.15 Self-gravity causes a molecular cloud to collapse, drawing parcels of gas toward a single point inside the cloud. The lighter blue arrows are examples of forces on the parcel due to other parcels of gas. The darker blue arrows show the sum of all those forces. That net force always points toward the center of mass of the cloud.

which acts to hold the cloud together, and sometimes is strong enough to cause the cloud to collapse.

Recall from our discussion of the Sun in Chapter 14 that hydrostatic equilibrium is the balance between gravity and pressure in a stable object. Interstellar clouds are not always in hydrostatic equilibrium—most interstellar clouds are large and very thin, so their self-gravity is weak. The internal pressure pushing out is much stronger than the self-gravity, so the cloud should expand. But the much hotter gas surrounding the clouds also exerts a pressure inward on a cloud. That external hot gas helps hold a cloud together and often serves as a trigger for collapse. If a cloud is massive enough and dense enough (or becomes so after a triggering event), self-gravity becomes stronger than the internal pressure, so the clouds collapse under their own weight, beginning a chain of events that will form a new generation of stars.

If self-gravity in a molecular cloud is much greater than internal pressure, gravity should win outright, and the cloud should rapidly collapse toward its center. In practice, the process goes very slowly because several other effects stand in the way of the collapse. One effect that slows a cloud's collapse is conservation of angular momentum (see Chapter 7). Others are turbulence, the effects of magnetic fields, and thermal pressure. Even though those effects may slow the collapse of a molecular cloud, in the end gravity will dominate. One part of the cloud can lose angular momentum to another part of the cloud, allowing the part of the cloud with less angular momentum to collapse further. Turbulence gradually fades away. Neutral matter crosses magnetic field lines, gradually increasing the gravitational pull toward the center until the force on the charged particles is large enough to drag the magnetic field toward the center as well. The details of those processes are complex and are the subject of much current research. The important point is that the effects that slow the collapse of a molecular cloud are temporary, whereas gravity is persistent. As the forces that oppose the cloud's self-gravity gradually fade, the cloud shrinks.

Molecular Clouds Fragment as They Collapse

Some regions within a molecular cloud are denser and collapse more rapidly than surrounding regions. **Figure 15.16** shows the process of collapse in a

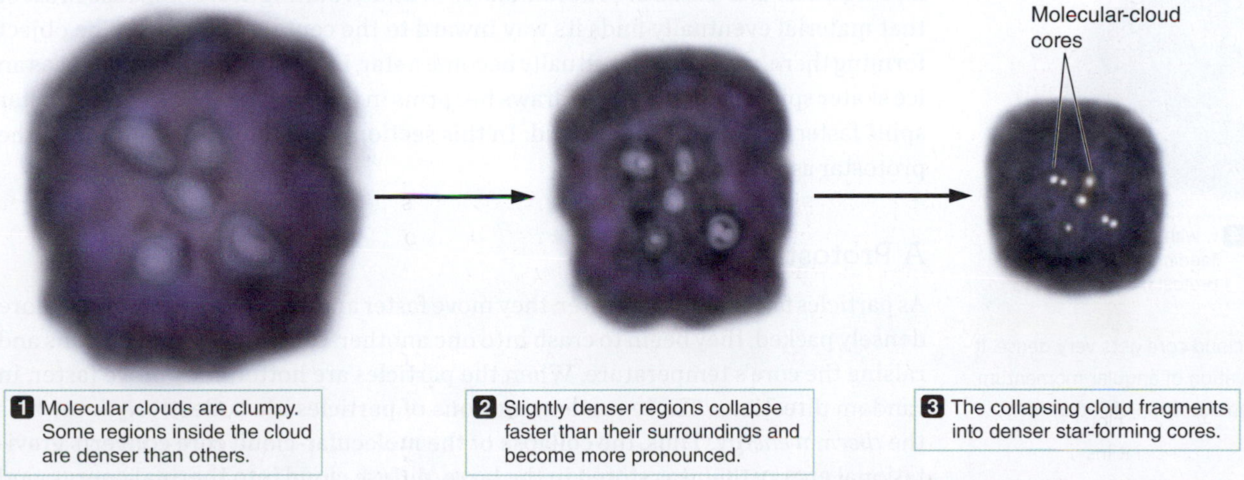

what if . . .

What if a molecular cloud has numerous nearby cloud cores that seemed destined to collapse to form O stars? How might the ionizing radiation from the first O stars affect the ability of the other cloud cores to form stars?

Astronomy in Action: Angular Momentum

AstroTour: Star Formation

Molecular-cloud cores

1 Molecular clouds are clumpy. Some regions inside the cloud are denser than others.

2 Slightly denser regions collapse faster than their surroundings and become more pronounced.

3 The collapsing cloud fragments into denser star-forming cores.

Figure 15.16 When a molecular cloud collapses, denser regions within the cloud collapse more rapidly than less dense regions. As that process continues, the cloud fragments into several very dense molecular-cloud cores embedded within the large cloud. Those cloud cores may go on to form stars.

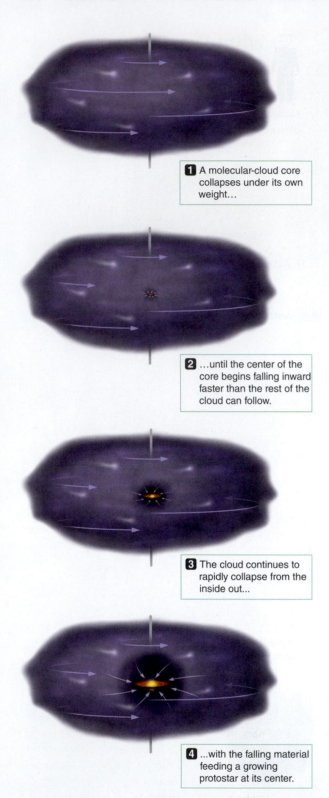

1 A molecular-cloud core collapses under its own weight…

2 …until the center of the core begins falling inward faster than the rest of the cloud can follow.

3 The cloud continues to rapidly collapse from the inside out…

4 …with the falling material feeding a growing protostar at its center.

Figure 15.17 When a molecular-cloud core gets very dense, it collapses from the inside out. Conservation of angular momentum causes the infalling material to form an accretion disk that feeds the growing protostar.

molecular cloud. Slight variations in the cloud's density become very dense, localized concentrations of gas. Instead of collapsing into a single object, the molecular cloud fragments into very dense **molecular-cloud cores**. A single molecular cloud may form hundreds or thousands of molecular-cloud cores, each of which is typically a few light-months in size. Some of those dense cores will eventually form stars.

As a molecular-cloud core collapses, the cloud's self-gravity grows stronger because the force of gravity is inversely proportional to the square of the radius. Suppose a cloud is 4 light-years across. When the cloud has collapsed to 2 light-years across, the different parts of the cloud are, on average, only half as far apart as when the collapse started. As a result, the self-gravity is 4 times stronger. When the cloud is one-fourth as large as it was at the beginning of the collapse, the self-gravity is 16 times stronger. As a core collapses, the self-gravity increases; as self-gravity increases, the collapse speeds up; as the collapse speeds up, the self-gravity increases even faster.

Eventually, gravity overwhelms the opposing forces due to pressure, magnetic fields, and turbulence. That happens first near the center of the molecular-cloud core, where the cloud material is densest. The inner parts of the molecular-cloud core start to fall rapidly inward. Those inner layers supported the weight of the layers farther out. Without the support of that inner material, the more distant material begins to fall freely toward the center. The molecular-cloud core collapses from the inside out, as shown in **Figure 15.17**.

CHECK YOUR UNDERSTANDING 15.2

Molecular clouds fragment as they collapse because: (a) the cloud's rotation throws some mass to the outer regions; (b) the density increases fastest in the cloud's center; (c) density variations from place to place grow larger as the cloud collapses; (d) the interstellar wind is stronger in some places than others.

15.3 Formation and Evolution of Protostars

Because of conservation of angular momentum, material that falls inward in a collapsing molecular-cloud core accumulates in a flat, rotating accretion disk. Most of that material eventually finds its way inward to the center of the disk. The object forming there, which will eventually become a star, is called a **protostar**. Just as an ice skater spins faster when she draws her arms in (recall Figure 7.5), the protostar spins faster than the original cloud. In this section, we follow the evolution of the protostar as it becomes a star.

A Protostar Forms

As particles fall toward the center, they move faster and faster. As they become more densely packed, they begin to crash into one another, causing random motions and raising the core's temperature. When the particles are hotter, they move faster, in random directions. Those random motions of particles are collectively known as the *thermal energy*. Thus, the collapse of the molecular-cloud core converts gravitational energy that was stored in the large, diffuse cloud into thermal energy, and the gas in the outer layers of the protostar is heated to thousands of kelvins, causing the protostar to shine.

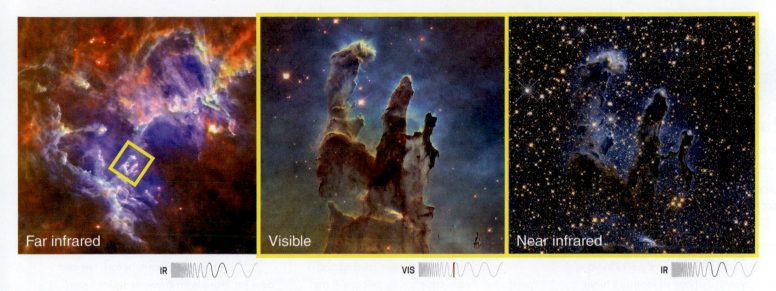

Figure 15.18 The Eagle Nebula contains dense columns of molecular gas and dust at the edge of an H II region. The yellow box in the left image identifies the region magnified in the middle and right images.

Due to the accumulation of thermal energy, the core gradually reaches more than 1 million K. At that temperature, deuterium (^{2}H) can fuse with hydrogen (^{1}H) to create helium-3. (Notice that this reaction occurs at a lower temperature than that at which hydrogen fusion occurs by the proton-proton chain; the object will not become a main-sequence star until the lone-proton version of hydrogen fusion begins.) Once deuterium fusion begins, it drives convection in the core. Recall from Chapter 14 that convection is the transport of energy by moving packets of gas. Convection temporarily keeps more gas from falling in and creates an apparent "surface"—more properly called a photosphere. The photosphere radiates away energy from the protostar. The hotter it gets, the more energy it radiates and the bluer that radiation becomes (see Working It Out 5.3). Because the protostar has not yet finished collapsing, the photosphere of a protostar is tens of thousands of times larger than the photosphere of the Sun today. Each square meter radiates away energy, so despite being thousands of times more luminous than the Sun, the protostar has a lower temperature.

Although the protostar is extremely luminous, astronomers often cannot observe it in visible light for two reasons. First, the photosphere of the protostar is relatively cool, so most of its radiation is in the infrared part of the spectrum. Second, and even more important, the protostar is buried deep in the heart of a dense and dusty molecular cloud. Instead, astronomers view protostars in the infrared part of the spectrum because much of the longer-wavelength infrared light from a protostar can escape through the cloud. Sometimes, as the dust absorbs the visible light, it warms up, and that heated dust also glows in the infrared. Even when astronomers cannot view a protostar directly, they can sometimes view that heated dust and know that a collapsing molecular-cloud core is hidden inside.

Sensitive infrared instruments developed since the 1980s have revolutionized the study of protostars and other young stellar objects. Clouds that appeared dark in the visible region of the spectrum, when viewed in the infrared, have revealed themselves to be clusters of dense cloud cores, young stellar objects, and glowing dust. Nearby hot stars sometimes blow away concealing dust and gas, so that molecular-cloud cores can be viewed more directly. For example, stars are forming in nodules located at the tops of the columns of dust and gas in the Eagle Nebula, shown in **Figure 15.18**.

unanswered questions

Many questions about star formation remain. For example, how must theories be modified to explain how binary stars or other multiple-star systems form? At what point during star formation is it determined that a collapsing cloud core will form several stars instead of just one? Some models suggest that this split may happen early in the process, during the fragmentation and collapse of the molecular cloud. The advantage of those ideas is that they provide a natural way of dealing with much of the cloud core's angular momentum: it goes into the orbital angular momentum of the stars around each other. Other models suggest that additional stars may form from the accretion disk around an initially single protostar.

Reading Astronomy News

Interstellar Dust Discovered inside NASA Spacecraft

Irene Klotz, *Discovery News*

Astronomers report on their search for dust from outside the Solar System.

Thanks to a massive effort by 30,716 volunteers, scientists have pinpointed what appear to be seven precious specks of dust from outside the Solar System, each bearing unique stories of exploded stars, cold interstellar clouds, and other past cosmic lives.

The Herculean effort began eight years ago after NASA's *Stardust* robotic probe flew by Earth to deposit a capsule containing samples from a comet and dust grains from what scientists hoped would be interstellar space. The spacecraft was outfitted with panels containing a smoke-like substance called aerogel that could trap and preserve fast-moving particles.

Stardust twice put itself into position to fish for interstellar grains, which are so small that a trillion of them would fit in a teaspoon. The only way scientists back on Earth would be able to find them was by the microscopic trails the grains made as they plowed into the aerogel.

"When we did the math we realized it would take us decades to do the search ourselves," physicist Andrew Westphal, with the

University of California, Berkeley, told *Discovery News*.

The team used an automated microscope to scan the collector and put out a call for volunteers.

"This whole approach was treated with pretty justifiable criticism by people in my community. They said, 'How can you trust total strangers to take on this project?'" Westphal said.

"We really didn't know how else to do it. We still don't," he added.

Recruits were trained and had to pass a test before they were given digital scans to peruse. Scientists sometimes inserted images with known trails just to see if the volunteers, known as "dusters," would spot them.

"We were very pleased to see that people are really good at finding these tracks, even really, really difficult things to find," Westphal said.

More than 50 candidate dust motes turned out to be bits of the spacecraft itself, but scientists found seven specks that bear chemical signs of interstellar origin and travel.

The grains are surprisingly diverse in shape, size, and chemical composition. The larger ones, for example, have a fluffy, snowflake-like structure.

Additional tests are needed to verify the grains' interstellar origins and ferret out their histories. But the grains are so tiny that with currently available technology, additional analysis would mean their demise.

"It'll probably be years before we can do a lot more with these samples," said space scientist Mike Zolensky, who oversees NASA's collection of cosmic dust, moon rocks, and other extraterrestrial samples at the Johnson Space Center in Houston.

"But we've got them safely tucked away and we can hang on to them until those techniques come along," Zolensky said.

QUESTIONS

1. Why do scientists want to identify interstellar dust grains?

2. How were those dust grains distinguished from Solar System grains?

3. Why can't the scientists do a complete analysis on those particles?

4. How did volunteers, known as "citizen scientists," assist with that project?

5. Use the Internet to explore: Is this project still continuing? What else has been discovered?

Source: https://www.seeker.com/interstellar-dust-discovered-inside-nasa-spacecraft-1768946686.html.

The Evolving Protostar

At any instant, the protostar is in balance—that is, the forces from hot gas pushing outward and the force of gravity pulling inward exactly oppose each other. However, that balance is constantly changing. Once the mode of energy transport in the core switches from convection to radiation, the deuterium in the core becomes depleted because it is no longer being replaced by material from the outer layers brought in by convection. The nuclear reactions slow down, reducing the pressure from the hot gas and allowing material to resume falling onto the protostar. That infalling material adds to the mass and self-gravity of

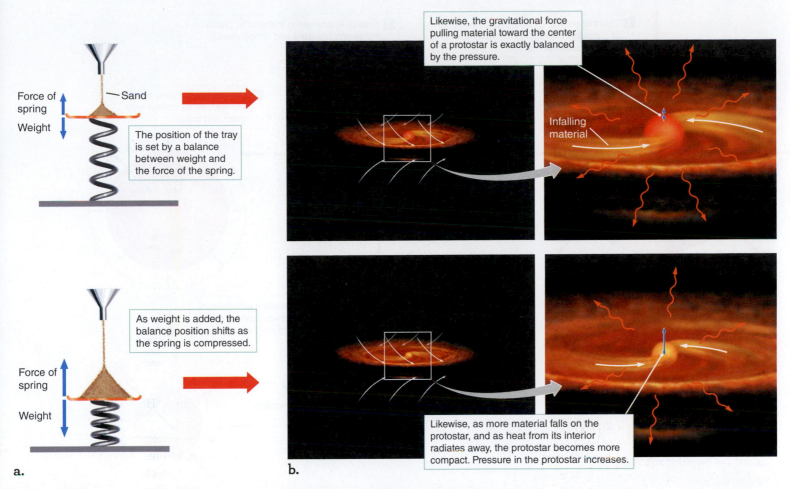

Likewise, the gravitational force pulling material toward the center of a protostar is exactly balanced by the pressure.

Force of spring
Weight
Sand

The position of the tray is set by a balance between weight and the force of the spring.

Infalling material

As weight is added, the balance position shifts as the spring is compressed.

Force of spring
Weight

Likewise, as more material falls on the protostar, and as heat from its interior radiates away, the protostar becomes more compact. Pressure in the protostar increases.

a.

b.

Figure 15.19 **a.** As weight is added to a pan on top of a spring, the pressure on the spring increases, and the spring compresses. As sand is added, the downward force of gravity is matched by the upward force of the compressed spring. **b.** Similarly, adding material to the surface of the protostar compresses the protostar, increasing the pressure inside. That balance between gravity and pressure determines the structure of a protostar.

the protostar and therefore increases the weight that inner layers of the protostar must support. The protostar also slowly loses its internal thermal energy by radiating it away.

How can an object be in perfect balance and yet be changing at the same time? Consider an Earth-bound example. **Figure 15.19a** shows a simple spring balance, which works on the principle that the more a spring is compressed, the harder it pushes back. You can measure the weight of an object by determining the point at which the pull of gravity and the push of the spring are equal.

When sand is poured slowly onto the spring balance, at any instant the downward weight of the sand is balanced by the upward force of the spring. As the weight of the sand increases, the spring is slowly compressed. The spring and the weight of the sand are always in balance, but that balance is changing as more sand is added. The situation is analogous to that of the protostar, shown in **Figure 15.19b**, in which the outward pressure of the gas behaves like the spring. The self-gravity is always matched by the internal pressure pushing out.

Material falls onto the protostar, adding to its self-gravity. While the protostar slowly loses internal thermal energy by radiating it away, the material that has fallen onto the protostar compresses the protostar and heats it up. The interior becomes

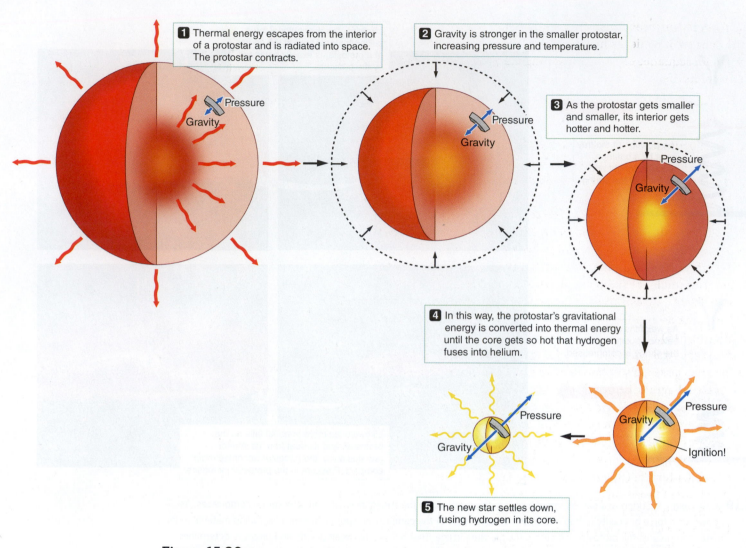

1 Thermal energy escapes from the interior of a protostar and is radiated into space. The protostar contracts.

2 Gravity is stronger in the smaller protostar, increasing pressure and temperature.

3 As the protostar gets smaller and smaller, its interior gets hotter and hotter.

4 In this way, the protostar's gravitational energy is converted into thermal energy until the core gets so hot that hydrogen fuses into helium.

5 The new star settles down, fusing hydrogen in its core.

Pressure
Gravity
Ignition!

Figure 15.20 A protostar's luminosity comes from gravitational collapse. As a protostar radiates away energy, the pressure drops and the protostar contracts. That contraction drives the interior pressure up. The counterintuitive result is that outward radiation of energy causes the interior of the protostar to grow hotter and hotter until nuclear reactions begin in its interior. The dashed line represents the previous size of the protostar.

unanswered questions

Do high-mass and low-mass stars form very differently? The smallest stars, spectral type M, are most likely to form as single stars, but a high fraction of medium-mass stars are formed in binary pairs. According to one theory, those binaries start out as triple systems, from which the smallest star is ejected, leading to a remaining pair and a single star. The highest-mass stars are less likely to form alone; instead, many form in larger groups of massive stars in which the formation of one large star may trigger the formation of another nearby in the molecular cloud.

denser and hotter, and the pressure rises—just enough to balance the increased weight of the material above it. Dynamic balance is always maintained as the protostar slowly contracts.

Figure 15.20 illustrates that chain of events as the protostar shrinks. Gravitational energy is converted to thermal energy, which heats the core, raising the pressure to oppose gravity. That process continues, with the protostar becoming smaller and smaller and its interior growing hotter and hotter. If the protostar is massive enough, its interior will eventually become so hot that nuclear fusion of hydrogen to helium can begin. That is when the transition from protostar to star takes place. The distinction between the two is that a protostar draws its energy from gravitational collapse, whereas a star draws its energy from thermonuclear reactions in its interior.

Whether the protostar will actually become a star depends on its mass. As the protostar slowly collapses, the temperature at its center rises. If the protostar's mass is greater than about 0.08 times the Sun's mass (0.08 M_{Sun}), the temperature in its core will eventually reach 10 million K and fusion of hydrogen into helium will begin. The newly born star will once again adjust its structure until it is radiating energy away from its surface at just the rate that energy is being liberated in its interior. As it does so, it achieves hydrostatic and thermal equilibrium and "settles" onto the main sequence of the Hertzsprung-Russell (H-R) diagram, where it will spend most of its life.

Brown Dwarfs

If the mass of the protostar is less than 0.08 M_{Sun}, it will never reach the point at which sustained nuclear fusion takes place. An object of roughly that mass is called a **brown dwarf** or sometimes a *substellar object*. Brown dwarfs form in the same way that a star forms, and like stars, they sometimes have binary companions and planets. In many other respects a brown dwarf is more like a giant planet Jupiter than like a star. The International Astronomical Union (IAU) has set the smallest mass for a brown dwarf at 13 Jupiter masses (13 M_{Jup}), although some astronomers think that 10 M_{Jup} is more likely the lower limit at which brief nuclear fusion can occur. The upper limit of brown dwarf masses is about 70 M_{Jup}. Despite that range of masses, brown dwarfs all have radii about the same as Jupiter's radius because more massive brown dwarfs are denser (Figure 7.16).

Brown dwarf spectral types L, T, and Y (**Figure 15.21**) have been added to the sequence of spectral classes to sort brown dwarfs by their temperature onto an H-R diagram. On an H-R diagram, brown dwarfs would be found at the far lower right, with temperatures below 1000 K and absolute magnitudes fainter than 15. A brown dwarf never grows hot enough to fuse the most common hydrogen nuclei consisting of a single proton (^{1}H), but instead glows primarily by continually releasing its own gravitational energy. The cores of brown dwarfs larger than 13 M_{Jup} can get hot enough to fuse deuterium (^{2}H), and those with a mass greater than 65 M_{Jup} can fuse lithium. But both of those energy sources are limited, and after a brief period of deuterium or lithium fusion, brown dwarfs shine only by the energy of their own gravitational contraction. A brown dwarf becomes progressively smaller and fainter. The coldest Y dwarfs observed with the WISE infrared space telescope are colder than the human body, which radiates at 310 K.

The cooler brown dwarfs have atmospheres full of methane and ammonia, like the giant planets of our Solar System. Brown dwarfs are thought to be convective all the way down to the center, so that the elemental composition of a brown dwarf's atmosphere reflects the composition in the deep interior. Winds on brown dwarfs can be very high, producing weather (including clouds) far more violent than storms observed in the atmospheres of the giant planets. Brown dwarfs are both convective and rapidly rotating, and so the magnetic fields at the surface can become tangled. Interactions between those tangled magnetic fields and the hot material beneath the surface can cause a brief lightning-like X-ray flare.

Since the first brown dwarfs were identified in the mid-1990s, more than a thousand have been found. Estimates from studies of young star clusters indicate that the Milky Way Galaxy contains between 25 and 100 billion brown dwarfs. The nearest brown dwarf to us is a binary brown dwarf system about 2 parsecs (6.5 light-years) away.

CHECK YOUR UNDERSTANDING 15.3

The energy required to begin nuclear fusion in a protostar originally came from: (a) the gravitational potential energy of the protostar; (b) the kinetic energy of the protostar; (c) the wind from nearby stars; (d) the pressure from the interstellar medium.

what if . . .

What if Jupiter were 15 times more massive than it is, thus qualifying as a small brown dwarf? How would Jupiter's appearance in the night sky be different?

Figure 15.21 This artist's conception shows the three types of brown dwarf stars: L dwarfs ($T \approx 1700$ K), T dwarfs ($T \approx 1200$ K), and Y dwarfs ($T \approx 500$ K).

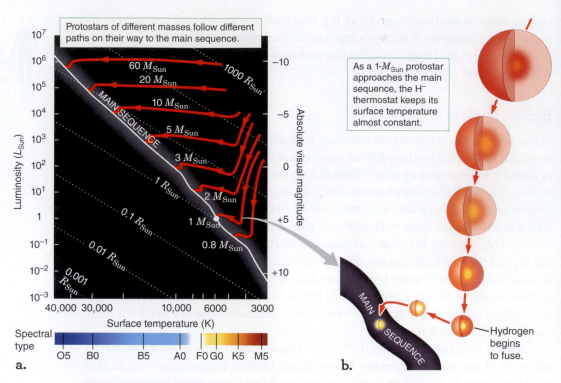

Figure 15.22 a. The evolution of pre-main-sequence stars can be followed on the H-R diagram. More massive protostars in the upper right portion of the diagram follow horizontal tracks. **b.** The roughly vertical, constant-temperature part of the evolutionary track of a low-mass protostar is called the Hayashi track.

15.4 Evolution before the Main Sequence

Protostars and young evolving stars change their location on the H-R diagram as they settle into the main sequence, where they will spend the bulk of their lives. In this section, we examine some of the early stages in the life of the new stars.

The Evolutionary Track of an Evolving Star

In Chapter 13, we introduced the H-R diagram (see Figure 13.15) and used it to help explain how the properties of stars differ. As stars evolve through their lifetimes, their position on the H-R diagram changes, creating a path known as an **evolutionary track**. Because a protostar is so large, it is more luminous than a star of the same temperature on the main sequence, so a protostar's evolutionary track is located above the main sequence on the H-R diagram. **Figure 15.22a** shows the evolutionary tracks of protostars of several different masses. For protostars with high masses, a radiative zone develops around the core, causing the surface temperature to increase. Because of that increase in surface temperature, the luminosity of these protostars stays roughly constant even as it contracts; on the H-R diagram, its track moves left horizontally to the main sequence (see the track for the $60\text{-}M_{\text{Sun}}$ protostar). Protostars with low masses are convective all the way to the core, so the surface temperature stays about the same as they collapse. As those protostars collapse, the luminosity decreases. On the H-R diagram, the track moves vertically down to the main sequence (see the track for the $1\text{-}M_{\text{Sun}}$ protostar); that path is called a **Hayashi track (Figure 15.22b)**.

In the 1960s, the theoretical physicist Chushiro Hayashi (1920–2010) explained the difference between the surface temperature of a star or protostar and the temperature deep in its interior. Hayashi showed that the atmospheres of stars and protostars contain a natural thermostat: the H⁻ ion. (A negative **ion**, such as H⁻, is an atom that has acquired an extra electron and therefore has a negative charge.) The amount of H⁻ in the atmosphere of a protostar is highly sensitive to the temperature at the protostar's surface. The cooler the atmosphere of a star, the more slowly atoms and electrons are moving and the easier it is for a hydrogen atom to hold on to an extra electron. As a result, the amount of H⁻ increases as the temperature of the atmosphere of the star decreases.

The H⁻ ion, in turn, helps control how much energy a star or protostar radiates away. The more H⁻ the atmosphere of the star or protostar has, the more opaque the atmosphere is and the more effectively the thermal energy of the protostar is trapped in its interior. Imagine that the surface of the protostar is "too cool," meaning that extra H⁻ forms in the atmosphere and makes the atmosphere of the protostar more opaque. The atmosphere thus traps more of the radiation trying to escape, and the trapped energy heats up the star. As the temperature increases, the H⁻ ions lose an electron to form neutral H atoms. Now imagine the other possibility—that the protostar is too hot. Then H⁻ in the protostar's atmosphere is destroyed, so the atmosphere becomes more transparent, allowing radiation to escape more freely from the interior. Because the protostar cannot hold on to enough of its energy to stay warm, the surface cools. In either case—too cold or too hot—H⁻ is formed or destroyed, respectively, until the star's atmosphere once again traps just the right amount of escaping radiation. The H⁻ ion is basically doing the same thing that you do with your bedcovers at night. If you get too cold, you pile on extra covers to trap your body's thermal energy and keep warm (corresponding to more H⁻ ions with the star). If you get too hot, you kick off some covers to cool down (fewer H⁻ ions).

The amount of H⁻ in the atmosphere keeps the surface temperature of protostars with low masses somewhere between about 3000 and 5000 K, depending on the protostar's mass and age. Because the protostar's surface temperature is not changing much, the amount of energy per unit time (**power**) radiated away by each square meter of the protostar's surface does not change much, either. Recall the Stefan-Boltzmann law from Chapter 5, which says that the amount radiated by each square meter of an object's surface depends on its temperature. As the protostar shrinks, the area of its surface shrinks as well. With fewer square meters of surface from which to radiate, the protostar's luminosity decreases. As viewed from outside, the protostar stays at nearly the same temperature and color but gradually gets fainter as it evolves toward its eventual life as a main-sequence star. The relationship among luminosity, surface temperature, and radius of protostars is further explored in **Working It Out 15.2.**

Bipolar Outflow

As shown in **Figure 15.23**, material falls onto the accretion disk around a young stellar object and moves inward toward the star's equator. Meanwhile, other material is blown away from the protostar and disk in two opposite directions from the

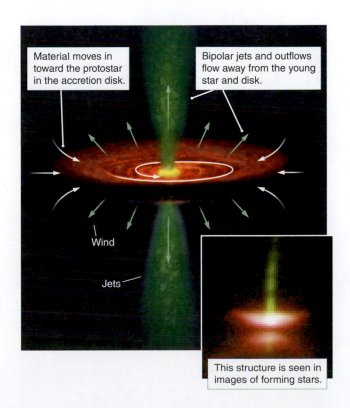

Material moves in toward the protostar in the accretion disk.

Bipolar jets and outflows flow away from the young star and disk.

Wind

Jets

This structure is seen in images of forming stars.

Figure 15.23 Material falls onto an accretion disk around a protostar and then moves inward, eventually falling onto the protostar. In the process, some of that material is driven away in powerful jets that stream perpendicular to the disk.

working it out 15.2

Luminosity, Surface Temperature, and Radius of Protostars

Recall from Chapter 13 that the luminosity, surface temperature, and radius of a star are related as follows:

$$L = 4\pi R^2 \sigma T^4$$

What does that equation reveal about the changing properties of the protostar as its radius decreases?

Suppose that when the Sun was a protostar, it had a radius 10 times what it is now and a surface temperature of 3300 K. What would its luminosity have been? The equations for each are

$$L_{\text{protostar}} = 4\pi R_{\text{protostar}}^2 \sigma T_{\text{protostar}}^4$$

and

$$L_{\text{Sun}} = 4\pi R_{\text{Sun}}^2 \sigma T_{\text{Sun}}^4$$

We can set that relationship up as a ratio, comparing the luminosity of the protostar Sun with its luminosity now, L_{Sun}:

$$\frac{L_{\text{protostar}}}{L_{\text{Sun}}} = \frac{4\pi R_{\text{protostar}}^2 \sigma T_{\text{protostar}}^4}{4\pi R_{\text{Sun}}^2 \sigma T_{\text{Sun}}^4}$$

We rewrite that expression as follows, grouping like terms together:

$$\frac{L_{\text{protostar}}}{L_{\text{Sun}}} = \frac{4\pi\sigma}{4\pi\sigma} \times \left(\frac{R_{\text{protostar}}}{R_{\text{Sun}}}\right)^2 \times \left(\frac{T_{\text{protostar}}}{T_{\text{Sun}}}\right)^4$$

Then we cancel out the constants, $4\pi\sigma$, and use the value for $T_{\text{Sun}} = 5780$ K from Chapter 14. We know that the protostar's radius is 10 times that of the Sun, so $R_{\text{protostar}}/R_{\text{Sun}} = 10$. Then the equation becomes

$$\frac{L_{\text{protostar}}}{L_{\text{Sun}}} = \left(\frac{10}{1}\right)^2 \times \left(\frac{3300}{5780}\right)^4 = 10^2 \times (0.57)^4 = 10.6$$

So the Sun was about 10.6 times more luminous as a protostar than it is now. We see that on the H-R diagram of protostars (see Figure 15.22). As a 1-M_{Sun} star approaches the main sequence on the diagram, it moves down (toward lower luminosity) and to the left (toward higher surface temperature).

what if . . .

What if you optically observe a T Tauri star having well-defined polar jets but no discernable accretion disk? What would you conclude about that star?

plane of the disk. The resulting stream of material away from the protostar is called a **bipolar outflow**. Powerful outflows can disrupt the molecular-cloud core and accretion disk from which the protostar formed, stopping material from falling onto the protostar.

Some bipolar outflows from young stellar objects are slow and fairly disordered, but others produce remarkable **jets** of material moving at hundreds of kilometers per second. The material in those jets flows out into the interstellar medium, where it heats, compresses, and pushes away surrounding interstellar gas. Knots of glowing gas accelerated by jets are called **Herbig-Haro objects** (or **HH objects** for short), named after the two astronomers who first identified them and associated them with star formation. An example is shown in **Figure 15.24**.

The origin of outflows from protostars is not well understood, but current models suggest that outflows are the result of magnetic interactions between the protostar and the disk. The interior of a protostar with low mass on a Hayashi track is convective. That convection, coupled with the protostar's rapid rotation, forms a dynamo, similar to the dynamo that drives the Sun's magnetic field, but much more powerful. The resulting strong magnetic field might cause the protostar to begin blowing a powerful wind. As the protostar's rotation drags the magnetic field lines around, the lines wind up like the fibers in a rope. Those tightly wound magnetic field lines collimate the outflow into jets.

Protostars with higher masses also show evidence of jets, and the mechanism may be similar, but because these protostars are not convective all the way to

IR

Figure 15.24 ★ WHAT AN ASTRONOMER SEES In this image of Herbig-Haro object 212, taken by ESO's Infrared Spectrometer And Array Camera (ISAAC), an astronomer will notice the accretion disk of the protostar, which is visible as a fuzzy dark band cutting between the two jets. She will know that the protostar is concealed inside that accretion disk and is only a few thousand years old. An astronomer will spend some time carefully comparing the two jets, noticing that the jets are remarkably symmetric, with several knots appearing at intervals. That symmetry suggests that the knots in both jets are produced by the same mechanism deep inside the accretion disk. The knots also indicate that the jet pulses regularly and over a short timescale. If the astronomer has velocity information and a distance, she can even determine that timescale, which may be as short as 30 years. An astronomer will also notice that at the ends of the jets, ejected gas collides with interstellar dust and gas, forming bow shocks, like the pileup of material in front of the bow of a boat. This means that material in the jet is moving very quickly—namely, several hundred kilometers per second. Finally, an astronomer will spend a moment or two considering the color choice for this infrared image; because it's not a visible image, the creators could have chosen any color, but this sepia tone really makes the contrast between the bright knots and the very faint dusty envelope near the center more noticeable.

Credit: ESO/M. McCaughrean, https://www.eso.org/public/images/potw1541a/. https://creativecommons.org /licenses/by/4.0/.

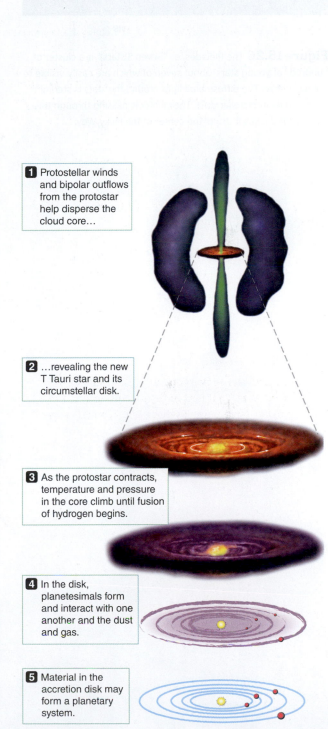

1 Protostellar winds and bipolar outflows from the protostar help disperse the cloud core…

2 …revealing the new T Tauri star and its circumstellar disk.

3 As the protostar contracts, temperature and pressure in the core climb until fusion of hydrogen begins.

4 In the disk, planetesimals form and interact with one another and the dust and gas.

5 Material in the accretion disk may form a planetary system.

the core, questions remain about how the magnetic field forms and influences the infalling material from the disk.

As the wind from the protostar disperses the remains of the dusty molecular-cloud core from which the protostar formed, the first direct, visible-light view of the protostar emerges. Some of those protostars of lower mass are called **T Tauri stars**. The name comes from the first recognized member of that class of objects, the star labeled T in the constellation Taurus. Higher-mass protostars of spectral type B or A are called Herbig Be or Ae stars. **Figure 15.25** summarizes the evolution of a protostar from the onset of winds to the formation of planets (recall that steps 4 and 5 were covered in Chapter 7).

The Influence of Mass

Astronomers are interested in how and why molecular clouds fragment to form stars with a range of masses. Astronomers do not understand why some cloud cores become 1-M_{Sun} stars, whereas others become 5- or 10-M_{Sun} stars. The details of that division—specifically, what fraction of newly formed stars will have which

Figure 15.25 An overview of how stars like the Sun form, beginning with the onset of stellar winds and ending with the ignition of a star sitting at the center of a revolving system of planets.

VIS

Figure 15.26 The Pleiades, or "Seven Sisters," is a cluster of hundreds of young stars, about seven of which are easily visible to the naked eye. The diffuse blue light around the stars is starlight scattered by interstellar dust. The cluster is passing through this interstellar dust as it orbits the center of the Milky Way.

masses—are crucial to understanding how observations of the stars near the Sun today relate to the history of star formation in our galaxy.

A look around the disk of our galaxy reveals a variety of stars—some very old and others very young. If those were the only stars available to study, scientists could not easily learn much about how stars evolve. But astronomers have long known that stars are often found close together in collections called **star clusters**. Star clusters are collections of stars that all formed in the same place, from the same material, and at about the same time. **Figure 15.26** shows one such star cluster: a group called the Pleiades, or "Seven Sisters." Clusters such as that one serve as extremely useful samples for studying star formation. Even though the few brightest and most massive stars in a cluster dominate any observation of a cluster, most of the stars in a cluster are less massive than the Sun. In fact, some star-forming regions seem to form no high-mass stars at all.

After a molecular-cloud core collapses, how a protostar evolves depends almost entirely on its mass. A star like the Sun takes about 10 million years or so to descend its Hayashi track and become a star on the main sequence. The total time for such a star to form—from initial fragmentation up to the ignition of hydrogen fusion—might be more like 30 million years. Because the self-gravity of a more massive core is stronger, more massive cores collapse to form stars more quickly. Thus, a 10-M_{Sun} star will form in only 100,000 years, and a 100-M_{Sun} star might take less than 10,000 years. By comparison, a 0.1-M_{Sun} star might take 100 million years to reach the main sequence.

The 30 million years that the Sun took to form is a long time, yet it is still just a tiny fraction of the 10 billion years during which the Sun will steadily fuse hydrogen into helium as a main-sequence star. Because stars spend so much longer on the main sequence than they do as protostars, it is no wonder that so few among the many stars visible in the sky are protostars. But every star was once a protostar, including the Sun.

CHECK YOUR UNDERSTANDING 15.4

What causes the wind from a protostar to form jets? (a) The accretion disk can supply material to the jet only along the equator of the star. (b) Magnetic fields wind up and direct the outflow. (c) The star is larger at the equator, so material leaves the poles more easily. (d) The star spins faster at the equator, so material leaves the poles more easily.

Origins: Star Formation, Planets, and Life

When astronomers consider the possibility of other life in the universe, one of the first things they think about is the formation of stars and planets. Life probably needs planets, and planets form along with stars. The conditions under which a star is born, and the mass and chemical composition that it has when it begins its nuclear fusion, set the stage for the rest of its life. In Chapter 7, we noted that most stars have planets that form at about the same time as the star. If the star is going to have rocky planets with hard surfaces (such as the planets of the inner Solar System) or gaseous planets with molten rocky cores and rocky moons (such as the planets of the outer Solar System), the material from which the star and planets form must be "enriched" with the heavy elements that make up those rocky surfaces.

Those enriched clouds would also provide elements essential to life on Earth. In addition to the presence of organic molecules mentioned previously in the chapter, astronomers have detected water in star-forming regions such as W3 IRS5 (**Figure 15.27**). Water exists as ice mixed with dust grains in the cool molecular clouds, or it exists as vapor when it is closer to a protostar and the dust grains and ice evaporate. In 2011, the Herschel Space Observatory detected oxygen molecules (O_2, the type we breathe) in a star-forming complex in Orion. Oxygen is the third-most-common element in the universe, yet it had not been decisively observed before in molecular form. That oxygen also may have come from the melting and evaporation of water ice on the tiny dust grains.

As noted in Chapter 13, astronomers had doubted that planets could exist in stable orbits in binary star systems, but now a few such systems have been found. Planets that form within associations of O and B stars may be too unstable to last very long. Isolated planets unattached to any star are moving through the Milky Way. Perhaps those rogue planets were gravitationally ejected soon after they formed in a multiple system. But those planets do not have a source of energy like Earth's Sun. Astronomers theorize that only planets that orbit stars can support life. So when trying to estimate the possibility of life in the galaxy, astronomers include estimations of the rate of star formation in the galaxy, along with the fraction of stars that have planets. Advances in the study of star formation and planet detection help astronomers understand better the conditions under which life might develop elsewhere.

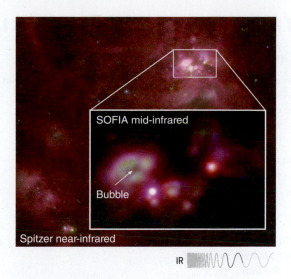

Figure 15.27 Water was detected in the W3 IRS5 star-forming complex. Here, W3 IRS5 is observed with the Spitzer Space Telescope in near infrared and with the Stratospheric Observatory for Infrared Astronomy (SOFIA) telescope in mid-infrared. A massive star has cleared the dust and gas from a small bubble, sweeping it into a dense shell (green).

SUMMARY

Astronomers have never watched the full process of a single star forming from beginning to end. Instead, they have observed many stars at different stages in their formation and evolution, at different wavelengths, and they have used their knowledge of physical laws to tie those observations together into a coherent, consistent description of how, why, and where stars form. Stars form from material in the interstellar medium, under the influence of gravity, often triggered by external factors such as the winds of hot stars nearby. The conditions under which a star is born determine whether it will have planets along with the chemical elements required by life as it exists on Earth. The star's mass at formation will determine how it evolves.

(1) **Categorize the types and states of material that exist in the space between the stars and describe how that material is detected.** The interstellar medium ranges from cold, relatively dense molecular clouds to hot, tenuous intercloud gas heated and ionized by energy from stars and stellar explosions. Dust and gas in the interstellar medium blocks much visible light but becomes more transparent at longer, infrared wavelengths. Different phases of the interstellar medium emit various types of radiation and can be observed at different wavelengths, ranging from radio waves to X-rays. Neutral hydrogen cannot be detected at visible and infrared wavelengths, but its presence is revealed by its 21-cm emission.

(2) **Explain the conditions under which a cloud of gas can contract into a stellar system and contrast the roles that gravity and angular momentum**

play in forming stars and planets. Star formation begins when the self-gravity of dense clouds exceeds outward pressure. The clouds collapse, heat up, and fragment to form stars. The conservation of angular momentum is important to the formation of disks during the collapse. Forming stars are detected directly from their infrared emission and indirectly from the dust around them, which is heated by the forming star.

(3) **List the steps in the evolution of a protostar, and summarize how the mass of a protostar affects its evolution.** Protostars collapse, radiating away their gravitational energy until fusion starts in their cores. When protostars reach hydrostatic and thermal equilibrium, they settle onto the main sequence. Stars form in clusters from dense cores buried within giant molecular clouds. A protostar must have a mass of at least 0.08 M_{Sun} to become a true star. Brown dwarfs are neither stars nor planets but have masses in between.

(4) **Reconstruct the track of a protostar as it evolves to a main-sequence star on the Hertzsprung-Russell (H-R) diagram.** Because star formation takes tens of thousands to millions of years, astronomers learn about the formation of stars from observations of many protostars at various stages of development. Higher-mass protostars move horizontally across the H-R diagram, at roughly constant luminosity, to the main sequence. Lower-mass protostars drop nearly straight down, at constant temperature, toward the main sequence. Once hydrogen fusion begins, the star moves nearly horizontally until it reaches the main sequence.

QUESTIONS AND PROBLEMS

TEST YOUR UNDERSTANDING

1. Phases of the interstellar medium include (choose all that apply)
   a. hot, low-density gas.
   b. cold, high-density gas.
   c. hot, high-density gas.
   d. cold, low-density gas.

2. Dust in the interstellar medium can be observed in
   a. visible light.
   b. infrared radiation.
   c. radio waves.
   d. X-rays.

3. The interstellar medium in the Sun's region of the galaxy is closest in composition to that of
   a. the Sun.
   b. Jupiter.
   c. Earth.
   d. comets in the Oort Cloud.

4. Interstellar dust is effective at blocking visible light because
   a. the dust is so dense.
   b. dust grains are so few.
   c. dust grains are so small.
   d. dust grains are so large.

5. Hot intercloud gas is heated primarily by
   a. starlight.
   b. protostars.
   c. neutrinos.
   d. supernova explosions.

6. Astronomers determined the composition of the interstellar medium from
   a. observing its emission and absorption lines.
   b. measuring the composition of the planets.
   c. samples returned from spacecraft.
   d. the composition of meteorites.

7. In astronomy, the term *bipolar* refers to outflows that
   a. point in opposite directions.
   b. alternate between expanding and collapsing.
   c. rotate about a polar axis.
   d. show spiral structure.

8. Which of the following has contributed most to our understanding of star formation?
   a. Astronomers have observed star formation as it happens for a few stars.
   b. Astronomers have observed star formation as it happens for many stars.
   c. Astronomers have observed many stars at various steps of the formation process.
   d. Theoretical models predict how stars form.

9. Cold neutral hydrogen can be detected because
   a. it emits light when electrons drop through energy levels.
   b. it blocks the light from more distant stars.
   c. it is always hot enough to glow in the radio and infrared wavelengths.
   d. the atoms in the gas change spin states.

10. The Hayashi track is a nearly vertical evolutionary track on the H-R diagram for low-mass protostars. Which of the following would you expect from a protostar moving along a vertical track?
    a. The star remains the same brightness.
    b. The star remains the same luminosity.
    c. The star remains the same color.
    d. The star remains the same size.

11. Which two forces establish hydrostatic equilibrium in an evolving protostar?
    a. the force from pressure and gravity
    b. the force from pressure and the strong nuclear force
    c. gravity and the strong nuclear force
    d. the energy emitted and the energy produced

12. Suppose you are studying a visible-light image of a distant galaxy, and you see a dark lane cutting across the bright disk. That dark line is most likely caused by
    a. gravitational instabilities that clear the area of stars.
    b. dust in the Milky Way blocking the view of the distant galaxy.
    c. dust in the distant galaxy blocking the view of stars in the disk.
    d. a flaw in the instrumentation.

13. What causes a hydrogen atom to radiate a photon of 21-cm radio emission?
    a. The electron drops down one energy level.
    b. The formerly free electron is captured by the proton.
    c. The electron flips to an aligned spin state.
    d. The electron flips to an unaligned spin state.

14. Astronomers know that dusty accretion disks exist around protostars because
    a. a dark band often appears across the protostar.
    b. a bright band often appears across the protostar.
    c. accretion disks should be there, according to theory.
    d. planets are in the Solar System.

15. What is the single most important property of a star that will determine its evolution?
    a. mass
    b. composition
    c. temperature
    d. radius

THINKING ABOUT THE CONCEPTS

16. The interstellar medium is approximately 99 percent gas and 1 percent dust. Why does dust and not gas block a visible-light view of the galactic center?

17. Explain why observations in the infrared are necessary for astronomers to study the detailed processes of star formation.

18. How does the material in interstellar clouds and intercloud gas differ in density and distribution?

19. When a star forms inside a molecular cloud, what happens to the cloud? Can a molecular cloud remain cold and dark with one or more stars inside it? Explain your answer.

20. If you placed your hand in boiling water (100°C) for even 1 second, you would get a very serious burn. If you placed your hand in a hot oven (200°C) for a second or two, you would hardly feel the heat. Explain that difference and how it relates to million-kelvin regions of the interstellar medium.

21. How do astronomers know that the Sun is located in a "local bubble" formed by a supernova?

22. Interstellar gas atoms typically cool by colliding with other gas atoms or grains of dust; during the collision, each gas atom loses energy and hence its temperature is lowered. How does that explain why very low-density gases are generally so hot, whereas dense gases tend to be so cold?

23. Explain how the 21-cm line discussed in the Process of Science Figure supports the cosmological principle (which states that the laws of physics must be the same everywhere). 👁

24. Molecular hydrogen is very difficult to detect from the ground, but astronomers can easily detect carbon monoxide (CO) by observing its 2.6-cm microwave emission. Describe how observations of CO might help astronomers infer the amounts and distribution of molecular hydrogen within giant molecular clouds.

25. The Milky Way contains several thousand giant molecular clouds. Describe a giant molecular cloud and its role in star formation.

26. As a cloud collapses to form a protostar, the forces of gravity experienced by all parts of the cloud (which follow an inverse square law) become stronger and stronger. One might argue that under those conditions, the cloud should keep collapsing until it becomes a single massive object. Why doesn't that happen?

27. ★ **WHAT AN ASTRONOMER SEES**   The image in Figure 15.24 was taken in the infrared part of the spectrum. If you could obtain an image at similar resolution in the radio and visible parts of the spectrum, which parts would you expect to be brighter in the radio image? Which parts would you expect to be brighter in the visible image? 👁

28. What are the similarities and differences between a brown dwarf and a giant planet such as Jupiter? Would you classify a brown dwarf as a supergiant planet? Explain your answer.

29. The H⁻ ion acts as a thermostat that controls the surface temperature of a protostar. Explain that process.

30. How does the composition of the molecular cloud affect the type of planets and stars that form within it?

APPLYING THE CONCEPTS

31. A typical temperature for intercloud gas is 8000 K. What is the peak wavelength at which that gas would radiate? 🟡—●—🟢
  a. Make a prediction: Is this gas "hot," "warm," or "cold"? What part of the electromagnetic spectrum do you expect the peak wavelength to be in?
  b. Calculate: Follow Working It Out 15.1 to calculate the peak wavelength of the radiation from this gas.
  c. Check your work: Verify that your answer has units of length, and compare your answer to your prediction.

32. Assume that a brown dwarf has a surface temperature of 1000 K and approximately the same radius as Jupiter. What is its luminosity compared with that of the Sun? 🟡—●—🟢
  a. Make a prediction: Based on what you have learned about brown dwarfs, do you expect a brown dwarf to be more or less luminous than the Sun? By a lot or by a little?
  b. Calculate: Follow Working It Out 15.2 to find the ratio of luminosity of the brown dwarf to the luminosity of the Sun. Move the luminosity of the Sun to the other side of the equation to solve for the luminosity of the brown dwarf.
  c. Check your work: Verify that your ratio is dimensionless (has no units), and then compare your final answer to your prediction.

33. Some parts of the Orion Nebula have a blackbody peak wavelength of 0.29 μm. What is the temperature of those parts of the nebula?
  a. Make a prediction: What part of the electromagnetic spectrum is this peak wavelength in? Do you expect to find that these parts of the nebula are "hot," "warm," or "cold"?
  b. Calculate: Follow Working It Out 15.1 backward to calculate the temperature of these parts of the Orion Nebula.
  c. Check your work: Verify that your answer has units of kelvin, and compare your answer to your prediction.

34. A protostar with the mass of the Sun starts out with a temperature of about 3500 K and a luminosity about 200 times larger than the Sun's present value. Estimate that protostar's radius in units of the radius of the Sun today (R_{Sun}). 👁
  a. Make a prediction: Does a protostar start out large or small compared to its final size? Is this a large difference or a small one?
  b. Calculate: Rearrange the equation in Working It Out 15.2 to solve for the radius of the protostar in units of the radius of the Sun.
  c. Check your work: Compare your answer to your prediction and to the radius lines in the H-R diagram in Figure 15.22.

35. A protostar starts out with a temperature of about 3500 K and a luminosity about a million times larger than the Sun's present value. Estimate that protostar's radius in units of the radius of the Sun today (R_{Sun}). 👁
  a. Make a prediction: Does a protostar start out large or small compared to its final size? Is this a large difference or a small one?
  b. Calculate: Rearrange the equation in Working It Out 15.2 to solve for the radius of the protostar in units of the radius of the Sun.
  c. Check your work: Compare your answer to your prediction and to the radius lines in the H-R diagram in Figure 15.22.

36. Recall from Chapter 13 that astronomers can measure the temperature of a star by comparing its brightness in blue and yellow light. Does reddening by interstellar dust affect a star's temperature measurement? If so, how?

37. When a hydrogen atom is ionized, it splits into two components.
   a. Identify the two components.
   b. If both components have the same kinetic energy, which moves faster?

38. Estimate the typical density of dust grains (grains per cubic centimeter) in the interstellar medium. A typical grain has a mass of about 10^{-17} kilogram (kg). (Hint: You know the typical density of gas and the fraction of the interstellar medium's mass that is made of dust.)

39. Use Figure 15.4 to estimate the effective blackbody temperature of the star as shown in part b. (without dust) and part c. (with dust). How significant are the effects of interstellar dust when observed data are used to determine the properties of a star? ★

40. Stellar radiation can convert atomic hydrogen (H I) to ionized hydrogen (H II).
   a. Why does a B8 main-sequence star ionize far more interstellar hydrogen in its vicinity than a K0 giant of the same luminosity?
   b. What properties of a star are important in determining whether it can ionize large amounts of nearby interstellar hydrogen?

41. A proton has 1850 times the mass of an electron. If a proton and an electron have the same kinetic energy ($E_K = 1/2\, mv^2$), how many times greater is the velocity of the electron than that of the proton?

42. If a typical hydrogen atom in a collapsing molecular-cloud core starts at a distance of 1.5×10^{12} km (10,000 AU) from the core's center and falls inward at an average velocity of 1.5 km/s, in how many years does it reach the newly forming protostar? Assume that a year is 3×10^7 seconds.

43. The ratio of hydrogen atoms (H) to carbon atoms (C) in the Sun's atmosphere is approximately 2400:1 (see Table 13.1). We can reasonably assume that this ratio also applies to molecular clouds. If 2.6-cm radio observations indicate 100 M_{Sun} of carbon monoxide (CO) in a giant molecular cloud, what is the implied mass of molecular hydrogen (H_2) in the cloud? (Carbon represents 3/7 of the mass of a CO molecule.)

44. The Sun took 30 million years to evolve from a collapsing cloud core to a star, with 10 million of those years spent on its Hayashi track. The Sun will spend 10 billion years on the main sequence. Suppose that the Sun's main-sequence lifetime were compressed into a single day.
   a. How long would the total collapse phase last?
   b. How long would the Sun spend on its Hayashi track?

45. The star-forming region 30 Doradus is 160,000 light-years away in the nearby galaxy called the Large Magellanic Cloud, and it appears about one-sixth as bright as the faintest stars visible to the naked eye. If it were located at the distance of the Orion Nebula (1300 light-years away), how much brighter than the faintest visible stars would it appear?

EXPLORATION   The Stellar Thermostat

In this Exploration, you will see how the H⁻ thermostat works in the formation of stars. You will need about 20 coins (they do not have to be all the same type).

Place your coins on a sheet of paper and draw a circle around them—the smallest possible circle that will fit all the coins. Then divide the circle into three parts, as shown in **Figure 15.28**. This circle represents a star with a changing temperature. The coins represent H⁻ ions. Removing a coin from the circle means that the H⁻ ion has turned into a neutral hydrogen atom. Placing a coin in the circle means that the neutral hydrogen atom has become an H⁻ ion.

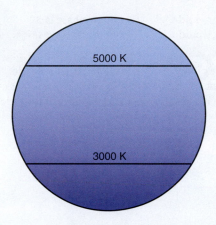

Figure 15.28 This circle represents a star with a changing temperature.

Place all the coins back on the circle.

1   How many "H⁻ ions" are now in the star?

_____

_____

The "blanket" of H⁻ ions holds heat in the star, so the star begins to heat up until it reaches about 5000 K. At that surface temperature, the H⁻ ions begin to be destroyed. Now that the star is hot, begin removing coins one at a time, starting from the top of the circle and working downward. When you see the line marking 3000 K, stop removing coins.

2   How many "H⁻ ions" are now in the star?

_____

_____

3   What will happen to the surface temperature of the star now that it has fewer ions?

_____

_____

When the star cools off to about 3000 K, H⁻ ions begin to form. Place the coins back on the circle, starting from the bottom and working your way up to the line at 5000 K.

4   How many "H⁻ ions" are now in the star?

_____

_____

5   What will happen to the surface temperature of the star now that it has more ions?

_____

_____

Now that the star is hot, begin removing coins one at a time, starting from the top of the circle and working downward. When you see the line marking 3000 K, stop removing coins.

6   What should happen next?

_____

_____

7   Make a circular flowchart that includes the following steps in the proper order: the star heats up; the star cools down; H⁻ is formed; H⁻ is destroyed.

_____

_____

Evolution of Low-Mass Stars

As stars begin the final stages of their evolution, they become both brighter and redder, as we explain here in Chapter 16. Stars are red, white, or blue, depending on the temperature of their outer layers. Predict whether you will see more red, white, or blue stars in the night sky. These colors can be subtle; some of the most prominent stars in the night sky are red giants or supergiants, but many people do not notice the color until it is pointed out to them! On a clear night, take a star chart out with you, and carefully observe the stars. When you find a red one, identify it with your star chart, and make a log of your observation, including the other bright stars of the constellation in which you found it. While not all red stars are red giants or supergiants, the vast majority of red stars are far too faint to see. If you see a bright red star in the night sky, it is most likely a dying star that is quickly passing through the giant or supergiant phase. While you are observing, carefully note the colors of other stars that you see.

EXPERIMENT SETUP

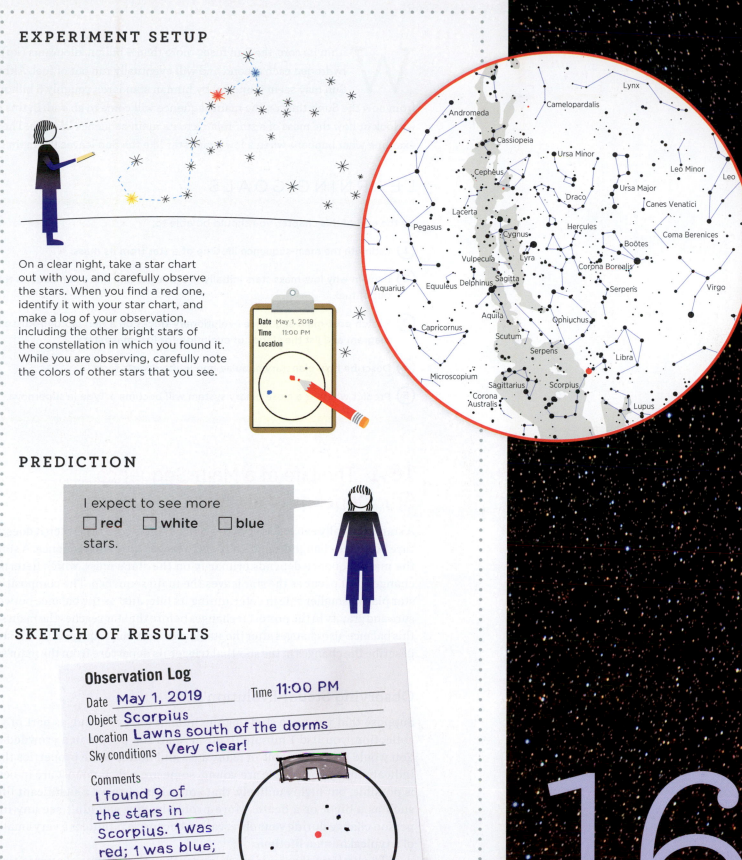

On a clear night, take a star chart out with you, and carefully observe the stars. When you find a red one, identify it with your star chart, and make a log of your observation, including the other bright stars of the constellation in which you found it. While you are observing, carefully note the colors of other stars that you see.

PREDICTION

I expect to see more
☐ red   ☐ white   ☐ blue
stars.

SKETCH OF RESULTS

Observation Log

Date May 1, 2019   Time 11:00 PM
Object Scorpius
Location Lawns south of the dorms
Sky conditions Very clear!

Comments
I found 9 of the stars in Scorpius. 1 was red; 1 was blue; the rest were white.

16

Within its core, the Sun fuses more than 4 billion kilograms (kg) of hydrogen each second and will eventually run out of fuel. Although the Sun may seem immortal by human standards, roughly 5 billion years from now the Sun's time on the main sequence will come to an end. In this chapter, we look at how the mass of a star relates to its main-sequence lifetime. Then we examine what happens when a low-mass star like the Sun leaves the main sequence.

LEARNING GOALS

By the end of this chapter, you should be able to:

(1) Estimate the main-sequence lifetime of a star from its mass.

(2) Explain why low-mass stars initially grow larger and more luminous as they run out of fuel.

(3) Sketch post-main-sequence evolutionary tracks on a Hertzsprung-Russell (H-R) diagram and list the stages of evolution for low-mass stars.

(4) Describe how planetary nebulae and white dwarfs form.

(5) Predict whether a close binary system will become a Type Ia supernova.

16.1 The Life of a Main-Sequence Star Depends on Its Mass

A star eventually exhausts the hydrogen fuel in its core, and when it does, its structure begins to change dramatically and it leaves the main sequence. A star's life on the main sequence depends primarily on the star's mass, which determines the changes that occur as the star leaves the main sequence. The composition of the star plays a smaller role in determining its fate. Just as the balance between pressure and gravity in the protostar changes before the star reaches the main sequence, this balance also changes after the star leaves the main sequence. In this section, we describe the changes in the star that trigger its departure from the main sequence.

Observing Stellar Evolution

Suppose that you were studying the human life cycle and, as part of your data collection, you had 1 minute to observe all the people in a crowded stadium. You would observe people of many ages and notice some properties that would indicate that some people are young, some are old, and most are in between. It is possible, but highly unlikely, that you might observe a significant life change such as a birth or a death. More probably, you wouldn't see any individual person change during your observation because a minute is a very small fraction of a typical human lifetime.

The lifetime of a star is much longer than the lifetime of a person, so astronomers would have to observe a star like the Sun for several hundred years to observe the equivalent of 1 minute in a human life span. Astronomers do not see individual stars "age." Instead, they observe many, many stars for a short time to piece together a stellar life cycle. Sometimes, just by chance, astronomers

Name	High-mass stars	Medium-mass stars	Low-mass stars	Very low-mass stars	Brown dwarfs
Spectral type	O, B	B	A, F, G, K	M	M, L, T, Y
Minimum mass	8 M_{Sun}	3 M_{Sun}	0.5 M_{Sun}	0.08 M_{Sun}	~0.01 M_{Sun} (~13 $M_{Jupiter}$)

Figure 16.1 This chart organizes main-sequence stars and brown dwarfs into five broad categories that depend on mass. Each mass category has a typical set of spectral types, which in turn means a typical temperature range, and a typical peak wavelength, or color. Spectral type B has stars in both the high-mass and medium-mass categories.

observe a star undergoing a dramatic change. From these observations, astronomers determine how stars change over time, more commonly known as **stellar evolution**.

Observers collect information about stellar properties and theorists model the nuclear reactions that take place inside stars of a given mass and chemical composition. Theorists use those models to predict how a star's radius and luminosity will evolve. Those predictions are compared with observations, then theorists adjust and improve their models. The study of stellar evolution through this interaction between observation and theory has led to a detailed understanding of how stars live and die. One key conclusion of these studies is that stars with different masses evolve differently.

Stars can be divided into five broad categories by mass, as laid out in **Figure 16.1**. The members of each group, from high-mass stars down to brown dwarfs, evolve in distinctly different ways. Stars above 8 M_{Sun} are called **high-mass stars**. Because few high-mass stars form initially, and because high-mass stars don't last very long, most stars are not high-mass stars. The fate of high-mass stars is discussed in Chapter 17. Here in Chapter 16, we describe the fate of stars with a mass less than 8 M_{Sun}. Stars with masses less than 8 M_{Sun} fall into four categories: **medium-mass stars** (for stars with mass between 3 and 8 M_{Sun}), **low-mass stars** (between 0.5 and 3 M_{Sun}), and **very low-mass stars** (between 0.08 and 0.5 M_{Sun}). **Brown dwarfs** are substellar objects with mass less than 0.08 M_{Sun}.

The Lowest-Mass Stars

Small red dwarf stars with masses lower than about 0.4–0.5 M_{Sun} may account for most stars in our galaxy, but they are small and faint and hard to detect. Those stars remain on the main sequence longer than the 13.8-billion-year age of the universe, so not enough time has passed for any of those stars to finish their lifetime on the main sequence. The very lowest-mass objects, those with masses less than 0.08 M_{Sun}, are known as brown dwarfs. These objects sit at the boundary between very large planets and very low-mass stars. They do not become main-sequence stars, so they do not follow post-main-sequence steps.

Main-Sequence Lifetime

The length of time you can drive your car before it runs out of gas depends on how much gas your tank holds (assuming a full tank) and the rate at which the engine uses the gas. Both factors are important. Even if a large car or truck has a larger tank, it may run out of fuel faster than a small car if it burns gas more quickly. How long the car runs before running out of gas can be found by dividing the amount of gas in the tank by how quickly the car uses it:

$$\text{Lifetime of tank of gas (hours)} = \frac{\text{Amount of fuel (gallons)}}{\text{Rate at which fuel is used (gallons/hour)}}$$

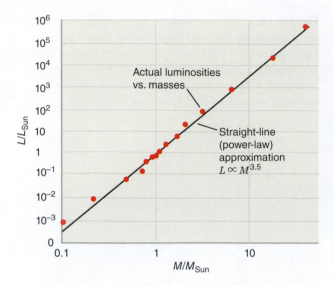

Figure 16.2 This graph plots the mass-luminosity relationship for main-sequence stars: $L \propto M^{3.5}$. The exponent can vary from 2.5 to 5.0, depending on the mass of the star. The average value, over the wide range of main-sequence star masses, is 3.5. Observational data show that how much stars deviate from the average relationship depends on their composition.

Table 16.1	Main-Sequence Lifetimes		
Spectral Type	Mass (M_{Sun})	Luminosity (L_{Sun})	Main-Sequence Lifetime (years)
O5	60	500,000	3.6×10^5
B0	17.5	32,500	7.8×10^6
B5	5.9	480	1.2×10^8
A0	2.9	39	7×10^8
A5	2.0	12.3	1.6×10^9
F0	1.6	5.2	3.1×10^9
F5	1.4	2.6	4.3×10^9
G0	1.05	1.25	8.9×10^9
G2 (Sun)	1.0	1.0	1.0×10^{10}
G5	0.92	0.8	1.2×10^{10}
K0	0.79	0.55	1.8×10^{10}
K5	0.67	0.32	2.7×10^{10}
M0	0.51	0.08	5.4×10^{10}
M5	0.14	0.008	4.9×10^{11}
M8	~0.08	0.0003	1.1×10^{12}

For example, if your car has a 15-gallon tank and your engine is burning fuel at a rate of 3 gallons each hour, your car will use up all the gas in 5 hours.

The same principle applies to main-sequence stars. Stars on the main sequence shine because they are fusing hydrogen into helium, releasing energy. The amount of fuel available is proportional to the mass of the star—that is, more massive stars have more hydrogen, so they have more "fuel in the tank." In order to remain in equilibrium, a main-sequence star must radiate the same amount of energy it produces in the core. Therefore, the rate at which the hydrogen "fuel" is used is proportional to the luminosity: If one main-sequence star is three times as luminous as another of the same size, it must fuse hydrogen at three times the rate.

The **main-sequence lifetime** of a star is the amount of time that it spends fusing hydrogen as its primary source of energy. A star's main-sequence lifetime is calculated using the same logic that was used to calculate the "lifetime" of a tank of gas:

$$\text{Lifetime of star} = \frac{\text{Amount of fuel } (\propto \text{ mass of star})}{\text{Rate at which fuel is used } (\propto \text{ luminosity of star})}$$

Surprisingly, stars with higher masses have shorter lifetimes, because even though the higher-mass star has more fuel, it also uses the fuel more quickly. Both the amount of fuel (the mass) and the rate at which the fuel is used (the luminosity) are larger in more massive stars. However, as the mass increases, the increase in luminosity is greater than the increase in the amount of fuel available. The graph in **Figure 16.2** shows how the luminosity depends on the mass.

A larger mass increases the luminosity of a star because more mass bearing down on the core causes a higher pressure there. This higher pressure affects the luminosity for two reasons. First, under higher pressure, the volume shrinks and nuclei are pushed closer together. Nuclei are more likely to collide under these conditions, and so the hydrogen burns faster and the star becomes more luminous. Second, at higher pressure, the temperature in the core also rises, so the atomic nuclei move faster. Those faster nuclei collide more violently, increasing the chances that they will overcome the electrostatic repulsion and actually fuse together. Therefore, increased temperature also increases the rate at which hydrogen burns in the core. Modest increases in mass dramatically increase the amount of energy released by nuclear fusion. Stars with higher mass use their fuel faster, and so their main-sequence lifetimes are shorter than those of stars with lower mass. The relationship between mass and main-sequence lifetime is developed further in **Working It Out 16.1**, and the results of similar calculations for different spectral types are listed in **Table 16.1**.

Changes on the Main Sequence

Hydrogen fuses into helium most rapidly at the center of the core of a main-sequence star because the temperature and pressure are highest there. Thus, helium accumulates most rapidly at the star's center. **Figure 16.3** shows how the chemical composition inside a star like the Sun changes throughout its main-sequence lifetime. When the Sun formed, it had a uniform composition of roughly 70 percent hydrogen and 27 percent helium by mass (Figure 16.3a). Since then, the Sun has converted hydrogen into helium, mostly via the proton-proton chain. As hydrogen fused into helium, the helium accumulated in the core of the Sun. Today, roughly 5 billion years later, about half the hydrogen in the core has been converted to helium; as a result, only about 35 percent of the mass in the Sun's core is hydrogen (Figure 16.3b). Five billion years from now, the Sun will begin to leave the main sequence, because the very center of the core will have no hydrogen left to fuse (Figure 16.3c).

While converting the fuel in its core from hydrogen to helium, a main-sequence star must maintain energy balance. Between the time the Sun was born and the time it will leave the main sequence, its luminosity will roughly double, with most of that change occurring during the last billion years of its main-sequence lifetime. That main-sequence evolution is slow and modest in comparison with the events that follow.

CHECK YOUR UNDERSTANDING 16.1

The main-sequence lifetime of a star depends on its mass because: (a) more massive stars fuse fuel faster and therefore have shorter lives; (b) more massive stars have more fuel and therefore have longer lives; (c) more massive stars fuse different fuels and therefore have longer lives; (d) more massive stars have different initial compositions and therefore have shorter lives.

Answers to Check Your Understanding questions are in the back of the book.

16.2 The Star Leaves the Main Sequence

As discussed in Chapter 14, two ^{3}He nuclei fuse to form one ^{4}He and two ^{1}H nuclei at the end of the proton-proton chain in main-sequence stars. At the temperatures in the centers of main-sequence stars, however, collisions are not energetic or frequent enough for the more massive ^{4}He nuclei to fuse into even heavier elements. Once the star has used up the hydrogen in the star's center core, thermal energy leaks out of the helium core into the surrounding layers of the star, but no energy is generated within the helium core to replace it. The balance that has maintained the structure of the star throughout its life is now disrupted. The star's life on the main sequence has come to an end, and its further evolution depends on core temperature changes, which govern fusion reactions.

Electron-Degenerate Matter in the Helium Core

Once a low-mass star like the Sun exhausts the hydrogen at its center, the star is no longer generating energy to keep it from collapsing. Gravity begins to "win," compared to the pressure from the hot gas, so the helium core collapses and becomes denser. Because of the rules of quantum physics, no two electrons can occupy the same state in the same place at the same time. This limits how densely electrons can be packed together, which in turn limits how densely hydrogen and helium atoms can be packed together. As the star's core collapses, it reaches that limit, in which the electrons are smashed as tightly together as possible. Matter in which electrons are packed as closely as possible is called **electron-degenerate** matter. Because the electrons cannot be packed more closely together, they produce a type of pressure, known as **degeneracy pressure**, which pushes outward against gravity and keeps the core from collapsing further. The presence of the electron-degenerate core triggers a chain of events that will dominate the evolution of a 1-M_{Sun} star for the next 50 million years, as it leaves the main sequence after the hydrogen in its core runs out.

what if . . .

What if stars were very turbulent in their interiors so that hydrogen and helium were constantly being well mixed? How would that affect stellar lifetimes?

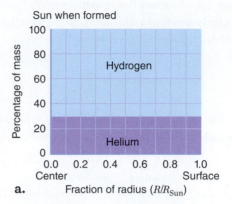

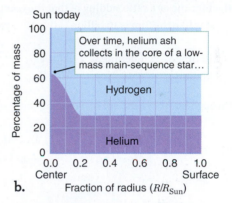

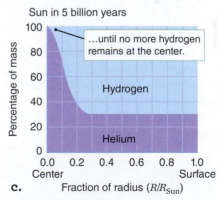

Figure 16.3 Chemical composition of the Sun is plotted here as a percentage of mass against distance from the center of the Sun. **a.** When the Sun formed 5 billion years ago, the composition was uniform throughout: about 27 percent of the Sun's mass was helium and 70 percent was hydrogen. **b.** Today, the material at the center of the Sun is about 65 percent helium and 35 percent hydrogen. **c.** The Sun's main-sequence life will end in about 5 billion years, when all the hydrogen at its center will be exhausted.

working it out 16.1

Estimating Main-Sequence Lifetimes

You can estimate the lifetime of main-sequence stars by combining the energy production rate of nuclear fusion in the core (the luminosity) with the fraction of its hydrogen that a star fuses (the mass available). The main-sequence lifetime, Lifetime_{MS}, can be expressed as:

$$\text{Lifetime}_{MS} \propto \frac{M_{MS}}{L_{MS}}$$

where M is mass (the amount of fuel available) and L is luminosity (the rate at which the fuel is used). The same equation would apply for the Sun:

$$\text{Lifetime}_{Sun} \propto \frac{M_{Sun}}{L_{Sun}}$$

We can express the lifetime as a ratio, adding in that the computed lifetime of a 1-M_{Sun} star like the Sun is 10 billion (1.0 × 10¹⁰) years:

$$\frac{\text{Lifetime}_{MS}}{\text{Lifetime}_{Sun}} = \frac{\text{Lifetime}_{MS}}{10^{10} \, \text{yr}} = \frac{M_{MS}/L_{MS}}{M_{Sun}/L_{Sun}}$$

Multiplying through by 10¹⁰ years and rearranging the fractions yields

$$\text{Lifetime}_{MS} = 10^{10} \, \text{yr} \times \frac{M_{MS}/L_{MS}}{M_{Sun}/L_{Sun}} = 10^{10} \times \frac{M_{MS}/M_{Sun}}{L_{MS}/L_{Sun}} \, \text{yr}$$

Now let's compare the lifetime of a star with that of the Sun. The relationship between the mass and the luminosity of stars is such that relatively small differences in the masses of stars result in large differences in their main-sequence luminosities. Figure 16.2 shows the **mass-luminosity relationship**, $L \propto M^{3.5}$, for main-sequence stars. If we express that relationship in units of the Sun's mass and luminosity, we find:

$$\frac{L_{MS}}{L_{Sun}} = \left(\frac{M_{MS}}{M_{Sun}}\right)^{3.5}$$

Substituting that relationship into the lifetime equation gives

$$\text{Lifetime}_{MS} = 10^{10} \times \frac{M_{MS}/M_{Sun}}{(M_{MS}/M_{Sun})^{3.5}} \, \text{yr} = 10^{10} \times \left(\frac{M_{MS}}{M_{Sun}}\right)^{-2.5} \, \text{yr}$$

Let's use a K5 main-sequence star as a specific example. According to Table 16.1, a K5 star has a mass of about 0.67 M_{Sun}:

$$\text{Lifetime}_{K5} = 10^{10} \times (0.67)^{-2.5} \, \text{yr} = 2.7 \times 10^{10} \, \text{yr}$$

Instead of the 10-billion-year life span of the Sun, a K5 star has a main-sequence lifetime of 27 billion years—2.7 times longer than the Sun's. Even though the K5 star starts out with less fuel than the Sun, it fuses that fuel so much more slowly that it lives longer.

what if . . .

What if the Earth were squeezed to a hundredth of its current radius, making Earth's density equal to that of a gas having degenerate electrons? How would that affect the orbit of the Moon, and what would it do to the acceleration of gravity on Earth's surface?

Electron-degenerate matter is not like ordinary matter that you interact with every day. Electron-degenerate matter is more than a hundred thousand times denser than lead—thus, a sugar-cube-sized block of electron-degenerate matter has a mass of at least 1000 kg. Oddly, a more massive electron-degenerate core *is smaller* than a less massive one. That trend is the *opposite* of that of ordinary matter, such as the cows that produce the milk you pour on your breakfast cereal—namely, more massive cows are bigger, not smaller. More massive stellar cores are smaller because gravity is stronger, so the electrons are smashed together into a smaller volume before their degeneracy pressure can stop the collapse.

Hydrogen Shell Fusion

After a low-mass star has converted all of the hydrogen in its core to helium, nuclear fusion in the core pauses. The core collapses to become electron degenerate. This more compact core has stronger gravity, so it produces higher pressures and temperatures in a shell around it. These pressures and temperatures become high enough for hydrogen to fuse in a layer around the inert core, in a process called **hydrogen shell fusion**. This hydrogen shell fusion is faster than the fusion in a main-sequence core, because the pressure and temperature are higher. Faster nuclear reactions in the shell release more energy, so the star's luminosity increases. At this time, the interior structure of the star is like that of a plum, with an internal seed (inert helium), a thin seed coat (hydrogen-fusing shell), and a large sphere of flesh (inert hydrogen).

Over time, the mass of the degenerate helium core grows as more and more hydrogen is converted into helium in the surrounding shell. As the mass of the degenerate helium core grows, so too does its gravitational pull, further increasing the rate of hydrogen fusion in the surrounding shell. The liberated energy drives the overlying layers of the star outward, so the star expands and grows larger. The gas of the outer layers of the star expands, so the star cools and therefore becomes redder. The star is red and very large, so it is called a **red giant**. The internal structure of the main-sequence star (**Figure 16.4a**) changes as the star evolves to become a red giant (**Figure 16.4b**). The core of the red giant is compact: much of the star's mass becomes concentrated into a volume only a few times the size of Earth. A red giant fuses hydrogen in a shell around a degenerate helium core and is larger, more luminous, and redder than it was on the main sequence.

The relation among radius, temperature, and luminosity, $L = 4\pi R^2 \sigma T^4$ (see Working It Out 13.3 and Chapter 15), still applies. The red giant may grow to have a luminosity hundreds of times that of the Sun and a radius of more than 50 solar radii ($50\ R_{Sun}$), but the surface temperature will fall only a few thousand degrees.

The Red Giant Branch

In Chapter 15, we used the Hertzsprung-Russell (H-R) diagram to show the changes in a protostar on its way to the main sequence; the diagram can also be used to track the changes in the star as it evolves away from the main sequence. As soon as the star exhausts the hydrogen in its core, it leaves the main sequence and becomes a **subgiant**: more luminous, larger, and cooler than it was on the main sequence. As it grows more luminous and its surface temperature decreases, its position on the H-R diagram moves upward and to the right. When the surface temperature of the subgiant star has dropped about 1000 kelvins (K) below its temperature on the main sequence, many H⁻ ions start to form in its atmosphere. Recall from Chapter 15 that H⁻ ions absorb and scatter outgoing radiation, trapping that radiation in the protostar and preventing it from cooling down. The H⁻ ions serve the same role in subgiants, regulating how much radiation can escape from the subgiant and preventing the temperature from dropping further.

Once this H⁻ "thermostat" begins working, the subgiant can cool no further. The red giant grows larger and therefore more luminous, but remains about the same temperature. As a result, the star's evolutionary track on the H-R diagram is nearly vertical. The path that a star follows on the H-R diagram as it leaves the main sequence is like a tree "branch" growing out of the "trunk" of the main sequence, as shown in **Figure 16.5**. The lower part of that track (which moves somewhat horizontally) is the **subgiant branch**. The vertical part is the **red giant branch**.

The evolutionary path of a red giant on the H-R diagram is similar to the path that it followed as a collapsing protostar, except in reverse: This time, the star is moving up that path instead of coming down it. That similarity is *not* a coincidence. The same physical processes (such as the H⁻ thermostat) that give rise to the vertical Hayashi track followed by a collapsing protostar also control the relationship of luminosity, size, and surface temperature in an expanding red giant.

As the star leaves the main sequence, the changes in its structure occur slowly at first, but then the star moves up the red giant branch faster and faster. A star like the Sun takes several hundred million years to go from the main sequence to the top of the red giant branch. Roughly half of that time is spent on the subgiant branch

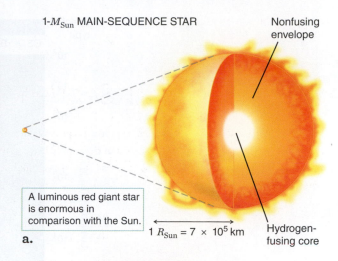

1-M_{Sun} MAIN-SEQUENCE STAR

Nonfusing envelope

A luminous red giant star is enormous in comparison with the Sun.

1 R_{Sun} = 7 × 10⁵ km

Hydrogen-fusing core

a.

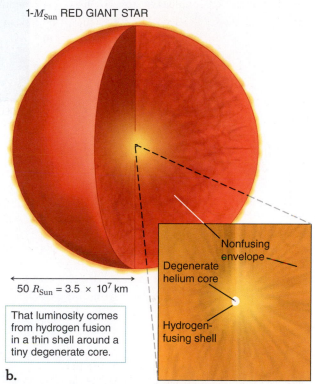

1-M_{Sun} RED GIANT STAR

50 R_{Sun} = 3.5 × 10⁷ km

That luminosity comes from hydrogen fusion in a thin shell around a tiny degenerate core.

Nonfusing envelope

Degenerate helium core

Hydrogen-fusing shell

b.

Figure 16.4 a. The size and structure of the Sun is compared with **b.** the size of a star near the top of the red giant branch of the Hertzsprung-Russell (H-R) diagram. The 1-M_{Sun} star is first shown on the left in a. to proper scale with the red giant star in b. It is necessary to "zoom in" to compare the inner structure in that main-sequence star to the structure of the red giant.

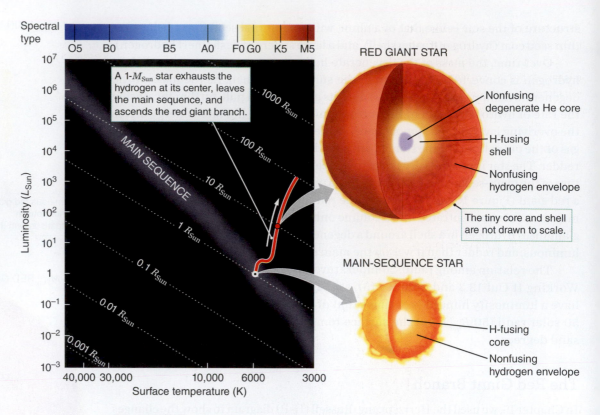

Figure 16.5 A red giant star consists of a degenerate core of helium surrounded by a hydrogen-fusing shell. Moving up the red giant branch, the star comes close to retracing the Hayashi track that it followed when it was a protostar collapsing toward the main sequence.

as the star's luminosity increases to about 10 times the luminosity of the Sun (that is, to 10 L_{Sun}). (During the subgiant and red giant phases, the luminosity increases but the mass does not, so the main-sequence mass–luminosity relation no longer applies.) During the second half of that time, the star climbs the red giant branch due to the feedback loop shown in **Figure 16.6**. Helium produced by hydrogen shell fusion adds mass to the helium core, a more massive core causes faster shell fusion, and faster shell fusion leads to faster core growth. As the core gains mass and the shell becomes more luminous, the energy released causes the outer layers to swell. The increased size causes the star's luminosity to climb from 10 to almost 1000 L_{Sun}, while the temperature remains about the same.

CHECK YOUR UNDERSTANDING 16.2

The red giant branch is nearly vertical on the H-R diagram because: (a) the surface temperature rises, but the luminosity is nearly constant; (b) the luminosity rises, but the temperature is nearly constant; (c) the luminosity and the temperature rise significantly; (d) the surface temperature and the luminosity both remain nearly constant.

16.3 Helium Fuses in the Degenerate Core

On the main sequence, a star fuses hydrogen in the core. On the red giant branch, the core consists of an inert helium core surrounded by a shell of fusing hydrogen. As the hydrogen in the shell is fused, the red giant's core contracts and heats.

Figure 16.6 As a star moves up the red giant branch in the H-R diagram, the luminosity of the star grows faster and faster. The fusion of hydrogen to helium in a shell surrounding a degenerate helium core feeds on itself, creating a cycle that speeds up as time goes on.

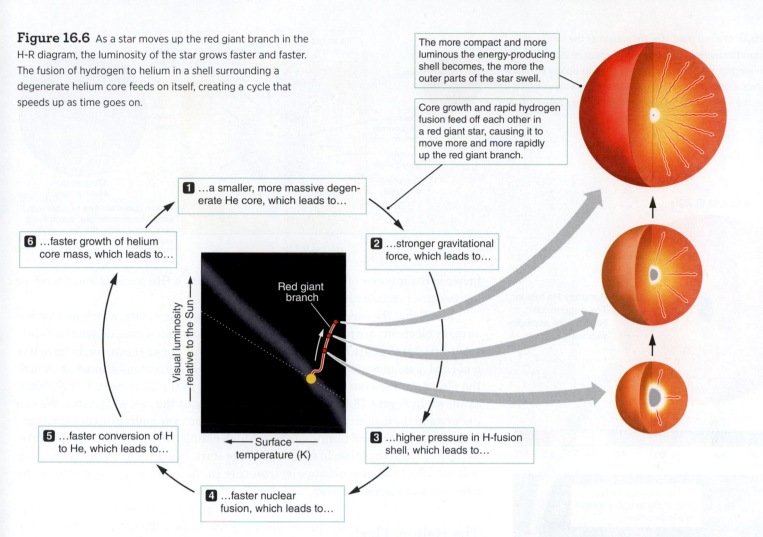

The more compact and more luminous the energy-producing shell becomes, the more the outer parts of the star swell.

Core growth and rapid hydrogen fusion feed off each other in a red giant star, causing it to move more and more rapidly up the red giant branch.

1 …a smaller, more massive degenerate He core, which leads to…

6 …faster growth of helium core mass, which leads to…

2 …stronger gravitational force, which leads to…

Red giant branch

Visual luminosity relative to the Sun →

← Surface temperature (K)

5 …faster conversion of H to He, which leads to…

3 …higher pressure in H-fusion shell, which leads to…

4 …faster nuclear fusion, which leads to…

Eventually, the temperature and pressure rise enough to start the next stage of stellar evolution: helium fusion in the core.

Helium Fusion and the Triple-Alpha Process

As the red giant evolves, its helium core becomes more massive, so it shrinks—remember that the core is electron degenerate, so it doesn't behave like ordinary matter. The temperature rises, due partly to the gravitational energy released as the core shrinks and due partly to the energy released by hydrogen fusion in the surrounding shell. The thermal motions of the atomic nuclei in the core become more and more energetic. Eventually, at a temperature of about 100 million (10^8) K, the ^{4}He nuclei are slammed together hard enough for helium fusion to begin.

Helium fuses in a two-stage sequence called the **triple-alpha process**, so named because it involves the fusion of three ^{4}He nuclei, which are sometimes called **alpha particles**. The process, illustrated in **Figure 16.7**, begins when two helium-4 (^{4}He) nuclei fuse to form a beryllium-8 (^{8}Be) nucleus consisting of four protons and four neutrons. The ^{8}Be nucleus is extremely unstable and decays after a short time. But if it collides with another ^{4}He nucleus before it decays, the two nuclei will fuse into a stable nucleus of carbon-12 (^{12}C) consisting of six protons and six neutrons. The reaction rate depends on the temperature: higher temperatures

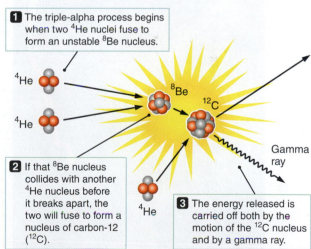

1 The triple-alpha process begins when two ^{4}He nuclei fuse to form an unstable ^{8}Be nucleus.

^{4}He
^{4}He
^{8}Be
^{12}C
Gamma ray

2 If that ^{8}Be nucleus collides with another ^{4}He nucleus before it breaks apart, the two will fuse to form a nucleus of carbon-12 (^{12}C).

^{4}He

3 The energy released is carried off both by the motion of the ^{12}C nucleus and by a gamma ray.

Figure 16.7 The triple-alpha process produces a stable nucleus of carbon-12. First, two helium-4 (^{4}He) nuclei fuse to form an unstable beryllium-8 (^{8}Be) nucleus. If that nucleus collides with another ^{4}He nucleus before breaking apart, the two will fuse to form a stable nucleus of carbon-12 (^{12}C). The energy produced is carried off by both the motion of the ^{12}C nucleus and a high-energy gamma ray emitted in the second step of the process.

Figure 16.8 In a red giant star, the weight of the overlying layers is supported by electron degeneracy pressure in the core arising from electrons that are packed together as tightly as quantum mechanics allows. Atomic nuclei in the core can move freely about within the sea of degenerate electrons, so they behave as a normal gas.

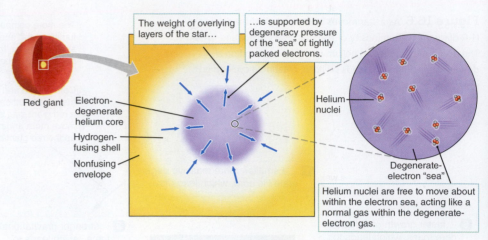

HELIUM FLASH

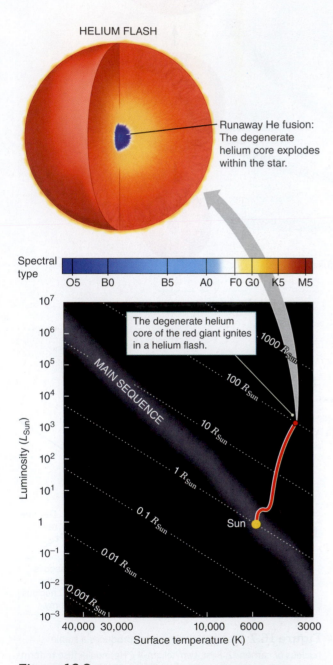

Runaway He fusion: The degenerate helium core explodes within the star.

The degenerate helium core of the red giant ignites in a helium flash.

Figure 16.9 At the end of its life, a low-mass star travels a complex path on the H-R diagram. The first part of that path takes it up the red giant branch to a point where helium ignites in a helium flash. After a few hours, the core of the star begins to inflate, ending the helium flash.

increase the number of ^{8}Be nuclei that collide with a ^{4}He nucleus and therefore increase the reaction rate.

Recall that the core of the red giant is electron-degenerate, which means that as many electrons are packed into that space as possible. Those degenerate electrons prevent the core from collapsing. The atomic nuclei, however, behave like a normal gas, moving through the sea of degenerate electrons almost as though the electrons were not there. The atomic nuclei in the core move freely about, as shown in **Figure 16.8**, just as they do throughout the rest of the star. We can understand the fusion of helium by treating the nuclei as matter in a normal state. Once the pressure and temperature are high enough, those nuclei begin to fuse, as the hydrogen nuclei do in main-sequence stars. However, the energy released will *not* affect the degenerate-electron core in the same way that it affects the core in a main-sequence star.

The Helium Flash

Degenerate material conducts thermal energy very well, so any differences in temperature within the core rapidly dissipate. As a result, when helium fusion begins at the center of the core, the energy released quickly heats the entire core. Within a few minutes, the entire core is fusing helium into carbon by the triple-alpha process. Some of the carbon will fuse with an additional ^{4}He nucleus to form stable oxygen-16 (^{16}O; each nucleus contains eight protons and eight neutrons).

In a normal gas, such as the air around you, the pressure of the gas comes from the random thermal motions of the atoms, which are faster at higher temperatures. Increasing the temperature increases the pressure of the gas. The core of a main-sequence star acts this way, so heating it causes it to expand; the temperature, density, and pressure then decrease; nuclear reactions slow; and the star settles into a new balance between gravity and pressure. Those are exactly the sorts of changes steadily occurring within the core of a main-sequence star like the Sun as the structure of the star shifts in response to the changing composition in the star's core.

However, the degenerate core of a red giant is not a normal gas. The pressure in a red giant's degenerate core comes primarily from how tightly the electrons in the core are packed together. Heating the core does not change the number of electrons that can be packed into its volume, so the core's pressure does not respond to changes in temperature. Because the pressure does not increase, the core does not expand when heated, as a normal gas would.

Although the higher temperature does not change the pressure, it does cause the helium nuclei to collide more often and with greater force, so the nuclear reactions become more vigorous. That process creates another feedback loop: More vigorous reactions increase the temperature, and higher temperature means even more vigorous reactions. Helium fusion in the degenerate core rapidly increases due to this feedback. As long as the degeneracy pressure from the electrons is greater than the thermal pressure from the nuclei, the feedback loop continues.

Helium fusion begins at a temperature of about 100 million K. By the time the temperature has climbed by just 10 percent, to 110 million K, the rate of helium fusion has increased to 40 times what it was at 100 million K. By the time the core's temperature reaches 200 million K, the core is fusing helium 460 million times faster than it was at 100 million K. As the temperature in the core rises, the thermal motions of the electrons and nuclei become more energetic, and the pressure increases. Within seconds of helium ignition, the thermal pressure grows larger than the degeneracy pressure. When this threshold is crossed, the helium core explodes in what is known as the **helium flash**. Because the explosion occurs deep within the star, however, it cannot be seen outside the star. Instead of appearing as light outside the star, the energy released lifts the intermediate layers of the star and expands the core. The electrons in the core spread out, becoming nondegenerate and the pressure falls. This slows the nuclear reactions and stops the helium flash after just a few hours. After the helium flash, helium fusion in the core keeps the core of the star puffed up. The luminosity falls, over 100,000 years or so, as the star settles into a new equilibrium of stable helium fusion. The helium flash marks the top of the red giant branch, shown in **Figure 16.9**, because it marks the moment when the star begins to drop in luminosity.

The Horizontal Branch

Once the star settles into its new equilibrium, it spends about 100 million years fusing helium into carbon in a normal, nondegenerate core while hydrogen fuses to helium in a surrounding shell. During this time, the star is about a hundred times less luminous than it was when the helium flash occurred. Slower energy production in the interior means that gravity becomes stronger than the outward pressure of the escaping radiation and this pulls the outer layers back in. The star shrinks, and its surface temperature climbs as gravitational energy is converted to thermal energy. The star moves horizontally to the left across the H-R diagram, remaining at the same luminosity but increasing in surface temperature. That portion of a star's path on the H-R diagram is called the **horizontal branch (Figure 16.10)**. At that point in their evolution, low-mass stars with chemical compositions similar to that of the Sun lie on the H-R diagram just to the left of the red giant branch. Stars that contain much less iron than the Sun tend to distribute themselves away from the red giant branch along the horizontal branch toward the hotter side of the diagram.

In a horizontal branch star, fusion takes place in the core and a shell, both surrounded by a nonfusing hydrogen envelope. The horizontal branch star fuses helium into carbon in the core, with a shell of hydrogen fusion around that core. The star is consuming fuel more rapidly, so it is more luminous than it was on the main sequence. In addition, helium fusion releases less energy than hydrogen fusion, so the helium fusion must take place even faster to maintain equilibrium. These factors mean that the star's time on the horizontal branch is much shorter than its time on the main sequence. The horizontal branch star remains stable for about 100 million years.

what if . . .
What if the intrinsic quantum phenomena of degeneracy didn't exist and the interiors of stars always behaved like normal gases. How might that effect the post-main-sequence evolution of a star?

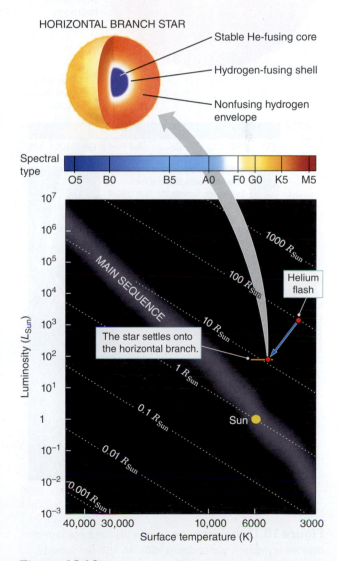

Figure 16.10 The star moves down from the red giant branch onto the horizontal branch. For about 100 million years or less, the star will remain on the horizontal branch and fuse helium in its core and hydrogen in a surrounding shell.

The temperature at the center of a horizontal branch star is not high enough for most of the carbon to fuse, so carbon builds up in the heart of the star. When the horizontal branch star has fused all the helium at its core, gravity once again begins to overwhelm the pressure of the escaping radiation. The carbon core is crushed by the weight of the star's layers above it until it becomes electron degenerate, with physical properties much like those of the degenerate helium core at the center of a red giant.

CHECK YOUR UNDERSTANDING 16.3

Stars begin fusing helium to carbon when the temperature rises in the core. That temperature increase is caused by (choose all that apply): (a) gravitational collapse; (b) fusion of hydrogen into helium in the core; (c) fusion of hydrogen into helium in a shell around the core; (d) electron degeneracy pressure.

16.4 Dying Stars Shed Their Outer Layers

After the horizontal branch, small changes in a star's properties—mass, chemical composition, magnetic field strength, and even the rotation rate—can lead to noticeable differences in how the star (especially its outer envelope) evolves. In this section, we follow a 1-M_{Sun} star with solar composition as it concludes its evolutionary stages, losing its outer layers and leaving behind a cooling carbon core.

Stellar-Mass Loss and the Asymptotic Giant Branch

After the horizontal branch, a small, dense, electron-degenerate carbon core remains. That core is very compact, causing the gravity in the inner parts of the star to be very high, which in turn drives up the pressure, which speeds up the nuclear reactions in the helium-burning shell, which causes the degenerate core to grow more rapidly. Those internal changes should sound familiar—they are similar to the changes that took place at the end of the star's main-sequence lifetime, and the path the star follows as it leaves the horizontal branch parallels that earlier phase of evolution. Just as the star accelerated up the red giant branch as its degenerate helium core grew, the star now leaves the horizontal branch and once again begins to grow larger, redder, and more luminous as its degenerate carbon core grows. As shown in **Figure 16.11**, the path that the star follows, called the **asymptotic giant branch (AGB)** of the H-R diagram, approaches the red giant branch as the star grows more luminous. An AGB star fuses helium and hydrogen in nested concentric shells around a degenerate carbon core as the star moves once again up the H-R diagram.

AGB stars are huge objects. When the Sun becomes an AGB star, its outer layers will engulf the orbits of the inner planets, possibly including Earth and maybe even Mars. When a star expands to such a size, the gravitational force at its surface is only 1/10,000 as strong as the gravity at the surface of the present-day Sun. Pushing surface material away from the star thus takes little extra energy. Before the temperature in the carbon core becomes high enough for carbon to fuse, the radiation pressure pushing outward on the outer layers of the star becomes greater than the gravitational pull inward. Those outer layers begin to drift away into space in a process called **stellar-mass loss**. Stellar-mass loss actually begins when the star is still on the red giant branch: By the time a 1-M_{Sun} main-sequence star reaches the horizontal branch, it may have lost 10–20 percent of its total mass. As the star ascends the asymptotic giant branch, it loses another 20 percent or even more of its total

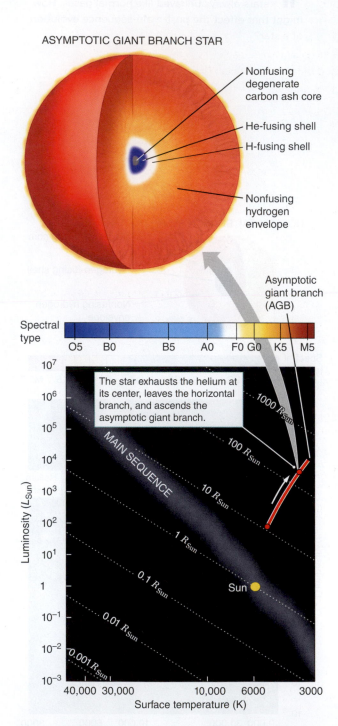

ASYMPTOTIC GIANT BRANCH STAR

Nonfusing degenerate carbon ash core

He-fusing shell

H-fusing shell

Nonfusing hydrogen envelope

Asymptotic giant branch (AGB)

The star exhausts the helium at its center, leaves the horizontal branch, and ascends the asymptotic giant branch.

Figure 16.11 The star moves up from the horizontal branch onto the asymptotic giant branch (AGB). An AGB star consists of a degenerate carbon core surrounded by helium-fusing and hydrogen-fusing shells. As the carbon core grows, the star brightens, accelerating up the AGB just as it earlier accelerated up the red giant branch while its degenerate helium core grew.

working it out 16.2

Escaping the Surface of an Evolved Star

Why are giant stars likely to lose mass? The escape velocity from the surface of a planet or star was given in Working It Out 4.2:

$$v_{esc} = \sqrt{\frac{2GM}{R}}$$

How does v_{esc} change when a star becomes a red giant? Let's look at the Sun as an example. When the Sun is on the main sequence, the escape velocity from its surface can be calculated by using $M_{Sun} = 1.99 \times 10^{30}$ kg, $R_{Sun} = 6.96 \times 10^5$ km, and $G = 6.67 \times 10^{-20}$ km³/(kg s²):

$$v_{esc} = \sqrt{\frac{2 \times [6.67 \times 10^{-20}\ km^3/(kg\ s^2)] \times (1.99 \times 10^{30}\ kg)}{6.96 \times 10^5\ km}}$$

$$v_{esc} = \sqrt{3.81 \times 10^5\ km^2/s^2} = 617\ km/s$$

What will the escape velocity be when the Sun becomes a red giant, with a radius 50 times greater than the radius it has today and a mass 0.9 times its present mass?

$$v_{esc} = \sqrt{\frac{2 \times [6.67 \times 10^{-20}\ km^3/(kg\ s^2)] \times 0.9 \times (1.99 \times 10^{30}\ kg)}{50 \times (6.96 \times 10^5\ km)}}$$

$$v_{esc} = \sqrt{6.86 \times 10^3\ km^2/s^2} = 83\ km/s$$

The escape velocity from the surface of a red giant star is only 13 percent that of a main-sequence star:

$$[(83\ km/s)/(617\ km/s)](100\%) = 13\%$$

That is part of the reason that red giant and AGB stars lose mass. The Sun may eventually lose half its mass.

---

mass. By the time the star is near the top of the asymptotic giant branch, a star that began as a 1-M_{Sun} star may have lost more than half its original mass. Stellar-mass loss is further explored in **Working It Out 16.2**.

Mass loss on the AGB can be spurred on by the star's unstable interior. The extreme sensitivity of the triple-alpha process to temperature in the core can lead to episodes of fast fusion and rapid energy release, which can provide the extra kick needed to expel material from the star's outer layers. Even stars that are initially similar can behave very differently when they reach that stage in their evolution.

Planetary Nebula

Toward the end of an AGB star's life, mass loss itself becomes a runaway process. When a star loses a bit of mass from its outermost layers, the weight pushing down on the underlying layers of the star is reduced. Without that weight holding them down, the remaining outer layers of the star puff up further. The post-AGB star, which is now both less massive and larger, is even less tightly bound by gravity, so less energy is needed to push its outer layers away. Mass loss leads to weaker gravity, which leads to faster mass loss, which leads to weaker gravity, and so on. By the time the last layers are lost, much of the remaining mass of the star is ejected into space, typically at speeds of 20–30 kilometers per second (km/s).

After ejection of its outer layers, all that is left of the low-mass star is a tiny, very hot, electron-degenerate carbon core surrounded by a thin envelope in which hydrogen and helium are still fusing. The star is now less luminous than when it was at the top of the AGB, but it is still much more luminous than a horizontal branch star. The remaining hydrogen and helium in the star rapidly fuse to carbon, and

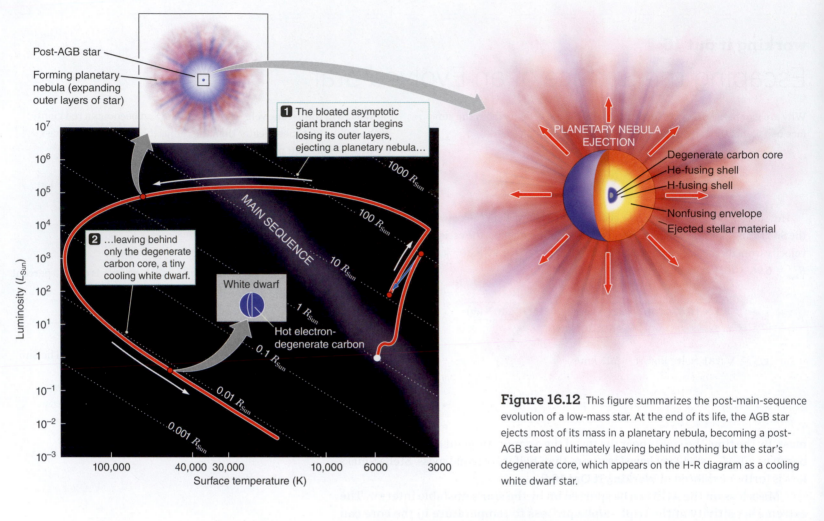

Post-AGB star

Forming planetary nebula (expanding outer layers of star)

1 The bloated asymptotic giant branch star begins losing its outer layers, ejecting a planetary nebula…

2 …leaving behind only the degenerate carbon core, a tiny cooling white dwarf.

White dwarf

Hot electron-degenerate carbon

MAIN SEQUENCE

1000 R_{Sun}
100 R_{Sun}
10 R_{Sun}
1 R_{Sun}
0.1 R_{Sun}
0.01 R_{Sun}
0.001 R_{Sun}

Luminosity (L_{Sun})

Surface temperature (K)

PLANETARY NEBULA EJECTION

Degenerate carbon core
He-fusing shell
H-fusing shell
Nonfusing envelope
Ejected stellar material

Figure 16.12 This figure summarizes the post-main-sequence evolution of a low-mass star. At the end of its life, the AGB star ejects most of its mass in a planetary nebula, becoming a post-AGB star and ultimately leaving behind nothing but the star's degenerate core, which appears on the H-R diagram as a cooling white dwarf star.

unanswered questions

Why do planetary nebulae have different shapes? Some are not simply chaotic but are well organized, with various types of symmetry. Some are spherically symmetric (like a ball), some have bipolar symmetry (like a long, hollow tube, pinched in the middle), and some are even point-symmetric (like the letter *S*). How can an essentially spherically symmetric object such as a star produce such beautifully organized outflows? Because stars are three-dimensional objects, and astronomers can view each object from only one direction, researchers can't easily determine how much of that variation is due to orientation and how much is due to actual differences in the object's shape. For example, a bipolar nebula, viewed from one end, would appear spherically symmetric. That orientation effect, among other problems, complicates efforts to understand how those shapes are formed. No single explanation has yet satisfactorily covered all the object types.

as more and more of the mass of the star ends up in the carbon core, the star itself shrinks and becomes hotter and hotter. Over the course of only about 30,000 years after the beginning of runaway mass loss, the star moves from right to left across the top of the H-R diagram, as shown in **Figure 16.12**.

The surface temperature of the star may eventually rise above 100,000 K. At such temperatures, the radiation's peak wavelength is in the high-energy ultraviolet (UV) part of the spectrum, as determined by Wien's law. That intense UV light heats and ionizes the ejected, expanding shell of gas, causing it to glow in the same way that UV light from an O star causes an H II region to glow. When those glowing shells were first observed in small telescopes, they were named "planetary nebulae" because they appeared fuzzy like nebular clouds of dust and gas, but they were approximately round, like planets. Later imagery, such as the images in **Figure 16.13**, showed that those objects are not like planets at all. Rather, a **planetary nebula** is the remaining outer layers of a star, ejected into space at the end of the star's ascent of the AGB. A planetary nebula may be visible for 50,000 years or so before the gas ejected by the star disperses so far that the nebula is too faint to be seen. Not all stars form planetary nebulae. Stars more massive than about 8 M_{Sun} pass through the post-AGB stage too quickly. Stars without enough mass take too long in the post-AGB stage, so their envelope evaporates before they can illuminate it. Astronomers do not know whether our own Sun will retain enough mass during its post-AGB phase to form a planetary nebula.

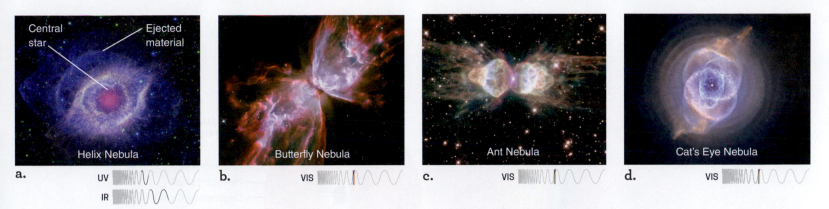

Figure 16.13 ★ WHAT AN ASTRONOMER SEES These images of planetary nebulae from the Hubble Space Telescope and the Spitzer Space Telescope show the wealth of structures that result from the complex processes by which low-mass stars eject their outer layers. An astronomer looking at these four images will be reminded that planetary nebulae come in a wide variety of shapes that reveal the details of the history of mass loss from the central star. In general, material that is further from the central star was emitted earlier, although sometimes later material travels faster and overtakes earlier emissions. An astronomer will know that the colors are due to atom- or ion-specific emission lines, and she will recognize that the colors can be used not only to find out what atoms and ions are present but also to further figure out what kinds of photons are reaching that area. She will further be distracted by trying to think about the objects in three dimensions. For example, if we could view the Butterfly Nebula in panel **b.** along its axis, from a direction near the top right of the image, it might look much like the Helix Nebula in panel **a.** An astronomer will also know that bright spots might mean more material, or they might mean that the material is better illuminated. Figuring out what's happening with a planetary nebula requires sorting out all of these effects.

The detailed structure of a planetary nebula may contain concentric rings of varying density, indicating that the rate of stellar-mass loss varied—sometimes faster, sometimes slower. In the image of the Helix nebula in Figure 16.13, the colored patches are emission lines from ions located at different places in the nebula. When that nebula was forming, a lot of mass was lost nearly all at once. Then the mass loss ceased, resulting in a hollow shell around the central star. As the light makes its way out from the central star, higher-energy light gets "used up" in ionizing the inner layers. Outer layers have emission lines from atoms that require less energy to ionize. In other planetary nebulae, the material may be concentrated parallel to the equator or poles of the star, indicating that the stellar-mass loss was blocked in some directions. You can see that effect in the images of the Butterfly, Ant, and Cat's Eye nebulae in Figure 16.13.

The gas in a planetary nebula carries the chemical elements from the star's outer layers off into interstellar space. Planetary nebulae often show a greater percentage of elements such as carbon, nitrogen, and oxygen than the percentage of those elements present in the outer layers of the Sun. Those elements are by-products of nuclear fusion either from the star that produced the planetary nebula or from the stars of earlier generations. Once that chemically enriched material leaves the star, it mixes with interstellar gas, increasing the chemical diversity of the interstellar medium.

White Dwarfs

Within about 50,000 years, a post-AGB star fuses all the fuel remaining on its surface, leaving behind a nonfusing ball of carbon with a mass less than 70 percent of the original star's mass. As that occurs, the post-AGB star becomes smaller

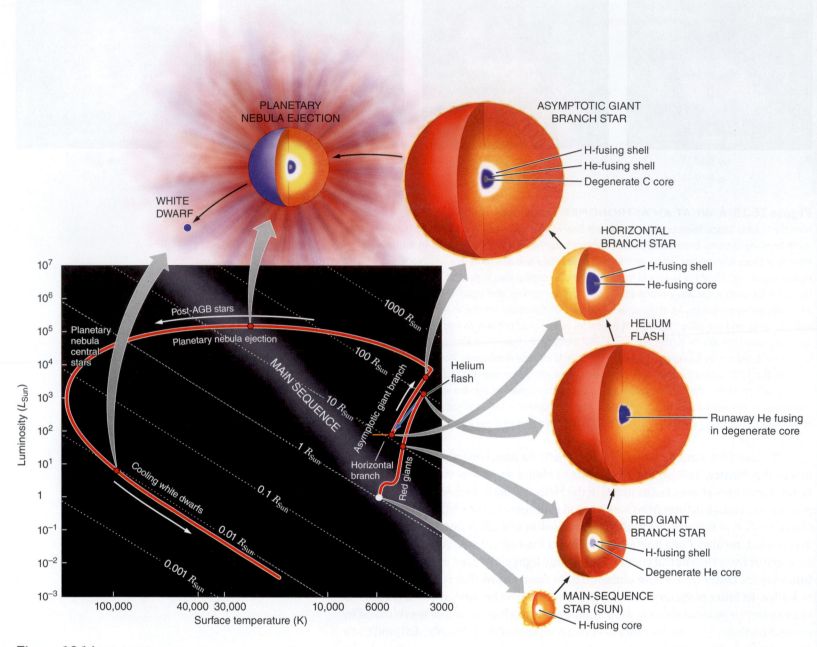

Figure 16.14 This H-R diagram summarizes the stages in the post-main-sequence evolution of a 1-M_{Sun} star.

and fainter but remains about the same temperature. Its position on the H-R diagram falls down the left side. Within a few thousand years, it shrinks to about the size of Earth, at which point it has become fully electron degenerate and can shrink no further. That remnant of stellar evolution is called a **white dwarf**.

The white dwarf, composed of nonfusing electron-degenerate carbon and maybe some oxygen, continues to radiate energy away into space. As it does so it cools, just like the heating coil on an electric stove once it is turned off. Because the white dwarf is electron degenerate, its size remains nearly constant as it cools, so it follows a line of constant radius down and to the right on the H-R diagram. Its tiny size means it is a thousand times less luminous than a main-sequence star like the Sun. Many white dwarfs are known, but none can be seen without a telescope. Sirius, the brightest star in Earth's sky, has a faint white dwarf as a binary companion.

Figure 16.14 summarizes the evolution of a solar-type, 1-M_{Sun} main-sequence star from the time it leaves the main sequence through its final state as a

~0.6-M_{Sun} white dwarf. The star leaves the main sequence, climbs the red giant branch, falls to the horizontal branch, climbs back up the AGB, takes a left across the top of the diagram while ejecting a planetary nebula, and finally falls to its final resting place in the bottom left of the diagram. That process is representative of the fate of low-mass stars. Although every low-mass star forms a white dwarf at the end point of its evolution, the precise path a low-mass star follows from core hydrogen fusion on the main sequence to white dwarf depends on many details particular to the star.

Some stars less massive than the Sun may become white dwarfs composed largely of helium rather than carbon. Conversely, temperatures in the cores of evolved 2- to 3-M_{Sun} stars are high enough to allow additional nuclear reactions to occur, leading to the formation of more massive white dwarfs composed of materials such as oxygen, neon, and magnesium. Differences in chemical composition of a star can also lead to differences in its post-main-sequence evolution.

The spectacle of a red giant or AGB star is ephemeral. Once it leaves the main sequence, the Sun will travel the path from red giant to white dwarf in less than one-tenth of the time it spent on the main sequence, steadily fusing hydrogen to helium in its core. Stars spend most of their luminous lifetimes on the main sequence, which is why most of the stars in the sky are main-sequence stars. The fainter white dwarfs constitute the final resting place for most of the stars that have been or ever will be formed.

The Fate of the Planets

What happens to the planets orbiting a low-mass star as it goes through those post-main-sequence stages? The first decade of exoplanet discoveries yielded primarily planets with orbits very close to their respective stars. Astronomers assumed that any planets closer than 1–2 astronomical units (AU) would not survive the post-main-sequence expansion of the star, so they did not expect to find planets in orbit around evolved stars. However, since that time, many planets have been discovered in orbit around red giants, AGB stars, and horizontal branch stars.

Astronomers cannot be sure whether the exoplanets they observe have remained in the same orbital locations as when their stars were on the main sequence or whether the planets migrated to different orbits. The surface gravity of a red giant is low, and some of the mass in its outermost layers will blow away. That decrease in mass reduces the gravitational force between the star and its planets, which could lead to planetary orbits evolving outward away from the star. In some models of planetary migration, the tidal forces between planets or between a star and a planet are significant factors, too, suggesting that some planetary orbits could evolve inward.

As a star loses mass during the evolutionary process, stellar or planetary companions may change their orbits. Rocky asteroids or smaller planets could migrate inward past the Roche limit (see Chapter 4) of the white dwarf and break up. Some of the material could remain in orbit around the white dwarf as a debris disk, similar to those seen in main-sequence stars with a planetary system. Those dusty debris disks have been observed around white dwarf stars with the Hubble Space Telescope and the Spitzer Space Telescope. Some of that dusty or rocky material may also fall onto the white dwarf, "polluting" its spectrum with heavy elements that were not produced in the stellar core.

what if . . .

What if the Sun becomes a red giant, and then undergoes a helium flash? How might that affect the planets in our Solar System?

Reading Astronomy News

Scientists Solve Riddle of Celestial Archaeology

University of Leicester Press Office

Scientists report that the spectra of some dead stars show evidence of rocky material left over from planetary systems.

A decades old space mystery has been solved by an international team of astronomers led by Professor Martin Barstow of the University of Leicester and President-elect of the Royal Astronomical Society.

Scientists from the University of Leicester and University of Arizona investigated hot, young, white dwarfs—the super-dense remains of Sun-like stars that ran out of fuel and collapsed to about the size of the Earth. Their research is featured in MNRAS—the *Monthly Notices of the Royal Astronomical Society*, published by Oxford University Press.

It has been known that many hot white dwarf atmospheres, essentially of pure hydrogen or pure helium, are contaminated by other elements—like carbon, silicon, and iron. What was not known, however, was the origins of these elements, known in astronomical terms as metals.

"The precise origin of the metals has remained a mystery and extreme differences in their abundance between stars could not be explained," said Professor Barstow, a Pro-Vice-Chancellor at the University of Leicester whose research was assisted by his daughter Jo, a coauthor of the paper, during a summer work placement in Leicester. She has now gone on to be an astronomer working in Oxford—on exoplanets.

"It was believed that this material was 'levitated' by the intense radiation from deeper layers in the star," said Professor Barstow.

Now the researchers have discovered that many of the stars show signs of contamination by rocky material, the leftovers from a planetary system.

The researchers surveyed 89 white dwarfs, using the Far Ultraviolet Spectroscopic Explorer to obtain their spectra (dispersing the light by color) in which the "fingerprints" of carbon, silicon, phosphorus, and sulfur can be seen when these elements are present in the atmosphere.

"We found that in stars with polluted atmospheres the ratio of silicon to carbon matched that seen in rocky material, much higher than found in stars or interstellar gas.

"The new work indicates that around one-third of all hot white dwarfs are contaminated in this way, with the debris most likely in the form of rocky minor planet analogs. This implies that a similar proportion of stars like our Sun, as well as stars that are a little more massive like Vega and Fomalhaut, build systems containing terrestrial planets. This work is a form of celestial archaeology where we are studying the 'ruins' of rocky planets and/or their building blocks, following the demise of the main star."

"The mystery of the composition of these stars is a problem we have been trying to solve for more than 20 years. It is exciting to realize that they are swallowing up the leftovers from planetary systems, perhaps like our own, with the prospect that more detailed follow-up work will be able to tell us about the composition of rocky planets orbiting other stars," said Professor Barstow.

The study also points to the ultimate fate of the Earth billions of years from now—ending up as a contamination within the white dwarf Sun.

QUESTIONS

1. The article refers to the white dwarfs in the study as "hot, young, white dwarfs." What does "young" mean in that context?

2. The spectra described are compared to fingerprints. In what ways are white dwarf spectra like fingerprints?

3. Why were scientists surprised to find elements other than hydrogen and helium in the atmospheres of white dwarf stars? Where do they think those other elements originate?

4. Why does a telescope need to be in space to observe far-UV wavelengths?

5. How common are contaminated white dwarfs in the sample in the study? Compare that with the percentage of Sun-like stars with planets (22 percent). Does the finding in the article seem like a sensible number?

Source: https://phys.org/news/2014-03-scientists-riddle-celestial-archaeology.html

What will happen to Earth? In Chapter 8, you learned that radioactive dating of meteorites indicates that the Solar System formed 4.6 billion years ago. In Chapter 14, you learned, by calculating its rate of hydrogen fusion, that the Sun might last about 10 billion years. The Sun's luminosity is about 30 percent higher now than it was early in the history of the Solar System, and it will continue to increase steadily over the rest of the Sun's main-sequence lifetime of another 5 billion years or so. The Sun's luminosity may increase enough—even while the Sun is a main-sequence star—that Earth will heat up to the point where the oceans evaporate, perhaps as soon as 1 billion to 2 billion years from now. By the time the Sun is a red giant, it might be impossible for any planet within the orbit of Jupiter to maintain liquid water on the surface.

Scientists are not certain whether the radius of the red giant Sun will extend past Earth to engulf the planet. The red giant Sun will have low surface gravity and will lose mass, which could cause Earth's orbit to expand, thereby enabling Earth to escape the encroaching solar surface. Alternatively, as the Sun expands in radius and its rotation rate slows (see Working It Out 7.1), tidal forces might pull Earth inward. The solar core will eventually become a white dwarf, perhaps with a dusty disk and a "polluted" atmosphere as the only remaining evidence of our rocky planet.

CHECK YOUR UNDERSTANDING 16.4

A planetary nebula forms from: (a) the ejection of mass from a low-mass star; (b) the collision of planets around a dying star; (c) the collapse of the magnetosphere of a high-mass star; (d) the remainders of the original star-forming nebula.

16.5 Binary Star Evolution

So far in this chapter, we have discussed the evolution of single stars in isolation. But many stars are members of binary systems. Those systems may result when a molecular-cloud core has too much angular momentum for a single star to form. Binary systems are common—as many as half of all stars may be in binary systems, and that fraction is largest among more massive stars. How is evolution different for stars in those systems? Sometimes, if the stars are close together and one star is more massive, the post-main-sequence evolution of both stars may be linked. In this section, we discuss the evolution of this type of system as it becomes a nova or supernova.

Mass Flows from an Evolving Star onto Its Companion

In a close system in which one star is more massive than the other, mass can transfer from the more massive star to the less massive star. To understand why, think for a moment about what would happen if you were to travel in a spacecraft from Earth toward the Moon. When you are still near Earth, the force of Earth's gravity is far stronger than that of the Moon. As you move away from Earth and closer to the Moon, Earth's gravitational attraction weakens, and the Moon's gravitational attraction becomes stronger. You eventually reach an intermediate zone where neither body has the stronger pull. If you continue beyond that point, the lunar gravity begins to dominate until you find yourself firmly in the gravitational grip of the Moon. The regions surrounding the two objects—their gravitational domains—are called the **Roche lobes** of the system.

what if . . .

What if astronomers discover a single white dwarf star surrounded by a system of seven planets? In what ways might that planetary system differ from "normal" planetary systems around main-sequence stars?

unanswered questions

Could Earth be moved farther from the Sun to accommodate the Sun's inevitable changes in luminosity, temperature, and radius? One proposal suggests that Earth could capture energy from a passing asteroid and migrate outward, thus staying in the habitable zone while moving farther from the Sun as the Sun ages. Or a huge, thin "solar sail" could be constructed so that radiation pressure from the Sun would slowly push Earth into a larger orbit. Such feats of "astronomical engineering" are not feasible anytime soon, but perhaps they could be accomplished by the time they are needed, hundreds of millions of years from now.

what if . . .

What if star 1 and star 2 are very widely separated, so that their Roche lobes are very large? In this case, star 1 may not get large enough to fill its Roche lobe. How would the evolution of this binary system be different from that of the system in Figure 16.15?

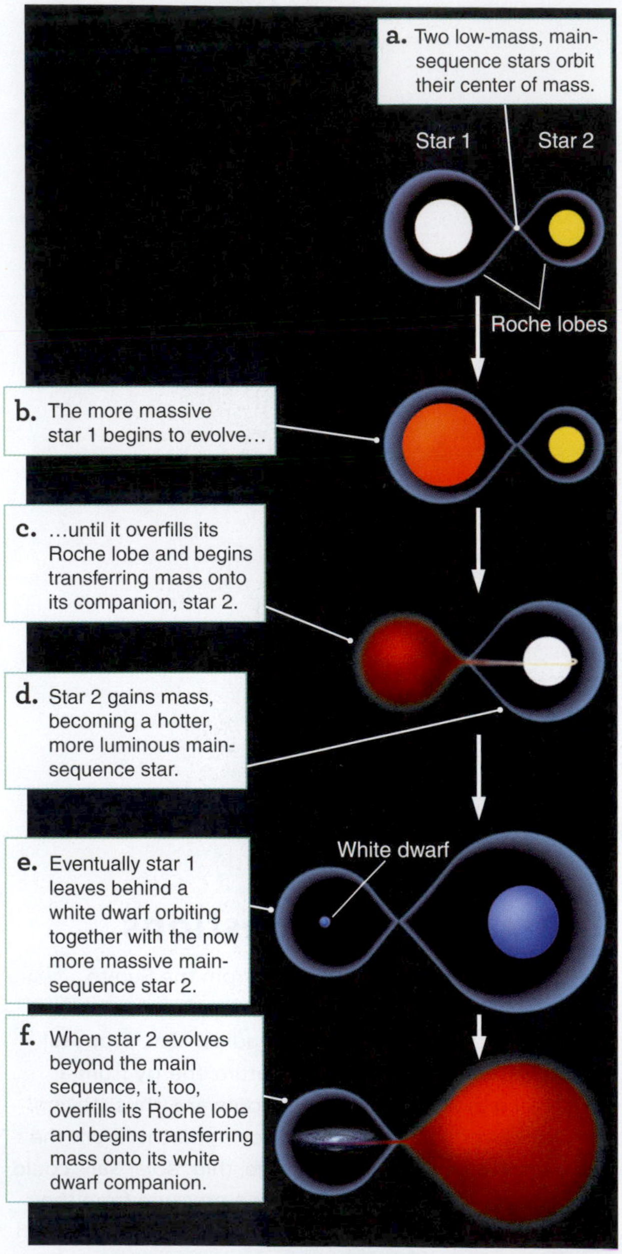

a. Two low-mass, main-sequence stars orbit their center of mass.

Star 1   Star 2

Roche lobes

b. The more massive star 1 begins to evolve…

c. …until it overfills its Roche lobe and begins transferring mass onto its companion, star 2.

d. Star 2 gains mass, becoming a hotter, more luminous main-sequence star.

White dwarf

e. Eventually star 1 leaves behind a white dwarf orbiting together with the now more massive main-sequence star 2.

f. When star 2 evolves beyond the main sequence, it, too, overfills its Roche lobe and begins transferring mass onto its white dwarf companion.

Figure 16.15 A compact binary system consisting of two low-mass stars that evolve through stages, transferring mass back and forth.

The same situation exists between two stars, as shown in **Figure 16.15**. Gas near each star clearly belongs to that star. When the more massive star leaves the main sequence and swells up, its outer layers may cross that gravitational dividing line separating the star from its companion. Once a star expands past the boundary of its Roche lobe, some of its material begins to fall onto the other star. That transfer of material from one star to the other is called **mass transfer**.

Evolution of a Close Binary System

The best way to understand how mass transfer affects the evolution of stars in a binary system is to apply what is known from studying the evolution of single low-mass stars. Figure 16.15a shows a binary system consisting of two low-mass stars: Star 1 is more massive, and star 2 is less massive. That is an ordinary binary system, and each star is an ordinary main-sequence star for most of the system's lifetime.

More massive main-sequence stars evolve more rapidly than less massive main-sequence stars. Therefore, star 1 will be the first to use up the hydrogen at its center and begin to evolve off the main sequence, as shown in Figure 16.15b. If the two stars are close enough to each other, star 1 will eventually grow to overfill its Roche lobe, and material will transfer onto star 2, as shown in Figure 16.15c. The transfer of mass between the two stars results in a "drag" that causes the orbits of the two stars to shrink, bringing the stars closer together and further enhancing mass loss. In addition, as star 1 loses mass, its Roche lobe shrinks, further enhancing the mass transfer. The two stars sometimes even reach the point at which they are effectively two cores sharing the same extended envelope of material, called a *common envelope*.

Despite those complexities, star 2 probably remains a basically normal main-sequence star throughout the process, fusing hydrogen in its core. Over time, however, the mass of star 2 increases because of the accumulation of material from its companion. As it does so, the structure of star 2 must change to accommodate its new status as a higher-mass star. If we plotted star 2's position on the H-R diagram during that period, we would see it move up and to the left along the main sequence, becoming larger, hotter, and more luminous.

Star 1, because it is losing mass to star 2, never grows larger than its Roche lobe, so it does not become an isolated red giant or AGB star at the top of the H-R diagram. Yet star 1 continues to evolve, fusing helium in its core on the horizontal branch, proceeding through helium shell fusion, and finally losing its outer layers and leaving behind a white dwarf. Figure 16.15e shows the binary system after star 1 has completed its evolution. All that remains of star 1 is a white dwarf, orbiting about star 2, its bloated main-sequence companion.

Novae

As star 2 begins to evolve off the main sequence, it expands to fill its Roche lobe, as shown in Figure 16.15f. Like star 1 before it, star 2 grows to fill its Roche lobe: material from star 2 begins to pour through the "neck" connecting the Roche lobes of the two stars. However, this time the mass is not being added to a normal star but is drawn toward the tiny white dwarf left behind by star 1. Because the system is revolving and the white dwarf is so small, the infalling material generally misses the star, instead landing on an accretion disk around the white dwarf. That disk is similar to the accretion disk that forms around a protostar. As with star formation, the accretion disk accumulates material that has too much angular momentum to hit the white dwarf directly.

A white dwarf has a mass comparable to that of the Sun but a size comparable to that of Earth. Having a large mass and a small radius means strong surface gravity. The material streaming toward the white dwarf in the binary system falls into an incredibly deep gravitational "well." The depth of that well affects the amount of energy with which matter impacts the white dwarf. A kilogram of material falling from space onto the surface of a white dwarf releases 100 times more energy than a kilogram of material falling from the outer Solar System onto the surface of the Sun. All that energy is turned into thermal energy. The spot where the stream of material from star 2 hits the accretion disk can be heated to millions of kelvins, where it glows in the far-UV and X-ray parts of the electromagnetic spectrum.

The infalling material accumulates on the surface of the white dwarf (**Figure 16.16**), where the enormous gravitational pull of the white dwarf compresses the material to a density close to that of the white dwarf itself. As more and more material builds up on the surface of the white dwarf, the white dwarf shrinks (just as the core of a red giant shrinks as it grows more massive). The density increases more and more while the release of gravitational energy drives the temperature of the white dwarf higher and higher. The infalling material comes from the outer, unfused layers of star 2, so it is composed mostly of hydrogen. That hydrogen is compressed to higher and higher densities and heated to higher and higher temperatures on the surface of the white dwarf.

Once the temperature at the base of the white dwarf's surface layer of hydrogen reaches about 10 million K, that hydrogen begins to fuse to helium. But that process is not the contained hydrogen fusion that takes place in the center of the Sun; instead, it is explosive hydrogen fusion in a degenerate gas. Energy released by hydrogen fusion drives up the temperature. Because the surface is degenerate, that rising temperature does not cause an expansion, as in normal matter, but instead drives up the rate of hydrogen fusion. That runaway thermonuclear reaction is much like the runaway helium fusion that takes place during the helium flash, except now no overlying layers of a star are there to absorb the energy liberated by fusion. The result is a tremendous explosion that blows part of the layer covering the white dwarf out into space at speeds of thousands of kilometers per second, as shown in Figure 16.16. An exploding white dwarf of that kind is called a **nova**.

The explosion of a nova does not destroy the underlying white dwarf star. In fact, much of the material that had built up on the white dwarf may remain behind after the explosion. Afterward, the binary system is in much the same configuration as before—namely, material from star 2 is still pouring onto the white dwarf. The nova can repeat many times as material builds up and ignites again and again on the surface of the white dwarf. If the underlying white dwarf is old and cooler, or the mass accumulates slowly on its surface, outbursts are separated by thousands of years, so most novae have been seen only once in historical times. But if the white dwarf is hotter, or mass is transferring quickly, such explosions can happen every few years or even every few months. Such recurring novae are called dwarf novae.

About 50 novae occur in our galaxy each year. Roughly 10 of those can be detected with telescopes, but very few become visible to the eye. The rest are blocked from view by dust in the disk of our galaxy. Novae reach their peak brightness in only a few hours, and for a short time they can be several hundred thousand times more luminous than the Sun. Although the brightness of a nova sharply drops in the weeks after the outburst, it can sometimes still be seen for years. During

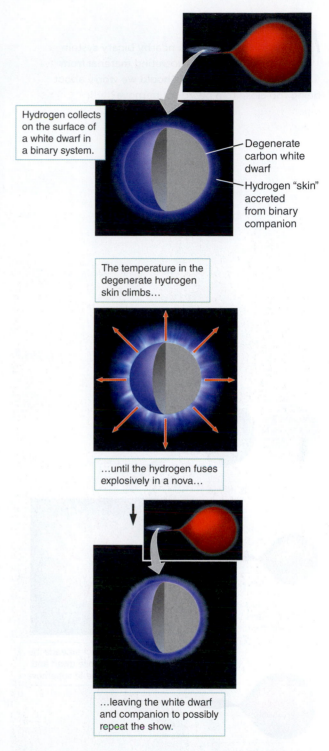

Hydrogen collects on the surface of a white dwarf in a binary system.

Degenerate carbon white dwarf

Hydrogen "skin" accreted from binary companion

The temperature in the degenerate hydrogen skin climbs…

…until the hydrogen fuses explosively in a nova…

…leaving the white dwarf and companion to possibly repeat the show.

Figure 16.16 In a binary system in which mass is transferred onto a white dwarf, a layer of hydrogen builds up on the surface. That hydrogen may ignite on the surface of the white dwarf, producing a nova.

what if . . .

What if there were a nearby binary system with a white dwarf accreting material from a star with 0.8 M_{Sun}? Should we worry about being too close to a nearby supernova?

that time, the glow from the expanding cloud of ejected material is caused by the decay of radioactive isotopes created in the explosion. Several novae have been accompanied by large bursts of gamma rays.

Supernovae

A white dwarf with a mass greater than 1.4 M_{Sun} cannot remain stable. This value is referred to as the **Chandrasekhar limit**, named for Subrahmanyan Chandrasekhar (1910–1995), who derived it. Above that mass, even the pressure supplied by degenerate electrons is no longer enough to balance gravity, so the white dwarf would collapse. A white dwarf that is accumulating mass, however, likely does not reach the Chandrasekhar limit. As the star reaches about 1.38 M_{Sun}, the core pressure and temperature rises enough to ignite carbon and begin a simmering phase that holds off thermonuclear runaway for a while. Once the temperature reaches about 1.0×10^8 K, the runaway carbon fusion involves the entire white dwarf. Within about a second, the whole white dwarf is consumed in the resulting explosion. In that instant, 100 times more energy is liberated than a star like the Sun emits over its 10-billion-year lifetime on the main sequence. Runaway fusion reactions convert a large fraction of the star's mass into elements such as iron and nickel, and the material in the explosion is ejected into space at speeds in excess of 20,000 km/s, enriching the interstellar medium with those heavier elements. This explosion of a white dwarf is known as a **Type Ia supernova**.

How might a white dwarf gain mass and explode in that way? Three options exist, as shown in **Figure 16.17**.

The first possibility, in the binary system shown in Figure 16.16, is that star 2 eventually may simply go on to form a white dwarf, leaving behind a stable binary system consisting of two white dwarfs, as in Figure 16.17a. Those two white dwarfs may then eventually spiral together and merge. If the sum of their masses is greater than 1.4 M_{Sun}, the resulting merged star will explode. That explosion destroys both of the original white dwarfs. The amount of mass involved in the explosion may range from 1.4 to 2.8 M_{Sun}. A majority of Type Ia supernovae may be of that type.

Second, while it is still a main-sequence star, star 2 may lose mass to the white dwarf, as in Figure 16.17b. Through millions of years of mass transfer from star 2 onto the white dwarf, and possibly through countless nova outbursts, the white dwarf's mass slowly increases—until it approaches the Chandrasekhar limit and later explodes. Here, the white dwarf's mass is always almost exactly 1.4 M_{Sun}. Only one object of that type has so far been observed, from a pulse of blue light emitted when the white dwarf exploded and heated up star 2.

Third, star 2 may evolve off the main sequence to become a red giant, filling its Roche lobe. The material from that star flows onto the white dwarf over millions of years, as shown in Figure 16.17c. As in the second case, the white dwarf approaches the Chandrasekhar limit and then explodes. Here, the white dwarf's mass is almost exactly 1.4 M_{Sun} every time. A red giant companion has not yet been directly observed in a Type Ia supernova.

In a galaxy the size of the Milky Way, Type Ia supernovae occur about once a century. They can briefly shine with a luminosity billions of times that of our Sun, possibly outshining the galaxy itself. Those objects are particularly useful to astronomers because their luminosities can be approximately determined from a careful study of their light curves. Because that type of supernova happens during the explosion of an object with a mass between 1.4 and 2.8 M_{Sun}, the total energy involved

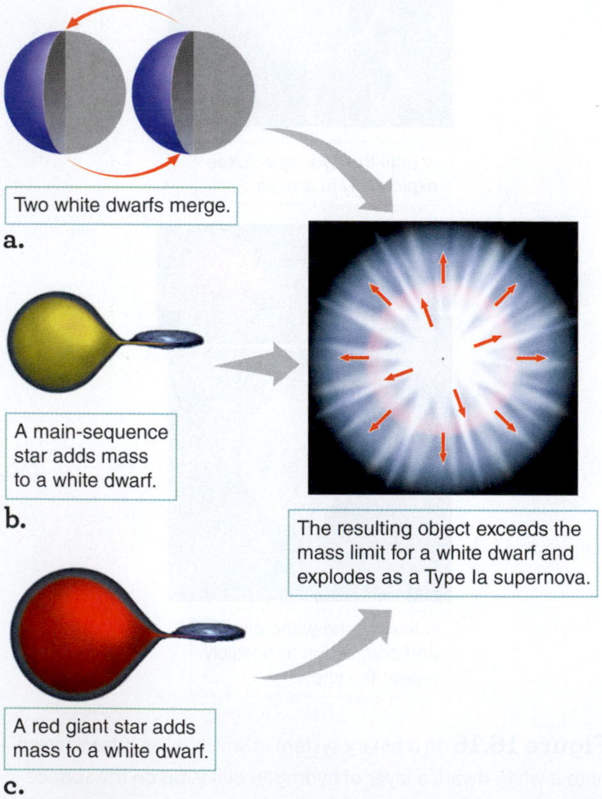

Figure 16.17 A Type Ia supernova results when a white dwarf exceeds the mass limit. That can happen because **a.** two white dwarfs merge; **b.** mass from a main-sequence companion falls onto the white dwarf, increasing its mass to the limit; or **c.** mass from an evolved companion falls onto the white dwarf, increasing its mass to the limit.

Two white dwarfs merge.
a.

A main-sequence star adds mass to a white dwarf.
b.

The resulting object exceeds the mass limit for a white dwarf and explodes as a Type Ia supernova.

A red giant star adds mass to a white dwarf.
c.

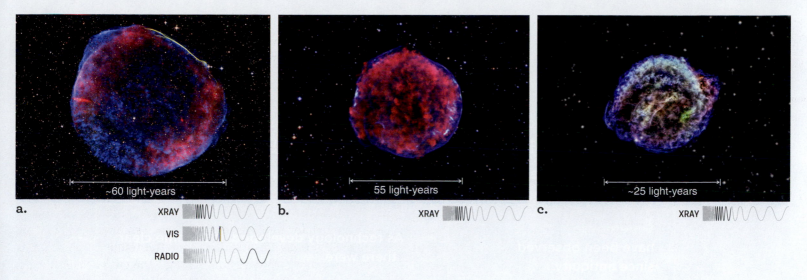

Figure 16.18 These images show the remnants of Type Ia supernovae. The material heated by the expanding blast wave from a supernova glows in the X-ray portion of the spectrum. **a.** SN 1006 is the brightest recorded supernova and was observed in China, Japan, Europe, and the Middle East in 1006. X-ray data are shown in blue, whereas radio data are shown in red. **b.** Tycho's supernova was observed in 1572. Low-energy X-rays are red, whereas high-energy X-rays are blue. **c.** Kepler's supernova, observed in 1604, is shown in five X-ray wavelengths.

will vary by at most a factor of 2. Therefore, the luminosity of Type Ia supernovae should vary by no more than a factor of 2. As we explain in the next few chapters, objects with known luminosity can be used to find distances to far away galaxies. Combining the luminosity of a Type Ia supernova with its apparent brightness gives the distance to the host galaxy. Because they are such an important distance indicator, Type Ia supernovae are a very active area of exploration in astronomy (see the **Process of Science Figure**).

Supernovae leave behind expanding shells of dust and gas called **supernova remnants**, as shown in **Figure 16.18**. The material is heated by the expanding blast wave from the supernova and glows in X-rays, as well as having line emission in the visible part of the spectrum. These remnants reach enormous sizes before gradually fading away as their material returns to the interstellar medium where some of it will eventually form new stars.

CHECK YOUR UNDERSTANDING **16.5**

A white dwarf will become a supernova: (a) if the original star was more than 1.38 M_{Sun}; (b) if it accretes an additional 1.38 M_{Sun} from a companion; (c) if some mass falls onto it from a companion; (d) if enough mass accretes from a companion to give the white dwarf a total mass of 1.38 M_{Sun}.

Origins: Stellar Lifetimes and Biological Evolution

From fossil records and DNA analysis, scientists estimate that life took hold on Earth within 1 billion years after the Solar System and Earth formed 4.6 billion years ago. It took another 1.5 billion years for more complex cells to develop and another billion years to develop multicellular life. The first animals didn't appear

Science Is Unfinished

Understanding Type Ia supernovae is at the foundation of measuring distances to the farthest galaxies and therefore critical for our conclusions about the universe as a whole. The observational evidence, though, remains incomplete.

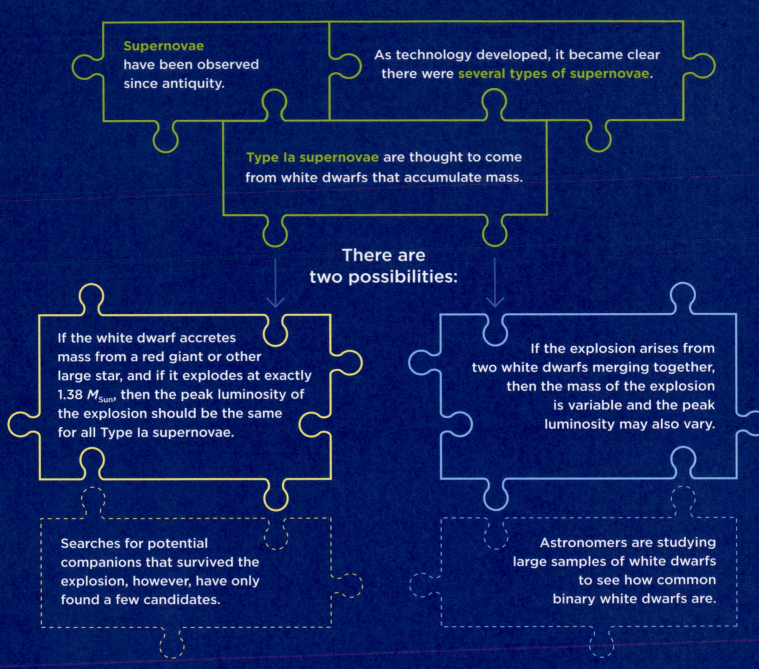

Supernovae have been observed since antiquity.

As technology developed, it became clear there were **several types of supernovae.**

Type Ia supernovae are thought to come from white dwarfs that accumulate mass.

There are two possibilities:

If the white dwarf accretes mass from a red giant or other large star, and if it explodes at exactly 1.38 M_{Sun}, then the peak luminosity of the explosion should be the same for all Type Ia supernovae.

If the explosion arises from two white dwarfs merging together, then the mass of the explosion is variable and the peak luminosity may also vary.

Searches for potential companions that survived the explosion, however, have only found a few candidates.

Astronomers are studying large samples of white dwarfs to see how common binary white dwarfs are.

When observational evidence is inconclusive, each scientist adds a piece to the puzzle. Some will conduct larger observational studies, whereas others will create new theoretical models. Eventually, these efforts bring clarity to a confusing situation.

on Earth until 600 million years ago, 4 billion years after the formation of the Sun and the Solar System.

The only known example of biology is the life on Earth. Extrapolating from one data point is always risky, and it is not known whether biology is widespread in the universe or whether Earth's biological timeline is typical. Still, reasoning from our one example is the only way to begin thinking about life in the universe. How does the preceding timeline of the evolution of life on Earth compare with the lifetimes of main-sequence stars? Table 16.1 indicates that the lifetime of an O5 star is less than half a million years, whereas the lifetime of a B5 star is about 120 million years, and the lifetime of an A0 star is about 700 million years. Those stars would have run out of hydrogen in the core and started post-main-sequence evolution in less time than Earth took to settle down after its periods of heavy bombardment by debris early in the Solar System's history.

The 1-billion-year main-sequence lifetime of an A2 star corresponds to the time the simplest life-forms took to develop on Earth. The 3-billion-year lifetime for F0 stars corresponds to the time multicellular life took to develop on Earth. Only stars cooler and less massive than F5 stars have main-sequence lifetimes longer than the 4 billion years life took to evolve into animals on Earth. Thus, searches for exoplanets survey stars that are F5 or cooler—because hotter and more massive stars probably don't live long enough on the main sequence for complex life to develop.

After a star leaves the main sequence, the helium-fusing red giant stage is estimated to last for about one-tenth of the main-sequence lifetime, so that doesn't help stars with short lifetimes to last long enough for complex biology to evolve. Could life survive the transition of its star to a red giant? As noted previously in the chapter, even if a planet is not destroyed, its orbit, temperature, and atmospheric conditions will drastically change, and any life might have to relocate if it is to survive.

SUMMARY

Stars with masses similar to that of the Sun form planetary nebulae like the ones shown in Figure 16.13. That is a likely fate for our own Sun at the end of its evolution. After leaving the main sequence, low-mass stars follow a convoluted path along the H-R diagram that includes the red giant branch, the horizontal branch, the asymptotic giant branch, and a path across the top and then down to the lower left of the diagram. Those stages of evolution are dominated by the balance between gravity and energy production from various fusion processes that "turn on and off" in the core of the star. That entire process takes much less time than the main-sequence lifetime of the star. Stars following that path have main-sequence lifetimes comparable to the evolutionary timescales of life on Earth. If life on Earth is typical, then more massive stars with shorter lifetimes will not be stable long enough for complex life to evolve.

(1) **Estimate the main-sequence lifetime of a star from its mass.** All stars eventually exhaust their nuclear fuel as hydrogen fuses to helium in the cores of main-sequence stars. Less massive stars exhaust their fuel more slowly and have longer lifetimes than more massive stars.

(2) **Explain why low-mass stars initially grow larger and more luminous as they run out of fuel.** When a low-mass star uses up the hydrogen in its core, it begins to fuse hydrogen in a shell around the core, heating the gaseous interior. The star expands and becomes a red giant.

(3) **Sketch post-main-sequence evolutionary tracks on a Hertzsprung-Russell (H-R) diagram and list the stages of evolution for low-mass stars.** After exhausting its core hydrogen, a low-mass star leaves the main sequence and swells to become a red giant, with a helium core made of electron-degenerate matter. The red giant fuses helium to carbon via the triple-alpha process, and quickly the core ignites in a helium flash. The star then

moves onto the horizontal branch. A horizontal branch star accumulates carbon and sometimes oxygen in its core and then moves up the asymptotic giant branch. The star becomes an AGB star and loses some of its mass.

④ **Describe how planetary nebulae and white dwarfs form.** In their dying stages, some stars eject their outer layers to form planetary nebulae. All low-mass stars eventually become white dwarfs, which are very hot but very small.

⑤ **Predict whether a close binary system will become a Type Ia supernova.** Transfer of mass within some binary systems can lead to a nuclear explosion. A nova occurs when hydrogen collects and ignites on the surface of a white dwarf in a binary system. If the mass of the white dwarf approaches 1.38 M_{Sun}, the entire star may become involved in a Type Ia supernova. That explosion can occur in several ways.

QUESTIONS AND PROBLEMS

TEST YOUR UNDERSTANDING

1. Place the main-sequence lifetimes of the following stars in order from shortest to longest.
   a. the Sun: mass, 1 M_{Sun}; luminosity, 1 L_{Sun}
   b. Capella Aa: mass, 3 M_{Sun}; luminosity, 76 L_{Sun}
   c. Rigel: mass, 24 M_{Sun}; luminosity, 85,000 L_{Sun}
   d. Sirius A: mass, 2 M_{Sun}; luminosity, 25 L_{Sun}
   e. Canopus: mass, 8.5 M_{Sun}; luminosity, 13,600 L_{Sun}
   f. Achernar: mass, 7 M_{Sun}; luminosity, 3150 L_{Sun}

2. Place the following steps in the evolution of a low-mass star in order.
   a. main-sequence star
   b. planetary nebula ejection
   c. horizontal branch
   d. helium flash
   e. red giant branch
   f. asymptotic giant branch
   g. white dwarf

3. If a star follows a horizontal path across the H-R diagram, the star
   a. maintains the same temperature.
   b. stays the same color.
   c. maintains the same luminosity.
   d. keeps the same spectral type.

4. Degenerate matter is different from normal matter because as the mass of degenerate material increases,
   a. the radius decreases.
   b. the temperature decreases.
   c. the density decreases.
   d. the luminosity decreases.

5. The most massive stars have the shortest lifetimes because
   a. the temperature is higher in the core, so they fuse faster.
   b. they have less fuel in the core when the star forms.
   c. their fuel is located farther from the core.
   d. the temperatures are lower in the core, so they fuse their fuel more slowly.

6. If a main-sequence star suddenly started fusing hydrogen at a faster rate in its core, it would become
   a. larger, hotter, and more luminous.
   b. larger, cooler, and more luminous.
   c. smaller, hotter, and more luminous.
   d. smaller, cooler, and more luminous.

7. A low-mass main-sequence star's climb up the red giant branch is halted by
   a. the end of hydrogen shell burning.
   b. the beginning of helium fusion in the core.
   c. electron-degeneracy pressure in the core.
   d. instabilities in the star's expanding outer layers.

8. Post-main-sequence stars lose up to half their mass because
   a. jets from the poles release material at an increasing rate.
   b. the mass of the star drops because of mass loss from fusion.
   c. the magnetic field causes increasing numbers of coronal mass ejections.
   d. the star swells until the surface gravity is too weak to hold material.

9. A planetary nebula glows because
   a. it is hot enough to emit UV radiation.
   b. fusion is happening in the nebula.
   c. it is heating up the interstellar medium around it.
   d. light from the central star causes emission lines.

10. As an AGB star evolves into a white dwarf, it runs out of nuclear fuel, and you might guess that the star should cool off and move to the right on the H-R diagram. Why does the star move to the left instead?
    a. It becomes larger.
    b. More of the star is involved in fusion.
    c. As outer layers are lost, deeper layers are exposed.
    d. The temperature of the core rises.

11. When compressed, ordinary gas heats up but degenerate gas does not. Why, then, does a degenerate core heat up as the star continues shell fusion around it?
    a. It is heated by the radiation from fusion.
    b. It is heated by the gravitational collapse of the shell.
    c. It is heated by the weight of helium falling on it.
    d. It is insulated by the shell.

12. All Type Ia supernovae
    a. are at the same distance from Earth.
    b. always involve two stars of identical mass.
    c. are extremely luminous.
    d. always release the same amount of energy in fusion.

13. In Latin, *nova* means "new." That word is used for novae and super-novae because they are
    a. newly formed stars.
    b. newly dead stars.
    c. newly visible stars.
    d. new main-sequence stars.

14. When the Sun runs out of hydrogen in its core, it will become larger and more luminous because
    a. it will start fusing hydrogen in a shell around a helium core.
    b. it will start fusing helium in a shell and hydrogen in the core.
    c. infalling material will rebound off the core and puff up the star.
    d. the energy balance will no longer hold, and the star will drift apart.

15. A white dwarf is located in the lower left of the H-R diagram. From that information alone, you can determine that the star is
    a. very massive.
    b. very dense.
    c. very hot.
    d. very bright.

THINKING ABOUT THE CONCEPTS

16. Can a star skip the main sequence and immediately begin fusing helium in its core? Explain your answer.

17. Suppose a main-sequence star suddenly started fusing hydrogen at a faster rate in its core. How would the star react? Discuss changes in size, temperature, and luminosity.

18. Describe some possible ways in which the temperature in the core of a star might increase while the density decreases.

19. Astronomers typically say that the mass of a newly formed star determines its destiny from birth to death. However, that statement is not true for one common environmental circumstance. Identify that circumstance and explain why the birth mass of a star might not fully account for the star's destiny.

20. Study the Process of Science Figure. Suppose that a new mechanism is found to explain Type Ia supernovae. In that mechanism, all Type Ia supernovae are more luminous than previously thought. Would the derived distances to galaxies be larger or smaller than we understand them to be now? 👁

21. Do stars change structure while on the main sequence? Why or why not?

22. Suppose Jupiter were not a planet but a G5 main-sequence star with a mass of 0.8 M_{Sun}.
    a. How would life on Earth be affected, if at all?
    b. How would the Sun be affected as it comes to the end of its life?

23. Explain the similarity in the paths that a star follows along the H-R diagram as it forms from a protostar and as it leaves the main sequence to climb the red giant branch.

24. Why is a horizontal branch star (which fuses helium at a high temperature) less luminous than a red giant branch star (which fuses hydrogen at a lower temperature)?

25. Suppose the core temperature of a star is high enough for the star to begin fusing oxygen. Predict how the star will continue to evolve, including its path on the H-R diagram.

26. ★ **WHAT AN ASTRONOMER SEES**   The Cat's Eye in Figure 16.13 has a bluish inner part, surrounded by concentric rings. Which was emitted from the star first: the blue inner part or the concentric rings? How do you know? 👁

27. Why does a white dwarf move down and to the right along the H-R diagram?

28. Suppose the more massive red giant star in a binary system engulfs its less massive main-sequence companion, and their nuclear cores combine. What structure will the new star have? Where will the star lie on the H-R diagram?

29. T Coronae Borealis is a well-known recurrent nova.
    a. Is it a single star or a binary system? Explain.
    b. What mechanism causes a nova to flare up?
    c. How can a nova flare-up happen more than once?

30. Why do astronomers prefer to search for planets around low-mass stars?

APPLYING THE CONCEPTS

31. What is the main-sequence lifetime for a 0.5-M_{Sun} star? 🟢🟢🟢
    a. Make a prediction: Do you expect this star's lifetime to be longer or shorter than the main-sequence lifetime of the Sun (10^{10} years)?
    b. Calculate: Follow Working It Out 16.1 to find the lifetime.
    c. Check your work: Compare your answer with your prediction, and compare your answer with the value in Table 16.1.

32. The Sun will eventually become an AGB star with a radius 200 times greater and a mass only 0.7 times that of today. What will the escape velocity from the surface be at that time? 🟡🟢🟢
    a. Make a prediction: Is the escape velocity from the surface of an AGB star more or less than the escape velocity from the star on the main sequence?
    b. Calculate: Follow Working It Out 16.2 to calculate the escape velocity from the Sun when it becomes an AGB star.
    c. Check your work: Verify that your answer has units of km/s, and compare it to both your prediction and the results of the calculations in Working It Out 16.2.

33. What is the main-sequence lifetime for a 6-M_{Sun} star?
    a. Make a prediction: Do you expect this star's lifetime to be longer or shorter than the main-sequence lifetime of the Sun (10^{10} years)?
    b. Calculate: Follow Working It Out 16.1 to find the lifetime.
    c. Check your work: Compare your answer with your prediction, and compare your answer with the value in Table 16.1.

34. What is the main-sequence lifetime for a 60-M_{Sun} star?
    a. Make a prediction: Do you expect this star's lifetime to be longer or shorter than the main-sequence lifetime of the Sun (10^{10} years)?
    b. Calculate: Follow Working It Out 16.1 to find the lifetime.
    c. Check your work: Compare your answer with your prediction, and compare your answer with the value in Table 16.1.

35. As the Sun climbs the red giant branch, it will grow to have a radius about 100 times greater than on the main sequence, but its mass will not have changed significantly. What will the escape velocity from the surface be at that time?

   a. Make a prediction: Is the escape velocity from the surface of red giant star more or less than the escape velocity from the star on the main sequence?

   b. Calculate: Follow Working It Out 16.2 to calculate the escape velocity from the Sun at this point on the red giant branch.

   c. Check your work: Verify that your answer has units of km/s, and compare it to both your prediction and the results of the calculations in Working It Out 16.2, and the answer to Question 34.

36. Figure 16.2 contains the label "Straight-line (power law) approximation." What does that label tell you about the axes on the graph—that is, are they linear or logarithmic? Explain why those data are plotted that way. ⊙

37. Use Figure 16.3 to estimate the percentage of the Sun's mass that is turned from hydrogen into helium over its lifetime. ⊙

38. Study Figure 16.14. How many times brighter is a star at the top of the giant branch than the same star (a) when it was on the main sequence and (b) when it was on the horizontal branch? ⊙

39. Study Figure 16.14. Make a graph of surface temperature versus time for the evolutionary track shown—from the time the star leaves the main sequence until it arrives at the dot showing that it is a white dwarf. Your time axis may be approximate, but it should show that the star spends different amounts of time in the different phases. ⊙

40. For most stars on the main sequence, luminosity scales with mass as $M^{3.5}$ (see Working It Out 16.1). What luminosity does that relationship predict for (a) 0.5-M_{Sun} stars, (b) 6-M_{Sun} stars, and (c) 60-M_{Sun} stars? Compare those numbers with the values in Table 16.1.

41. Each form of energy generation in stars depends on temperature.

   a. The rate of hydrogen fusion (proton-proton chain) near 10^7 K increases with temperature as T^4. If the temperature of the hydrogen-fusing core is raised by 10 percent, by how much does the hydrogen fusion energy increase?

   b. Helium fusion (the triple-alpha process) at 10^8 K increases with an increase in temperature at a rate of T^{40}. If the temperature of the helium-fusing core is raised by 10 percent, by how much does the helium fusion energy increase?

42. A planetary nebula has an expansion rate of 20 km/s and a lifetime of 50,000 years. Roughly how large will that planetary nebula grow before it disperses?

43. Suppose a companion star transferred mass onto a white dwarf at a rate of about 10^{-9} M_{Sun} per year. Roughly how long after mass transfer begins will the white dwarf explode as a Type Ia supernova? How does that length of time compare with the typical lifetime of a low-mass star? Assume that the white dwarf started with a mass of 0.6 M_{Sun}.

44. Use Kepler's third law to estimate how fast material in an accretion disk (size = 2×10^5 km) orbits around a 0.6 M_{Sun} white dwarf.

45. A white dwarf has a density of approximately 10^9 kilograms per cubic meter (kg/m^3). Earth has an average density of 5500 kg/m^3 and a diameter of 12,700 km. If compressed to the same density as a white dwarf, what would Earth's radius be?

EXPLORATION   Evolution of Low-Mass Stars

digital.wwnorton.com/astro7

The evolution of a low-mass star, as discussed here in Chapter 16, corresponds to many twists and turns on the H-R diagram. In this exploration, we return to the H-R Diagram Interactive Simulation to investigate how these twists and turns affect the appearance of the star.

Visit the Student Site on the Digital Resources page and open the H-R Diagram Interactive Simulation in Chapter 16. The box labeled "Comparison to Sun" shows an image of both the Sun and the test star. Initially, these two stars have identical properties: the same temperature, the same luminosity, and the same size.

Examine the box labeled "Test Star Properties." This box shows the temperature, luminosity, and radius of a test star located at the X in the H-R diagram. Before you change anything, answer questions 1–4.

1. What is the temperature of the test star?

_____

2. What is the luminosity of the test star?

_____

3. What is the radius of the test star?

_____

4. What do you predict will happen to the temperature, luminosity, and radius of the test star if it moves up and to the right on the H-R diagram?

_____

As a star leaves the main sequence, it moves up and to the right on the H-R diagram. Grab the cursor (the X on the H-R diagram) and move it up and to the right.

5. What changes about the image of the test star next to the Sun?

_____

6. What is the test star's temperature? What property of the image of the test star indicates that its temperature has changed?

_____

7. What is the test star's luminosity?

_____

8. What is the test star's radius?

_____

9. Ordinarily, the hotter an object is, the more luminous it is. In this case, the temperature has decreased, but the luminosity has increased. How can this be?

_____

The star then moves around quite a lot in that part of the H-R diagram. Look at the H-R diagrams in Chapter 16 (Figures 16.5–16.14), and then use the cursor to approximate the motion of the star as it moves up the red giant branch, back down and onto the horizontal branch, and then back to the right and up the asymptotic giant branch.

10. Are the changes you observe in the image of the star as dramatic as the ones you observed for question 5?

_____

11. What is the most noticeable change in the star as it moves through this portion of its evolution?

_____

Next, the star begins moving across the H-R diagram to the left, maintaining almost the same luminosity.

12. Predict how the temperature, luminosity, and radius of the star will change as it moves across the top of the diagram toward the left.

_____

Drag the cursor across the top of the H-R diagram to the left, and study what happens to the image of the star in the "Comparison to Sun" box.

13. What changed about the star as you dragged it across the H-R diagram?

_____

14. How does its size now compare to that of the Sun?

_____

Finally, the star drops to the bottom of the H-R diagram and then begins moving to the right.

15. Predict how the temperature, luminosity, and radius of the star will change as it drops to the bottom of the H-R diagram and moves to the right.

_____

Move the cursor toward the bottom of the H-R diagram, where the star becomes a white dwarf.

16. What changed about the star as you dragged it down the H-R diagram?

_____

17. How does its radius now compare to that of the Sun?

_____

To solidify your understanding of stellar evolution, press the "Reset" button and then move the star from main sequence to white dwarf several times. This will help you remember how this part of a star's life appears on the H-R diagram.

Evolution of High-Mass Stars

Some massive stars end their lives as neutron stars that rapidly spin. In some cases, the star has a strong magnetic field with an axis that does not line up with the rotation axis. This makes the star "pulse" on and off, as the beam of light that comes from the magnetic field axis passes in front of the observer. For each rotation of the neutron star, predict how many "pulses" an observer will see. To see why a star pulses, build a model of a pulsar with two pencils, a large wad of paper, some tape, and some string or ribbon (optional). Tape the two pencils together so that they form an X. One of these pencils is the rotation axis, and one is the axis of the magnetic field. Wad up some paper around the place where the pencils cross, to represent the neutron star. Choose one of the pencils to be the axis of the magnetic field, and tape pieces of string or ribbon to each end (the optional string or ribbon represents the beam of light that comes from the magnetic field axis). Hold the other pencil, which represents the rotation axis, and spin the model quickly. Notice how sometimes the magnetic field points toward you, and sometimes it does not. How many times does the magnetic field point toward you during each rotation of the model?

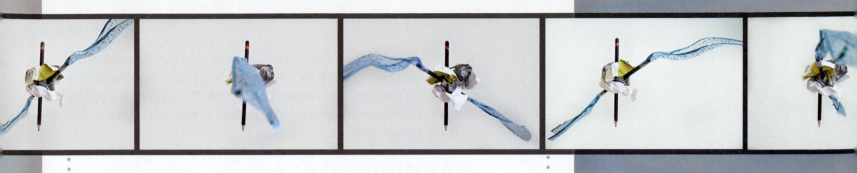

EXPERIMENT SETUP

To see why a star pulses, build a model of a pulsar with two pencils, a large wad of paper, some tape, and some string or ribbon (optional).

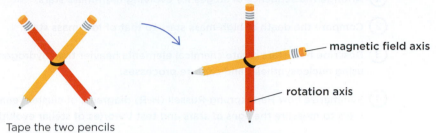

magnetic field axis

rotation axis

Tape the two pencils
so that they form an "X."

One of these pencils is the rotation axis,
and one is the axis of the magnetic field.

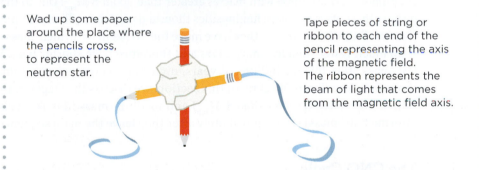

Wad up some paper
around the place where
the pencils cross,
to represent the
neutron star.

Tape pieces of string or
ribbon to each end of the
pencil representing the axis
of the magnetic field.
The ribbon represents the
beam of light that comes
from the magnetic field axis.

Hold the other pencil, which represents the rotation axis, and spin
the model quickly. Notice how sometimes the magnetic field points
towards you, and sometimes it does not. How many times does the
magnetic field point toward you during each rotation of the model?

PREDICTION

I predict that for each rotation
the observer will see
☐ 0   ☐ 2   ☐ 3   ☐ 4
pulse/pulses.

17

Most stars are smaller and less massive than the Sun and live a relatively long time. O and B stars, which are much more massive than the Sun, are rarer and live a much shorter time. Those massive stars die more quickly and explosively than stars like the Sun. In this chapter, we explore the life and death of high-mass stars.

LEARNING GOALS

By the end of this chapter, you should be able to:

(1) Arrange the sequence of stages for evolving high-mass stars.

(2) Compare the death of high-mass stars to that of low-mass stars.

(3) Describe how stars create chemical elements heavier than hydrogen and helium using nucleosynthesis and capture processes.

(4) Summarize how Hertzsprung-Russell (H-R) diagrams of clusters enable astronomers to measure the ages of stars and test theories of stellar evolution.

17.1 High-Mass Stars Follow Their Own Path

High-mass stars are those with masses greater than about 8 M_{Sun}. Due to their higher masses, these stars also have luminosities thousands or even millions of times greater than the Sun's. Even though they have more fuel to begin with, high-mass stars use it up much faster than low-mass stars and therefore have shorter lives. Whereas low-mass stars live for billions of years, high-mass stars live only hundreds of thousands to millions of years. In this section, we discuss the stages in the evolution of high-mass (greater than 8 M_{Sun}) and medium-mass (3–8 M_{Sun}; also called intermediate-mass) stars, immediately after they leave the main sequence.

The CNO Cycle

A high-mass star has greater gravitational force pressing down on the interior than a low-mass star does. That greater force leads to higher temperature and pressure in the core. Those two factors increase the rate of nuclear fusion, which therefore makes the stars more luminous. Those two factors also mean that additional nuclear reactions become possible beyond the fusion of hydrogen into helium. Recall from Chapter 14 that the hydrogen nucleus has only one proton—a single positive charge—so hydrogen fuses at lower temperatures (at minimum 10 million kelvins) than any other atomic nucleus. However, the probability that any two hydrogen atoms will fuse is low. The low probability of that first step in the proton-proton chain limits how rapidly the entire process can move forward.

If the star formed with elements from a previous generation of stars, carbon and other elements will be mixed with the hydrogen. In the core of a massive star, the temperature and pressure are high enough that the **carbon-nitrogen-oxygen (CNO) cycle**, a nuclear fusion process that converts hydrogen to helium in the presence of carbon, occurs. That process is illustrated in **Figure 17.1a**, which shows each step. First, a hydrogen nucleus (^{1}H) fuses with a carbon-12 (^{12}C) nucleus to form nitrogen-13 (^{13}N). Second, a proton in that ^{13}N nucleus is converted to a neutron, so that the atom is once again carbon. However, the "extra" neutron in that carbon

▶▶ **Interactive Simulation:** CNO Cycle

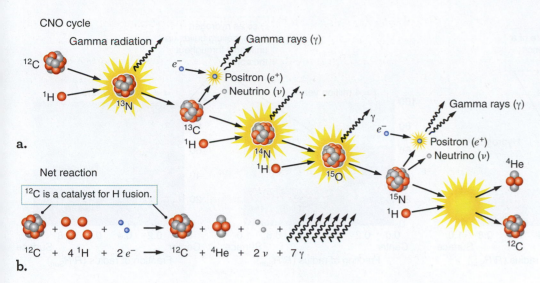

Figure 17.1 a. In high-mass stars, carbon serves as a catalyst for the fusion of hydrogen to helium. That process is the carbon-nitrogen-oxygen (CNO) cycle. **b.** The CNO cycle takes carbon-12, hydrogen, and electrons as inputs and produces carbon-12, helium, neutrinos, and gamma rays.

nucleus makes it carbon-13 (^{13}C) now, not carbon-12. Third and fourth, two more hydrogen nuclei then fuse with that ^{13}C nucleus, creating nitrogen-14 (^{14}N) and then oxygen-15 (^{15}O). Fifth, a proton in the oxygen nucleus is converted to a neutron, creating nitrogen-15 (^{15}N). One more proton enters the nucleus, causing the ejection of a helium-4 (^{4}He) nucleus and leaving behind a ^{12}C nucleus, which can participate in the cycle again. This cycle begins with a ^{12}C nucleus and one is regenerated at the end, so ^{12}C is considered to be a catalyst in the overall fusion of four ^{1}H to form one ^{4}He.

Notice that only one of two reactions occurs at most steps in that process: either a hydrogen nucleus fuses with another nucleus to create a new element with a higher atomic number or a proton in the nucleus spontaneously decays to a neutron to create a new element with a lower atomic number. Along the way, several by-products are formed: a positron and a neutrino are ejected each time a proton decays to form a neutron. Each positron then annihilates with an electron to produce a gamma ray. Additional gamma rays are released each time fusion occurs. **Figure 17.1b** shows the net reaction: a carbon-12 nucleus and four hydrogen nuclei combine with two electrons to produce a carbon-12 nucleus, a helium-4 nucleus, two neutrinos, and seven gamma rays. Getting a hydrogen nucleus past the electric barrier set up by a carbon nucleus, with its six protons, takes a lot of energy. But when that barrier can be overcome at high pressure and temperature, fusion is much more probable than it is in the interaction between two hydrogen nuclei. The CNO cycle is far more efficient than the proton-proton chain in stars more massive than about 1.3–1.5 M_{Sun}.

In a high-mass star, the temperature difference between the center of the star and the outside of the core is so large that convection occurs within the core, "stirring" the core like the water in a boiling pot. As shown in **Figure 17.2**, helium spreads uniformly throughout the core of a high-mass star as the star consumes its hydrogen. To see how this differs from low-mass stars, in which the helium builds up from the center outward, compare the graphs in Figure 17.2 with the ones in Figure 16.3. The difference results from the different processes in which helium fusion occurs in high-mass and low-mass stars.

The High-Mass Star Leaves the Main Sequence

As a high-mass star runs out of hydrogen in its core, the weight of the overlying star compresses the core, just as in a low-mass star. In the core of a high-mass star, however, the pressure and temperature (10^8 K) become high enough for helium

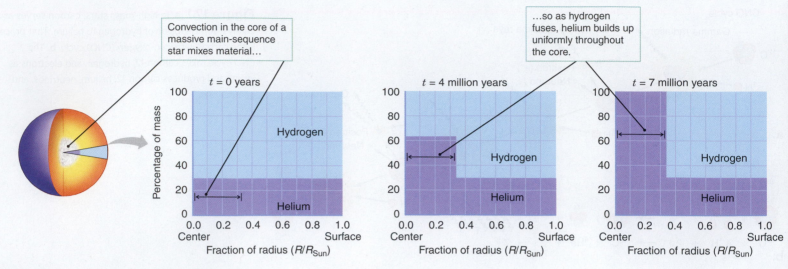

Figure 17.2 Convection keeps the core of a high-mass main-sequence star well mixed, so the composition remains uniform throughout the core as it evolves from zero age in the first graph to age 7 million years in the last graph. (Evolution times are for a 25-M_{Sun} star.)

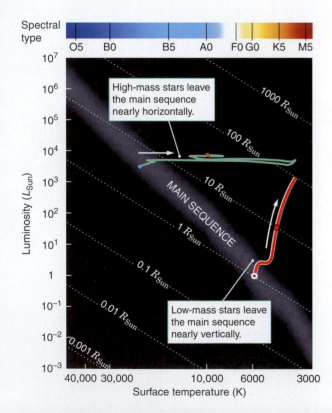

Figure 17.3 When high-mass stars leave the main sequence, they move horizontally across the H-R diagram, unlike low-mass stars, which move nearly vertically.

fusion to begin before the high-mass star forms an electron-degenerate core. As a result, the star makes a fairly smooth transition from hydrogen fusion to helium fusion as it leaves the main sequence.

When a low-mass star leaves the main sequence, the path it follows on the Hertzsprung-Russell (H-R) diagram is nearly vertical (**Figure 17.3**) as the star becomes more luminous at roughly constant temperature. In contrast, as a high-mass star leaves the main sequence, the radius increases while its surface temperature decreases, so its evolutionary track is nearly horizontal on the H-R diagram (Figure 17.3). The massive star fuses helium in its core and hydrogen in a surrounding shell. Recall that this is the same structure as a low-mass horizontal branch star. Stars more massive than 10 M_{Sun} become red supergiants during their helium-fusing phase. They have very cool surface temperatures (about 4000 K) and radii as much as 1000 times that of the Sun.

The next stage in the evolution of a high-mass star has no analog in low-mass stars. When the high-mass star exhausts the helium in its core, the core begins to collapse again. Carbon fusion begins when the core reaches temperatures of 8×10^8 K or higher. Carbon fusion produces more massive elements (sometimes referred to as "heavy elements"), including oxygen (O), neon (Ne), sodium (Na), and magnesium (Mg). The star then consists of a carbon-fusing core surrounded by a helium-fusing shell surrounded by a hydrogen-fusing shell, which moves outward as fusion products accumulate in the core. When carbon is exhausted as a nuclear fuel at the center of the star, neon either breaks down or fuses to magnesium; and when neon is exhausted, oxygen begins to fuse. The structure of the evolving high-mass star, shown in **Figure 17.4**, is like that of an onion, with many concentric layers.

Medium-mass stars fuse hydrogen via the CNO cycle like high-mass stars. Stars with medium mass leave the main sequence as high-mass stars do, fusing helium in their cores immediately after their hydrogen is exhausted and skipping the helium flash phase of low-mass star evolution. When helium fusion in the core is complete, however, the temperature at the center of a medium-mass star is too low for carbon to fuse. From that point on, the star evolves more like a low-mass star, ascending the asymptotic giant branch (AGB), fusing helium and hydrogen in shells around a degenerate core, then ejecting its outer layers and leaving behind a white dwarf.

Stars on the Instability Strip

While undergoing post-main-sequence evolution, a star may make one or more passes through a region of the H-R diagram known as the **instability strip**, which consists of the dashed-line region in **Figure 17.5**. As stars pass through the instability strip, they grow and shrink repeatedly, so they appear to pulsate, regularly growing bright and faint and then bright again. The time for one pulsation is called the **period**. Such stars are called **pulsating variable stars**. A pulsating variable star does not achieve a steady balance between pressure and gravity; instead, it repeatedly overshoots the equilibrium radius, shrinking too far before being pushed back out by pressure or expanding too far before being pulled back in by gravity. Pulsating variable stars are one type of **variable star**, which is a more general name for a star that varies in brightness as time passes.

The most luminous pulsating variable stars are **Cepheid variables**, named after the prototype star Delta Cephei. Type I, or classical, Cepheids are massive and luminous yellow supergiants. A Cepheid variable takes between 1 and 100 days to complete one pulsation period. The period correlates with the luminosity, creating a **period-luminosity relationship** in which stars with longer periods are more

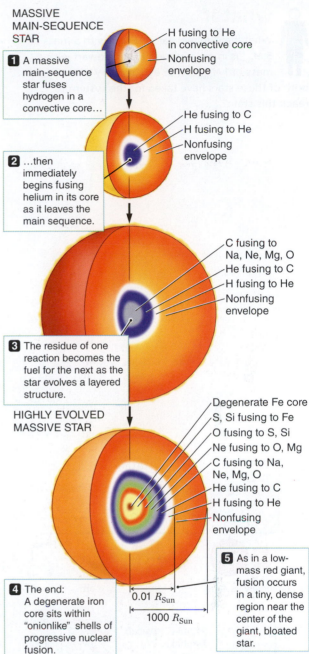

Figure 17.4 As a high-mass star evolves, it builds up a layered structure like that of an onion, with progressively more advanced stages of nuclear fusion found deeper and deeper within the star. The bottom image has been reduced to fit on the page.

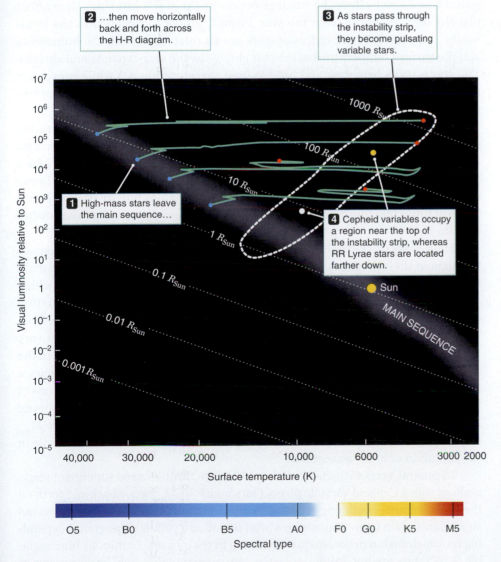

Figure 17.5 The paths of high-mass stars along the H-R diagram take them through a region known as the instability strip, shown by the region surrounded by the white dashed line. Pulsating variable stars such as Cepheid variables and RR Lyrae stars are found in that region of the H-R diagram.

what if . . .

What if a Cepheid variable star with a mass of 5 M_{Sun} is in orbit around a white dwarf with a mass of 1 M_{Sun}? What evolutionary steps might both of these stars have taken for the system to reach this state?

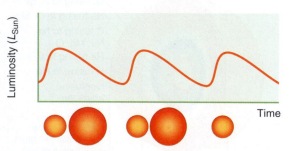

a. Luminosity vs. time

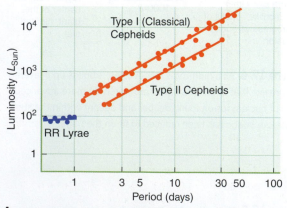

b. Period-luminosity relationship

Figure 17.6 a. Cepheid variable stars pulsate in size, so their luminosity changes periodically. The shape of the resulting curve is distinctive. **b.** The length of the period of pulsation is related to the star's luminosity at maximum.

luminous, as shown in **Figure 17.6**. Henrietta Leavitt (1868–1921) experimentally determined the period-luminosity relationship for Cepheid variables in 1912, now called the Leavitt law. The relationship is important because it allows astronomers to find the distance to a Cepheid variable star. The process has three steps: First, astronomers observe the period of the Cepheid variable star. Then they use that period, combined with the period-luminosity relationship, to find the luminosity. Finally, they combine that luminosity with the observed brightness in the sky to find the distance to the star. If the Cepheid variable is in a distant galaxy, astronomers then know the distance to that galaxy.

Thermal energy powers the pulsations of stars like Cepheid variables. An ionized gas is nearly opaque because ions can interact with light of all wavelengths, scattering many of the photons back toward the interior. In contrast, a neutral gas can absorb only light with energies that match its electron transitions. As the helium atoms in the atmosphere of the star alternate between ionized and neutral states, the atmosphere of the star alternates between opaque and transparent. That in turn alternately traps and releases light from the star. The star's atmosphere is much like a lid on a pot of boiling water. The pot builds up pressure enough to pop open the lid and let steam escape. Then gravity pulls the lid back down, and pressure builds again to repeat the process. Those pulsations do not affect the nuclear fusion in the star's interior—they affect just the light escaping from the star. From Chapter 13, recall that the luminosity, temperature, and radius of stars are all related, so both the luminosity and the surface temperature of the star change as the star expands and shrinks. The star is at its brightest and hottest (and therefore bluest) while it expands through its equilibrium size and is at its faintest and coolest (and therefore reddest) while it falls back inward.

High-mass Type I Cepheid variables are not the only type of variable star. The low-mass horizontal branch also passes through the instability strip on the H-R diagram. Low-mass (~0.8-M_{Sun}) stars can be Type II Cepheid variables or **RR Lyrae variables**, which have periods of less than a day. RR Lyrae stars pulsate by the same mechanism as Cepheid variables but are typically hundreds of times less luminous. They, too, follow a period-luminosity relationship (Figure 17.6b).

High-mass stars expel a significant percentage of their mass back into space throughout their lifetimes, changing their composition. Even while on the main sequence, massive O and B stars have low-density winds with velocities as high as 3000 kilometers per second (km/s). The pressure of the intense radiation generated in the core overcomes the star's gravity and drives away material from its outermost layers. Main-sequence O and B stars lose mass at rates that vary from 10^{-7} up to 10^{-5} M_{Sun} of material per year. The fastest mass loss occurs in the most massive stars. Those numbers may sound tiny, but over millions of years, a high-mass star may lose many solar masses of material. An O star with a mass of 20 M_{Sun} may lose about 20 percent of its mass (4 M_{Sun}) while on the main sequence and possibly more than half its mass (10 M_{Sun}) over its lifetime. Even an 8-M_{Sun} star may lose 5–10 percent of its mass. That mass loss plays a prominent role in the evolution of high-mass stars.

In general, stars with masses below 15 M_{Sun} go through a red supergiant stage, perhaps with a Cepheid variable phase. Stars under 30 M_{Sun} move back and forth on the H-R diagram—becoming red supergiants, then blue supergiants, and then red supergiants again (sometimes with a brief period as a yellow supergiant), depending on which fusion processes are occurring in their cores. Luminous blue variable (LBV) stars are hot, luminous, extremely rare stars that may be as massive as

150 M_{Sun}. An example is Eta Carinae (**Figure 17.7**), a binary system with at least 120 times the mass of the Sun and a luminosity (summed over all wavelengths) of 5 million L_{Sun}. Eta Carinae is now losing one solar mass every 1000 years. During a 19th century eruption, however, when Eta Carinae became the second-brightest star in the sky, it shed ~10 M_{Sun} of material in only 20 years. Eta Carinae is expected to explode in the astronomically near future.

CHECK YOUR UNDERSTANDING 17.1

How does energy production in a high-mass, main-sequence star differ from energy production in the Sun? (Choose all that apply.) (a) High-mass stars get a lot of energy through non-nuclear processes. (b) High-mass stars produce energy at a faster rate. (c) High-mass stars fuse carbon on the main sequence. (d) High-mass stars use up all their carbon in a process that fuses hydrogen into helium.

Answers to Check Your Understanding questions are in the back of the book.

17.2   High-Mass Stars Go Out with a Bang

A low-mass star approaches the end of its life relatively slowly and gently, ejecting its outer parts into nearby space and leaving behind a degenerate core. In contrast, a high-mass star ends its life suddenly and explosively. In this section, we explain how the energy involved in binding atomic nuclei affects nuclear fusion in the core and results in an explosive end for a high-mass star.

Binding Energy

An evolving high-mass star builds up an onionlike structure as nuclear fusion in its interior proceeds to more and more massive elements (see Figure 17.4). Hydrogen fuses to helium, helium fuses to carbon and oxygen, carbon fuses to magnesium, oxygen fuses to sulfur and silicon, and then silicon and sulfur fuse to iron. Many types of nuclear reactions occur up to that point, forming almost all the stable isotopes of elements less massive than iron. But the chain of nuclear fusion stops with iron.

The **binding energy** of an atomic nucleus is the energy required to break the nucleus into its constituent parts. A nuclear reaction releases energy if it increases the binding energy in a nucleus. Conversely, a nuclear reaction absorbs energy if it decreases the binding energy in the nucleus. The graph in **Figure 17.8** shows the binding energy per nucleon (that is, per each proton or neutron in the nucleus) for different atomic nuclei. Moving from hydrogen to helium increases the binding energy, so fusing hydrogen releases energy, sustaining the nuclear reaction. Moving from helium to carbon also increases the binding energy, so that process also helps sustain the nuclear reactions in the core (**Working It Out 17.1**). Iron is at the peak of the binding-energy curve, so fusing lighter elements up to iron (Fe) also releases energy; however, fusing iron to make heavier elements decreases the binding energy. Because that process absorbs energy from the core, it lowers the temperature and the pressure. Iron fusion, therefore, cannot sustain itself. Once the star's core is filled with iron, the star will not last much longer.

The End of a Massive Star

Per reaction, helium fusion produces about one-quarter as much energy as hydrogen fusion, so when a star is fusing helium into carbon, the pressure inside the

Figure 17.7 In this image of the luminous blue variable star Eta Carinae, an expanding cloud of ejected dusty material is seen in visible (blue) and X-ray (yellow) light. The star itself, largely hidden by the surrounding dust, has a luminosity of 5 million L_{Sun} and a mass probably in excess of 120 M_{Sun}. Dust is created when volatile material ejected from the star condenses.

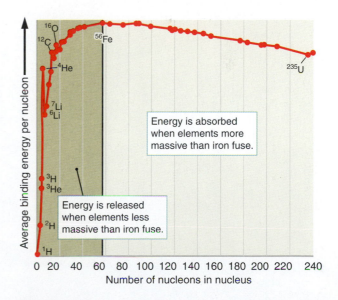

Figure 17.8 The binding energy per nucleon is plotted against the number of nucleons for each element. That is the energy required to break the atomic nucleus apart into protons and neutrons. Energy is released by nuclear fusion only if the resulting element is higher on the curve.

Binding Energy of Atomic Nuclei

The net energy released by a nuclear reaction is the difference between the binding energy of the products and the binding energy of the reactants:

$$\text{Net energy} = \left(\begin{array}{c}\text{Binding energy}\\\text{of products}\end{array}\right) - \left(\begin{array}{c}\text{Binding energy}\\\text{of reactants}\end{array}\right)$$

In the triple-alpha process, the binding energy of the initial helium nuclei is 6.824×10^{14} joules (J) per kilogram (kg) of helium, and the binding energy of the produced ^{12}C nuclei is 7.402×10^{14} J per kg of carbon. The amount of energy available from fusing 1 kg of helium nuclei into carbon is given by

$$\left(\begin{array}{c}\text{Net energy from}\\\text{fusing 1 kg of He}\end{array}\right) = \left(\begin{array}{c}\text{Binding energy}\\\text{of C formed}\end{array}\right) - \left(\begin{array}{c}\text{Binding energy}\\\text{of He fused}\end{array}\right)$$

$$= (7.402 \times 10^{14}\,\text{J}) - (6.824 \times 10^{14}\,\text{J})$$

$$= 5.780 \times 10^{13}\,\text{J}$$

Thus, fusing He to C releases energy, which indicates that helium is a good nuclear fuel, as Figure 17.8 also shows.

What about fusing iron into more massive elements? Because iron is at the peak of the binding-energy curve, the products of iron fusion will have less binding energy than the initial reactants. Going from iron to more massive elements means moving down on the binding-energy curve in Figure 17.8, so the net energy in the reaction will be negative. Instead of producing energy, iron fusion absorbs energy.

star is lower than when it was fusing hydrogen into helium on the main sequence. The star compresses, increasing the reaction rate so the helium is used up more quickly. Although hydrogen fusion can supply the energy needed to support the high-mass star on the main sequence for millions of years, helium fusion can support the star for only a few hundred thousand years.

Maintaining the balance in a star is like trying to keep a leaky balloon inflated. The larger the leak, the more rapidly air must be pumped into the balloon to keep it inflated. A star fusing hydrogen or helium is like a balloon with a slow leak. At the temperatures of hydrogen or helium fusion, energy leaks out of a star's interior primarily by radiation and convection. Neither process is very efficient because the outer layers of the star act like a thick, warm blanket. Much of the energy is kept in the star, so nuclear fuels do not need to fuse very fast to support the weight of the outer layers of the star while keeping up with the energy escaping from the surface.

Once carbon fusion begins, that balance shifts. Energy is carried away primarily by neutrinos—referred to as **neutrino cooling**—rather than by radiation and convection. Recall from your study of the Sun in Chapter 14 that neutrinos escape easily, carrying energy away from the core. Like air pouring out through a huge hole in the side of a balloon, neutrinos produced inside the star stream through the overlying layers of the star as though they were not even there, carrying the energy from the stellar interior out into space. As thermal energy pours out of the interior of the star, the outer layers of the star fall inward, driving up the density and temperature and increasing the rate of nuclear reactions.

Table 17.1	Fusion Stages in High-Mass Stars		
Core Fusing Stage	Duration for a 15-M_{Sun} Star	Duration for a 25-M_{Sun} Star	Typical Core Temperature (K)
Hydrogen (H) fusion	11 million years	7 million years	$(3–10) \times 10^7$
Helium (He) fusion	2 million years	800,000 years	$(1–7.5) \times 10^8$
Carbon (C) fusion	2000 years	500 years	$(0.8–1.4) \times 10^9$
Neon (Ne) fusion	8 months	11 months	$(1.4–1.7) \times 10^9$
Oxygen (O) fusion	2.6 years	5 months	$(1.8–2.8) \times 10^9$
Silicon (Si) fusion	18 days	0.7 day	$(2.8–4) \times 10^9$

Once neutrino cooling becomes significant, the star begins evolving much more rapidly. **Table 17.1** shows that the star fuses faster at each stage of nuclear fusion, so that the entire time spent fusing all the elements beyond hydrogen is very short in comparison with the main-sequence lifetime. A star that fuses silicon into iron in the core is not much more luminous than it was while fusing helium. But because of neutrino cooling, the silicon-fusing star actually releases about 200 million times more energy per second than it did while it was fusing helium—most of that energy is released in the form of neutrinos.

The Collapse of the Core and Subsequent Explosion

After silicon fuses to form an iron core in the star, the end comes suddenly and dramatically because no source of nuclear energy remains to replenish the energy emitted with the neutrinos. No longer supported by thermonuclear fusion, the iron core of the massive star begins to collapse.

Figure 17.9 shows the stages a high-mass star passes through at the end of its life. The early stages of collapse of the iron core of an evolved massive star are much the same as those in the collapse of a nonfusing core in a low-mass star (Figure 16.14). As the core collapses, the force of gravity increases and the density and temperature skyrocket. The core becomes electron-degenerate when it is about the size of Earth. Unlike the electron-degenerate core of a low-mass red giant, however, the weight bearing down on the iron core is too great to be held up by electron degeneracy pressure (step 1 in Figure 17.9). As the core collapses, the core temperature climbs to 10 billion K (10^{10} K) and higher while the density exceeds 10^{10} kilograms per cubic meter (kg/m³)—10 times the density of an electron-degenerate white dwarf.

The phenomenal temperatures and pressures trigger fundamental changes in the core. At those temperatures, the nucleus of the star is filled with thermal radiation so energetic that it is in the gamma-ray part of the spectrum. Those gamma-ray photons have enough energy to break iron nuclei apart into helium nuclei (step 2 in Figure 17.9) in a process called **photodisintegration**. Photodisintegration absorbs thermal energy and reverses the results of nuclear fusion. Also, the pressure in the core is now so great that electrons are forced into atomic nuclei, where they combine with protons to produce neutrons and neutrinos (step 3 in Figure 17.9) in a process called **charge destruction**. Both charge destruction

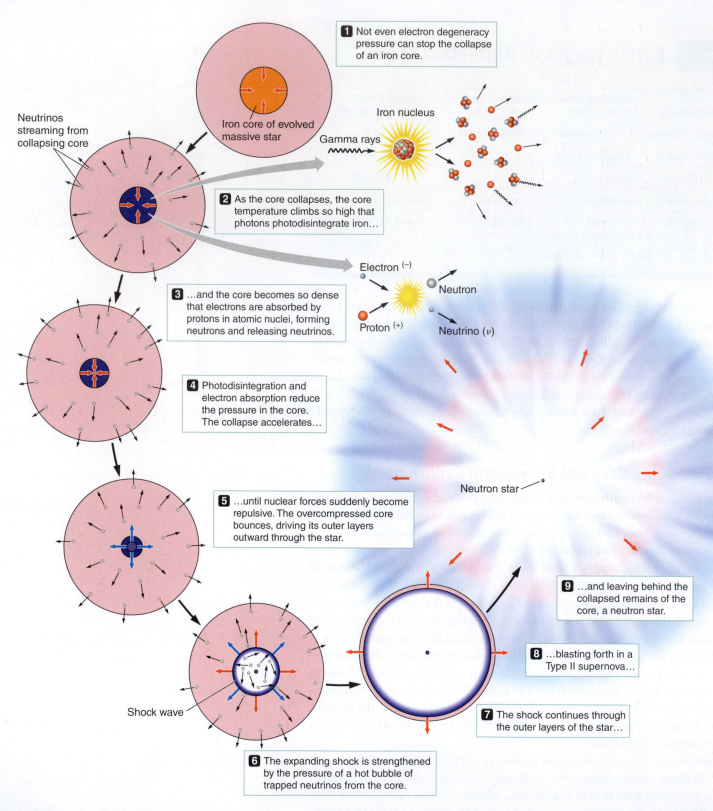

1 Not even electron degeneracy pressure can stop the collapse of an iron core.

Iron core of evolved massive star

Neutrinos streaming from collapsing core

Iron nucleus

Gamma rays

2 As the core collapses, the core temperature climbs so high that photons photodisintegrate iron…

Electron $^{(-)}$

Neutron

Proton $^{(+)}$

Neutrino (ν)

3 …and the core becomes so dense that electrons are absorbed by protons in atomic nuclei, forming neutrons and releasing neutrinos.

4 Photodisintegration and electron absorption reduce the pressure in the core. The collapse accelerates…

Neutron star

5 …until nuclear forces suddenly become repulsive. The overcompressed core bounces, driving its outer layers outward through the star.

9 …and leaving behind the collapsed remains of the core, a neutron star.

8 …blasting forth in a Type II supernova…

7 The shock continues through the outer layers of the star…

Shock wave

6 The expanding shock is strengthened by the pressure of a hot bubble of trapped neutrinos from the core.

Figure 17.9 A high-mass star goes through several stages at the end of its life as its core collapses and the star explodes as a Type II supernova.

and photodisintegration absorb much of the energy that was holding up the dying star. Neutrinos continue to take enormous amounts of energy with them as they leave the star. The collapse of the core accelerates, reaching a speed of 70,000 km/s, or almost one-fourth the speed of light (step 4 in Figure 17.9). Together, all those events—photodisintegration, charge destruction, and collapse—take place in less than a second.

As material in the collapsing core exceeds the density of an atomic nucleus, the strong nuclear force becomes repulsive (step 5 in Figure 17.9). About half of the collapsing core suddenly slows its inward fall. The remaining half slams into the innermost part of the star at a significant fraction of the speed of light and "bounces," sending a shock wave back out through the outer layers of the star (step 6 in Figure 17.9).

Over the next second or so, almost one-fifth of the core mass is converted into neutrinos. Most of those neutrinos travel immediately outward through the star; but at the extreme densities found in the collapsing core of the massive star, not even neutrinos pass with complete freedom. The dense material traps a few tenths of a percent of the energy of the neutrinos streaming out of the core of the dying star. The energy of those trapped neutrinos drives the pressure and temperature in that region higher, inflating a bubble of extremely hot gas and intense radiation around the core of the star. The pressure of that bubble adds to the strength of the shock wave moving outward through the star. Within about a minute, the shock wave has pushed its way out through the helium shell within the star. Within a few hours it reaches the surface of the star itself, heating the stellar surface to 500,000 K and blasting material outward at velocities of up to about 30,000 km/s. The evolved massive star explodes, becoming more than a billion times as luminous as the Sun and leaving behind a cloud of dust and gas (**Figure 17.10**). That type of supernova, triggered by the collapse of the core, is called a **Type II supernova**, or sometimes a "core-collapse" supernova.

The difference in appearance between a Type Ia supernova (see Figure 16.18 and its accompanying discussion) and a Type II supernova is subtle. Both become suddenly very luminous. Both leave behind expanding clouds of dust and gas. In the first year or so after the explosion, however, it is possible to distinguish between the types by their light curves, as shown in **Figure 17.11**. Type II supernovae have more varied light curves, with a peak luminosity less than that of Type Ia supernovae and a light curve that falls off less rapidly. The spectra of those objects also have observable differences.

In 1987, astronomers observed the explosion of a massive star in the Large Magellanic Cloud (LMC; a companion galaxy to the Milky Way 160,000 light-years

what if . . .

What if a Type II supernova is observed using both neutrinos and optical light? In what order would you expect to see these signals arrive at Earth?

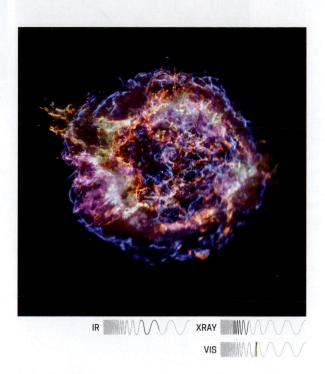

IR   XRAY   VIS

Figure 17.10 When an evolved star explodes, it forms a supernova remnant like Cassiopeia A. X-ray observations such as these suggest that the star may have turned itself inside out as it spit out elements from its core.

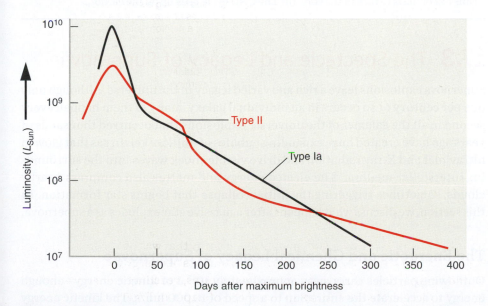

Figure 17.11 These light curves show the changes in brightness of average Type Ia and Type II supernovae.

Figure 17.12 Supernova 1987A exploded in a small companion galaxy of the Milky Way called the Large Magellanic Cloud (LMC). These images show the LMC **a.** before the explosion and **b.** while the supernova was near its peak. Notice the "new" bright star at lower right. **c.** An X-ray image from *Chandra* is combined with Hubble Space Telescope optical and ground-based infrared images to show freshly formed dust inside the glowing rings of gas.

Astronomy in Action: Type II Supernova

away). Astronomers working in all parts of the electromagnetic spectrum pointed their telescopes at Supernova 1987A (**Figure 17.12**)—the first supernova visible to the naked eye since the invention of the telescope. Astronomers were ultimately surprised to discover from looking at older photographs that the star that blew up was not a red supergiant but rather a 20-M_{Sun} B3 I blue supergiant now classified as a luminous blue variable star. Neutrino telescopes recorded a burst of neutrinos passing through Earth from that tremendous stellar explosion that had occurred in the LMC. The detection of neutrinos from SN 1987A gave astronomers a rare glimpse of the very heart of a massive star at the moment of its death, confirming a fundamental prediction of theories about the collapse of the core and its effects. Because this supernova is so close, astronomers have been able to observe its evolution in more detail than any other (**Figure 17.13**).

CHECK YOUR UNDERSTANDING **17.2**

What causes a high-mass star to explode as a Type II supernova? (a) The high-mass star merges with another star. (b) Iron absorbs energy when it fuses. (c) The high-mass star runs out of mass to fuse in the core. (d) The CNO cycle uses up all the carbon.

17.3 The Spectacle and Legacy of Supernovae

Supernova explosions leave a rich and varied legacy in the universe. Although only one per century or so occurs in an individual galaxy, many of them happen every second in all the galaxies of the universe. Explosions that occurred thousands of years ago have created huge expanding bubbles of million-kelvin gas that glow in ultraviolet and X-ray radiation and drive visible shock waves into the surrounding interstellar medium. The ejected energy and matter also compress nearby clouds, sometimes triggering the initial collapse that begins star formation. In this section, we discuss what happens after a massive star explodes as a supernova.

The Energetic and Chemical Legacy of Supernovae

Outflowing particles carry away approximately 10^{47} J of kinetic energy—enough energy to accelerate the entire Sun to a speed of 10,000 km/s. The kinetic energy released from both Type Ia and Type II supernovae heats the hottest phases of the

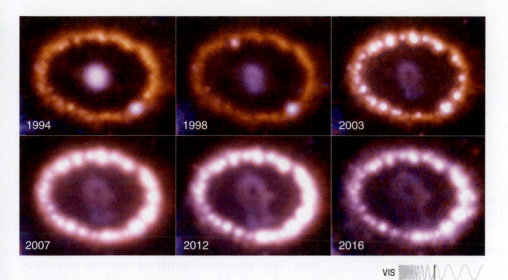

VIS

Figure 17.13 ★ WHAT AN ASTRONOMER SEES In general, astronomical objects change very slowly, so it is a little bit unusual that an astronomer has the chance to see such dramatic changes in celestial objects. This series of Hubble Space Telescope images shows how the innermost region of the supernova explosion SN1987A changed from 1994 to 2016. The ring of gas was produced late in the life of the star, at least 20,000 years before it exploded, and then was lit up by a blast of ultraviolet light from the explosion, causing it to glow for at least 30 years. An astronomer will notice that in 1994, seven years after the supernova occurred, the material leaving the star is still very compact, near the center of the ring. By 2012, that material had blasted out along an axis, nearly reaching the ring. She might look up the scale of the image to find that, in 2016, the central structure measured more than half a light-year across. From images like these, astronomers can observe the evolution of the supernova remnant, which has an organized structure reminiscent of some planetary nebulae.

interstellar medium and stirs the clouds in the interstellar medium. That kinetic energy is about 100 times as much energy as is carried away by light. Yet even that amount of energy is less than the energy carried away from the supernova explosion by neutrinos—an amount of energy at least another 100 times larger.

The chemical changes in the galaxy caused by supernova explosions may be even more important than the energy released. Only very low-mass chemical elements were present at the beginning of the universe: hydrogen, helium, and trace amounts of lithium and possibly beryllium. All the rest of the naturally occurring elements on the periodic table were released to the interstellar medium when stars died. The process of forming more massive atomic nuclei from less massive nuclei is called **nucleosynthesis**. Nucleosynthesis is responsible for the buildup of massive elements in the universe, forming those elements either during nuclear fusion in the core of a star or during the rapid nuclear reactions that occur during violent events like a supernova explosion.

Elements up to carbon, oxygen, and small amounts of neon and magnesium form from nuclear fusion in the cores of low-mass stars and are released to the interstellar medium as the star loses its outer layers. Recall that fusion up to iron creates energy, but fusion at and beyond iron absorbs energy (see Figure 17.8). A look at the periodic table of the elements (see Appendix 3) shows that many naturally occurring elements are more massive than iron. Elements heavier than iron fuse only under conditions in which abundant energy is available to be absorbed—as in supernova explosions. Thus, the naturally occurring elements heavier than iron were all produced in the deaths of high-mass stars.

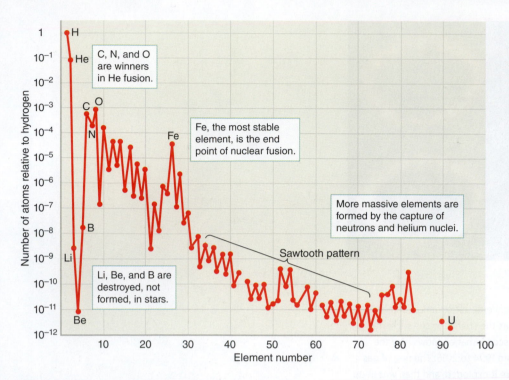

Figure 17.14 Observed relative abundances of different elements in the Solar System are plotted against the element number of each element's nucleus. That pattern can be understood as a result of nucleosynthesis in stars. The periodic table of the elements in Appendix 3 identifies individual elements by their element number (that is, by their number of protons).

Because like charges repel one another, electrostatic repulsion normally keeps atomic nuclei far apart. Extreme temperatures and pressures are needed to slam nuclei together hard enough to overcome that electrostatic repulsion. Neutrons, however, have no net electric charge, so free neutrons are not repelled by an atomic nucleus. Under normal conditions, free neutrons are rare. In the interiors of evolved stars, however, nuclear reactions occur that produce free neutrons. Under the conditions present shortly before and during a supernova, free neutrons are produced in very large numbers. Those free neutrons are easily captured by atomic nuclei and later decay to become protons. That process of **neutron capture** and decay forms the elements with atomic numbers and masses higher than those of iron. The abundance of helium in the outflowing material means that **helium capture**, in which a helium nucleus is captured by another nucleus, also is relatively common. Each time it occurs, helium capture increases the atomic number of the capturing atom by two protons. Proton capture also occurs, in which a free proton (a hydrogen nucleus, ^{1}H) is captured into a larger nucleus. That type of capture is most probable if the capturing nucleus has an odd number of protons.

Nuclear physics predicts how the abundances of the elements change as a result of those energetic explosions. Those predictions agree with abundances measured on Earth, in the Solar System (**Figure 17.14**), and in the atmospheres of stars and their remains. In general, less massive elements are more abundant than more massive elements because more massive elements are progressively built up from less massive elements. An exception to that pattern is the dip in the abundances of the light elements lithium (Li), beryllium (Be), and boron (B). Nuclear fusion easily destroys those elements, and they are not produced by the common reactions involved in fusing hydrogen (H) and helium (He). Conversely, carbon (C), nitrogen (N), and oxygen (O) are produced in quantity in the triple-alpha process, so they are abundant. The relatively high abundances of elements near iron is evidence of processes that favor those tightly bound nuclei. The sawtooth pattern observed in the figure's plot results from both how elements form in stars and how they persist in space. Atoms with odd numbers of protons are more likely to capture another proton, and atoms with even numbers of protons are more stable, so even-numbered elements are more abundant in the current universe. By comparing the predictions of nuclear physics with observations of elemental abundances, astronomers repeatedly test the theory of stellar evolution.

Neutron Stars and Pulsars

In the explosion of a Type II supernova, the star's outer parts are blasted back into interstellar space—but what remains of the collapsing core left behind? For cores less than about 3 M_{Sun}, the collapse is halted when neutrons are packed as tightly together as the rules of quantum mechanics permit. That *neutron-degenerate* core left behind by the explosion of a Type II supernova is called a **neutron star**. It

Reading Astronomy News

Supergiant Betelgeuse Smaller, Closer Than First Thought

Australian National University

Everyone hopes that a nearby supernova happens in their lifetime—but not too nearby.

It may be another 100,000 years until the giant red star Betelgeuse dies in a fiery explosion, according to a new study by an international team of researchers.

The study, led by Dr Meridith Joyce from The Australian National University (ANU), not only gives Betelgeuse a new lease on life, but shows it is both smaller and closer to Earth than previously thought.

Dr Joyce says the supergiant—which is part of the Orion constellation—has long fascinated scientists. But lately, it's been behaving strangely.

"It's normally one of the brightest stars in the sky, but we've observed two drops in the brightness of Betelgeuse since late 2019," Dr Joyce said.

"This prompted speculation it could be about to explode. But our study offers a different explanation.

"We know the first dimming event involved a dust cloud. We found the second smaller event was likely due to the pulsations of the star."

The researchers were able to use hydrodynamic and seismic modelling to learn more about the physics driving these pulsations—and get a clearer idea of what phase of its life Betelgeuse is in.

According to co-author Dr Shing-Chi Leung from The University of Tokyo, the analysis "confirmed that pressure waves—essentially, sound waves—were the cause of Betelgeuse's pulsation."

"It's burning helium in its core at the moment, which means it's nowhere near exploding," Dr Joyce said.

"We could be looking at around 100,000 years before an explosion happens."

Co-author Dr László Molnár from the Konkoly Observatory in Budapest says the study also revealed how big Betelgeuse is, and its distance from Earth.

"The actual physical size of Betelgeuse has been a bit of a mystery—earlier studies suggested it could be bigger than the orbit of Jupiter. Our results say Betelgeuse only extends out to two thirds of that, with a radius 750 times the radius of the Sun," Dr Molnár said.

"Once we had the physical size of the star, we were able to determine the distance from Earth. Our results show it's a mere 530 light years from us—25 percent closer than previous thought."

The good news is Betelgeuse is still too far from Earth for the eventual explosion to have significant impact here.

"It's still a really big deal when a supernova goes off. And this is our closest candidate. It gives us a rare opportunity to study what happens to stars like this before they explode," Dr Joyce said.

The study was funded by The Kavli Institute for the Physics and Mathematics of the Universe (WPI), The University of Tokyo, and facilitated by the ANU Distinguished Visitor's program. It involved researchers from the United States, Hungary, Hong Kong, and the United Kingdom, as well as Australia and Japan.

The study has been published in *The Astrophysical Journal*.

Questions

1. Betelgeuse dimmed twice since late 2019. What was the cause of each event?

2. Betelgeuse is "burning helium in its core at the moment." Where should Betelgeuse be placed on the H-R diagram?

3. Betelgeuse is very nearly unique, because it is large enough and close enough that astronomers can measure its angular size on the sky. It's larger than a point. If a star's angular size is known, what else do you need to know in order to figure out how far away it is?

4. Make a sketch to show how to find distance from the angular size and your answer to question 3.

Source: https://www.anu.edu.au/about/global-engagement/southeast-asia-liaison-office/supergiant-betelgeuse-smaller-closer-than.

---

has a radius of 10–15 km, making it roughly the size of a small city, but it contains between 1.4 and around 2 M_{Sun}. (Cores greater than ~2 M_{Sun} have too much gravity to be supported by neutron degeneracy and become black holes (see Chapter 18). Combining the mass and the size shows that the neutron star's density is about 10^{18} kg/m^3—about a billion times denser than a white dwarf and roughly equivalent to the density of an atomic nucleus. If Earth were crushed down to the size

working it out 17.2

Gravity on a Neutron Star

Neutron stars are incredibly dense objects. As a result, the surface gravity and escape velocity of neutron stars are very high. For example, let's look at a typical neutron star, which has a radius of 15 km and a mass of $2\,M_{Sun}$.

Recall from Working It Out 4.1 that the acceleration due to gravity on the surface—here, the surface of a neutron star (NS)—is given by

$$g = \frac{GM_{NS}}{R^2_{NS}}$$

$$g = 6.67 \times 10^{-20}\,\frac{km^3}{kg\,s^2} \times \frac{2.0 \times (1.99 \times 10^{30}\,kg)}{(15\,km)^2}$$

$$g = 1.2 \times 10^9\,\frac{km}{s^2}$$

Dividing that number by the gravitational acceleration on Earth, $9.8\,m/s^2 = 0.0098\,km/s^2$, shows that the gravitational acceleration on a neutron star is more than 100 billion times as large as that on Earth.

What about the escape velocity from a neutron star? From Working It Out 16.2, we know that the escape velocity is given by

$$v_{esc} = \sqrt{\frac{2GM}{R}}$$

Putting in the above numbers for a typical neutron star yields

$$v_{esc} = \sqrt{\frac{2 \times [6.67 \times 10^{-20}\,km^3/(kg\,s^2)] \times 2.0 \times (1.99 \times 10^{30}\,kg)}{(15\,km)}}$$

$$v_{esc} = 190{,}000\,km/s$$

Dividing that result by the speed of light gives

$$\frac{v_{esc}}{c} = \frac{190{,}000\,km/s}{300{,}000\,km/s} = 0.63$$

The escape velocity from that neutron star is more than 60 percent of the speed of light and almost 17,000 times greater than the escape velocity from Earth (11.2 km/s). The physicist Albert Einstein showed that strange things happen at velocities near the speed of light (including modifications to Newton's equations—see Chapter 18).

of a football stadium, the planet would then have about the same density as a neutron star. Neutron stars are so compact that the acceleration due to gravity on the surface is more than 100 billion times the acceleration on Earth, as shown in **Working It Out 17.2**. That extremely high surface gravity implies a very large escape velocity. A spacecraft would need to be traveling at 0.63c to escape from the surface of a typical neutron star.

Many of the unusual objects discussed in Chapters 16 and 17—such as pulsating stars, supernovae, and planetary nebulae—puzzled astronomers when they were first observed but were later understood to be associated with the end points of stellar evolution. In contrast, neutron stars were predicted in 1934, not long after neutrons themselves were discovered. Astronomers Walter Baade and Fritz Zwicky proposed that supernova explosions could lead to the formation of a neutron star. But neutron stars were not actually observed for another 30 years. Rapidly pulsing objects were first discovered in 1967 by people observing radio wavelengths (see the **Process of Science Figure**).

Those objects, which blinked like very fast, regularly ticking clocks, puzzled astronomers. Today, those rotating neutron stars are called **pulsars**. More than 2000 pulsars are known, and more are being discovered all the time.

Recall from Chapter 15 that a collapsing molecular cloud begins to spin faster as it shrinks into a protostar because angular momentum is conserved. The core of a massive star spins faster as it collapses, too, for the same reason. A massive main-sequence O star may rotate once every few days. As a neutron star, it might rotate closer to a thousand times each second.

unanswered questions

What creates magnetars, a class of pulsars that are characterized by extremely large magnetic fields? Magnetars are observed to produce bursts of lower-energy gamma rays. The origin of their huge magnetic fields is not well understood. Those fields may originate from a dynamo in the interior of a superconducting region of the neutron star, but we do not know whether ordinary pulsars go through a magnetar phase.

Occam's Razor

Occam's razor is a guiding principle in science: when scientists consider two hypotheses that explain a phenomenon equally well, they should give preference to the "simpler" theory and prioritize testing it. "Simpler" does not mean that the math is easier, or even that the concept is easy to understand. It means that the fewest number of other, new assumptions need to be made.

In 1967, Jocelyn Bell, a student at Cambridge, discovered a "mystery signal" in her data. Her adviser, Antony Hewish, half-jokingly suggested "little green men" caused the signal.

Bell and Hewish find three more such signals. It was unlikely that the same "little green men" would be sending the same signal from four separate locations in the sky.

Bell and Hewish suggested the signals were from pulsating white dwarfs or neutron stars.

Meanwhile, Franco Pacini and Thomas Gold developed a detailed explanation involving rotating neutron stars. This explanation relied entirely on previously understood physical phenomena: rotation, magnetic fields, and neutron stars. It did not require assumptions about the existence of extraterrestrials.

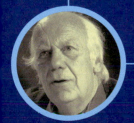  

The neutron star explanation is "simpler" and has become the accepted explanation for Bell's observations.

Simpler theories are easier to rule out, so if they survive testing, then scientists are more likely to consider them more seriously.

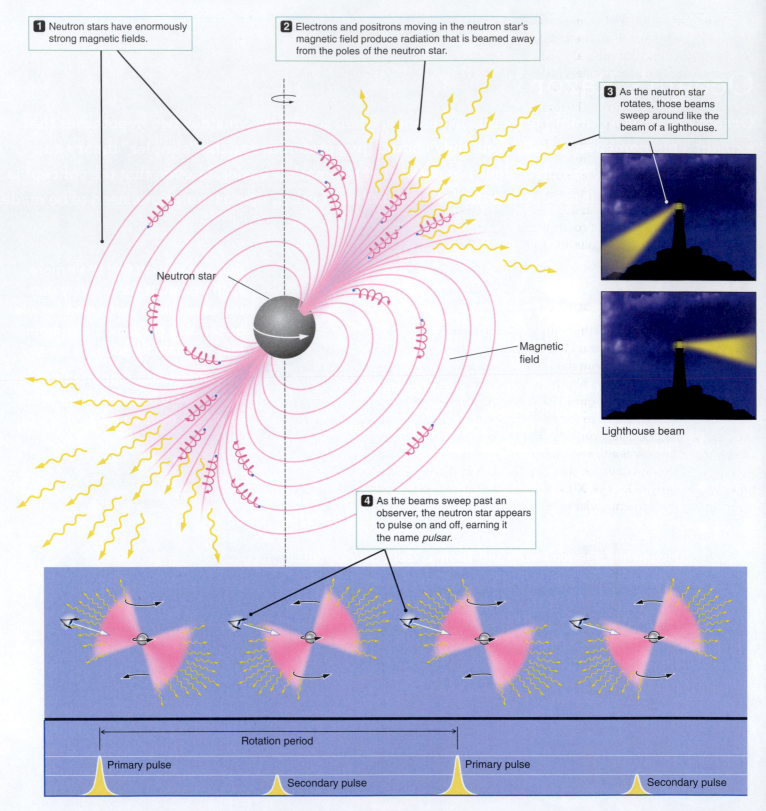

1 Neutron stars have enormously strong magnetic fields.

2 Electrons and positrons moving in the neutron star's magnetic field produce radiation that is beamed away from the poles of the neutron star.

3 As the neutron star rotates, those beams sweep around like the beam of a lighthouse.

Lighthouse beam

Neutron star

Magnetic field

4 As the beams sweep past an observer, the neutron star appears to pulse on and off, earning it the name *pulsar*.

Rotation period

Primary pulse

Secondary pulse

Primary pulse

Secondary pulse

Figure 17.15 When a highly magnetized neutron star rotates rapidly, light is given off, much like the beams from a rotating lighthouse lamp. As those beams sweep past Earth, the star will appear to pulse on and off, earning it the name *pulsar*.

The collapsing star also concentrates the magnetic field to strengths trillions of times greater than the magnetic field at Earth's surface. That magnetic field does not necessarily have its poles aligned with the rotation axis of the neutron star. A neutron star has a magnetosphere just like Earth does, except that the neutron star's magnetosphere is much stronger and rotates very fast.

Electrons and positrons become trapped in the magnetic-field lines of the neutron star and are "funneled" along the field toward the magnetic poles of the system. Any accelerating charged particle produces radiation, so those particles produce beams of radiation along the magnetic poles of the neutron star, as shown in **Figure 17.15**. As the neutron star rotates, those beams sweep through space like the rotating beams of a lighthouse. When Earth is in the paths of those beams, the neutron star appears to flash on and off regularly with a period equal to either the period of rotation of the star (if one beam is seen) or half the rotation period (if both beams are seen). That flashing light is the determining characteristic of a pulsar.

Binary Systems and Short Gamma-Ray Bursts

The massive star from which the neutron star formed may have been part of a binary system. Unless the massive star loses so much mass that gravity no longer holds the two stars together, the neutron star is left with a binary companion. Processes such as those in the white dwarf binary systems responsible for novae and Type Ia supernovae are possible. **Figure 17.16** illustrates an **X-ray binary**, a binary system in which mass from an evolving star spills over onto a collapsed companion such as a white dwarf, neutron star, or black hole. As the lower-mass star in such a binary system evolves and overfills its Roche lobe, matter falls toward the accretion disk around the neutron star, heating it to millions of kelvins and causing it to glow brightly in X-rays. X-ray binaries sometimes develop powerful jets of material that are perpendicular to the accretion disk and that carry material away

🎥 **Astronomy in Action:** Pulsar Rotation

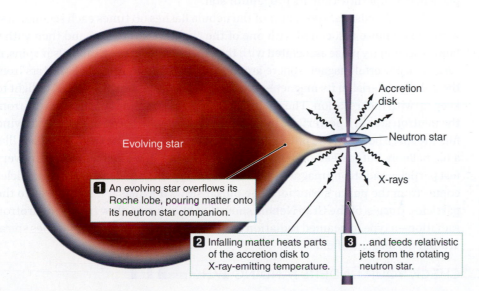

Accretion disk

Evolving star

Neutron star

X-rays

1 An evolving star overflows its Roche lobe, pouring matter onto its neutron star companion.

2 Infalling matter heats parts of the accretion disk to X-ray-emitting temperature.

3 ...and feeds relativistic jets from the rotating neutron star.

Figure 17.16 X-ray binaries are systems consisting of a normal evolving star with a white dwarf, a neutron star, or a black hole. As the evolving star overflows its Roche lobe, mass falls toward the collapsed object. The gravitational well of the collapsed object is so deep that when the material hits the accretion disk, the disk is heated to such high temperatures that it radiates away most of its energy as X-rays.

what if . . .

What if you could observe the Crab Nebula a million years from now? How do you think the nebula would appear, and what will be the fate of the rotating pulsar?

VIS   GAMMA
XRAY

Figure 17.17 This image of the Crab Nebula is a composite of images from across the electromagnetic spectrum, gathered from multiple space observatories. The outer layers show primarily line emission in the visible part of the spectrum, while the glowing purplish disk near the center is emitting X-rays from extremely hot gas. Gamma rays have also been detected coming from within this region. A pulsar lies at the center of the nebula.

at speeds close to the speed of light. If the neutron star or white dwarf accretes enough matter to become a nova, once or more than once, it is often called a "cataclysmic variable star" to describe how the observed brightness changes.

In 2013, the Hubble Space Telescope followed up on a gamma-ray observation, detecting infrared radiation from a "kilonova." Astronomers hypothesized that those kilonovae and the associated gamma-ray bursts were caused by the merger of two neutron stars. In August 2017, the Laser Interferometer Gravitational-Wave Observatory (LIGO) and a plethora of other observatories detected the collision of two neutron stars. That was the first time that astronomers observed the same object both in gravitational waves and in electromagnetic waves. That "multi-messenger" observation allowed astronomers to confirm that neutron star mergers are a cause of **short gamma-ray bursts**, bright gamma-ray events that last between a few milliseconds and two seconds. Follow-up observations have suggested that neutron star mergers may produce most of the gold and silver in the universe.

The Crab Nebula

In 1054, Chinese astronomers noticed a "guest star" in the direction of the constellation Taurus. The new star was so bright that it could be seen during the daytime for 3 weeks, and it did not fade away altogether for many months. From the Chinese description of the changing brightness and color of the object, modern astronomers have concluded that the guest star of 1054 was a fairly typical Type II supernova. Today, an expanding cloud of debris from the explosion occupies that place in the sky—forming an object called the Crab Nebula (**Figure 17.17**).

The Crab Nebula has filaments of glowing gas expanding away from the central star at 1500 km/s—50 times faster than the expansion rate of a planetary nebula. Those filaments contain unusually high abundances of helium and other more massive chemical elements—the result of the nucleosynthesis that took place in the supernova and its progenitor star.

The Crab pulsar at the center of the nebula flashes 60 times each second: first with a main pulse associated with one of the "lighthouse" beams and then with a fainter secondary pulse associated with the other beam. As the Crab pulsar spins, it carries its powerful magnetosphere around with it. A few thousand kilometers from the pulsar, material in its magnetosphere must move at almost the speed of light to keep up with that rotation. The rotating magnetosphere flings particles away from the neutron star in a powerful wind moving at nearly the speed of light. That wind fills the space between the pulsar and the expanding shell. The Crab Nebula is like a big balloon, but instead of being filled with hot air, it is filled with a mix of very fast particles and strong magnetic fields. The energy that accelerates those particles comes from the pulsar's rotation. The pulsar slows down as it loses energy to the particles. Images of the Crab Nebula show that bubble as a glow from synchrotron radiation—a type of beamed radiation emitted as very fast-moving particles spiral around the magnetic field.

CHECK YOUR UNDERSTANDING **17.3**

One reason astronomers think neutron stars were formed in supernova explosions is that: (a) all supernova remnants contain pulsars; (b) pulsars are made of heavy elements, such as those produced in supernova explosions; (c) pulsars spin very rapidly, as did the massive star just before it exploded; (d) pulsars sometimes have material around them that looks like the ejecta from supernovae.

17.4 Star Clusters Are Snapshots of Stellar Evolution

Recall from Chapter 15 that when an interstellar cloud collapses, it breaks into pieces, forming many stars of different masses. Those groups of gravitationally bound stars are called star clusters. **Globular clusters** are densely packed collections of tens of thousands to a few million stars (**Figure 17.18a**). **Open clusters** are much less tightly bound collections of a few dozen to a few thousand stars (**Figure 17.18b**). Because stars in a cluster form out of the same cloud at nearly the same time, observations of star clusters can be used to determine how stars of different masses evolve.

Cluster Distances and Ages

In the 1920s, astronomers plotted the observed brightness versus the spectral type for as many stars as possible in each cluster. The resulting cluster H-R diagrams showed stars of all the categories in the H-R diagram (see Figure 13.15). Because all the stars in a given cluster are approximately the same distance from Earth, the effect of distance on the brightness of each star is the same. Shifting the main sequence of the cluster up so that it overlaps the main sequence of the "textbook" H-R diagram indicates how much the brightness of the cluster has been diminished by distance. That is one method, called "main-sequence fitting," that astronomers can use to determine the distance to a cluster.

The H-R diagrams of clusters also offer snapshots of stellar evolution. All the stars in a cluster formed together at nearly the same time, so by the time the cluster is 10 million years old, all the stars in the cluster are 10 million years old. Comparing clusters of different ages shows which stars arrive on the main sequence first and which evolve off it first. **Figure 17.19a** shows an H-R diagram of a very young cluster called NGC 6530. O, B, and A stars are on the main sequence, whereas F through M stars with lower masses are still evolving *to* the main sequence. No

Figure 17.18 a. Globular clusters can have tens of thousands to a few million stars. **b.** Open clusters have up to hundreds or thousands of stars. These images are from the Hubble Space Telescope. Credit (part a.): ESA/Hubble & NASA, https://esahubble.org/images /potw1140a/. https://creativecommons .org/licenses/by/4.0/.

a.   VIS

b.   VIS

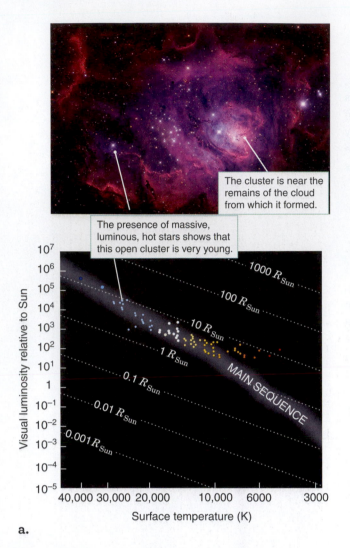

a.

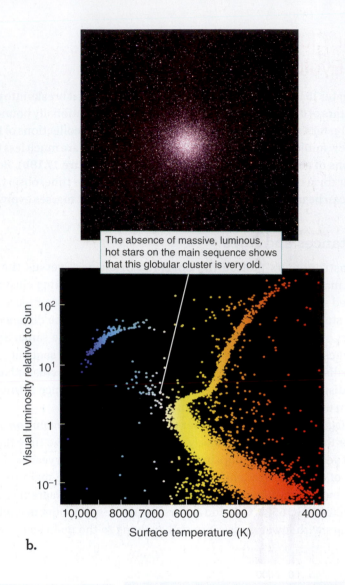

b.

Figure 17.19 H-R diagrams of **a.** a very young cluster (2 million years old) called NGC 6530 and **b.** a very old cluster (12 billion years old) called M55. In a, some of the stars haven't yet arrived on the main sequence. In b, more than the top half of the main sequence has already evolved. While a few "blue stragglers" remain, the vast majority of hot stars have left the main sequence. The vertical scales are logarithmic; b. is zoomed in compared with a.
Credit (part b. photo): ESO, https://www.eso.org/public/images/m55 -3point6-m_copy/. https://creativecommons.org/licenses /by/4.0/.

red giants or white dwarfs are present. In contrast, **Figure 17.19b** shows the H-R diagram of a very old cluster called M55. No high-mass (O and B) stars are on the main sequence because they have evolved off it, but stars are on the horizontal, red giant, and asymptotic giant branches and in the lower part of the main sequence. M55 is about 12 billion years old.

Astronomers cannot watch an individual cluster age over millions of years, but they can observe clusters of different ages. Astronomers explore cluster evolution by examining H-R diagrams of a *simulated* cluster of 40,000 stars as it would appear at several ages and then comparing it with the observed H-R diagrams of actual clusters. Star formation in a molecular cloud is spread over several million years, and it takes considerable time for lower-mass stars to contract to reach the main sequence. The H-R diagram of a very young cluster typically shows many lower-mass stars located well above the main sequence; eventually, they move onto the main sequence. In **Figure 17.20a**, we have ignored that complication to more clearly define the main sequence itself.

The more massive a star is, the shorter its life on the main sequence will be. **Figure 17.20b** shows that after only 4 million years, all stars with masses greater than about 20 M_{Sun} have evolved off the main sequence and are now spread across the top of the H-R diagram. The most massive stars have already disappeared from the H-R diagram, having vanished in supernovae. As time goes on, stars of lower

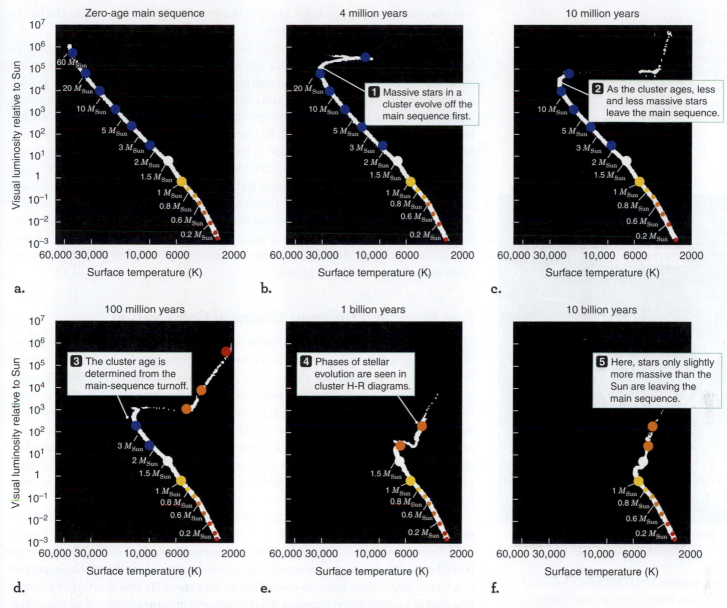

Figure 17.20 H-R diagrams of star clusters are snapshots of stellar evolution. These H-R diagrams of a simulated cluster of 40,000 stars of solar composition are shown at different times after the birth of the cluster. Note the progression of the main-sequence turnoff to lower and lower masses. In the simulation, the stars are all placed on the main sequence at zero age. In reality, however, the lowest-mass stars have not yet reached the main sequence by the time the most massive stars have left it.

and lower mass evolve off the main sequence, and the top of the main sequence moves toward the bottom right in the H-R diagram. By the time the cluster is 10 million years old (**Figure 17.20c**), only stars with masses less than about $15\,M_{Sun}$ remain on the main sequence. The location of the most massive star that remains on the main sequence is called the **main-sequence turnoff**. As the cluster ages, the main-sequence turnoff moves farther and farther down the main sequence to stars of lower and lower mass.

As a cluster ages further (**Figures 17.20d and 17.20e**), we see the details of all stages of stellar evolution. By the time the star cluster is 10 billion years old (**Figure 17.20f**), stars with masses of only $1\,M_{Sun}$ are beginning to die. Stars slightly more massive than that are seen as giant stars of various types. Note how few supergiant and giant stars are present in any of the cluster H-R diagrams. The supergiant, giant, horizontal, and asymptotic giant branch phases in the evolution of stars pass so quickly in comparison with a star's main-sequence lifetime that even though that simulated cluster started with 40,000 stars, only a handful of stars are seen in those phases of evolution. Similarly, even though most evolved stars in an old cluster are white dwarfs, all but a few of those stars will have cooled and faded

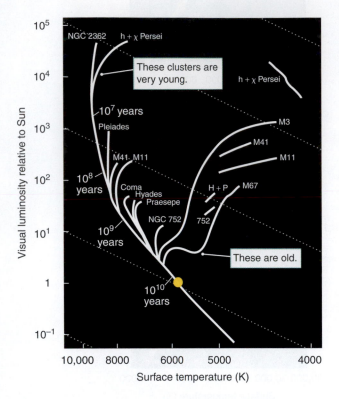

Figure 17.21 H-R diagrams for clusters having a range of ages. The ages associated with the different main-sequence turnoffs are indicated.

unanswered questions

Blue straggler stars, found in clusters, are bluer and brighter than the stars at the main-sequence turnoff point. How do they fit into the picture of stellar evolution? They may have resulted from mass transfer in a binary pair or from the merger of two single or two binary stars, either of which could have resulted in a more massive star than what might be expected from the age of the cluster. Astronomers study the environments of those stars by estimating the likelihood of collisions and the number of binary systems, which may be different in clusters where the density of stars is high.

into obscurity at any given time. More of those evolved stars are seen in the larger globular cluster M55 in Figure 17.19b.

To astronomers observing a star cluster, the location of the main-sequence turnoff immediately indicates the age of the cluster. **Figure 17.21** traces the observed H-R diagrams for several real star clusters. Once you know what to look for, the difference between young and old clusters is obvious. NGC 2362 is a young cluster. Its complement of massive, young stars on the main sequence shows it to be only a few million years old. In contrast, cluster M3 has a main-sequence turnoff that indicates its cluster age is about 11 billion years. When the H-R diagrams of open clusters are studied, a wide range of ages is observed. Some open clusters contain the short-lived O and B stars and are therefore very young. Other open clusters contain stars older than the Sun. But even the youngest globular clusters are several billion years older than the oldest open clusters. Open clusters tend to be young because their stars are loosely bound and gradually leave the cluster. Globular clusters, in contrast, are tightly bound by gravity, allowing them to survive for billions of years.

Age, Color, and Different Chemical Composition

That understanding of stellar evolution applies even when groups of stars are so far away that individual stars cannot be seen. The light from a star cluster is dominated by its most luminous stars—namely, by massive, short-lived blue O and B main-sequence stars and evolved post-main-sequence supergiant and giant stars. If the cluster is young, most of the light comes from luminous, hot, blue stars and some red supergiants. If the cluster is old, the light from the cluster has the color of red giants and red dwarf stars. However, that relationship can be complicated by chemical composition—stars with smaller amounts of massive elements in their atmospheres often look significantly bluer than stars that formed from chemically enriched material.

A group of stars with similar ages and other shared characteristics is called a **stellar population**. The link between color and specific characteristics becomes useful when we begin to discuss the much larger collections of stars called galaxies (see Chapter 19). Astronomers usually can figure out something about the properties of a stellar population from its overall color. An especially bluish color to a galaxy or a part of a galaxy often signifies that the galaxy contains a young stellar population that still includes hot, luminous, blue stars that formed recently. In contrast, a galaxy or part of a galaxy that has a reddish color is usually composed primarily of an old stellar population.

All elements more massive than boron are formed in stars and are expelled into the interstellar medium when a star dies. In main-sequence stars, material from the core does not mix with material in the atmosphere, so the abundances of chemical elements inferred from the spectrum of a star are the same as the abundances in the interstellar gas from which the star formed. The abundance of massive elements in the interstellar medium serves as a record of the cumulative amount of star formation that has taken place so far. Gas that shows large abundances of massive elements has gone through a great deal of stellar processing—so it contains more "recycled" material. Gas with low abundances of massive elements is more pristine.

Stars in globular clusters contain only very small amounts of massive elements; some globular-cluster stars contain only 0.5 percent as much of those massive elements as the Sun does, indicating that they were among the earliest stars to form.

Open clusters are younger and formed from a more enriched interstellar medium. Therefore, they have larger amounts of the more massive elements.

Even the very oldest globular-cluster stars contain massive chemical elements. Therefore, at least one generation of massive stars must have lived and died, ejecting newly synthesized massive elements into space, before even those oldest globular clusters formed. Most of the luminous matter in the universe is still composed of hydrogen and helium formed before the first stars. Even a chemically rich star like the Sun, which is made of gas processed through approximately 9 billion years of previous generations of stars, is composed of less than 2 percent massive elements. In upcoming chapters, we explain that those variations in the chemical content of stars indicate a lot about the chemical evolution of galaxies.

CHECK YOUR UNDERSTANDING **17.4**

If the main-sequence turnoff of a globular cluster occurs near the very bottom of the main sequence, the cluster is: (a) very old; (b) very young; (c) very hot; (d) very dense.

Origins: Seeding the Universe with New Chemical Elements

Massive elements are everywhere on Earth. The surfaces of rocky planets contain silicon, oxygen, magnesium, and sodium. The iron-and-nickel solid inner core and liquid outer core of Earth are responsible for Earth's magnetic field. The most common chemical elements in biological molecules are carbon (C), hydrogen (H), nitrogen (N), oxygen (O), phosphorus (P), and sulfur (S)—all but hydrogen were most likely created in a dying star. The fact that those elements are here means that the Sun is not a first-generation star. Instead, it formed from material in the interstellar medium enriched by material from dying massive stars.

In supergiant, giant, and AGB stars, more massive chemical elements that formed from nuclear fusion deep within their interiors are carried upward and mixed with material in the outer parts of the star. As a star ages, its core grows hotter, and the temperature gradient within the star grows steeper. Under certain circumstances, convection can spread so deep into a star that chemical elements formed by nuclear fusion within the star are dredged up and carried to the star's surface. For example, the spectra of some AGB stars show an excess of carbon and other by-products of nuclear fusion. That extra carbon originated in each star's helium-fusing shell and was carried to the surface by convection. For stars with lower masses, stellar winds and planetary nebulae carry the enriched outer layers off into interstellar space. Nuclear fusion goes beyond the formation of elements such as carbon. Supernova explosions and merging neutron stars seed the universe with much more massive atoms, from iron and nickel up to uranium.

The oxygen atoms in the air you breathe and the water you drink were created by nucleosynthesis in dying stars. The iron atoms that are a key element of hemoglobin, which makes up the red blood cells that carry oxygen from your lungs to the rest of your body, formed in the explosions of massive stars. The nickel, copper, and zinc atoms in the coins in your pocket were formed in supernovae, and the rare-earth atoms in your electronics were created in the mergers of neutron stars. The Sun, the planets (including Earth), and all life on Earth are made of recycled stars. Dying stars are in you.

SUMMARY

As high-mass stars evolve, their interiors form concentric shells of progressive nuclear fusion. Once those stars leave the main sequence, they may pass through the instability strip and become pulsating variable stars. High-mass stars eventually explode as Type II supernovae, which eject newly formed massive elements into interstellar space. The supernova explosion that ends the life of a massive star leaves behind a neutron star that contains between about 1 and 2 M_{Sun} of neutron-degenerate matter packed into a sphere 10–15 km in diameter. Pulsars are rapidly spinning, magnetized neutron stars. Accretion of mass onto neutron stars produces X-rays in some binary systems. Binary neutron stars sometimes merge, producing short gamma-ray bursts. The Sun, the Solar System, Earth, and all life on Earth contain heavy elements created in earlier generations of short-lived massive stars.

1 **Arrange the sequence of stages for evolving high-mass stars.** Evolving high-mass stars leave the main sequence as they fuse heavier elements. Once an iron core is produced, the star becomes unstable and the core collapses, heating the material to cause photodisintegration of the iron nuclei and the merging of protons and electrons into neutrons. The outer layers bounce off the dense core and produce a shock wave that travels outward. That shock wave causes neutrons to fuse with atomic nuclei and form more massive elements.

2 **Compare the death of high-mass stars to that of low-mass stars.** The larger masses of high-mass stars allow them to fuse heavier elements than those produced in low-mass stars. That process leads to a more violent death that leaves massive cores behind.

3 **Describe how stars create chemical elements heavier than hydrogen and helium using nucleosynthesis and capture processes.** The chain of nuclear fusion reactions consists of increasingly shorter stages, resulting in more massive elements up to iron. However, those elements are destroyed in the set of processes that define the core collapse. Heavy elements in the universe today were formed during the rebound explosion of massive stars as high-energy neutrons penetrated atomic nuclei. Those neutrons then decay to protons, creating new elements with higher atomic numbers.

4 **Summarize how Hertzsprung-Russell (H-R) diagrams of clusters enable astronomers to measure the ages of stars and test theories of stellar evolution.** Clusters are groups of stars born together, so they are all at about the same distance from Earth. H-R diagrams of clusters show stars leaving the main sequence in a progression from the highest-mass stars to the lowest-mass stars, confirming theories of stellar evolution. The location of the main-sequence turnoff indicates the age of the cluster.

QUESTIONS AND PROBLEMS

TEST YOUR UNDERSTANDING

1. Why does the interior of an evolved high-mass star have layers like those of an onion?
   a. Heavier nuclei sink to the bottom because stars are not solid.
   b. Before the star formed, heavier atoms accumulated in the centers of clouds because of gravity.
   c. Heavier nuclei fuse closer to the center because the temperature and pressure are higher there.
   d. Different energy transport mechanisms occur at different densities.

2. Arrange the following elements in the order they fuse inside the nucleus of a high-mass star during the star's evolution.
   a. helium
   b. neon
   c. oxygen
   d. silicon
   e. hydrogen
   f. carbon

3. Elements heavier than iron originated
   a. in the Big Bang.
   b. in the cores of low-mass stars.
   c. in the cores of high-mass stars.
   d. in the supernova explosions of high-mass stars.

4. A pulsar pulses because
   a. its spin axis crosses Earth's line of sight.
   b. it spins.
   c. it has a strong magnetic field.
   d. its magnetic axis crosses Earth's line of sight.

5. Study the H-R diagram in Figure 17.5. If you could watch a high-mass star move to the right, along one of those post-main-sequence lines, what would you observe happening to the star's color? ★
   a. It would become redder.
   b. It would become bluer.
   c. It would remain the same.

6. Study the H-R diagram in Figure 17.5. If you could watch a high-mass star move to the right, along the topmost of those post-main-sequence lines, what would you observe happening to the star's size? ★
   a. It would become much larger.
   b. It would become much smaller.
   c. It would remain the same.

7. Study Figure 17.20. If the Sun were a member of a globular cluster, that cluster's H-R diagram would fall between ★
   a. (a) and (b).
   b. (b) and (c).
   c. (c) and (d).
   d. (d) and (e).
   e. (e) and (f).

8. In a high-mass star, the dominant hydrogen fusion process is
   a. the proton-proton chain.
   b. the CNO cycle.
   c. gravitational collapse.
   d. spin-spin interaction.

9. The layers in a high-mass star occur roughly in order of
   a. atomic number.
   b. decay rate.
   c. magnetic field strength.
   d. spin state.

10. Eta Carinae is an extreme example of
    a. a massive star.
    b. a planetary nebula.
    c. a supernova remnant.
    d. an ancient star.

11. Iron fusion cannot support a star because
    a. iron oxidizes too quickly.
    b. iron absorbs energy when it fuses.
    c. iron emits energy when it fuses.
    d. iron is not dense enough to hold up the layers.

12. The start of photodisintegration of iron in a star sets off a process that *always* results in a
    a. supernova.
    b. neutron star.
    c. supergiant.
    d. pulsar.

13. The Crab Nebula serves as a test of our ideas about supernova explosions because
    a. the system contains an X-ray binary.
    b. the nebula is slowly expanding.
    c. the supernova was observed in 1054 and now astronomers see a pulsar in the nebula.
    d. the original star was like the Sun before exploding.

14. What mechanism supplies the pressure inside a neutron star?
    a. ordinary pressure from hydrogen and helium gas
    b. degeneracy pressure from neutrons
    c. degeneracy pressure from electrons
    d. neutrino pressure

15. Very young star clusters have main-sequence turnoffs
    a. that drop below the main sequence.
    b. at the top left of the main sequence.
    c. at the bottom right of the main sequence.
    d. in the middle of the main sequence.

THINKING ABOUT THE CONCEPTS

16. ★ **WHAT AN ASTRONOMER SEES**   All six of the images in Figure 17.13 are the same size. Study the bumps in the large bright ring. Do these bumps change location over the two decades shown here, or do they simply change in brightness? Does the ring change size during this time? ★

17. Explain the differences between the ways that hydrogen is converted to helium in a low-mass star (proton-proton chain) and in a high-mass star (CNO cycle). What is the catalyst in the CNO cycle, and how does it take part in the reaction?

18. How does a high-mass star begin fusing helium in its core? How is that process different from what happens in low-mass stars?

19. Why does the core of a high-mass star not become degenerate, as the core of a low-mass star does?

20. List the two reasons why each post-helium-fusion cycle for high-mass stars (carbon, neon, oxygen, silicon, and sulfur) becomes shorter than the preceding cycle.

21. Cepheids are highly luminous, variable stars in which the period of variability is directly related to luminosity. Why are Cepheids good indicators for determining stellar distances?

22. Identify and explain two important ways in which supernovae influence the formation and evolution of new stars.

23. Study the Process of Science Figure. Why is that pulsar explanation "simpler"? ★

24. Describe what an observer on Earth will witness when Eta Carinae explodes.

25. Recordings show that neutrinos from SN 1987A were detected on February 23, 1987. About 3 hours later the supernova was detected in optical light. What caused the delay?

26. Why can the accretion disk around a neutron star release so much more energy than the accretion disk around a white dwarf, even though the two stars have approximately the same mass?

27. In Section 17.2, you learned that Type II supernovae blast material outward at up to 30,000 km/s. The material in the Crab Nebula described in Section 17.3 is expanding at only 1500 km/s. What explains the difference?

28. An experienced astronomer can take one look at the H-R diagram of a star cluster and immediately estimate its age. How is that possible?

29. Explain how astronomers know that an even earlier generation of stars existed before the oldest observed stars.

30. What is the binding energy of an atomic nucleus? How does that quantity help astronomers calculate the energy given off in nuclear fusion reactions?

APPLYING THE CONCEPTS

31. In the proton-proton chain, the binding energy of the final helium nucleus is 6.824×10^{14} joules (J) per kilogram (kg) of helium. The binding energy of the protons is 0 J/kg. How much energy is available from fusing protons to form 1 kg of helium nuclei?
    a. Make a prediction: Study the graph in Figure 17.8. Do you expect the energy available through fusing hydrogen into helium to be more or less than the amount of energy from fusing helium into carbon? ★
    b. Calculate: Follow Working It Out 17.1 to find the energy produced when fusing protons to form 1 kg of helium nuclei.
    c. Check your work: Compare your answer to your prediction. ★

32. What is the surface gravity on a neutron star with a radius of 10 km and a mass of 2.8 M_{Sun}? ●–●–●

   a. Make a prediction: Do you expect the surface gravity to be higher or lower than the surface gravity on Earth (9.8 m/s²)? Do you expect the surface gravity to be higher or lower than the example in Working It Out 17.2?

   b. Calculate: Follow the first half of Working It Out 17.2 to find the surface gravity on this neutron star.

   c. Check your work: Verify that your answer has the correct units. Compare your answer to your prediction and to the value in Working It Out 17.2 to evaluate the "size" of your answer.

33. What is the escape velocity on a neutron star with a radius of 10 km and a mass of 2.8 M_{Sun}?

   a. Make a prediction: Do you expect the escape velocity to be higher or lower than the example in Working It Out 17.2?

   b. Calculate: Follow the second half of Working It Out 17.2 to find the escape velocity from this neutron star.

   c. Check your work: Verify that your answer has the correct units. Compare your answer to the value in Working It Out 17.2 to evaluate the "size" of your answer.

34. It can be instructive to work the escape velocity problem backward. Imagine an object that has a mass of 2 M_{Sun} and an escape velocity equal to the speed of light (c = 300,000 km/s). What is the radius of this object?

   a. Make a prediction: Do you expect this object's radius to be larger or smaller than the 15-km radius of the neutron star in Working It Out 17.2?

   b. Calculate: Use the formula from Working It Out 17.2 to calculate the radius of this object.

   c. Check your work: Verify that your answer has units of km, and compare your answer to your prediction.

   d. Can an object located at that radius ever escape from the object? Why or why not?

35. Suppose you observe a classical Cepheid variable with a period of 10 days. What is the luminosity of that star? What other piece of information would you need to find out how far away the star is?

   a. Make a prediction: Where are Cepheid variables located on the H-R diagram? Do you expect to find a luminosity that is comparable to the Sun, much brighter, or much fainter?

   b. Calculate: The approximate relationship between the luminosity and the period of Cepheid variables is L_{star} (in L_{Sun}) = 335 P (in days). Find the luminosity of this Cepheid variable.

   c. Check your work: Compare the luminosity you found with your prediction and with the approximate luminosity of a Cepheid variable with a 10-day period in the graph in Figure 17.6b. ✦

36. Study the graphs in Figure 17.2. What fractions of the star are helium at t = 0 years and at t = 7 million years? ◉

37. Study the H-R diagram in Figure 17.5. How much hotter, larger, and more luminous than the Sun is the uppermost main-sequence star? ★

38. Study the sequence in Figure 17.4. Are the radius of the core and the radius of the star represented to scale? What fraction of the star's radius is the core's radius? ★

39. If the Crab Nebula has been expanding at an average velocity of 1500 km/s since the year 1054, what was its average radius in the year 2019? (Note: A year has approximately 3 × 10⁷ seconds.)

40. Figure 17.14 shows the relative abundances of the elements. Is that graph a log or a linear plot? Explain what it means that oxygen lies on the y-axis at 10⁻³. ★

41. The Milky Way has about 50,000 stars of average mass (0.5 M_{Sun}) for every main-sequence star of 20 M_{Sun}. But 20-M_{Sun} stars are about 10,000 times as luminous as the Sun, and 0.5-M_{Sun} stars are only 0.08 times as luminous as the Sun.

   a. How much more luminous is a single massive star than the total luminosity of the 50,000 less massive stars?

   b. How much more mass is in the lower-mass stars than the single high-mass star?

   c. Which stars—lower-mass or higher-mass stars—contain more mass in the galaxy, and which produce more light?

42. In a large outburst in 1841, the 120-M_{Sun} star Eta Carinae was losing mass at a rate of 0.1 M_{Sun} per year.

   a. The mass of the Sun is 2 × 10³⁰ kg. How much mass (in kilograms) was Eta Carinae losing each minute?

   b. The mass of the Moon is 7.35 × 10²² kg. How does Eta Carinae's mass loss per minute compare with the mass of the Moon?

43. An O star can lose 20 percent of its mass during its main-sequence lifetime. Estimate the average mass loss rate (in solar masses per year) of a 25-M_{Sun} O star with a main-sequence lifetime of 7 million years.

44. The approximate relationship between the luminosity and the period of Cepheid variables is L_{star} (in L_{Sun}) = 335 P (in days). Delta Cephei has a cycle period of 5.4 days and a parallax of 0.0033 arcsecond (arcsec). A more distant Cepheid variable appears 1/1000 as bright as Delta Cephei and has a period of 54 days.

   a. How far away (in parsecs) is the more distant Cepheid variable?

   b. Could the distance of the more distant Cepheid variable be measured by parallax? Explain.

45. Estimate the size of a neutron star that has the mass of the Sun.

EXPLORATION The CNO Cycle

digital.wwnorton.com/astro7

Nuclear reactions usually involve many steps. In the Exploration for Chapter 14, you investigated the proton-proton chain. In this Exploration, you will study the CNO cycle, which is even more complex. Visit the Digital Resources Page and on the Student Site open the "CNO Cycle" Interactive Simulation in Chapter 17.

First, press "Play" and watch the animation all the way through. Next, press "Reset" to clear the screen, and then press "Play" again, allowing the animation to proceed past the first collision before pressing "Pause."

1 Which atomic nuclei are involved in that first collision?

_____

2 What color is used to represent the proton (hydrogen nucleus, ^1H)?

_____

3 What does the blue squiggle represent?

_____

4 What atomic nucleus is created in the collision?

_____

5 The resulting nucleus is not the same type of element as either of the two that entered the collision. Why not?

_____

_____

Press "Play" again, and then pause as soon as the yellow ball and the dashed line appear.

6 Is that a collision or a spontaneous decay?

_____

7 What does the yellow ball represent?

_____

8 What does the dashed line represent?

_____

9 The resulting nucleus has the same number of nucleons (13), but it is a different element. (Remember that both protons and neutrons are nucleons.) What happened to the proton that was in the nitrogen nucleus but is not in the carbon nucleus?

_____

Proceed past the next two collisions, to "^{15}O."

10 Study the pattern that is forming. When a blue ball comes in, what happens to the number of nucleons and the type of the nucleus (that is, what happens to the "12" and the "C," or the "14" and the "N")?

_____

11 What is emitted in those collisions?

_____

Proceed until "^{15}N" appears.

12 Is that a collision or a spontaneous decay?

_____

13 Which previous reaction is that most like?

_____

Now proceed to the end of the animation.

14 After the final collision, a line is drawn back to the beginning, telling you what type of nucleus the upper red ball represents. What is that nucleus?

_____

15 How many nucleons are not accounted for by the upper red ball? (Hint: Don't forget the ^1H that came into the collision.) Those nucleons must be in the nucleus represented by the bottom red ball.

_____

16 Carbon has six protons. Nitrogen has seven. How many protons are in the nucleus represented by the bottom red ball?

_____

17 How many neutrons are in the nucleus represented by the bottom red ball?

_____

18 What element does the bottom red ball represent?

_____

19 What is the net reaction of the CNO cycle? That is, which nuclei are combined and turned into the resulting nucleus?

_____

20 Why is ^{12}C not considered part of the net reaction?

_____

_____

Relativity and Black Holes

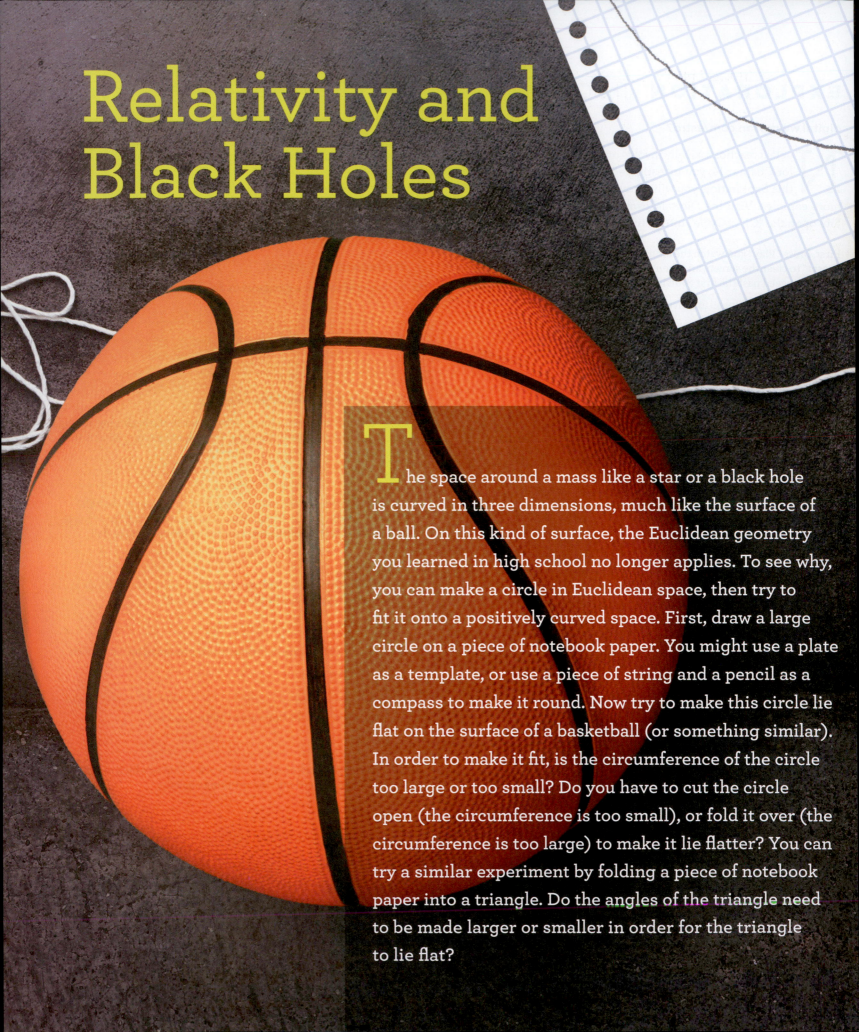

The space around a mass like a star or a black hole is curved in three dimensions, much like the surface of a ball. On this kind of surface, the Euclidean geometry you learned in high school no longer applies. To see why, you can make a circle in Euclidean space, then try to fit it onto a positively curved space. First, draw a large circle on a piece of notebook paper. You might use a plate as a template, or use a piece of string and a pencil as a compass to make it round. Now try to make this circle lie flat on the surface of a basketball (or something similar). In order to make it fit, is the circumference of the circle too large or too small? Do you have to cut the circle open (the circumference is too small), or fold it over (the circumference is too large) to make it lie flatter? You can try a similar experiment by folding a piece of notebook paper into a triangle. Do the angles of the triangle need to be made larger or smaller in order for the triangle to lie flat?

EXPERIMENT SETUP

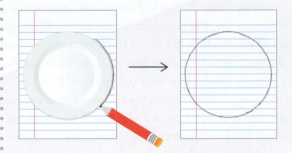

First, draw a large circle on a piece of notebook paper. Use something like a plate as a template, or use a piece of string and a pencil as a compass to make it round.

Cut the circle out from the piece of paper.

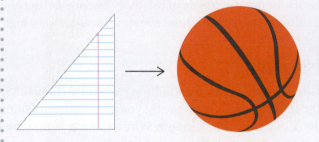

Now try to make this circle lie flat on the surface of a basketball (or something similar). In order to make it fit, is the circumference of the circle too large or too small?
Do you have to cut the circle open (the circumference is too small), or fold it over (the circumference is too large) to make it lie flatter?

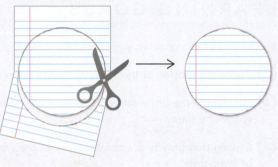

You can try a similar experiment by folding a piece of notebook paper into a triangle. Do the angles of the triangle need to be made larger or smaller in order for the triangle to lie flat?

PREDICTION

I predict that the circumference of the circle will be
☐ **too large**   ☐ **too small**
for the circle to lie flat on the ball.

18

Some stars leave behind a black hole at the end of their lives. Black holes have conditions so extreme that the laws of Newtonian physics can't describe them. To discuss black holes, we must understand how Albert Einstein changed the way physicists thought about the nature of space and time in the early 20th century. According to Einstein's *special* theory of relativity, matter behaves very differently when it is traveling near the speed of light. According to his *general* theory of relativity, space itself is very warped near very massive objects. That warping of space is so extreme at a black hole that nothing can travel fast enough to escape. Here in Chapter 18, we move beyond Newtonian ideas of space and time to understand black holes.

LEARNING GOALS

By the end of this chapter, you should be able to:

1. Relate the motion of the observer to the observed motion of other objects.

2. Summarize the observable consequences of the relationship between space and time.

3. Explain that gravity is a consequence of the way mass distorts the very shape of spacetime.

4. Explain the formation of black holes from the most massive stars, and describe the key properties and observational consequences of those stellar black holes.

18.1 Relative Motion Affects Measured Velocities

All observers, whatever their motion, measure the same speed of light. That observed fact has profound implications for relative motion, space, and time. In this section, we lay the groundwork for our discussion of relativity by considering how relative motion affects measurements and how those effects increase at very high speeds.

Aberration of Starlight

Imagine that you are sitting in a car in a windless rainstorm (**Figure 18.1**). If the car is stationary and the rain is falling vertically, you see raindrops falling straight down when you look out your side window. When the car is moving forward, however, the situation is different. Between the time a raindrop appears at the top of your window and the time it disappears beneath the bottom of your window, the car has moved forward. The raindrop disappears beneath the window behind the point at which it appeared at the top of the window, which means the raindrop looks as though it is falling at an angle, even though in reality it is falling straight down. As you go faster, the apparent front-to-back slant of the raindrops increases, and their apparent paths become more angled. An observer by the side of the road would say the raindrops are coming from directly overhead, but to you in the moving car they are coming from a direction in front of the car. You are observing that apparent motion of the raindrops from within your own unique frame of reference. (The observer by the side of the road has her own unique frame of reference, too.)

a.

b.

Figure 18.1 On a windless day, the direction in which rain falls depends on the frame of reference in which it is viewed. **a.** From outside the car, the rain is seen to fall vertically downward whether the car is stationary or moving. **b.** From inside the car, the rain is seen to fall vertically downward if the car is stationary; but if the car is moving, the rain is seen to fall at an angle determined by the speed and direction of the car's motion.

The light from a distant star arrives at Earth from the direction of the star (**Figure 18.2**). However, just as the raindrops appeared to be coming from in front of the moving car in Figure 18.1b, an observer on the moving Earth sees the starlight coming from a slightly different direction. Because the direction of Earth's motion around the Sun continuously changes during the year, the apparent position of a star in the sky moves in a small loop, a phenomenon known as the **aberration of starlight**. That shift in apparent position was first detected in the 1720s by the astronomers Samuel Molyneux and James Bradley. Measurement of the aberration of starlight shows that Earth moves on its path about the Sun with an average speed of just under 30 kilometers per second (km/s). Because distance equals speed multiplied by time, astronomers used that measurement to determine the circumference, and therefore the radius, of Earth's orbit. The speed of Earth (29.8 km/s) multiplied by the number of seconds in 1 year (3.16×10^7 seconds) gives a circumference of 9.42×10^8 km. Astronomers used that circumference to estimate the radius of Earth's orbit (1.5×10^8 km). The aberration of starlight is an astronomical example of how relative motion affects a measurement.

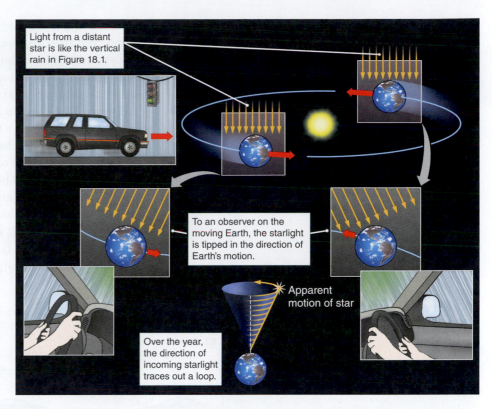

Figure 18.2 When we observe stars, their apparent positions are deflected slightly toward the direction in which Earth is moving. As Earth orbits the Sun, stars appear to trace out small loops in the sky. That effect is called *aberration of starlight*.

Relative Speeds Close to the Speed of Light

Every observer inhabits a reference frame, a set of coordinates in which the observer measures distances and speeds. A nonaccelerating reference frame, in which no net force is acting, is known as an **inertial reference frame**. Newton's laws of motion predict how the motion of an object will be measured by observers in different inertial reference frames. For example, the observed motion of a ball thrown from a moving car depends on the reference frame of the observer (recall the analogy for the Coriolis effect in Figure 2.12). If light behaved like other objects, the speed of light should differ from one observer to the next as a result of the observer's motion, just like the speed of a ball thrown from a moving car. However, the results of laboratory experiments with light in the late 19th century showed something very different. Scientists found instead that *all* observers measure the same value for the speed of light, regardless of the observer's frame of reference.

Imagine that you are in a red car traveling at 100 kilometers per hour (km/h) relative to the ground (**Figure 18.3a**), and you throw a ball at a speed of 50 km/h out the window at an oncoming green car, also traveling at 100 km/h relative to the ground. An observer standing by the side of the road watches the entire event. In *your* reference frame (Figure 18.3a, top panel), the red car is stationary; that is, you and the car are moving together, so the car does not move relative to you. In your reference frame, then, you measure the ball traveling at 50 km/h. To the observer standing by the road (Figure 18.3a, middle panel), however, the ball is moving at 150 km/h (50 km/h from throwing it plus 100 km/h from the motion of the car). To passengers in the oncoming green car moving at 100 km/h (Figure 18.3a, bottom panel), the speed of the

what if . . .

What if the speed of light were much smaller, say 200 km/hr. How might that affect a game of soccer?

ball is 250 km/h because the ball approaches them at 150 km/h relative to the ground, and they approach the ball at 100 km/h. As a result, the speed of the ball depends on how the observer, the cars, and the ball are moving relative to one another. The velocities add together to yield the velocity of one object relative to another. That is Galilean relativity, which you use in everyday life.

Figure 18.3b shows how light differs from the ball in the previous example. Imagine that you are riding in the yellow spaceship at half the speed of light (0.5c) and you shine a beam of laser light forward (Figure 18.3b, top panel). You measure the speed of the beam of light to be c, or 3×10^8 meters per second (m/s)—as expected because you are holding the source of the light. But the observer on the planet also measures the speed of the passing beam of light to be c, not 1.5c, which would be the sum of the speed of light plus the speed of the spaceship (Figure 18.3b, middle panel). Even a passenger in an oncoming blue spacecraft traveling at 0.5c (Figure 18.3b, bottom panel) finds that the beam from your light is traveling at exactly c in her own reference frame, and not

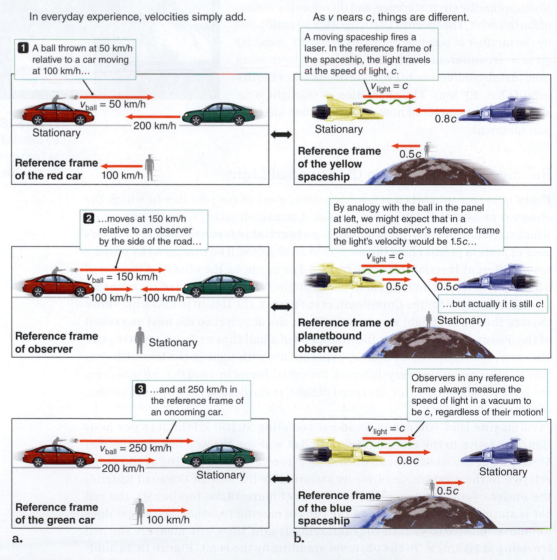

Figure 18.3 a. The Newtonian rules of motion apply in daily life; however, those rules break down when speeds approach the speed of light. **b.** Light itself always travels at the same speed for any observer, and that fact is the basis of special relativity, which implies that velocities don't simply add together, as they do in the Newtonian world.

at 2.0*c*, which would be the sum of the speeds of the spacecraft and the laser light. At speeds close to the speed of light, speeds do not add simply. That statement is true not only for light but also for all ordinary objects moving at nearly the speed of light. According to Einstein's relativistic formulas, the relative speed between the two spacecraft in the top and bottom panels of Figure 18.3b (0.5*c* + 0.5*c*) adds to 0.8*c*, not 1.0*c*! At **relativistic speeds** (speeds close to the speed of light), everyday experience no longer holds true. Every observer always finds that light in a vacuum travels at the same speed, *c*, regardless of his or her own motion or the motion of the light source. That observation directly conflicts with Newtonian theory and Galilean relativity.

CHECK YOUR UNDERSTANDING 18.1

Which beam of light is moving faster? (a) the one from the headlight of a parked car; (b) the one from the headlight of a moving car; (c) the one from the headlight of a moving spaceship; (d) all the beams are moving at *c*.

Answers to Check Your Understanding questions are in the back of the book.

18.2 Special Relativity Explains How Time and Space Are Related

Albert Einstein's first scientific paper, written when he was a 16-year-old student, was about traveling along with a light wave, moving in a straight line at constant speed. Einstein reasoned that according to Newton's laws of motion, you should be able to "keep up" with light so that you are moving right along with it. In that inertial reference frame, the light is stationary—it is an oscillating electric and magnetic wave that does not move. However, Maxwell's equations for electromagnetic waves (discussed in Chapter 5) are actually the same in all inertial frames. So Einstein took a radical new approach. Instead of starting with preconceived ideas about space and time, Einstein started with the observed fact that light always travels at the same speed, and then he reasoned backward to find out what that must imply about space and time. That reasoning led to the 1905 publication of his **special theory of relativity**, sometimes called *special relativity*, which describes the effects of traveling at constant speeds close to the speed of light. In this section, we explore special relativity and some of its consequences.

Time and Relativity

In developing special relativity, Einstein focused his thinking on pairs of *events*. An **event** is something that happens at a particular location in space at a particular time. Snapping your fingers is an event because that action has both a time and a place. Everyday experience indicates that the distance between any two events depends on the reference frame of the observer. Imagine you are sitting in a car that is traveling on the highway in a straight line at a constant speed of 60 km/h. You snap your fingers (event 1), and a minute later you snap your fingers again (event 2). In your reference frame you are stationary, so the two events happened at the same place—in the car. The events were, however, separated by a minute in time. What you saw is very different from what happens in the reference frame of an observer sitting by the road. That observer agrees that the second snap of your fingers (event 2) occurred a minute after the first snap of your fingers (event 1), but to that observer the two events were separated from each other in space by a kilometer, the distance your car traveled in the minute between snaps. In that "Newtonian" view, the distance between two events depends on the motion of the observer, but the *time* between the two events does not. Special relativity says instead that both the distance *and* the time between events vary depending on the motion of the observer.

The notion that different observers will measure time differently is a *very* counterintuitive idea, but it is central to special relativity and therefore to our scientific understanding of the universe as well. To see how Einstein arrived at the concept of relative time, consider his thought experiment known as the boxcar experiment. In that experiment, observer 1 is in a boxcar of a train moving to the right. Observer 1 has a lamp, a mirror (mounted on the roof of the boxcar), and a clock. Observer 2 is standing on the ground outside. The clock is based on a value that everyone can agree on—such as the speed of light.

Figure 18.4a shows the experimental setup as seen by observer 1, who is stationary with respect to the clock. At time t_1, event 1 happens: the lamp gives off a pulse of light. The light bounces off a mirror at a distance ℓ meters away and then heads back toward its source. At time t_2, event 2 happens: the light arrives at the clock and is recorded by a photon detector. (Note that the light beam is shown leaving and arriving at two locations. The artist did that so that you can see both events in the figure. Both events actually occur at the same location.) The time between events 1 and 2 is just the distance the light travels (2ℓ meters) divided by the speed of light: $t_2 - t_1 = 2\ell/c$.

Figure 18.4b shows the experiment as seen by observer 2, who is stationary on the ground outside the train, which is moving at speed v. In observer 2's reference

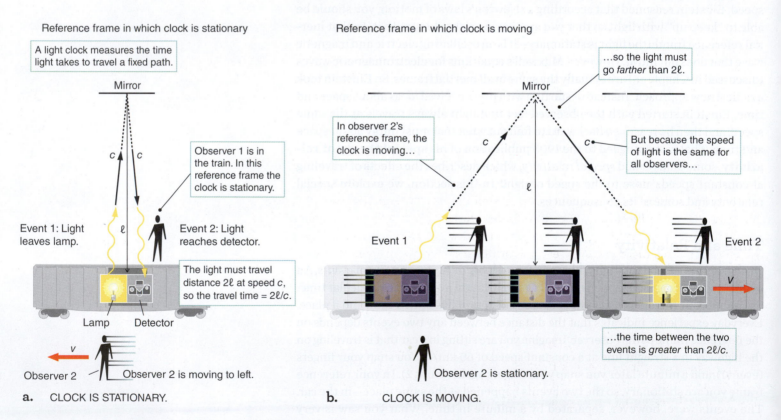

Figure 18.4 The "tick" of a light clock is different when seen in two reference frames: **a.** stationary, as in the reference frame of observer 1 on the boxcar, and **b.** moving, as in the reference frame of observer 2 on the tracks. As Einstein's thought experiment demonstrates, if the speed of light is the same for every observer, moving clocks *must* run more slowly than stationary clocks.

frame, the clock moves to the right between the two events, so the light has farther to go because of the horizontal distance. The time between the two events is still the distance traveled divided by the speed of light, but now that distance is *longer* than 2ℓ meters. Because the speed of light is the same for all observers, the time between the two events must be longer as well.

The two events are the same two events, regardless of the reference frame from which they are observed. Because the speed of light is the same for all observers, more time must pass between the two events when viewed from a reference frame in which the clock is moving (observer 2). The seconds of a moving clock are stretched. That is, a moving clock takes more time than a stationary clock to complete one "tick." Therefore, the passage of time must depend on an observer's frame of reference. Because both frames of reference are equally good places to do physics, both time measurements are valid in their own frames, even though they differ from each other.

In that experiment, light travels farther between events in a moving boxcar than between events in a stationary boxcar and consequently takes longer to travel between the events in the moving boxcar. Einstein realized that the *only* way the speed of light can be the same for all observers is *if the passage of time is different from one observer to the next*. For moving observers, the time is stretched out, so that the time interval between events is longer, a phenomenon known as **time dilation**.

The Newtonian view of the world describes a three-dimensional space through which time marches steadily onward. Einstein discovered, however, that time flows differently for different observers. He reshaped the three dimensions of space and the one dimension of time into a four-dimensional combination called **spacetime**. Events occur at specific locations within that four-dimensional spacetime, but how much of a spacetime distance is measured in space and how much is measured in time depends on the observer's reference frame.

Einstein did not "disprove" Newtonian physics. At speeds much less than the speed of light, Einstein's equations become identical to the equations of Newtonian physics, so that Newtonian physics is contained within special relativity. Only when objects approach the speed of light do our observations begin to depart measurably from the predictions of Newtonian physics. Those departures are called **relativistic** effects. In our everyday lives, we never encounter relativistic effects directly because we never travel at speeds that approach the speed of light. Even the fastest object ever made by humans, the *Helios II* spacecraft, traveled at only about $0.00023c$.

Einstein's ideas remained controversial well into the 20th century, and his 1921 Nobel Prize in Physics was awarded for his work on the photoelectric effect (see Chapter 5), not for his work on relativity, because not all physicists were yet convinced that his theory was correct. But as one experiment after another confirmed the strange and counterintuitive predictions of relativity, scientists came to accept its validity.

The Implications of Relativity

Today, special relativity shapes our thinking about the motions of both the tiniest subatomic particles and the most distant galaxies. In this subsection, we discuss only a few of the essential insights that come from Einstein's work.

Mass and Energy   What we think of as "mass" and what we think of as "energy" are actually closely related. The energy of an object depends on its speed: The faster it moves, the more energy it has. According to Einstein's equation, $E = mc^2$,

what if . . .

What if you could measure mass in a laboratory with great precision? In an experiment that mixes two chemicals together inside a closed container, you observe that the chemical reaction produces light. Would you expect the mass of the container full of chemicals to change? Explain why or why not.

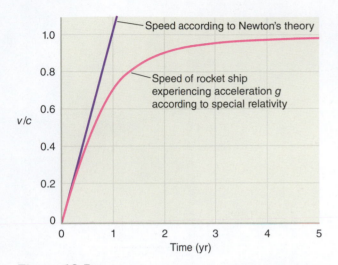

Figure 18.5 The speed of a rocket ship experiencing an acceleration equal to Earth's acceleration of gravity (g) increases with time. The purple curve shows the speed according to Newtonian theory. The pink curve shows the effects of special relativity, where the rocket ship approaches, but never reaches, the speed of light. The pink line correctly accounts for relativistic effects, whereas the purple line does not.

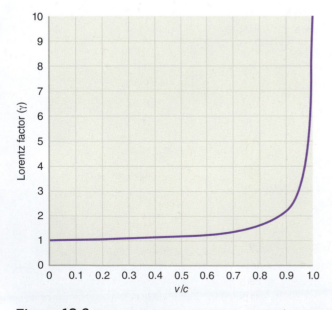

Figure 18.6 The Lorentz factor, γ, is plotted against v/c. It is not much different from one until velocities are about 50 percent of the speed of light.

though, even a stationary object has an intrinsic "rest" energy that equals the mass (m) of the object multiplied by the speed of light (c) squared. The speed of light is a very large number, so a small mass has a very large rest energy. A single tablespoon of water, for example, has a rest energy equal to the energy released in the explosion of more than 300,000 tons of TNT. We used that relationship between mass and energy in Chapter 14 when discussing the nuclear fusion that makes stars shine.

In Chapter 3, we connected the *mass* of an object to its inertia—its resistance to changes in motion. At relativistic speeds, adding to the energy of motion of an object increases its inertia. For example, a proton in a high-energy particle accelerator may travel so close to the speed of light that its total energy is 1000 times greater than its rest energy. Such an energetic proton is harder to "push around" (in other words, it has more inertia) than a proton at rest. It also more strongly attracts other masses through gravity.

The Ultimate Speed Limit   We already discussed Einstein's insight that if traveling at the speed of light were possible, then in that reference frame light would cease to be a traveling wave, and the laws of electromagnetism wouldn't be valid. You can also think about that limit in terms of the equivalence of mass and energy. As the speed of an object gets closer and closer to the speed of light, the energy of that object, and therefore its mass, becomes greater and greater, so it becomes increasingly resistant to further changes in its motion. Only a photon or other massless particle can travel at the speed of light. We rely on the fact that light travels at a constant speed in a vacuum whenever we use the travel time of light to describe astronomical distances.

Adding energy to an object will cause its velocity to get closer and closer to the speed of light, but the velocity will never actually reach the speed of light. To accelerate an object with a nonzero rest mass to the speed of light would take an *infinite* amount of energy. Not enough energy exists in the entire universe to accelerate even one electron to the speed of light. The electron can get arbitrarily close to that number—0.999999999999999999999 . . . × c is possible (at least in principle)—but not enough energy is available to accelerate the electron beyond that to the speed of light. **Figure 18.5** shows how a rocket ship, which experiences a constant acceleration equal to that of gravity on Earth (so that its occupants will feel "normal" gravity), moves faster and faster but never reaches the speed of light. In the Newtonian view shown by the purple curve, that limit is not present, so the speed of the spaceship would continuously increase at the same rate.

Time   Time passes more slowly in a moving reference frame: For moving objects, the seconds are stretched out by time dilation. No inertial reference frame is special. If you compared clocks with an observer moving at nine-tenths the speed of light ($0.9c$) relative to you, you would find that the other observer's clock was running 2.29 times slower than your clock. The other observer would find instead that *your* clock was running 2.29 times slower. To you, the other observer may be moving at $0.9c$, but to the other observer, *you* are the one who is moving. Both frames of reference are equally valid, so you would each find the other's clock to be slower than your own. That time dilation effect increases with speed, and that symmetry holds as long as neither frame accelerates. **Figure 18.6** and **Table 18.1** show that how much the time is stretched depends on the object's speed. The factor of 2.29 by which time is stretched in the preceding example is called the Lorentz factor and is usually denoted by the symbol γ.

A scientific observation demonstrates time dilation in nature. Fast particles called cosmic-ray muons (**Figure 18.7**) are produced 15 km up in Earth's atmosphere when high-energy **cosmic rays**—elementary particles moving at nearly the speed of light—strike atmospheric atoms or molecules. Muons at rest decay very rapidly into other particles. That decay happens so quickly that even if they could move at the speed of light, virtually all muons would have decayed long before traveling the 15 km to reach Earth's surface. Time dilation slows the muons' clocks, however, so the particles live longer and can travel farther and reach the ground. As muons move faster and faster, their clocks run slower and slower, and more of them can reach the ground. The same general principle is observed in particle accelerators, in which particles traveling at speeds near the speed of light live longer before decaying. **Working It Out 18.1** shows some examples of time dilation.

Length An object appears shorter in motion than it is at rest. Moving objects are compressed in the direction of their motion by a factor of $1/\gamma$, where γ is the same Lorentz factor introduced in the discussion of time dilation. That phenomenon is called **length contraction**. A meterstick moving at $0.9c$ appears to be 0.44 meters long. That finding also explains our muon experiment from the perspective of the muons themselves. In the reference frame of the fast-moving muon produced at a height of 15 km, Earth's atmosphere is moving fast and appears to be much shorter than 15 km; indeed, it is so compressed from the muon's perspective that the muon may be able to reach the ground before decaying. That length contraction effect also increases with speed.

Twin Paradox Suppose you take a trip into space and leave your identical twin back on Earth. You accelerate to nearly the speed of light as you leave Earth. After you arrive at your destination, you return to Earth, again traveling at a speed close to c. To your twin on Earth, you were the one in a moving reference frame, so your twin measures your time as running much slower (see Working It Out 18.1), in which case you should return younger than your twin. However, from your perspective, the spaceship didn't move. Instead, Earth receded from you, stopped, and returned. Your twin is the one who moved at just under the speed of light, and your twin's time ran more slowly than yours, so your twin has aged much less. Both you and your twin cannot be correct, and that is the nature of the paradox.

One key difference between you and your twin resolves the paradox. You experienced acceleration during your trip, whereas your twin did not. Accelerated motion is not uniform motion. As a result, you changed reference frames during your trip. You changed reference frames when you left Earth, changed again when you stopped at your destination, changed a third time when you left your destination to return home, and changed reference frames one final time when you arrived back at Earth. Your twin, however, remained in Earth's reference frame. Upon your return, you would find that more time has passed for your twin on Earth than for you and that your twin has aged more than you have.

Space Travel What is the actual reality of human travel in space? Some U.S. astronauts have been to the Moon and back, and several robotic spacecraft have been sent to explore objects throughout the Solar System. (To date, only a few robotic spacecraft are traveling at a high enough velocity to leave the Solar System.) But humans could not easily visit and explore other planetary systems in the Milky Way Galaxy because of the constraints of energy and the ultimate speed limit of light. With current technology, engineers can construct rockets that can travel at speeds of up to 20,000 m/s. At such a speed, a one-way trip to Earth's nearest neighbor star,

Table 18.1	Lorentz Factor	
v/c		**γ**
0.10		1.005
0.20		1.02
0.30		1.05
0.40		1.09
0.50		1.15
0.60		1.25
0.70		1.40
0.80		1.67
0.90		2.29
0.95		3.20
0.99		7.09
0.995		10.01
0.999		22.37
0.9999		70.71

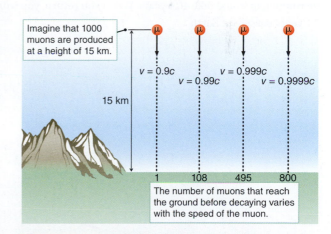

Figure 18.7 Muons created by cosmic rays high in Earth's atmosphere decay long before reaching the ground if they are not traveling at nearly the speed of light. Here, we show what happens to 1000 muons produced at an altitude of 15 km for a variety of speeds. Faster muons have slower clocks, so more of them survive long enough to reach the ground—many more than would be expected simply because of the faster speed.

working it out 18.1

Time Dilation

Physicist Hendrik Lorentz (1853–1928) derived the equation for how much time is dilated and how much space is contracted when something is traveling at velocities near the speed of light. That *Lorentz factor* (abbreviated γ) is given by

$$\gamma = \frac{1}{\sqrt{1 - \frac{v^2}{c^2}}}$$

Figure 18.6 shows the Lorentz factor plotted against velocity, and Table 18.1 gives the calculated value of γ for different values of velocities v/c. For something moving at half the speed of light, the Lorentz factor is 1.15. For something moving at 90 percent of c, however, the factor is 2.29, and it goes up quickly from there, becoming arbitrarily large as the velocity approaches, but never quite reaches, the speed of light.

Suppose you take a trip to a star 20 light-years away to study the "super-Earth" planets in orbit around it. You travel at 0.99c, whereas your twin stays behind on Earth. Our common Newtonian experience on Earth would tell you how long the trip takes, using the familiar equation that time equals distance divided by speed:

$$\text{Time passed} = \frac{20\,\text{light-years}}{0.99c} = 20.2\,\text{years}$$

The round trip would take 40.4 years. When you return, you will find that 40.4 years passed on Earth.

To you on the spaceship moving at 0.99c, however, time will pass more slowly. For you, the time in the spaceship is equal to the time on Earth divided by the Lorentz factor: $t_{\text{ship}} = \frac{1}{\gamma}t_{\text{Earth}}$. Your trip to the star would take (1/7.09) × 20.2 years = 2.8 years and another 2.8 years for the return trip. So, as you traveled to that star and back, 5.6 years would have passed for you in the spaceship, but *40.4* years would have passed on Earth. Your twin on Earth would be almost 35 years older than you are!

For a more practical experiment that can be performed on Earth, we've noted that time dilation can be seen with subatomic particles. Suppose one type of particle, called a pion, can "live" for 20 nanoseconds (ns) before it decays into different particles. If pions are produced in a particle accelerator at a speed of 0.999c, how long will physicists see the pions before they decay? Here $v = 0.999c$, so from Table 18.1, $\gamma = 22.37$. In the reference frame of the moving pions, they are still living for 20 ns: they are like the twin who traveled into space at high speed. In the reference frame of the physicists, though, special relativity predicts that the particle will last longer—namely, 20 ns × 22.37 = 447 ns. Indeed, physicists observe that pions moving at nearly the speed of light "live" longer and travel farther before decay (like the muons in Figure 18.7).

The same factor γ applies for mass and length. If physicists measured a ruler moving at 0.999c, the ruler would be 22.37 times shorter than its length when at rest. Similarly, the pions traveling at 0.999c will behave as though their mass were 22.37 times larger. If the high-speed pions collided with other particles, for example, the energy of the collision would be as though they had the higher mass.

Proxima Centauri, at a distance of 4.2 light-years, would take well over 50,000 years. Travel to more distant stars would take even longer.

In principle, travel just under the speed limit c is possible, so one could take advantage of relativistic time dilation to make such adventures well within the lifetime of a space traveler. In the example in Working It Out 18.1, an astronaut experiences a round-trip travel time to planets 20 light-years away in just 5.6 years at a speed of 0.99c. Or an astronaut could travel to the center of the Milky Way Galaxy and back in just 2 years by traveling at 0.9999999992c. Unfortunately, for a spacecraft traveling at those speeds, any impact with any object, even a quite small one, would be catastrophic, tearing a hole straight through the spacecraft. Much stronger shielding will need to be invented before such a trip can be survived.

Although theoretically possible, travel at those speeds in practice would require an impractical amount of energy. If M is the mass of the astronauts, rocket ship, and fuel, just accelerating the rocket ship to such a high speed would take γMc^2 of energy, or $10Mc^2$ in the Proxima Centauri example and $25,000Mc^2$ in the example of the trip to the center of the Milky Way. For the second example, the energy to accelerate just the astronaut (not even including the spaceship) to such energies is more than that contained in 10 billion nuclear weapons. So, although not theoretically impossible, visits to other stars in our galaxy will not take place anytime soon.

CHECK YOUR UNDERSTANDING 18.2

Suppose that your friend flies past you in a spaceship moving at 0.9c, and both of you measure the time the spaceship takes to pass your location. Which of the following is true? (a) The time you measure is *greater* than the time your friend measures. (b) The time you measure is *less* than the time your friend measures. (c) You both measure the same amount of time.

18.3 Gravity Is a Distortion of Spacetime

Our exploration of special relativity began with the observation that the speed of light is always the same regardless of the motion of an observer or the motion of the source of the light. We have seen that three-dimensional space and time are actually just the result of a particular, limited perspective on a four-dimensional spacetime that is different for each observer. That four-dimensional spacetime is itself warped and distorted by the masses it contains. As we discuss the properties of black holes—indeed, of all massive objects in the universe—the concepts of space and time will diverge even further from the absolutes of Newtonian physics. In this section, we explore the general theory of relativity, which describes how mass affects space and time.

The Equivalence Principle

We have already discussed one fundamental connection between gravity and spacetime in Chapter 4 and showed that the *inertial mass* of an object—the mass appearing in Newton's equation $F = ma$—is *exactly* the same as the object's gravitational mass. In addition, any two objects at the same location and moving with the same velocity will follow the same path through spacetime, regardless of their masses. The astronaut in an orbiting spaceship falls around Earth, moving in lockstep with the spaceship itself. A feather dropped by an *Apollo* astronaut standing on the Moon falls toward the surface of the Moon at the same rate as a dropped hammer does. Instead of thinking of gravity as a force that acts on objects, it is more accurate to think of gravity as a consequence of the warping of spacetime in the presence of a mass. *Gravitation is the result of the shape of spacetime that objects move through.* That is one of the key insights of the **general theory of relativity**, Einstein's theory of gravity.

According to special relativity, any inertial reference frame is as good as any other. Thus, no experiment can distinguish between sitting in an enclosed spaceship floating "stationary" in deep space (**Figure 18.8a**) and sitting in an enclosed spaceship traveling at 0.9999c (**Figure 18.8b**). Those two situations feel the same because neither observer feels an acceleration. Both are equally valid inertial reference frames. As long as nothing accelerates either spaceship, neither observer can distinguish between the two spacecraft.

But what if an acceleration occurs? An acceleration can change either the speed or the direction of an object. Consider an astronaut inside a spaceship orbiting

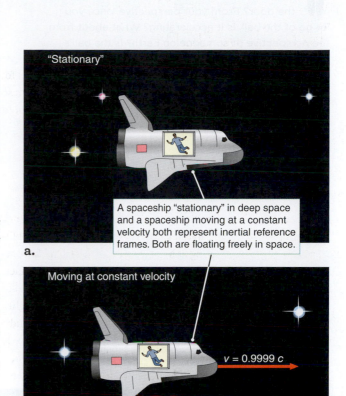

"Stationary"

A spaceship "stationary" in deep space and a spaceship moving at a constant velocity both represent inertial reference frames. Both are floating freely in space.

a.

Moving at constant velocity

$v = 0.9999\ c$

b.

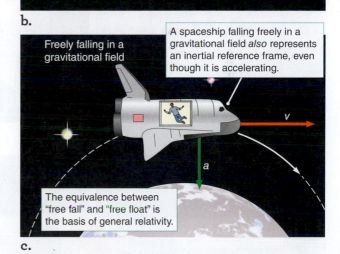

Freely falling in a gravitational field

A spaceship falling freely in a gravitational field *also* represents an inertial reference frame, even though it is accelerating.

v

a

The equivalence between "free fall" and "free float" is the basis of general relativity.

c.

Figure 18.8 According to special relativity, no difference exists between **a.** a reference frame floating stationary in space and **b.** one moving through the galaxy at constant velocity. General relativity adds that no difference exists between those inertial reference frames and **c.** an inertial reference frame falling freely in a gravitational field. Free fall is the same as free float, as far as the laws of physics are concerned.

what if . . .

What if you are orbiting Earth in the International Space Station and you throw a ball out the door? From your perspective, once you let go of the ball, is it accelerating? What about from the perspective of someone on Earth? Is the ball's reference frame inertial?

Figure 18.9 According to the equivalence principle, **a.** an object falling freely in a gravitational field is in an inertial reference frame, and **b.** an object at rest in a gravitational field is in an accelerated reference frame. As a result, sitting in a spaceship accelerating at 9.8 m/s² feels the same as sitting still on Earth.

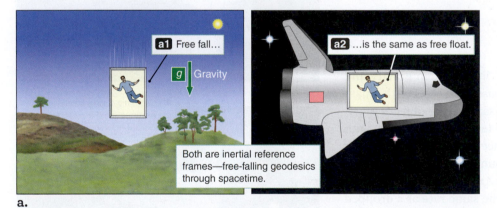

Both are inertial reference frames—free-falling geodesics through spacetime.

a.

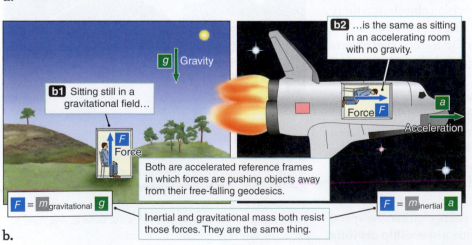

Both are accelerated reference frames in which forces are pushing objects away from their free-falling geodesics.

Inertial and gravitational mass both resist those forces. They are the same thing.

b.

Earth, as shown in **Figure 18.8c**. That spaceship is accelerating because the direction of its velocity is constantly changing as it orbits Earth. The astronaut is accelerating, too. Because he feels weightless, the astronaut has no way to tell the difference between being inside the spaceship as it falls around Earth and being inside a spaceship floating through interstellar space. Even though the ship's velocity is constantly changing as it falls, the inside of a spaceship orbiting Earth is an inertial frame of reference just as an object drifting along a straight line through interstellar space is an inertial frame of reference.

The idea that a freely falling reference frame is equivalent to a freely floating reference frame is called the **equivalence principle**. If you close your eyes and jump off a diving board, for the brief time that you are falling freely through Earth's gravitational field, the sensation you feel is the same as the sensation you would feel floating in interstellar space.

The natural path that an object will follow through spacetime in the absence of other forces is called the object's **geodesic**. In the absence of a gravitational field, the geodesic of an object is a straight line, in accordance with Newton's first law: An object will move at a constant speed in a constant direction unless acted on by a net external force. However, the shape of spacetime becomes distorted in the presence of mass, so an object's geodesic becomes curved.

Figure 18.9a shows two examples of inertial frames: an astronaut in a spaceship coasting through space and a person in a box falling toward the ground with an acceleration g. Both people are following their geodesics, so those two inertial reference frames are equivalent, and neither observer can distinguish between them. The equivalence principle also applies in cases of accelerations that result in a change in speed. Imagine an observer sitting in a closed box on Earth's surface (**Figure 18.9b**). The floor of the box pushes on him to keep him from following his geodesic, and he feels that force. Now imagine the box is inside a spaceship accelerating through deep space at a rate of 9.8 meters per second per second (m/s²) in the direction of the arrow shown in Figure 18.9b. The floor of the box pushes on the observer to overcome his inertia and cause him to accelerate at 9.8 m/s², so he feels as though he is being pushed into the floor of the box. In both cases, a force acts on the observer so that he does not follow the same free-falling geodesic. The observer feels as though he is being pushed into the floor of the box, so he feels the acceleration, and his frame of reference is not inertial. According to the equivalence principle, sitting in an armchair in a spaceship traveling with an acceleration of 9.8 m/s² is equivalent to sitting in an armchair on the surface of Earth reading this book. In each case, it is the same mass—the mass that gives an object inertia, from a Newtonian perspective—that resists the change. Gravitational mass and inertial mass are the same thing.

The equivalence principle has an important caveat. In an accelerated reference frame such as an accelerating spaceship, the same acceleration (both magnitude and direction) is experienced everywhere. In contrast, the curvature of space by a massive object is weaker farther from the object. The effects of gravity and acceleration are equivalent only locally; that is, the equivalence principle is valid only as long as attention is restricted to small enough volumes of space so that differences in gravity within that volume can be ignored.

Mass Distorts Spacetime

The general theory of relativity describes how mass distorts the geometry of spacetime. Imagine the surface of a tightly stretched, flat rubber sheet. A marble will roll in a straight line across the sheet. Euclidean geometry—the geometry of everyday life—applies on the surface of the sheet: If you draw a circle, its circumference is equal to 2π times its radius, r. If you draw a triangle, the angles add up to 180 degrees. Lines that are parallel anywhere are parallel everywhere.

Now place a bowling ball in the middle of the rubber sheet, creating a deep depression, or "well," as in **Figure 18.10**. The surface of the sheet is no longer flat, and Euclidean geometry no longer applies. If you roll a marble across the sheet (Figure 18.10a), its path dips and curves. You can roll the marble so that it moves around and around the bowling ball, like a planet orbiting about the Sun. If you draw a circle around the bowling ball (Figure 18.10b), the circumference of the circle is less than $2\pi r$. If you draw a triangle (Figure 18.10c), the angles add up to more than 180 degrees.

Mass affects the fabric of spacetime like the bowling ball affects the fabric of the rubber sheet. The bowling ball stretches the sheet, changing the distances between any two points on the surface of the sheet. Similarly, mass distorts spacetime, changing the distance between any two locations or events. Larger masses produce larger distortions in spacetime. We can see how a rubber sheet with a bowling ball on it is stretched through a third spatial dimension, but most people cannot picture what a curved four-dimensional spacetime would "look like." Yet experiments verify that the geometry of four-dimensional spacetime is distorted much like the rubber sheet, whether or not it can be easily pictured.

When One Physical Law Supplants Another

Previously in this book, we described gravity as a force that obeys Newton's universal law of gravitation: $F = Gm_1m_2/r^2$. Here in Chapter 18, we have introduced the ideas of general relativity and asked you to change the way you think about gravity. If general relativity is correct, does that mean Newton's formulation of gravity is wrong? If so, then why does Newton's law continue to be used?

Those questions go to the heart of how science progresses. As long as a gravitational field is not too strong, Newton's law of gravitation is a very close *approximation* to the results of a calculation using general relativity. In that context, even the gravitational field near the core of a massive main-sequence star would be

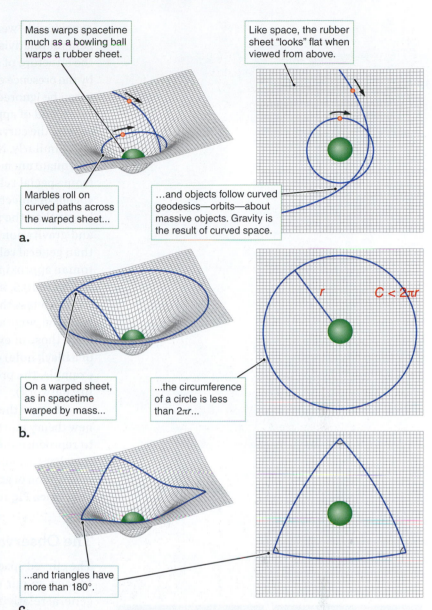

Mass warps spacetime much as a bowling ball warps a rubber sheet.

Like space, the rubber sheet "looks" flat when viewed from above.

Marbles roll on curved paths across the warped sheet...

...and objects follow curved geodesics—orbits—about massive objects. Gravity is the result of curved space.

a.

On a warped sheet, as in spacetime warped by mass...

...the circumference of a circle is less than $2\pi r$...

r $C < 2\pi r$

b.

...and triangles have more than 180°.

c.

Figure 18.10 Mass warps the geometry of spacetime much as a bowling ball warps the surface of a stretched rubber sheet. That distortion of spacetime has many consequences; for example, **a.** objects follow curved paths, or geodesics, through curved spacetime; **b.** the circumference of a circle around a massive object is less than 2π times the radius of the circle; and **c.** angles in triangles add up to more than 180 degrees.

considered "weak." An astronomer will obtain the same results if she uses a general relativistic formulation of gravity instead of Newton's laws to calculate the structure of a main-sequence star. Similarly, even though spacetime is curved by the presence of mass, that curvature near Earth is so slight that over small regions it can be ignored entirely, and flat (Euclidean) geometry can be used. That is the same kind of approximation people use when they navigate with a flat road map, despite the curvature of Earth.

Similarly, Newton's laws of motion are *approximations* of special relativity and quantum mechanics. In fact, Newton's laws can be mathematically derived from special relativity and quantum mechanics by using the assumptions that speeds of objects are much less than the speed of light and that objects are much larger than the particles from which atoms are made. Newton's laws of motion and gravitation are used most of the time because they are far easier to apply than general relativity and because any inaccuracies introduced by using Newtonian approximations are usually far too tiny to measure. That was illustrated in Figure 18.5, in which the behavior of an accelerated object moving at a velocity much less than c is the same whether or not relativity is used. Calculations based on general relativity are required only when conditions are very different from those of everyday life (for example, the behavior of the gravitational field of a black hole) or in special cases when extremely high accuracy is needed (for example, the precise timing used by the global positioning system [GPS] satellite network).

If a new theory is to replace an earlier, highly successful scientific theory, the new theory must hold the old theory within it—that is, the new theory must be able to reproduce the successes of the earlier theory. Special relativity contains Newton's laws of motion, and general relativity holds within it the successful Newtonian description of gravity that we have relied on throughout this book (see the **Process of Science Figure**).

The Observable Consequences of General Relativity

The curved spacetime of general relativity does have observable consequences. Indeed, in our own Solar System, observations can be made that distinguish general relativity from Newtonian physics. In Newton's theory, orbits are elliptical and fixed in space. In contrast, general relativity predicts that the long axis of an elliptical orbit slowly rotates, or precesses. **Figure 18.11** illustrates the difference between an orbit in Newton's theory, which remains stationary (left panel), and one in general relativity, which precesses (right panel). In our Solar System, even after one accounts for the effects of other planets on Mercury, a very small shift in its axis remains, equal to 43 arcseconds (arcsec) per century—which cannot be explained by Newton's laws alone. General relativity predicts exactly that precession for Mercury.

General relativity has other unique implications. A beam of light moving through empty space travels in a straight line, but a beam of light moving through the distorted spacetime around a massive object is bent by gravity, just as the lines in Figure 18.10 are bent by the curvature of the sheet. That bending of the light path by curved spacetime is called **gravitational lensing** because optical lenses also bend light paths. The first measurement of gravitational lensing came during the total solar eclipse of 1919. Several months before the eclipse, astrophysicist Sir Arthur Stanley Eddington (1882–1944) measured the positions of several stars in the direction of the sky where the eclipse would occur. He then repeated the measurements

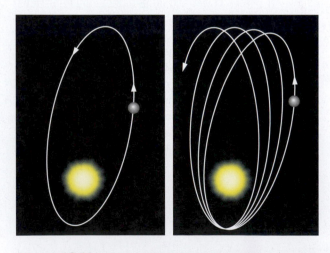

Figure 18.11 The left panel denotes an elliptical orbit about the Sun. In the Newtonian view, that elliptical orbit remains stationary. In the right panel, Mercury's orbit precesses as a result of the warped spacetime near the Sun.

New Science Can Encompass the Old

General relativity is more accurate and broadly applicable than Newton's laws. It also explains why gravity acts as it does. Still, for ordinary objects on Earth, Newton's and Einstein's theories agree.

General relativity is needed when masses are large and distances are small, where the pull of gravity is large.

Far from a mass, however, where gravity is weak and spacetime is relatively flat, general relativity gives the same result that Newton found.

One way that scientists check new theories is by considering the limits. What happens at great distances? What happens if the mass is very small? In these limits, the new and more complete theories must give the same results as old theories.

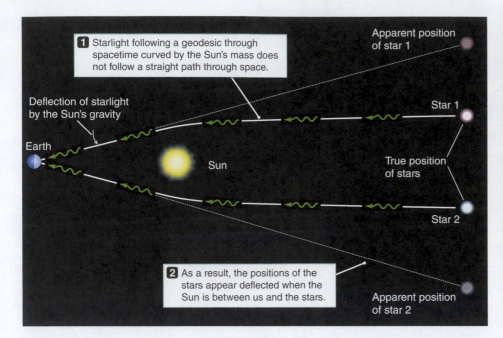

Figure 18.12 According to measurements taken during the total solar eclipse of 1919, the gravity of the Sun bends the light from distant stars by the amount predicted by Einstein's general theory of relativity. That is an example of gravitational lensing. Note that the "triangle" formed by Earth and the two stars contains more than 180 degrees, just like the triangle in Figure 18.10c.

Figure 18.13 This Hubble Space Telescope image shows four images of the same distant supernova (arrows), where the images arise from the gravitational lensing due to a massive foreground galaxy.

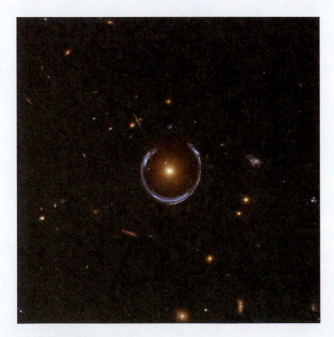

Figure 18.14 An Einstein ring created by the gravity of a luminous red galaxy gravitationally lensing the light from a much more distant blue galaxy.
Credit: ESA/Hubble & NASA, https://apod.nasa.gov/apod/ap111221.html. https://creativecommons.org/licenses/by/4.0/.

during the eclipse. **Figure 18.12** shows how the light from distant stars curved as it passed the Sun, causing the measured positions of the stars to shift outward. The stars appeared farther apart in Eddington's second measurement than in his first by the amount predicted by general relativity. That is another example of the effects of general relativity—in curved space, parallel lines actually can intersect.

Images of galaxies are sometimes distorted by gravitational lensing by other galaxies or clusters of galaxies. In the Hubble Space Telescope image shown in **Figure 18.13**, a supernova in a distant galaxy is lensed into four images by a nearby massive galaxy. What makes that case particularly interesting is that those four images appeared at different times because the light for each image took a different path, and some paths were longer than others. In an extreme case, a lensed galaxy image can be distorted into an **Einstein ring** (**Figure 18.14**).

Mass distorts the geometry not only of space but also of time. The deeper one descends into the gravitational field of a massive object, the more slowly clocks appear to run from the perspective of a distant observer—an effect called **general relativistic time dilation**. Suppose a light is attached to a clock sitting on the surface of a neutron star. The light is adjusted so that, to a person on the neutron star, the light flashes once each second. Because time near the surface of the star is dilated as a result of the depth of the gravitational field, an observer far from the neutron star perceives the light to be pulsing with a lower frequency—less than once a second. Now suppose an emission line source is on the surface of the neutron star. Because time is running slowly on the surface of the neutron star, the light that reaches the distant observer will have a lower frequency than when it was emitted. It is as though photons of light have lost energy in their escape from the star's surface. A lower frequency means a longer wavelength, so the light from the source will be seen at a longer, redder wavelength than the wavelength at which it was emitted. That shift in the wavelengths of light from objects deep within a gravitational well is called the **gravitational redshift** (Figure 18.15). The effect of gravitational redshift is similar to that of the Doppler redshift discussed in Chapter 5. In fact, no way exists to tell the difference between light that has been redshifted by gravity and light from an object moving away from you that has been Doppler shifted.

The following example brings the phenomenon of general relativistic time dilation a bit closer to home. A clock on the top of Mount Everest runs faster, gaining about 80 ns a day compared with a clock at sea level. The difference between an object on the surface of Earth and an object in orbit is even greater. A GPS receiver uses the results of sophisticated calculations of the effects of general relativistic gravitational redshift to help you accurately find your position on the surface of Earth. Satellites in orbit travel quickly enough that the effects of special relativity are measurable. Even after one allows for slowing due to special relativity, the clocks on the satellites that make up the GPS run faster than clocks on the surface of Earth. If the satellite clocks *and* your GPS receiver did not correct for that and other effects of general relativity, the position your GPS receiver reported would be in error by up to half a kilometer. The fact that the GPS can be accurate to a few meters provides strong experimental confirmation of two predictions of general relativity—gravitational redshift and general relativistic time dilation.

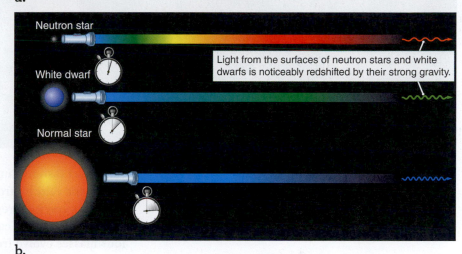

Figure 18.15 Time passes more slowly near massive objects because of the curvature of spacetime. To a distant observer, then, light from near a massive object will have a lower frequency and longer wavelength. **a.** The closer the source of radiation is to the object or **b.** the more massive and compact the object is, the greater the gravitational redshift will be.

CHECK YOUR UNDERSTANDING **18.3**

In general relativity, what causes gravity? (a) It is a result of Newton's law of gravitation. (b) It is a consequence of time dilation and length contraction. (c) It is the result of masses having greater effect when they move at high speeds. (d) It is the result of the distortion in spacetime around a massive object.

18.4 Black Holes

General relativity also predicts the existence of black holes. The Newtonian world has no well-formulated theory of black holes. When placed on the surface of a rubber sheet, a mass causes a funnel-shaped distortion analogous to the distortion of spacetime by a mass. Now imagine the limit in which the funnel is *infinitely* deep—it gets narrower as it goes deeper but it has no bottom. That is the rubber-sheet analog to a black hole. The mathematics describing a black hole approaches infinity, just as the mathematical expression $1/x^2$ approaches infinity as x approaches zero. Such a mathematical anomaly is called a **singularity**. Black holes contain singularities in spacetime, and that mathematical complication indicates that extreme conditions exist near (and inside) a black hole.

How Black Holes Form

We have seen how stellar evolution can lead to the formation of compact stellar remnants such as white dwarfs and neutron stars. However, an even more

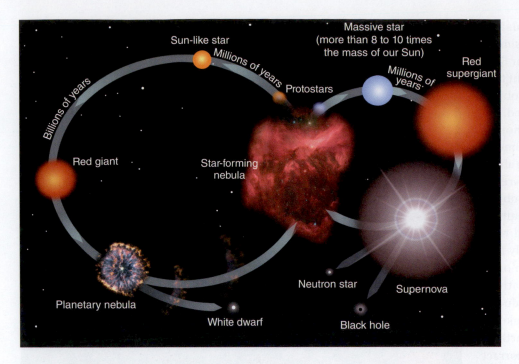

Figure 18.16 Stars of various masses evolve from protostars to stellar remnants and in the process may recycle material back into the interstellar medium to produce new stars. (This figure is not to scale.)

extreme fate awaits some massive stars at the end of their evolution. Recall from Chapter 16 that a white dwarf can have a mass of no more than the Chandrasekhar limit—about 1.4 M_{Sun}. If the mass of the object exceeds that limit, gravity can overcome electron degeneracy pressure, and the white dwarf will begin collapsing again.

The physics of a neutron star is much like the physics of a white dwarf, except that neutrons rather than electrons are what cause a neutron star to be degenerate. Analogous to what happens with a white dwarf, if the mass of a neutron star exceeds ~2.5 M_{Sun}, gravity begins to win out over pressure once again. The neutron star grows smaller, and gravity at the star's surface becomes stronger and stronger at an ever-accelerating pace. Recall from Chapter 4 that the escape velocity from a planet or moon depends on its surface gravity. Now imagine surface gravity so strong that the escape velocity approaches the speed of light. Nothing, though, not even light, can travel faster than *c*. So when the neutron star crosses the threshold where the escape velocity from its surface exceeds the speed of light, not even light can escape its gravity. A region of space where neither matter nor radiation can escape the pull of gravity is called a **black hole**. **Figure 18.16** illustrates how stars of various masses evolve from protostars to their ultimate end, including sometimes as black holes. Material released when stars die is recycled into the interstellar medium to form new stars.

A neutron star in a binary system can collapse to become a black hole if it accretes enough matter from its companion to push it over the ~2.5-M_{Sun} limit, similar to the way some Type Ia supernovae accrete matter from a companion and approach the Chandrasekhar limit of 1.4 M_{Sun}. Regardless of how it forms, any stellar remnant with a mass greater than ~2.5 M_{Sun} must be a black hole.

Properties of Black Holes

From the outside, you can never actually "see" a black hole. The closer an object approaches a black hole, the greater is its escape velocity (the speed that the object would need to escape from the gravity of the black hole). Shortly after Einstein's theory of general relativity was announced, Karl Schwarzschild (1873–1916) explored

Figure 18.17 **a.** A black hole's size is determined by the Schwarzschild radius and the corresponding event horizon. This image is a two-dimensional analog for a black hole. In reality, the event horizon is a sphere. **b.** If the object that formed the black hole was spinning, its angular momentum would be conserved, and the black hole would twist the spacetime around it.

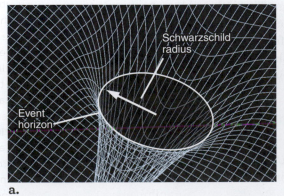

a.

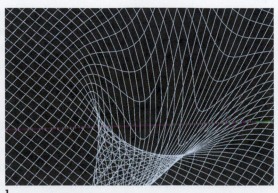

b.

the limits of Einstein's equations, and predicted the existence of the objects we now know as black holes. Schwarzschild calculated the radius at which the escape velocity is equal to the speed of light. That radius is called the **Schwarzschild radius**, and it is proportional to the mass of the black hole:

$$R_S = \frac{2GM_{BH}}{c^2} = 3\,\text{km} \times \frac{M_{BH}}{M_{Sun}}$$

where R_S is the Schwarzschild radius, G is the universal gravitational constant, M_{BH} is the mass of the black hole, and c is the speed of light. The sphere around the black hole at that distance is called its **event horizon**—a surface from which nothing, not even light, can escape. A black hole with a mass of $1\,M_{Sun}$ has a Schwarzschild radius of about 3 km. A black hole with a mass of $5\,M_{Sun}$ has a Schwarzschild radius 5 times that, or about 15 km. If Earth were squeezed into a black hole, it would have a Schwarzschild radius of only about a centimeter. All the mass of a black hole is concentrated at its very center, but that fact is unobservable outside the black hole. **Figure 18.17a** shows a rubber-sheet analog to a black hole, with the Schwarzschild radius and the event horizon indicated.

A black hole has only three observable properties: mass, electric charge, and angular momentum. The mass of a black hole determines the Schwarzschild radius. The electric charge of a black hole is the net electric charge of the matter that fell into it. The angular momentum of a rotating black hole twists the space-time around it, as shown in **Figure 18.17b**. Apart from those three properties, all information about the material that fell into the black hole is lost. Nothing of its former composition, structure, or history survives.

Imagine that an adventurer falls into a black hole (**Figure 18.18**). From our perspective far from the event horizon of the black hole, the adventurer would appear to fall toward the event horizon. As she fell, her watch would appear to run more and more slowly, and her progress toward the event horizon would slow as well. At the event horizon, the gravitational redshift becomes infinite, and clocks appear to stop altogether. She would approach the event horizon, but from our perspective she would never quite make it. The adventurer's experience, however, would be very different. From her perspective, she would see nothing special about the event horizon. She would fall past the event horizon and continue deeper into the black hole's gravitational well. She would now have entered a region of spacetime cut off from the rest of the universe. The event horizon is like a one-way door: after the adventurer has passed through, she can never again pass back into the larger universe she once belonged to.

Actually, we have overlooked an important detail—the adventurer would have been torn to shreds long before she reached the black hole. Near the event horizon of a 3-M_{Sun} black hole, the difference in gravitational acceleration between the adventurer's feet and her head would be about a billion times her gravitational acceleration on the surface of Earth. In other words, her feet would be accelerating a billion times faster than her head. That would be a most unpleasant experience. Although scientific theories must produce testable predictions, not all individual predictions have to be tested directly.

"Seeing" Black Holes

In 1974, the physicist Stephen Hawking realized that black holes should be sources of radiation. In the ordinary vacuum of empty space, particles and their antiparticles spontaneously appear and then, within about 10^{-21} second, annihilate each

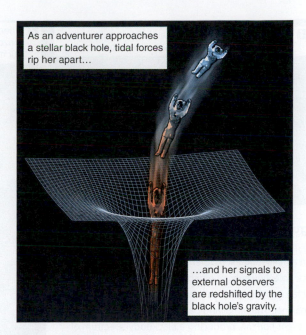

As an adventurer approaches a stellar black hole, tidal forces rip her apart...

...and her signals to external observers are redshifted by the black hole's gravity.

Figure 18.18 An adventurer falling into a black hole would be torn apart by the extreme tidal forces.

unanswered questions

What happens to the information that falls into a black hole? We said that a black hole is characterized by only three properties: mass, angular momentum, and electric charge. Where did all the other information go? To a distant observer, material takes an infinitely long time to fall into a black hole, so although the observer sees less and less radiation from the material, the properties of the material seem to be the same for all time. From the perspective of the infalling material, however, crossing the event horizon takes a finite amount of time, so information other than mass, angular momentum, and charge can no longer be shared with the outside world. Today, some scientists argue that as a black hole radiates and decays, the emitted Hawking radiation encodes the information that was "lost" in the black hole. Observing Hawking radiation is far beyond current technology.

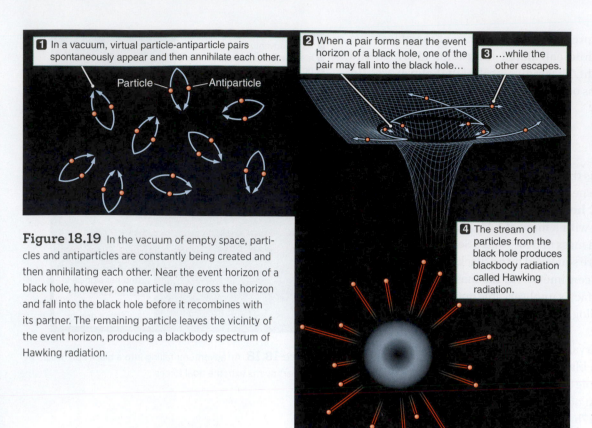

1 In a vacuum, virtual particle-antiparticle pairs spontaneously appear and then annihilate each other.

Particle — Antiparticle

2 When a pair forms near the event horizon of a black hole, one of the pair may fall into the black hole...

3 ...while the other escapes.

4 The stream of particles from the black hole produces blackbody radiation called Hawking radiation.

Figure 18.19 In the vacuum of empty space, particles and antiparticles are constantly being created and then annihilating each other. Near the event horizon of a black hole, however, one particle may cross the horizon and fall into the black hole before it recombines with its partner. The remaining particle leaves the vicinity of the event horizon, producing a blackbody spectrum of Hawking radiation.

other and disappear. Such particles are called **virtual particles** because they exist for only a very short time. If a pair of virtual particles comes into existence near the event horizon of a black hole, one of the particles might fall into the black hole while the other particle escapes (**Figure 18.19**). Some of the gravitational energy of the black hole will have been used up in making one of the pair of virtual particles real. Hawking showed that through that process, a black hole should emit a Planck blackbody spectrum and that the effective temperature of that spectrum would increase as the black hole became smaller through that "evaporation" process. After a very, very long time (on the order of 10^{61} years for a black hole with a mass of the Sun), the black hole would become small enough that it would become unstable and explode. Although the light that emerges, called **Hawking radiation**, is of considerable interest to physicists, the low intensity of Hawking radiation means it is not a likely way to see a black hole.

The first strong direct evidence for black holes that result from supernovae came from X-ray binary stars. The radio emission from Cygnus X-1 (an object originally identified in X-rays) flickers rapidly, changing in as little as 0.01 second. That means the source of the X-rays must be smaller than the distance that light travels in 0.01 second, or 3000 km—smaller than Earth. Cygnus X-1 was also identified as both a radio source and an already cataloged star called HD 226868. The spectrum of HD 226868 shows that it is a normal O9.7 I supergiant star with a mass of about 19 M_{Sun}—far too cool to produce the observed X-ray emission by itself. The wavelengths of absorption lines in the spectrum of HD 226868 are Doppler-shifted back and forth with a period of 5.6 days, indicating that HD 226868 is part of a binary system. Using the same techniques we discussed to measure the masses of binary stars in Chapter 13, astronomers found that the mass of the unseen compact companion of HD 226868 must be about 15 M_{Sun} (**Working It Out 18.2**). The companion to HD 226868 is too compact to be a normal star, yet it is much more massive than the Chandrasekhar limit for a white dwarf or the upper mass limit of a neutron star. Such an object must be a black hole. The X-ray emission from Cygnus X-1 arises when material from the O9.7 I supergiant falls onto an accretion disk surrounding the black hole, as illustrated in **Figure 18.20**.

In some similar systems, winds have been observed blowing off the disk around the black hole. Those winds are probably caused by magnetic fields in the disk. The fastest winds observed are in the binary system IGR J17091, where the wind speeds are as high as 32 million km/h (about 3 percent of the speed of light). That wind is blowing in many directions, and it

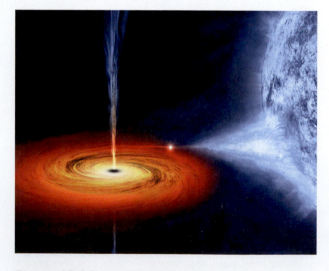

Figure 18.20 As material from a companion falls toward a black hole, some of it will impact the accretion disk. X-ray emission will be produced from this "hot spot," which moves as the companion orbits the more massive black hole. This artist's depiction illustrates the process at one moment in time.

working it out 18.2

Masses in X-ray Binaries

Cygnus X-1 is part of a binary system with a blue supergiant star O9.7 I (very close to B0 I) and an unseen compact object located about 0.2 astronomical unit (AU) away from it. The blue supergiant and the compact object orbit a common center of mass every 5.6 days. We can use the simple formula from Working It Out 13.4 to calculate the sum of the masses:

$$\frac{M_{blue}}{M_{Sun}} + \frac{M_{compact}}{M_{Sun}} = \frac{A^3}{P^2}$$

with A in astronomical units and P in years. Here, $A = 0.2$ AU and $P = 5.6/365.24$ years, so

$$\frac{M_{blue}}{M_{Sun}} + \frac{M_{compact}}{M_{Sun}} = \frac{0.2^3}{\left(\dfrac{5.6}{365.24}\right)^2} = 34$$

Thus, the sum of the masses of the two stars is 34 M_{Sun}.

To find the values of the two individual masses from their orbits, we need to know the velocities of the two stars or the distance of each star to the center of mass and the orbital inclination of the system. Obtaining such information is difficult when one star is compact and not observed separately. However, the mass of the blue supergiant star can be estimated from spectroscopic and photometric data at many wavelengths, as long as the distance to the system is known. When that is done, the mass of the supergiant is estimated at 19 M_{Sun}. If we subtract that from 34, the mass of the compact object is 15 M_{Sun}—well over the mass limit for a neutron star. Therefore, Cygnus X-1 is assumed to be a black hole. A recent study with data from several X-ray telescopes confirmed that mass.

---

may be carrying away more mass than is being captured by the black hole (**Figure 18.21**).

Astronomers have modeled the observational data of dozens of good candidates for stellar-mass black holes in X-ray binary systems in the Milky Way. Those findings indicate that the black hole masses are greater than 4.5–5 M_{Sun}, which is not very close to the limit of ~2.5 M_{Sun} for a neutron star. That gap in mass between the most massive neutron stars and the least massive stellar black holes is not yet understood, and it is assumed to be a result of the mass transfer processes between the stars.

In 2019, astronomers took advantage of gravitational lensing to obtain the first picture of a black hole's "shadow" (**Figure 18.22**). Widely advertised as "the first image of a black hole," it is the image of the shadow left behind when the path of light is severely bent by a mass of some 6.5 billion times the mass of the Sun.

Detection of Gravitational Waves

General relativity also predicts the existence of a wholly new form of radiation. If you exert a force on the surface of a rubber sheet, accelerating it downward, waves will move away from where you struck it, like ripples spreading out over the surface of a pond. Similarly, the equations of general relativity predict that if you accelerate the fabric of spacetime (for example, through the catastrophic asymmetrical collapse of a high-mass star), ripples in spacetime, or **gravitational waves**, will move outward at the speed of light. Those gravitational waves are like electromagnetic waves in some respects. Accelerating an electrically charged particle gives rise to an electromagnetic wave. Deforming a massive object gives rise to gravitational waves. Both types of waves can travel through a vacuum, but gravitational waves

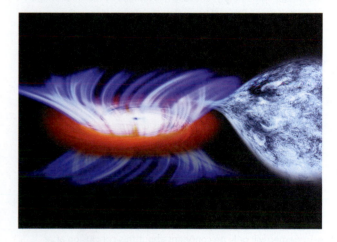

Figure 18.21 This artist's model shows strong winds being emitted from the disk around a stellar black hole. Those winds can remove more material than the amount that actually falls into the hole.

what if . . .

What if there were a nearby isolated black hole with a mass of 10 M_{sun}? How might astronomers be able to detect that black hole?

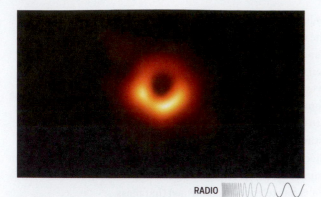

RADIO

Figure 18.22 ★ WHAT AN ASTRONOMER SEES

This "image" of a black hole was obtained by the Event Horizon Telescope Collaboration, a global network of telescopes that take data simultaneously. The image shows a bright ring formed as light bends in the intense gravity field around a black hole with a mass of 6.5 billion solar masses. An astronomer will recognize this image immediately, because it is famous as the "first image of a black hole." She will know that the dark region in the center is the "shadow" of the black hole, caused by the capture of light by the black hole. This dark region is about 2.5 times larger than the event horizon of the black hole. The orange ring of material around the outside of the shadow is the result of distorted paths of light passing through the surrounding material. An astronomer's attention will be caught by the asymmetry in the picture—namely, the bottom half of the ring is brighter than the upper half—and an astronomer will immediately wonder why. The material in the bottom half is moving toward the observer, so it benefits from an effect called "relativistic beaming," in which emitting objects look brighter when moving toward the observer at high speed. Conversely, the material in the upper half of the ring is moving away. This conclusion, however, is not obvious; an astronomer will only be confident about this conclusion after comparing the image to the output of a large number of computer simulations.

do not involve electric or magnetic fields. Instead, gravitational waves are ripples in the very fabric of spacetime.

Scientists have long had indirect evidence for the existence of gravitational waves. General relativity predicts that the orbit of binary neutron stars should lose energy, which will be carried away as gravitational waves. In 1974, astronomers discovered a binary system of two neutron stars, one of which is an observable pulsar. Using the pulsar as a precise clock, astronomers accurately measured the orbits of both stars. The orbits are gradually losing energy at the rate predicted by general relativity. Other similar binary pairs have been found with an orbital energy loss consistent with the radiation of gravitational waves.

On 14 September 2015, almost exactly 100 years after the prediction of gravitational waves by Albert Einstein, the Laser Interferometer Gravitational-Wave Observatory (LIGO) detected the gravitational waves from the merger of two black holes in a distant galaxy. The signal LIGO recorded precisely agreed with the prediction of the gravitational wave signal from such a merger, and the three leaders of the LIGO team, Kip Thorne, Rainer Weiss, and Barry Barrish, won the 2017 Nobel Prize in Physics for that discovery. Since then, more than 50 mergers have been observed by LIGO and its European counterpart, Virgo (**Figure 18.23**). They also detected the merger of two neutron stars, as discussed in Chapter 17.

The black holes we discussed here in Chapter 18 came from collapsing massive stars, but that is not the only type of black hole. In Chapters 19 and 20, we explain that supermassive black holes can be found at the centers of galaxies, including the Milky Way.

CHECK YOUR UNDERSTANDING 18.4

If a black hole suddenly doubled in mass, the event horizon would become _____ its original radius. (a) one-quarter; (b) one-half; (c) twice; (d) 3 times; (e) 4 times

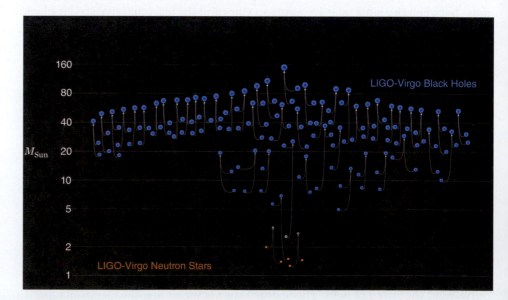

Figure 18.23 An illustration of the black hole and neutron star mergers seen by LIGO and Virgo through 2020. Objects higher on the chart are more massive; each merging pair results in one higher-mass object. The objects are spread horizontally simply so you can see them. (This is not a time axis.)

Reading Astronomy News

LIGO Just Detected the Oldest Gravitational Waves Ever Discovered

Karla Lant, Futurism.com

The first few gravitational wave detections were big news! It's difficult to overstate the excitement of the early days of gravitational wave detection, when every detection was something brand new.

Gravitational Waves Revealing the Universe

The Laser Interferometer Gravitational-wave Observatory (LIGO) just detected gravitational waves, ripples in time and space, for the third time. Two black holes collided, forming a huge black hole 49 times more massive than our sun, and this generated the waves. This kind of collision was also the cause of the waves detected previously by LIGO, although the masses of the black holes varied. This repetition of the discovery confirms that a new area of astronomy now exists.

"We have further confirmation of the existence of stellar-mass black holes that are larger than 20 solar masses—these are objects we didn't know existed before LIGO detected them," MIT's David Shoemaker, a LIGO spokesperson, said in a press release. "It is remarkable that humans can put together a story, and test it, for such strange and extreme events that took place billions of years ago and billions of light-years distant from us. The entire LIGO and Virgo scientific collaborations worked to put all these pieces together."

In September 2015, LIGO first directly observed these gravitational waves during its first run since receiving Advanced LIGO upgrades. The second detection followed in December 2015, and this latest detection, called GW170104, followed in January of this year [2017]. In each case, both of LIGO's twin detectors perceived gravitational waves from the collisions of the black holes, but this latest observation does offer a few new pieces of information.

For example, it suggests which directions the black holes might be spinning in, and indicates that at least one of the black holes in the pair may not be aligned with the overall orbital motion. Scientists are hoping that they can learn more about how binary black holes form by making more LIGO observations.

LIGO's Future

This work is testing, and thus far providing proof for, the theories proposed by Albert Einstein. For example, the theory of relativity says that dispersion, the effect that happens as light waves in a physical medium travel at different speeds, cannot happen in gravitational waves. LIGO has not found any evidence of dispersion in gravitational waves, as predicted by relativity.

"It looks like Einstein was right—even for this new event, which is about two times farther away than our first detection," Georgia Tech's Laura Cadonati, the Deputy Spokesperson of the LIGO Scientific Collaboration (LSC), said in the press release. "We can see no deviation from the predictions of general relativity, and this greater distance helps us to make that statement with more confidence."

Moving forward, the LIGO-Virgo team will keep searching LIGO data for any hint of gravitational waves emanating from the far corners of the Universe. The sensitivity of the detector will improve during the next run starting in late 2018 after researchers apply technical upgrades, hoping to see even more. Caltech's David Reitze, the LIGO Laboratory's executive director, said in the press release, "While LIGO is uniquely suited to observing these types of events, we hope to see other types of astrophysical events soon, such as the violent collision of two neutron stars."

QUESTIONS

1. How do these results help confirm the theory of general relativity?

2. Why were astronomers surprised about the masses of the two black holes?

3. What was different about that particular detection?

4. Go to the website for LIGO detections (http://ligo.org/detections.php). What was found in the 2018–2019 (or later) observing runs?

Source: https://futurism.com/ligo-detected-the-oldest-gravitational-waves-ever-discovered.

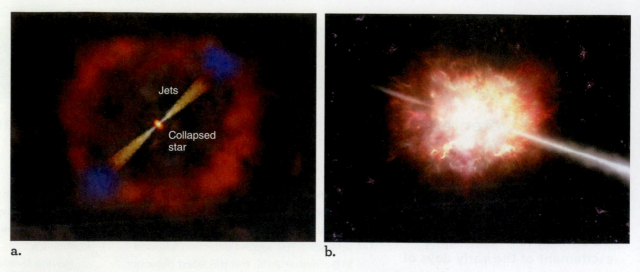

Figure 18.24 An artist's model of a gamma-ray burst (GRB). **a.** Narrow beams of intense energy are sent in two opposite directions. **b.** If one beam is pointed toward the observer, the GRB will appear bright.
Credit (part b.): ESO/A. Roquette, https://www.eso.org/public/images/eso0917a/. https://creativecommons.org/licenses/by/4.0/.

unanswered questions

Do wormholes exist in spacetime, connecting one region with another, perhaps through black holes? Wormholes are a mathematical solution to the equations of general relativity. The idea is that when something goes into a black hole, it travels through a wormhole and emerges in a different part of the universe. In that way a wormhole acts as a shortcut through spacetime. In science fiction, wormholes are a popular means of traveling large distances by exploiting the strange geometry of spacetime. But many scientists doubt that wormholes can exist in nature, and even if they do exist, strong tidal forces would pull apart anything that falls into a black hole before it emerged from a wormhole.

Origins: Gamma-Ray Bursts

The most energetic explosions in the universe are probably related to stellar black holes or merging neutron stars. **Gamma-ray bursts (GRBs)** are intense bursts of gamma rays. The bursts are followed by a weaker "afterglow" that is observed at many wavelengths. GRBs were first observed in the 1960s by satellites designed to look for radiation from nuclear weapons being tested in space after such tests were banned on Earth. In the 1990s, gamma-ray astronomy satellites discovered that those bursts were coming from all directions in the sky and that they might be associated with supernova explosions in distant galaxies. Short-duration GRBs, which last less than 2 seconds, probably originate from the merging of two neutron stars or a neutron star and a black hole in a close binary system that collapses into a single black hole. The more common long-duration GRBs are easier to study because they have a longer afterglow. Astronomers think that those originate in the collapse of a very high-mass, rapidly spinning star to a black hole or a neutron star after a supernova explosion.

Unlike regular supernovae, which radiate equally in all directions, GRBs are beamed events, so most of their enormous energies are concentrated into two opposite jets of emission (**Figure 18.24a**). In addition to the electromagnetic radiation, relativistic jets of cosmic rays are emitted. Astronomers have not observed any GRBs in the Milky Way: A massive supernova has not occurred in our galaxy for at least a century. But the energy of GRBs is so intense that people have wondered what might happen to Earth if one went off nearby with its radiation beamed in Earth's direction, a possibility imagined in **Figure 18.24b**. A leading candidate for a future GRB in our galaxy is the massive star Eta Carinae (see Figure 17.7). That star is 7500 light-years away—a neighbor, astronomically speaking. Its rotation axis, however, is such that it is unlikely to form a GRB that beams toward Earth.

Some scientists wonder whether past supernova and GRB events could have affected the history of life on Earth. Supernova "archaeologists" may have found evidence on Earth of past supernovae. In one study, rocks deep in the Pacific Ocean were found to have amounts of a radioactive isotope of iron, iron-60, that is too short-lived to be left over from the formation of Earth. That iron-60 could have been deposited 2.8 million years ago on Earth after a supernova explosion. In another study, high concentrations of nitrates were found in some layers in Antarctic ice cores. Gamma radiation from supernovae can produce excess nitrogen oxides in the atmosphere, which then become converted to nitrates that are trapped in snowfall. Nitrate spikes correlate to 1006 and 1054 CE—two years when bright supernovae are known to have appeared in the Milky Way.

What about more drastic effects on Earth from a nearby supernova or a more distant but beamed GRB? Normally, Earth is protected from cosmic radiation and cosmic-ray particles by its ozone layer and magnetic field. Cosmic-ray particles might not be a major problem if they arose very close to Earth, but high-energy gamma-ray radiation could have a more serious effect on Earth. The excess nitrogen oxides they produce in the atmosphere can absorb sunlight, which would cool Earth. The gamma radiation could ionize Earth's atmosphere, shrinking or destroying the ozone layer that protects life from ultraviolet radiation. Even a burst of a few seconds could lead to ozone damage lasting for decades. The gamma rays could trigger a burst of solar ultraviolet radiation at Earth's surface, which could damage the DNA of phytoplankton a few hundred meters deep in the ocean, affecting their ability to photosynthesize. Phytoplankton are the base of Earth's food chain, so a drastic reduction in phytoplankton could upset the entire biosphere. It has been hypothesized that such an event may have happened in Earth's history.

Statistically speaking, GRBs that beam to Earth may be rare, and some astronomers have argued that they are less likely to be produced in a Milky Way–type galaxy than in other types of galaxies. A lot of uncertainty surrounds any estimate of how close the supernova or GRB must be and how often those explosive events must occur to seriously affect Earth. According to one estimate, a supernova or GRB explosion close enough to alter Earth's biosphere occurs a few times every billion years, possibly leading to mass extinction events. In Chapter 17, we noted that the chemical elements that make up life were created in supernova explosions. The discussion here suggests that supernovae may have had some effect on the *evolution* of life on Earth as well.

SUMMARY

The highest mass stars leave behind black holes. In the environment surrounding black holes, relativistic effects become important. A black hole's mass determines its Schwarzschild radius: the boundary from which light cannot escape. Gamma-ray bursts are beamed high-energy explosions that result from the merger of two compact objects or the rapid collapse of a high-mass star into a black hole. The radiation from those bursts could affect life on Earth.

(1) **Relate the motion of the observer to the observed motion of other objects.** Even at low relative speeds, the motion of the observer can affect the direction of the measured velocity, as in the aberration of starlight. At higher speeds, the magnitude of the measured velocity also is affected. The speed of light in a vacuum, c, is the ultimate speed limit. Observers in all inertial reference frames will measure the same speed of light.

(2) **Summarize the observable consequences of the relationship between space and time.** Special relativity connects space and time into four-dimensional spacetime. How the spacetime distance between events is divided between space and time depends on the observer's motion. As velocities approach the speed of light, observers detect that moving clocks run more slowly than their own clocks, that moving objects are contracted in length, and that moving objects behave as though they were more massive.

(3) **Explain that gravity is a consequence of the way mass distorts the very shape of spacetime.** Inertial mass and gravitational mass are the same, leading to the principle of equivalence, in which acceleration cannot be distinguished from gravity if the acceleration is small enough. In general relativity, mass warps the fabric of spacetime so that objects move on the shortest path in that warped geometry. Gravity is a consequence of the spacetime warping. Time runs more slowly near massive objects, and radiation from any light source near a black hole's event horizon is redshifted.

(4) **Explain the formation of black holes from the most massive stars, and describe the key properties and observational consequences of those stellar black holes.** The masses for both white dwarfs and neutron stars have an upper limit. Dense stellar remnants more massive than about 2.5 M_{Sun} collapse to form black holes. The supernova explosion that ends the life of a massive star leaves behind a neutron star or a black hole. The mathematical singularity at the center of a black hole is still a mystery to science. However, scientists have now observed the mergers of many black holes through the gravitational waves emanating from such events. Black holes may, after a very, very long time, be destroyed by evaporation through Hawking radiation.

QUESTIONS AND PROBLEMS

TEST YOUR UNDERSTANDING

1. Rank the following (from least massive to most massive) in terms of the mass of the star that produces each.
   a. neutron star
   b. black hole
   c. white dwarf

2. A car approaches you at 50 km/h. A fly inside the car is flying toward the back of the car at 7 km/h. From your point of view by the side of the road, the fly is moving at _____ km/h.
   a. 7
   b. 28.5
   c. 43
   d. 57

3. A car approaches you at 50 km/h. The driver turns on the headlights. From your point of view, the light from the headlights is moving at
   a. $c + 50$ km/h.
   b. $c - 50$ km/h.
   c. $(c + 50$ km/h$)/2$.
   d. c.

4. Imagine that you are on a spaceship. A second spaceship rockets past yours at $0.5c$. You start a stopwatch and stop it 10 seconds later. For an astronaut in the other spaceship, the number of seconds that have ticked by during the 10 seconds on your stopwatch is
   a. more than 10 seconds.
   b. equal to 10 seconds.
   c. less than 10 seconds.

5. The International Space Station flies overhead. Using a telescope, you take a picture and measure its length to be _____ than its length as it would be measured if it were sitting on the ground.
   a. much greater
   b. slightly greater
   c. slightly less
   d. much less

6. Astronauts in the International Space Station
   a. have no mass.
   b. have no energy.
   c. are outside Earth's gravitational field.
   d. are in free fall.

7. Einstein's formulation of gravity
   a. is approximately equal to Newton's universal law of gravitation for small gravitational fields.
   b. is always used to calculate gravitational effects in modern times.
   c. explains why Newton's universal law of gravitation describes the motions of masses.
   d. both a and c

8. As the mass of a black hole increases, its Schwarzschild radius
   a. increases as the square of the mass.
   b. increases proportionately.
   c. stays the same.
   d. decreases proportionately.
   e. decreases as the square of the mass.

9. If a neutron star is more than ~2.5 times as massive as the Sun, it collapses because
   a. the force of electron degeneracy is stronger than gravity.
   b. gravity overpowers the force of electron degeneracy.
   c. gravity overpowers the force of neutron degeneracy.
   d. the force of neutron degeneracy is stronger than gravity.

10. Relative motion between two objects is apparent
    a. even at everyday speeds, such as 10 km/h.
    b. only at very large speeds, such as $0.8c$.
    c. only near very large masses.
    d. only when both objects are in the same reference frame.

11. If a spaceship approaches you at $0.5c$, and a light on the spaceship is turned on pointing in your direction, how fast will the light be traveling when it reaches you?
    a. $1.5c$
    b. between $1.0c$ and $1.5c$
    c. exactly c
    d. between $0.5c$ and $1.0c$

12. Imagine two protons traveling past each other at a distance d, with relative speed $0.9c$. Compared with two *stationary* protons a distance d apart, the gravitational force between those two protons when at their closest will be
    a. smaller because they interact for less time.
    b. smaller because the moving proton acts as though it has less mass.
    c. the same because the particles have the same mass.
    d. larger because the moving proton acts as though it has more mass.

13. If two spaceships approach each other, each traveling at 0.5c relative to an outside observer, spaceship 1 will measure spaceship 2 to be traveling
    a. much faster than c.
    b. slightly faster than c.
    c. at c.
    d. more slowly than c.

14. Some black holes are rapidly flickering X-ray sources. That observation indicates that
    a. X-rays are not light, because even light cannot escape from a black hole.
    b. the X-rays are coming from a very small source—so small that it could be a black hole.
    c. black holes are (at least sometimes) surrounded by hot gas.
    d. black holes have very high temperatures.

15. The current model of long-duration GRBs includes jets from the collapsed star. You have seen jets like that before, when studying (choose all that apply)
    a. giant planets.
    b. white dwarfs.
    c. pulsars.
    d. young stellar objects.
    e. supernovae.
    f. planetary nebulae.

THINKING ABOUT THE CONCEPTS

16. An astronomer sees a redshift in the spectrum of an object. Without any other information, can she determine whether it is an extremely dense object (exhibiting gravitational redshift) or one that is receding from her (exhibiting Doppler redshift)? Explain your answer.

17. Imagine you are traveling in a spacecraft at 0.9999999c. You point your laser pointer out the back window of the spacecraft. At what speed does the light from the laser pointer travel away from the spacecraft? What speed would be observed by someone on a planet traveling at 0.000001c?

18. According to Einstein's special theory of relativity, no object with mass can travel faster than, or even at, the speed of light. Recall that light is both an electromagnetic wave and a particle called a photon. If light acts as a particle, how can a photon travel at the speed of light?

19. Twin A takes a long trip in a spacecraft and returns younger than twin B, who stayed behind. Could twin A ever return before twin B was born? Explain.

20. In one frame of reference, event A occurs before event B. In another frame of reference, could the two events be reversed, so that B occurs before A? Explain.

21. ★ WHAT AN ASTRONOMER SEES   In Figure 18.22, suppose that the ring of light around the black hole's shadow was equally bright all the way around. What would you conclude about the rotation of the material around the black hole? Would this surprise you? Explain why or why not. 👁

22. Suppose you had a density meter that could instantly measure the density of an object. You point the meter at a person in a spacecraft zipping by at very high speed. Is that person's density value larger or smaller than an average person's? Explain.

23. Imagine a future astronaut traveling in a spaceship at 0.866c. According to special relativity, the spaceship is only half as long along the direction of flight as it was when at rest on Earth. The astronaut checks that prediction with a meterstick that he brought with him. Will his measurement confirm the contracted length of his spaceship? Explain your answer.

24. Within a speedy spacecraft, astronauts are playing soccer with a spherical ball. How would the shape of the ball appear to an observer watching the spacecraft speed by?

25. You observe a meterstick traveling past you at 0.9999c. You measure the meterstick to be 1 meter long. How is the meterstick oriented relative to you?

26. Suppose astronomers discover a 3-M_{Sun} black hole located a few light-years from Earth. Should they be concerned that its tremendous gravitational pull will lead to Earth's untimely demise?

27. If you could watch a star falling into a black hole, how would the color of the star change as it approached the event horizon?

28. Why don't people detect the effects of special and general relativity in their everyday lives here on Earth?

29. Many movies and television programs (such as *Star Wars*, *Star Trek*, and *Battlestar Galactica*) are premised on faster-than-light travel. How likely is it that such technology will be developed in the near future?

30. How could a gamma-ray burst in our galaxy potentially affect life on Earth?

APPLYING THE CONCEPTS

31. How much younger than your twin would you be if you made the journey described in Working It Out 18.1 at 0.5c? ●–●–●
    a. Make a prediction: Is 0.5c more or less than the speed in the example of Working It Out 18.1? So, then, do you expect your age and your twin's age to be closer together or further apart?
    b. Calculate: Follow Working It Out 18.1 to calculate how much younger you would be than your twin.
    c. Check your work: Compare your answer to your prediction.

32. What is the Schwarzschild radius of a black hole that has a mass of 1.4 M_{Sun}? ●–●–●
    a. Make a prediction: Review the paragraph after the Schwarzschild equation in Section 18.4. Do you expect this Schwarzschild radius to be comparable to the examples presented there? Much smaller? Much larger?
    b. Calculate: Solve the Schwarzschild equation to find the radius of this black hole.
    c. Check your work: Verify that your answer has units of length, and compare your result to your prediction.

33. How much younger than your twin would you be if you made the journey described in Working It Out 18.1 at 0.999c?
    a. Make a prediction: Is 0.999c more or less than the speed in the example of Working It Out 18.1? So, then, do you expect your age and your twin's age to be closer together or further apart?
    b. Calculate: Follow Working It Out 18.1 to calculate how much younger you would be than your twin.
    c. Check your work: Compare your answer to your prediction.

34. What is the Schwarzschild radius of a black hole that has a mass equal to the average mass of a person (~70 kilograms)?
    a. Make a prediction: Review the paragraph after the Schwarzschild equation in Section 18.4. Do you expect this Schwarzschild radius to be comparable to the examples presented there? Much smaller? Much larger?
    b. Calculate: Solve the Schwarzschild equation to find the radius of a 70-kg black hole.
    c. Check your work: Verify that your answer has units of length, and compare your result to your prediction.

35. What is the mass of a black hole with a Schwarzschild radius of 1.5 km?
    a. Make a prediction: Review the paragraph after the Schwarzschild equation in Section 18.4. Do you expect this black hole's mass to be comparable to the examples presented there? Much smaller? Much larger?
    b. Calculate: Solve the Schwarzschild equation to find the mass of this black hole.
    c. Check your work: Verify that your answer has units of solar masses, and compare your result to your prediction.

36. Does the angle of the rain falling outside the car in Figure 18.1b depend on the speed of the car? Knowing only that angle and the information on your speedometer, how could you determine the speed of the falling rain? ⊙★

37. Compare Figure 18.1 with Figure 18.2. If you knew the speed of Earth in its orbit from a prior experiment, how could you determine the speed of light from the angle of the aberration of starlight? ⊙★

38. According to Einstein, mass and energy are equivalent. So, which weighs more on Earth—a cup of hot coffee or a cup of iced coffee? Why? Do you think the difference is measurable?

39. As explained in the Process of Science Figure, a new theory should contain the old theory within it. Study Figure 18.5, which compares two imaginary rocket ships experiencing the same acceleration, g. ⊙★
    a. Approximately how fast (as a fraction of the speed of light) are the two spaceships going when the effects of relativity begin to be significant?
    b. Convert that speed to kilometers per hour. How does that speed compare with the speeds at which you usually travel? Why do you not usually see relativistic effects in your life?

40. Figure 18.6 shows how the Lorentz factor depends on speed. At about what speed (in terms of c) does the Lorentz factor begin to differ noticeably from 1? What happens to the Lorentz factor as the speed of an object approaches the speed of light? ⊙★

41. Imagine that the perihelion of Mercury advances 2 degrees per century. How many arcseconds does the perihelion advance in a year? (Recall that there are 60 arcseconds in an arcminute and 60 arcminutes in a degree.) Can Mercury's position be measured well enough to detect the advance of perihelion in 1 year?

42. The Moon has a mass equal to $3.7 \times 10^{-8} \, M_{Sun}$. Suppose the Moon suddenly collapsed into a black hole.
    a. What would be the Schwarzschild radius of the black-hole Moon?
    b. How would that collapse affect tides raised by the Moon on Earth? Explain.
    c. Would that event generate gravitational waves? Explain.

43. If a spaceship approaching Earth at $0.9c$ shines a laser beam at Earth, how fast will the photons in the beam be moving when they arrive at Earth?

44. Suppose you discover signals from an alien civilization coming from a star 25 light-years away, and you go to visit it by using the spaceship described in the discussion of the twin paradox in Working It Out 18.1.
    a. How long will you take to reach that planet, according to your clock? According to a clock on Earth? According to the aliens on the other planet?
    b. How likely is it that someone you know will be here to greet you when you return to Earth?

45. Working It Out 18.2 relates the mass of a binary pair to the period and the size of the orbit. Suppose that a spaceship orbited a black hole at a distance of 1 AU, with a period of 0.5 year. What assumptions could you make that would allow you to calculate the mass of the black hole from that information? Make those assumptions, and calculate the mass.

EXPLORATION Black Holes

Because grabbing a black hole and bringing it into the lab is not possible, and because Earth has never actually been close to one, astronomers can conduct only thought experiments to explore the properties of black holes. The following are a few thought experiments to help you better understand what's happening near a black hole.

Imagine a big rubber sheet. It is very stiff and not easily stretched, but it does have some "give" to it. At the moment, it is perfectly flat. Imagine rolling some golf balls across it.

1 Describe the path of the golf balls across the sheet.

_____

Now imagine putting a bowling ball (very much heavier than a golf ball) in the middle of the sheet, so that it makes a big, slope-sided pit. Roll some more golf balls.

2 What happens to the path of the golf balls when they are very far from the bowling ball?

_____

_____

3 What happens to the path of the golf balls when they come just inside the edge of the dip?

_____

_____

4 What happens to the path of the golf balls when they go directly toward the bowling ball?

_____

_____

5 How do each of the three cases in questions 2–4 change if the golf balls are moving very, very fast? What if they are moving very slowly?

_____

_____

6 What happens to the depth and width of the pit as the golf balls fall into the center near the bowling ball? (Imagine putting *lots* of golf balls in.)

_____

_____

_____

All of the preceding thought experiments relate to ordinary stuff. Stars, people, planets—everything interacts in that way because of gravity. With black holes, things are different. Here, it is more accurate to think of the bowling ball as a hole in the sheet that pulls it down rather than as an object that sits on it. But the bowling ball still affects the sheet in the same way. The hole is a good analogy for the event horizon of a black hole. Objects outside the event horizon will know that the black hole is there because the sheet is sloping, but they won't be captured unless they come within the event horizon. Think about light for a moment as though it were, say, grains of sand rolling across the sheet.

7 What happens to the light as it passes far from the pit? What happens if it reaches the hole?

_____

_____

Now suppose you roll another bowling ball across the sheet.

8 What happens to the sheet when the second bowling ball falls in after the first? Would that change affect your golf balls and grains of sand? How? What happens to the hole? What happens to the size of the pit?

_____

_____

_____

_____

None of those thought experiments take into account relativistic effects (length contraction and time dilation). Imagine for a moment that you are traveling close to the black hole.

9 Look out into the galaxy and describe what you see. Consider the lifetimes of stars, the distances between them, their motions in your sky, and how they die. Add anything else that occurs to you.

_____

_____

_____

_____

Galaxies

Astronomers use many different methods for determining distance, as we discuss here in Chapter 19. In Chapter 13, you learned how astronomers use parallax to find the distance to the closest stars. To go further, astronomers need a "standard candle"—a type of object that has a known luminosity. Street lights along a single street typically use the same bulbs—they all have the same luminosity, so they make good standard candles. If a street light is fainter, it is farther away. Find a straight street that is well lit by regularly placed street lights—the longer, the better. Look at the nearest street light, and then look down the line of lights until you find one that is half as bright. Repeat this experiment as many times as you can, finding the next light that is half as bright as the previous one. If you do this carefully and well, you can make an accurate graph of the brightness of the lights versus the distance, measured as the number of street lights.

EXPERIMENT SETUP

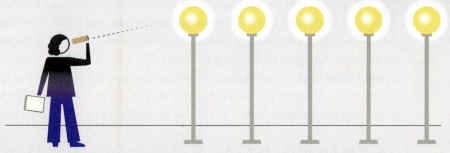

Take a toilet paper tube out with you at night and find a straight street that is well-lighted by regularly placed street lights—the longer, the better. The brightness of each light tells you how near it is to you.

Use your toilet paper tube to restrict your vision to just one street light at a time. Look at the nearest street light.

Move down the line of lights until you find one that is half as bright. How many street lights away is the light that is half as bright as the nearest one?

Repeat this experiment as many times as you can, finding the next light that is half as bright as the previous one. If you do this carefully and well, you can make an accurate graph of the brightness of the lights versus the distance, measured as the number of street lights.

PREDICTION

I predict that for each drop in brightness by half, I will count ☐ **the same** ☐ **a different** number of street lights.

SKETCH OF RESULTS (in progress)

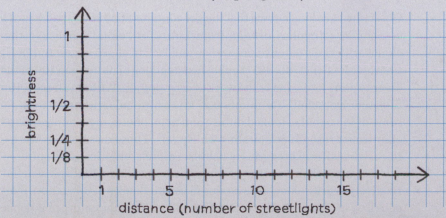

19

Nearly a century has passed since astronomers realized that the universe is filled with huge collections of stars, gas, and dust called galaxies. Just as stars vary in their mass or their stage of evolution, galaxies come in many forms. Chapter 19 begins our discussion of galaxies with a survey of the types of galaxies and their basic properties to understand better the differences among them. Chapter 20 then looks in detail at our galaxy, the Milky Way, and in later chapters we describe the evolution of galaxies and of the universe itself.

LEARNING GOALS

By the end of this chapter, you should be able to:

① Determine a galaxy's type from its appearance and describe the motions of its stars.

② Organize the steps of the distance ladder and explain how distances to galaxies are measured at each step.

③ Describe the evidence suggesting that galaxies are composed mostly of dark matter.

④ Discuss the evidence indicating that most—perhaps all—large galaxies have supermassive black holes at their centers.

19.1 Galaxies Come in Different Shapes and Sizes

A **galaxy** is a gravitationally bound collection of a million to hundreds of billions of stars, dust, gas, and dark matter. Even though the number of stars in the Milky Way Galaxy is very large—a few hundred billion—the number of galaxies in the universe is even larger; astronomers estimate there are about two trillion galaxies in the observable universe. Most of those galaxies are so far away that they appear too small and faint to detect with any but the most powerful telescopes. In this section, we explain how astronomers concluded that galaxies were separate from our Milky Way, and how the types of galaxies are different from one another.

The Discovery of Galaxies in the 20th Century

Observers have long known that the sky contains faint, misty patches of light, known as nebulae (singular: *nebula*, from Latin for "cloud") because they resemble clouds with their patchy appearance and lack of sharp edges. In 1784, astronomer Charles Messier (1730–1817) published a catalog of more than 100 of these nebulae. Twenty years later, due to the work of astronomers William Herschel and his sister, Caroline Herschel, the number of known nebulae jumped to 2500. Although some of the nebulae looked diffuse and amorphous, most appeared to be round or elliptical or resembled spiraling whirlpools. Those distinctions were the basis for the original three categories of nebulae: diffuse, elliptical, and spiral.

For more than a century, astronomers speculated that spiral nebulae might be relatively nearby planetary systems in various stages of formation, like the systems described in Chapter 7. Alternatively, the influential 18th century philosopher Immanuel Kant (1724–1804) proposed that spiral nebulae were instead "island universes"—

separate from the Milky Way Galaxy. In order to test these hypotheses, astronomers needed to figure out how far away the nebulae were, as well as determine the size of the Milky Way. If the nebulae were close enough to be inside the Milky Way, they might be planetary systems in formation. If the nebulae were much farther than the edge of the Milky Way, the "island universe" hypothesis would be supported instead.

The presence of interstellar dust complicated early attempts to understand the size of the Milky Way. Because early astronomers did not know that the dust exists or that it blocks the passage of visible light through the Milky Way, they seriously underestimated the size of the Milky Way, concluding that the Milky Way was only 1800 parsecs (pc) across (recall from Chapter 13 that 1 parsec = 3.26 light-years). In the beginning of the 20th century, astronomer Harlow Shapley (1885–1972) observed globular clusters, which tend to be found most often in the outer reaches of the Milky Way. His observations "grew" the size of the Milky Way by more than 50 times, to 92,000 pc in diameter. Because there were more globular clusters in one direction than another, Shapley was also able to determine that the Sun was not at the center of the Milky Way.

Shapley thought his far larger estimate for the size of the Milky Way meant that it was big enough to encompass everything in the universe, and therefore he thought that the spiral and elliptical nebulae were inside the Milky Way. Astronomer Heber D. Curtis (1872–1942) preferred the earlier, smaller model of the Milky Way. He also favored the idea that the spiral nebulae were in fact galaxies separate from the Milky Way and that therefore the whole universe was larger than the Milky Way. In 1920, Shapley and Curtis met in Washington, D.C., to debate publicly their interpretations of the nature of spiral nebulae. Historians call that meeting astronomy's *Great Debate*. Although that debate did not resolve the issue at the time, it set the stage and gave direction to the later work of Edwin P. Hubble (1889–1953), who not only settled this issue, but also fundamentally changed the modern understanding of the universe.

Using the newly finished 100-inch telescope on Mount Wilson, high above the then-small city of Los Angeles, Hubble found some variable stars in the spiral nebula Andromeda (**Figure 19.1**). Those stars were very similar to but appeared fainter than other known Cepheid variable stars. Hubble used Henrietta Swan Leavitt's period-luminosity relation for Cepheid variable stars (discussed in Chapter 17) to find the luminosity of these stars. He compared this luminosity to the observed average brightness to find the distance. The results showed that the distances to those stars were far greater than Shapley's size of the Milky Way. Therefore, the distance to the spiral nebula Andromeda was also far greater than the size of the Milky Way. Hubble concluded that Kant was correct: the Milky Way is one of many island universes. Further work showed that most diffuse nebulae are clouds of gas and dust within the Milky Way Galaxy, but Messier's elliptical and spiral nebulae are instead galaxies, which are similar in size to the Milky Way but located at truly immense distances.

Types of Galaxies

Imagine taking a handful of coins and throwing them in the air, as shown in **Figure 19.2a**. You know that all of the coins are flat and circular. When you look at them falling through the air, however, they do not appear all the same. Some coins look circular because you see them "face on." Others look like thin lines, because you see them "edge on." Most of the coins, though, are seen from an angle between those two extremes, so they appear with various degrees of ellipticity, or flattening. Even if that single image of many coins was the only information you had, you could use it to figure out that the three-dimensional shape of a coin is flat and circular.

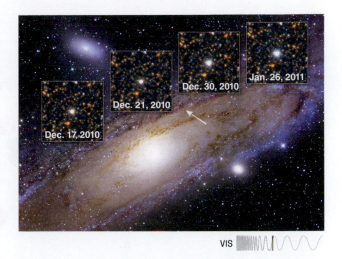

Figure 19.1 The Andromeda Galaxy, the nearest large galactic neighbor to the Milky Way, is about 2.5 million light-years (780,000 pc) away. The arrow points to the Cepheid variable star V1, a standard candle that Hubble used to estimate the distance to Andromeda. The insets show the variability in the light from V1. Hubble's measurement provided the first observational evidence of the vastness of the universe.

Astronomy in Action: Galaxy Shapes and Orientation

Figure 19.2 **a.** A handful of coins thrown in the air illustrates the difficulties in identifying the shapes of certain types of galaxies. Like the coins in this picture, galaxies are seen in various orientations—some face on, some edge on, and most somewhere in between. **b.** These disk-shaped galaxies are seen from various perspectives or angles, corresponding to the range of perspectives for the coins. Credit (part b, inset 3): ESO, https://www.eso.org/public/hungary/images/eso0902c/. https://creativecommons.org/licenses/by/4.0/.

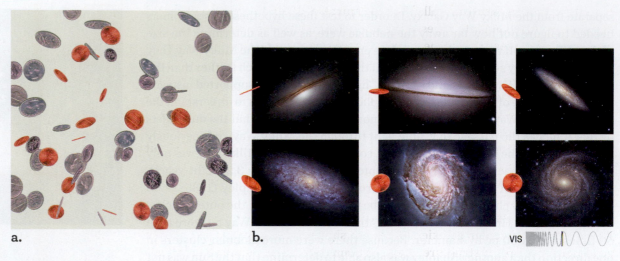

a.

b.

VIS

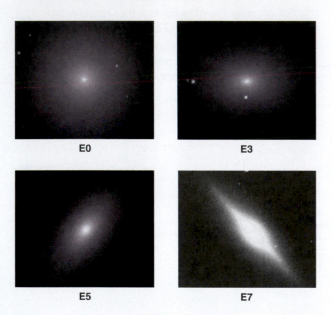

E0

E3

E5

E7

Figure 19.3 Elliptical galaxies are classified on the basis of how elliptical they appear to be from our point of view. The most circular are E0 galaxies, whereas the most elliptical are classified as E7.

Similarly, by studying galaxies in their random orientations on the sky, astronomers can reconstruct their three-dimensional shapes. **Figure 19.2b** shows a set of galaxies, viewed from our perspective at various viewing angles. The similarities in appearance among these galaxies indicate that they share common features and can be grouped into one classification. In the 1930s, Edwin Hubble sorted galaxies into three categories, based on their appearance in visible light: ellipticals, spirals, and irregulars.

Elliptical galaxies have either spherical or ellipsoidal shapes in three dimensions, as shown in **Figure 19.3**. Elliptical galaxies are further divided into categories based on their roundness (from our point of view): they have numbered subtypes ranging from nearly spherical (E0) to flattened (E7). The appearance of an elliptical galaxy in our sky does not necessarily tell us its true shape. For example, a galaxy might actually be shaped like a rugby ball (which is elongated, but has rounded ends), but if viewed end on, it looks round like a soccer ball instead. Modern telescopic observations have revealed that many, if not most, elliptical galaxies contain small rotating disks at their centers.

Spiral galaxies (**Figure 19.4**) have spiral arms in a flattened, rotating disk. They also have a central **bulge**, which extends above and below the disk. In about 2/3 of spiral galaxies, this bulge is stretched out, forming a bar across the center of

Sa

Sb

Sc

SBa

SBb

SBc

Figure 19.4 Spiral and barred spiral galaxies are classified by the openness of the arms and the prominence of the bulge.

the galaxy: those galaxies are called **barred spirals (SB)**. Both spirals and barred spirals are subdivided into types a, b, and c according to the prominence of the central bulge and how tightly the spiral arms are wound. For example, Sa and SBa galaxies have the largest bulges and display tightly wound and smooth spiral arms. Sc and SBc galaxies have small central bulges and more loosely wound spiral arms that are often very knotty in appearance. The Milky Way Galaxy is a barred spiral, classified SBbc (between SBb and SBc).

Some galaxies, known as **S0 galaxies**, have a shape that combines properties of both spiral and elliptical galaxies. The S0 galaxies have stellar disks but no spiral arms, so the disk is smooth in appearance, like an elliptical galaxy, as shown in **Figure 19.5**. Hubble differentiated S0 galaxies as either barred (SB0) or unbarred (S0).

Irregular galaxies do not fit neatly into any of these categories. As their name implies, irregular galaxies like those shown in **Figure 19.6** often lack symmetry in shape or structure. Many of them once were spirals or ellipticals that became distorted by the gravity of another galaxy. **Table 19.1** summarizes the criteria that Hubble used to classify galaxies.

Using larger telescopes than those available to Hubble, as well as improved camera technology and more extensive deep-sky surveys, astronomers have identified many small, faint galaxies. **Dwarf galaxies** are usually 1/10 or less the size of our Milky Way, with lower total luminosities than the types discussed previously. Dwarf galaxies can be elliptical, spheroidal, irregular, or more rarely spiral. Some are very compact with high density and are like the cores of ellipticals; some are very compact and blue with clusters of hot, young stars. When objects (like galaxies and nebulae) are large in the sky, astronomers consider not only their total brightness (all the light from the object added together), but also the **surface brightness**—how bright each area is. The surface brightness is important because it often limits whether the object can be seen at all (that is, the object must be brighter than the surrounding sky in order to be observed). An object that is very bright in total but also very extended might not be observable because the light is too spread out. Many known dwarf galaxies have a surface brightness that is only a few percent brighter than the background sky; there are almost certainly many more dwarf galaxies that have even lower surface brightness, so they cannot be observed. Dwarf galaxies are more difficult to detect than the other types, but are probably the most common kind of galaxy in the universe.

VIS

Figure 19.5 An S0 galaxy has a disk structure, but lacks the dust and gas of a spiral galaxy, so there are no clear dark lanes cutting through the disk. The tiny dots surrounding NGC 5308 in this image are not image defects; they are large globular clusters in orbit around the galaxy.
Credit: ESA/Hubble & NASA Acknowledgement: Judy Schmidt (Geckzilla), https://esahubble.org/images/potw1620a/. https://creativecommons.org/licenses/by/4.0/.

VIS

Figure 19.6 Irregular galaxies are often the result of collisions. This montage of six examples shows how these collisions destroy the tidy structure of spiral galaxies as they collide and, eventually, merge.
Credit: ESA/Hubble & NASA, A. Adamo et al., https://esahubble.org/images/heic2101a/. https://creativecommons.org/licenses/by/4.0/.

Table 19.1	The Hubble Classification of Galaxies

A Classification Scheme Based on the Properties of Galaxies

Category	Criteria	Abbreviation	Range of Features			
Elliptical	Mostly bulge Old, red stellar population Smooth-appearing	E0 ↕ E7	More spherical ↕ More elongated			
S0 (unbarred/barred)	Bulge and disk with no arms and with mostly old, red stars	S0/SB0	Smooth disk and bulge			
Spiral (unbarred/barred)	Bulge and disk with arms Bulge has old, red stars Disk has both old, red stars and young, blue stars Spirals (S) have roundish bulges Barred spirals (SB) have elongated or barred bulges	Sa/SBa Sb/SBb Sc/SBc	More bulge ↕ Little bulge	Tightly wound arms ↕ Open arms	Smooth arms ↕ Knotty arms	
Irregular	No arms, no bulge Some old stars, but mostly young stars, gas, and dust, giving a knotty appearance	Irr				

Stellar Motions and Galaxy Shape

Stellar motions determine galaxy shapes. A galaxy is not a solid object like a coin but rather a collection of stars, gas, particles, and dust. In an elliptical galaxy, stars move in all possible directions, following orbits with a wide range of shapes and orientations (**Figure 19.7**). Those orbits are more complex than the elliptical orbits of planets about a star. Taken together, all those stellar orbits give an elliptical galaxy its shape. The faster the stars are moving, the more spread out the galaxy is. If the stars in an elliptical galaxy are moving in truly random directions, the galaxy will have a spherical shape. If stars are more likely to have certain directions of motion than others at each location, however, the galaxy will be more spread out in that direction, giving it an elliptical shape.

The orbits of stars in spiral galaxy disks are different from those of stars in elliptical galaxies. The components of a barred spiral galaxy are shown in **Figure 19.8**. The defining feature of a spiral galaxy is that it has a flattened, rotating disk. Most stars in the disk of a spiral galaxy follow nearly circular orbits and travel in the same direction around the bulge at the center of the galaxy. But the stellar orbits

Figure 19.7 Elliptical galaxies take their shape from the orbits of the stars they contain. The colored lines superimposed on the galaxy represent the complex orbits of its stars.
Credit: NOIRLab/NSF/AURA, https://noirlab.edu/public/images /noao-m59/. https://creativecommons.org/licenses/by/4.0/.

VIS

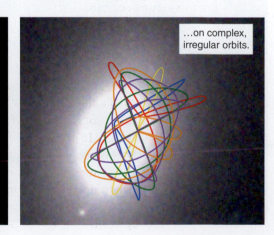

Stars in elliptical galaxies move in all directions…

…on complex, irregular orbits.

in a spiral galaxy's central bulge are different from those in the galaxy's disk. As with elliptical galaxies, the stars in the bulge follow orbits in random directions. The bulges of unbarred spiral galaxies are thus roughly spherical.

CHECK YOUR UNDERSTANDING 19.1a

Galaxies are classified according to: (a) mass; (b) color; (c) density; (d) shape.

Answers to Check Your Understanding questions are in the back of the book.

Other Differences among Galaxies

In addition to the differences in their stellar orbits, other important distinctions exist between spiral and elliptical galaxies. Those distinctions carry information about how they evolved in the past and how they will evolve in the future.

Gas and Dust   Most spiral galaxies contain large amounts of dust and cold, dense molecular gas concentrated in the midplanes of their disks, in addition to neutral hydrogen gas. Just as the dust in the Milky Way's disk can be seen on a clear summer night as a dark band slicing the galaxy in two (see Figure 15.2), the dust in an edge-on spiral galaxy appears as a dark, obscuring band running down the midplane of the disk (**Figure 19.9**), sometimes called a "dust lane." The cold molecular gas that accompanies the dust can be seen in radio observations of spiral galaxies. In contrast, giant elliptical galaxies contain large amounts of very hot gas that astronomers see primarily by observing the X-rays emitted by the gas, but they contain very little cold dust and gas.

The difference in shape between elliptical and spiral galaxies offers some insight into why the gas in giant ellipticals is hot, whereas the gas in spirals is cold and dense. Conservation of angular momentum causes cold gas to settle into the disk of a spiral galaxy, just as gas settles into a disk around a forming star. In contrast, elliptical galaxies do not have an overall rotation—the orbits of stars or dust and gas particles are in random directions, so the gas does not settle into a disk. The only place in an elliptical galaxy where cold gas could collect is at the center. However, the density of stars in elliptical galaxies is so high that evolving stars and Type Ia supernovae continually reheat that gas, thus preventing most of it from cooling off and forming cold interstellar clouds.

Color   The colors of spiral and elliptical galaxies reveal a great deal about their star formation histories and their present star formation. Recall from Chapter 15 that stars form from dense clouds of cold molecular gas. Because the gas seen in elliptical galaxies is very hot, active star formation is not taking place in those galaxies today. The reddish colors of elliptical and S0 galaxies confirm that little or no star formation has occurred there for some time. The stars in those galaxies are an older population of lower-mass stars. In contrast, the bluish colors of the disks of spiral galaxies indicate that massive, young, hot stars are forming in the cold molecular clouds contained within the disk. Even though most of the stars in a spiral disk are old, the massive, young stars are so luminous that their blue light dominates. Star formation in most irregular galaxies is like that in spiral galaxies. Some irregular galaxies form stars at prodigious rates, given their relatively small sizes. Irregular and disk galaxies that are experiencing intense bursts of star formation are called starburst galaxies.

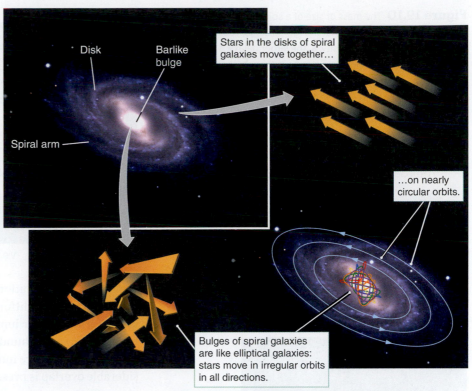

Figure 19.8 The components of a barred spiral galaxy include a barlike bulge, a disk, and spiral arms. The orbits of stars in the rotating disk are different from the orbits of stars in the elliptically shaped bulge.
Credit: NOIRLab/NSF/AURA, https://noirlab.edu/public/images /noao-m109/. https://creativecommons.org/licenses/by/4.0/.

VIS

Figure 19.9 This Hubble Space Telescope (HST) image shows the nearly edge-on spiral galaxy M104 (the Sombrero Galaxy; type Sa). The dust in the plane is seen as a dark, obscuring band in the midplane of the galaxy. Note the bright halo made up of stars and globular clusters. Compare this image with Figure 15.2, which shows the dust in the plane of the Milky Way.

Figure 19.10 The mass or size of a spiral galaxy does not determine its appearance. Even though these galaxies *appear* to be similar in size and luminosity, **a.** the larger galaxy is 4 times more distant and 10 times more luminous than **b.** the smaller galaxy.
Credit (part b): ESO, https://www.eso.org/public/hungary/images/eso9949a/. https://creativecommons.org/licenses/by/4.0/.

Large, distant spiral

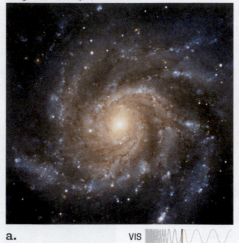

Smaller, nearby spiral

a.   VIS

b.   VIS

Luminosity   The relationship between luminosity and radius among the types of galaxies is not straightforward. Galaxies range in luminosity from tens of thousands up to a trillion solar luminosities (10^4 to 10^{12} L_{Sun}). They range in size from a few hundred parsecs to hundreds of thousands of parsecs. Although the most luminous elliptical galaxies are more luminous than the most luminous spiral galaxies, considerable overlap is present in the range of luminosities among all galaxy types.

Mass   For a star, mass is the single most important parameter in determining its properties and evolution. In contrast, differences in mass (and also size) typically do not lead to obvious differences among galaxies. Even when a larger, more distant spiral galaxy (**Figure 19.10a**) is seen next to a smaller, nearby spiral galaxy (**Figure 19.10b**), telling which is which by appearance alone can be hard. Galaxies that have relatively low luminosity (less than 1 billion L_{Sun}) are the dwarf galaxies, and those that are more than 1 billion L_{Sun} are called giant galaxies, because the luminosity indicates the number of stars and therefore the total amount of stellar mass. (For comparison, the Milky Way has a luminosity of about 100 billion L_{Sun}.) The difference between a dwarf elliptical galaxy and a giant elliptical galaxy is shown in **Figure 19.11**. The giant elliptical galaxy has a much higher density of stars, and they are more centrally concentrated than stars in the dwarf elliptical galaxy.

Dwarf elliptical galaxy

a.   VIS

Giant elliptical galaxy

b.   VIS

Figure 19.11 a. Dwarf elliptical galaxies differ in appearance from **b.** giant elliptical galaxies. Stars in giant elliptical galaxies are more centrally concentrated than those in dwarf elliptical galaxies.
Credit (part a): NOIRLab/NSF/AURA, https://noirlab.edu/public/images/noao-m110/. https://creativecommons.org/licenses/by/4.0/.

CHECK YOUR UNDERSTANDING 19.1b

Currently, star formation rates are highest in (a) elliptical galaxies; (b) S0 galaxies; (c) spiral galaxies; (d) dwarf galaxies.

19.2 Astronomers Use Several Methods to Find Distances to Galaxies

To determine the distances to galaxies, astronomers started with the closest objects and looked for patterns that would help them find distances to the farthest objects—just as they did for stars. Distances are measured in a series of different methods called the **distance ladder**, which relates distances on a variety of overlapping scales, each method building on the previous one. In this section, we discuss those distance methods and how they led to the surprising discovery that the galaxies are moving away from us.

The Distance Ladder

The modern distance ladder is summarized in **Figure 19.12**. Since the 1960s, distances within the Solar System have been found using radar and signals from space probes. Using the distance from Earth to the Sun (it was estimated long before it was measured by radar), trigonometric parallax (see Chapter 13) was used to determine the distances to nearby stars. The distances to those nearby stars can then be combined with their brightness in our sky to find their luminosity. From their luminosity and the spectral classifications of nearby stars, astronomers created the H-R diagram. For more distant stars, astronomers use the spectral classification of a main-sequence star to determine its position on the H-R diagram. That position provides a star's luminosity, enabling astronomers to then estimate its distance by comparing its apparent brightness with its luminosity through using spectroscopic parallax (see Chapter 13 and Appendix 7).

In order to measure distances to objects in the outer reaches of the Milky Way and beyond, astronomers use *standard candles*, a term borrowed from a method of measuring the brightness of actual candles. **Standard candles** are objects of a particular type, such as Type Ia supernovae, that have a known luminosity. Astronomers combine the known luminosity and measured brightness of the object to find its distance. The objects must be bright enough as viewed from here on Earth to be recognizable in the distant galaxy.

Objects that can be used as standard candles include main-sequence O stars, globular clusters, planetary nebulae, novae, Cepheid variable stars, and supernovae. For example, Hubble Space Telescope (HST) observations of Cepheid variables enable astronomers to measure distances accurately to galaxies as far away as 30 million parsecs (also called 30 **megaparsecs [Mpc]**). Type Ia supernovae are even more luminous than the Cepheids, and thus detectable at greater distances.

Recall from Chapter 16 that Type Ia supernovae can occur when gas flows from an evolved star onto its white dwarf companion, pushing the white dwarf up toward $1.4\,M_{Sun}$, which is the Chandrasekhar limit for the mass of an electron-degenerate object. When that happens, the white dwarf burns carbon,

what if . . .

What if two giant spiral galaxies merge together into a more giant galaxy? What do you think the result will look like, and will it depend on the relative orientation of the two galaxies' angular momentum?

Figure 19.12 The distance ladder, indicating how the distances to remote objects are estimated through a series of methods beginning with relatively nearby objects. Distances are given in parsecs (pc). The scale is logarithmic, so each blue arc is 10 times farther away than the one before it. That means that radar is useful for only a tiny portion of the universe, and the methods using distant objects such as Type Ia supernovae cover most of the universe.

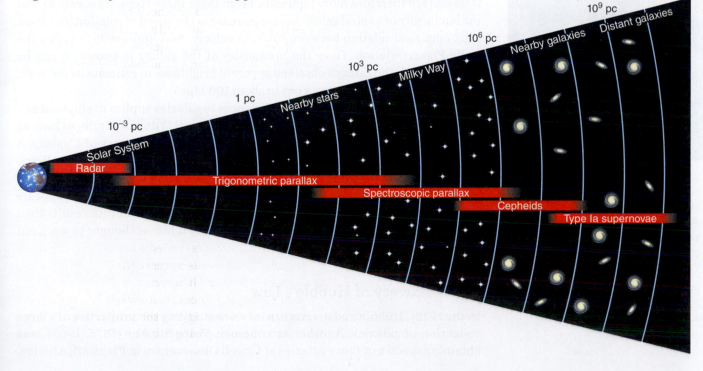

a.   UV

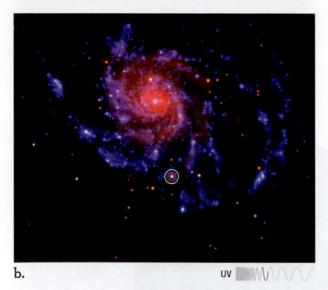

b.   UV

Figure 19.13 Type Ia supernovae are extremely luminous standard candles. These ultraviolet images show the Pinwheel Galaxy, located 6.4 Mpc away, **a.** before and **b.** after Supernova 2011fe.

▶❚❚ **AstroTour:** Hubble's Law

collapses, and then explodes. At first, astronomers thought that those Type Ia supernovae all occurred in white dwarfs of just below 1.4 M_{Sun}. In that situation, all such explosions would occur at the same mass and have similar luminosity, with some calibration adjustment for the rate at which the brightness declines after it peaks. But now astronomers estimate that 80 percent of Type Ia supernovae come from double-degenerate systems—for example, two white dwarfs that merge and then explode. The double white dwarf supernova could have up to twice the mass of a single white dwarf supernova. Because the amount of mass involved may range from one to two times the Chandrasekhar limit, the luminosity might vary by as much as a factor of two. In addition, Type Ia supernovae at large distances may have different colors or compositions than nearby ones, because they formed when galaxies were younger. To test whether all Type Ia supernovae have about the same luminosity, astronomers observe nearby Type Ia supernovae in galaxies with distances determined from other methods, such as by Cepheid variables. So far, these Type Ia supernovae *do* have similar luminosities to one another, but it is an ongoing project to patiently wait for more nearby Type Ia supernovae so that we can determine the full range of luminosities involved. With a peak luminosity that can outshine a billion Suns (**Figure 19.13**), Type Ia supernovae can be seen and measured with modern telescopes at very large distances (**Working It Out 19.1**).

There are two other methods for roughly estimating the distance to a galaxy. First, the Tully-Fisher relation connects the rotation of spiral galaxy disks to the luminosity. In a rotating spiral galaxy, half of the disk is approaching Earth, so the light it emits is blueshifted, whereas the other half of the disk is moving away from Earth, so the light it emits is redshifted. In most cases, an observed spectral line, for example the 21-cm line of hydrogen, will not split into two (a redshifted part and a blueshifted part); instead, the line will be broadened, or made wider in the spectrum. There are three logical steps required to use this observation to find the luminosity of the galaxy: (1) faster rotations make broader lines; (2) faster rotating disks also require more mass to hold them together; (3) more massive galaxies have more stars and are therefore more luminous. From these three steps, we conclude that the luminosity of a spiral galaxy should increase as the speed of rotation increases. That empirical relation between rotation velocity and luminosity is called the Tully-Fisher relation. Once the luminosity of the galaxy is known, it can be compared with the galaxy's observed apparent brightness to estimate its distance. This method is thought to work out to about 100 Mpc.

The second method to estimate distances to galaxies applies to elliptical galaxies. Elliptical galaxies (and the bulges of S0 galaxies) do not rotate, so instead astronomers look at the distribution of the surface brightness of a galaxy. A closer galaxy shows more variations in surface brightness because the distribution of stars throughout the galaxy isn't perfectly uniform. For more distant galaxies, those variations are less noticeable, and the surface brightness appears more uniform across the galaxy. That method yields less accurate results than the Tully-Fisher method for spirals, but generally it also is thought to work out to about 100 Mpc.

The Discovery of Hubble's Law

In the 1920s, Hubble and his coworkers were studying the properties of a large collection of galaxies. Another astronomer, Vesto Slipher (1875–1969), was obtaining spectra of those galaxies at Lowell Observatory in Flagstaff, Arizona.

working it out 19.1

Finding the Distance from a Type Ia Supernova

How do astronomers use a standard candle to estimate distance? Figure 19.13 shows the Pinwheel Galaxy (M101), with a supernova that was observed in 2011. That supernova was detected before its brightness reached its peak. Astronomers can then compare the peak observed brightness of this supernova with the peak luminosity for that type of supernova to compute the distance.

In Section 13.1, we related brightness, luminosity, and distance using the following equation:

$$\text{Brightness} = \frac{\text{Luminosity}}{4\pi d^2}$$

Rearranging to solve for distance gives:

$$d = \sqrt{\frac{\text{Luminosity}}{4\pi \times \text{Brightness}}}$$

The maximum observed brightness of the supernova was 7.5×10^{-12} watts per square meter (W/m^2). The graph in Figure 17.11 shows that the typical maximum luminosity L of a Type Ia supernova is 9.5×10^9 times the luminosity of the Sun:

$$L = 9.5 \times 10^9 \times L_{\text{Sun}}$$

$$L = 9.5 \times 10^9 \times (3.9 \times 10^{26}\,\text{W}) = 3.7 \times 10^{36}\,\text{W}$$

Thus, we can solve the equation:

$$d = \sqrt{\frac{3.7 \times 10^{36}\,\text{W}}{4\pi \times 7.5 \times 10^{-12}\,\text{W/m}^2}} = 2.0 \times 10^{23}\,\text{m}$$

We can convert that value from meters to megaparsecs as follows:

$$d = \frac{2.0 \times 10^{23}\,\text{m}}{3.1 \times 10^{22}\,\text{m/Mpc}} = 6.4\,\text{Mpc}$$

The distance is 6.4 Mpc. Because the Pinwheel Galaxy is relatively close, other standard candles can be observed in that galaxy to help calibrate the distance. Finally, note that 6.4 Mpc = 21 million light-years, so that supernova explosion took place 21 million years ago.

Slipher's galaxy spectra looked like the combined spectra of many stars with a bit of glowing interstellar gas mixed in. He was surprised to find, however, that the emission and absorption lines in the spectra of those galaxies were rarely seen at the same wavelengths as in the spectra of stars observed in the Milky Way Galaxy. Instead, the lines were almost always shifted to longer wavelengths, as seen in **Figure 19.14**.

Slipher characterized most of the observed shifts in galaxy spectra as redshifts because the light from those galaxies is shifted to longer (redder) wavelengths. Hubble interpreted Slipher's redshifts as Doppler shifts, concluding that almost all the galaxies in the universe are moving away from the Milky Way. Recall from Chapter 5 that objects with larger Doppler redshifts are moving away more quickly than those with smaller redshifts. When Hubble combined the measurements of galaxy velocities with his own estimates of the distances to those galaxies, he found that distant galaxies are moving away from Earth more rapidly than are nearby galaxies. Specifically, *the velocity at which a galaxy is moving away from an observer is proportional to the distance of that galaxy*. That relationship between distance and recession velocity is known as **Hubble's law** (also known as the Hubble-Lemaître law). Hubble's law is foundational to modern astrophysics, and we explain more about it not only in this chapter, but also in chapters to come.

The data that support Hubble's law are shown in **Figure 19.15**, which plots the measured recession velocities of galaxies against their measured distances. The points lie along a line on the graph with a slope equal to the proportionality constant H_0, called the **Hubble constant**. Notice how well the data line up along the line,

unanswered questions

How standard are "standard candles"? For example, Cepheid variable light curves are slightly different, depending on the amount of heavy elements in the stars, so that variation must be calibrated. Most Type Ia supernovae may originate from the merging of two compact objects rather than from one white dwarf accreting mass and exploding when it gets close to $1.4\,M_{\text{Sun}}$. Astronomers try to address those difficulties by using multiple methods to find distances to galaxies—for example, observing many Cepheids, bright O stars, and Type Ia supernovae in the same galaxy—to check that their calculated distances agree. In one recent study, astronomers used the HST to observe more than 600 Cepheid variable stars in eight galaxies in which Type Ia supernovae had been detected, and they reduced the uncertainty in their distances. The Type Ia supernovae calibrated in that way are consistent with earlier-calibrated maximum luminosities. However, those supernovae in galaxies that are far enough away that no other standard candles exist may have slightly different maximum luminosities, leading to a less precise distance estimate.

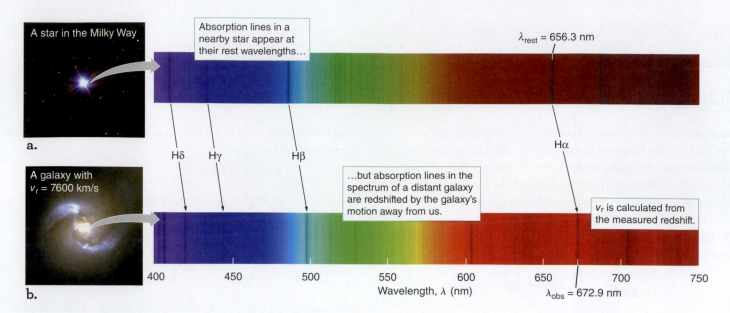

Figure 19.14 a. The spectrum of a star in our galaxy shows absorption lines, which here lie at the rest wavelength. **b.** A distant galaxy, shown with its spectrum at the same scale as that of the star, has lines redshifted to longer wavelengths. v_r is recession velocity, or radial velocity.

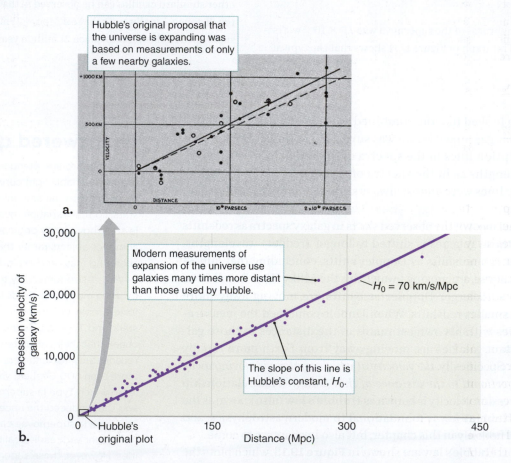

Figure 19.15 a. Hubble's original graph shows that more distant galaxies are receding faster than less distant galaxies. **b.** Modern data on galaxies many times farther away than those studied by Hubble show that recession velocity is proportional to distance.

working it out 19.2

Redshift—Calculating the Recession Velocity and Distance of Galaxies

Recall from Working It Out 5.2 that the Doppler equation for the radial velocity of spectral lines is

$$v_r = \frac{\lambda_{obs} - \lambda_{rest}}{\lambda_{rest}} \times c$$

Recall, too, that the radial velocity is the part of the motion that is either toward you or away from you. The fraction in front of the c is equal to z, the redshift. Substituting z for the fraction, we get

$$v_r = z \times c$$

(Note: That correspondence between velocity and redshift requires a correction as velocities approach the speed of light.)

Because spectral lines from distant galaxies have wavelengths shifted to the red, the galaxies must be moving away from Earth. Suppose astronomers observe a spectral line with a rest wavelength of 373 nanometers (nm) in the spectrum of a distant galaxy. If the observed wavelength of the spectral line is 379 nm, its redshift (z) is

$$z = \frac{\lambda_{obs} - \lambda_{rest}}{\lambda_{rest}}$$

$$z = \frac{379\ nm - 373\ nm}{373\ nm} = 0.0161$$

The value of the redshift of a galaxy is independent of the wavelength of the line used to measure it: The same result would have been calculated if a different line had been observed.

We can now calculate the recession velocity from that redshift as follows:

$$v_r = z \times c = 0.0161 \times 300,000\ km/s = 4830\ km/s$$

How far away is that distant galaxy? Here is where *Hubble's law* and the *Hubble constant* apply. Hubble's law relates a galaxy's recession velocity to its distance as

$$v_r = H_0 \times d_G$$

where d_G is the distance to a galaxy measured in megaparsecs. Dividing through by $H_0 = 70$ (km/s)/Mpc yields

$$d_G = \frac{v_r}{H_0} = \frac{4,830\ km/s}{70\ km/s/Mpc} = 69\ Mpc$$

From a measurement of the wavelength of a spectral line, we see that the distant galaxy is approximately 69 Mpc away.

showing that velocity is proportional to distance. That strong correlation indicates that the nearby universe follows Hubble's law very closely; for example, a galaxy 30 Mpc from Earth moves away twice as fast as a galaxy 15 Mpc distant. Over time, the accuracy of the measurement of the Hubble constant has improved, as more galaxy distances are measured, and as astronomers learn to better understand Cepheid variables and other standard candles. Today, astronomers have measured the Hubble constant to an accuracy of a few percent, and continue to refine it as they obtain more data on more types of standard candles. In this text, we use a value of 70 (km/s)/Mpc as an approximation to the best current measured values of 67 to 74 (km/s)/Mpc. These current measurements have uncertainties of 1–3 (km/s)/Mpc, depending on the measurement. The unit means that for every additional megaparsec of distance, the galaxy's motion is an additional 70 km/s faster.

Hubble's law is written mathematically as $v_r = H_0 \times d_G$, where d_G is the distance to the galaxy, H_0 is the Hubble constant, and v_r is the galaxy's recession velocity (**Working It Out 19.2**). Once H_0 is known, Hubble's law makes the once-difficult task of measuring distances in the universe relatively straightforward. Astronomers measure the redshift of a galaxy to find its distance. In fact, astronomers often use the redshift interchangeably with the distance, stating the redshift of a galaxy instead of its distance to indicate how far away it is. Astronomers have used Hubble's law to map the structure of the observable universe. We return to Hubble's law and its implications for understanding the universe as a whole in Chapter 21.

CHECK YOUR UNDERSTANDING **19.2**

Hubble's law displays the relationship between which two properties of a galaxy: (a) size; mass (b) distance; rotation speed (c) distance; recession velocity (d) size; recession velocity?

19.3 Galaxies Are Mostly Dark Matter

Efforts to measure the masses of galaxies during the 20th century led to the discovery of dark matter—mass that does not interact with light and cannot be detected via the light it emits. To understand that discovery, you first need to understand how astronomers measure the mass of a galaxy and then see how they concluded that much of the mass in a galaxy is dark matter.

Finding the Mass of a Galaxy

▶❙❙ **AstroTour:** Dark Matter

One way to measure the mass of a galaxy is to add up the mass of all of the observed stars, dust, and gas. Because a galaxy's spectrum is composed primarily of starlight, once astronomers know what types of stars are in the galaxy, they can estimate the total mass of stars in the galaxy from the galaxy's luminosity. Astronomers then estimate the mass of the dust and gas from observations of interstellar gas at X-ray, infrared, and radio wavelengths. Together, the stars, gas, and dust in a galaxy are called **luminous matter** (or sometimes **normal matter**) because that matter emits or scatters electromagnetic radiation.

A second way to measure the mass of a galaxy is to measure its gravitational pull. Stars in galactic disks follow orbits that are much like the orbits of planets around their parent stars and binary stars around each other (see Working It Out 13.4). Newton's version of Kepler's third law (see Working It Out 4.3) connects the speed of stars in the disk of a galaxy to the mass inside their orbits, just as it does for those other systems.

Figure 19.16 **a.** The profile of visible light in a typical spiral galaxy drops off with distance from the center. **b.** The predicted mass density of stars and gas located at a given distance from the galaxy's center follows the light profile. If stars and gas accounted for all of the mass of the galaxy, then the galaxy's rotation curve would be as shown in **c.** However, observed galaxy rotation curves look more like the curve shown in **d.**

Dark Matter

Astronomers originally assumed that all the mass in a galaxy was luminous matter. They observed that the light of all galaxies, including spiral galaxies, was highly concentrated toward the center (**Figure 19.16a**). On the basis of the observed location of light, astronomers predicted that nearly all mass in a spiral galaxy was concentrated toward its center (**Figure 19.16b**). The situation is much like that in the Solar System, where nearly all the mass is in the Sun—at the center of the Solar System. Therefore, they predicted that the velocities of *individual* stars in a spiral galaxy's disk should vary like the velocities of planets in our Solar System. A **rotation curve** plots the orbital velocity on the vertical axis and distance on the horizontal axis, and shows the predicted relationship: faster orbital velocities near the

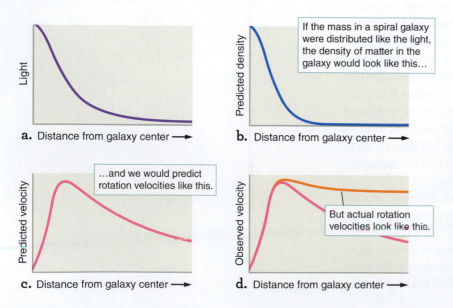

a. Distance from galaxy center ⟶ (vertical axis: Light)

b. Distance from galaxy center ⟶ (vertical axis: Predicted density)
If the mass in a spiral galaxy were distributed like the light, the density of matter in the galaxy would look like this…

c. Distance from galaxy center ⟶ (vertical axis: Predicted velocity)
…and we would predict rotation velocities like this.

d. Distance from galaxy center ⟶ (vertical axis: Observed velocity)
But actual rotation velocities look like this.

center of the spiral galaxy and slower orbital velocities farther out (**Figure 19.16c**). The measurement required to test that prediction is similar to that required for the Tully-Fisher relation, described in the previous section. In this case, however, astronomers measure the redshift and blueshift of *individual* stars and clumps of gas at various distances from a galaxy's center. Combining these data into a rotation curve enables astronomers to determine directly how the mass in a galaxy is distributed.

Vera Rubin (1928–2016) pioneered work on galaxy rotation rates in the 1970s. She discovered that, contrary to the earlier prediction (Figure 19.16c), the rotation velocities of spiral galaxies remain about the same out to the most distant measured parts of the galaxies (**Figure 19.16d**). Observations of 21-cm radiation from neutral hydrogen show that the rotation curves appear level, or "flat," in their outer parts even well outside the extent of the visible disks. The observations did not confirm the prediction, forcing astronomers to consider that the mass in a galaxy does *not* have the same distribution as the light.

Astronomers turned to the mathematics of gravity to figure out what mass distribution would cause that unexpected rotation curve. Recall from Chapter 4 that only the mass inside a given radius contributes to the net gravitational force experienced by an orbiting object. Therefore, from the rotation velocity, you can calculate the mass within the orbit of the object. **Figure 19.17** shows the results of a calculation for a particular galaxy. The black line shows the observed rotation curve of the galaxy. The pink line indicates how much luminous mass is observed inside a given radius. To produce a rotation curve like the one shown in black, that galaxy must have a second, larger component consisting of matter that is not luminous, indicated by the purple line. Furthermore, most of this matter must be in the dark outer regions of the galaxy, implying that the matter neither emits nor absorbs light. That material, which does not interact with light, and reveals itself only by the influence of its gravity, is called **dark matter**.

Astronomers currently estimate that as much as 95 percent of the total mass in some spiral galaxies consists of a **dark matter halo** (**Figure 19.18**), which can extend up to 10 times farther than the visible spiral portion of the galaxy located at the galaxy's center. That is a startling statement: The luminous part of a spiral galaxy is only a small part of a much larger distribution of mass dominated by some type of invisible dark matter.

What about elliptical galaxies? Again, astronomers need to compare the luminous mass measured from the light they can see with the gravitational mass measured from the effects of gravity. Because stars in elliptical galaxies have random orbits, astronomers cannot use Kepler's laws to measure the gravitational mass from the orbits of stars. Instead, astronomers noticed that an elliptical galaxy's ability to hold on to its hot, X-ray–emitting gas depends on its mass. If the galaxy is not massive enough, then the hot atoms and molecules will escape into intergalactic space. To find the mass of an elliptical galaxy, astronomers first infer the total amount of gas from X-ray images, such as the (false color) blue and purple halo seen in **Figure 19.19**. Then they calculate the total mass needed to hold on to the gas and compare that gravitational mass with the luminous mass. The amount of dark matter is the difference between what is needed to hold on to the inferred amount of gas and the observed amount of luminous matter.

Some elliptical galaxies contain up to 20 times as much mass as can be accounted for by their stars and gas alone, so they must be dominated by dark matter, just like spiral galaxies. As with spirals, the luminous matter in ellipticals is more centrally

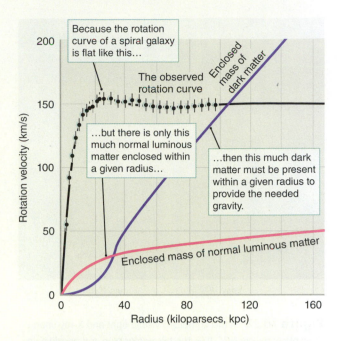

Figure 19.17 The flat rotation curve of the spiral galaxy NGC 3198 can be used to determine the total mass within a given radius. Notice that the normal mass that can be accounted for by stars and gas provides only part of the needed gravity. Extra dark matter is needed to explain the rotation curve. (1 kiloparsec [kpc] = 1000 parsecs. The mass curves have different units.)

Figure 19.18 In addition to the matter that is visible, galaxies are surrounded by halos containing a large amount of dark matter.

XRAY VIS

Figure 19.19 In this combined visible-light and X-ray image of elliptical galaxy NGC 1132, the false-color blue and purple halo is X-ray emission from hot gas surrounding the galaxy. The hot gas extends well beyond the visible light from stars.

concentrated than the dark matter. The transition from the inner parts of galaxies (where luminous matter dominates) to the outer parts (where dark matter dominates) is remarkably smooth. Some galaxies may contain less dark matter than others, but 90–95 percent of the total mass in a typical galaxy is in the form of dark matter.

The high percentage of dark matter distinguishes smaller dwarf galaxies from globular clusters, which do not have dark matter. That difference is an important observation that guides astronomers in the development of a theory of the evolution of galaxies.

The Composition of Dark Matter

What is the dark matter that makes up most of a galaxy? Several candidates have been investigated, including objects such as large planets, compact stars, black holes, and exotic unknown elementary particles, but the answer remains unknown.

At first, astronomers considered small main-sequence M stars, Jupiter-sized planets, white dwarfs, neutron stars, or black holes as dark matter candidates. These objects are collectively referred to as **MaCHOs**, which stands for *massive compact halo objects*. If the dark matter in a galaxy consists of MaCHOs, there must be a truly enormous number of them, and they must each exert gravitational force. These properties provide an avenue of investigation through gravitational lensing. If a MaCHO passed between Earth and a distant star, the star's light would be deflected by the MaCHO's gravity. Astronomers monitored tens of millions of stars in two small companion galaxies of the Milky Way for several years. While they did observe a few of these "micro-lensing" events, there were too few to account for the dark matter in the halo of our galaxy. Thus, MaCHOs with mass less than 100 M_{Sun} cannot account for the dark matter in the galaxy. In recent years, the discovery of black holes with masses approaching 100 M_{Sun} has revived the MaCHO hypothesis, but this version of the hypothesis has yet to be tested observationally.

Astronomers have also considered the hypothesis that unknown elementary particles, called **WIMPs**, which stands for *weakly interacting massive particles*, make up the dark matter in the galaxy. Like neutrinos, the hypothesized WIMPs barely interact with ordinary matter, yet are more massive and move more slowly. Many current particle physics experiments are seeking these unknown WIMPs. There are experiments at the Large Hadron Collider to produce and detect a WIMP. There are also multiple experiments in laboratories and in deep mines to detect halo dark matter particles as they pass through Earth. Astronomers are attempting to detect the (very weak) interactions of dark matter particles in the halos of galaxies. So far, there have been no confirmed direct or indirect detections of dark matter particles; however, the hypothesis has not been ruled out.

Other hypotheses have also received attention. Other types of exotic elementary particles, for example, might be found. Some scientists are exploring theoretical alternatives to dark matter, in which Newtonian dynamics or Newtonian gravity behave differently under conditions like those found in the outer regions of galaxies. Some of those alternative models of Newtonian gravity can produce the observed rotation curves in spiral galaxies, but they cannot replace dark matter as an explanation for other aspects of galaxy evolution, as we explain in subsequent chapters.

Despite scientists' best efforts, the composition of dark matter remains completely unknown. As a result, it has been said that "most of the mass in the universe is missing." That's not really true, though, because astronomers know *where* it is . . . they just don't know *what* it is. Knowing that discovery is right around the corner—if we can just figure out the right experiment to run or observation to make—builds a feeling of suspense and excitement in the entire astronomical community. Astronomers are following the latest dark matter news, hoping for a big announcement, sometime soon.

CHECK YOUR UNDERSTANDING 19.3

Astronomers detect dark matter: (a) by comparing luminous mass with gravitational mass; (b) because it blocks background light; (c) because more distant galaxies move away faster; (d) because it emits lots of X-rays.

19.4 Most Galaxies Have a Supermassive Black Hole at the Center

Studying the centers of galaxies is difficult because so many stars and so much dust and gas are in the way that astronomers cannot get a clear picture of the center, even for nearby galaxies. Observations of the most distant objects in the universe were the first to yield clues about what lies in the centers of massive galaxies.

The Discovery of Quasars

In the late 1950s, radio surveys detected several bright, compact objects that at first seemed to have no optical counterparts. Improved radio positions revealed that the radio sources coincided with faint, very blue, starlike objects. Unaware of the true nature of those objects, astronomers called them "radio stars." Obtaining spectra of the first two radio stars was a laborious task, requiring 10-hour exposures. Those spectra did not display the expected absorption lines characteristic of blue stars. Instead, the spectra showed only a single pair of emission lines that were broad—indicating very rapid motions within the objects—and that did not seem to correspond to the spectral lines of any known substances.

For several years, astronomers believed they had discovered a new type of star until astronomer Maarten Schmidt realized that those broad spectral lines, shown in **Figure 19.20**, were the highly redshifted lines of ordinary hydrogen. The implications were surprising: the objects were not stars—they were extraordinarily luminous objects located at enormous distances. Those "quasi-stellar radio sources" were named **quasars**. Other quasars were soon found by the same techniques, and astronomers began cataloging them.

Quasars are phenomenally powerful, shining with the luminosity of a trillion to a thousand trillion (10^{12} to 10^{15}) Suns. They are all very far away: hundreds or thousands of megaparsecs. Billions of galaxies are closer to Earth than the nearest quasar is. Recall that the distance to an object also indicates how much time has passed since the light from that object left its source, so when we observe to great distances, we are seeing earlier times in the universe. The fact that no quasars are close to Earth implies that they are rare in the universe now but were once much more common. The discovery that quasars existed in the distant and therefore earlier universe is compelling evidence that the universe has evolved.

Since their discovery, astronomers have come to understand that quasars are a specific type of **active galactic nucleus** (or **AGN**, where the "N" can stand for either *nucleus* or *nuclei*) (**Figure 19.21**). The distinct types of active nuclei are identified from the galaxy spectrum. A "normal" galaxy has an absorption spectrum that is a composite of the light from its billions of stars. A galaxy with an AGN exhibits emis-

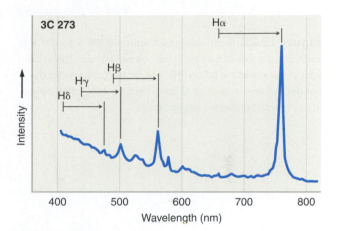

Figure 19.20 This graph shows the spectrum of quasar 3C 273, one of the closest and most luminous known quasars. The emission lines are redshifted by $z = 0.16$ from marked rest wavelengths, indicating that the quasar is at a distance of about 685 Mpc. (More detailed calculations, that take into account other complicating factors that will be introduced in Chapter 22, yield a distance of 750 Mpc.)

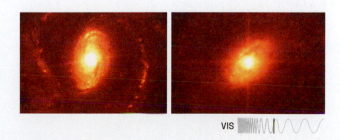

Figure 19.21 These HST images show bright quasars embedded in the centers of galaxies.
Credit: ESO, https://www.eso.org/public/hungary/news/eso0529/. https://creativecommons.org/licenses/by/4.0/.

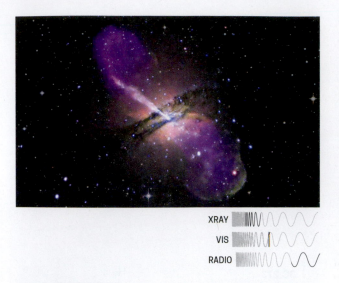

Figure 19.22 Radio galaxy Centaurus A is the closest AGN to Earth, at a distance of 3.4 Mpc. In this composite image, the visible-light image shows the galaxy, the X-ray image (pink) shows the hot gas and an energetic jet blasting from the AGN, and the radio image (purple) shows the jets and lobes.

▶❙❙ **AstroTour:** Active Galactic Nuclei

📷 **Astronomy in Action:** Size of Active Galactic Nuclei

sion lines in addition to the stellar absorption spectrum. Thus, AGN are identified by the emission lines in their spectra, which distinguishes them from normal galaxies.

AGN come in several types and can occur in spiral or elliptical galaxies. **Seyfert galaxies**, named after Carl Seyfert (1911–1960), who discovered them in 1943, are spiral galaxies whose centers contain AGN. The luminosity of a typical Seyfert nucleus can be 10 billion to 100 billion L_{Sun}, comparable to the luminosity of the rest of the host galaxy as a whole. Similarly, **radio galaxies** are elliptical galaxies whose centers contain AGN; their emission is usually most prominent in radio wavelengths. Radio galaxies and the more distant and luminous quasars are often the sources of slender jets that extend outward millions of light-years from the galaxy, powering twin lobes of radio emission (**Figure 19.22**). Quasars have the most extreme form of activity that can occur in the nuclei of galaxies, often resulting from interactions with other galaxies.

Much of the light from AGN is synchrotron radiation—the same type that comes from extreme environments such as the Crab Nebula supernova remnant (see Figure 17.17). Synchrotron radiation comes from relativistic charged particles spiraling around magnetic field lines. AGN accelerate large amounts of material to nearly the speed of light in the presence of strong magnetic fields, so they emit a lot of synchrotron radiation. The spectra of many quasars and Seyfert nuclei also show emission lines that are smeared out by the Doppler effect across a wide range of wavelengths. That observation implies that gas in AGN is swirling around the centers of those galaxies at speeds of thousands or even tens of thousands of kilometers per second.

AGN Are the Size of the Solar System

The enormous luminosity and mechanical energy of AGN are made even more spectacular by the fact that they are no larger than a light-day or so across—comparable in size to the Solar System. Although the HST and large, ground-based telescopes show faint fuzz—light from the surrounding galaxy—around the images of some quasars and other AGN, the objects themselves remain as unresolved points of light.

To understand how astronomers determine that AGN are compact objects, think about the halftime show at a football game. **Figure 19.23** illustrates a problem faced by every director of a marching band. When a band is all together in a tight formation at the center of the field, the notes you hear in the stands are clear and crisp; the band plays together beautifully. But, as the band spreads out across the field, its sound begins to get mushy. That is not because the marchers are poor musicians. Instead, it is because sound travels at a finite speed of about 340 meters per second (m/s). At that speed, sound takes approximately 1/3 of a second to travel from one end of the football field to the other. Even if every musician on the field plays a note at exactly the same instant in response to the director's cue, in the stands you hear the instruments close to you first but have to wait longer for the sound from the far end of the field to arrive.

If the band is spread from one end of the field to the other, the beginning of a note will be smeared out over about 1/3 of a second, the time sound takes to travel from one end of the field to the other. If the band were spread out over two football fields, the sound from the most distant musician would take about 2/3 of a second to arrive at your ear. If the marching band were spread out over a kilometer, it would be roughly 3 seconds—the time sound takes to travel a kilometer—before you heard a crisply played note start and stop. Even with your eyes closed, you could easily tell whether the band was in a tight group or spread out across the field.

The same principle applies to the light observed from AGN. Quasars and other AGN change their brightness dramatically over only a day or two—and sometimes as quickly as in a few hours. That rapid variability sets an upper limit on the size of the

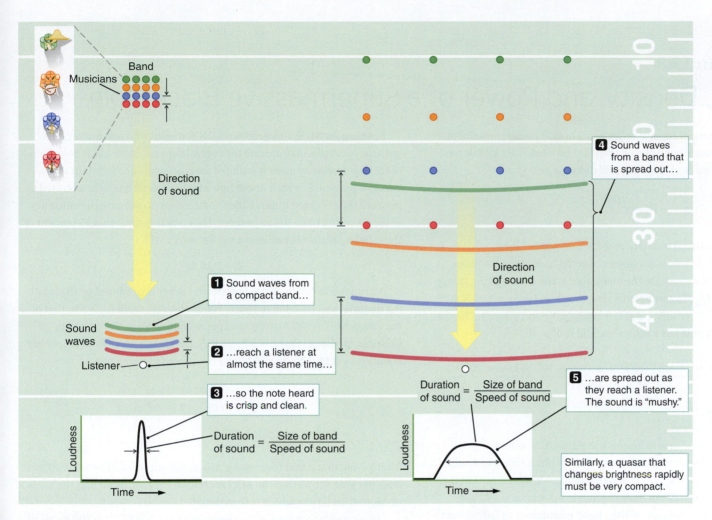

Figure 19.23 A marching band spread out across a field cannot play a clean note. Similarly, AGN must be very compact to explain their rapid variability.

AGN, just as hearing clear music from a marching band indicates that the musicians are close together. An AGN must be smaller than a light-day or so across, because if it were larger, its brightness could not possibly change in a day or two. This is a remarkable conclusion: A region of an AGN roughly the size of the orbit of Neptune may be emitting as much light as 10,000 galaxies.

Supermassive Black Holes and Accretion Disks

The conclusion that AGN emit so much energy from such a small region had to be explained. In thinking about what type of object could be very small yet very energetic, astronomers hypothesized that galaxies with an AGN contain **supermassive black holes**—black holes with masses from thousands to tens of billions of solar masses. Violent accretion disks surround those supermassive black holes. As matter in the disk falls inward, gravitational energy is converted to heat, providing the luminous energy of the AGN. You have already learned about accretion disks surrounding young stars, white dwarfs, neutron stars, and stellar-mass black holes, all of which are fed by small amounts of material from a cloud or star each year. The accretion disks around supermassive black holes are fed by several solar masses every year (**Working It Out 19.3**).

unanswered questions

Where do supermassive black holes come from? To explore that question, astronomers are studying computer models in which a supermassive black hole grew from many stellar-mass black holes, or grew along with a galaxy by swallowing large amounts of central gas, or increased after the merger of two or more galaxies. We return to that question when we discuss galaxy evolution in Chapter 23.

working it out 19.3

The Size, Density, and Power of a Supermassive Black Hole

Size. What are the sizes of supermassive black holes? Recall from Chapter 18 that the Schwarzschild radii of stellar-mass black holes are kilometers in size. The formula for the Schwarzschild radius is given by

$$R_S = \frac{2GM_{BH}}{c^2}$$

where G is the gravitational constant and c is the speed of light.

The largest supermassive black holes observed have about 10 billion solar masses (M_{Sun}). For example, the black hole at the center of the galaxy M87 is 6.69 billion M_{Sun}. To compute its size, recall that $M_{Sun} = 1.99 \times 10^{30}$ kilograms (kg), $c = 3 \times 10^5$ km/s, and $G = 6.67 \times 10^{-20}$ km³/(kg s²). Then, a 6.6 billion-M_{Sun} black hole has a Schwarzschild radius of

$$R_S = \frac{2 \times [6.67 \times 10^{-20} \, km^3/(kg\,s^2)] \times (6.6 \times 10^9 \times 1.99 \times 10^{30} \, kg)}{(3 \times 10^5 \, km/s)^2}$$

$$R_S = 1.9 \times 10^{10} \, km$$

We can convert that value into astronomical units (AU), where 1 AU = 1.5 × 10⁸ km. Therefore, that supermassive black hole has a radius of 130 AU—about 4 times the radius of Neptune's orbit. We know that light takes 8.3 minutes to reach Earth from the Sun at a distance of 1 AU, so 130 AU corresponds to a distance of 1080 light-minutes, or 18 light-hours.

Density. What is the average density of that object inside of the event horizon? The mass of the black hole divided by the volume within the Schwarzschild radius is

$$\text{Density} = \frac{\text{Mass}}{\text{Volume}} = \frac{(6.69 \times 10^9) \times (1.99 \times 10^{30} \, kg)}{4/3 \times \pi \times (1.9 \times 10^{10} \, km)^3}$$

$$\text{Density} = 4.0 \times 10^8 \, kg/km^3 = 0.40 \, kg/m^3$$

That is about 1/3 of the density of air at sea level, or 1/2500 the density of water. Supermassive black holes do not have the extremely high densities of stellar-mass black holes, which are at around 10²⁰ times denser.

Feeding an AGN. Power for an AGN is produced when matter falls onto the accretion disk around the central supermassive black hole. Some of that high-velocity mass is radiated away according to Einstein's mass-energy equation: $E = mc^2$. About how much material has to be accreted to produce the observed luminosities? Astronomers estimate the efficiency of the accretion to be 10–20 percent. Here, we'll assume that 15 percent of the infalling matter is radiated away as energy, or

$$E = 0.15 \, mc^2$$

Astronomers can measure how much energy is produced by the infalling material and radiated to space. For a relatively weak AGN like that of the galaxy M87, $L = 5 \times 10^{35}$ joules per second (J/s), or 5×10^{35} kg m²/s² each second. Recall that $c = 3 \times 10^8$ m/s. Dividing both sides of Einstein's equation by 0.15 c^2 gives us the mass consumed each second:

$$m = \frac{E}{0.15\,c^2} = \frac{5 \times 10^{35} \, kg\,m^2/s^2}{0.15 \times (3 \times 10^8 \, m/s)^2}$$

$$m = 3.7 \times 10^{19} \, kg$$

Multiplying that result (the mass consumed each second) by 3.2 × 10⁷ seconds per year shows that the AGN accretes 10²⁷ kg, or about half the mass of Jupiter, each year, which is then radiated away as energy.

If we consider a quasar with a luminosity (L) of 10³⁹ J/s = 10³⁹ kg m²/s² each second (= 2.5 trillion L_{Sun}), the mass accreted each second is given by

$$m = \frac{10^{39} \, kg\,m^2/s^2}{0.15 \times (3 \times 10^8 \, m/s)^2}$$

$$m = 7.4 \times 10^{22} \, kg$$

Multiplying by 3.2 × 10⁷ seconds per year yields a mass of 2.4 × 10³⁰ kg per year. The mass of the Sun is 1.99 × 10³⁰ kg, so to radiate that much energy, the supermassive black hole is accreting about 1.2 M_{Sun} each year. A quasar with 10 times that luminosity would be accreting 10 times the mass.

In the discussion of star formation in Chapter 15, you learned that gravitational energy is converted to thermal energy as material moves inward toward a growing protostar. In an AGN, as material moves inward toward the supermassive black hole, conversion of gravitational energy heats the accretion disk to hundreds of thousands of kelvins, causing it to glow brightly in visible, ultraviolet, and X-ray light. Conversion of gravitational energy to thermal energy as material falls onto the accretion disk also is a source of energetic emission. As much as 20 percent of the mass of infalling material around a supermassive black hole is converted to luminous energy. The rest of that mass is pulled into the black hole itself, causing it to grow even more massive.

The interaction of the accretion disk with the black hole creates powerful radio jets that emerge perpendicular to the disk (as in the jet in the upper left of Figure 19.22). Throughout, twisted magnetic fields accelerate charged particles such as electrons and protons to relativistic speeds, producing synchrotron emission. Gas in the accretion disk or in nearby clouds orbiting the central black hole at high speeds produces emission lines smeared out by the Doppler effect into the broad lines seen in AGN spectra. That accretion disk surrounding a supermassive black hole is the "central engine" that powers AGN.

The Unified Model of AGN

Astronomers have developed the basic picture of a supermassive black hole surrounded by an accretion disk into a more complete AGN model. The **unified model of AGN** attempts to explain all types of AGN—quasars, Seyfert galaxies, and radio galaxies. **Figure 19.24** shows the various components of that AGN model, in which an accretion disk surrounds a supermassive black hole. Much farther out from the accretion disk lies a large **torus**, or "doughnut" of gas and dust consisting of material feeding the central engine. Located far from the inner turmoil of the accretion disk, and far larger than the central engine, some of that torus is ionized by UV light from the AGN.

In the unified model of AGN, the various AGN observed from Earth are partly explained by astronomers' view of the central engine. The torus of gas and dust obscures that view in different ways, depending on the viewing angle. Variation in that angle, in the mass of the black hole, and in the rate at which it is being fed accounts for a wide range of observed AGN properties. When the AGN is viewed edge on, astronomers see emission lines from the surrounding torus and other surrounding gas. Astronomers also can sometimes see the torus in absorption against the background of the galaxy. From that nearly edge-on orientation, they cannot see the accretion disk itself, so they do not see the Doppler-smeared lines that originate closer to the supermassive black hole. If jets are present in the AGN, though, they should be visible emerging from the center of the galaxy. This unified model is much simpler than a model in which each type of AGN is unrelated to every other type (see the **Process of Science Figure**).

If astronomers observe the accretion disk more face on, they can see over the edge of the torus and thus get a more direct look at the accretion disk and the location of the black hole. Then they see more of the synchrotron emission from the region around the black hole and the Doppler-broadened lines produced in and around the accretion disk. **Figure 19.25** shows an image of one such object, the galaxy M87, at an intermediate inclination. M87 is a source of powerful jets that continue outward for 100,000 light-years but originate in the tiny engine at the heart of the galaxy. Spectra of the disk at the center of that galaxy show the rapid rotation of material around a central black hole that has a mass of 3 billion (3×10^9) M_{Sun}.

The material in an AGN jet travels very close to the speed of light, so what astronomers see is strongly influenced by relativistic effects. One of those effects is called **relativistic beaming**: Matter traveling at close to the speed of light concentrates any radiation it emits into a tight beam pointed in the direction in which it is moving. So, astronomers often observe only one side of the jets from AGN, even though the radio lobes of radio galaxies are usually two-sided. The jet moving away is just too faint to observe.

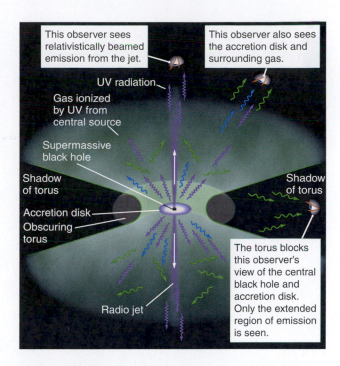

Figure 19.24 The basic model of an active galactic nucleus, with a supermassive black hole surrounded by an accretion disk at the center. A larger, dusty torus sometimes blocks the view of the black hole. The mass of the central black hole, the rate at which it is being fed, and the viewing angle determine the observational properties of an AGN.

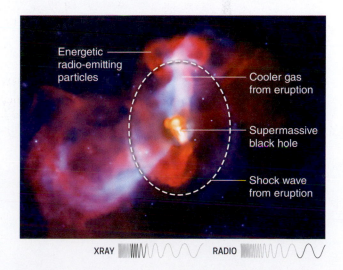

Figure 19.25 This image of M87 in radio and X-rays shows the location of the supermassive black hole.

what if . . .

What if you observed two AGN in the process of merging, with their nuclei quite close together? Would you expect to observe gravitational waves from this system?

In rare instances when the accretion disk in a quasar or radio galaxy is viewed almost directly face on, relativistic beaming dominates the observations. In those *blazars*, emission lines and other light coming from hot gas in the accretion disk are overwhelmed by the bright glare of the jet emission beamed directly at Earth.

Normal Galaxies and AGN

The essential elements of an AGN are a central engine (an accretion disk surrounding a supermassive black hole) and a source of fuel (gas and stars flowing onto the accretion disk). Without a source of matter falling onto the black hole, an AGN would no longer be an active nucleus. Astronomers looking at such an object would observe a normal galaxy with a supermassive black hole sitting in its center.

Only a few percent of present-day galaxies contain AGN as luminous as the host galaxy. But when astronomers look at more distant galaxies (and therefore look further back in time), the percentage of galaxies with AGN is much larger. These observations show that when the universe was younger, many more AGN existed than are there today. If astronomers' understanding of AGN is correct, all the supermassive black holes that powered those dead AGN should still be around. Astronomers hypothesize that many massive galaxies today contain supermassive black holes.

There are several observations that support this hypothesis. Such a concentration of mass at the center of a galaxy should draw surrounding stars close to it. The central region of such a galaxy should be much brighter than could be explained if stars alone were responsible for the gravitational field in the inner part of the galaxy. Stars experiencing the gravitational pull of a supermassive black hole in the center of a galaxy should also orbit at very high velocities and therefore show large Doppler shifts. Astronomers have found those large Doppler shifts in every normal galaxy with a substantial bulge in which a careful search has been conducted. The masses inferred for those black holes range from 10,000 to 20 billion M_{Sun}. The mass of the supermassive black hole seems to be related to the mass of the bulge in which it is found. Astronomers conclude that most large galaxies, whether elliptical or spiral, probably contain supermassive black holes.

Apparently, the only difference between a normal galaxy and an active galaxy is whether the supermassive black hole at its center is being fed when we see that galaxy. The rarity of present-day galaxies with very luminous AGN does not indicate which galaxies have the potential for AGN activity. Rather, it indicates which galaxy centers are being lit up at the moment. If a large amount of gas and dust were dropped directly into the center of any large galaxy, that material would fall inward toward the central black hole, forming an accretion disk and a surrounding torus. That process would change the nucleus of that galaxy into an AGN.

In Chapter 23, we discuss galaxy evolution and note that many of the observed properties of galaxies discussed here in Chapter 19—including the type of galaxy that forms, spiral structure, star formation, and AGN—depend on the interactions and mergers between galaxies. To account for the many large galaxies visible today, interactions and mergers must have been much more prevalent in the past when the universe was younger; that is one explanation for the larger number of AGN that existed in the past. Computer models show that galaxy-galaxy interactions can cause gas located thousands of parsecs from the center of a galaxy to fall inward toward the galaxy's center, where the gas can provide fuel for an AGN. During mergers, a significant fraction of a galaxy might wind up being cannibalized. HST images of quasars often show that quasar host galaxies are tidally distorted or are surrounded

Finding the Common Thread

When first discovered, active galactic nuclei (AGN) seemed to come in many different types, with dramatically different spectra. Later, it was realized that this could be an orientation effect: the many different types of objects could all be explained with one type of object, viewed from different angles.

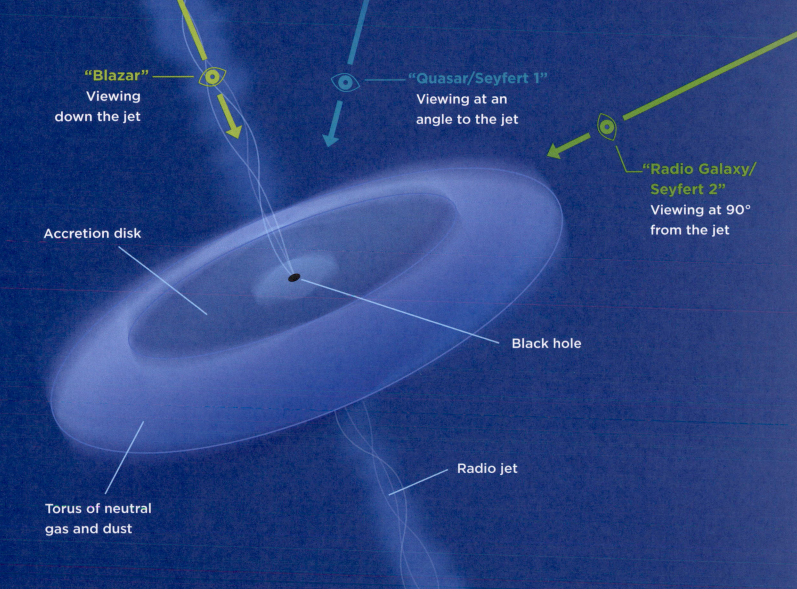

"Blazar"
Viewing
down the jet

"Quasar/Seyfert 1"
Viewing at an
angle to the jet

**"Radio Galaxy/
Seyfert 2"**
Viewing at 90°
from the jet

Accretion disk

Black hole

Radio jet

Torus of neutral
gas and dust

Scientists seek underlying principles that explain more than one phenomenon. In the case of AGN, scientists developed one model to explain many types of objects. This model encompasses the idea that our classification can be affected by our viewing angle, which is true for galaxy shapes generally. A unified model is much simpler than having a different model for every type of object.

Figure 19.26 ★ WHAT AN ASTRONOMER SEES

This gallery of interacting galaxies shows the distorted shapes that sometimes form as the gravitational interaction warps and twists the galaxies. Interactions and mergers often disrupt the dust and gas in a galaxy, triggering many stars to form at once. An astronomer will notice the color differences, from one region to another. Young stars are hot and blue, and so very blue regions often indicate recent star formation. An astronomer will look for dark lanes of dust and gas, as well as reddish glow from old stars. Tidal tails of stars, dust, and gas are often left behind as the galaxies accelerate toward each other, showing the direction each galaxy came from. An astronomer looking at images like these sees evidence of a dynamic, ever-changing universe.

VIS

by other visible matter that is probably still falling into the galaxies (**Figure 19.26**). Galaxies that show evidence of recent interactions with other galaxies are more likely to house AGN in their centers (**Figure 19.27**). Any large galaxy might be only an encounter away from becoming an AGN.

CHECK YOUR UNDERSTANDING 19.4

Supermassive black holes: (a) are extremely rare—only a handful exist in the universe; (b) are completely hypothetical; (c) occur in most, perhaps all, large galaxies; (d) occur only in the space between galaxies.

Origins: Habitability in Galaxies

Now that we know something about the types of galaxies that have been observed, what can we say about their potential for life? No solid information is available because those galaxies are too far away for astronomers to have detected any planets around their stars. Instead, all we can do is speculate about the habitability of other galaxies. Two key requirements are the presence of heavy elements to form planets (and life) and an environment without too much radiation that might be damaging to life.

Figure 19.27 The Swift Gamma-Ray observatory has detected active black holes (circles) in these merging galaxies.

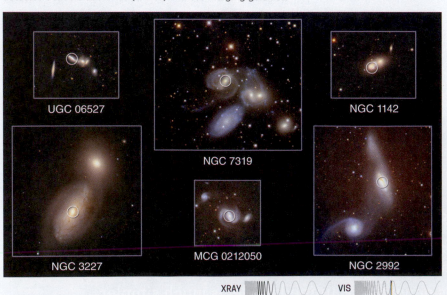

UGC 06527

NGC 7319

NGC 1142

NGC 3227

MCG 0212050

NGC 2992

XRAY   VIS

A study of the host stars of exoplanet candidates discovered by the Kepler telescope suggests that stars with a higher percentage of heavier elements may be more likely to have planets. (That finding fits with the core accretion models of planet formation discussed in Chapter 7.) The first generation of stars made from the hydrogen and helium of Big Bang nucleosynthesis do not have heavy elements. Recall from the discussion of stellar evolution that elements heavier than helium are created in the cores of dying stars and then are scattered into the galactic environment through planetary nebulae, stellar winds, and supernova explosions. So, the amount of heavier elements in a star depends on the cosmic history of the material from which the star formed. Therefore, astronomers must consider the galactic environment of the star, which varies among types of galaxies and locations within the galaxies.

Spiral galaxies have had more continual star formation in their disks throughout their history. They contain more stars

Reading Astronomy News

Earth Faster, Closer to Black Hole in New Map of Galaxy

National Astronomical Observatory of Japan, *November 26, 2020*

Scientists constantly refine and improve measurements of our place in the universe.

Earth just got 7 km/s faster and about 2000 light-years closer to the supermassive black hole in the center of the Milky Way Galaxy. But don't worry, this doesn't mean that our planet is plunging toward the black hole. Instead the changes are results of a better model of the Milky Way Galaxy based on new observation data, including a catalog of objects observed over the course of more than 15 years by the Japanese radio astronomy project VERA.

VERA (VLBI Exploration of Radio Astrometry, by the way "VLBI" stands for Very Long Baseline Interferometry) started in 2000 to map three-dimensional velocity and spatial structures in the Milky Way. VERA uses a technique known as interferometry to combine data from radio telescopes scattered across the Japanese archipelago in order to achieve the same resolution as a 2300 km diameter telescope would have. Measurement accuracy achieved with this resolution, 10 micro-arcseconds, is sharp enough in theory to resolve a United States penny placed on the surface of the Moon.

Because Earth is located inside the Milky Way Galaxy, we can't step back and see what the galaxy looks like from the outside. Astrometry, accurate measurement of the positions and motions of objects, is a vital tool to understand the overall structure of the galaxy and our place in it. This year, the First VERA Astrometry Catalog was published containing data for 99 objects.

Based on the VERA Astrometry Catalog and recent observations by other groups, astronomers constructed a position and velocity map. From this map they calculated the center of the galaxy, the point that everything revolves around. The map suggests that the center of the galaxy, and the supermassive black hole which resides there, is located 25,800 light-years from Earth. This is closer than the official value of 27,700 light-years adopted by the International Astronomical Union in 1985. The velocity component of the map indicates that Earth is traveling at 227 km/s as it orbits around the Galactic Center. This is faster than the official value of 220 km/s.

Now VERA hopes to observe more objects, particularly ones close to the central supermassive black hole, to better characterize the structure and motion of the galaxy. As part of these efforts VERA will participate in EAVN (East Asian VLBI Network), comprised of radio telescopes located in Japan, South Korea, and China. By increasing the number of telescopes and the maximum separation between telescopes, EAVN can achieve even higher accuracy.

QUESTIONS

1. In which part of the electromagnetic spectrum were these observations made? From that information, would you conclude that the astronomers were observing stars or gas clouds in the Milky Way?

2. Does the article actually state what objects were observed to make this determination?

3. Calculate the percent change in the velocity measurement: Divide the difference between the new velocity and the old one (7 km/s) by the old velocity (220 km/s), and then convert that decimal to a percentage. Is this a large change or a refinement of only a few percent?

4. Similarly, calculate the percent change in the distance to the center of the Milky Way. (Take a moment to make a prediction: Do you expect this percent change to be about the same, larger, or smaller than the percent change in question 3?) Is this a large change or a refinement of only a few percent?

5. Is this supermassive black hole at the center of the Milky Way part of an AGN? How do you know?

Source: https://www.nao.ac.jp/en/news/science/2020/20201126-mizusawa.html.

born from recycled material and therefore more stars with a higher fraction of heavy elements. Elliptical (and S0) galaxies have older, redder populations of stars and very little current star formation. Old, massive ellipticals have a larger percentage of lower-mass stars than that of smaller ellipticals or spirals. Astronomers had previously thought that difference meant that large elliptical galaxies would not be good environments for planet formation. But the Kepler telescope has found many planets around small, red, main-sequence stars like the ones that populate elliptical galaxies. One study of two elliptical galaxies showed that both had some fraction of stars with a heavy-element fraction similar to that of the stars hosting Kepler exoplanets in the Milky Way, so ellipticals are not ruled out.

Another issue is the presence of radiation that might be hazardous to life. That radiation would most likely come from the center of the galaxy. Galaxies in an active AGN state might have too much radiation in regions close to their centers to be conducive to life. Stars whose orbits cross spiral arms many times also might be exposed to higher-than-average levels of radiation, but that is not the case for most stars in a galaxy.

The conditions in those galaxies may also change as the galaxies evolve. Galaxy mergers can shake up stellar orbits and move stars and their planets to different locations. Mergers also may affect the growth and activity level of supermassive black holes and thus the presence of radiation. Some galactic environments just may not remain habitable for the length of time—billions of years—that life took to evolve from bacteria to human intelligence on Earth.

SUMMARY

Galaxies are classified on the basis of their shape and the types of orbits of their stars. Most of the mass of a galaxy is dark matter, which interacts with light very weakly, if at all. The nature of that matter is not yet known. Most (non-dwarf) galaxies have a supermassive black hole at the center, which may become an AGN if gas accretes onto it. In thinking about the potential habitability of galaxies, astronomers consider the activity state of the galaxy, including the presence or absence of an AGN and mergers, and the amount of heavy elements in the stars in the galaxy, which is related to the star formation rate and galaxy type.

① **Determine a galaxy's type from its appearance and describe the motions of its stars.** Spiral galaxies are distinguished by their flat disk and spiral arms. The stars in that disk all orbit the center of the galaxy in the same direction. Elliptical galaxies are shaped like a rugby ball, and the stars orbit in all directions. Irregular galaxies are galaxies that fit neither classification, usually because they are interacting with another galaxy.

② **Organize the steps of the distance ladder and explain how distances to galaxies are measured at each step.** Astronomers build a distance ladder to galaxies by observing objects of known luminosity, such as Cepheid variable stars and Type Ia supernovae, in distant galaxies. Observations of the distance and the velocity of galaxies show that the two factors are related: More distant galaxies move away from us faster. Hubble's law, $v = H_0 d$, offers a method to find the distances of the most remote objects.

③ **Describe the evidence suggesting that galaxies are composed mostly of dark matter.** Most of the mass in galaxies does not reside in gas, dust, or stars; instead, galaxy rotation curves indicate that about 90 percent of a galaxy's mass is in the form of dark matter, which does not emit or absorb light to any significant degree. Dark matter is so far identified by its gravitational interaction with ordinary matter. After many searches and experiments, astronomers still do not know what constitutes dark matter.

④ **Discuss the evidence indicating that most—perhaps all—large galaxies have supermassive black holes at their centers.** Observations of distant quasars reveal extremely luminous, compact sources near the centers of galaxies. Those active galactic nuclei (AGN) are best explained as supermassive black holes, surrounded by an accretion disk and a torus of dust and gas. AGN can emit as much as 1000 times the light of the whole galaxy, all coming from a region the size of the Solar System. Surveys of the centers of large galaxies show that most have black holes at their centers.

QUESTIONS AND PROBLEMS

TEST YOUR UNDERSTANDING

1. Which of the following contributes the largest percentage to the total mass of a spiral galaxy?
   a. dark matter
   b. central black hole
   c. stars
   d. dust and gas

2. In the context of spiral galaxies, Kepler's laws could be used to estimate
   a. P, the period of galactic rotation.
   b. A, the radius of the galaxy.
   c. M, the mass of the galaxy.
   d. v, the rotation speed of the galaxy.

3. Astronomers determine the radius of an AGN by measuring
   a. how much light comes from it.
   b. how hard it pulls on stars nearby.
   c. how quickly its light varies.
   d. how quickly it rotates.

4. If you observed a galaxy with an Hα emission line that had a wavelength of 756.3 nm, what would be the galaxy's redshift? Note that the rest wavelength of the Hα emission line is 656.3 nm.
   a. 0.01
   b. 0.05
   c. 0.10
   d. 0.15

5. As astronomers extend their distance ladder beyond 30 Mpc, they change their measuring standard from Cepheid variable stars to Type Ia supernovae. Why is that change necessary?
   a. Type Ia supernovae are more luminous than Cepheid variables.
   b. Type Ia supernovae are less luminous than Cepheid variables.
   c. Type Ia supernovae vary more slowly than do Cepheid variables.
   d. Type Ia supernovae vary more quickly than do Cepheid variables.

6. Which galaxy types have a spherical bulge and a well-defined disk?
   a. spiral
   b. barred spiral
   c. elliptical
   d. irregular

7. Which galaxy type is shaped like a rugby ball?
   a. Sb
   b. SBb
   c. irregular
   d. E5

8. For a variable star to be useful as a standard candle, its luminosity must be related to its
   a. period of variation.
   b. mass.
   c. temperature.
   d. radius.

9. If all the stars in an elliptical galaxy traveled in random directions in their orbits, the elliptical galaxy would be type
   a. E0.
   b. E2.
   c. E5.
   d. E7.

10. The flat rotation curves of spiral galaxies imply that the distribution of mass
    a. is like the Solar System; most mass is concentrated in the center.
    b. is a wheel; the density remains the same as the radius increases.
    c. resembles the light distribution of the galaxy; a large concentration occurs in the middle, but significant mass exists quite far out.
    d. extends farther out than the visible galaxy.

11. Astronomers observe two galaxies, A and B. Galaxy A has a recession velocity of 2500 km/s, whereas galaxy B has a recession velocity of 5000 km/s. According to those data,
    a. galaxy A is 4 times as far away as galaxy B.
    b. galaxy A is twice as far away as galaxy B.
    c. galaxy B is twice as far away as galaxy A.
    d. galaxy B is 4 times as far away as galaxy A.

12. The Hubble constant is found from
    a. the slope of the line fit to the data in Hubble's law.
    b. the y-intercept of the line fit to the data in Hubble's law.
    c. the spread in the data in Hubble's law.
    d. the inverse of the slope of the line fit to the data in Hubble's law.

13. Astronomers know that dark matter is present in galactic halos because the speeds of orbiting stars _____ far from the center of the galaxy.
    a. decrease
    b. increase
    c. remain about constant
    d. fluctuate dramatically

14. What accounts for the observed differences among types of AGN?
    a. the type of the host galaxy
    b. the size of the central black hole
    c. the amount of dark matter in the galaxy's halo
    d. the luminosity and our viewing angle

15. If a Seyfert galaxy's nucleus varies in brightness on the timescale of 10 hours, approximately what is the size of the emitting region?
    a. 20 AU
    b. 70 AU
    c. 90 AU
    d. 140 AU

THINKING ABOUT THE CONCEPTS

16. ★ **WHAT AN ASTRONOMER SEES**   Figure 19.26 shows 8 different pairs of interacting galaxies. How many of these interactions are between two spiral galaxies? How many are between two ellipticals? How many are between an elliptical and a spiral? 👁

17. What was the subject of the Great Debate, and why was it important to astronomers' understanding of the scale of the universe?

18. How did observations of Cepheid variable stars finally settle the Great Debate?

19. Why is observing more than one type of standard candle in a distant galaxy better?

20. Explain what astronomers mean by *distance ladder*.

21. Why is knowing the type of progenitor of a Type Ia supernova in a distant galaxy important?

22. Some galaxies have regions that are relatively blue; other regions appear redder. What does that variation indicate about the differences between those regions?

23. Describe how elliptical galaxies and spiral bulges are similar.

24. Which is more luminous: a quasar or a galaxy with 100 billion solar-type stars? Explain your answer.

25. The nearest observed quasar is about 750 Mpc away. Why don't astronomers observe many that are closer?

26. What distinguishes a normal galaxy from one that contains an AGN?

27. Compare the size of a typical AGN with the size of the Solar System. How do astronomers know the size of an AGN?

28. The Process of Science Figure depicts how astronomers unify the types of AGN so that all have the same fundamental components. Which types of AGN are seen when the accretion disk is visible? 👁

29. Study the distance ladder in Figure 19.12. The different "rungs" of the distance ladder overlap in the distances that they measure. Why is that important? 👁

30. Why do astronomers think that planets around stars near the centers of galaxies would not be good locations for the formation of life?

APPLYING THE CONCEPTS

31. Suppose that you observe a Type Ia supernova with a maximum brightness of 2.5×10^{-12} watts per square meter (W/m²). The typical maximum luminosity L of a Type Ia supernova is 9.5×10^9 times the luminosity of the Sun. How far away is the galaxy that hosts this supernova? 🟢●●
   a. Make a prediction: Study Working It Out 19.1. Was that supernova brighter or fainter than this one? Would you expect this supernova to be closer or farther away?
   b. Calculate: Follow Working It Out 19.1 to find the distance to this supernova.
   c. Check your work: Verify that your answer has units of distance and compare the numerical value with your prediction.

32. The spectrum of a distant galaxy shows the Hα line of hydrogen ($\lambda_{rest} = 656.28$ nm) at a wavelength of 750 nm. Assume that $H_0 = 70$ km/s/Mpc. What is the distance of the galaxy in megaparsecs? 🟢●●
   a. Make a prediction: Compare the redshift of this galaxy to the redshift in Working It Out 19.2. Do you expect this galaxy to be much farther away or much closer than that galaxy?
   b. Calculate: Follow Working It Out 19.2 to find the recession velocity and then the distance to this galaxy.
   c. Check your work: Verify that your answer has units of distance and compare the numerical result to your prediction.

33. If a luminous quasar has a luminosity of 2×10^{41} W, or J/s, how many solar masses ($M_{Sun} = 2 \times 10^{30}$ kg) per year does that quasar consume to maintain its average energy output? 🟢●●
   a. Make a prediction: Compare the luminosity of this quasar to the luminosity of the quasar in Working It Out 19.3. Do you expect that this quasar will be consuming more or less mass than the quasar calculated there? About how much more or less?
   b. Calculate: Calculate how much mass is required to maintain this quasar's luminosity each year.
   c. Check your work: Compare your answer to your prediction.

34. Suppose that you are working on a project to verify whether Type Ia supernovae really do make good standard candles. You observe a Type Ia supernova with a maximum brightness of 6.2×10^{-12} watts per square meter (W/m²). Other astronomers have used other standard candles to determine that the distance to the host galaxy is 2.0 Mpc. What is the luminosity of this Type Ia supernova?

35. One of the closer known quasars is 3C 273. It is located in the constellation Virgo and is bright enough to be seen in a medium-sized amateur telescope. With a redshift of 0.158, what is the distance to 3C 273 in parsecs?

36. In Figure 19.17, the small vertical bars (known as error bars) on the data points indicate the size of the measurement error. 👁
   a. At a radius of 25,000 parsecs (pc), what is the approximate measurement error in the rotation velocity?
   b. What is that value as a percentage of the measured velocity?
   c. Error bars are important because they show how wrong the measurement could possibly be. One way to think about this is that the black line could be as high as the top of the error bars or as low as the bottom of the error bars. In either case, would shifting the black line change the overall conclusion about redshift and distance? Why or why not?

37. Suppose the number density of galaxies in the universe is, on average, 3×10^{-68} galaxies/m³. If astronomers could observe all galaxies out to a distance of 10^{10} pc, how many galaxies would they find?

38. The quasar 3C 273 has a luminosity of $10^{12} L_{Sun}$. Assuming that the total luminosity of a large galaxy, such as the Andromeda Galaxy, is 10 billion times that of the Sun, compare the luminosity of 3C 273 with that of the entire Andromeda Galaxy.

39. A quasar has the same brightness as a galaxy seen in the foreground 2 Mpc distant. If the quasar is 1 million times more luminous than the galaxy, what is the distance of the quasar?

40. Estimate the Schwarzschild radius for a supermassive black hole with a mass of 26 billion M_{Sun}.

41. You read on the Internet that astronomers have discovered a "new" cosmological object that appears to be flickering with a period of 83 minutes. Because you have read *21st Century Astronomy*, you can estimate quickly the maximum size of that object. How large can it be?

42. A quasar has a luminosity of 10^{41} W, or J/s, and $10^8 M_{Sun}$ to feed it. If you assume constant luminosity and 20 percent conversion efficiency, what is your estimate of the quasar's lifetime?

43. A solar-type star ($M = 2 \times 10^{30}$ kg) approaches a supermassive black hole. Half of its mass falls into the black hole, whereas the other half is completely converted to energy in the form of light. How much energy does that dying star send out to the rest of the universe?

44. Suppose a Type Ia supernova is found in a distant galaxy. The measured supernova brightness is 10^{-17} W/m². What is the distance of the galaxy?

45. Suppose that an object with the mass of Earth ($M_{Earth} = 5.97 \times 10^{24}$ kg) fell into a supermassive black hole with a 10 percent energy conversion.
   a. How much energy (in joules) would be radiated by the black hole?
   b. Compare your answer with the energy radiated by the Sun each second: 3.85×10^{26} J.

EXPLORATION Galaxy Classification

Galaxy classification sounds simple, but it can become complicated when you actually attempt it. **Figure 19.28**, taken by the Hubble Space Telescope, shows a small portion of the Coma Cluster of galaxies. The Coma Cluster contains thousands of galaxies, each containing billions of stars. Some of the objects in that image (the ones with a bright cross) are foreground stars in the Milky Way. Some of the galaxies are far behind the Coma Cluster. Working with a partner, in this Exploration you will classify the 20 or so brightest galaxies in that cluster.

First, make a map by laying a piece of paper over the image and numbering the 20 or so brightest (or largest) galaxies in the image (label them "galaxy 1," "galaxy 2," and so on). Copy the map so that you and your partner each have a list of the same galaxies.

Separately, classify each galaxy by type. If it is a spiral galaxy, is its subtype a, b, or c? If it is an elliptical, how elliptical is it? Make a table that contains the galaxy number, the type you have assigned it, and any comments that will help you remember why you made that choice. When you are done classifying, compare your list with your partner's. Now comes the fun part! Argue about the classifications until you agree—or until you agree to disagree.

1 Which galaxy type was easiest to classify?

_____

_____

2 Which galaxy type was hardest to classify?

_____

_____

3 What makes classifying some of the galaxies hard?

_____

_____

4 Which galaxy type did you and your partner agree about most often?

_____

_____

VIS

Figure 19.28 This Hubble Space Telescope image of the Coma Cluster shows a diversity of shapes.

5 Which galaxy type did you and your partner disagree about most often?

_____

_____

6 How might you improve your classification technique?

_____

_____

If you found that activity interesting and rewarding, astronomers can use your help: go to https://galaxyzoo.org to get involved in a citizen science project to classify galaxies, some of which have never been viewed before by human eyes.

The Milky Way— A Normal Spiral Galaxy

The stars in a spiral galaxy are not uniformly distributed. This is also true for our Milky Way, a barred spiral galaxy. Here in Chapter 20, you will learn why this is so. Before you begin this experiment, predict whether you will see more stars in the plane of the Milky Way or far from it. Then, find a dark sky that is clear in all directions. Bring a star chart for the date and time of your observation, as well as a paper towel tube. Point your paper towel tube at the plane of the Milky Way, and count the stars that you see through the tube. Move the tube so that it points as far from the plane of the Milky Way as possible (90°, if you can!). Count the stars again. Halfway between these two points, make a third star count.

EXPERIMENT SETUP

Find as dark a site as possible, where you can see lots of stars. Bring a star chart for the date and time of your observation, as well as a paper towel tube.

1 Point your paper towel tube at the plane of the Milky Way, and count the stars that you see through the tube.

2 Move the tube so that it points as far from the plane of the Milky Way as possible (90°, if you can!). Count the stars again.

3 Halfway between these two points, make a third star count.

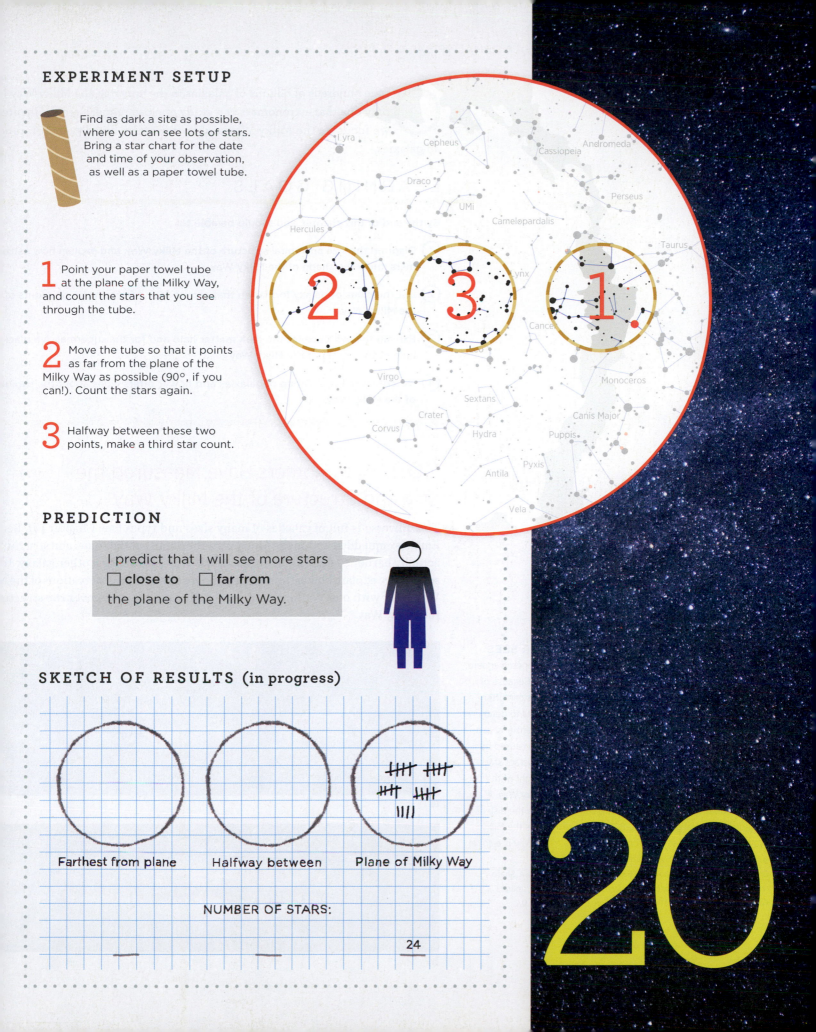

PREDICTION

I predict that I will see more stars
☐ **close to**   ☐ **far from**
the plane of the Milky Way.

SKETCH OF RESULTS (in progress)

Farthest from plane   Halfway between   Plane of Milky Way

NUMBER OF STARS:

24

20

Of the hundreds of billions of galaxies in the universe, the Milky Way is the only one that astronomers can study at close range. Here in Chapter 20, we focus on the Milky Way and how it offers clues to understanding all galaxies.

LEARNING GOALS

By the end of this chapter, you should be able to:

(1) Diagram the size and spiral structure of the Milky Way, and explain how astronomers know the shape of the Milky Way.

(2) List the clues of galaxy formation that can be found from the components of the Milky Way.

(3) Explain the evidence for the dark matter halo and for the supermassive black hole at the center of the Milky Way.

(4) Describe the Local Group of galaxies and how it offers clues about the evolution of the Milky Way.

20.1 Astronomers Have Measured the Size and Structure of the Milky Way

The universe is full of galaxies of many sizes and types (see Chapter 19). Because Earth is embedded within the Milky Way, the details of the shape and structure are actually harder to determine for the Milky Way than for any other galaxy. In this section, we explain how astronomers compare our limited observations of the Milky Way itself with observations of more distant galaxies to discover the structure of the Milky Way.

Figure 20.1 ★ WHAT AN ASTRONOMER SEES
Most astronomical objects have relatively simple shapes: a sphere, an ellipse, a disk, or a combination of these shapes. Because of this small set of shapes to choose from, astronomers can often compare two-dimensional images of different objects to determine whether they are the same shape in three dimensions. In **a**, an astronomer would note the prominent dark lanes caused by interstellar dust that obscures the light from more distant stars. She would identify the brighter yellow area to the right of the center of the image as the galactic center of the Milky Way. A comparison of this image to the image in **b**, in which the disk of the edge-on spiral galaxy NGC 891 greatly resembles the Milky Way, would convince an astronomer that the Milky Way is a spiral galaxy and the Solar System resides within the disk.
Credit (part b): C.Howk (JHU), B.Savage (U. Wisconsin), N.A.Sharp (NOAO)/WIYN/NOIRLab/NSF, https://noirlab.edu/public/images/noao-n891/. https://creativecommons.org/licenses/by/4.0/.

a.

b.   VIS

Spiral Structure in the Milky Way

From a dark location at night, you can see dark bands of interstellar gas and dust that obscure much of the central plane of the Milky Way Galaxy (**Figure 20.1a**). Compare this view of the Milky Way from Earth with **Figure 20.1b**, which shows a spiral galaxy, viewed edge-on. The similarities between those images suggest that the Milky Way is a spiral galaxy and that we are viewing it edge on, from a position in the disk.

In 2005, Spitzer Space Telescope observations of the distribution and motions of stars in the inner part of the galaxy confirmed that the Milky Way has a substantial bar with a modest bulge at its center. **Figure 20.2** shows an artist's rendering of the major features of the Milky Way. Two major spiral arms—Scutum-Centaurus and Perseus—connect to the ends of the central bar and sweep through the galaxy's disk, just like the arms observed in external spiral galaxies. Our galaxy has several smaller arm segments, including the Orion Spur, which contains the Sun and Solar System. Astronomers conclude that the Milky Way is a giant barred spiral that is more luminous than an average spiral. Viewed from the outside, the Milky Way would look much like the barred spiral galaxy M109 (**Figure 20.3**). Observations of ionized hydrogen gas in visible light reveal concentrations of young, hot O and B stars that confirm that the disk of the Milky Way contains at least two spiral arms.

Finding further details about the size and shape of the Milky Way requires extensive observations in the visible, infrared, and radio regions of the electromagnetic spectrum. Recall from Chapter 15 that neutral hydrogen emits radiation at a wavelength of 21 centimeters (cm) in the radio region of the spectrum, and that molecules emit in the infrared and radio regions of the electromagnetic spectrum as well. Maps of that radiation show spiral structure in other galaxies and suggest spiral structure in the Milky Way. Because these observations typically also contain velocity information, from the redshifts or blueshifts of the lines, astronomers can infer not only the direction to clouds of neutral hydrogen, but also their distance (**Figure 20.4**). Each cloud orbits the center of the Milky Way and so has a part of its velocity along the line of sight. Astronomers use Newton's version of Kepler's third law, along with geometry, to turn images like the one shown in Figure 20.4 into three-dimensional maps of the gas in the Milky Way. Those observations not only confirm that the Milky Way is a spiral galaxy, but trace out the specific locations of the spiral arms.

Figure 20.2 Infrared and radio observations contribute to an artist's model of the Milky Way Galaxy. The galaxy's two major arms (Scutum-Centaurus and Perseus) are seen attached to the ends of a thick central bar.

VIS

Figure 20.3 From the outside, the Milky Way would look much like this barred spiral galaxy, M109.
Credit: NOIRLab/NSF/AURA, https://noirlab.edu/public/images /noao-m109/. https://creativecommons.org/licenses/by/4.0/.

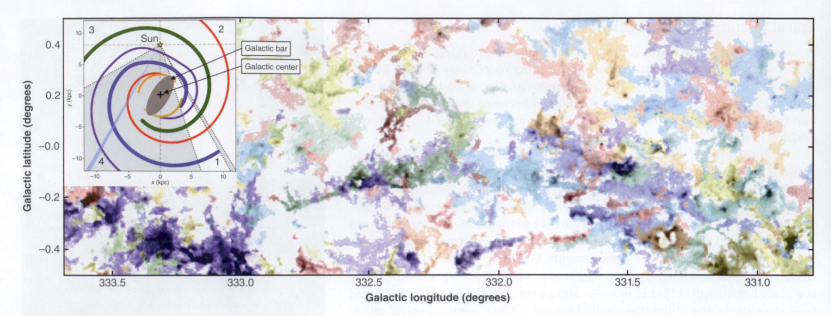

Figure 20.4 Astronomers use observations of cold molecular gas in the Milky Way to infer the detailed structure of its spiral arms. The blue segment in the inset shows the direction the telescope was pointing to obtain the larger image. Various colors in the larger image indicate different clouds, which are known to be separate because of their different velocities. Astronomers use these velocities to determine how far away a particular cloud is, and therefore map out the locations of the clouds and the gaps between them. This image is a snapshot of a much larger survey that covers the entire gray region shown in the inset sketch of the structure of the Milky Way.

Spiral Arms and Star Formation

In pictures of spiral galaxies, the arms are often the most prominent feature, as in the Andromeda Galaxy (**Figure 20.5**). The spiral arms are prominent in the ultraviolet image (Figure 20.5a), and although they are less prominent in visible light (Figure 20.5b), they are still clearly defined. You might then conclude that most stars in the disk of a spiral galaxy are concentrated in the spiral arms. That turns out *not* to be the case: Although stars are slightly concentrated in spiral arms, the concentration is not strong enough to account for their prominence. Structures associated with star formation, however, such as molecular clouds and associations of luminous O and B stars, are all concentrated in spiral arms. Spiral arms are prominent because star formation is occurring there, so the arms contain significant concentrations of young, massive, hot, and therefore luminous stars that emit strongly in blue light.

Stars form when dense interstellar clouds become so dense that they begin to collapse under the force of their own gravity (see Chapter 15). Because stars form

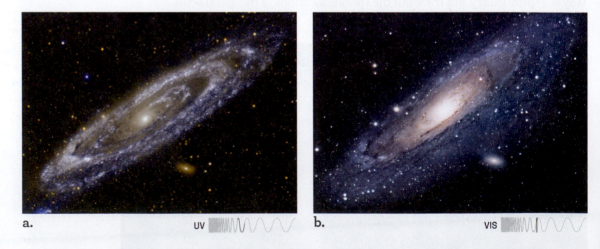

Figure 20.5 The Andromeda Galaxy in **a.** ultraviolet light and **b.** visible light. The spiral arms, dominated by hot, young stars, are most prominent in ultraviolet light. The spiral arms are less prominent in visible light.

Reading Astronomy News

Scientists Peer into the 3-D Structure of the Milky Way

Science Daily; Cardiff University

Because astronomers cannot leave the Milky Way, and look back to take a picture, they have to be clever in order to figure out the details of the Milky Way's shape.

Scientists from Cardiff University have helped produce a brand-new, three-dimensional survey of our galaxy, allowing them to peer into the inner structure and observe its star-forming processes in unprecedented detail.

The large-scale survey, called SEDIGISM (Structure, Excitation and Dynamics of the Inner Galactic Interstellar Medium), has revealed a wide range of structures within the Milky Way, from individual star-forming clumps to giant molecular clouds and complexes, that will allow astronomers to start pushing the boundaries of what we know about the structure of our galaxy.

SEDIGISM has been unveiled today through the publication of three separate papers in the *Monthly Notices of the Royal Astronomical Society*, authored by an international team of over 50 astronomers.

"With the publication of this unprecedentedly detailed map of cold clouds in our Milky Way, a huge observational effort comes to fruition," says Frederic Schuller from the Max Planck Institute for Radio Astronomy (MPIfR), lead author of one of the three publications, presenting the data release.

Dr Ana Duarte Cabral, a Royal Society University Research Fellow from Cardiff University's School of Physics and Astronomy, was lead author on one of the papers and has provided a catalogue of over 10,000 clouds of molecular gas in our Milky Way.

The Milky Way, named after its hazy appearance from Earth, is a spiral galaxy with an estimated diameter between 170,000 and 200,000 light-years which contains between 100–400 billion stars.

The Milky Way consists of a core region that is surrounded by a warped disk of gas and dust that provides the raw materials from which new stars are formed.

For Dr Duarte Cabral, the new catalogue of gas clouds will allow scientists to probe exactly how the spiral structure of our own Milky Way affects the life cycle of clouds, their properties, and ultimately the star formation that goes on within them.

"What is most exciting about this survey is that it can really help pin down the global galactic structure of the Milky Way, providing an astounding 3-D view of the inner galaxy," she said. "With this survey we really have the ability to start pushing the boundaries of what we know about the global effects of the galactic structures and dynamics, in the distribution of molecular gas and star formation, because of the improved sensitivity, resolution, and the 3D view."

The catalogue of molecular gas clouds was created by measuring the rare isotope of the carbon monoxide molecule, ^{13}CO, using the extremely sensitive 12-metre Atacama Pathfinder Experiment telescope on the Chajnantor plateau in Chile.

This allowed the team to produce more precise estimates of the mass of the gas clouds and discern information about their velocity, therefore providing a truly three-dimensional picture of the galaxy.

Dr Duarte Cabral and colleagues are already beginning to tease out information from the vast amount of data at their disposal.

"The survey revealed that only a small proportion, roughly 10%, of these clouds have dense gas with ongoing star formation," said James Urquhart from the University of Kent, the lead author of the third publication.

Similarly, the results from the work led by Dr Duarte Cabral suggest that the structure of the Milky Way is not that well defined and that the spiral arms are not that clear.

They have also shown that the properties of clouds do not seem to be dependent on whether a cloud is located in a spiral arm or an inter-arm region, where they expected very different physics to be playing a role.

"Our results are already showing us that the Milky Way may not be a strong grand design type of spiral galaxy as we thought, but perhaps more flocculent in nature," Dr Duarte Cabral continued.

"This survey can be used by anyone that wants to study the kinematics or physical properties of individual molecular clouds or even make statistical studies of larger samples of clouds, and so in itself has a huge legacy value for the star formation community."

QUESTIONS

1. The article does not say, but in what part of the electromagnetic spectrum were these observations of molecular clouds most likely made: radio, infrared, visible, or X-ray?

2. About how many gas clouds were catalogued in this survey? According to Dr. Urquhart, 10 percent of those have dense gas with ongoing star formation. How many clouds in this survey have ongoing star formation?

3. Why does knowing the velocity of a cloud in the disk of the Milky Way help astronomers figure out how far away it is?

4. Dr. Cabral states that "the Milky Way may not be a strong grand design type of spiral galaxy as we thought, but perhaps more flocculent in nature." Look up the word "flocculent." What is Dr. Cabral saying about the disk of the Milky Way?

Source: https://www.sciencedaily.com/releases/2020/12/201203122257.htm.

a.

VIS ⩍⩍⩍⩍⩍

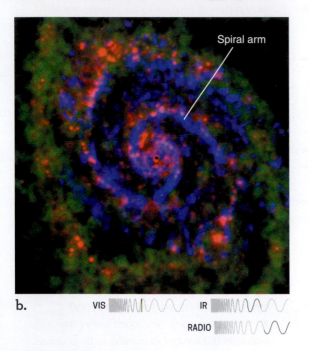

b.

VIS ⩍⩍⩍⩍⩍ IR ⩍⩍⩍⩍⩍

RADIO ⩍⩍⩍⩍⩍

Figure 20.6 These two images of a face-on spiral galaxy show the spiral arms. **a.** This visible-light image also shows dust absorption. **b.** This image shows the distribution of neutral interstellar hydrogen (green), carbon monoxide (CO) emission from cold molecular clouds (blue), and hydrogen alpha (Hα) emission from ionized gas (red).
Credit (part a): Todd Boroson/NOIRLab/NSF/AURA/, https://noirlab.edu /public/images/noao-m74/. https://creativecommons.org/licenses/by/4.0/.

in spiral arms, spiral arms must be places where clouds of interstellar gas and dust pile up and are compressed. Those clouds can be observed where the dust and gas block starlight (**Figure 20.6a**) or where gases such as neutral hydrogen or carbon monoxide emit at various wavelengths (**Figure 20.6b**).

Disks of galaxies do not rotate like a solid body. Instead, material close to the center takes less time to travel around the galaxy than material farther out, and so the inner part of the disk gets ahead of the outer part. This means that any disturbance in the disk of a spiral galaxy will cause a spiral pattern because the disk rotates. **Figure 20.7** illustrates the point: in the second frame, the outer part of the line is trailing behind the inner part. As time passes, a straight line through the center becomes a spiral. This concept cannot fully explain spiral arms, however, because orbital periods are relatively short (the Sun takes about 230 million years to orbit the center of the Milky Way). This kind of spiral structure tightens very quickly and disappears.

A spiral pattern may form if a spiral galaxy is disturbed. For example, star formation itself can create spiral structure. Regions of star formation release considerable energy into their surroundings through UV radiation, stellar winds, and subsequent supernova explosions. That energy compresses clouds of gas and triggers more star formation. Typically, many massive stars form in the same region at about the same time, and their combined mass outflows and supernova explosions after short lifetimes occur one after another in the same region of space over only a few million years. The result can be large, expanding bubbles of hot gas that sweep out cavities in the interstellar medium and concentrate the swept-up gas into dense, star-forming clouds, much like the snow that piles up in front of a snowplow. In that way, star formation can propagate through the disk of a galaxy. Rotation bends the resulting strings of star-forming regions into spiral structures. However, as with the winding process shown in Figure 20.7, a single disturbance will also not produce a stable spiral-arm pattern. Spiral arms produced from a single disturbance will wind themselves up completely in two or three rotations of the disk and then disappear.

Stable spiral arms require repeated disturbances. In barred spirals, for example, the elongated bulge gravitationally tugs on the disk differently in different areas of the disk. As the disk rotates past the elongated bulge, it is repeatedly disturbed by the changing gravitational pull. Repeated episodes of star formation occur, and stable spiral arms form. Many galaxies show clear evidence of a relationship between the shapes of their bulges and the structure of their spiral arms. Barred spirals have a characteristic two-armed spiral pattern that is connected to the elongated bulge, as seen in Figure 20.3. Even the bulges of galaxies that are not obviously barred may be elongated enough to contribute to the formation of a two-armed spiral structure. Smaller galaxies in orbit about larger galaxies also can give rise to a periodic gravitational disturbance, triggering a similar two-armed structure.

Stars move in and out of spiral arms as they orbit the center of a galaxy. The conditions are roughly analogous to a traffic jam on a busy highway. The cars in the jam are changing all the time, yet the traffic jam persists as a place of higher density—a place with more cars than usual. The traffic jam itself moves slowly backward, even as the cars move forward and pass through it. Just like the traffic jam, the disturbance of the spiral arm also moves at a different speed from that of the individual stars. Those disturbances in the disks of spiral galaxies are called **spiral density waves** because they are waves of greater mass density and increased pressure in the galaxy's interstellar medium. Those waves move around a disk in the pattern of a two-armed spiral that does not rotate at the same rate as the stars, gas, or dust. As material in the disk orbits the center of the galaxy, it passes through those spiral density waves.

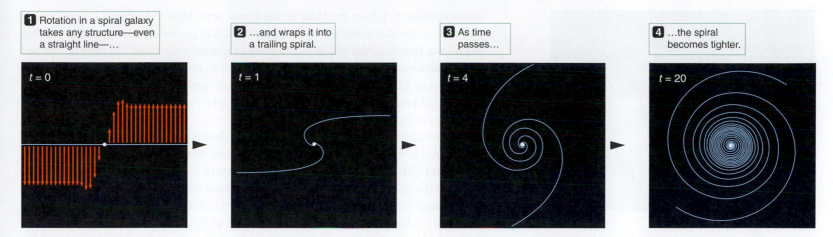

1 Rotation in a spiral galaxy takes any structure—even a straight line—...

2 ...and wraps it into a trailing spiral.

3 As time passes...

4 ...the spiral becomes tighter.

t = 0

t = 1

t = 4

t = 20

Figure 20.7 As the disk of a spiral galaxy rotates in the direction indicated by the orange arrows, it will naturally take even an originally linear structure and wrap it into a progressively tighter spiral as time (*t*) goes by.

A spiral density wave has very little effect on the motions of stars as they pass through it, but it does compress the gas that flows through it. Gas flows into the spiral density wave and piles up. Stars form in the resulting compressed gas. Massive stars are concentrated in the arms because they have such short lives (typically 10 million years or so) that they never have the chance to drift far from the spiral arms where they were born. Less massive stars, however, have plenty of time to move away from their places of birth, so they form a smooth underlying disk.

The Size of the Milky Way Galaxy

If you go out on a dark night, away from any street lights, and look toward the center of the Milky Way—located in the constellation Sagittarius—you will see the dark lane of dusty clouds shown in Figure 20.1a. Because the Solar System is inside the dusty disk of the Milky Way, the visible-light view of the galaxy itself is badly obscured, so how do astronomers know that the center of the Milky Way lies in that direction? And how do they know how far away it is?

In the 1920s, Harlow Shapley made a map of globular clusters in the sky. Recall from Chapter 17 that globular clusters are large, spheroidal groups of stars held together by gravity. The Milky Way contains more than 150 cataloged globular clusters, and dust in the disk hides many more. Globular clusters are very luminous (as much as 1 million L_{Sun}), so the ones that lie outside the dusty disk can be easily seen as round, fuzzy blobs even through small telescopes and even at great distances. From his map, Shapley found that there were more globular clusters in the direction of Sagittarius than in any other direction. Imagine that you are sitting in a full movie theater, and you see more people to your left than your right. You would immediately know that you are not at the center, even if you can't see the walls. Similarly, Shapley concluded that the Sun was not at the center of the Milky Way, and that the center is in the direction of Sagittarius.

Shapley was able to go one step further by finding the distances to the globular clusters. In a Hertzsprung-Russell (H-R) diagram of an old cluster, the horizontal branch crosses the instability strip, which contains pulsating stars such as RR Lyrae stars and Cepheid variables. RR Lyrae stars are easy to spot in globular clusters because they are relatively luminous and have a distinctive light curve. As

what if . . .

What if astronomers observed only K- and M-type stars in the spiral arms of galaxies? What would you expect to find out about the concentration of dust and gas in these arms?

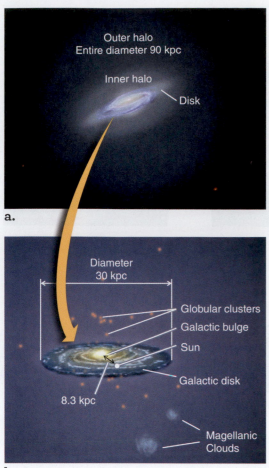

a.

b.

Figure 20.8 Parts of the Milky Way include **a.** the disk and the inner and outer halos and **b.** the galactic bulge that crosses the center of the disk. Globular clusters are located in the halo, and the Sun is located in the disk, about 8.3 kpc from the center. (1 kpc = 1000 pc.) The Magellanic Clouds are nearby dwarf galaxies, companions to the Milky Way.

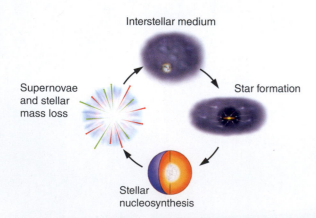

Figure 20.9 Matter moves from the interstellar medium into stars and back again in a progressive cycle that has enriched today's universe with massive elements.

with Cepheid variables, the time an RR Lyrae star takes to undergo one pulsation is related to the star's luminosity. Shapley used that period-luminosity relationship to find the luminosities of RR Lyrae stars in globular clusters. He then combined those luminosities with measured brightnesses to determine the distances to globular clusters. Finally, Shapley cross-checked his results by noting that more distant clusters (as measured by the RR Lyrae stars) also tended to appear smaller in the sky, as expected.

Those globular clusters trace out the luminous part of the galactic halo of the Milky Way Galaxy, a large spherical volume of space surrounding the disk and bulge. The center of the distribution of globular clusters coincides with the gravitational center of the galaxy. Shapley realized that because he could determine the distance to the center of that distribution, he had actually determined the Sun's distance from the center of the Milky Way, as well as the size of the galaxy itself. His map showed that globular clusters occupy a roughly spherical region of space with a diameter of about 90 kiloparsecs (kpc), or 90,000 parsecs (pc). Shapley did not know about gas and dust, however, so he overestimated the distance to the globular clusters. A modern determination indicates that the Sun is located about 8300 pc (27,000 light-years) from the center of the galaxy, or roughly halfway out toward the edge of the disk, as shown in **Figure 20.8**.

CHECK YOUR UNDERSTANDING **20.1**

List three pieces of evidence supporting the theory that the Milky Way is a spiral galaxy.

Answers to Check Your Understanding questions are in the back of the book.

20.2 The Components of the Milky Way Reveal Its Evolution

The Sun is a middle-aged star located among other middle-aged stars that orbit around the galaxy within the galactic disk, as do the gas and dust in the disk. The stars in the halo move in random orbits similar to those of stars in elliptical galaxies, sometimes at high velocities. Some of those stars can be observed near the Sun as their orbits carry them swiftly through the disk. Most of the halo stars are much older than the Sun. The bar in the galactic bulge of the Milky Way is shaped primarily by stars and gas moving not only in highly elongated orbits up and down the long axis of the bar but also in short orbits aligned perpendicular to the bar. All those stellar orbits determine the shapes of the various parts of the galaxy, and stellar orbits are easier to measure in the Milky Way than in other galaxies. Using the ages, chemical abundances, and motions of nearby stars, astronomers can differentiate between disk and halo stars to learn more about the galaxy's structure. In this section, we describe how astronomers study the constituents of the Milky Way to find clues about how spiral galaxies form.

Age and Chemical Compositions of Stars

Over time, as stars are born, live, and die, they add massive elements to the interstellar medium (**Figure 20.9**). The interstellar medium therefore reflects all the stellar evolution that has taken place up to the present time. As this gas forms new stars, it carries those massive elements along, so more recently formed stars have a greater abundance of massive elements. As illustrated in **Figure 20.10**, the chemical composition of a star's atmosphere reflects all of the stellar birth and death that occurred before that star formed. Although the details are complex, several clear and important lessons can be learned from the observed patterns in the amounts of massive elements in the stars of the galaxy.

The relationship between age and abundances of massive elements is evident throughout much of the galaxy. Stars in globular clusters contain only very small amounts of massive elements, indicating their great age. Some globular-cluster stars contain only 0.5 percent as much of those massive elements as the Sun. All the stars in the galaxy's halo have smaller amounts of heavy elements. Within the disk, younger stars typically have higher abundances of massive elements than those of older stars. Similarly, older stars in the outer parts of the galaxy's bulge have lower massive-element abundances than those of young stars in the disk.

There are two types of star clusters—globular and open. Globular clusters orbit in the halo of the Milky Way and are tightly bound by gravity. Open clusters orbit in the disk of the Milky Way and are so loosely bound by gravity that the stars eventually drift away from one another. As with globular clusters, the stars in an open cluster all formed at about the same time, so astronomers can use the H-R diagram to find the cluster's age. Some open clusters are very young, containing the very youngest stars known, whereas other open clusters contain stars older than the Sun; but all open clusters are younger than the youngest globular clusters. Because the globular clusters are old and located in the halo, astronomers infer that stars in the halo formed before stars in the disk. That epoch of halo star formation did not last long. Star formation in the disk started later and has been continuing ever since.

Within the galaxy's disk, astronomers observe differences in abundances of massive elements from place to place, related to the rate of star formation in different regions. Star formation is generally more active in the denser, inner part of the Milky Way than in the outer parts. Observations of chemical abundances in the interstellar medium, based both on interstellar absorption lines in the spectra of stars and on emission lines in glowing clouds of neutral hydrogen gas known as H II regions, confirm that prediction by showing a smooth decline in abundances of massive elements from the inner to the outer parts of the disk. Astronomers have observed similar trends in other galaxies. Within a galactic disk, relatively old stars near the center of a galaxy often have greater massive-element abundances than those of young stars in the outer parts of the disk.

Even the very oldest globular-cluster stars contain some chemical elements fused in previous generations of more massive stars. That observation implies that globular-cluster stars and other halo stars were not the first stars in the Milky Way to form. At least one generation of massive stars lived and died, ejecting newly synthesized massive elements into space, before even the oldest globular clusters formed. (We return to these first stars in Chapter 23.) Every star less massive than about $0.8\ M_{\text{Sun}}$ that ever formed is still around as a main-sequence star today. The oldest known star is a red dwarf in this mass category (forgettably named 2MASS J18082002-5104378 B). It is estimated to be about 13.5 billion years old, a bare 200 million years younger than the universe itself. Even that ancient star has a small abundance of massive elements. Observations of ancient stars like this one indicate that the gas that wound up in the disk of the Milky Way must have seen a significant amount of star formation before it made stars. Still, even a chemically "rich" star like the Sun, which is made of gas processed through approximately 9 billion years of previous generations of stars, is composed of less than 2 percent massive elements.

what if . . .

What if you observe a nearby star that shows absolutely no evidence of elements heavier than helium in its spectrum? What could you conclude about the mass of this star, and how fast do you expect it to be moving relative to the Solar System?

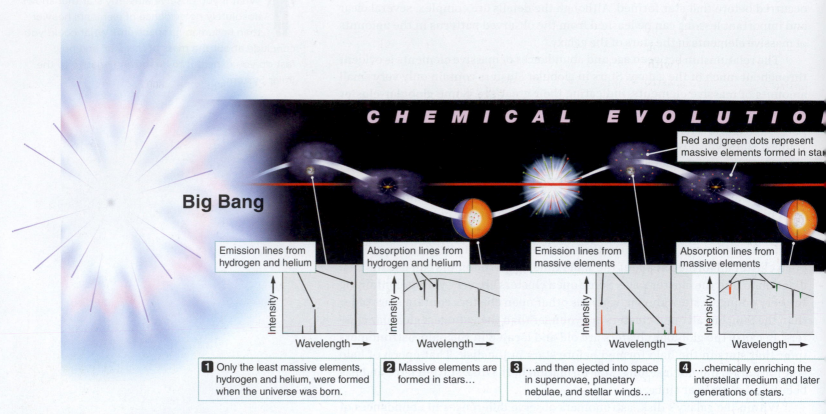

CHEMICAL EVOLUTION

Big Bang

Red and green dots represent massive elements formed in stars

Emission lines from hydrogen and helium

Absorption lines from hydrogen and helium

Emission lines from massive elements

Absorption lines from massive elements

Intensity

Wavelength →

Intensity

Wavelength →

Intensity

Wavelength →

Intensity

Wavelength →

1 Only the least massive elements, hydrogen and helium, were formed when the universe was born.

2 Massive elements are formed in stars…

3 …and then ejected into space in supernovae, planetary nebulae, and stellar winds…

4 …chemically enriching the interstellar medium and later generations of stars.

Figure 20.10 As subsequent generations of stars form, live, and die, they enrich the interstellar medium with massive elements—the products of stellar nucleosynthesis. The chemical evolution of the Milky Way and other galaxies can be traced in many ways, including by the strength of interstellar emission lines and stellar absorption lines.

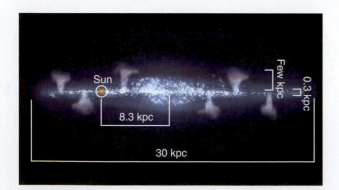

Few kpc

0.3 kpc

Sun

8.3 kpc

30 kpc

Figure 20.11 Artist's concept of a "galactic fountain," in which gas is pushed away from the disk of the galaxy by energy released by young stars and supernovae and then falls back onto the disk. The distance of the Sun from the center of the galaxy and the diameter of the disk are shown here for scale.

Luminous matter in the universe is still dominated by hydrogen and helium formed just after the Big Bang, long before the first stars.

Higher massive-element abundances tend to correlate with the regions of more active star formation in the inner galaxy, but there are complicating factors. New material falling into the galaxy adds more hydrogen and helium, reducing the abundance of massive elements in the interstellar medium. The energy of massive stars propels enriched material from the inner disk into the halo in great "fountains," as shown in the plumes of the artist's depiction in **Figure 20.11**. This material falls back onto the disk elsewhere. Past interactions with other galaxies stirred the Milky Way's interstellar medium, mixing gas from those other galaxies with gas already there. Deciphering the history of star formation from the variations of chemical abundances within the Milky Way and other galaxies remains an active topic of research.

Components of the Disk

Astronomers divide the disk of the Milky Way into a **thin disk** and a **thick disk** (**Figure 20.12a**). The youngest stars in the galaxy are most strongly concentrated in the middle of the disk, defining a thin disk about 300 pc (1000 light-years) thick but more than 30,000 pc (100,000 light-years) across. That ratio of the thickness to the diameter of the disk is similar to that of a DVD. The youngest stars are concentrated in the thin disk because that is where the molecular clouds and gas are most concentrated. Dust is also concentrated in the thin disk of the galaxy, as seen in a recent image from the *Gaia* mission (**Figure 20.12b**). Stars

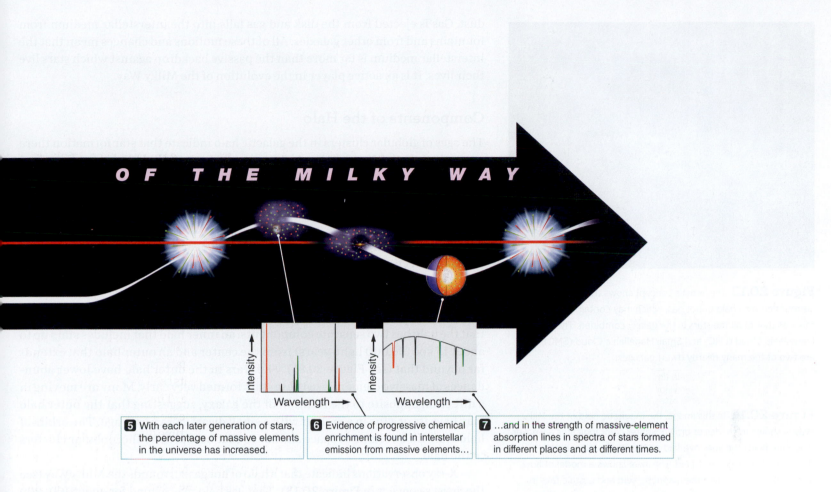

5 With each later generation of stars, the percentage of massive elements in the universe has increased.

6 Evidence of progressive chemical enrichment is found in interstellar emission from massive elements...

7 ...and in the strength of massive-element absorption lines in spectra of stars formed in different places and at different times.

in the thin disk show a decrease in massive-element abundances as the distance from the galactic center increases, implying that the youngest stars are closer to the galactic center.

The thick disk is about 3700 pc (12,000 light-years) thick, and it contains the older population of disk stars, distinguishable by lower abundances of massive elements. The oldest stars are closer to the galactic center. Astronomers are still debating how distinct those two disks are and how they originated.

The interstellar medium in the disk is a dynamic place—energy from star-forming regions can form gigantic and interesting structures in the interstellar medium. The death of stars constantly changes the composition of the gas and the

Figure 20.12 **a.** This illustration shows the disks, bulge, and inner halo of the Milky Way and the location of globular clusters. **b.** An all-sky view of the interstellar dust that fills the Milky Way Galaxy, from observations by ESA's *Gaia* satellite.

Credit (part b): © ESA/Gaia/DPAC, CC BY-SA 3.0 IGO, https://sci.esa.int/web/gaia/-/60171-gaia-s-view-of-dust-in-the-milky-way. https://creativecommons.org/licenses/by/3.0/igo/.

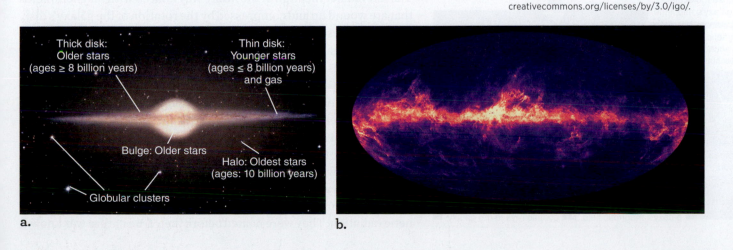

a.

b.

Figure 20.13 This artist's concept shows the Milky Way surrounded by a halo of hot gas, which may contain as much mass as that of all the stars in the galaxy combined. The Large Magellanic Cloud (LMC) and Small Magellanic Cloud (SMC) are two of the many nearby dwarf galaxies.

Figure 20.14 In this image, the magnetic field of the Milky Way is shown in shades of brown; lighter color indicates a weaker magnetic field. The lines overlaid on the image indicate the direction of the magnetic field. The inset shows a model of how interstellar clouds affect the magnetic field and cosmic rays in the galaxy.

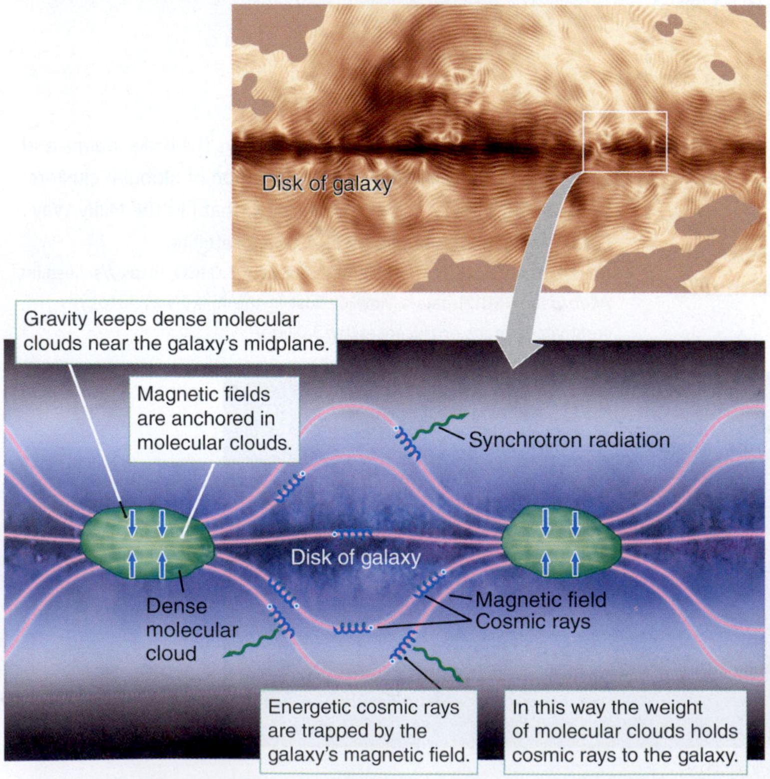

dust. Gas is ejected from the disk and gas falls into the interstellar medium from fountains and from other galaxies. All of these motions and changes mean that the interstellar medium is far more than the passive backdrop against which stars live their lives. It is an active player in the evolution of the Milky Way.

Components of the Halo

The ages of globular clusters in the galactic halo indicate that star formation there was early and brief. Yet globular clusters account for only about 1 percent of the total mass of stars in the halo. As halo stars fall through the disk of the Milky Way, some pass close to the Sun, providing a sample of the halo that can be studied at closer range. Astronomers can distinguish nearby halo stars because halo stars appear to be moving at higher relative velocities than disk stars. Most stars near the Sun are disk stars, so they orbit the center of the galaxy at nearly the same speed, in roughly the same direction as the Sun. In contrast, halo stars orbit the center of the galaxy in random directions, so the relative velocity between the halo stars and the Sun tends to be high. Those stars are known as high-velocity stars.

By studying the orbits of high-velocity stars, astronomers have determined that the halo has two separate components: an inner halo that includes stars up to about 15 kpc (50,000 light-years) from the center and an outer halo that extends far beyond that (see Figure 20.8a). The stars in the outer halo have lower abundances of massive elements, so these stars formed very early. Many are moving in a direction opposite to the rotation of the galaxy, suggesting that the outer halo may have its origins in a merger with a small dwarf galaxy long ago. The orbits of halo stars fill a volume of space similar to that occupied by the globular clusters in the halo.

X-ray observations indicate that a halo of hot gas surrounds the Milky Way (see the artist's concept in **Figure 20.13**). That gas halo may extend for about 100–200 kpc from the galactic center, encompassing two nearby small galaxies and containing as much mass as that of all the stars in our galaxy. The halo's temperature is about 2 million kelvins (K), so the gas particles are ionized and moving very quickly. The gas is extremely diffuse, however, so the particles rarely collide with one another to transfer energy. Much like the gas in the solar corona, this halo gas wouldn't "feel" hot.

Magnetic Fields and Cosmic Rays Fill the Galaxy

The interstellar medium of the Milky Way is laced with magnetic fields that are wound up and compressed by the rotation of the galaxy's disk. The total interstellar magnetic field, however, is about a hundred thousand times weaker than Earth's magnetic field. Charged particles and magnetic fields interact strongly; the particles spiral around magnetic fields, moving along the field rather than across it. Conversely, magnetic fields cannot freely escape from a cloud of gas containing even a small number of charged particles. The dense clouds of interstellar gas in the thin disk of the Milky Way (**Figure 20.14**) anchor the galaxy's magnetic field to the disk, in turn anchoring high-energy charged particles, known as cosmic rays, to the galaxy.

Cosmic rays are charged particles that travel close to the speed of light. Despite their name, cosmic rays are not a form of electromagnetic radiation: They were named before their true nature was known.

Figure 20.15 The Pierre Auger Observatory in Argentina is an array of stations designed to catch the particles that shower from collisions of cosmic rays with the upper atmosphere. Each station in the array is equipped with its own particle collectors, carefully protected from the elements.

Most cosmic-ray particles are protons, but some are nuclei of helium, carbon, and other elements produced by nucleosynthesis. A few are high-energy electrons and other subatomic particles.

Cosmic rays span an enormous range in particle energy. Astronomers observe the lowest-energy cosmic rays with interplanetary spacecraft. Those cosmic rays have energies as low as about 10^{-11} joule (J), which corresponds to the energy of a proton moving at a velocity of 1/3 the speed of light. In contrast, the most energetic cosmic rays are 10 trillion (10^{13}) times as energetic as the lowest-energy cosmic rays, and they move at $0.999999c$—very close to the speed of light. High-energy cosmic rays are detected from the showers of elementary particles that they cause when crashing through Earth's atmosphere. Those particle showers are observed by special telescopes such as the High Energy Stereoscopic System (HESS) imaging telescopes in Namibia, or the Pierre Auger Observatory in Argentina (**Figure 20.15**).

Astronomers hypothesize that most cosmic rays are accelerated to those incredible energies by supernova explosions. The very highest-energy cosmic rays are as much as a hundred million times more energetic than any particle ever produced in a particle accelerator on Earth. Those extremely high energies make the highest-energy cosmic rays much more difficult to explain than those with lower energies.

Cosmic rays spiraling around the magnetic field in the Milky Way's disk produce synchrotron radiation (see Chapter 17). Such synchrotron emission is seen in other spiral galaxies as well, indicating that they, too, have magnetic fields and populations of energetic cosmic rays. The very highest-energy cosmic rays are moving much too fast to be confined by the gravitational force of their originating galaxy, or the galactic magnetic field. Any such cosmic rays that formed in the Milky Way would soon stream away from the galaxy into intergalactic space. Thus, some of the energetic cosmic rays reaching Earth probably originated in energetic events outside the Milky Way Galaxy.

The total energy of all the cosmic rays in the galactic disk can be estimated from the energy of the cosmic rays reaching Earth. The strength of the interstellar magnetic field can be measured by observing how the field affects the properties of radio waves passing through the interstellar medium. Those measurements indicate that in the Milky Way Galaxy, the magnetic-field energy and the cosmic-ray energy are about equal. Both are comparable to the energy present in other energetic components of the galaxy, including the motions of interstellar gas and the total energy of electromagnetic radiation within the galaxy.

CHECK YOUR UNDERSTANDING 20.2

What parts of the Milky Way contain old stars, and what parts contain young stars?

20.3 Most of the Milky Way Is Unseen

As in other galaxies, the most interesting parts of the Milky Way may be the parts that can't be seen directly but are detected only by their gravitational influence on the stars around them. Dark matter accounts for most of the mass in a galaxy and extends far beyond a galaxy's visible boundary. As in all other spiral galaxies, compelling evidence indicates that dark matter dominates the Milky Way. From radio and infrared observations, astronomers determine how the disk of the Milky Way moves, and from that motion, they can then determine its mass.

The supermassive black hole at the center of the Milky Way poses a different kind of observation problem. That object also is detected by its gravitational effects on the stars nearby but cannot be seen directly. In this section, we describe what can be inferred about those two components of the Milky Way from how their gravity affects other objects.

Dark Matter in the Milky Way

The rotation of the disk of the Milky Way can be determined from observations of the relative velocities of interstellar hydrogen measured from 21-cm radiation. **Figure 20.16** shows how those velocities vary with viewing direction from the Sun. Looking toward the center of the galaxy, we see that the gas is stationary relative to the Sun. When we look in the direction of the Sun's motion around the galactic center, hydrogen clouds appear to be moving toward Earth, whereas in the opposite direction, clouds are moving away from Earth. In other directions, the measured velocities are complicated by Earth's moving vantage point within the disk and so are more difficult to interpret at a glance. That pattern of the rotation velocity of gas in a disk is like those you learned about in Chapter 19. The only difference is that instead of looking at it from outside, we see the Milky Way Galaxy's rotation curve from a vantage point located within— and rotating with—the galaxy. Even so, observed velocities of neutral hydrogen enable astronomers to measure the Milky Way Galaxy's rotation curve and even determine the structure present throughout its disk.

Figure 20.16 a. From Earth's perspective within the Solar System, the signature of a rotating disk is clear from views of either side of the galactic center. **b.** Those velocities vary between redshift and blueshift as an observer looks around in the disk of the Milky Way.
Credit (part b): ESA/Gaia/DPAC, CC BY-SA 3.0 IGO, https://sci.esa.int/web/gaia/-/60223-gaia-s-all-sky-map-of-radial-velocities. https://creativecommons.org/licenses/by/3.0/igo/.

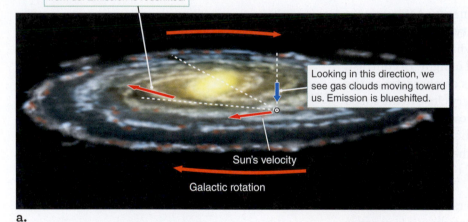

a.

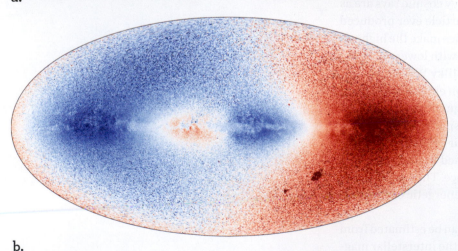

b.

working it out 20.1

The Mass of the Milky Way inside the Sun's Orbit

The Sun orbits about the center of the Milky Way Galaxy. Even though the gravitational pull on the Sun comes from all the material inside that orbit (see Chapter 4), Newton showed that we could treat the system as though all the mass were concentrated at the center. Thus, we can apply Newton's and Kepler's laws to calculate the mass of the Milky Way inside the Sun's orbit. Newton's version of Kepler's third law relates the period of the orbit to the orbital radius and the masses of the objects. But here, the mass of the galaxy is much larger than the mass of the Sun, so the Sun's mass is negligible by comparison. In addition, what astronomers can *measure* is the orbital speed of the Sun or other stars about the galactic center, rather than the orbital period. Thus, we can use a rearranged form of the same equation from Working It Out 4.3 that we used to estimate the mass of the Sun from the orbit of Earth:

$$M = \frac{r v_{circ}^2}{G}$$

The Sun orbits the center of the galaxy at 240 kilometers per second (km/s). The distance of the Sun from the center of the galaxy is 8300 pc, which converts to kilometers as follows:

$$8300 \text{ pc} \times (3.09 \times 10^{13} \text{ km/pc}) = 2.56 \times 10^{17} \text{ km}$$

Because we know the value of the gravitational constant, $G = 6.67 \times 10^{-20}$ km^3/(kg s^2), we can calculate the mass of the portion of the Milky Way inside the Sun's orbit:

$$M = \frac{(2.56 \times 10^{17} \text{ km}) \times (240 \text{ km/s})^2}{6.67 \times 10^{-20} \text{ km}^3/(\text{kg s}^2)}$$

$$M = 2.21 \times 10^{41} \text{ kg}$$

To put that result in units of the Sun's mass, we divide the answer by $M_{Sun} = 1.99 \times 10^{30}$ kg, yielding

$$M = \frac{2.21 \times 10^{41} \text{ kg}}{1.99 \times 10^{30} \text{ kg}/M_{Sun}} = 1.11 \times 10^{11} M_{Sun}$$

The mass of the Milky Way inside the Sun's orbit is about 111 billion times the mass of the Sun.

Recall (Chapter 19) that observations of rotation curves led astronomers to conclude that the masses of spiral galaxies consist mostly of dark matter. **Figure 20.17** shows the rotation curve of the Milky Way as inferred primarily from 21-cm observations. The orbital motion of the nearby dwarf galaxy called the Large Magellanic Cloud provides data for the outermost point in the rotation curve, at a distance of roughly 50,000 pc (160,000 light-years) from the center of the galaxy. Like other spiral galaxies, the Milky Way has a fairly flat rotation curve, which indicates the presence of large amounts of dark matter. The mass outside the Sun's orbit does not greatly affect the Sun's orbit (**Working It Out 20.1**).

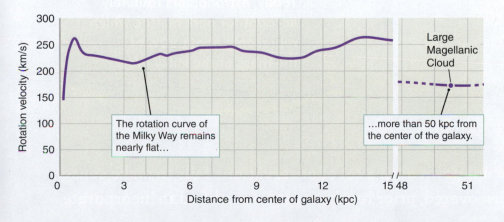

Figure 20.17 Rotation velocity is plotted against distance from the center of the Milky Way. The most distant point comes from measurements of the orbit of the Large Magellanic Cloud. The nearly flat rotation curve indicates that dark matter dominates the outer parts of the Milky Way. Part of this graph has been hidden to fit the Large Magellanic Cloud into the figure. The broken x-axis, marked by the diagonal hatch marks, indicates the hidden portion.

Unknown Unknowns

Initial efforts to measure the Milky Way did not include dark matter, because astronomers did not know about it.

Fritz Zwicky
1898–1974

Fritz Zwicky was the first to predict that dark matter existed, based on observations of clusters of galaxies.

Vera Rubin
1928–2016

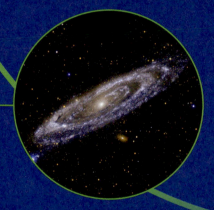

Vera Rubin subsequently discovered dark matter in the Andromeda Galaxy. This galaxy rotated faster than could be accounted for by the observed mass. Following up with observations of other galaxies, she confirmed that dark matter was an important component in nearly every galaxy.

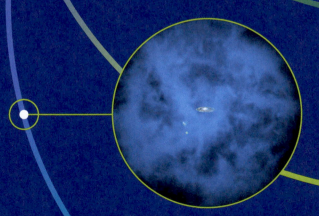

Today, astronomers routinely account for dark matter, which represents about 90% of the mass of the Milky Way.

Often, scientists don't know what is unknown until a set of observations fails to make sense. Once this "unknown unknown" is discovered, prior results must be modified to incorporate the new knowledge.

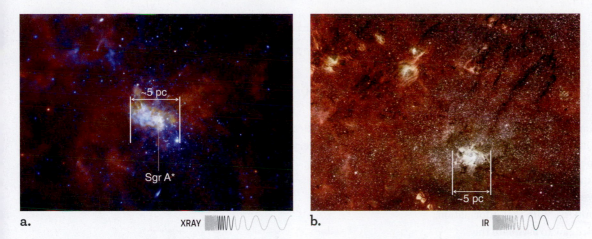

Figure 20.18 a. This X-ray view of the Milky Way's central region shows the active source, Sagittarius A* (Sgr A*), as the brightest spot at the middle of the image. Lobes of superheated gas (shown in red) are evidence of recent, violent explosions happening near Sgr A*. **b.** This infrared view of the central core of the Milky Way shows hundreds of thousands of stars. The bright white spot at the lower right marks Sgr A*, the location of the supermassive black hole.

The total mass of the Milky Way Galaxy is currently estimated to be about 1.0 trillion to 1.5 trillion times the mass of the Sun. The luminous mass, however, estimated by adding the masses of stars, dust, and gas, is only about one-tenth as much. Astronomers conclude that, like other spiral galaxies, the Milky Way's mass consists mainly of dark matter (see the **Process of Science Figure**). The spatial distribution of dark and normal matter within the Milky Way is also much like that of other galaxies, with dark matter dominating its outer parts.

The Supermassive Black Hole

Figure 20.18 shows images of the Milky Way's center taken with the Chandra X-ray Observatory and the Spitzer Space Telescope. The X-ray view (Figure 20.18a) shows the location of a strong radio source called Sagittarius A* (abbreviated Sgr A*), which lies at the center of the Milky Way. The infrared image (Figure 20.18b) cuts through the dust to reveal the galaxy's crowded, dense core containing hundreds of thousands of stars.

Studies of the motions of stars closest to the Sgr A* source suggest a central mass very much greater than that of the few hundred stars orbiting there. Furthermore, observations of the galaxy's rotation curve show rapid rotation velocities very close to the galactic center. Stars closer than 0.1 light-year from the galactic center follow Kepler's laws, indicating that their motion is dominated by mass within their orbit. The closest stars studied are only about 0.01 light-year from the center of the galaxy—so close that their orbital periods are only about a dozen years. The positions of those stars change noticeably over time, and astronomers can see them speed up as they whip around what can only be a supermassive black hole at the focus of their elliptical orbits (**Figure 20.19**). Using Newton's version of Kepler's third law, we can then estimate that the black hole at the center of the Milky Way Galaxy is a relative lightweight, having a mass of "only" 4 million times the mass of the Sun (**Working It Out 20.2**).

Clouds of interstellar gas at the galaxy's center are heated to millions of degrees by supernova explosions and stellar winds from young, massive stars. Superheated gas produces X-rays, and the Chandra X-ray Observatory has detected more than 9000 X-ray sources within the central region of the galaxy. Those sources include frequent, short-lived X-ray flares near Sgr A* (see Figure 20.18a), which provide

what if . . .

What if one of the Milky Way's companions, such as the Large Magellanic Cloud, were to merge with our galaxy? What effect might this have on the black hole in the nucleus of our galaxy?

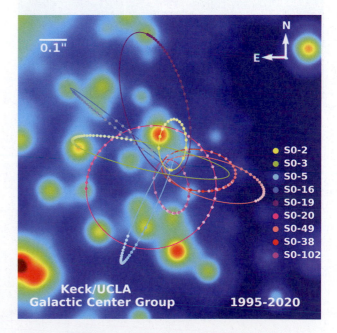

Figure 20.19 The orbits of eight stars within 0.03 pc (0.1 light-year, or about 6000 astronomical units [AU]) of the Milky Way's center. Applying Kepler's laws to the motions of those stars reveals the presence of a 4-million-M_{Sun} supermassive black hole at the galaxy's center. Colored dots show the measured positions of each star over many years: the dots progress from lighter in 1995 to darker in 2016.

working it out 20.2

The Mass of the Milky Way's Central Black Hole

Figure 20.19 illustrates data points for the stars in the central region orbiting closely to the central black hole of the Milky Way Galaxy. Those stars have highly elliptical orbits with changing speeds, but the orbital periods are short enough that they can be observed and measured. Star S0-2 in the figure has a measured orbital period of 15.8 years. The semimajor axis of its orbit is estimated to be 1.5×10^{11} km = 1000 AU. With that information, we can use Newton's version of Kepler's third law to estimate the mass inside S0-2's orbit. Setting up the equation as we did in Working It Out 13.4:

$$\frac{m_{BH}}{M_{Sun}} + \frac{m_{S0\text{-}2}}{M_{Sun}} = \frac{A_{AU}^3}{P_{years}^2}$$

The mass of star S0-2 is much less than the mass of the black hole, so the sum of the two is very close to the mass of the black hole. Therefore, we can write

$$\frac{m_{BH}}{M_{Sun}} = \frac{A_{AU}^3}{P_{years}^2} = \frac{1000^3}{15.8^2} = 4.0 \times 10^6$$

$$m_{BH} = 4.0 \times 10^6 \, M_{Sun}$$

The supermassive black hole at the center of the Milky Way has a mass 4 million times that of the Sun. That value is considerably less than the billion-solar-mass black holes in some of the active galactic nuclei discussed in Chapter 19.

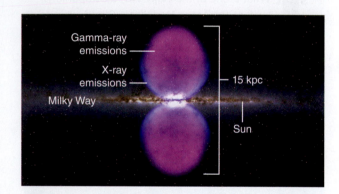

Figure 20.20 The Fermi Gamma-ray Space Telescope observed gamma-ray bubbles (purple) extending 8 kpc above and below the galactic plane. Hints of the edges of the bubbles were first observed in X-rays (blue) in the 1990s. In this artist's conceptual view from outside the galaxy, the gamma-ray jets are in magenta.

direct evidence that matter falling toward the supermassive black hole fuels the energetic activity at the galaxy's center.

The Fermi Gamma-ray Space Telescope has observed gamma-ray-emitting bubbles that extend 8 kpc (25,000 light-years) above and below the galactic disk. The bubbles may have formed after a burst of star formation a few million years ago produced massive star clusters near the center of the galaxy. If some of the gas formed stars and about 2000 M_{Sun} of material fell into the supermassive black hole, then enough energy could have been released to power the bubbles. More recently, faint gamma-ray signals were observed that look like jets coming from the center, within the bubbles (see the artist's depiction in **Figure 20.20**). Using the Hubble Space Telescope, astronomers have estimated that those jets originated from material falling into the supermassive black hole about 6 million to 9 million years ago. Some astronomers predict that gas clouds are heading toward the center and will soon be accreted by the black hole. Currently, the observed activity is not as intense as that seen in active galactic nuclei with central, supermassive black holes. The inner Milky Way is a reminder that it was almost certainly "active" in the past and could become active once again.

CHECK YOUR UNDERSTANDING 20.3

Which property is detectable for both dark matter and the supermassive black hole at the center of the Milky Way? (a) luminosity; (b) temperature; (c) gravity; (d) composition

20.4 The History and Future of the Milky Way

A fundamental goal of stellar astronomy is to understand the life cycle of stars, including how they form from clouds of interstellar gas. In Chapter 15, we described stellar evolution, and tied that story strongly to observations of Earth's galactic neighborhood. Galactic astronomy has a similar basic goal. Astronomers would like to have a complete and well-tested theory of how the Milky Way formed and to be able to make predictions about its future. The distribution of stars of different ages with different amounts of heavy elements is one clue. Additional clues come from studying other galaxies at different distances (and therefore of different ages), their supermassive black holes, and their merger history. In this section, we explore the history and the future of the Milky Way.

The Local Group

Galaxies do not exist in isolation. Most galaxies are parts of gravitationally bound collections of galaxies, the smallest and most common of which are called **galaxy groups**. A galaxy group contains as many as several dozen galaxies, most of them dwarf galaxies. The Milky Way is a member of the **Local Group** (see Chapter 1), first identified by Edwin Hubble in 1936. Hubble labeled 12 galaxies as part of the Local Group, but now astronomers count at least 50. The Local Group (**Figure 20.21**) includes the two giant barred spirals—the Milky Way Galaxy and the Andromeda Galaxy—along with a few ellipticals and irregulars and at least 30 smaller dwarf galaxies in a volume of space about 3 million pc (10 million light-years) in diameter. Almost 98 percent of all the galaxy mass in the Local Group resides in just those two giant galaxies. The third-largest galaxy, Triangulum, is an unbarred spiral with a few percent the mass of the Milky Way or Andromeda. Most, but not all, of the dwarf elliptical and dwarf spheroidal galaxies in the group are satellites of the Milky Way or Andromeda. The Local Group interacts with a few nearby groups, which is discussed further in Chapter 23.

Many of the dwarf galaxies in the Local Group are gravitationally bound to the Milky Way. Observations of the motions and speeds of the dwarf galaxies about the Milky Way may lead to new estimates of the dark matter mass within the Milky Way itself.

Some of the fainter dwarf galaxies were discovered only very recently because of their low luminosity. The dwarf galaxies are the lowest-mass galaxies observed, and they are dominated by an even greater percentage of invisible dark matter than are other known galaxies. They also contain stars very low in elements more massive than helium. Those old ultrafaint dwarf galaxies offer clues to the formation of the Local Group.

The Formation of the Milky Way

We have seen that globular clusters and high-velocity stars must have been among the first stars formed in the Milky Way that still exist. The fact that they are not concentrated in the disk or bulge of the galaxy indicates that they formed from clouds of gas well before those clouds settled into the galaxy's disk. That hypothesis is supported by observations that globular clusters are very old and that the youngest

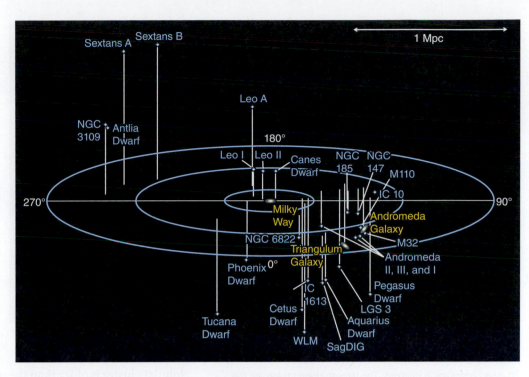

Figure 20.21 This graphical map shows some of the members of the Local Group of galaxies. Most are dwarf galaxies. Spiral galaxies are shown in yellow. The closest galaxies to the Milky Way (such as the Large Magellanic Cloud and Small Magellanic Cloud) are not seen on this scale. (1 Mpc = 1 megaparsec = 1 million parsecs = 3.26 million light-years.)

unanswered questions

Do many more ultrafaint dwarf galaxies exist than have so far been detected? Those types of dwarf galaxies are so faint that they are hard to detect even when close to the Milky Way. As we discuss in Chapter 23, the giant spirals such as the Milky Way were built up by mergers of those small, faint galaxies, so models predict that hundreds or even thousands of them should exist. They could be so dominated by dark matter that they are not at all visible, or too small ever to have formed stars, or they might have merged with the Milky Way or other Local Group members long ago.

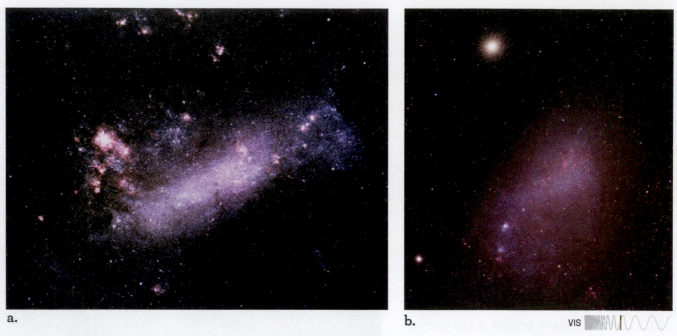

a.   b.   VIS

Figure 20.22 The Milky Way is surrounded by more than 20 dwarf companion galaxies. The largest among them are the **a.** Large Magellanic Cloud (LMC) and **b.** Small Magellanic Cloud (SMC) (see Figure 20.13).

globular cluster is older than the oldest disk stars. The presence of extremely small amounts of massive elements in the atmospheres of halo stars also indicates that at least one generation of stars must have lived and died *before* the formation of the halo stars visible today. That line of reasoning implies that the Milky Way formed from the merger of several smaller clumps of matter, which included both stars and clouds of dust and gas.

Combining that conclusion with the presence of the central, supermassive black hole and the number of nearby dwarf galaxies, astronomers conclude that the Milky Way must have formed when the gas within a huge "clump" of dark matter collapsed into many small protogalaxies. Some of those protogalaxies then merged to form the large, barred spiral galaxies in the Local Group, but some of those smaller protogalaxies are still around today in the form of the small, satellite dwarf galaxies near the Milky Way. The largest among them are the Large Magellanic Cloud and the Small Magellanic Cloud (**Figure 20.22**), which are easily seen in the Southern Hemisphere without the aid of a telescope. The Magellanic Clouds were named for Ferdinand Magellan (1480–1521), who headed a European expedition that ventured far enough into the Southern Hemisphere to see them.

The Future of the Milky Way

Mergers and collisions of Local Group galaxies continue today. Among the closest companions to the Milky Way is the Sagittarius Dwarf Galaxy, which is plowing through the disk of the Milky Way on the other side of the bulge. Astronomers have observed streams of stars, as sketched in **Figure 20.23**, from Sagittarius Dwarf and some of the other dwarf galaxies that are being tidally disrupted by the Milky Way. Those dwarf galaxies will be incorporated into the Milky Way—an indication that the galaxy is still growing. Computer simulations suggest that the spiral-arm structure could be the result of such mergers.

The Andromeda Galaxy appears to violate Hubble's law because its spectrum shows blueshifts, indicating that the galaxy is moving toward, not away from, the Milky Way at 110 km/s (400,000 km/h). Andromeda and the Milky Way are the two

Figure 20.23 This artist's impression shows tidal tails of stars from the Sagittarius Dwarf elliptical galaxy (reddish-orange). Those stars have been stripped from the dwarf galaxy by the much more massive Milky Way, and in billions of years the two galaxies will eventually merge.

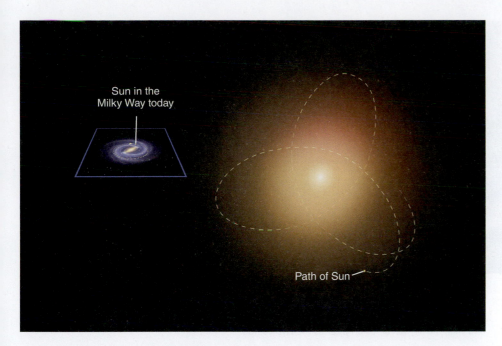

Figure 20.24 A computer simulation of the new orbit of the Sun within the Milky Way–Andromeda merger remnant 10 billion years from now.

largest galaxies in the Local Group, about 770 kpc (2.5 million light-years) apart. (If each galaxy were the size of a quarter, they would be about an arm's length apart.) They are so massive that their mutual gravitational attraction is strong and dominates over cosmological expansion, so they are moving toward each other and we see blueshifts.

Astronomers used the Hubble Space Telescope to measure the motions of Andromeda and determine whether a collision will be head-on, partial, or a total miss. They concluded that the two galaxies will "collide" head-on in about 4 billion years. Most of a galaxy is space between the stars, however, so actual collisions between the stars themselves are unlikely: although diffuse gas in the interstellar medium will collide, sweeping up and heating the gas, most of the material of one galaxy will pass through the other. The two galaxies will take another 2 billion years to merge completely and form one giant elliptical galaxy. The third spiral in the Local Group, the Triangulum Galaxy, may merge first with the Milky Way, or with Andromeda, or be ejected from the Local Group.

Note the timing here. This first close encounter with Andromeda in 4 billion years will occur *before* the Sun runs out of hydrogen in its core, although the Sun will probably have increased its luminosity enough by then that Earth's habitability will have been affected. In 6 billion years, the Sun and its planets could end up near the center of the merged galaxy, or more probably they will have a new location farther from the center with a different orbit and in a different stellar neighborhood (**Figure 20.24**). If by chance the encounter leads to a star passing close to the Sun, the orbits of the Solar System planets could be disrupted, causing them to be at different distances from the Sun. **Figure 20.25** shows how that merged galaxy might look in Earth's sky—if anyone is still around to see it!

CHECK YOUR UNDERSTANDING 20.4

Why is Andromeda now moving toward us?

what if . . .

What if instead of being born in the Local Group, our Milky Way Galaxy was formed in a cluster of galaxies having a thousand giant galaxies? How might our past and future be different?

unanswered questions

Will the merger of the Milky Way and Andromeda form a quasar at the center of the new elliptical galaxy? If the dust and gas at the center of the galaxy did not block the view from Earth, such a quasar could be brighter than the full Moon. The supermassive black hole at the center of Andromeda is thought to be 25–50 times larger than the one in the Milky Way, but both black holes combined still would be on the lower end of the masses of the black holes in the active galactic nuclei discussed in Chapter 19.

Figure 20.25 This computer simulation depicts the view in the sky on Earth when the Andromeda and Milky Way galaxies collide. The spiral Andromeda appears larger in the sky as it gets closer, and then the sky becomes brighter as the two collide.

Origins: The Galactic Habitable Zone

In Chapter 19, we discussed the concept of galactic habitable zones in general; in this "Origins," we focus more specifically on ideas about the habitable zone of the Milky Way. Stars situated too far from the galactic center may have protoplanetary disks with insufficient quantities of heavy elements—such as oxygen, silicon, iron, and nickel (to make up rocky planets like Earth), or carbon, nitrogen, and oxygen (to make up the molecules of life). Stars too close to the galactic center may have planets too strongly affected by its high-energy radiation environment (X-rays and gamma rays from the supermassive black hole) and by supernova explosions and gamma-ray bursts. The bulge has a higher density of stars, creating a strong radiation field, and the halo and thick disk have stars with smaller amounts of heavy elements, so perhaps only stars in the thin disk of the galaxy are candidates for residence in a galactic habitable zone.

Astronomers also must compare stellar lifetimes against the 4 billion years after the formation of Earth that life took to evolve into land animals—so only stars with masses low enough that they will live at least 4 billion years on the main sequence are considered potential hosts of more complex life. In that simple model based on the evolution times for life on Earth, the galactic habitable zone could be a doughnut-shaped region around the galactic center. In one version of the model, that zone is estimated to contain stars born 4 billion to 8 billion years ago located 7–9 kpc from the galactic center, with the Sun exactly in the middle of the doughnut. The doughnut would grow larger as heavier-element formation spread outward from the galactic center.

Some researchers have conducted additional studies and proposed more complex models. For example, one group searched for the molecule formaldehyde (H_2CO)—a key "prebiotic" molecule—by observing molecular clouds in the outer parts of the Milky Way, 12–23.5 kpc from the galactic center. Formaldehyde was detected in two-thirds of the group's sample of 69 molecular clouds, suggesting that at least one important prebiotic molecule is available far from the galactic center.

Another computer model took into account details of the evolution of individual stars within the Milky Way, including birth rates, locations, distribution within the galaxy, abundances of heavy elements, stellar masses, main-sequence lifetimes, and the likelihood that stars became or will become supernovae. The model also assumed that the development of complex life would take 4 billion years. One result of that model is that stars in the inner part of the Milky Way are more likely to be affected by supernova explosions, but those stars are even more likely to have the heavier elements for the formation of planets. In that model, then, the inner part of the galaxy, about 2.5–4 kpc from the center, in and near the middle of the thin disk, is the most likely place for habitable planets. Here, the Sun is *not* in the middle of the most probable zone for habitable planets. As more exoplanets are discovered, astronomers will have a better idea of their distribution throughout the Milky Way.

Mergers with other galaxies could cause stars to migrate into or out of the galactic habitable zone. The uncertainties increase as the assumptions in those models move from the astronomical to the biological. For example, maybe life evolved faster on other planets, so stars of higher mass and shorter lifetimes should be included. Or maybe life evolved slower elsewhere, in which case the older stars would be the best candidates. Scientists do not know whether intense radiation from a supernova or a gamma-ray burst would permanently sterilize a planet or only affect evolution for a while. For example, if Earth's ozone layer were temporarily destroyed, life on land might die out, but life in the oceans would continue. Habitability in the Milky Way Galaxy is complicated, with many unanswered questions.

SUMMARY

Astronomers compare images and spectra of other galaxies with observations of the Milky Way. From that, they determine that the Milky Way is a barred spiral of type SBbc. The Milky Way formed from a collection of smaller protogalaxies that collapsed out of a halo of dark matter. The idea of a galactic habitable zone is that certain parts of the Milky Way may be more suitable for the existence of habitable planets. That zone would have enough heavy elements for the formation of rocky planets and organic molecules, but not so much radiation that it would damage any life.

(1) **Diagram the size and spiral structure of the Milky Way, and explain how astronomers know the shape of the Milky Way.** The Milky Way is a spiral galaxy with several spiral arms, a central bulge, and globular clusters spread through a large spherical halo. The Sun is located about 8300 pc (27,000 light-years) from the Milky Way's center, and the Milky Way's disk is 30,000 pc (100,000 light-years) across. The distances to globular clusters can be found from the luminosity of variable stars within them. Because those globular clusters are symmetrically distributed around the center of the Milky Way, the center of the distribution is located at the center of the galaxy. Radio observations of neutral hydrogen gas and observations of star-forming nebulae provide evidence for the Milky Way's spiral structure.

(2) **List the clues of galaxy formation that can be found from the components of the Milky Way.** The chemical composition of the Milky Way has evolved as material has cycled between stars and the interstellar medium. A generation of stars must have existed before the formation of the oldest halo and globular-cluster stars we see today. The Milky Way has a disk consisting of two parts, the thick disk of old stars and the thin disk of young stars, implying that gas and dust from merging galaxies settled onto the disk while the stars passed through. The galactic halo consists of an inner halo and outer halo of stars and globular clusters, as well as a large, hot gas halo. The abundance of heavy elements in the Milky Way has increased as each generation of stars has produced more of those elements during the final phases of the stars' lives.

(3) **Explain the evidence for the dark matter halo and for the supermassive black hole at the center of the Milky Way.** The Doppler velocities of radio spectral lines show that the rotation curve of the Milky Way is flat, like those of other galaxies. But the inferred mass cannot be accounted for by the mass that is observed directly. That finding indicates that the Milky Way's mass is mostly in the form of dark matter. Evidence for the black hole at the center of the galaxy includes rapid orbital velocities of nearby stars and symmetric X-ray and gamma-ray outflows of material.

④ **Describe the Local Group of galaxies and how it offers clues about the evolution of the Milky Way.** The Milky Way is part of the Local Group of galaxies, which consists of two large, barred spirals and several dozen smaller galaxies. Collisions and mergers between those galaxies probably happened in the past, and a merger with the Andromeda Galaxy may be part of the Milky Way's future. The dwarf satellites and other neighbors in the Local Group are evidence that the Milky Way is growing through accretion.

QUESTIONS AND PROBLEMS

TEST YOUR UNDERSTANDING

1. The size of the Milky Way is determined from studying _____ stars in globular clusters.
   a. Cepheid variable
   b. blue supergiant
   c. RR Lyrae
   d. Sun-like

2. Detailed observations of the structure of the Milky Way are difficult because
   a. the Solar System is embedded in the dust and gas of the disk.
   b. the Milky Way is mostly dark matter.
   c. too many stars are in the way.
   d. the galaxy is rotating too fast (about 200 km/s).

3. In general, as time passes, the Milky Way
   a. has the same chemical composition.
   b. has more abundant hydrogen.
   c. has more abundant heavy elements.
   d. has less abundant heavy elements.

4. The magnetic field of the Milky Way has been detected by
   a. synchrotron radiation from cosmic rays.
   b. direct observation of the field.
   c. the field's interaction with Earth's magnetic field.
   d. studying molecular clouds.

5. Evidence of a supermassive black hole at the center of the Milky Way comes from
   a. direct observations of stars that orbit it.
   b. visible light from material that is falling in.
   c. strong radio emission from the black hole itself.
   d. streams of cosmic rays from the center of the galaxy.

6. The Large Magellanic Cloud and Small Magellanic Cloud will probably
   a. become part of the Milky Way.
   b. remain orbiting forever.
   c. become attached to another passing galaxy.
   d. escape from the gravity of the Milky Way.

7. Globular clusters are important to understanding the Milky Way because
   a. they are so young that they provide information about current star formation.
   b. they provide information about dwarf ellipticals from which the Milky Way formed.
   c. they reveal the size of the Milky Way and Earth's location in it.
   d. the stars in them are highly enhanced in metals.

8. A globular cluster with no variable stars would have been left out of Shapley's study because
   a. the cluster would have been too far away.
   b. the distance to the cluster could not have been determined.
   c. the cluster would have been too faint to see.
   d. the cluster would have been too young to determine its evolutionary state.

9. The best evidence for the presence of dark matter in the Milky Way comes from the observation that the rotation curve
   a. is flat at great distances from the center.
   b. rises swiftly in the interior.
   c. falls off and then rises again.
   d. has a peak at about 2000 light-years from the center.

10. Cosmic rays are
    a. a form of electromagnetic radiation.
    b. high-energy particles.
    c. high-energy dark matter.
    d. high-energy photons.

11. What kind of galaxy is the Milky Way?
    a. elliptical
    b. spiral
    c. barred spiral
    d. irregular

12. Where are the youngest stars in the Milky Way Galaxy?
    a. in the core
    b. in the bulge
    c. in the disk
    d. in the halo

13. Halo stars are found near the Sun. What observational evidence distinguishes them from disk stars? (Choose all that apply.)
    a. their direction of motion
    b. their speed
    c. their composition
    d. their temperature

14. Why are most of the Milky Way's satellite galaxies so difficult to detect?
    a. They are very small.
    b. They are very far away.
    c. The halo of the Milky Way blocks the view.
    d. They are very faint.

15. Which of the following are considered in the concept of a galactic habitable zone? (Choose all that apply.)
    a. the radiation field.
    b. the ages of stars.
    c. the amount of heavy elements.
    d. the distance of a planet from its central star.

THINKING ABOUT THE CONCEPTS

16. ★ **WHAT AN ASTRONOMER SEES**   On the basis of images like Figure 20.1a, explain the logic that leads us to determine that the Milky Way is a spiral galaxy. How would these images be different if we lived in an elliptical galaxy? ⊙

17. Describe the distribution of globular clusters within the Milky Way, and explain what that distribution implies about the size of the galaxy and our distance from its center.

18. In which parts of the Milky Way do astronomers find open clusters? In which parts do they find globular clusters?

19. Old stars in the inner disk of the Milky Way have higher abundances of massive elements than those of young stars in the outer disk. Explain how that difference might have developed.

20. How do 21-cm radio observations reveal the rotation of the Milky Way Galaxy?

21. Halo stars are found near the Sun. What observational evidence distinguishes them from disk stars?

22. What is one source of synchrotron radiation in the Milky Way, and where is it found?

23. Why must astronomers use X-ray, infrared, and 21-cm radio observations to probe the center of the galaxy?

24. What is Sgr A*, and how was it detected?

25. Explain the evidence for a supermassive black hole at the center of the Milky Way. How does the mass of the supermassive black hole at the center of our galaxy compare with that found in most other spiral galaxies?

26. To observers in Earth's Southern Hemisphere, the Large Magellanic Cloud and Small Magellanic Cloud look like detached pieces of the Milky Way. What are those "clouds," and why is it not surprising that they look so much like pieces of the Milky Way?

27. What is the origin of the Milky Way's satellite galaxies? What has been the fate of most of the Milky Way's satellite galaxies? Why are most of the Milky Way's satellite galaxies so difficult to detect?

28. Use your imagination to describe how Earth's skies might appear if the Sun and Solar System were located (a) near the center of the galaxy; (b) near the center of a large globular cluster; (c) near the center of a large, dense molecular cloud.

29. What factors do astronomers consider when thinking about a galactic habitable zone in the Milky Way?

30. Scientists can never know everything, especially at the beginning of a set of research programs. For example, Shapley did not know about dust in the Milky Way. Yet, scientists must often set aside that problem of the unknowns they don't know and do the best they can with the knowledge they have. Explain why that is a necessary step toward scientific understanding.

APPLYING THE CONCEPTS

31. Use Figure 20.17 to find the rotation velocity of a star 12 kpc from the center of the Milky Way. How much mass must be interior to that radius? ⊙ ●–●–●
   a. Make a prediction: Working It Out 20.1 demonstrates this calculation for the Sun. Do you expect there to be more or less mass interior to 12 kpc than is found in Working It Out 20.1? By a lot or by a little?
   b. Calculate: Follow Working It Out 20.1 to calculate the mass in the Milky Way interior to a radius of 12 kpc.
   c. Check your work: Compare your result to the answer in Working It Out 20.1.

32. A star is observed in a circular orbit about a black hole with an orbital radius of 1.5×10^{11} km and an average speed of 2000 km/s. What is the mass of that black hole in solar masses? ●–●–●
   a. Make a prediction: About how many solar masses is the black hole at the center of the Milky Way? Do you expect that this black hole will be something like that, much smaller, or much larger?
   b. Calculate: Follow Working It Out 20.1 to calculate the mass of the black hole.
   c. Check your work: Verify that your answer has units of solar masses, and compare the mass to your prediction.

33. A star in a circular orbit about the black hole at the center of the Milky Way (whose mass $M_{BH} = 8 \times 10^{36}$ kg) has an orbital radius of 0.0131 light-year (1.24×10^{14} meters). What is the average speed of that star in its orbit?
   a. Make a prediction: Study Working It Out 20.1. Compare the masses and the distances of the orbit of this star and the Sun. Do you expect this star to be moving faster or slower than the Sun does in its orbit? By a lot or by a little?
   b. Calculate: Rearrange the equation in Working It Out 20.1 to solve for v_{circ}. Then substitute the values for r and M in this problem to solve for the speed of the star.
   c. Check your work: Verify that your answer has units of km/s, and compare your answer to your prediction.

34. Star S0-19 in Figure 20.19 has a semimajor axis that is approximately three times as large as the orbit of S0-2. What is the period of this orbit? ⊙
   a. Make a prediction: Working It Out 20.2 gives an algebraic relationship among the semimajor axis, the period, and the mass of the black hole. Do you expect the period of S0-19 to be larger or smaller than the period of S0-2? What will the units of the period be when you have solved for it using the relationship in Working It Out 20.2?
   b. Calculate: Rearrange the equation in Working It Out 20.2 to solve for the period. Plug in the numbers from this problem to find the period of S0-19.
   c. Check your work: Verify that your answer has the correct units, and compare your answer to your prediction.

35. Suppose that a star near the central black hole of the Milky Way has a period 3 times as long as the period of S0-2. What is the semimajor axis of its orbit?
   a. Make a prediction: Do you expect the semimajor axis to be longer or shorter than the semimajor axis of S0-2?
   b. Calculate: Set up a ratio using the equation in Working It Out 20.2 to solve for the semimajor axis of the orbit of this star.
   c. Check your work: Compare your answer to your prediction.

36. From Figure 20.8, estimate the ratio between the radius of the Milky Way's outer halo and the radius of the disk. ⊛

37. According to Figure 20.17, what is the rotation velocity of a disk star located 6000 pc from the center of the Milky Way? If you assume a circular orbit, how long does that star take to orbit once? ⊛

38. From the data in Figure 20.17, estimate the time the Large Magellanic Cloud would take to orbit the Milky Way if the Large Magellanic Cloud were on a circular orbit. ⊛

39. The Sun completes one trip around the center of the galaxy in approximately 230 million years. How many times has the Solar System made the circuit since its formation 4.6 billion years ago?

40. The Sun is located about 8300 pc from the center of the galaxy, and the galaxy's disk probably extends another 9000 pc farther out from the center. Assume that the Sun's orbit takes 230 million years to complete.
   a. With a truly flat rotation curve, how long would a globular cluster located near the edge of the disk take to complete one trip around the center of the galaxy?
   b. How many times has that globular cluster made the circuit since its formation about 13 billion years ago?

41. Parallax measurements of the variable star RR Lyrae indicate that it is located 230 pc from the Sun. A similar star observed in a globular cluster located far above the galactic disk appears 160,000 times fainter than RR Lyrae.
   a. How far from the Sun is that globular cluster?
   b. What does your answer to part (a) tell you about the size of the galaxy's halo in comparison with the size of its disk?

42. One of the fastest cosmic rays ever observed had a speed of $(1.0 - [1.0 \times 10^{-24}]) \times c$ (that is, very, very close to c). Assume that the cosmic ray and a photon left a source at the same instant. To a stationary observer, how far behind the photon would the cosmic ray be after traveling for 100 million years?

43. Imagine that a black hole has a mass of 5 million M_{Sun}. If a star's orbit about the black hole has a semimajor axis of 0.02 light-year (1.9×10^{14} meters), then what is the star's orbital period? (Hint: You may want to refer to Chapter 4.)

44. What is the Schwarzschild radius of the black hole at the center of the Milky Way? What is its density? How does that value compare with the density of a stellar black hole?

45. One model of the galactic habitable zone contains stars in a doughnut-shaped region in the disk between 7 and 9 kpc from the center of the galaxy. If you assume that this doughnut is as thick as the disk itself, what fraction of the disk of the Milky Way lies in that habitable zone?

EXPLORATION　The Center of the Milky Way

Astronomers once thought that the Sun was at the center of the Milky Way. In this Exploration, you will repeat Shapley's globular-cluster experiment that led to a more accurate picture of the size and shape of the Milky Way.

Imagine that the disk of the Milky Way is a flat, round plane, like a pizza. Globular clusters are arranged in a rough sphere around that plane. To map globular clusters on **Figure 20.26**, imagine that a line is drawn straight "down" from a globular cluster to the disk of the Milky Way. The "projected distance" in kiloparsecs is the distance from the Sun to the place where the line hits the plane. The galactic longitude indicates the direction toward that point; it is marked around the outside of the graph, along with the several constellations.

Adapted from *Learning Astronomy by Doing Astronomy*, by Ana Larson

Make a dot at the location of each globular cluster by finding the galactic longitude indicated outside the circle and then coming in toward the center to the projected distance. The two globular clusters in boldface in **Table 20.1** have been plotted for you as examples. After plotting all the globular clusters, estimate the center of their distribution and mark it with an X. That is the center of the Milky Way.

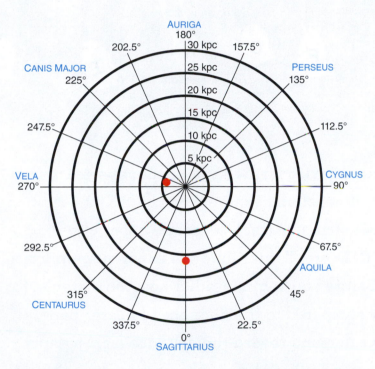

Figure 20.26 This polar graph can be used to plot distance and direction.

| Table 20.1 | Globular Cluster Data |

Cluster	Galactic Longitude	Projected Distance (kpc)	Cluster	Galactic Longitude	Projected Distance (kpc)
104	306	3.5	6273	357	7
362	302	6.6	**6287**	**0**	**16.6**
2808	283	8.9	6333	5	12.6
4147	**251**	**4.2**	6356	7	18.8
5024	333	3.4	6397	339	2.8
5139	309	5	6535	27	15.3
5634	342	17.6	6712	27	5.7
Pal 5	1	24.8	6723	0	7
5904	4	5.5	6760	36	8.4
6121	351	4.1	Pal 10	53	8.3
O 1276	22	25	Pal 11	32	27.2
6638	8	15.1	6864	20	31.5
6171	3	15.7	6981	35	17.7
6218	15	6.7	7089	54	9.9
6235	359	18.9	Pal 12	31	25.4
6266	353	11.6	288	147	0.3
6284	358	16.1	1904	228	14.4
6293	357	9.7	Pal 4	202	30.9
6341	68	6.5	4590	299	11.2
6366	18	16.7	5053	335	3.1
6402	21	14.1	5272	42	2.2
6656	9	3	5694	331	27.4
6717	13	14.4	5897	343	12.6
6752	337	4.8	6093	353	11.9
6779	62	10.4	6541	349	3.9
6809	9	5.5	6626	7	4.8
6838	56	2.6	6144	352	16.3
6934	52	17.3	6205	59	4.8
7078	65	9.4	6229	73	18.9
7099	27	9.1	6254	15	5.7

1 What is the approximate distance from the Sun to the center of the Milky Way?

_____

2 What is the galactic longitude of the center of the Milky Way?

_____

3 How do astronomers know that the Sun is not at the center of the Milky Way?

_____

The Expanding Universe

The universe expands, carrying galaxies farther and farther apart. This can be hard to visualize, but a one-dimensional example can help. Cut a rubber band in one place, making a "rubber string." Attach at least 6 paper clips to the rubber band, and label them A, B, C, D, etc. To simulate the expanding universe, you will stretch the rubber band. Before you begin, predict how the distance to more distant paper clips will grow compared to the distance to nearer paper clips. Lay the rubber band along a ruler, and record the distance from a paper clip near the center (the "home" paper clip) to all the other paper clips. Stretch the rubber band, keeping the home paper clip at the same location on the ruler. Measure and record the distance to all the other paper clips along the stretched rubber band. (You may need the help of a friend.) Select a new "home" paper clip near the center, and repeat the experiment, first measuring distances along the unstretched rubber band, then measuring distances along the stretched rubber band. Make sure to stretch the rubber band the same amount in both experiments. Make a graph of "stretched distance" versus "original distance" for each experiment, and compare these graphs. Is the shape of the graph the same for each experiment?

EXPERIMENT SETUP

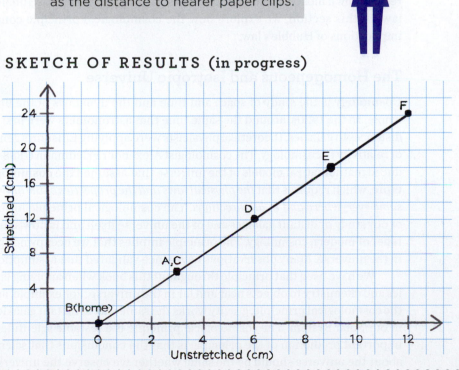

A **B** C D E F

home = B

Lay the rubber band along a ruler and measure the distance from "home" to the other paper clips.

A **B** C D E F

Stretch the rubber band and measure again.

A B C **D** E F

home = D

Repeat the experiment with a new "home" paper clip, stretching the rubber band the same amount.

B C **D** E F

PREDICTION

I predict that the distance to more distant paper clips will grow
☐ **more** ☐ **less** ☐ **the same**
as the distance to nearer paper clips.

SKETCH OF RESULTS (in progress)

Stretched (cm) vs Unstretched (cm)

F at (12, 24)
E at (9, 18)
D at (6, 12)
A,C at (3, 6)
B(home) at (0, 0)

The cosmological principle lies at the center of astronomers' conceptual understanding of the universe. The cosmological principle states that the conclusions we reach about our universe should be more or less the same, regardless of whether we live in the Milky Way or in a galaxy billions of light-years away. The cosmological principle, combined with observations of the universe that only became possible in the 20th century, leads logically to the concept that the universe had a hot, dense beginning known as the Big Bang. Here in Chapter 21, we explain more about the cosmological principle and explore the observational evidence for the Big Bang.

LEARNING GOALS

By the end of this chapter, you should be able to:

① Explain in detail the cosmological principle.

② Demonstrate how the Hubble constant can be used to estimate the age of the universe.

③ Describe the observational evidence for the Big Bang.

④ List the chemical elements that were created in the early hot universe, and explain why no other elements were created at that time.

21.1 The Cosmological Principle

In Chapter 19, you learned about observations of galaxy motions and distances, made in the early 20th century. From those observations, Edwin Hubble created a graph that showed the galaxies moving apart, with the more distant ones moving faster—a result known most commonly as Hubble's law, but also called the Hubble-Lemaître law. In this section, we explore how the cosmological principle constrains the implications of Hubble's law.

The Homogeneous and Isotropic Universe

Cosmology is the study of space and time and the dynamics of the universe as a whole. In the 1920s, around the same time that astronomers were first measuring distances to galaxies, theoretical physicists were applying Einstein's general theory of relativity to cosmology. The cosmologist Alexander Friedmann (1888–1925) produced mathematical models of the universe that assumed the **cosmological principle**, which requires that *the physical laws and the properties of the universe are the same everywhere and in all directions*. The cosmological principle forms the basis of our study of distant galaxies and of the universe itself and is now a fundamental tenet of modern cosmology. The principle implies that we occupy no special place in the universe. Many observations have since validated that assumption.

We have used that principle throughout this book when applying laws of physics and chemistry to objects near and far in the universe. For example, according to the cosmological principle, gravity works the same way in distant galaxies as it does in our Solar System. The cosmological principle is a testable scientific theory. An important prediction of the principle is that the conclusions that scientists reach about the universe should be the same whether we observe the universe from the

Milky Way or from a galaxy billions of parsecs away. In other words, if the cosmological principle is correct, the universe is **homogeneous**, having the same composition and properties at all places.

In an absolute sense, the universe is *not* homogeneous because the conditions on Earth are very different from those in the center of the Sun or in deep space. In cosmology, homogeneity of the universe means that stars and galaxies in Earth's part of the universe are much the same, and behave in the same manner, as stars and galaxies in remote corners of the universe. Homogeneity also means that stars and galaxies everywhere are distributed in space in much the same way that they are distributed in Earth's cosmic neighborhood and that observers in those galaxies would see the same properties for the universe that astronomers see from here. However, homogeneity does not mean that the universe is unchanging over time; it was different in the past and will be different in the future.

Directly verifying the prediction of homogeneity is not easy. Scientists cannot travel from the Milky Way to a galaxy in the remote universe to see whether conditions are the same there. They can, however, compare light arriving from closer and farther locations in the distant universe and see the ways in which features look the same or different. For example, astronomers can test homogeneity by looking at how galaxies are distributed in distant space and seeing whether that distribution is similar to the distribution nearby.

In addition to predicting that the universe is homogeneous, the cosmological principle requires that all observers measure the same properties of the universe, regardless of the direction in which they are looking. If something is the same in all directions, it is **isotropic**. The prediction that the universe is isotropic is much easier to test directly than is homogeneity. If galaxies were lined up in rows, like the shelves of a supermarket, astronomers would measure very different properties of the universe if they looked in a direction along the rows than in a direction perpendicular to them. The universe would still be homogeneous, but not isotropic—so it would not satisfy the cosmological principle.

Usually, isotropy goes together with homogeneity, and the cosmological principle requires both. **Figure 21.1** shows examples of how the universe could have violated the cosmological principle by not being homogeneous or isotropic, as well as examples of how the universe might satisfy the cosmological principle. All observations of the actual universe show that the properties of the universe are basically the same, regardless of the direction in which observers are looking. When averaged over very large scales (thousands of millions of parsecs), the universe appears homogeneous as well.

The Hubble Expansion

According to Hubble's law (Chapter 19), the distance of a galaxy (d_G) is proportional to its recession velocity (v_r): $d_G = v_r/H_0$. Hubble's law helps astronomers investigate whether the universe is homogeneous and isotropic. They can confirm its isotropy by observing that galaxies in one direction in the sky obey the same Hubble law as galaxies in other directions in the sky. Hubble's law says that Earth is located in an expanding universe and that the expansion looks the same regardless of the location of the observer. To help you visualize that uniformity, the chapter-opening experiment gives a useful model that you can build for yourself with materials you probably already own.

In that experiment, the paper clips located along the rubber band obey a Hubble-like law, with more widely separated paper clips moving apart "more quickly" than

what if . . .

What if the universe were homogeneous but not isotropic? Would that mean that we live in a special place in the universe?

unanswered questions

Does the cosmological principle hold everywhere and at all times? Occasionally, observations have been made that call the cosmological principle into question. So far, there has turned out to be another explanation for the apparent discrepancy, and the overwhelming preponderance of evidence upholds the cosmological principle. Still, the process of science is to always question the assumptions; this question will likely never be fully answered.

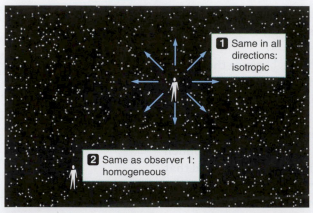

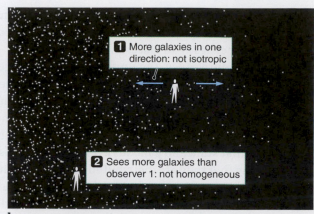

a. Homogeneous and isotropic

b. Neither homogeneous nor isotropic

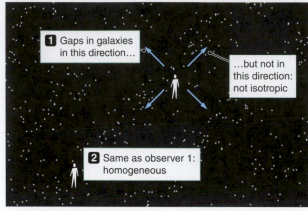

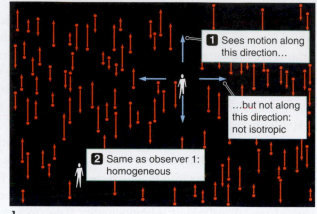

c. Homogeneous, not isotropic

d. Homogeneous, not isotropic

Figure 21.1 Homogeneity and isotropy in four theoretical models of a universe. Blue arrows indicate the direction of view. **a.** The distribution of galaxies is uniform, so this universe is both homogeneous and isotropic. **b.** The density of galaxies is decreasing in one direction, so this universe is neither homogeneous nor isotropic. **c.** The bands of galaxies lie along a unique axis, making this universe not isotropic. **d.** The distribution of galaxies is uniform, but galaxies move along only one direction, so this universe also is not isotropic.

🎥 **Astronomy in Action:** Infinity and the Number Line

paper clips near one another. The key insight to this metaphor for the universe comes from realizing that nothing is special about which paper clip you choose to measure from. If you repeat the experiment for any paper clip along the rubber band, you will arrive at the same result. The stretching rubber band, like the universe, is "homogeneous." The same Hubble-like law applies, regardless of where the observer is located.

This insight helps explain why an isotropic expansion does *not* imply that Earth is at the center of the expanding universe. In the rubber-band experiment, the observation that nearby paper clips move away slowly and distant paper clips move away more rapidly does not mean that the paper clip selected as a vantage point is at the center of anything. Instead, it means that the rubber band is being stretched uniformly along its length. Similarly, Hubble's law shows that nearby galaxies are carried away slowly by expanding space, and distant galaxies are carried away more rapidly. Any observer in any galaxy sees nearby galaxies moving away slowly and more distant galaxies moving away more rapidly. The same Hubble's law applies from their vantage point as applies from our vantage point on Earth in our galaxy; the expansion of the universe is homogeneous and isotropic. The motion of galaxies as a result of the expansion of the universe is called the **Hubble flow**.

For galaxies that are close together, gravitational attraction dominates over the expansion of space. As discussed in Chapter 20, the Andromeda Galaxy and the Milky Way are being pulled together by their gravity. The Andromeda Galaxy is approaching the Milky Way at about 110 kilometers per second (km/s), so the light from the Andromeda Galaxy is blueshifted, not redshifted. Gravity pulls the two galaxies together more quickly than the Hubble flow would carry them apart. The fact that gravitational or electromagnetic forces can overwhelm the expansion of space also explains why the Solar System is not expanding, and neither are you.

The Universe in Space and Time

Hubble's law gives astronomers a practical tool for measuring distances to remote objects. Once astronomers know the value of the Hubble constant, H_0, they can use a straightforward measurement of the redshift of a galaxy to find its distance. In other words, once H_0 is known for relatively nearby galaxies, Hubble's law makes the once-difficult task of measuring large distances in the universe relatively straightforward. Astronomers need only measure the redshift of very distant galaxies, and then use Hubble's law to determine the distance. This works well for redshifts less than one. There are complications, such as interactions among galaxies that are close to one another, or for redshifts larger than one (which we discuss in Chapter 22), but otherwise Hubble's law provides remarkably accurate measurements of distance.

Hubble's law does more than place galaxies in space; it also places galaxies in time. Light travels at a huge but finite speed. Because it takes time for light to travel from one place to another, you observe the Sun as it existed 8 minutes ago (see Chapter 1). When you look at Alpha Centauri, the nearest stellar system beyond the Sun, you see it as it existed 4.3 years ago. If you look at the center of the Milky Way, the light you see is 27,000 years old. The **look-back time** to a distant object is the time the light from that object has taken to reach a telescope on Earth. As astronomers look into the distant universe, look-back times become very large. The distance to a galaxy with a redshift of $z = 0.1$ is 1.4 billion light-years (assuming $H_0 = 70$ km/s/Mpc), so the look-back time to that galaxy is 1.4 billion years. The look-back time to a galaxy in which $z = 0.2$ is 2.7 billion years. The look-back time of the most distant galaxies observed, at about $z = 10$, is 13.2 billion years. As astronomers observe objects with very large redshifts, they are observing the universe when it was very young.

CHECK YOUR UNDERSTANDING 21.1

In astronomy, *isotropic* means that the universe is the same _____, and *homogeneous* means that the universe is the same _____. (Choose all that apply for each blank.) (a) in all locations; (b) in all directions; (c) at all times; (d) at all size scales.

Answers to Check Your Understanding questions are in the back of the book.

21.2 The Universe Began in the Big Bang

Imagine watching a video of the universe, with the galaxies moving apart. Now reverse the video and run it backward in time. The galaxies become closer and closer together as the universe becomes younger and younger. In that way, the observation of the expansion of the universe leads to the idea of a beginning to the universe at a time that can be estimated from Hubble's law. In this section, we explore that implication of Hubble's law.

Expansion and the Age of the Universe

Hubble's law gives an estimate of the age of the universe. For the simplest estimate, we assume that the expansion speed has always been constant, and the age of the

▶❙❙ **AstroTour:** Hubble's Law

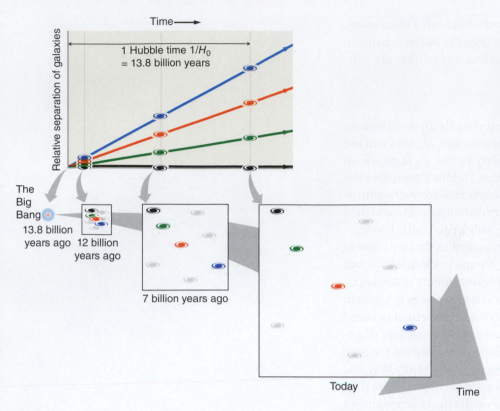

Figure 21.2 Looking backward in time, the distance between any two galaxies is smaller and smaller, until all matter in the universe is concentrated at the same point: the Big Bang.

Astronomy in Action: Expanding Balloon Universe

universe can be estimated from the slope of the line in a graph of the velocity of galaxies plotted against their distance. The slope has units that reduce to 1/time, so its inverse has units of time. That slope is the Hubble constant, H_0, and its inverse (1 divided by H_0) is the **Hubble time**. The Hubble time is an estimate of the universe's age: 13.8 billion years (**Working It Out 21.1**). If the expansion were faster, the Hubble constant would be larger, and the universe would be younger, because the galaxies would have moved farther apart in less time; similarly, a slower expansion would yield a smaller Hubble constant and an older universe.

If the universe expanded uniformly, then when it was about half its current age, about 6.9 billion years ago, all the galaxies in the universe were half as far apart as they are now, and 12.4 billion years ago, all the galaxies were about a tenth as far apart. If we assume that galaxies have been moving apart all that time at the same speed as they do today, then a little less than one Hubble time ago—13.8 billion years ago—almost no space existed between the particles that constitute today's universe. All such matter, as well as energy in the universe, must then have been unimaginably dense. Because expanding gases cool down, the universe then must have been much hotter than it is today in its expanded state. That hot, dense beginning, 13.8 billion years ago, is called the **Big Bang** (**Figure 21.2**).

Georges Lemaître (1894–1966) was the first to propose the theory of the Big Bang. The idea greatly troubled many astronomers in the early and middle years of the 20th century. Several suggestions were put forward to explain the observed fact of Hubble expansion without resorting to the idea that the universe came into existence in an extraordinarily dense "fireball" billions of years ago. As more and more distant galaxies have been observed, however, and as more discoveries about the structure of the universe have been made, the Big Bang theory has grown stronger. The major predictions of the Big Bang theory have proven to be correct.

The implications of Hubble's law forever changed the scientific concepts of the origin, history, and possible future of the universe. At the same time, Hubble's law has pointed to many new questions about the universe. To address them, we next need to consider more precisely what the term *expanding universe* means.

In this context, it's important to be very clear about the distinction between the universe and the observable universe. The **observable universe** is the part of the universe that we can see. The observable universe extends 13.8 billion light-years in every direction. That limit exists because it is the length of time the universe has been around. The light from more distant regions has not yet had time to travel to us, and so we cannot see it yet. As far as astronomers have been able to determine, the universe is much, much larger than the observable universe—so large that it might as well be infinitely large because any boundary that exists has no effect on the part of the universe we can observe.

Galaxies and the Expansion

At this point in our discussion, you may be picturing the expanding universe as a cloud of debris from an explosion flying outward through surrounding

working it out 21.1

Expansion and the Age of the Universe

We can use Hubble's law to estimate the age of the universe. Consider two galaxies located 30 Mpc ($d_G = 9.3 \times 10^{20}$ km) away from each other (**Figure 21.3**). If those two galaxies are moving apart, then at some time in the past they must have been together in the same place at the same time. According to Hubble's law, and on the assumption that $H_0 = 70$ km/s/Mpc, the distance (d_G) between those two galaxies is increasing at the following rate:

$$v_r = H_0 \times d_G$$
$$v_r = 70 \text{ km/s/Mpc} \times 30 \text{ Mpc}$$
$$v_r = 2100 \text{ km/s}$$

Knowing the velocity (v_r) at which they are traveling, we can calculate the time the two galaxies took to become separated by 30 Mpc:

$$\text{Time} = \frac{\text{Distance}}{\text{Velocity}} = \frac{9.3 \times 10^{20} \text{ km}}{2100 \text{ km/s}} = 4.4 \times 10^{17} \text{ s}$$

Dividing by the number of seconds in a year (about 3.16×10^7 s/yr) gives

$$\text{Time} = 1.4 \times 10^{10} \text{ yr} = 14 \text{ billion yr}$$

In other words, *if* expansion of the universe has been constant, two galaxies that today are 30 Mpc apart started out at the same place about 14 billion years ago.

Now let's do the same calculation with two galaxies 60 Mpc (18.6×10^{20} km) apart. Those two galaxies are twice as far apart, but the distance between them is increasing twice as rapidly:

$$v_r = H_0 \times d_G = 70 \text{ km/s/Mpc} \times 60 \text{ Mpc} = 4200 \text{ km/s}$$

Therefore,

$$\text{Time} = \frac{18.6 \times 10^{20} \text{ km}}{4200 \text{ km/s}} = 4.4 \times 10^{17} \text{ s} = 1.4 \times 10^{10} \text{ yr}$$

Again, we calculate time as distance divided by velocity (twice the distance divided by twice the velocity) to find that those galaxies also took about 14 billion years to reach their current locations. We can do that calculation again and again for any pair of galaxies in the universe today. The farther apart the two galaxies are, the faster they are moving. But all galaxies took the *same* amount of time to get to where they are today. (Small differences in the intermediate steps of those calculations are the result of rounding and are not significant to the argument.) The most precise measurements give a result of 13.8 billion years.

Working out the example with words instead of numbers makes it clear why the answer is always the same. Because the velocity we are calculating comes from Hubble's law, velocity equals the Hubble constant multiplied by distance. Writing that out as an equation, we get

$$\text{Time} = \frac{\text{Distance}}{\text{Velocity}} = \frac{\text{Distance}}{H_0 \times \text{Distance}}$$

Distance divides out to give

$$\text{Time} = \frac{1}{H_0}$$

where $1/H_0$ is the Hubble time. That approach is one way to estimate the age of the universe.

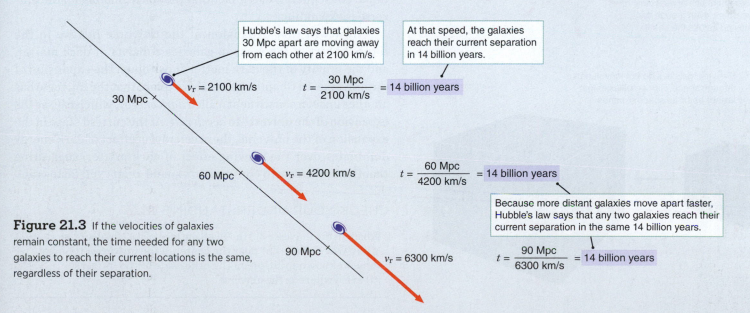

Hubble's law says that galaxies 30 Mpc apart are moving away from each other at 2100 km/s.

At that speed, the galaxies reach their current separation in 14 billion years.

$v_r = 2100$ km/s $t = \dfrac{30 \text{ Mpc}}{2100 \text{ km/s}} = 14$ billion years

30 Mpc

$v_r = 4200$ km/s $t = \dfrac{60 \text{ Mpc}}{4200 \text{ km/s}} = 14$ billion years

60 Mpc

Because more distant galaxies move apart faster, Hubble's law says that any two galaxies reach their current separation in the same 14 billion years.

$v_r = 6300$ km/s $t = \dfrac{90 \text{ Mpc}}{6300 \text{ km/s}} = 14$ billion years

90 Mpc

Figure 21.3 If the velocities of galaxies remain constant, the time needed for any two galaxies to reach their current locations is the same, regardless of their separation.

what if . . .

What if Hubble had discovered that the recession velocity increases as the square of the distance, so that $v_r = H_0 d^2$? What would that imply about Earth's location in the universe?

Astronomy in Action: Observable vs. Actual Universe

Figure 21.4 a. As a rubber sheet is stretched, coins on its surface move farther apart, even though they are not moving with respect to the sheet itself. Any coin on the surface of the sheet observes a Hubble-like law in every direction. **b.** Galaxies in an expanding universe are not flying apart through space. Rather, space itself is stretching.

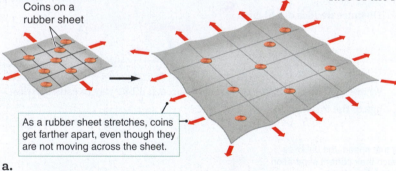

Coins on a rubber sheet

As a rubber sheet stretches, coins get farther apart, even though they are not moving across the sheet.

a.

Similarly, galaxies do not move apart through space. Rather, galaxies get farther apart as space expands.

Galaxies in space

b.

space. That is a common depiction of the Big Bang in movies and television shows, which portray a tiny bright spot that explodes to fill the screen. However, the Big Bang was not an explosion in the usual sense of the word, in which an event occurred at a location in preexisting space, and then its effects spread outward through that space. Instead, the Big Bang took place *everywhere*, because space itself began to exist at the moment of the Big Bang. Wherever anything is in the universe today, it is at the site of the Big Bang. The implication is that, like the paper clips on the rubber band in the chapter-opening experiment, galaxies are not flying through space at all. Instead, space, like the rubber band, is itself expanding, carrying the stars and galaxies that populate the universe along with it.

We have already dealt with the basic ideas that explain the expansion of space. In our discussion of black holes (Chapter 18), you encountered Einstein's general theory of relativity (see the **Process of Science Figure**). General relativity says that space is distorted by the presence of mass and that the consequence of that distortion is gravity. For example, the mass of the Sun, like any other object, distorts the geometry of spacetime around it; so Earth, coasting along in its inertial frame of reference, follows a curved path around the Sun. We illustrated that phenomenon in Figure 18.10 with the analogy of a ball placed on a stretched rubber sheet, showing how the ball distorted the surface of the sheet.

The surface of a rubber sheet can be distorted in other ways as well. Imagine several coins placed on a rubber sheet (**Figure 21.4**). Then imagine grabbing the edges of the sheet and beginning to pull them outward. As the rubber sheet stretches, each coin remains at the same location on the surface of the sheet, but the distances between the coins increase. Two coins sitting close to each other move apart only slowly, whereas coins farther apart move away from each other more rapidly. The distances and relative motions of the coins on the surface of the rubber sheet obey a Hubble-like relationship as the sheet is stretched.

That movement is analogous to what is happening in the universe, with galaxies taking the place of the coins and space itself taking the place of the rubber sheet. For a rubber sheet, the sheet can be stretched only so far before it breaks. With space and the real universe, no such limit exists. The fabric of space can, in principle, go on expanding forever. Hubble's law is the observational consequence of the fact that the space making up the universe is expanding.

How will that expansion of the universe behave in the future? Most of the mass in galaxies consists of dark matter, and the gravity of the dark matter slows down the expansion of the universe. In Chapter 22, we also explain that the universe has another unseen constituent, called *dark energy*, which causes the expansion of the universe to *accelerate*. At the current stage in the expansion of the universe, the accelerating effect of dark energy dominates over the slowing effect of dark matter, suggesting that the universe will continue to expand at an ever-faster rate.

CHECK YOUR UNDERSTANDING 21.2

Where in the universe did the Big Bang take place? (a) near the Milky Way Galaxy; (b) near the center of the universe; (c) near some unknown location on the other side of the universe; (d) everywhere in the universe

Data Are the Ultimate Authority

The genius of Albert Einstein is recognized by just about everyone, but even he had to change his mind in the face of new data.

1920

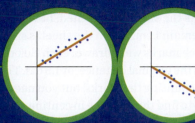

Λ

Einstein developed general relativity throughout the late 1910s. His equations predicted that the universe must be expanding or contracting.

Einstein inserted an extra term, indicated by the symbol Λ, into his equations to make the universe stationary, because he believed it should not expand or contract.

1930

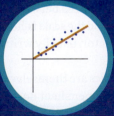

In 1929, Edwin Hubble combined observations of galaxies to show that nearly all galaxies moved away from Earth, with farther galaxies receding more quickly. These observations showed that the universe is expanding!

Einstein changed his mind and described his reluctance to accept the prediction of his original theory as "his biggest blunder."

Even the most brilliant scientists must adjust when the data contradict their conclusions.

21.3 Expansion Is Described with a Scale Factor

As the universe expands, the distance between any two objects increases because of the stretching of space. Astronomers find it useful to discuss that expansion in terms of the *scale factor* of the universe.

Scale Factor

Let's return to the analogy of the rubber sheet. Suppose you place a ruler on the rubber sheet and draw a tick mark every centimeter (**Figure 21.5a**). To measure the distance between two points on the sheet, you can count the marks between the two points and multiply by 1 centimeter (cm) per tick mark.

As the sheet is stretched, however, the distance between the tick marks does not remain 1 cm. When the sheet is stretched to 150 percent of the size it had when the tick marks were drawn, each tick mark is separated from its neighbors by 1 1/2 times the original distance, or 1.5 cm. The distance between two points can still come from counting the marks, but you need to scale up the distance between tick marks by 1.5 to find the distance in centimeters. If the sheet were twice the size it was when the tick marks were drawn (**Figure 21.5b**), each mark would correspond to 2 cm of actual distance. Astronomers use the term **scale factor (R_U)** to indicate the size of the sheet relative to its size when the tick marks were drawn. The scale factor also indicates how much the distance between points on the sheet has changed. In the first example, the scale factor of the sheet is 1.5; in the second, the scale factor is 2.

Suppose astronomers choose today to lay out a "cosmic ruler" on the fabric of space, placing an imaginary tick mark every 10 Mpc. The scale factor of the universe at this time is defined to be 1. In the past, when the universe was smaller, distances between the points in space marked by the cosmic ruler would have been less than 10 Mpc. The scale factor of that younger, smaller universe would have been less than today's scale factor and therefore less than 1. In the future, as the universe continues to expand, the distances between the tick marks on the cosmic ruler will grow to more than 10 Mpc, and the scale factor of the universe will be greater than 1. Astronomers use the scale factor, usually written as R_U, to keep track of the changing scale of the universe.

The laws of physics are themselves unchanged by the expansion of the universe, just as stretching a rubber sheet does not change the properties of the coins on its surface. At scales smaller than the Local Group, the nuclear and electromagnetic forces within and between atoms, as well as the gravitational forces between relatively close objects, dominate over the expansion. As the universe expands, the sizes and other physical properties of atoms, stars, and galaxies also remain unchanged.

As we look back in time, the scale factor of the universe gets smaller and smaller, approaching zero as it comes closer and closer to the Big Bang. The fabric of space that today spans billions of parsecs spanned much smaller distances when the universe was young. When the universe was only a day old, all the space visible today amounted to a region only a few times the size of the Solar System. When the universe was 1/50 of a second old, the vast expanse of space that makes up today's observable universe (and all the matter in it) occupied a volume only the size of today's Earth. As we go backward in time and approach the Big Bang itself, the space that makes up today's observable universe becomes smaller and smaller—the size of a grapefruit, a marble, an atom, a proton. Every point in the fabric of space that makes up today's universe was right there at the beginning, a part of that unimaginably tiny, dense universe that emerged from the Big Bang.

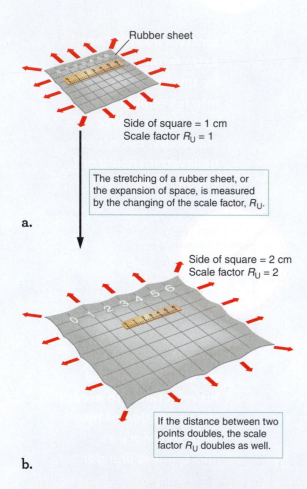

Rubber sheet

Side of square = 1 cm
Scale factor $R_U = 1$

The stretching of a rubber sheet, or the expansion of space, is measured by the changing of the scale factor, R_U.

a.

Side of square = 2 cm
Scale factor $R_U = 2$

If the distance between two points doubles, the scale factor R_U doubles as well.

b.

Figure 21.5 a. On a rubber sheet, tick marks are drawn 1 cm apart. As the sheet is stretched, the tick marks move farther apart. **b.** When the spacing between the tick marks is 2 cm, or twice the original value, the scale factor of the sheet, R_U, is said to have doubled. A similar scale factor, R_U, is used to describe the expansion of the universe.

It's worth repeating that the Big Bang did *not* occur at a specific point in space because space itself came into existence with the Big Bang. Instead, the Big Bang happened everywhere; no particular point in today's universe marked the site of the Big Bang. A Big Bang universe is homogeneous and isotropic, consistent with the cosmological principle.

CHECK YOUR UNDERSTANDING 21.3a

The scale factor keeps track of: (a) the movement of galaxies through space; (b) the current distances between many galaxies; (c) the changing distance between any two galaxies; (d) the location of the center of the universe.

Redshift Is Due to the Changing Scale Factor of the Universe

The ideas of general relativity discussed in Chapter 18 are powerful tools for interpreting Hubble's great discovery of the relationship between the velocity and distance of galaxies and of the expanding universe. Those velocities were determined from the redshift of the galaxies by using the Doppler effect. Although it is true that the distance between galaxies is increasing as a result of the expansion of the universe, and that we can use the equation for Doppler shifts to measure the redshifts of galaxies, those redshifts are not due to Doppler shifts in the same way that we described for a moving star. Light that comes from very distant objects was emitted when the universe was younger and therefore smaller. As that light comes toward Earth from distant galaxies, the scale factor of the space through which the light travels is constantly increasing, and as it does, the distance between adjacent light-wave crests increases as well. The light is "stretched out" as the space it travels through expands.

Figure 21.6 uses the rubber-sheet analogy to explain why the increasing distance between galaxies causes the light to be redshifted. If you draw a series of bands on the rubber sheet to represent the crests of an electromagnetic wave, you will see these crests move farther apart as the sheet expands. By the time the sheet is stretched to twice its original size—that is, by the time the scale factor of the sheet is 2—the distance between wave crests has doubled. When the sheet has been stretched to 3 times its original size (a scale factor of 3), the wavelength of the wave will be 3 times what it was originally (**Working It Out 21.2**).

When light left a distant galaxy, the scale factor of the universe was smaller than it is today. As light comes toward us from distant galaxies, the space through which the light travels is stretching, and the light is also "stretched out" as the space through which it travels expands. The universe expanded while the light was in transit, and as it did so, the wavelength of the light grew longer in proportion to the increasing scale factor of the universe. The redshift of light from distant galaxies is therefore a direct measure of how much the universe has expanded since the radiation left its source. Redshift measures how much the scale factor of the universe, R_U, has changed since the light was emitted. Unlike for ordinary redshifts, caused by an object moving through space, cosmological redshifts can be greater than 1 if the emitting galaxy is far enough away.

CHECK YOUR UNDERSTANDING 21.3b

What is the interpretation of a redshift larger than 1? (a) The object is moving faster than the speed of light. (b) The universe has more than doubled in size since the light from that object was emitted. (c) The light was shifted to longer wavelengths from gravitational radiation. (d) The rate of expansion of the universe is increasing.

what if . . .

What if the universe were contracting, rather than expanding, with a negative Hubble constant? How would this affect the Doppler shift of other galaxies?

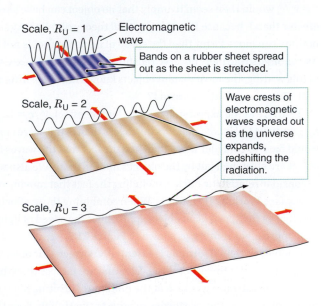

Scale, $R_U = 1$ Electromagnetic wave

Bands on a rubber sheet spread out as the sheet is stretched.

Scale, $R_U = 2$

Wave crests of electromagnetic waves spread out as the universe expands, redshifting the radiation.

Scale, $R_U = 3$

Figure 21.6 Bands drawn on a rubber sheet represent the positions of the crests of a light wave in space. As the rubber sheet is stretched—that is, as the universe expands—the wave crests get farther apart. The light is redshifted.

unanswered questions

What existed before the Big Bang? The usual (and less than satisfying) answer is that the Big Bang was the beginning of space and *time*, so time could not exist before it happened. A more recent answer is that this universe may be just one of many universes, and the Big Bang was the beginning of this universe only. We return to that topic in Chapter 22.

working it out 21.2

When Redshift Exceeds 1

From our discussion of the Doppler shift (Chapter 5), recall that $v_r/c = (\lambda_{obs} - \lambda_{rest})/\lambda_{rest}$, where v_r is the velocity of an object moving away and c is the speed of light. Edwin Hubble used that result to interpret the observed redshifts of galaxies as evidence that galaxies throughout the universe were moving away from the Milky Way. Hubble initially assumed that redshifts were due to the Doppler effect. The resulting relation, $z = v_r/c$, would then seem to imply that no object can have a redshift (z) greater than 1, because Einstein's special theory of relativity says that nothing can move faster than the speed of light. Yet that is not the case. Astronomers routinely observe redshifts significantly in excess of 1. As of this writing, the most distant objects known have redshifts as large as 9–11. How can redshifts exceed 1?

To arrive at the expression for the Doppler effect, $v_r/c = (\lambda_{obs} - \lambda_{rest})/\lambda_{rest}$, we have to *assume* that v_r is much less than c. If v_r were close to c, we would have to consider more than just the fact that the waves from an object are stretched out by the object's motion away. We also would have to consider relativistic effects, including the fact that moving clocks run slowly (see Working It Out 18.1). When combining those effects, we would find that as the speed of an object approaches the speed of light, its redshift approaches infinity, as shown in **Figure 21.7**.

Doppler's original formula is essentially correct—for objects close enough to us that their measured velocities are far less than the speed of light. When astronomers look at the motions of orbiting binary stars or the peculiar velocities of galaxies relative to the Hubble flow, that equation works just fine. But for any redshift of 0.4 or greater, relativity must be taken into account.

Another source of redshift is the gravitational redshift discussed in Chapter 18. As light escapes from deep within a gravitational well, it loses energy, so photons are shifted to longer and longer wavelengths. If the gravitational well is deep enough, the observed redshift of that radiation can be boundlessly large. In fact, the event horizon of a black hole—that is, the surface around the black hole from which not even light can escape—is where the gravitational redshift becomes infinite.

Cosmological redshift, which is most relevant to this chapter, results from the amount of "stretching" that space has undergone while the light from its original source has been en route to Earth. The amount of stretching is given by the factor $1 + z$. When astronomers observe light from a distant galaxy whose redshift of $z = 1$, the wavelength of that light is twice as long as when it left the galaxy. When the light left its source, the universe was half the size that it is today. When they see light from a galaxy with $z = 2$, the wavelength of the radiation is 3 times its original wavelength, and

they are seeing the universe when it was one-third its current size. That direct relationship enables astronomers to use the observed redshift of the galaxy to calculate the size of the universe at the look-back time to that galaxy. Nearby, that means that distance and look-back time are proportional to z. As astronomers look back closer and closer to the Big Bang, however, redshift climbs more and more rapidly, approaching infinity as the look-back time approaches the age of the universe.

Written as an equation, the scale factor of the universe (R_U) that astronomers see when looking at a distant galaxy is equal to 1 divided by 1 plus the redshift of the galaxy:

$$R_U = \frac{1}{1 + z}$$

For example, when astronomers report they have observed a galaxy with a redshift of 9, the scale factor when the light was emitted was

$$R_U = 1/(1 + 9) = 1/10.$$

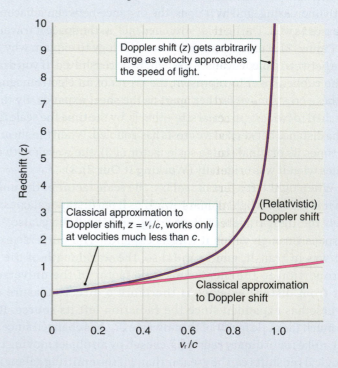

Figure 21.7 This graph shows the plot of the redshift (z) of an object versus its recession velocity (v_r) as a fraction of the speed of light. According to special relativity, the redshift becomes large without limit as v_r approaches c.

21.4 Astronomers Observe Cosmic Microwave Background Radiation

The Big Bang is an extraordinary conclusion to draw from the evidence of Hubble's law. Immediately, astronomers began asking: What predictions does the Big Bang theory make? Is there other evidence that the Big Bang actually took place? One piece of evidence comes from observations of the early universe across the entire sky. In this section we discuss the cosmic microwave background radiation, one of the major confirming observations of the theory that the universe had a beginning.

Radiation from the Big Bang

In the late 1940s, cosmologists Ralph Alpher (1921–2007), Robert Herman (1914–1997), and George Gamow (1904–1968) reasoned that because a compressed gas cools as it expands, the universe should also be cooling as it expands. When the universe was very young and small, it was filled with an extraordinarily hot, dense gas. That hot, dense gas should have been filled with radiation that exhibits a high-temperature Planck blackbody spectrum (see Chapter 5).

Gamow and Alpher took that idea a step further, noting that as the universe expanded, that radiation would have been redshifted to longer and longer wavelengths. According to Wien's law (Chapter 5), the temperature associated with Planck blackbody radiation is inversely proportional to the peak wavelength: $T = (2,900,000 \text{ nm K})/\lambda_{peak}$. Shifting the Planck radiation to longer and longer wavelengths would cause the peak of the radiation to shift to longer wavelengths, so that over time, the spectrum would appear to come from a progressively colder blackbody. As illustrated in **Figure 21.8**, doubling the wavelength of the photons in a Planck blackbody spectrum by stretching space and doubling the scale factor of the universe is equivalent to cutting the temperature of the Planck spectrum in half. As a result, the radiation from the early universe's hot, dense gas should still be detectable today and should have a Planck blackbody spectrum with a temperature of 5–50 kelvins (K). Alpher searched for the signal, but the technology of the late 1940s and early 1950s was not advanced enough for him to find it. A decade later, physicist Robert Dicke (1916–1997) and his colleagues at Princeton University also predicted a hot early universe, arriving independently at the same basic conclusions that Alpher and Gamow had reached earlier.

Measuring the Temperature of the Cosmic Microwave Background Radiation

In the early 1960s, two physicists at Bell Laboratories—Arno Penzias and Robert Wilson (**Figure 21.9**)—detected a faint signal in the microwave part of the electromagnetic spectrum. Within the limits of their equipment, this signal appeared uniform from all parts of the sky. The signal was identified as the radiation left behind by the hot early universe. The strength of the detected signal was consistent with the glow from a blackbody with a temperature of about 3 K, very close to the predicted value for the radiation from the early universe. Their results,

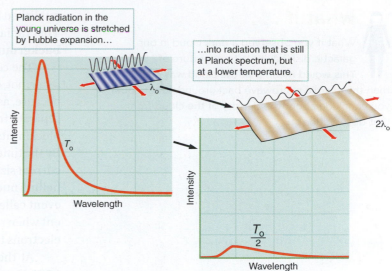

Figure 21.8 As the universe expanded, Planck radiation left over from the hot young universe was redshifted to longer wavelengths. Redshifting a Planck spectrum is equivalent to lowering its temperature.

Figure 21.9 Arno Penzias (left) and Robert Wilson next to the Bell Labs radio telescope antenna with which they discovered the cosmic microwave background radiation. That antenna is now a U.S. National Historic Landmark.

what if . . .

What if you observed a dark cloud in our galactic disk having a temperature of 1.5 K? This would mean that the cloud was colder than the cosmic microwave background. What could you conclude about the density of the cloud from that observation?

published in 1965, reported the discovery of the "glow" left behind by the Big Bang. That radiation left over from the early universe is called the **cosmic microwave background radiation (CMB)**.

The conditions within the early universe were much like the conditions within a star: hot, dense, and opaque (**Figure 21.10a**). At that time, the universe was opaque because, as in the interior of a star, all the atoms were ionized—the electrons were separate from the atomic nuclei. Free electrons interact strongly with radiation, blocking its progress, so the gas is opaque. As the universe expanded, the gas cooled. By the time the universe was about 380,000 years old, and about a thousandth of its current size, the temperature had dropped to a few thousand kelvins. Hydrogen and helium nuclei combined with electrons to form neutral atoms for the first time—an event called the **recombination** of the universe. The universe became transparent when recombination occurred (**Figure 21.10b**), because there were so few free electrons to impede the travel of radiation.

At the time of recombination, the temperature of the universe was about 3000 K, and the peak wavelength of that radiation was about 1 micron (μm). As it traveled through the expanding universe, that radiation was redshifted. Today, the scale of the universe has increased a thousandfold since recombination, and the peak

Figure 21.10 The cosmic microwave background radiation we see originated at the moment the universe became transparent. **a.** Before recombination, the universe was like a foggy day, except that the "fog" was a sea of electrons and protons. Radiation interacted strongly with free electrons and so could not travel far. The trapped radiation had a Planck blackbody spectrum. The tree in the picture is analogous to a source emitting just before recombination. **b.** When the constituents of the universe recombined to form neutral hydrogen atoms, the fog cleared and that radiation was free to travel unimpeded.

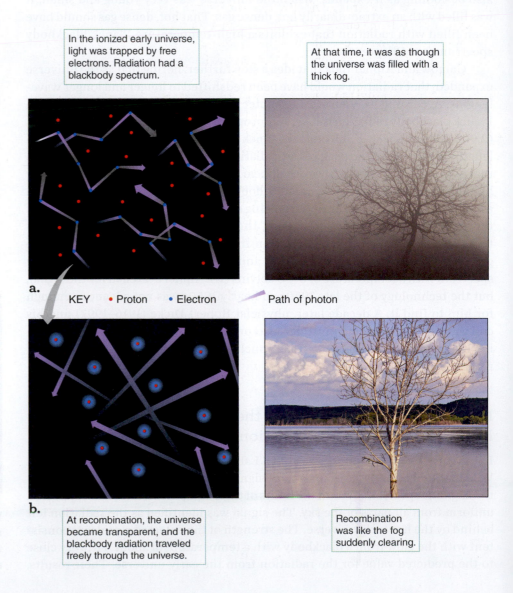

In the ionized early universe, light was trapped by free electrons. Radiation had a blackbody spectrum.

At that time, it was as though the universe was filled with a thick fog.

a.

KEY   • Proton   • Electron   Path of photon

At recombination, the universe became transparent, and the blackbody radiation traveled freely through the universe.

Recombination was like the fog suddenly clearing.

b.

wavelength of the cosmic background radiation has increased by a thousandfold as well, to a value close to 1 millimeter (mm). The spectrum of the CMB still has the shape of a Planck blackbody spectrum, but with a characteristic temperature of only about 3 K—a thousandth of what it was at the time of recombination.

Variations in the CMB

The presence of cosmic background radiation with a Planck blackbody spectrum is a strong prediction of the Big Bang theory. Penzias and Wilson had confirmed that a signal with the correct strength was there, but they could not say for certain whether the signal they saw had the spectral shape of a Planck blackbody spectrum. From the late 1960s to the 1990s, experiments at different wavelengths provided inconsistent results. The Cosmic Background Explorer (COBE) satellite made extremely precise measurements of the CMB at many wavelengths, from a few microns out to 1 cm. The COBE measurements showed the CMB to be a Planck blackbody spectrum with a temperature of 2.73 K (**Figure 21.11**). The observed spectrum perfectly matches the one predicted by Big Bang cosmology. This sort of evidence, in which a hypothesis is used to make a prediction that is later verified, is very strong evidence that the hypothesis is correct. In this case, the evidence is strong enough that the Big Bang theory is now the fundamental theory of the origin and evolution of the universe. Astrophysicists John Mather and George Smoot won the Nobel Prize in Physics in 2006 for their work on COBE.

COBE data included much more than a measurement of the spectrum of the cosmic background radiation. **Figure 21.12a** shows a map obtained by COBE of the CMB from the entire sky. The different colors in the map correspond to variations of about 0.1 percent in the peak wavelength of the CMB. Most of that variation in wavelengths is present because the Sun is moving at a velocity of 370 km/s in the direction of the constellation Crater. Radiation coming from the direction in which Earth is moving is slightly blueshifted (and thus appears to be shifted to a higher temperature) by that motion, whereas radiation coming from the opposite direction is slightly redshifted (cooler temperature). The Sun's motion is due to a combination of factors, including its orbit around the center of the Milky Way Galaxy and the motion of the Milky Way relative to the CMB.

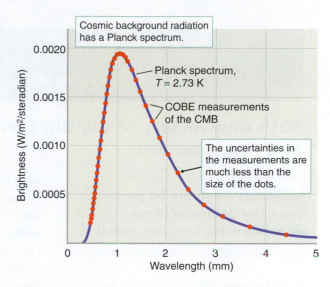

Figure 21.11 The spectrum of the CMB, as measured by the Cosmic Background Explorer (COBE) satellite, is shown by the red dots. The uncertainty in the measurement at each wavelength is much less than the size of a dot. The line running through the data is a blackbody spectrum with a temperature of 2.73 K.

Figure 21.12 ★ **WHAT AN ASTRONOMER SEES**
a. The COBE satellite mapped the temperature of the CMB. The CMB is slightly hotter (by about 0.003 K) in one direction in the sky than in the other direction. This difference is due to Earth's motion relative to the CMB. **b.** With Earth's motion removed, tiny ripples remain in the CMB. **c.** *WMAP* confirmed the fundamentals of cosmological theory at small and intermediate scales. **d.** The Planck mission has provided the highest resolution yet of the CMB and has detected some surprises, such as the "cold spot." The radiation seen in this image was emitted less than 400,000 years after the Big Bang. Comparing these four panels, an astronomer would be very aware of the improvement in technology over time. Panels a. and b. were imaged by COBE in 1989. The angular resolution in panel b. is about 10° (20 times larger than the full Moon). By the time *WMAP* launched in 2001, it was able to achieve an angular resolution that was 33 times better: about 0.25° (half the diameter of the full Moon). An astronomer would notice this distinct improvement in panel c. Planck launched in 2009, and panel d. has an angular resolution of about 5 arcminutes (0.083°). An astronomer would notice that large patterns remain the same, even when more detail is apparent, which lends confidence to the accuracy and interpretation of the images.

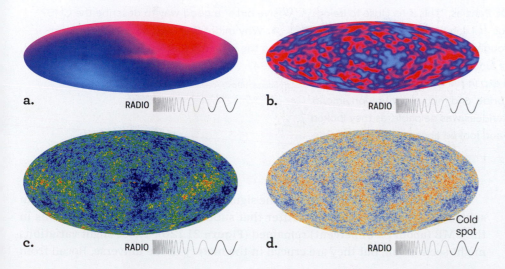

Reading Astronomy News

50th Anniversary of the Big Bang Discovery

Joanne Colella, *The Journal* (NJ)

Scientists celebrate 50 years since the detection of the cosmic microwave background radiation.

A unique gathering of some of the most brilliant minds in the world was held last month on Holmdel's Crawfords Hill on a beautiful spring day, celebrating the 50th anniversary of a truly stellar discovery: the detection of cosmic microwave background radiation (CMB), the thermal echo of the universe's explosive birth, and the evidence that proved the famed Big Bang theory, which would have taken place about 13.8 billion years ago. On May 20, 1964, American radio astronomers Robert Wilson and Arno Penzias confirmed that discovery, admittedly by accident but only after an exhaustive amount of investigative research to rule out every possible explanation for an odd buzzing sound that came from all parts of the sky at all times of day and night. The hum was detected by the enormous Horn Antenna at the Bell Labs site, now a national landmark. Puzzled by the noise, but initially not suspecting its significance, the pair went to great lengths to determine, or rule out, any possible source—including some pigeons that had nested in the antenna and were determined to return, even after being shipped to a distant location.

In 1978, Dr. Wilson and Dr. Penzias won the esteemed Nobel Prize in Physics for their work. Now 78 and 81 years old, respectively, the two came together again at the Horn Antenna site to celebrate the momentous anniversary with current and former Bell Labs colleagues. The event was headed up by Bell Labs President and Corporate CTO Marcus Weldon, who lauded their achievement and spoke about the company initiative to return "back to the future"—back to the classic model of Bell Labs, the research arm of Alcatel-Lucent, working to invent the future. Regarding the Big Bang theory, Mr. Weldon brought chuckles by stating, "In the beginning there was nothing and then it exploded . . . and then there was Arno and Bob." Subtle humor was a recurring theme by some speakers, who even poked fun at myths about Bell Labs, which has produced 12 Nobel Prize laureates. . . . Dr. Wilson and Dr. Penzias each spoke personally about their backgrounds, their work together, and the meaning of their discovery. The pair's distinct styles and personalities complemented each other perfectly. "It is very satisfying to look back and see we did our job right," said Dr. Wilson quietly and humbly. "This was pretty heady stuff," stated the talkative Dr. Penzias. "This is as close to being religious as I can be."

Throughout the presentations and celebratory luncheon, held under oversized tents on the expansive grounds, the air of excitement, pride, and mutual admiration among the attendees was palpable, as they looked back—and looked ahead—to the incredible legacy and pool of talent shared by Bell Labs personnel over the years. Robert Wilson and his wife still reside in Holmdel, as do many other Bell Labs employees, and the company has been an integral component of Holmdel and surrounding communities.

During the event, details were also announced about the establishment of the Bell Labs Prize, an annual competition to give scientists around the globe the chance to introduce their ideas in the fields of information and communications technology. The challenge offers a grand prize of $100,000, second prize of $50,000, and third prize of $25,000. Winners may also get the chance to develop their ideas at Bell Labs. The program is intended to inspire world-changing discoveries and innovations by young researchers.

QUESTIONS

1. What other main pieces of evidence supporting the Big Bang were known before the discovery of the CMB?

2. Why was that discovery seen as the final confirmation of the Big Bang?

3. Is "thermal echo of the universe's explosive birth" a good way to describe the CMB?

4. Why might Bell Labs have been supporting the research that led to that discovery in 1964?

Source: https://colellacommunications.com/50th -anniversary-of-the-big-bang-discovery/.

Astronomers know the shape of the signal from the Sun's velocity, so they can subtract it from the COBE map. After that subtraction, only slight variations in the CMB (about 0.001 percent) remained (**Figure 21.12b**). Those slight variations might seem small, but they are crucial in the history of the universe. Recall from

Chapter 18 that gravity itself can create a redshift. Those tiny fluctuations in the cosmic background radiation are the result of gravitational redshifts caused by concentrations of mass that existed in the early universe. Those concentrations later gave rise to galaxies and the rest of the structure that is evident in the universe today.

From 2001 to 2010, NASA's *WMAP* (Wilkinson Microwave Anisotropy Probe) satellite made more precise measurements of the variations of the CMB (**Figure 21.12c**). The European Space Agency's Planck space observatory collected data of even higher resolution from 2009 to 2013 (**Figure 21.12d**). The much higher-resolution maps obtained by *WMAP* and Planck enable astronomers to refine their ideas about the development of structure in the early universe, which we discuss in Chapters 22 and 23.

CHECK YOUR UNDERSTANDING **21.4**

The existence of the cosmic microwave background radiation tells us that the early universe was: (a) much hotter than it is today; (b) much colder than it is today; (c) about the same temperature as today but much more dense.

Origins: Big Bang Nucleosynthesis

The expansion of the universe and the cosmic microwave background radiation are two of the key pieces of observational evidence supporting the Big Bang theory. The third major piece of supporting evidence comes from observations of the number and types of chemical elements in the universe. For a short time after the Big Bang, the temperature and density of the universe were high enough for nuclear reactions to take place. Collisions between protons and neutrons in the early universe built up low-mass nuclei, including deuterium (heavy hydrogen, ^2H) and isotopes of helium, lithium, and beryllium. That process of element creation, called **Big Bang nucleosynthesis**, determined the final chemical composition of the matter that emerged from the hot phase of the Big Bang. Because the universe was rapidly expanding and cooling, the density and temperature of the universe fell too low for fusion to heavier elements such as carbon to occur. Therefore, all elements more massive than beryllium, including most of the atoms that make up Earth and its life, must have formed in later generations of stars.

Figure 21.13 shows the observed and calculated predictions of the amounts of deuterium, helium, and lithium from Big Bang nucleosynthesis, plotted as a function of the observed present-day density of normal (luminous) matter in the universe. Observations of current abundances are shown as horizontal bands. Theoretical predictions, which depend on the density of the universe, are shown as darker, thick lines. Big Bang nucleosynthesis predicts that about 24 percent of the mass of the normal matter formed in the early universe should have ended up in the form of the very stable isotope ^4He, regardless of the total density of matter in the

▶‖ **AstroTour:** Big Bang Nucleosynthesis

Figure 21.13 Observed and calculated abundances of the products of Big Bang nucleosynthesis, plotted against the density of normal matter in today's universe. Big Bang nucleosynthesis correctly predicts the amounts of those isotopes found in the universe today. (Note the two scale breaks on the *y*-axis.)

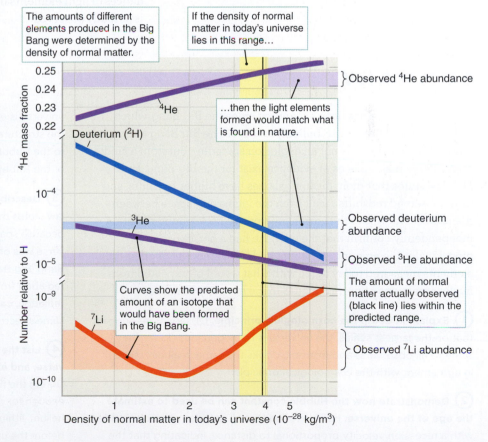

The amounts of different elements produced in the Big Bang were determined by the density of normal matter.

If the density of normal matter in today's universe lies in this range...

...then the light elements formed would match what is found in nature.

Curves show the predicted amount of an isotope that would have been formed in the Big Bang.

The amount of normal matter actually observed (black line) lies within the predicted range.

Observed ^4He abundance

Observed deuterium abundance

Observed ^3He abundance

Observed ^7Li abundance

^4He mass fraction

Number relative to H

Density of normal matter in today's universe (10^{-28} kg/m³)

universe. Indeed, that is what is observed. That agreement between theoretical predictions and observation serves as powerful evidence that the universe began in a Big Bang.

Unlike with helium, most other isotope abundances depend on the density of normal matter in the universe, so comparing current abundances with models of isotope formation in the Big Bang helps pin down the density of the early universe. Beginning with the abundances of isotopes, such as ^{2}H (deuterium) and ^{3}He found in the universe today (shown as the horizontal, light-colored bands in Figure 21.13), and comparing them with predictions of how abundant isotopes *should* be when formed at different densities (dark-colored curves in Figure 21.13), cosmologists can find the density of normal matter (the vertical yellow band in Figure 21.13). The best current measurements give a value of about 3.9×10^{-28} kg/m^3 for the average density of normal matter in the universe today. For comparison, the density of air at sea level on Earth is about 1.2 kg/m^3. That value for the universe lies well within the range predicted by the observations shown in Figure 21.13. The agreement is remarkable, and it holds for many isotopes.

Turning that process around, cosmologists can begin with an observation of the amount of normal matter in and around galaxies and then compare that with calculations of what the chemical composition emerging from the Big Bang should have been. The observations agree remarkably well with the amounts of those elements actually found in nature. The idea that light elements originated in the Big Bang is resoundingly confirmed. This is extremely strong evidence supporting the Big Bang model.

That agreement also creates a powerful constraint on the nature of dark matter, which dominates the mass in the universe. Dark matter cannot consist of normal matter made up of neutrons and protons; if it did, the density of neutrons and protons in the early universe would have been much higher, and the resulting abundances of light elements in the universe would have been much different from what is actually observed.

SUMMARY

The universe has been expanding since the Big Bang, which occurred nearly 13.8 billion years ago. The Big Bang happened everywhere—it is not an explosion spreading out from a single point. Three major pieces of evidence exist for the Big Bang. Hubble's law states that more distant galaxies have higher redshifts, and those observed redshifts result from the increasing space between them. Observations of the cosmic microwave background radiation independently confirm that the universe had a hot, dense beginning. Finally, the amounts of helium and trace amounts of other light elements measured today agree with what would be expected from nuclear reactions of normal matter in the hot early universe.

(1) **Explain in detail the cosmological principle.** Observations suggest that on the largest scales, the universe is homogeneous (looks the same to all observers) and isotropic (looks the same in all directions), in agreement with the cosmological principle.

(2) **Demonstrate how the Hubble constant can be used to estimate the age of the universe.** Hubble found that all galaxies are moving away, with a recession velocity proportional to distance, indicating that the universe is expanding. The expansion does not affect local physics and the structure of objects. Running the Hubble expansion backward leads to the Hubble time, and the age of the universe, found from the inverse of the Hubble constant.

(3) **Describe the observational evidence for the Big Bang.** Hubble's law states that light from distant galaxies is redshifted, which occurs because space itself is expanding. Hubble's law suggests that the universe was once very hot and very dense, a beginning known as the Big Bang. Big Bang theory predicts that we should be able to observe the radiation from a few hundred thousand years after the Big Bang. That radiation, called the cosmic microwave background radiation, has the same spectrum in every direction.

(4) **List the chemical elements that were created in the early hot universe, and explain why no other elements were created at that time.** During the first few minutes after the Big Bang, the universe was hot enough for nucleosynthesis to take place through fusion. Deuterium, helium, lithium, and beryllium, but not heavier elements, were created before the universe cooled too quickly for that fusion to progress further.

QUESTIONS AND PROBLEMS

TEST YOUR UNDERSTANDING

1. What do astronomers mean when they say that the universe is homogeneous?
   a. The universe looks the same from every perspective.
   b. Galaxies are generally distributed evenly throughout the universe.
   c. All stars in all galaxies have planetary systems just like ours.
   d. The universe has looked the same at all times in its history.

2. What do astronomers mean when they say that the universe is isotropic?
   a. More distant parts of the universe look just like nearby parts.
   b. Intergalactic gas has the same density everywhere in the universe.
   c. The laws of physics apply everywhere in the universe.
   d. The universe looks the same in every direction.

3. Cosmological redshifts are calculated from observations of spectral lines from
   a. individual stars in distant galaxies.
   b. clouds of dust and gas in distant galaxies.
   c. spectra of entire galaxies.
   d. rotations of the disks of distant galaxies.

4. Astronomers observe that all galaxies are moving away from the Milky Way. According to the cosmological principle, that observation suggests that
   a. the Milky Way is at the center of the universe.
   b. the Milky Way must be at the center of the expansion.
   c. the Big Bang occurred at the current location of the Milky Way.
   d. an observer in a distant galaxy would make the same observation.

5. Some galaxies have redshifts z that if equated to v_r/c correspond to velocities greater than the speed of light. Special relativity is not violated
   a. because of relativistic beaming.
   b. because it's a trick of the measurement angle.
   c. because redshifts carry no information.
   d. because those velocities do not measure motion through space.

6. The Big Bang theory predicted (select all that apply)
   a. the Hubble law.
   b. the cosmic microwave background radiation.
   c. the cosmological principle.
   d. the abundance of helium.
   e. the period-luminosity relationship of Cepheid variables.

7. The simplest way to estimate the age of the universe is from
   a. using the slope of Hubble's law.
   b. the age of Moon rocks.
   c. models of stellar evolution.
   d. measurements of the abundances of elements.

8. The CMB includes information about (select all that apply)
   a. the age of the universe.
   b. the temperature of the early universe.
   c. the density of the early universe.
   d. density fluctuations in the early universe.
   e. the motion of Earth around the center of the Milky Way.

9. Repeated measurements showing that the current helium abundance is much less than the value predicted by the Big Bang would imply that
   a. some part of the Big Bang theory is incorrect or incomplete.
   b. the current helium abundance is wrong.
   c. scientists don't know how to measure helium abundances.

10. According to the cosmological principle,
    a. the universe is expanding.
    b. the universe began in the Big Bang.
    c. the rules that govern the universe are the same everywhere.
    d. the early universe was 1000 times hotter than the characteristic temperature of the CMB.

11. Why is the Milky Way Galaxy not expanding together with the rest of the universe?
    a. Because it is at the center of the expansion.
    b. It is expanding, but the expansion is too small to measure.
    c. The Milky Way is a special location in the universe.
    d. Local gravity dominates over the expansion of the universe.

12. The scale factor keeps track of
    a. the movement of galaxies through space.
    b. the current distances between many galaxies.
    c. the changing distance between any two galaxies.
    d. the location of the center of the universe.

13. The Big Bang is
    a. the giant supernova explosion that triggered the formation of the Solar System.
    b. the explosion of a supermassive black hole.
    c. the eventual demise of the Sun.
    d. the beginning of space and time.

14. The CMB is essentially uniform in all directions in the sky. That fact is an example of
    a. anisotropy.
    b. isotropy.
    c. thermal fluctuations.
    d. Wien's law.

15. Which of the following was created as a result of Big Bang nucleosynthesis? (Choose all that apply.)
    a. helium   d. deuterium
    b. lithium   e. carbon
    c. hydrogen

THINKING ABOUT THE CONCEPTS

16. ★ **WHAT AN ASTRONOMER SEES**   Figure 21.12 shows four different views of the cosmic microwave background. Even without knowing any specifics, an astronomer would recognize that technology had improved between the times that images b, c, and d were taken. How would an astronomer know that the images were different because of a technology improvement, rather than because of a change in the cosmic microwave background? ★

17. Imagine that you are standing in the middle of a dense fog.
    a. Would you describe your environment as isotropic? Why or why not?
    b. Would you describe it as homogeneous? Why or why not?

18. As the universe expands from the Big Bang, galaxies are not actually flying apart from one another. What is really happening?

19. We see the universe around us expanding, which gives distant galaxies an apparent velocity of 70 km/s/Mpc. If you were an astronomer living today in a galaxy located 1 billion light-years away from us, at what rate would you see the galaxies moving away from you?

20. Does Hubble's law imply that our galaxy is sitting at the center of the universe? Explain.

21. What does the value of R_U, the scale factor of the universe, tell us?

22. Does the expansion of the universe make the Sun bigger? What about the Milky Way? Why or why not?

23. Science's greatest strength is the self-correcting nature of scientific inquiry: Minds can be changed in the face of new data, as described in the Process of Science Figure. Consider a modern paradigm of science such as the Big Bang, climate change, or the power source of stars. In a short paragraph, describe your current viewpoint and give a piece of evidence that would make you change your mind if it were true. For example, Einstein believed the universe was static. Measurements of the movement of galaxies caused him to change his mind. ★

24. Name two predictions of the standard Big Bang theory that have been verified by observations.

25. The general relationship between recession velocity (v_r) and redshift (z) is $v_r = cz$. That simple relationship fails, however, for very distant galaxies with large redshifts. Explain why.

26. Why is it significant that the CMB displays a Planck blackbody spectrum?

27. What is the significance of the tiny brightness variations observed in the CMB?

28. What important characteristics of the early universe are revealed by today's observed abundances of various isotopes, such as ^2H and ^3He?

29. Why were only a few of the chemical elements created in the Big Bang?

30. How do astronomers know that some of the observed helium is left over from the Big Bang?

APPLYING THE CONCEPTS

31. The Hubble time ($1/H_0$) represents the age of a universe that has been expanding at a constant rate since the Big Bang. Calculate the age of the universe in years if H_0 equals 74 km/s/Mpc. (Note: 1 year = 3.16×10^7 seconds, and 1 Mpc = 3.09×10^{19} km.) ●━●━●
    a. Make a prediction: Compare this value for H_0 to the value for H_0 in Working It Out 21.1. Do you expect that your answer will be close to or far from the age calculated for that value? Do you expect your answer to be larger or smaller than the answer in Working It Out 21.1?
    b. Calculate: Use unit analysis to calculate the age of the universe in years from this value of H_0.
    c. Check your work: Verify that your answer has units of years, and compare it to your prediction.

32. How much has the universe expanded since light was emitted from a galaxy with a redshift of $z = 8$? ●━●━●
    a. Make a prediction: Do you expect the scale factor to be large (the universe was much bigger then than now) or small (the universe was smaller then than now)?
    b. Calculate: Use the equation in Working It Out 21.2 to find R_U at the time the light from this galaxy was emitted. By what factor has the universe expanded since that time?
    c. Check your work: Compare your answer to your prediction.

33. Throughout the latter half of the 20th century, estimates of H_0 ranged from 50 to 100 km/s/Mpc. Calculate the age of the universe in years for both of those estimated values of H_0.

34. What was the scale factor, R_U, of the universe (compared with the present) when the CMB was emitted, at $z = 1000$?

35. What is the redshift of a galaxy observed at a time when R_U was 1/4 of its current value?

36. In Figure 21.5, a rubber sheet is shown as an analogy to help you think about the scale factor. Between the moments shown in parts (a) and (b), each square doubles in size on every edge. How does the area of a square change? Imagine that the sheet is now a block of rubber, expanding in three dimensions instead of two. How would the volume of a cube change between the moment shown in part (a) and the moment shown in part (b)? ◉

37. Error bars have not been plotted in Figure 21.11. Why not? Was that a very precise measurement or a very imprecise measurement? How does the precision of the measurement affect your confidence in the conclusions drawn from it? ◉

38. Figure 21.13 includes both predictions and observations. ◉
    a. What do the vertical yellow bar and the slanted lines and curves represent: theory or observation?
    b. What do the pastel horizontal lines and the vertical black line represent: predictions or observations?
    c. Do the predictions and observations match? Choose one example, and explain how you know.

39. Suppose a galaxy is observed with a redshift of $z = 2$. How much has the universe expanded since that light was emitted from those galaxies?

40. You observe a distant quasar in which a spectral line of hydrogen with rest wavelength $\lambda_{rest} = 121.6$ nm is found at a wavelength of 547.2 nm. What is its redshift? When the light from that quasar was emitted, how large was the universe in comparison with its current size?

41. A distant galaxy has a redshift of $z = 5.82$ and a recession velocity $v_r = 287,000$ km/s (about 96 percent of the speed of light).
    a. If $H_0 = 70$ km/s/Mpc and if Hubble's law remains valid out to such a large distance, how far away is that galaxy?
    b. Assuming a Hubble time of 13.8 billion years, how old was the universe at the look-back time of that galaxy?
    c. What was the scale factor of the universe at that time?

42. The spectrum of the CMB is shown as the red dots in Figure 21.11, along with a blackbody spectrum for a blackbody at a temperature of 2.73 K. From the graph, determine the peak wavelength of the CMB spectrum. Use Wien's law to find the temperature of the CMB. How does that rough measurement that you just made compare with the accepted temperature of the CMB? ★

43. COBE observations show that the Solar System is moving in the direction of the constellation Crater at a speed of 368 km/s relative to the cosmic reference frame. What is the blueshift (negative value of z) associated with that motion?

44. To get a feeling for the emptiness of the universe, compare its density (4×10^{-28} kg/m³) with that of Earth's atmosphere at sea level (1.2 kg/m³). How much denser is Earth's atmosphere? Write that ratio in standard notation.

45. Assume that the most distant galaxies have a redshift of $z = 10$. The average density of normal matter in the universe today is 4×10^{-28} kg/m³. What was its density when light was leaving those distant galaxies? (Hint: Keep in mind that volume is proportional to the *cube* of the scale factor.)

EXPLORATION Hubble's Law for Balloons

The expansion of the universe is extremely difficult to visualize, even for professional astronomers. In this Exploration, you will use the surface of a balloon to get a feel for how an "expansion" changes distances between objects. Throughout this Exploration, remember to think of the surface of the balloon as a two-dimensional object, much as the surface of Earth is a two-dimensional object for most people. The average person can move east or west, or north or south, but moving into Earth and out to space are not options. For this Exploration you will need a balloon, 11 small stickers, a piece of string, and a ruler. A partner is helpful as well. **Figure 21.14** shows some of the steps involved.

Blow up the balloon partially and hold it closed, but *do not tie it shut*. Stick the 11 stickers on the balloon (each represents a different galaxy) and number them. Galaxy 1 is the reference galaxy.

Measure the distance between the reference galaxy and each of the galaxies numbered 2–10. The easiest way to do that is to use your piece of string. Lay it along the balloon between the two galaxies and then measure the length of the string. Record those data in the "Distance 1" column of a table like the one shown here.

Simulate the expansion of your balloon universe by *slowly* blowing up the balloon the rest of the way. Have your partner count the seconds you take to do that, and record that number in the "Time Elapsed" column of the table (each row has the same time elapsed because the expansion occurred for the same amount of time for each galaxy). Tie the balloon shut. Measure the distance between the reference galaxy and each numbered galaxy again. Record those data under "Distance 2."

Subtract the first measurement (distance 1) from the second (distance 2). Record the difference in the table.

Divide that difference, which represents the distance traveled by the galaxy, by the time that blowing up the balloon took. Distance divided by time gives an average speed.

Make a graph with velocity on the *y*-axis and distance 2 on the *x*-axis to get "Hubble's law for balloons." You may want to roughly fit a line to those data to clarify the trend.

1 Describe your data. If you fit a line to them, is it horizontal or does it trend upward or downward?

_____

2 Is anything special about your reference galaxy? Is it different in any way from the others?

_____

3 If you had picked a different reference galaxy, would the trend of your line be different? If you are not sure of the answer, get another balloon and try it.

_____

4 The expansion of the universe behaves similarly to the movement of the galaxies on the balloon. We don't want to carry the analogy too far, but let's think about one more thing. In your balloon, some areas

Figure 21.14 Measure the distance around a curved balloon by using a string.

Galaxy Number	Distance 1	Distance 2	Difference (Distance 2 – Distance 1)	Time Elapsed	Velocity
1 (reference)	0	0	0		0
2					
3					
4					
5					
6					
7					
8					
9					
10					
11					

probably expanded less than others because the material was thicker; more "balloon stuff" was holding it together. How is that analogy similar to some places in the actual universe?

_____

_____

_____

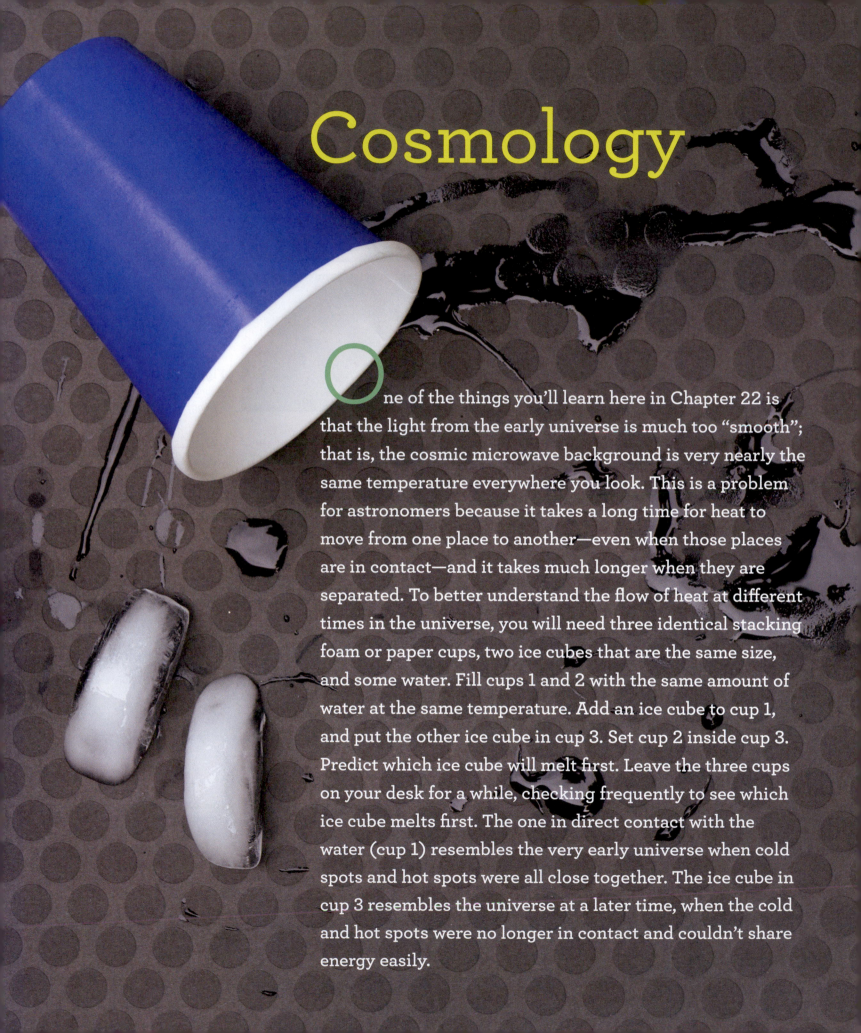

Cosmology

One of the things you'll learn here in Chapter 22 is that the light from the early universe is much too "smooth"; that is, the cosmic microwave background is very nearly the same temperature everywhere you look. This is a problem for astronomers because it takes a long time for heat to move from one place to another—even when those places are in contact—and it takes much longer when they are separated. To better understand the flow of heat at different times in the universe, you will need three identical stacking foam or paper cups, two ice cubes that are the same size, and some water. Fill cups 1 and 2 with the same amount of water at the same temperature. Add an ice cube to cup 1, and put the other ice cube in cup 3. Set cup 2 inside cup 3. Predict which ice cube will melt first. Leave the three cups on your desk for a while, checking frequently to see which ice cube melts first. The one in direct contact with the water (cup 1) resembles the very early universe when cold spots and hot spots were all close together. The ice cube in cup 3 resembles the universe at a later time, when the cold and hot spots were no longer in contact and couldn't share energy easily.

EXPERIMENT SETUP

Fill cups 1 and 2 with
the same amount of water
at the same temperature.

Add an ice cube to cup 1,
and put the other ice cube in cup 3.

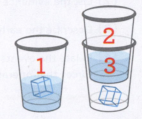

Set cup 2 inside cup 3.
Predict which ice cube will melt first.

Leave the three cups on your desk
for a while, checking frequently
to see which ice cube melts first.

PREDICTION

I predict that the ice cube in
☐ cup 1   ☐ cup 3
will melt first.

SKETCH OF RESULTS (in progress)

Time	Cup 1	Cup 3
0 min.	No melting.	No melting.
4 min.	Edges slightly more rounded.	No melting.
6 min.		
8 min.		

22

C osmology is the study of the large-scale universe, including its nature, origin, evolution, and ultimate destiny. Here in Chapter 22, we take a closer look at the nature of the universe, how it has evolved, and its ultimate fate. We also discuss the physics of the smallest particles, which is necessary for describing the earliest moments of the universe.

LEARNING GOALS

By the end of this chapter, you should be able to:

(1) Connect the mass within the universe and the gravitational force it produces to the history, shape, and fate of the universe.

(2) List the evidence for the accelerating expansion of the universe.

(3) Describe the early period of rapid expansion of the universe known as inflation.

(4) Relate the events that occurred in the earliest moments of the universe to the forces that operate in the modern universe.

22.1 Gravity and the Expansion of the Universe

What is the fate of the universe? That is a central question of modern cosmology, and the simplest answer depends in part on the average mass in the universe within a fixed volume: If the universe is dense enough, gravity will eventually cause it to collapse. Several factors determine the way mass is distributed on large scales across the universe. In this section, we explain how gravity affects the expansion of the universe.

Mass Distribution

Recall the effects of Earth's gravity on the motion of projectiles fired upward from the surface of Earth (Chapter 4). If the projectile's speed is *less* than Earth's escape velocity, gravity eventually stops the rise of the projectile and pulls it back to the ground. But if the speed is a bit *greater* than the escape velocity—fast enough to overcome both air resistance and Earth's gravity—the projectile will slow down, but it will never stop. It will escape from Earth entirely.

The escape velocity from a planet's surface depends on both the mass and the radius of the planet. If the planet is massive, then its gravity is strong, and the escape velocity is high. If the planet is less massive, however, then its gravity is weak, and the escape velocity is low. Thus, even quite slow projectiles might escape from an asteroid, for example.

How does the escape velocity depend on the radius of the planet? If two planets have the same mass, the one with the smaller radius is denser; the surface is closer to the planet's center, the gravitational pull is stronger, so the escape velocity is higher. Conversely, the planet with the larger radius is less dense: The surface from which the projectile is fired is farther from the center of the planet, the gravitational pull is weaker, and the escape velocity is lower. Whether a projectile fired with a given speed will escape depends on both the planet's mass and the planet's radius.

Just as the gravity slows the climb of a projectile, the mass distributed across the universe slows its expansion. If there is enough mass in the universe,

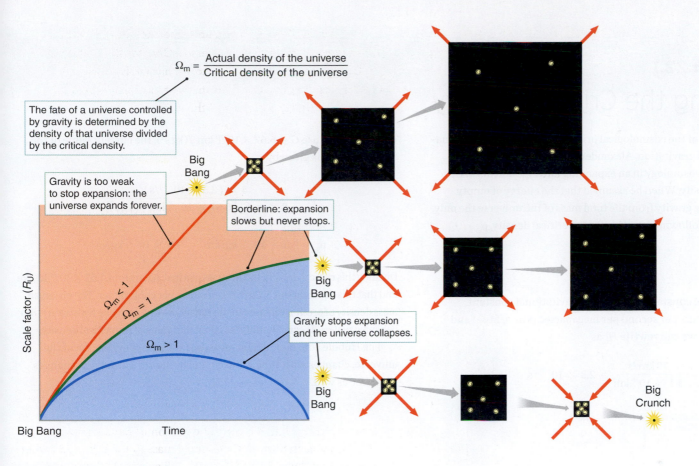

$$\Omega_m = \frac{\text{Actual density of the universe}}{\text{Critical density of the universe}}$$

The fate of a universe controlled by gravity is determined by the density of that universe divided by the critical density.

Gravity is too weak to stop expansion: the universe expands forever.

$\Omega_m < 1$

$\Omega_m = 1$

$\Omega_m > 1$

Big Bang

Borderline: expansion slows but never stops.

Gravity stops expansion and the universe collapses.

Big Bang

Big Crunch

Scale factor (R_U)

Time

Big Bang

Figure 22.1 The universe has three possible fates if gravity is the only important factor.

then gravity will be strong enough to stop the expansion. However, if the radius of the universe is large, then the escape velocity will be small, and the universe will continue to expand. Thus, the density of the universe is critical in determining its fate. If the average density is high, the expansion will slow, stop, and reverse. If the average density is low, the universe will expand forever.

Critical Density

The dividing point between an expansion that eventually reverses, and one that never stops depends on the **critical density**: the density of the universe that would cause the expansion to just barely slow to a stop after a very, very long time. If the universe is less dense than the critical density, gravity is too weak to stop the expansion, and the universe will expand forever. If the density of the universe is greater than the critical density, gravity is strong enough to stop and reverse the expansion eventually.

Astronomers use an uppercase omega (Ω), the last letter in the Greek alphabet, to characterize densities and the ultimate fate of the universe. The ratio of the actual density of mass to the critical density is called the *matter density parameter*, written as Ω_m and pronounced "omega matter." It is a ratio of two densities, so it has no units.

In a universe in which the expansion is only affected by gravity, there are three possible outcomes, depending on the value of Ω_m (**Figure 22.1**). If Ω_m is greater than 1, the density of the universe is greater than the critical density, so gravity is strong enough that the universe will eventually collapse. Conversely, if Ω_m is less than 1, the universe has a density lower than the critical density. The expansion of the universe

working it out 22.1

Calculating the Critical Density

Using the assumption of the cosmological principle and using a set of equations from Einstein on gravitation, Alexander Friedmann (1888–1925) derived equations for cosmology in an expanding universe and discussed the idea of critical density. When he assumed that the energy of empty space was zero, making gravity from the total mass of the universe the only factor, he derived the following equation for the critical density, ρ_c:

$$\rho_c = \frac{3H_0^2}{8\pi G}$$

where H_0 is the Hubble constant and G is the gravitational constant. Using $H_0 = 70$ kilometers per second per megaparsec (km/s/Mpc) and 1 Mpc $= 3.1 \times 10^{19}$ km, we can rewrite H_0 as

$$H_0 = \frac{70 \text{ km/s}}{3.1 \times 10^{19} \text{ km}} = 2.3 \times 10^{-18}/\text{s}$$

Then, when we use $G = 6.67 \times 10^{-20}$ km³/(kg s²), the current value of the critical density is given by

$$\rho_c = \frac{3 \times (2.3 \times 10^{-18}/\text{s})^2}{8 \times \pi \times [6.67 \times 10^{-20} \text{ km}^3/(\text{kg s}^2)]}$$

$$\rho_c = 9.5 \times 10^{-18} \text{ kg/km}^3 = 9.5 \times 10^{-27} \text{ kg/m}^3$$

Dividing this critical density by the mass of a hydrogen atom, 1.67×10^{-27} kg, we find that this equals about 5.7 hydrogen atoms per cubic meter. The observed mass density of ordinary matter in the universe—less than one hydrogen atom per cubic meter—is considerably less than that value of the critical density.

The Hubble constant, which measures the rate of expansion of the universe, changes with time. Consequently, the critical density changes with time as well.

will slow, but the universe will still expand forever. If Ω_m equals 1, the density of the universe is exactly equal to the critical density, and the universe is on the dividing line. The expansion will slow down, but never quite stop until an infinite amount of time has passed.

Just as a planet must be very massive to slow a very fast projectile, the universe must be very massive to slow a very fast expansion. Therefore, the critical density today depends on the Hubble constant, H_0. If gravity is the only factor affecting the expansion, the critical density of the universe is less than the mass of six hydrogen atoms in every cubic meter (**Working It Out 22.1**). Note that the parameters that describe both the expansion and the density change with time; in this book, we always refer to today's values (for H_0 and the current value of Ω_m) when discussing these parameters.

Until the closing years of the 20th century, most astronomers thought that this straightforward application of gravity was all that was needed to understand the expansion and fate of the universe. Researchers carefully measured the masses of galaxies, galaxy groups, and galaxy clusters in the expectation that those data would reveal the density and therefore the fate of the universe. The luminous matter seen in galaxies and groups of galaxies gives an Ω_m value of about 0.02. But galaxies contain about 10 times as much dark matter as normal luminous matter, so adding in the dark matter in galaxies pushes the value of Ω_m up to about 0.2. When the mass of dark matter *between* galaxies is included (a subject we return to in Chapter 23), Ω_m could increase to 0.3 or higher. By that accounting, the universe has at most only a third as much mass as is needed to stop the expansion. Many astronomers were convinced that the expansion of the universe would continue but that it would slow down because of gravity. They were wrong.

what if . . .

What if we live in a universe with Ω greater than one. After a long enough time, how would future astronomers perceive the universe?

22.2 The Accelerating Universe

Edwin Hubble showed that the universe is expanding (see Chapter 19). If the expansion of the universe is slowing down because of gravity, the expansion must have been faster in the past. If so, objects very far away—so that we see them as they were long ago—were moving faster. The most distant galaxies should therefore have larger redshifts than predicted by Hubble's law, which is derived from observations of local galaxies. However, if the expansion of the universe is accelerating, then very distant galaxies were moving slower at that time in the universe's history, and those galaxies should have *smaller* redshifts than predicted by Hubble's law. In this section, we examine the evidence that the expansion of the universe is accelerating.

The Cosmological Constant

During the 1990s, astronomers measured the brightness of Type Ia supernovae in very distant galaxies and compared the brightness of those supernovae with their expected brightness, which was based on the redshift distances of those galaxies. (Recall from Chapter 19 that Type Ia supernovae have a very high peak luminosity that can be determined independently from their brightness in our sky, so they are very useful for measuring distances to extremely distant galaxies.) The findings of those studies are shown as black dots in **Figure 22.2**. Toward the left side of the graph, the data lie nearer to the purple line than the red one, implying that their redshifts are smaller than expected. Astronomers were shocked. Instead of showing that the expansion of the universe has slowed down, the data indicated that the expansion is *speeding up—* accelerating. An "accelerating universe" does *not* mean that the universe is zooming through space faster and faster like a car along a road. It means instead that the expansion is happening faster and faster. For that to be true, a previously unknown force stronger than gravity must be pushing the *entire universe* apart in opposition to gravity. Results from the *WMAP* (Wilkinson Microwave Anisotropy Probe) spacecraft early in the 21st century confirmed the result independently, and in 2011, the Nobel Prize in Physics was awarded to Saul Perlmutter, Brian P. Schmidt, and Adam G. Riess for their observations of Type Ia supernovae and the discovery of the accelerating universe.

The idea of a repulsive force that opposes the attractive force of gravity is not entirely new. In the early 20th century, Einstein used his newly formulated equations of general relativity to calculate the structure of spacetime in the universe. The equations clearly indicated that any universe containing mass could not be static, any more than a ball can hang motionless in the air. He found the same result that Figure 22.1 illustrates—namely, that gravity always makes the expansion of the

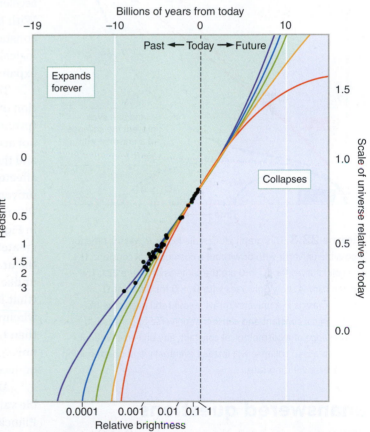

Figure 22.2 This graph plots the observed brightness of Type Ia supernovae as a function of their redshift. The different-colored lines represent different values of the cosmological constant. The observations (data points) indicate that the redshifts are too small for their distances: they best fit the line for an accelerating universe—a universe expanding faster today than it did in the past. The colored lines are more distinguishable at higher redshifts, so astronomers look for higher-redshift supernovae to differentiate the data better.

what if . . .

What if the cosmological constant Λ were 50 times larger than it is? How might that change the properties of the universe?

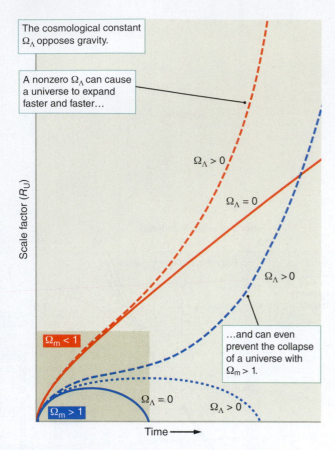

The cosmological constant Ω_Λ opposes gravity.

A nonzero Ω_Λ can cause a universe to expand faster and faster…

$\Omega_\Lambda > 0$

$\Omega_\Lambda = 0$

$\Omega_\Lambda > 0$

Scale factor (R_U)

$\Omega_m < 1$

…and can even prevent the collapse of a universe with $\Omega_m > 1$.

$\Omega_\Lambda = 0$

$\Omega_\Lambda > 0$

$\Omega_m > 1$

Time ⟶

Figure 22.3 This graph plots scale factor R_U versus time for a possible universe with or without a cosmological constant, Ω_Λ. The solid curves for $\Omega_\Lambda = 0$ are the same as in Figure 22.1. The dashed curves show universes with $\Omega_\Lambda > 0$ for different Ω_m. If enough mass is in a universe, gravity could still overcome the cosmological constant and cause that universe to collapse. In the presence of a cosmological constant, any universe without enough mass to collapse will instead eventually end up expanding at an ever-increasing rate.

unanswered questions

What is the origin of dark energy, and why is $\Omega_\Lambda + \Omega_\mu$ so close to 1 now? Is that a coincidence? We have already said that dark energy is a form of vacuum energy, and explaining how big Ω_Λ really is is one of the grand challenges of cosmology. The simplest estimates for the size of Ω_Λ yield results much larger than the observed value—by a factor of about 10^{120}. An as yet undetermined mechanism must exist that affects the size and evolution of Ω_Λ, and that mechanism is one of the biggest questions in modern cosmology.

universe slow down or even stop and reverse. However, Einstein's formulation of spacetime came more than a decade before Hubble discovered the expansion of the universe, and the conventional wisdom then was that the universe was static—that it neither expands nor collapses.

To force his new general theory of relativity to allow for a static universe, Einstein inserted a "fudge factor" called the **cosmological constant** into his equations. When Einstein added the cosmological constant to his equations of general relativity, he considered it a new fundamental constant, similar to Newton's universal gravitational constant G. Einstein's cosmological constant, which is constant in both space and time, acts as a repulsive force that opposes gravity. If it has just the right value, the cosmological constant can lead to a static universe in which galaxies remain stationary despite their mutual gravitational attraction.

When Hubble announced his discovery that the universe is expanding, Einstein realized his mistake. Einstein could have predicted that the universe must be either expanding or contracting with time, but instead he forced his equations to comply with conventional wisdom. He called the introduction of the cosmological constant the "biggest blunder" of his scientific career. The much more recent discovery of an accelerating universe, however, restores the credibility of the cosmological constant. With the results on the brightness of Type Ia supernovae, Einstein's cosmological constant was revived. A repulsive force, like the force associated with the cosmological constant in Einstein's equations, is just what is needed to describe a universe expanding at an ever-faster rate (see the **Process of Science Figure**).

Today, we write the cosmological constant as Λ (uppercase lambda). The fraction of the critical density provided by the cosmological constant is written as Ω_Λ (pronounced "omega lambda"). In Section 22.1, we considered a universe that does not accelerate; in that case Ω_Λ was zero. In an accelerating universe, Ω_Λ is not zero, and the fate of the universe is no longer controlled exclusively by Ω_m. Instead, it is affected by *both* Ω_Λ and Ω_m. The mass density needed to halt the expansion of the universe will be greater than the critical mass density discussed in Section 22.1. The fate of the universe depends on the balance between Ω_Λ and Ω_m, as illustrated in **Figure 22.3**. If the mass density is greater than the critical density, so that Ω_m is greater than 1 (solid and dotted blue lines), then gravity may stop the expansion, and the universe may collapse back on itself, even if Ω_Λ is not zero. Even in a universe unlike ours, where the density of the universe is greater than the critical density (that is, Ω_m is greater than 1), a large enough cosmological constant could overwhelm gravity and make the universe expand forever (dashed blue line). If Ω_m is less than 1, however, then it is not large enough to stop the expansion. In this case, the universe expands forever, and whether the expansion of the universe will accelerate or decelerate depends on the value of Ω_Λ (orange lines).

Astronomers have data from several experiments showing the range of possible values for Ω_m and Ω_Λ. The data from Type Ia supernovae—from *WMAP*, from Planck, and from clusters of galaxies (discussed in Chapter 23)—are all consistent with values for Ω_m and Ω_Λ of about 0.3 and 0.7, respectively. Thus, the expansion of the universe is apparently accelerating under the dominant effect of the cosmological constant and has been doing so for 5 billion to 6 billion years. This universe is represented by the dashed orange line in Figure 22.3.

So far, we have treated the cosmological constant as a mathematical construct—a thing we plug into the mathematics to match our equations of the universe to our observations of it. But what, physically, is the cosmological constant? Where does it come from? One idea is that it comes from the energy of empty space. Today, physicists call "empty space" the **vacuum** and recognize that it has some distinct

Never Throw Anything Away

The accelerating universe is a good example of a scientist being right for the wrong reason. Einstein's famous mathematical constant has become useful in a way he never would have imagined.

1930

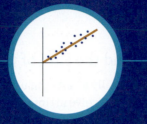

Almost 75 years after Hubble proved Einstein wrong . . .

. . . astronomers were working to measure the slowing down of the expansion of the universe, and to their surprise, they found that the expansion rate was accelerating.

2000

Einstein's "fudge factor" was revived as the cosmological constant, which accounts for the acceleration.

Science has a memory, so old ideas can be revisited in the light of new data, often saving a great deal of effort, even when the interpretation has changed.

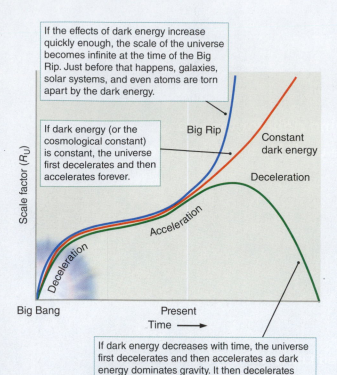

If the effects of dark energy increase quickly enough, the scale of the universe becomes infinite at the time of the Big Rip. Just before that happens, galaxies, solar systems, and even atoms are torn apart by the dark energy.

If dark energy (or the cosmological constant) is constant, the universe first decelerates and then accelerates forever.

Big Rip

Constant dark energy

Deceleration

Acceleration

Deceleration

Deceleration

Big Bang

Present
Time →

If dark energy decreases with time, the universe first decelerates and then accelerates as dark energy dominates gravity. It then decelerates again as dark energy subsides. If dense enough, the universe could collapse to a Big Crunch.

Figure 22.4 The scale factor R_U of the universe varies depending on how dark energy changes.

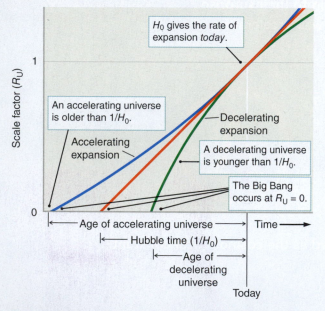

H_0 gives the rate of expansion *today*.

An accelerating universe is older than $1/H_0$.

Decelerating expansion

Accelerating expansion

A decelerating universe is younger than $1/H_0$.

The Big Bang occurs at $R_U = 0$.

Age of accelerating universe — Time →

Hubble time ($1/H_0$)

Age of decelerating universe

Today

Figure 22.5 This graph shows plots of the scale factor R_U versus time for three possible universes. If the universe has expanded at a constant rate, its age is equal to the Hubble time, $1/H_0$. If the expansion of the universe has slowed, the universe is younger than the Hubble time. If the expansion has sped up, the universe is older than $1/H_0$.

physical properties; for example, the vacuum may have nonzero energy even in the total absence of matter. That energy of empty space is called **dark energy**, and it produces exactly the kind of repulsive force that could accelerate the expansion of the universe.

Although the cosmological constant does not evolve with time, other versions of dark energy can evolve. Dark energy is a very active area of study, as scientists work to figure out where that energy comes from, its form and properties, and whether the amount changes. Scientists in an international collaboration called the Dark Energy Survey are working to measure the amount of dark energy and whether it changes with time. That effort combines multiple approaches to measuring dark energy, including finding many Type Ia supernovae, measuring the observed distortions of galaxies from dark matter fluctuations, and using other phenomena sensitive to dark energy.

Dark Energy and the Fate of the Universe

What fate, then, actually awaits the universe? If the cosmological constant is truly constant, and does not evolve in time, the answer is relatively straightforward. When the universe was young and compact, all of the mass was close together, so the effect of gravity was stronger than the effect of the cosmological constant. As the universe expands, the mass spreads out, so gravity grows progressively weaker relative to the cosmological constant. The effect of the cosmological constant becomes increasingly larger than the force of gravity over time, and the expansion of the universe will continue accelerating forever.

Because scientists do not yet understand the origin of dark energy, it is possible that it is not really a constant of nature and instead could be either increasing or decreasing. A changing cosmological constant would significantly change the future of the universe (**Figure 22.4**). For example, if dark energy were to decrease rapidly in the future, mass would once again dominate over dark energy, and turn the current acceleration into a *deceleration*. As mass becomes more dominant, the expansion might eventually reverse, and the universe could collapse to what astronomers call the **Big Crunch**. In contrast, if the effect of dark energy grows stronger with time, the universe would accelerate its expansion at an ever-increasing rate. Ultimately, expansion could be so rapid that the scale factor would become infinite within a finite period—a phenomenon called the **Big Rip**. In the Big Rip, the repulsive force of dark energy would become so dominant that the entire universe would come apart. First, gravity would no longer keep groups of galaxies together; then, gravity would no longer be able to hold individual galaxies together; and so on. Just before the end, the Solar System would come apart, and even atoms would be ripped into their constituent components. Don't worry too much about the Big Crunch or the Big Rip, though: The best observational data seem consistent with constant dark energy. In the most likely outcome, the universe continues expanding, and after 100 trillion years or so, the universe is cold and dark, filled with black holes, dead stars, and dead planets (sometimes called the *Big Chill*).

The Age of the Universe

The values for Ω_m and Ω_Λ not only affect predictions for the future of the universe but also influence how astronomers interpret the past. **Figure 22.5** shows plots of the scale factor of the universe versus time. Measurement of the Hubble constant (H_0) indicates how fast the universe is expanding *today*. That is, the Hubble constant

Reading Astronomy News

Astronomers Find Half of the Missing Matter in the Universe

Hannah Devlin, *The Guardian*

Scientists produce indirect evidence of gaseous filaments and sheets known as Whims linking clusters of galaxies in the cosmic web.

It is one of cosmology's more perplexing problems: that up to 90% of the ordinary matter in the universe appears to have gone missing.

Now astronomers have detected about half of this missing content for the first time, in a discovery that could resolve a long-standing paradox.

The conundrum first arose from measurements of radiation left over from the Big Bang, which allowed scientists to calculate how much matter there is in the universe and what form it takes. This showed that about 5% of the mass in the universe comes in the form of ordinary matter, with the rest being accounted for by dark matter and dark energy.

Dark matter has never been directly observed and the nature of dark energy is almost completely mysterious, but even tracking down the 5% of ordinary stuff has proved more complicated than expected. When scientists have counted up all the observable objects in the sky—stars, planets, galaxies and so on—this only seems to account for between a tenth and a fifth of what ought to be out there.

The deficit is known as the "missing baryon problem," baryons being ordinary sub-atomic particles like protons and neutrons.

Richard Ellis, a professor of astrophysics at the University College London, said: "People agree that there's a lot missing, raising the question where is it?"

The distribution of galaxies in the universe follows a web-like pattern and scientists have speculated that the missing baryons could be floating in diffuse gaseous filaments and sheets linking the galaxy clusters in the cosmic web.

Theoretical calculations suggest these gaseous threads, known as the warm–hot intergalactic medium, or the Whim, ought to be around a million degrees Celsius. A mist of gas at this temperature is too cold to emit X-rays that could be spotted by ordinary telescopes from the Earth—but not cold enough to absorb significant amounts of light passing through it.

"The trouble is, it's in this unusual temperature regime where we can't see it," said Ellis.

Now two separate teams of scientists, one at the University of Edinburgh, the other at the Institute of Space Astrophysics in Orsay, France, have produced compelling indirect evidence for the Whim. Both teams relied on the fact that when radiation travels through a hot gas, it is scattered, meaning that the Whim ought to appear as a dim outline in the cosmic microwave background.

The scientists overlaid observations of the cosmic background radiation, made by the Planck space observatory, and the most detailed three-dimensional map of the cosmic web, created by the Sloane Digital Sky Survey (SDSS). They hypothesized that if

there were gas threads linking galaxy clusters, these should show up in the Planck data.

The Edinburgh team found the regions between galaxies appeared to be about six times as dense as the surrounding bits of space and when summed up, these gaseous threads could amount to about 30% of the ordinary matter in the universe. The French teams' calculation came out at slightly less than this, but the numbers are consistent.

Ellis, who was not involved in either project, describes the findings as "inspirational." "These two papers have been very prominently discussed and people are excited," he added. "The Whim is out there."

The initial measurements still do not account for all the ordinary matter, and some believe the remaining portion could be made up by exotic unobserved objects such as black holes or dark stars. Cosmologists are also still yet to discover the nature of dark matter, which makes up even more of the universe.

QUESTIONS

1. What is "missing"?

2. What is a Whim? How is it detected?

3. Why can't the Whim be seen with X-ray telescopes?

4. Why is it important that two separate teams are reporting that result?

Source: https://www.theguardian.com/science/2017/oct/12/astronomers-find-half-of-the-missing-matter-in-the-universe.

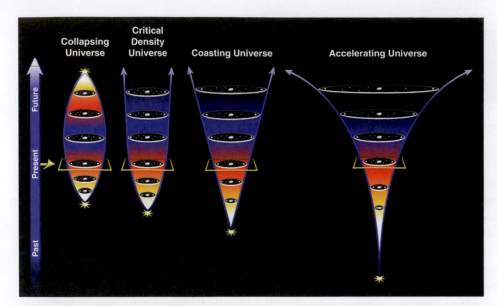

Figure 22.6 Past, present, and future for different scenarios. All models start in the past with a Big Bang (yellow starburst shape), expand to the present, and then have different possible futures. The present age of the universe is different in each model, as in Figure 22.5.

Figure 22.7 ★ WHAT AN ASTRONOMER SEES
Near a massive object, space is positively curved; as light passes through the bent space, its path is warped and twisted. In this image, an astronomer will notice that the foreground cluster of mostly large reddish galaxies are causing long arcs to form from the light coming from background galaxies. An astronomer will know from this gravitational lensing that an enormous amount of mass resides there. But she will also know that, as large as this galaxy cluster is, it still counts as "local" when compared to the universe as a whole. These positively curved spaces that exist "locally" are embedded in a space that may have its own curvature on larger scales.

indicates the slope of the curves in Figure 22.5 at the current time. If the expansion of the universe has not changed, the plot of the scale factor versus time is the straight red line in Figure 22.5. The age of the universe in that case is equal to the Hubble time: $1/H_0$. If the expansion of the universe has been slowing down (green line in Figure 22.5), the universe is actually younger than the Hubble time. If the expansion of the universe has been speeding up (blue line), the true age of the universe is greater than the Hubble time. That scenario is also illustrated in **Figure 22.6**, where looking down the blue time line from the present to the past shows that the different models take different amounts of time to go from the Big Bang to the present.

Recall that a Hubble constant of $H_0 = 70$ km/s/Mpc corresponds to a Hubble time ($1/H_0$) of about 13.8 billion years. If the expansion of the universe has slowed, the universe is actually younger than 13.8 billion years. Having a younger universe is a problem if the measured ages of globular clusters—13 billion years—is correct because globular clusters cannot be older than the universe that contains them. But if the expansion of the universe has sped up, as suggested by the observations of Type Ia supernovae and of the 2.7-K cosmic background radiation, the universe is at least 13.8 billion years old—comfortably older than globular clusters. The effects from gravity and dark energy on the age of the universe have nearly canceled out now.

The Shape of the Universe

We have already discussed such properties of the universe as density, dark energy, and age. The universe also has another key property: its shape. In Chapter 21, we used the metaphor that space is a "rubber sheet" that has stretched outward from the Big Bang. In Chapter 18, we used the same rubber sheet metaphor to show how the mass of a star, planet, or black hole causes a distortion in the shape of space. You saw how the shape of space around a massive object is detected through changes in geometric relationships, such as the ratio of the circumference of a circle to its radius or the sum of the angles in a triangle or whether parallel lines stay parallel. **Figure 22.7** reminds you how the distortion of space affects the light passing through it. Similarly, the mass of everything in the universe—including galaxies, dark matter, and dark energy—distorts the shape of the universe *as a whole*.

Three basic shapes are possible for the universe (**Figure 22.8**). Which shape actually describes the universe is determined by the total amount of mass and energy—in other words, the sum of Ω_m and Ω_Λ. The shapes are easiest to visualize in only two dimensions, as in the rubber sheet metaphor we have been using all along.

The first possibility is a **flat universe** (Figure 22.8a), corresponding to $\Omega_m + \Omega_\Lambda = 1$. A flat universe is described overall by the rules of basic Euclidean geometry. That is, circles in a flat universe have a circumference of 2π times their radius ($2\pi r$), and triangles contain angles whose sum is 180 degrees. A flat universe stretches on forever.

The second possibility, an **open universe** (Figure 22.8b), corresponds to $\Omega_m + \Omega_\Lambda < 1$ and is shaped something like a saddle. In an open universe, the

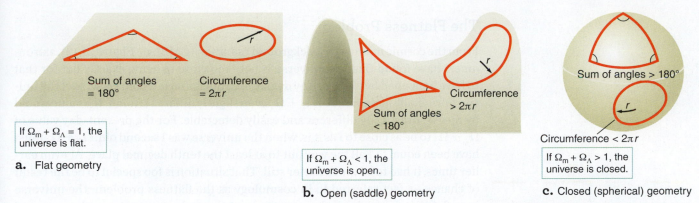

If $\Omega_m + \Omega_\Lambda = 1$, the universe is flat.

a. Flat geometry

If $\Omega_m + \Omega_\Lambda < 1$, the universe is open.

b. Open (saddle) geometry

If $\Omega_m + \Omega_\Lambda > 1$, the universe is closed.

c. Closed (spherical) geometry

Figure 22.8 These two-dimensional representations show possible geometries that space can have in a universe. **a.** In a flat universe, Euclidean geometry holds, so triangles have angles that sum to 180 degrees, and the circumference of a circle equals 2π times the radius. In **b.** an open universe or **c.** a closed universe, those relationships are no longer correct over very large distances.

circumference of a circle is greater than $2\pi r$, and triangles contain less than 180 degrees. An open universe, like a flat universe, is infinite.

The third possibility, a **closed universe** (Figure 22.8c) corresponds to $\Omega_m + \Omega_\Lambda > 1$ and is like the surface of a sphere. The geometric relationships on a sphere are similar to those near a massive object (see Chapter 18). Thus, the circumference of a circle on a sphere is less than $2\pi r$, and triangles contain more than 180 degrees. A closed universe is finite. The universe closes back on itself like the two spatial dimensions on the surface of a sphere. A closed universe is finite, but it has no boundary; it is much like the surface of Earth in that you could walk around and around on Earth forever, and never hit an "edge." The cosmological principle is satisfied by the finite yet unbounded nature of a closed universe. On scales smaller than the size of the observable universe, but larger than the size of superclusters, the space in a closed universe may be flat, and the expansion of the universe maintains both the cosmological principle and the apparent flatness of underlying space.

The measurements to estimate directly which of those shapes describes the universe are difficult. As noted earlier, the data suggest that $\Omega_m = 0.3$ and $\Omega_\Lambda = 0.7$, so $\Omega_m + \Omega_\Lambda$ is close to 1, in which case the universe is very nearly flat.

CHECK YOUR UNDERSTANDING 22.2

Dark energy has been hypothesized to solve which problem? (a) The universe is expanding. (b) The cosmic microwave background radiation is too smooth. (c) The expansion of the universe is accelerating. (d) Stars orbit the centers of galaxies too fast.

22.3 Inflation Solves Several Problems in Cosmology

A century ago, astronomers were struggling to understand the size of the universe. Today scientists have a comprehensive theory that ties together many diverse facts about nature: the constancy of the speed of light, the properties of gravity, the motions of galaxies, and even the origins of the atoms that make up planets and life. The case for the Big Bang is compelling. Even so, improved observations of the cosmic background radiation and measurements of the expansion of the universe have raised some questions about how the universe expanded when it was very young. In this section, we look at those questions and some potential answers.

The Flatness Problem

From the cosmic microwave background radiation (CMB; see Figure 21.12), astronomers have found that the universe is flat—too close to being exactly flat for that to have happened by chance. Any deviation from flatness would grow, so if the universe originally had a value of $\Omega_m + \Omega_\Lambda$ even slightly different from 1, the value would by now be drastically different and easily detectable. For the present-day value of $\Omega_m + \Omega_\Lambda$ to be as close to 1 as it is, when the universe was 1 second old, $\Omega_m + \Omega_\Lambda$ must have been equal to 1 all the way out to at least the tenth decimal place. At even earlier times, it had to be much flatter still. That situation is too special to be the result of chance—a fact referred to in cosmology as the flatness problem: the universe is so flat that some physical property about the early universe must have caused $\Omega_m + \Omega_\Lambda$ to have a value incredibly close to 1.

A universe that contains mass and does not start out perfectly flat has a very different fate. If a universe started out with Ω_m even slightly greater than 1, its expansion would slow more rapidly than that of the flat universe, meaning that less and less density would be required to stop the expansion. Meanwhile, the actual density would be falling less rapidly than in the flat universe. That disparity between the actual density of the universe and the critical density would increase, causing the ratio between the two, Ω_m, to skyrocket, so that gravity would quickly overwhelm the expansion. A universe that starts out even slightly closed rapidly becomes obviously closed and would collapse long before stars could form. Conversely, if a universe started with Ω_m even a tiny bit less than 1, the expansion would slow less rapidly than in a flat universe. As time passed, more and more mass would be required for gravity to stop the too-rapidly-expanding universe. Meanwhile, the actual density of the universe would be dropping faster than in a flat universe. In that case, Ω_m would plummet.

Adding Ω_Λ to the picture makes the math more complex but does not change the basic results. Try balancing a razor blade on its edge. If the blade is tipped just a tiny bit in one direction, it quickly falls that way. If the blade is tipped just a tiny bit in the other direction, it quickly falls in the other direction instead. It would seem that the actual universe should be obviously open or obviously closed—analogous to the tipped razor blade. Instead, the universe has $\Omega_m + \Omega_\Lambda$ so close to 1 that telling which way the razor blade is tipped is difficult—if it is tipped at all. Discovering that $\Omega_m + \Omega_\Lambda$ is extremely close to 1 after more than 13 billion years is like balancing a razor blade on its edge and coming back 10 years later to find that it still has not tipped over.

The Horizon Problem

Another problem faced by cosmological models is that the CMB is surprisingly smooth. After its discovery in the 1960s, many observational cosmologists turned their attention to mapping that background radiation. At first, result after result showed that the temperature of the CMB is remarkably constant, with variations of less than one part in 3000, regardless of where you look in the sky. Over time, though, that strong confirmation of Big Bang cosmology challenged cosmologists' view of the early universe. Once Earth's motion relative to the CMB is removed from the picture, the CMB is not just smooth—it is *too* smooth.

In Chapter 5, we discussed the bizarre world of quantum mechanics that shapes the world of atoms, light, and elementary particles. When the universe was extremely young, it was so small that quantum mechanical effects played a role in shaping the structure of the universe as a whole. The early universe was subject to the quantum mechanical **uncertainty principle**, which says that as a system is studied at extremely small scales, the properties of that system become less and less well determined. That principle applies to the properties of an electron in an atomic orbital or to the entire universe when it was very young and would have fit within the size of an atom.

Consider a simple analogy of how the uncertainty principle applies to the universe. Imagine sitting on the beach looking out across the ocean. Off in the distance, you see more total ocean and average the surface over larger scales; therefore, the surface of the ocean appears smooth and flat. The horizon looks almost like a geometric straight line. Yet the apparent smoothness of the ocean as a whole hides the tumultuous structure present at smaller scales, where waves and ripples fluctuate dramatically from place to place. Similarly, while the universe on large scales seems steady and smooth, quantum mechanics says that conditions must fluctuate unpredictably at smaller and smaller scales in the universe. In particular, quantum mechanics says that those fluctuations are more dramatic earlier and earlier in the history of the universe. When the universe was young, it could not have been smooth. Dramatic variations ("ripples") must have existed in the density and temperature of the universe from place to place.

If the universe had expanded slowly, those ripples would have smoothed themselves out, but the universe expanded much too rapidly for such smoothing to be possible. After the Big Bang, not enough time was available for a smoothing signal to travel from one region to the other. So when cosmologists look at the universe today, they should see the fingerprint of those early ripples imprinted on the cosmic background radiation—but they do not. The fact that the CMB is so smooth is called the *horizon problem* in cosmology. The **horizon problem** states that different parts of the universe are too much like other parts of the universe that should have been "over their horizon" and beyond the reach of any signals that might have smoothed out the early quantum fluctuations. In essence, the horizon problem is as follows: How can different parts of the universe that underwent different fluctuations and were never able to communicate with one another still show the same temperature in the cosmic background radiation to an accuracy of better than one part in 100,000?

Inflation: Early, Rapid Expansion

In the early 1980s, physicist Alan Guth (1947–) offered a solution to the flatness and horizon problems of cosmology. Guth suggested that the universe has not expanded at a steady pace but that instead it started out much more compact than steady expansion would predict. Then, briefly, the young universe expanded at a rate *far* in excess of the speed of light. That rapid expansion of the universe is called **inflation**. In the first 10^{-33} second of the universe, the distance between points in space increased by a factor of at least 10^{30} and perhaps very much more (**Figure 22.9**). In that incomprehensibly brief instant, the size of the observable universe grew from ten-trillionths the size of the nucleus of an atom to a region about 1 meter across. That is like a grain of very fine sand growing to the size of today's observable universe—all in a billionth of the time that light takes to cross the nucleus of an atom. During inflation, space itself expanded so rapidly that the distances between points in space increased faster than the speed of light. Inflation does *not* violate the rule that no signal can travel through space faster than the speed of light because the space itself was expanding.

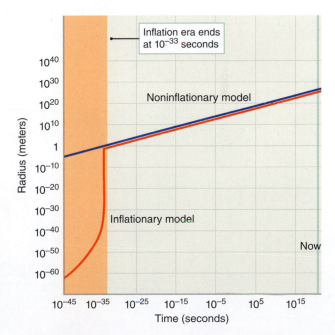

Figure 22.9 This graph illustrates expansion of the observable universe. In an inflationary model, prior to 10^{-33} second after the Big Bang, the universe expanded greatly.

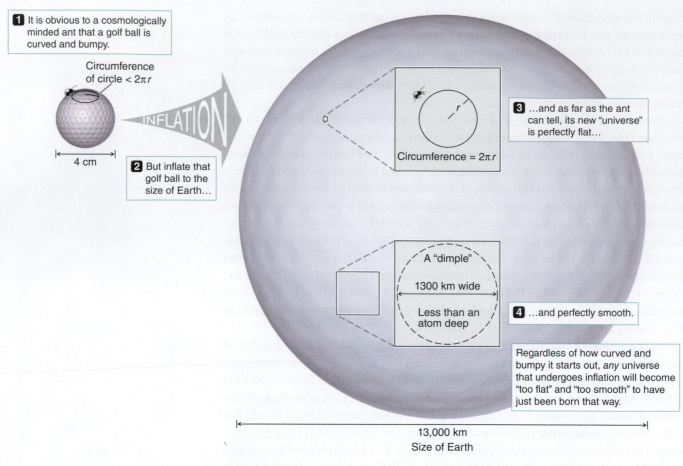

1 It is obvious to a cosmologically minded ant that a golf ball is curved and bumpy.

Circumference of circle < 2πr

4 cm

INFLATION

2 But inflate that golf ball to the size of Earth...

r

Circumference = 2πr

3 ...and as far as the ant can tell, its new "universe" is perfectly flat...

A "dimple"

1300 km wide

Less than an atom deep

4 ...and perfectly smooth.

Regardless of how curved and bumpy it starts out, *any* universe that undergoes inflation will become "too flat" and "too smooth" to have just been born that way.

13,000 km
Size of Earth

Figure 22.10 If a round, lumpy golf ball were suddenly inflated to the size of Earth, the ball would seem flat and smooth to an ant on its surface. Similarly, after inflation, any universe would seem both extremely flat and extremely smooth, regardless of the exact geometry and irregularities it started with.

To understand how inflation solves the flatness and horizon problems of cosmology, imagine that you are an ant living in the two-dimensional universe defined by the surface of a golf ball (**Figure 22.10**). This universe is positively curved, like the surface of Earth. If you were to walk around the circumference of a circle in your two-dimensional universe and then measure the radius of the circle, you would find the circumference to be less than $2\pi r$. If you were to draw a triangle in your universe, the sum of its angles would be greater than 180 degrees. Another obvious characteristic would be the dimples, approximately a half-millimeter deep, on the surface of the golf ball.

Now imagine that the golf-ball universe suddenly grew to the size of Earth. The curvature of the universe would no longer be apparent. An ant (or person) walking along the surface of the golf-ball universe would think the universe is flat. The circumference of a circle would be $2\pi r$, and a triangle would have 180 degrees. (In fact, it took most of human history for people to realize that Earth, which is not very large cosmologically speaking, is a sphere.) For inflationary cosmology, the universe after inflation would be extraordinarily flat (that is, having $\Omega_m + \Omega_\Lambda$ extraordinarily close to 1) *regardless* of what the geometry of the universe was before inflation. Because the universe was inflated by a factor of at least 10^{30}, $\Omega_m + \Omega_\Lambda$ immediately after inflation must have been equal to 1 within one part in 10^{60}, which is flat enough for $\Omega_m + \Omega_\Lambda$ to remain close to 1 today. If inflation occurred, today's universe is not flat by chance. It is flat because any universe that underwent inflation would become flat.

What about the horizon problem? When the golf-ball universe inflates to the size of Earth, the dimples that covered the surface of the golf ball stretch out as well. Instead of being a half millimeter or so deep and a few millimeters across, those dimples now are only an atom deep but are hundreds of kilometers across. The ant would not detect any dimples at all. For our universe, inflation took the large fluctuations in conditions caused by quantum uncertainty in the preinflationary universe and stretched them out so much that they are not measurable in today's postinflationary local universe. The slight irregularities observed in the CMB are the faint ghosts of quantum fluctuations that occurred as the universe inflated.

An early era of inflation in the history of the universe offers a way to solve the horizon and flatness problems, but the idea that the universe should have undergone a period during which it expanded at such a high rate seems remarkable. The cause of inflation must lie in the fundamental physics that governed the behavior of matter and energy at the earliest moments of the universe, but current physical theories have not yet been able to predict the details of an early inflationary phase of the universe. Although the existence of an inflationary epoch is difficult to test, it is not impossible, and astronomers are devising ways to test whether inflation occurred in the early universe. Note that one prediction of inflation is that the universe will be filled with a background of gravitational radiation, just as it is filled with the cosmic background radiation.

CHECK YOUR UNDERSTANDING 22.3

Identify the two problems of cosmology that inflation solves.

what if . . .

What if inflation never happened? What could you conclude about conditions right after the Big Bang?

22.4   The Earliest Moments of the Universe Connect the Very Largest Size Scales to the Very Smallest

At first glance, particle physics and cosmology might seem to have almost nothing in common. Particle physics is the study of subatomic particles, which are smaller than atoms. Whereas particle physics looks at the quantum mechanical world that exists on the tiniest scales imaginable, cosmology is the study of the changing structure of a universe that extends for billions of parsecs and probably much farther. Yet cosmologists and particle physicists have come to realize that the structure of the universe and the fundamental nature of matter are related. In this section, we describe the earliest moments of the universe, when atoms had not yet formed, and how the interaction among those subatomic particles governed the conditions and events that took place.

The Forces of Nature

Understanding the universe requires understanding the forces that govern the behavior of all matter and energy in it. Nature has four fundamental forces, and

Table 22.1   The Four Fundamental Forces of Nature

Force	Relative Strength	Range of Force	Particles That Can Carry the Force	Example of What the Force Does
Strong nuclear	1	10^{-15} m	Gluons	Holds protons and neutrons together in atomic nuclei.
Electromagnetic	10^{-2}	Infinite	Photons	Binds the electrons in an atom to the nucleus.
Weak nuclear	10^{-4}	10^{-16} m	W^+, W^-, and Z^0	Responsible for beta decay, in which a positron or electron is emitted from the nucleus.
Gravitational	10^{-38}	Infinite		Holds you to Earth; binds planetary systems, stars, galaxies, clusters of galaxies, and so forth.

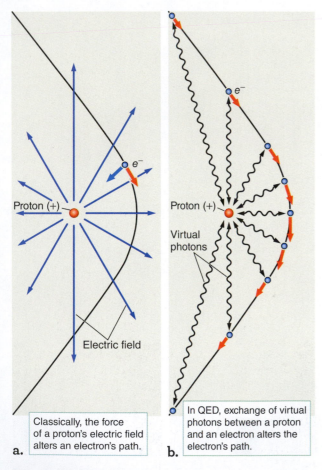

Proton (+)

Electric field

Proton (+)

Virtual photons

Classically, the force of a proton's electric field alters an electron's path.

a.

In QED, exchange of virtual photons between a proton and an electron alters the electron's path.

b.

Figure 22.11 a. This figure shows the classical view of an electron being deflected from its course (black line) by the electric field from a proton. **b.** According to quantum electrodynamics models, the interaction is viewed as an ongoing exchange of virtual photons between the two particles.

everything in the universe is a result of their action (**Table 22.1**). The **electromagnetic force**, which includes both electric and magnetic interactions, acts on charged particles such as protons and electrons. That force governs not only chemistry but also light. The **strong nuclear force** that binds the protons and neutrons in the nuclei of atoms governs reactions such as the fusion of hydrogen in the heart of the Sun (see Chapter 14). The **weak nuclear force** governs the radioactive decay of unstable nuclei. Finally, *gravity*, which plays a major role throughout astronomy, governs how matter affects the geometry of spacetime. In models of particle physics, the first three forces combine in a single force, leaving only gravity to stand alone. To understand how those four forces formed and came to govern the universe today, we must explore backward in time, toward the Big Bang itself.

The underlying concept of the standard model is that forces between particles are caused by the exchange of carrier particles. Light can be described either as an electromagnetic wave resulting from electric and magnetic fields or as a stream of particles called photons (see Chapter 5). Those two descriptions of electromagnetism coexist. The branch of physics that deals with a quantum description of radiation is called **quantum electrodynamics (QED)**.

QED treats charged particles almost as though they were baseball players engaged in an endless game of catch. As the players throw and catch baseballs, they experience forces. Similarly, in QED, charged particles "throw" and "catch" an endless stream of "virtual photons" (**Figure 22.11**). Quantum mechanics is a science of probabilities rather than certainties. The QED description of the electromagnetic interaction between two charged particles is an average of all the possible ways that the particles could throw photons back and forth. The resulting force acts, over large scales, like the classical electromagnetic force that we observe in chemistry.

The central idea of QED—that forces are really caused by the exchange of carrier particles—serves as a template for understanding two of the other three fundamental forces in nature. The electromagnetic and weak nuclear forces have been combined into a single theory called **electroweak theory**. That theory predicts the existence of three particles—labeled W^+, W^-, and Z^0—that mediate the weak nuclear force. Sheldon Glashow, Abdus Salam, and Steven Weinberg received the 1979 Nobel Prize in Physics for their work on the theory of the unified weak and electromagnetic forces. In the 1980s, physicists identified those particles in laboratory experiments and confirmed the essential predictions of electroweak theory.

The strong nuclear force is described by a distinct theory called **quantum chromodynamics (QCD)**. QCD states that particles such as protons and neutrons are composed of more fundamental building blocks called **quarks**, which are bound together by the exchange of another type of carrier particle, dubbed **gluons**. Together, electroweak theory and QCD comprise the **standard model** of particle physics. Excluding gravity, the standard model explains all the currently observed interactions of matter and has made many predictions that laboratory experiments later confirmed. Despite that great success, the standard model leaves many questions unanswered, such as whether neutrinos have mass or why strong interactions are so much stronger than weak interactions.

At high temperatures, the different forces are indistinguishable, because at those temperatures, the different carrier particles have such high energy. Therefore, our universe started out with all the forces unified into one force, as described by one (as yet unknown) theory of everything. As the universe expanded and cooled, the particles cooled, lost energy, and became distinguishable from one another. Physicists refer to this as "breaking the symmetry" of these particles. In the standard model of particle physics, a special particle called the Higgs boson creates the Higgs field that breaks the symmetry between the different types of carrier particles.

The Higgs boson is the particle that all other particles must interact with to have mass. In the standard model, all particles are created without mass. When the electroweak symmetry breaks (as expansion and cooling continue), that "special" particle is created throughout the universe. All existing particles interact with the Higgs field and gain their mass in the process. The existence of that particle was predicted in 1964, and it was finally detected at the Large Hadron Collider in Europe in 2012. In 2013, Peter Higgs and François Englert shared the Nobel Prize for their work predicting that particle. As the universe cooled, the carrier particles (and other particles, such as electrons) gained mass due to the Higgs field, and so the forces also began to be distinguishable.

A Universe of Particles and Antiparticles

In the standard model, every particle in nature has an **antiparticle** that is identical in mass. For example, the positron emitted in the proton-proton chain (see Chapter 14) is the antiparticle of an electron. A positron is identical to an electron except that it has a positive charge instead of a negative charge. For the proton, there is the antiproton; for the neutron, the antineutron; and so on down the list. Collectively, those antiparticles are called **antimatter**.

One property of particle-antiparticle pairs is that if you bring such a pair together, the two particles annihilate each other. When a particle-antiparticle pair annihilates, the mass of the two particles is converted into energy in accordance with Einstein's equation $E = mc^2$. In **Figure 22.12a**, for example, an electron and a positron annihilate each other, and the energy is carried away by a pair of gamma-ray photons. Particle-antiparticle pairs were produced when two high-energy photons collided with each other (**Figure 22.12b**), creating in their place an electron-positron pair. **Pair production**—when an energetic collision creates a particle and its corresponding antiparticle—has been observed in particle accelerators.

In principle, *any* type of particle and its antiparticle can be created by pair production. The only limitation comes when not enough energy is available to supply the mass of the particles being created (**Working It Out 22.2**). If two gamma-ray photons with a combined energy greater than the rest mass energy of an electron-positron pair collide, the two photons may disappear and leave an electron-positron

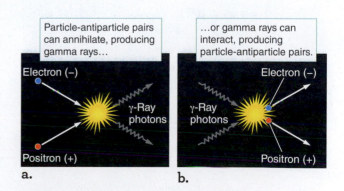

Figure 22.12 a. An electron and a positron annihilate, creating two gamma-ray photons that carry away the energy of the particles. **b.** In the reverse process, pair production, two gamma-ray photons collide to create an electron-positron pair.

working it out 22.2

Pair Production in the Early Universe

The early universe was hot and awash in a bath of Planck blackbody radiation. How did that radiation become particles? Recall Einstein's equation for the conversion of mass to energy and energy to mass:

$$E = mc^2$$

To produce a pair consisting of a particle plus an antiparticle of a certain rest mass, a minimum amount of energy is required. The formula shows that producing a higher-mass particle plus antiparticle requires more energy than does producing a lower-mass pair. For example, a proton has 1836 times the mass of an electron, so it will require 1836 times as much energy to produce a proton-antiproton pair of particles than to produce an electron-positron pair.

We can relate that amount of energy to the average energy of particles at a given temperature through the following equation:

$$E = (3/2)kT$$

where k is Boltzmann's constant: 1.38×10^{-23} kg m^2/s^2/K (which is joules per kelvin; J/K). Equating the two energies yields

$$mc^2 = (3/2)kT$$

Rearranging,

$$T = \frac{2mc^2}{3k}$$

The proton and antiproton each have a mass of 1.67×10^{-27} kg. What temperature would the radiation have to be to produce that proton-antiproton pair?

$$T = \frac{2 \times (2 \times 1.67 \times 10^{-27}\,\text{kg}) \times (3.0 \times 10^8\,\text{m/s})^2}{3 \times (1.38 \times 10^{-23}\,\text{kg m}^2/\text{s}^2/\text{K})}$$

$$T = 1.45 \times 10^{13}\,\text{K}$$

Such high temperatures are thought to have existed only during the first few seconds after the Big Bang.

Similarly, to produce an electron-positron pair, each of which has a mass of 9.11×10^{-31} kg, the temperature of the CMB would have to be

$$T = \frac{2 \times (2 \times 9.11 \times 10^{-31}\,\text{kg}) \times (3.0 \times 10^8\,\text{m/s})^2}{3 \times (1.38 \times 10^{-23}\,\text{kg m}^2/\text{s}^2/\text{K})}$$

$$T = 7.9 \times 10^9\,\text{K}$$

That is still very hot, but less than the temperature required to form a proton-antiproton pair, so electron-positron production lasted several minutes, a longer time after the Big Bang than did proton-antiproton production.

We also can think of that in terms of the energy of the photons involved in creating those particles. From Chapter 5, recall that the energy of a photon is related to its wavelength by

$$E = \frac{hc}{\lambda} \quad \text{or} \quad \lambda = \frac{hc}{E}$$

where h is Planck's constant ($h = 6.63 \times 10^{-34}$ kg m^2/s). We then use $E = (3/2)kT$ with our value of T for the electron-positron production above, yielding

$$\lambda = \frac{2hc}{3kT} = \frac{2 \times (6.63 \times 10^{-34}\,\text{kg m}^2/\text{s}) \times (3.0 \times 10^8\,\text{m/s})}{3 \times (1.38 \times 10^{-23}\,\text{kg m}^2/\text{s}^2/\text{K}) \times (7.92 \times 10^9\,\text{K})}$$

$$\lambda = 1.21 \times 10^{-12}\,\text{m}$$

According to the electromagnetic spectrum shown in Figure 5.6, photons with that wavelength are high-energy gamma-rays.

pair behind in their place. If the photons have more than the necessary energy, the extra energy goes into the kinetic energy of the two newly formed particles.

Now let's apply that idea to a hot universe full of blackbody radiation. When the universe was less than about 100 seconds old and had a temperature greater than a billion kelvins, it was filled with energetic photons constantly colliding, creating electron-positron pairs, and those electron-positron pairs were constantly annihilating each other, creating pairs of gamma-ray photons. Those two processes reached equilibrium, determined strictly by temperature, in which pair creation and pair annihilation exactly balanced each other. Instead of being filled only with a swarm of photons, the universe then was filled with a swarm of photons, electrons, and positrons. Earlier, when the universe was even hotter, photons would have produced a swarm of protons and antiprotons. Still earlier, a swarm of quarks/antiquarks and gluons existed, called a "quark-gluon plasma," as has been observed in some heavy-nucleus accelerators on Earth.

Grand Unified Theories

In pair production, a symmetry exists between matter and antimatter: for every particle created, its antiparticle is created as well. As the universe cooled, there was no longer enough energy to create particle pairs, so the swarm of particles and antiparticles that filled the early universe annihilated each other and were not replaced. When that cooling happened, first every proton should have been annihilated by an antiproton. Then at still cooler temperatures, every electron should have been annihilated by a positron. That was almost the case, but not quite. For every proton and electron in the universe today, 10 billion and one protons and electrons existed in the early universe, but only 10 billion antiprotons and positrons. That one-part-in-10-billion excess of electrons over positrons meant that when electron-positron pairs finished annihilating each other, some electrons were left over—enough to account for all the electrons in all the atoms in the universe today (**Figure 22.13**). Similarly, the early universe had an excess of protons over antiprotons, and the protons observed today are all that is left from the annihilation of proton-antiproton pairs.

If the standard model of particle physics were a complete description of nature, the imbalance of one part in 10 billion between matter and antimatter would not have been present in the early universe. The symmetry between matter and antimatter would have been complete. No matter at all would have survived into today's universe, and galaxies, stars, and planets would not exist. The fact that you are reading this page demonstrates that something more needs to be added to the model.

Several competing ideas seek to explain why there was more matter than antimatter in the early universe. One set of ideas is called **grand unified theories (GUTs)** because they join the electromagnetic force, weak nuclear force, and strong nuclear force together into a single force. Such theories do not include gravity, however. GUTs explain why the universe is composed of matter rather than antimatter. When the universe was very young (younger than about 10^{-35} second) and very hot (hotter than about 10^{27} K), enough energy was available for particles associated with a GUT to be freely created. During that time, the distinction among the electromagnetic, weak nuclear, and strong nuclear forces had not yet taken place. Only the one unified force existed. During that era of GUTs, the apparent size of the entire observable universe was less than a trillionth the size of a single proton.

Many possible GUTs exist, and they make many predictions about the universe. Unfortunately, most of those predictions are impossible to test with even the largest of today's particle colliders. The problem is that the particles carrying the grand unification forces are so massive that creating them takes enormous amounts of energy—roughly a trillion times as much energy as can be achieved in today's particle accelerators. Even so, some predictions of GUTs can be tested with current technology. For example, GUTs predict that protons should be unstable particles that, given enough time, will decay into other types of elementary particles. That process is *very* slow. Over 100 years, GUTs predict that as much as a 1 percent chance may exist that *one* of the 10^{28} or so protons in your body will decay. Experiments have given a lower limit on the lifetime of a proton of 10^{34} years, but as of this writing, proton decay has yet to be observed.

what if . . .

What if dark matter particles have dark matter antiparticles associated with them? Is our universe consistent with there having been an excess of dark matter particles over dark matter antiparticles, just as for ordinary matter?

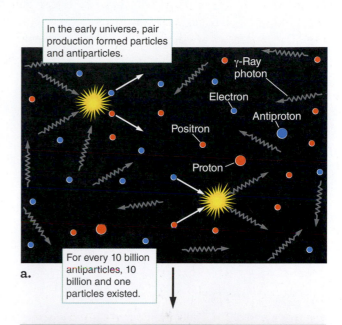

In the early universe, pair production formed particles and antiparticles.

γ-Ray photon

Electron

Antiproton

Positron

Proton

a.

For every 10 billion antiparticles, 10 billion and one particles existed.

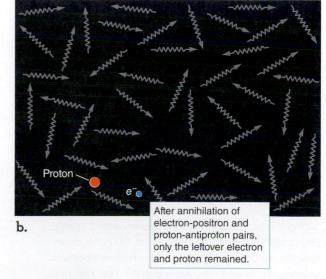

Proton

e^-

After annihilation of electron-positron and proton-antiproton pairs, only the leftover electron and proton remained.

b.

Figure 22.13 a. For every 10 billion antiparticles in the early universe, 10 billion and one particles existed. **b.** After those particles annihilated, only one electron and one proton remained.

How does gravity fit into that scheme? General relativity provides a description of gravity that correctly predicts the orbits of planets, describes the ultimate collapse of stars, and enables astronomers to calculate the structure of the universe. Yet general relativity's description of gravity is very different from the theories of the other three forces. Instead of talking about the exchange of photons or gluons or other mediating particles, general relativity talks about the large, smooth, continuous canvas of spacetime that events are painted on. The era of GUTs is described perfectly if gravity is treated as a force separate from the other three forces. However, a time even closer to the Big Bang exists when a description of the universe requires that gravity be treated as a quantum phenomenon.

Toward a Theory of Everything

Even earlier in the universe than when the three forces were unified, when the universe was younger than about 10^{-43} second, its density was incomprehensibly high. The observable universe was so small that 10^{60} universes would have fit into the volume of a single proton. Under those extreme conditions, general relativity can no longer describe spacetime: quantum physics is needed to describe not only particles but also spacetime itself. Rather than a smooth sheet, spacetime was a quantum mechanical "foam." The failure of general relativity to describe that early universe is much like the failure of Newtonian mechanics to describe the structure of atoms. An electron in an atom must be thought of in terms of probabilities rather than certainties. Similarly, no deterministic history is available for the earliest moments after the Big Bang. That era in the history of the universe is called the **Planck era**, signifying that physicists can understand the structure of the universe during that period only by using the ideas of quantum mechanics.

The conflict between general relativity and quantum mechanics is at the current limits of human knowledge. Known physics can explain things back to when the universe was a ten-millionth of a trillionth of a trillionth of a trillionth of a second old, but to push back any earlier, something new is needed. To understand the earliest moments of the universe, physicists need a theory that combines general relativity and quantum mechanics into a single theoretical framework unifying all *four* fundamental forces—a **theory of everything (TOE)**.

A successful TOE would do more than unify general relativity with quantum mechanics. It would suggest which of the possible GUTs is correct and would tell us the nature of dark matter. A successful TOE would also explain the how, when, and why of inflation and the underlying physics of the dark energy accelerating the expansion of the universe. Physicists are grappling with what a TOE might look like. One leading contender is superstring theory (see Section 22.5).

The Forces Separated in the Cooling Universe

To understand the very earliest moments in the history of the universe, physicists look backward to earlier and earlier times and, consequently, to higher and higher energies. Now let's organize the events the other way, beginning at the beginning.

In the early universe, the TOE could unite all four forces, and as the universe expanded and cooled, the various forces emerged separately. **Figure 22.14** illustrates how the four fundamental forces emerged in the evolving universe. In the first 10^{-43} second after the Big Bang, as described by the TOE, the physics of elementary particles and the physics of spacetime were one and the same. As the universe expanded and cooled, gravity separated from the forces described by the GUT. Spacetime

unanswered questions

What mechanism leads to an extra electron for every 10 billion electron-positron pairs in the early universe? Here in Chapter 22, we mentioned that grand unified theories predict an asymmetry between particles and antiparticles, as well as proton decay. But physicists do not know which GUT is the correct one, and so far they have not actually observed proton decay. Measuring the actual lifetime of a proton would enable physicists to determine the correct GUT and therefore the mechanism leading to particle-antiparticle asymmetry. In addition, if the correct TOE was really understood, that theory could predict which GUT describes the universe and therefore the mechanism and amount of asymmetry.

took on the properties described by general relativity. Inflation also may have been taking place then.

As the universe continued to expand and its temperature decreased further, less and less energy was available to create particle-antiparticle pairs. When the particles responsible for mediating GUT interactions could no longer form, the strong force split off from the others. As the unity of the original TOE was lost, the symmetry between matter and antimatter was broken. As a result, the universe ended up with more matter than antimatter.

The next big change took place at one 10-trillionth of a second when the temperature of the universe had fallen to 10^{16} K. At that time, particles responsible for unifying the electromagnetic and weak nuclear forces separated, leaving those two forces independent of one another. All four fundamental forces of nature that govern today's universe were then separate. It was a full minute or two later before the universe cooled to the billion-kelvin mark, below which not even pairs of electrons and positrons could form.

The universe then was too cool to form additional particles and their antiparticles. It was still hot enough, though, for the fast-moving protons to overcome the electric barriers between them, allowing nuclear reactions to take place. Those reactions formed the least massive elements and isotopes, including deuterium, helium, and lithium. Recall from stellar evolution that increasingly high temperatures are needed for the nucleosynthesis of increasingly heavy elements. So as the universe continued to expand, it soon became too cool for the nucleosynthesis of more massive elements.

The Big Bang nucleosynthesis discussed at the end of Chapter 21 ended by the time the universe was about 15–20 minutes old and the temperature of the universe had dropped below about 800 million K. By then the density of the universe had fallen to only about a tenth that of water. Normal matter consisted of atomic nuclei and electrons, but the universe was still dominated by the radiation from the Big Bang. After several hundred thousand years, the temperature dropped so low that electrons could combine with atomic nuclei to form neutral atoms. That was the era of recombination, which is seen directly when astronomers observe the CMB. At that stage, the radiation background could no longer dominate over matter: the universe became transparent, and light could move freely through it for the first time. In addition, gravity began playing its role in forming the vast structure of the universe that is now observed.

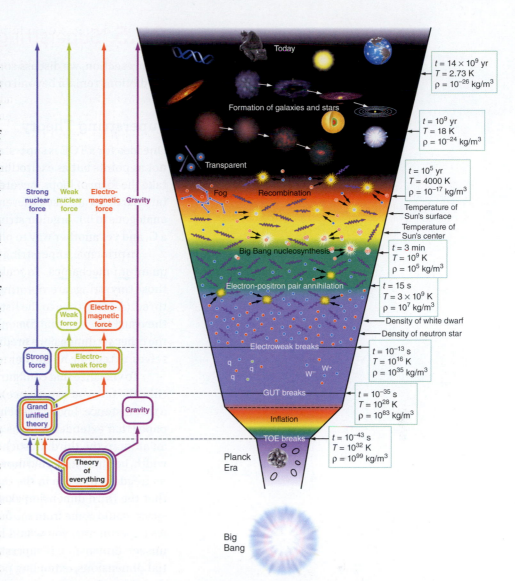

Figure 22.14 This figure conceptualizes eras in the evolution of the universe. The left side shows the four forces separating in stages after the Big Bang. The right side shows the temperature and density of the universe at different key times. As it expanded and cooled after the Big Bang, the universe went through phases determined by what types of particles could be created freely at that temperature. Later, the structure of the universe was set by the gravitational collapse of material to form galaxies and stars and by the chemistry made possible by elements formed in stars.

CHECK YOUR UNDERSTANDING 22.4

Most antimatter in the early universe: (a) is still around today, filling the space between galaxies; (b) became dark matter; (c) formed antimatter galaxies and stars; (d) annihilated with matter.

22.5 Superstring Theory and Multiverses

In this section, we discuss some speculative ideas in theoretical cosmology whose predictions remain beyond our current abilities to test.

Superstring Theory

One idea for a TOE is superstring theory, in which elementary particles are viewed not as points but as excitations or waves traveling along tiny "strings," which may take the form of loops. According to superstring theory, different types of elementary particles are like different "notes" played by vibrating loops of string. That is analogous to how a guitar string vibrates in one way to play an F, another way to play a G, and yet another way to play an A.

In principle, superstring theory offers a way to reconcile general relativity and quantum mechanics. To make superstring theory work, physicists imagine that those tiny strings are vibrating in a universe with nine spatial dimensions instead of three. (Adding time to the list would make the universe 10-dimensional.) Whereas the usual three spatial dimensions spread out across the vastness of the universe, the other six dimensions predicted by superstring theory wrap tightly around themselves, extending no further today than they did a brief instant after the Big Bang.

To visualize that configuration, imagine living in a three-dimensional universe (like the one you experience) in which one of those dimensions extended for only a tiny distance. Living in such a universe would be like living within a thin sheet of paper that extended billions of parsecs in two directions but was far smaller than an atom in the third. In such a universe, you would easily be aware of length and width, because you could move in those directions at will. However, you would have *no* freedom to move in the third dimension at all, and you might not even realize that the third dimension existed. Perhaps your only inkling of the true nature of space would come from the fact that in order to explain the results of particle physics experiments, you would have to assume that particles extended into a third, unseen dimension. If superstring theory is correct, everyone now sees three spatial dimensions, extending possibly forever, but doesn't notice that each point in that three-dimensional space actually has a tiny but finite extent in the six other dimensions at the same time.

Although called a "theory," superstring theory is not like the well-tested theories that have been discussed throughout this book. Instead, it is no more than a promising idea that gave direction to theorists searching for a TOE. Therefore, calling superstring theory a "string hypothesis" or a "string idea" would be more consistent with the definitions of idea, hypothesis, and theory given in Chapter 1. Physicists will probably never be able to build particle accelerators that enable them to search directly for the most fundamental particles predicted by a TOE: The energies required are simply too high. Other ways to test the theory may be available, however, and some progress may be made by studying the ultimate particle accelerator: the Big Bang itself.

Multiverses

Is our universe the only one? Because we've defined the universe as "everything," what does it mean to say "multiple universes"? Do parallel universes—either separated in space or even occupying exactly the same space as "our" universe—exist? Can experiments or observations test the idea of multiple universes, or **multiverses**?

Those ideas are speculative, but some cosmologists think seriously about the idea of collections of parallel universes.

The simplest example of such parallel universes is illustrated in **Figure 22.15**. The age of our universe—that is, the time that has passed since the Big Bang—is 13.8 billion years. The universe has been expanding since then; the current distance to an object that emitted light 13.8 billion years ago is about 47 billion light-years. Imagine freezing the expansion of the universe today. Then the distance between us and a source of light emitted at the time of the Big Bang that we see today is the comoving size of the universe. That observable universe—everything astronomers can possibly observe today—is within a sphere of radius of ~47 billion light-years. Anything farther than that is outside the observable universe and cannot be seen. As we discussed previously in the chapter, the observational evidence suggests that the geometry of space is flat. A flat universe is infinite in size and must therefore contain an infinite number of similar spheres. As dark energy causes the universe to expand faster and faster, the separate observable universes move farther apart and will never overlap. Those parallel universes are simply too far away to ever be observed from Earth, and they are moving farther away all the time.

What are these other parallel universes like? Physicists suggest several possibilities from what has been learned about the observable universe. First, if the cosmological principle holds, on large scales each of those observable universes should look pretty much like our own, although the details could be very different. In a truly infinite universe, an infinite number of observable universes exactly like this one must exist, with exact copies of you reading identical versions of *21st Century Astronomy*. The argument is that if our own observable universe is cooler than about 10^8 K everywhere, no more than 10^{118} particles can be in the observable universe, and those particles can be distributed in only so many ways. If you then ask how far you must go before you will find an observable universe *identical to* our own, the answer is about $10^{10^{118}}$ m. (Yes, that's 10 raised to the power 10^{118}. This is so large that the units don't actually matter. The number in light-years is the same as the number in meters for far more digits than your calculator holds.) So, in an infinite universe—as enormous as it might be—the nearest identical universe must be at about that distance.

Other types of multiverses include those in which a universe undergoes **eternal inflation**, with no beginning or end to the inflation. If such a universe exists, small quantum fluctuations may cause some regions to expand more slowly than the rest of the universe. As a result, such a region may form a bubble whose inflating phase ends quickly. In that scenario, Earth is inside such a region, and "our" Big Bang would just be the condensation of our bubble within the eternally inflating universe. That type of multiverse neatly answers the question of what existed before the Big Bang. Because the universe has been inflating and will continue to inflate forever, it has no beginning or end. "Our" own bubble or parallel universe separated from the rest of the universe at a time called the Big Bang, but other bubbles are constantly separating and becoming their own parallel-universe big bangs.

In another type of multiverse derived from quantum physics, which describes a probabilistic universe, each event in the universe spawns multiple universes in which each possible outcome of the event exists. Although cosmologists consider that phenomenon at the level of particle interactions, it is more simply explained at the human scale: You made a decision about having breakfast this morning. That event caused two universes—one in which you did have breakfast and one in which you didn't. Which one are you in while you read this book? Yet another type of

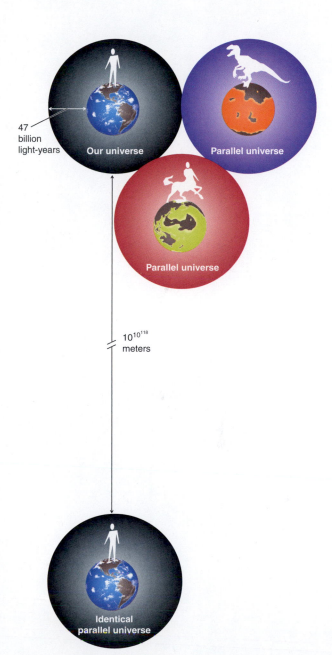

Figure 22.15 The observable universe is a sphere with a radius equal to the distance light has traveled since the Big Bang (47 billion light-years). Because the universe is infinite, an infinite number of similar spheres must exist. The rules of probability dictate that some of those spheres are *exactly* like our own.

multiverse is characterized by parallel universes that have different mathematical structure to describe the different physics within those universes. In that all-encompassing case, almost any behavior for the universe is possible.

Multiverses are common in science fiction and in popular science books and form a type of common mythology about alternative realities that is explored with enthusiasm in popular culture. Many astronomers, though, ask whether the idea of parallel universes, or multiverses, is really science. Throughout this book, we have emphasized that any legitimate scientific theory must be testable and ultimately falsifiable. Do tests exist that can prove those multiverse ideas to be wrong? Possibly. For example, the first multiverse we described, involving distinct observable bubbles, is tested when astronomers measure the isotropy of the CMB, the flatness of the universe, or the large-scale distribution of galaxies. The eternal inflation model is difficult to test because observing parallel universes directly is impossible, though such a theory could produce a feature in the cosmic background radiation. But if physicists obtain a TOE that predicts eternal inflation, and if that theory is itself falsifiable, a connection exists between eternal inflation and observation.

Considerable debate occurs within the scientific community as to whether the multiverse hypotheses, except for the first type, can be tested and falsified. Even if some tests are possible, the concern is that they are not truly meaningful. For example, you might test Newton's theory of gravity by releasing an apple and watching whether it falls upward or downward. But that test is not very discerning: It might not always distinguish between two or more sensible theories. Similarly, for the multiverse hypothesis, whether the tests that seem possible would be meaningful tests of the theory is still debatable.

CHECK YOUR UNDERSTANDING **22.5**

Which of the following statements about multiverses are true? (Choose all that apply.) (a) They represent a scientific theory only if they can be tested and falsified. (b) They might explain what happened before the Big Bang. (c) They are an inevitable consequence of modern observational cosmology. (d) Two identical multiverses can exist.

Origins: Our Own Universe Must Support Life

In some models of an inflationary universe, each bubble could contain different values of the fundamental constants of physics. In some bubble universes, for example, the strength of the nuclear force might be larger, the electric charge might be smaller, and the gravitational constant G might be much smaller than in others. To address what are the properties of our universe, scientists invoke the **anthropic principle**, which states that our universe (or our bubble in the universe) must have physical properties that allow intelligent life to develop. Because humans exist, are intelligent, and can observe the surrounding universe, our universe must have the properties that would allow intelligent life to evolve. That is, this universe must have had the right physical properties and existed long enough for atoms, stars, galaxies, planets, and life to have formed; otherwise, we wouldn't be here to observe it.

For a multiverse that contains bubbles with different physical constants in each of them, the anthropic principle provides information about the values of those physical constants. Consider a few examples. In a bubble universe where the gravitational constant G was much bigger than G as measured in our universe, stellar evolution would occur much faster, and intelligent life might not have time to evolve on a planet before its star burned itself out. Similarly, the anthropic principle provides a relatively narrow range for the strength of the strong nuclear force that holds nuclei together. If that force were much weaker than what physicists now measure, nuclei would not be able to overcome their electric repulsion in order to fuse. Without nuclear fusion, stars could not shine, heavy elements would not form, and planets and life as we know it could not evolve. Alternatively, if the nuclear force were stronger, two protons would more easily fuse in the early universe; thus, most hydrogen would fuse to helium, and less hydrogen would exist to form water and organic molecules that are necessary for life as we know it.

As another example of the anthropic principle applied to multiverses, recall the fraction of the critical density provided by the cosmological constant, Ω_Λ. If Ω_Λ were 20 times larger than what is observed now, the universe would have begun accelerating much earlier, and galaxies would not have had time to form. Without galaxies, little star formation would occur. Therefore, stellar nucleosynthesis would not have taken place, and rocky planets and life could not have evolved. Alternatively, if the cosmological constant were negative, this entire bubble universe would have reached a maximum size and begun to collapse even before galaxies and stars could have evolved. A universe with an intermediate value might last only a few billion years—enough time for stars and galaxies to be established, but perhaps not enough time for sufficient amounts of heavy elements to form or for life to evolve to intelligence. As those examples illustrate, the cosmological constant must be within a particular range of values to allow the intelligent life that exists on Earth to evolve.

SUMMARY

Both gravity and the cosmological constant (or dark energy) determine the fate of the universe. Observations indicate that instead of slowing down, the expansion of the universe is accelerating. The best explanation for that phenomenon is dark energy. The very early universe may have gone through a brief but dramatic period of exceptionally rapid expansion, known as inflation. If so, inflation would explain both the flatness and the uniformity of the universe we see today. During the very earliest moments in the universe, the four fundamental forces of nature were unified. Ideas about multiple universes have been proposed, but none are yet directly testable. Our observable universe must be one in which physics can support the formation of life.

① Connect the mass within the universe and the gravitational force it produces to the history, shape, and fate of the universe. The rate of expansion of the universe is affected by its density (mass/volume). Mass can slow and even stop the expansion of the universe. The density also affects the shape of the universe: If density is high, the universe is shaped more like a sphere; if density is low, the shape is more like a saddle. In between, at the critical density, the universe is flat.

② List the evidence for the accelerating expansion of the universe. Recent observations of distant Type Ia supernovae suggest that the expansion of the universe was slower in the past. That means the expansion is speeding up, implying that a force must be acting to increase the expansion rate. That force may come from dark energy: the energy of empty space. Because the rate of expansion of the universe is increasing, the universe will probably expand forever.

③ Describe the early period of rapid expansion of the universe known as inflation. Inflationary models, which state that the universe expanded very rapidly in a very short time soon after the Big Bang, were proposed to solve several problems in cosmology. Astronomers are looking for an observational signature of inflation.

④ Relate the events that occurred in the earliest moments of the universe to the forces that operate in the modern universe. During the earliest moments of the universe, the four forces (gravity, electromagnetism, strong nuclear, and weak nuclear) split off, each becoming separate at a different time. Modern physicists search for the combined theory of all four forces: the theory of everything.

QUESTIONS AND PROBLEMS

TEST YOUR UNDERSTANDING

1. If astronomers ignored any cosmological constant (or dark energy), the future of the universe could be determined solely from
   a. the mass of the universe.
   b. the volume of the universe.
   c. the amount of light in the universe.
   d. the density of the universe.

2. The cosmological constant accounts for the effects of
   a. dark matter.
   b. the Big Bang.
   c. dark energy.
   d. gravity.

3. The cosmic microwave background radiation (CMB) indicates that the early universe
   a. was uniform.
   b. varied greatly in density from one place to another.
   c. varied greatly in temperature from one place to another.
   d. was shaped differently from the modern universe.

4. According to the definitions of these terms in Chapter 1, superstring theory is
   a. a hypothesis.
   b. a theory.
   c. a law.
   d. a principle.

5. Place in order the following events in the history of the universe.
   a. Planck era
   b. grand unified theory breaks
   c. today
   d. Big Bang nucleosynthesis
   e. electroweak breaks
   f. theory of everything breaks
   g. electron-positron pair annihilation
   h. formation of galaxies and stars
   i. recombination
   j. inflation

6. Astronomers will never directly observe the first few minutes of the universe because
   a. the universe was opaque then.
   b. the universe is too large now.
   c. no particles or other matter existed for astronomers to see.
   d. no photons existed.

7. As applied to the universe, what is the meaning of *critical density*?
   a. Above that density, nebulae collapse to form stars.
   b. Above that density, dark matter becomes important.
   c. Above that density, the universe will eventually collapse.
   d. Above that density, matter becomes degenerate.

8. Of the four fundamental forces in nature, which two become unified at the lowest energy?
   a. gravitational force
   b. electromagnetic force
   c. strong nuclear force
   d. weak nuclear force

9. Suppose you measure the angles of a triangle and find that they add to 185 degrees. From that you can determine that the space the triangle occupies is
   a. flat.
   b. positively curved.
   c. negatively curved.
   d. filled with dark matter.

10. Which of the following is the correct order for these objects to form after the Big Bang?
    a. neutral atoms, protons, nuclei
    b. protons, nuclei, neutral atoms
    c. nuclei, neutral atoms, protons
    d. protons, neutral atoms, nuclei

11. Quarks are
    a. virtual particles.
    b. massless particles.
    c. candidates for dark matter.
    d. building blocks of larger particles.

12. Current understanding indicates that the universe (choose all that apply)
    a. is closed.
    b. is flat.
    c. is open.
    d. is inflating.
    e. has accelerating expansion.

13. When a particle and an antiparticle come together, they
    a. annihilate each other, releasing photons.
    b. create a black hole.
    c. release enormous amounts of energy.
    d. create new particles.

14. Observations of Type I supernovae in distant galaxies have shown that
    a. the rate of star formation in galaxies decreases with increasing redshift.
    b. the expansion rate of the universe is increasing.
    c. the cosmological constant is zero.
    d. dark energy is negligible now.

15. The anthropic principle states that
    a. the universe was created so that life exists.
    b. life exists, so the universe must be such that life can exist.
    c. if the universe were otherwise, life would not exist.
    d. life has made the universe the way it is.

THINKING ABOUT THE CONCEPTS

16. What set of circumstances would cause an expanding universe to reverse its expansion and end up in a "Big Crunch"?

17. Describe the observational evidence suggesting that Einstein's cosmological constant (a repulsive force) may be needed to explain the historical expansion of the universe. Explain how Einstein was "right for the wrong reason."

18. What do astronomers mean by *dark energy*?

19. If the universe is being forced apart by dark energy, why isn't the Milky Way Galaxy, the Solar System, or the planet Earth being torn apart?

20. In Chapter 21, we said we could estimate the age of the universe with Hubble time ($1/H_0$). Why does that method not give the best answer?

21. ★ **WHAT AN ASTRONOMER SEES**   Figure 22.7 includes a large number of arcs formed when the light from distant galaxies is bent as it passes through this foreground cluster. Explain how these arcs might be different if the space near the cluster were flat (not curved at all), or open (curved like a saddle). 👁★

22. During the period of inflation, the universe may have briefly expanded at 10^{30} (a million trillion trillion) or more times the speed of light. Why did that ultra-rapid expansion not violate Einstein's special theory of relativity, which says that neither matter nor communication can travel faster than the speed of light?

23. Why is particle physics important for understanding the early universe?

24. The fundamental forces of the universe are generally assumed not to change.
   a. How would the fate of the universe be affected if Newton's gravitational constant changed with time?
   b. What if, instead, the electric force between charged particles changed with time?

25. The standard model cannot explain why neutrinos have mass or why electron-positron asymmetry existed in the early universe. Do those failings make it an incomplete theory? Should all of its predictions be ignored until the theory can resolve those remaining issues?

26. Explain pair production.

27. Describe the Planck era.

28. What are the basic differences between a grand unified theory (GUT) and a theory of everything (TOE)?

29. Consider the term *superstring theory* in light of the discussion of scientific theory in Chapter 1. Some scientists object to using the word *theory* to describe superstring theory. Why?

30. Suggest another example of how a different value of a physical constant would affect conditions in a "parallel" universe.

APPLYING THE CONCEPTS

31. The range of reasonable values for H_0 runs from 67 km/s/Mpc to 74 km/s/Mpc. Using the highest reasonable number ($H_0 = 74$ km/s/Mpc), calculate the critical density. 🟢—●—🟢
   a. Make a prediction: Do you expect your answer to be larger or smaller than the answer given in Working It Out 22.1, where the critical density was calculated using $H_0 = 70$ km/s/Mpc?
   b. Calculate: Follow Working It Out 22.1 to find the critical density that corresponds to $H_0 = 74$ km/s/Mpc.
   c. Check your work: Verify that your answer has units of density, and compare your result to your prediction.

32. The proton and antiproton each have the same mass, $m_p = 1.67 \times 10^{-27}$ kg. What is the energy (in joules) of each of the two gamma rays created in a proton-antiproton annihilation? 🟢—●—🟢
   a. Make a prediction: You will need to use the mass-energy relation: $E = mc^2$. The speed of light is 3×10^8. Estimate the order of magnitude of your answer (the power of ten), by adding −27 and 16 (8×2).
   b. Calculate: Use $E = mc^2$ to accurately calculate the energy of each gamma ray created in a proton-antiproton annihilation.
   c. Check your work: Compare the order of magnitude of your answer to your prediction.

33. The range of reasonable values for H_0 runs from 67 km/s/Mpc to 74 km/s/Mpc. Using the lowest reasonable number ($H_0 = 67$ km/s/Mpc), calculate the critical density.

34. The electron and antielectron each have the same mass, $m_p = 9.11 \times 10^{-31}$ kg. What is the energy (in joules) of each of the two gamma rays created in an electron-antielectron annihilation?

35. About 500 million CMB photons are in the universe for every hydrogen atom. Assume that the average wavelength of a CMB photon is 1 mm. Using $E = \frac{hc}{\lambda}$ and $E = mc^2$, what is the equivalent mass of those 500 million photons?

36. Consider Figure 22.2. ★
   a. Is the vertical axis linear or logarithmic?
   b. The horizontal axes have two labels. The top label is measured in billions of years. Is that axis linear or logarithmic?
   c. The bottom label for the horizontal axis is measured in relative brightness. Is that axis linear or logarithmic?
   d. What is the relationship between billions of years and relative brightness?

37. The solid blue curve in Figure 22.3 indicates a model in which the universe first decelerated and then accelerated, whereas the solid orange curve indicates continual deceleration.
   a. How are the two curves different? ★
   b. What is it about one of those curves that indicates deceleration? What indicates acceleration?
   c. If a straight line were plotted on that graph, what would the model that the new line represents indicate about the expansion of the universe?

38. On the graph in Figure 22.2, the colored lines represent various models, and the black dots represent data taken in the actual universe. ★
   a. Why are there no data points on the right-hand side of the graph?
   b. Which models are excluded by the data?
   c. Roughly how far back in time do the data go?
   d. What fraction of the age of the universe is the answer to part (c) (if you assume an age of 13.8 billion years)?

39. Is the plot in Figure 22.9 linear or logarithmic? How much did the universe increase in size during the time of inflation? How much smaller was the universe at the beginning in an inflationary model than in the noninflationary model? 👁

40. Consider Figure 22.14. 👁⭐
   a. Is the time axis (the vertical dimension of the figure) approximately linear or approximately logarithmic?
   b. By how many orders of magnitude (factors of 10) has the density ρ of the universe decreased since earliest time?
   c. By how many orders of magnitude has the temperature decreased since earliest time?

41. Currently, the Hubble constant has an uncertainty of about 4 percent. What are the corresponding maximum and minimum ages allowed for the universe?

42. Estimate the average density of the Milky Way. How does it compare to the critical density?

43. The universe today has an average density $\rho_0 = 9.9 \times 10^{-27}$ kg/m³. If you assume that the average density depends on the scale factor, as $\rho = \rho_0/R_U^3$, what was the scale factor of the universe when its average density was about the same as Earth's atmosphere at sea level ($\rho = 1.2$ kg/m³)?

44. Suppose you brought together a gram of ordinary-matter hydrogen atoms (each composed of a proton and an electron) and a gram of antimatter hydrogen atoms (each composed of an antiproton and a positron). Keeping in mind that 2 grams is less than the mass of a dime:
   a. Calculate how much energy (in joules) would be released as the ordinary-matter and antimatter hydrogen atoms annihilated one another.
   b. Compare that amount of energy with the energy released by a 1-megaton hydrogen bomb (4.2×10^{15} J).

45. One GUT predicts that a proton will decay in about 10^{31} years, which means if you have 10^{31} protons, you should see one decay per year. The Super-Kamiokande observatory in Japan holds about 20 million kg of water in its main detector, and it did not see any decays in 5 years of continuous operation. What limit does that observation place on proton decay and on the GUT described here?

EXPLORATION Studying Particles

Just as astronomers use images to study distant clouds of dust and gas, so too do particle physicists use images to study the tiny particles. Those images capture the tracks the particles make. As the particle passes through vapor in a chamber, it ionizes some of the molecules of the vapor. The vapor then condenses around the ionized molecules, leaving a thin trail of cloud. A strong magnetic field fills the chamber, so that the charged particles will turn as they travel through it.

Figure 22.16 shows tracks of particles moving through a bubble chamber. Studying that figure in more detail will give you a feel for how scientists have gained an understanding of those tiny particles.

1 Estimate the number of particle tracks that you can see in Figure 22.16.

2 Recall that charged particles spiral around magnetic field lines. From the image of Figure 22.16, determine which way the magnetic field points—that is, perpendicular or parallel to the image.

3 Identify each of the following types of tracks in Figure 22.16:
   a. a track that looks straight
   b. a track that spirals tightly (a small spiral) and a track that spirals loosely (a large spiral)
   c. a bright track and a fainter track
   d. two tracks that form a "V" shape
   e. a track with an obvious beginning and ending
   f. a track with a kink, or abrupt change of direction

Oppositely charged particles spiral in opposite directions. For example, if the positive particles spiral clockwise, the negative ones spiral counterclockwise. More massive particles are harder to turn than less massive particles. Faster-moving particles also are harder to turn. The fact that both the mass and the velocity of a particle affect how easily it turns makes separating the two properties difficult.

4 Only charged particles interact with the vapor in the cloud chamber to produce tracks, so all the tracks should be curved in the magnetic field. But some of those tracks are very nearly straight. What can you say about the velocity of particles that leave straight tracks?

5 Consider the tight spiral and the loose spiral tracks.
   a. If both tracks were made by electrons, which particle was traveling faster?
   b. If both tracks were *not* made by electrons, what can you say about the total energy of each particle, including its mass?

6 The brightness of the track tells you how long the particle spent in each position: a brighter track is produced when the particle stays near the same spot longer.
   a. Consider the bright track and the faint track. Which track shows the faster particle?
   b. Now apply that reasoning to the tight and loose spiral tracks. Can you tell which particle was moving faster?
   c. If the looser track is also brighter, were both the loose and the tight tracks made by the same type of particle?

7 A track that is shaped like a "V" indicates that two charged particles were produced at the base of the "V" and then proceeded onward. That can occur, for example, if a neutral particle decays to produce a positive and a negative charge. If the two particles were a proton and an electron, but both arms of the "V" curved the same (possibly very small) amount, what can you say about their relative velocities?

8 A track with an obvious beginning and endpoint shows an event initiated by a neutral particle. Because neutral particles do not leave tracks in the chamber, that type of track must indicate a charged particle that suddenly begins moving. How might a neutral particle initiate one of those tracks?

9 A kink occurs when a charged particle decays, emitting a particle and in the process becoming a different charged particle.
   a. Is the particle emitted in the decay neutral or charged? How do you know?
   b. Does the remaining charged particle have more or less mass than before it decayed?
   c. Sketch the kink that you are studying, and indicate the direction the particle was traveling along the track. (Hint: Consider the tightness of the spiral!)

To learn more, you would have to know more about the experimental setup—the strength of the magnetic field, the speed of the particles as they enter the chamber, and so on. By studying images like those, and "crunching the numbers," scientists determine the properties and interactions of the fundamental building blocks of the universe.

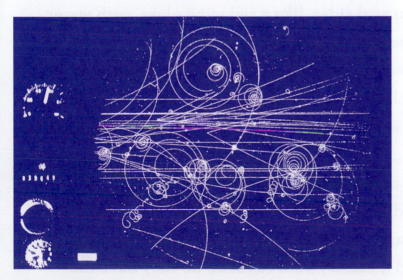

Figure 22.16 The decay of a particle in a hydrogen bubble chamber.

Large-Scale Structure in the Universe

A foreground cluster of galaxies will warp and twist the light coming from a background galaxy. A similar phenomenon happens when using an irregularly shaped lens of any kind. Fill a clear glass with curved sides, like a red wine glass, with water, and use one eye to look through the glass at a lamp or other light source across the room. Move the glass closer and farther from you. Move it up and down. Rotate it to one side as far as you can without spilling. Rotate it away from you. Observe how the image of the lamp changes, warping and splitting, depending on the relative location of the "lens," the lamp, and your eye. If you have a second, differently shaped glass, try both of them; do they have the same effect on the image of the lamp?

EXPERIMENT SETUP

Take a clear glass with curved sides, and fill it with water. Use one eye to look through the glass at a lamp or other light source across the room.

Move the glass closer and farther from you.

Move it up and down.

Rotate it to one side as far as you can without spilling. Rotate it away from you.

Observe how your view of the lamp changes, warping and splitting, depending on the relative location of the "lens," the lamp, and your eye.

If you have differently shaped glasses, try them; do they have the same effect on the image of the lamp?

PREDICTION

I predict that each of the items checked will affect the shape of the image of the lamp, while the others will not.

- ☐ **Distance of the glass from my eye**
- ☐ **Distance of the glass from the lamp**
- ☐ **Moving the glass left and right**
- ☐ **Moving the glass up and down**
- ☐ **Rotating the glass to the side**
- ☐ **Rotating the glass away from me**

SKETCH OF RESULTS

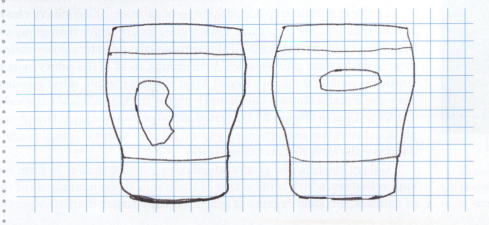

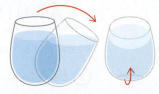

23

The universe that emerged from the Big Bang was incredibly uniform—wholly unlike today's universe of galaxies, stars, and planets. Here in Chapter 23, we investigate the origin of the current structure of the universe and find that complex structure is a natural consequence of the action of physical laws in an evolving universe.

LEARNING GOALS

By the end of this chapter, you should be able to:

(1) Describe the distribution of galaxies in the universe.

(2) Explain how the large-scale structure of today's universe evolved from the structure that began to form shortly after the Big Bang.

(3) Describe the formation of the first stars and the first galaxies.

(4) Relate the observations of galaxies at different redshifts to the evolution of the large-scale structure of the universe.

23.1 Galaxies Form Groups, Clusters, and Larger Structures

Just as stars and clouds of glowing gas reveal the structure of the Milky Way, the distribution of galaxies indicates the structure of the universe. And just as gravity holds stars together to form galaxies, gravity holds galaxies together to form larger structures. In this section, we look at how galaxies cluster into larger structures.

Types of Galaxy Structures

Most galaxies are part of a larger structure: a group, a cluster, or a supercluster. Galaxy groups are the smallest and most common collection of galaxies and contain up to several dozen galaxies, most of which are dwarf galaxies. The Milky Way is a member of the Local Group (see Chapter 20), which consists of three large spiral galaxies—the Milky Way Galaxy, the Andromeda Galaxy, and the Triangulum Galaxy—along with more than 50 smaller dwarf elliptical or dwarf spheroidal galaxies in a volume of space roughly 3 megaparsecs (Mpc) in diameter. Most of the mass in the Local Group, both dark matter and luminous matter, resides in the three spiral galaxies.

Larger gravitationally bound systems of galaxies, called **galaxy clusters**, can consist of thousands of galaxies. Galaxy clusters are larger than groups, typically occupying a volume of space 2–10 Mpc across. Galaxy clusters also contain far more dwarf galaxies than giant galaxies. As in groups, most of the galaxy mass in galaxy clusters resides in the giant galaxies. Most clusters contain spiral galaxies, though the more massive clusters have a much higher fraction of giant elliptical galaxies. The Virgo Cluster (**Figure 23.1**), located 16.5 Mpc from the Local Group, is an example of a cluster that contains mostly spiral galaxies. Giant elliptical and S0 galaxies dominate the more distant Coma Cluster.

Galaxy clusters and groups of galaxies bunch together to form enormous **superclusters**, which contain tens of thousands or even hundreds of thousands of galaxies and span regions of space typically larger than 30 Mpc. Our Local

VIS

Figure 23.1 The Virgo Cluster of galaxies, at a distance of about 16.5 Mpc (54 million light-years), is the closest cluster to the Milky Way. This image shows the center of the Virgo Cluster.

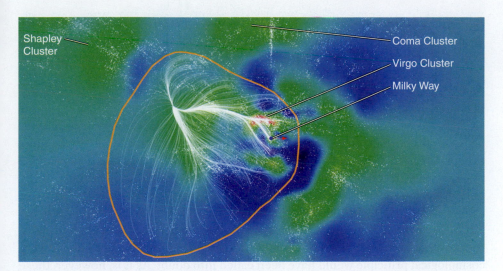

Figure 23.2 This computer-generated visualization shows a slice of the Laniakea Supercluster. Individual galaxies are white dots, blue areas are voids, red dots indicate the Virgo Cluster, and green areas contain many galaxies. White lines indicate the movement of galaxies toward the center of the supercluster. The orange contour encloses the outer limits of those galaxies streaming to the center. The dark blue dot shows the location of the Milky Way.

Group is part of the Laniakea Supercluster, which also includes the Virgo Cluster (**Figure 23.2**).

Mapping the Universe

Hubble's law for the expansion of the universe ($v_r = H_0 \times d_G$) is a powerful tool for mapping the distribution of galaxies, groups, clusters, and superclusters in space. Astronomers measure the redshift in the galaxy's spectrum, and then determine the distance to a galaxy (d_G) using Hubble's law. The first redshifts were measured from spectra recorded on photographic plates, which required exposures of several hours to capture the faint signal. Now astronomers can measure the redshifts of galaxies by using larger telescopes equipped with electronic detectors and spectrographs that can observe many galaxies at once. We now know the redshifts of several million galaxies, so we now know the approximate distances to those galaxies. From that information, we can develop a three-dimensional map of the structure of the universe on the largest scales.

The results of large redshift surveys are often presented as a two-dimensional "slice of the universe," which follows an arc across part of the sky and plots the distances of galaxies along that arc (**Figure 23.3a**). This leads to a shape much like a piece of pizza; with Earth's position located at the point of the wedge. The most distant galaxies are located where the pizza slice outer crust would be. These maps show that clusters and superclusters of galaxies are not scattered randomly throughout space but instead are linked in an intricate network of relatively thin structures known as filaments and walls. The Sloan Digital Sky Survey (SDSS) includes the Sloan Great Wall (**Figure 23.3b**), a string of galaxies 400 Mpc long. Filaments and walls surround some of the largest structures known: large regions of space with very few galaxies, known as **voids**. Though the voids may seem empty, we do not know that they are devoid of matter—only that they have very few observable galaxies. On large scales, the universe has a structure much like that of a sponge or a pile of soap bubbles. Filaments, voids, and walls, together, form the **large-scale structure** of the universe.

what if . . .

What if you observe the effect of gravitational lensing by galaxy structures that contain only dark matter, located in voids? How would this change our understanding of the universe if the voids were full of such dark matter structures as opposed to having no such galaxy structures at all?

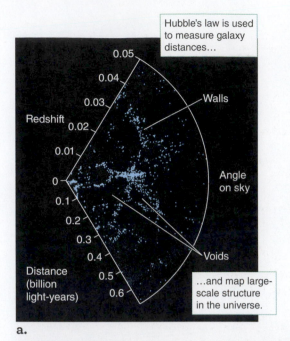

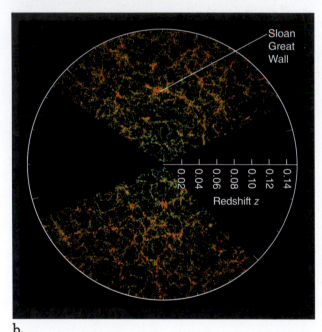

a.

b.

Figure 23.3 Redshift surveys use Hubble's law to map the universe. **a.** In 1986, the Harvard-Smithsonian Center for Astrophysics redshift survey, called "A Slice of the Universe," first showed that clusters and superclusters of galaxies are part of even larger-scale structures. **b.** The 2008 Sloan Digital Sky Survey map of the universe extends outward to a distance of about 600 Mpc. Shown here is a sample of 67,000 galaxies colored according to the ages of their stars, with the redder, more strongly clustered points being galaxies made of older stars.

The **peculiar velocity** of a galaxy is its motion relative to the Hubble flow. Peculiar velocities often result from local gravitational attraction to other galaxies. Observations of a galaxy's peculiar velocity reveal the distribution of mass near that galaxy. For example, the Local Group is pulled toward the center of mass of the large supercluster called Laniakea, with the Local Group and the Milky Way located at the outer edge (Figure 23.2).

The large-scale structure of the universe provides direct observational evidence for the cosmological principle—that the universe is homogeneous (the same everywhere) and isotropic (the same in every direction)—because the observations of large-scale structure show that on that very largest scale, the structure of the universe is the same everywhere, in every direction. If, for example, the universe in the top half of the multicolored image of Figure 23.3b showed a uniform distribution of galaxies, whereas the universe in the bottom half showed walls and voids, the universe would not be homogeneous or isotropic. All conclusions based on the cosmological principle would have been called into doubt. As it is, however, these observations support that underlying principle of cosmology.

Dark Matter in Galaxy Groups and Clusters

Just as dark matter dominates galaxies, it also dominates galaxy groups and clusters. Astronomers use the motion of a small satellite galaxy orbiting the central dominant galaxy of a cluster to estimate the mass of the cluster inside that orbit—similar to the way they measure dark matter in spiral galaxies. Alternatively, they look at the motions of all of a cluster's galaxies and calculate how strong gravity must be to hold the cluster together. The total mass of the clusters, including dark matter, must be about 8–10 times greater than the normal matter they contain (**Working It Out 23.1**) in order for the cluster to hold together.

Another piece of evidence for the presence of dark matter is that the space between galaxies in a cluster is filled with extremely hot gas that is 10 million to 100 million kelvins (K), making it bright in X-rays (**Figure 23.4**). Even though that gas is of extremely low density, the volume of space that it occupies is

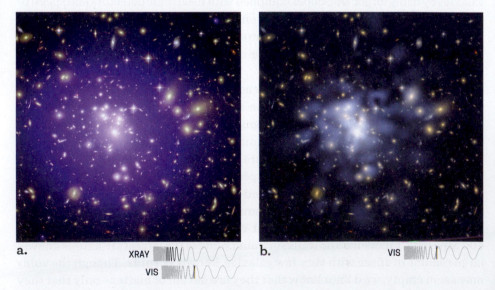

a.   XRAY   VIS

b.   VIS

Figure 23.4 a. Galaxy clusters, such as Abell 1689 shown here, are rich in hot, X-ray-emitting gas (shown in purple overlaying the image taken in the visible part of the spectrum). **b.** Abell 1689, with the inferred dark matter distribution shown in blue-white.

working it out 23.1

Mass of a Cluster of Galaxies

Recall from Working It Out 20.1 that astronomers use the orbit of a star about the center of a galaxy to estimate the mass of the galaxy within the star's orbit. A similar estimation is made with groups or clusters of galaxies. Those groups or clusters are often dominated by one giant galaxy at the center, with smaller ones orbiting around it. The orbital velocities of the smaller galaxies are measured from the Doppler shifts of the lines in their spectra. The distance between the central and orbiting galaxies (the radius of a circular orbit) is estimated. The equation then looks like the one from Working It Out 20.1:

$$M = \frac{r v_{\text{circ}}^2}{G}$$

Suppose a smaller galaxy is orbiting the large galaxy at the center of the cluster at a speed of 1000 km/s at a distance of about 3 Mpc. The gravitational constant is $G = 6.67 \times 10^{-20}$ km³/(kg s²). As in Chapter 20, we must convert the orbital radius (3 Mpc) to kilometers:

$$(3 \times 10^6 \text{ pc}) \times (3.09 \times 10^{13} \text{ km/pc}) = 9.3 \times 10^{19} \text{ km}$$

The mass of the cluster core is given by

$$M = \frac{(9.3 \times 10^{19} \text{ km}) \times (10^3 \text{ km/s})^2}{6.67 \times 10^{-20} \text{ km}^3/(\text{kg s}^2)}$$

$$M = 1.4 \times 10^{45} \text{ kg}$$

We can divide that result by the mass of the Sun, 2.0×10^{30} kg, to get a cluster mass of $7.0 \times 10^{14} M_{\text{Sun}}$. If we divide that core cluster mass by the mass of the Milky Way Galaxy, $10^{12} M_{\text{Sun}}$, then the cluster core has a mass of about 700 Milky Way galaxies. That value includes the dark matter inside the orbit.

In 1933, astronomer Fritz Zwicky (1898–1974) measured the velocities of many galaxies within a collection of clusters, and in each case he found the entire cluster contained more mass than could reside in the stars that produce the cluster's visible light. He concluded there was a considerable amount of extra and yet unseen mass in the clusters. In the 1970s, Vera Rubin (1928–2016) and colleagues measured rotation curves of individual galaxies and discovered evidence of dark matter in those galaxies, too.

enormous: The mass of that hot gas can be up to 5 times the mass of all the stars in that cluster. X-ray spectra show that the gas contains significant amounts of massive elements that must have formed in stars. That chemically enriched gas has been either blown out of galaxies in winds driven by the energy of massive stars or stripped from galaxies during encounters with neighboring galaxies. The amount of luminous mass is not enough to keep that hot gas from escaping the cluster, so it would have dispersed long ago if not for the gravity of the dark matter filling the volume of the cluster.

Another way to look for dark matter relies on the predictions of Einstein's general theory of relativity (see Chapter 18), which states that mass distorts the geometry of spacetime, causing even light to bend near a massive object. In particular, light from a distant object is bent by the gravity of a galaxy or cluster of galaxies, so that images of the distant object can be seen magnified on both sides of the intervening galaxy or cluster. The result is a gravitational lens (**Figure 23.5a**). Recall from the discussion of massive compact halo objects (MaCHOs) in Chapter 19 that lenses can make background objects appear brighter. Lenses can also show multiple images of background objects, and those magnified images are often drawn out into arcs. The amount of gravitational lensing increases as the amount of mass in the cluster increases. **Figure 23.5b** shows a galaxy cluster acting as a gravitational lens for several background galaxies. Analysis of such images reveals the mass of the lensing cluster.

Regardless of how astronomers measure the masses of galaxy clusters—by looking at the motions of their galaxies, by measuring their hot gas, or by using them as

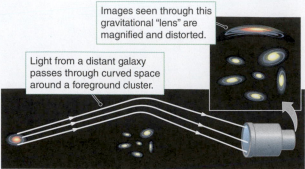

Images seen through this gravitational "lens" are magnified and distorted.

Light from a distant galaxy passes through curved space around a foreground cluster.

a.

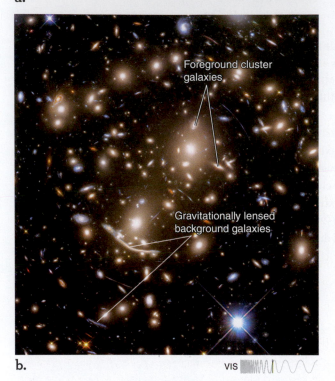

Foreground cluster galaxies

Gravitationally lensed background galaxies

b.   VIS

Figure 23.5 a. This illustration shows the geometry of a gravitational lens. A mass can gravitationally focus the light from a distant object, thereby magnifying and distorting the image. **b.** A Hubble Space Telescope image of the cluster Abell 370 shows many gravitationally lensed galaxies, seen as arcs.

gravitational lenses—the results are the same. Dark matter dominates the mass of galaxy clusters and superclusters.

CHECK YOUR UNDERSTANDING 23.1

Place the following types of galaxy collections in order of increasing size: (a) wall; (b) cluster; (c) group; (d) supercluster

Answers to Check Your Understanding questions are in the back of the book.

23.2 Gravity Forms Large-Scale Structure

How did the universe evolve from a very smooth distribution of radiation and matter after the Big Bang to the large-scale structure of filaments and voids seen today? Astronomers approach that question both observationally and theoretically: Observers use telescopes to study the most distant objects and the cosmic microwave background radiation (CMB), whereas theorists use the largest supercomputers to simulate the growth of small- and large-scale structure. In this section, we describe the role of gravity and dark matter in the formation of structure.

Gravitational Instabilities

In Chapter 15, you learned about gravitational instabilities involved in star formation. Star formation can begin in a molecular cloud with clumps inside it. Gravity causes those clumps to collapse faster than their surroundings; thus, gravity can turn density variations within clouds into stars. The same gravitational instability can turn density variations within the universe into galaxies.

The end of this galaxy formation process can be characterized by looking for the first galaxies. More distant galaxies have higher measured values of redshift (z), so the observed light was emitted when the universe was *younger*, closer to the time of the Big Bang (**Table 23.1**). In this way, astronomers can use observations of galaxies with different redshifts to determine how the universe has changed over time. Astronomers diligently search for galaxies at higher and higher redshifts. However, a gap of several hundred million years remains between the CMB maps of the universe at age 380,000 years and the highest-redshift galaxies observed by Hubble Space Telescope (**Figure 23.6**). The international Atacama Large Millimeter/submillimeter Array (ALMA) in Chile has observed even younger galaxies. The future James Webb Space Telescope (JWST) may detect galaxies at even higher redshifts. Filling in this gap in time, between recombination (when the CMB was emitted) and the earliest known galaxy, will answer many questions about the formation of structure.

As discussed in Chapter 21, space missions such as the Cosmic Background Explorer (COBE), Wilkinson Microwave Anisotropy Probe (*WMAP*), and the Planck space observatory have revealed variations in the CMB. Those tiny variations reflect the imprinted structure of the early universe at the time of inflation, providing the "clumps" or "seeds" from which galaxies and collections of galaxies grew. Over time, gravity amplified those clumps by pulling the slightly dense regions closer together, increasing the difference in density to more than twice the average. Smaller structures such as dwarf galaxies formed first, whereas larger structures such as clusters, superclusters, filaments, and voids formed later. That process is called **hierarchical clustering** because the structure forms in a "bottom up" hierarchy. Hierarchical clustering is supported by observations and is fundamental to how structure formed in the universe.

The initial variations are essential because gravity cannot produce structure in a perfectly uniform universe, in which every particle would be pulled equally in every direction by gravity. Where did they come from? Those initial seeds are thought to have formed from quantum fluctuations during inflation in the early universe. That model has a profound implication: The seeds leading to galaxies, clusters, and superclusters (the largest structures in the universe) arose from the same quantum physics that describes the smallest structures in the universe (atoms, nuclei, and elementary particles). All of science is connected.

A problem remains with the model as we've explained it so far. Variations in the CMB are found at a level of about one part in 100,000. The theoretical models clearly show that such tiny variations at the time of recombination (when the universe was about 380,000 years old) are far too small to explain the structure observed in today's universe. Gravity is not strong enough for galaxies and clusters of galaxies to have grown from such small clumps. Those models indicate that for today's galaxies to have formed, the density of those clumps must have been a few tenths of a per-cent greater than the average density of the universe at the time of recombination. But if normal luminous matter in the early universe had clumps with that higher density, the variations in the CMB today would be at least 30 times larger than they are. How do astronomers reconcile that problem? Dark matter holds the key.

Galaxies Formed because of Dark Matter

Dark matter is an essential ingredient in the formation of the structure we observe. The comparison of Big Bang nucleosynthesis models with observed amounts of light elements predicts a current density of normal matter in the universe very close to what we observe. That agreement in the observed density of normal mat-ter powerfully constrains the nature of dark matter, which dominates the mass in the universe. Dark matter cannot consist of normal matter made up of neutrons and protons. If it did, the density of neutrons and protons in the early universe would have been much higher, and the resulting amounts of light elements in the universe from Big Bang nucleosynthesis would have been very different from what we actually observe (see Chapter 21).

Table 23.1	Redshift and Age
Observed z	**Age of Universe (years)**
1100	380,000 (recombination)
30	100 million
20	200 million
15	270 million
10	480 million
9	560 million
8	650 million
7	750 million
6	900 million
5	1.2 billion
4	1.6 billion
3	2.2 billion
2	3.3 billion
1	5.9 billion
0.5	8.6 billion
0.25	10.5 billion
0	13.8 billion

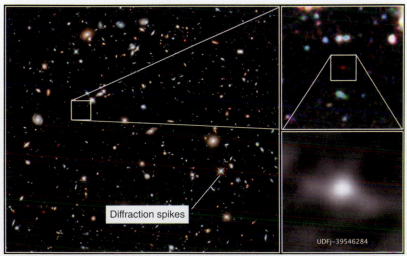

Diffraction spikes

UDFj-39546284

IR

Figure 23.6 ★ WHAT AN ASTRONOMER SEES
An astronomer will be fascinated by a Hubble Space Telescope image like this one, which was taken over a very long time. This patch of sky was chosen because it appeared to be empty of stars in all previous images. Knowing that nearly every blob in this image is a galaxy, composed of billions or trillions of stars, will give her pause. Perhaps she will take a moment to appre-ciate that the size of this image on the sky is about the size of a basketball, 6 kilometers (about 3.78 miles) away. She may take a moment to think about the fact that every tiny patch of sky in every direction is as dense with galaxies as this image. Naturally, she will immediately look for the few specks that are *not* galaxies but instead are nearby stars. She will identify these from the diffraction spikes, which make an X across the star's image. She will focus in on the colors of the galaxies in the field and search for at least one galaxy of every Hubble type (spiral, barred spiral, elliptical, and irregular). The earliest, highest-redshifted galaxies are observable in infrared light; observations of these galaxies are foundational to tests of Big Bang cosmology. Contained within this image is a compact faint galaxy as it existed about 480 million years after the Big Bang. At least 100 of these small galaxies would be needed to build up a galaxy like the Milky Way.

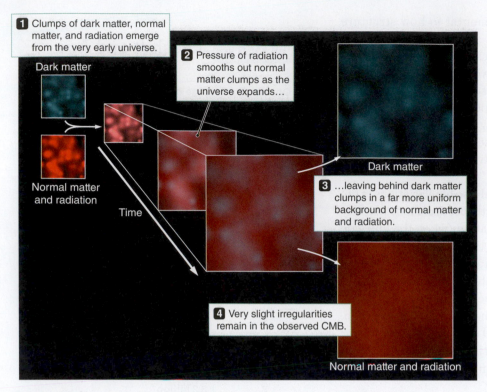

1 Clumps of dark matter, normal matter, and radiation emerge from the very early universe.

Dark matter

Normal matter and radiation

Time

2 Pressure of radiation smooths out normal matter clumps as the universe expands…

3 …leaving behind dark matter clumps in a far more uniform background of normal matter and radiation.

4 Very slight irregularities remain in the observed CMB.

Dark matter

Normal matter and radiation

Figure 23.7 Radiation pressure and other processes in the early universe smoothed out variations in normal matter, but clumps in the dark matter survived and later pulled in normal matter to form galaxies.

Dark matter interacts only weakly with normal matter and does not interact with electromagnetic radiation. Dark matter and normal matter behaved differently in the early universe. After the Big Bang, radiation pressure smoothed out ripples in the distribution of normal matter. However, feebly interacting dark matter is immune to those processes, so whereas normal matter smoothed out, dark matter remained clumpy. Because clumps of such dark matter in the early universe did not interact with radiation or normal matter, astronomers would not see them directly when looking at the CMB.

Figure 23.7 shows how the distributions of normal and dark matter differ in the early universe. Although those clumps of dark matter do cause slight gravitational redshifts in the light coming from normal matter, the resulting variations fit well with current observations of the CMB. Dark matter solves the problems of the formation of galaxies and clusters of galaxies by producing a stronger gravitational attraction that remains clumpy in the early universe, unlike normal matter, which is smoothed out by radiation pressure.

Hot and Cold Dark Matter

Dark matter in the early universe was much more strongly clumped than normal matter. Within a few million years after recombination, those dark matter clumps pulled in the surrounding normal matter. Later, gravitational instabilities caused those clumps to collapse. The normal matter in the clumps went on to form visible galaxies. The details of how that happened depend greatly on the properties of the dark matter itself. Even though we do not yet know exactly what dark matter is made of, we can distinguish two broad classes, called cold dark matter and hot dark matter, based on how dark matter behaves.

Cold dark matter consists of feebly interacting particles moving about relatively slowly, like the slow-moving atoms and molecules in a cold gas. Several candidates are possible for the composition of cold dark matter. Most likely, cold dark matter consists of an unknown **elementary particle**. One candidate is the **axion**, a hypothetical particle first proposed to explain some observed properties of neutrons. Axions should have very low mass, and they would have been produced in great abundance in the Big Bang. Another candidate is the **photino**, a massive particle—perhaps 1000 times the mass of a proton—predicted in certain particle theories such as string theory. Physicists are using particle accelerators such as the Large Hadron Collider to look for those types of particles, and several experiments are under way to search for axions and photinos trapped in the dark matter halo of the Milky Way.

Hot dark matter consists of particles moving so rapidly that gravity cannot confine them to the same region as the luminous matter in the galaxy. Neutrinos are one example of hot dark matter. Neutrinos interact with matter so weakly (see Chapter 14) that they can flow freely outward from the center of the Sun, passing through the overlying layers of matter as though they were not there. The universe is filled with neutrinos, which might account for a few percent of its mass. Although that percentage is not high enough to account for most of the dark matter in the universe, it may still have noticeably affected the formation of structure.

Cold and hot dark matter affect structure formation differently because of how they respond to a gravitational field. Slow-moving particles are more easily held by gravity than are fast-moving particles, so particles of cold dark matter clump together more easily into galaxy-sized structures than do particles of hot dark matter. As a result, theoretical models show that on the largest scales of massive superclusters, both hot dark matter and cold dark matter can form the kinds of structures observed; on much smaller scales, however, only cold dark matter can clump enough to produce structures like the galaxies filling the universe. To account for the formation of today's galaxies, we need cold dark matter.

Forming a Galaxy from an Instability

We can best see how models of galaxy formation work by following the events predicted by the models step by step. Consider a universe made up primarily of cold dark matter, clumped together with normal matter in a manner consistent with observations of the CMB. On the scale of an individual galaxy, the effect of the cosmological constant is negligible.

Figure 23.8a shows the first step of a model simulation of one clump of dark matter at the time of recombination. In this model, the dark matter was slightly clumpier than normal matter, but overall the distribution of matter was uniform. **Figure 23.8b** shows that by a few million years after recombination, the universe in the simulation had expanded severalfold; spacetime is expanding, so the clumps of dark matter also are expanding. However, the clumps of dark matter did not expand as rapidly as their surroundings because their self-gravity slowed down their expansion. The clumps of dark matter stood out more with respect to their surroundings. The gravity of the dark matter clumps began to pull in normal matter. Eventually, the clumps of dark matter stopped expanding when their own self-gravity slowed and then stopped their initial expansion. By the stage shown in **Figure 23.8c**, normal matter clumped in much the same way as dark matter. Unlike dark matter, which cannot emit radiation, the normal matter in the clumps radiated away energy and cooled, collapsing toward the center of the dark matter clumps.

Those clumps did not exist in isolation: They were tugged on by the gravity of neighboring clumps, and they were pushed around by the pressure waves that ran through the young universe, smoothing out its structure. As a result, each clump may have been rotating slightly when it began to collapse. As normal matter fell inward toward the center of a rotating dark matter clump, that rotation forced much of the gas to settle into a rotating disk, just as the collapsing clouds of protostars settled first into an accretion disk. Later, that rotating disk became the disk of a spiral galaxy (**Figures 23.8d–f**). That model has succeeded because it yields the right masses for observed galaxies. For example, a system too low in mass would not cool fast enough to separate the visible matter from the dark matter. The dark matter we have been talking about as a major

Figure 23.8 A spiral galaxy passes through roughly six stages as it forms from the collapse of a clump of cold dark matter.

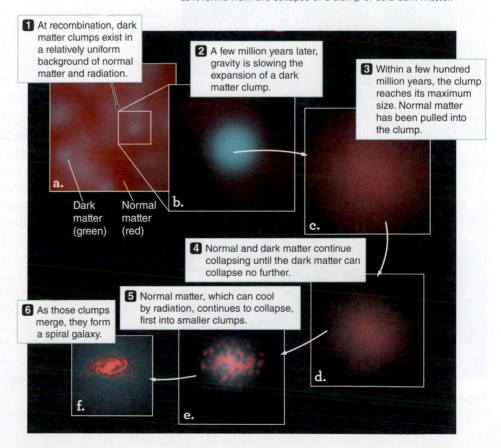

1 At recombination, dark matter clumps exist in a relatively uniform background of normal matter and radiation.

2 A few million years later, gravity is slowing the expansion of a dark matter clump.

3 Within a few hundred million years, the clump reaches its maximum size. Normal matter has been pulled into the clump.

Dark matter (green)   Normal matter (red)

4 Normal and dark matter continue collapsing until the dark matter can collapse no further.

5 Normal matter, which can cool by radiation, continues to collapse, first into smaller clumps.

6 As those clumps merge, they form a spiral galaxy.

a.
b.
c.
d.
e.
f.

○ what if . . .

What if the universe were composed entirely of ordinary (normal) matter but with the same CMB variations 18 billion years after the Big Bang? How might that affect the time at which intelligent life like us would first appear?

component of galaxies turns out to be responsible for their formation as well (see the **Process of Science Figure**).

By the beginning of the 21st century, a "standard model" of Big Bang cosmology had broad support for explaining the large-scale structure and accelerated expansion of the universe, as well as the CMB and the amounts of the light elements from Big Bang nucleosynthesis (discussed in Chapters 21 and 22). That model is called **Lambda-CDM**—from *lambda* for the cosmological constant Ω_Λ and *CDM* for the cold dark matter that played a vital role in structure formation.

CHECK YOUR UNDERSTANDING 23.2

The dominant factor in the formation of galaxies is the distribution of _____ in the early universe. (a) ordinary matter; (b) dark matter; (c) energy; (d) dark energy

23.3 First Light of Stars and Galaxies

Recombination occurred approximately 380,000 years after the Big Bang (see Chapter 21), when the universe cooled enough for electrons and protons to combine to form hydrogen and helium atoms. Before that time the universe was opaque; after recombination it became transparent. The CMB was emitted at that moment, and we can observe it today. That began the epoch in the history of the universe called the **Dark Ages** because no visible "light" came from astronomical objects. The only light available then was from the cooling CMB, and some amount of 21-centimeter (21-cm) radio radiation from transitions occurring within hydrogen atoms. Several projects are ongoing to detect that 21-cm radio radiation, which has been redshifted to much longer radio wavelengths today.

The Dark Ages lasted from about 380,000 to several hundred million years after the Big Bang. By several hundred million years after the Big Bang, the first stars were forming from the elements created in the Big Bang. As those stars formed, they heated up until they emitted ultraviolet (UV) photons with enough energy to reionize neutral hydrogen in interstellar space. During that **reionization** stage, starting about 500 million years after the Big Bang, the UV light from the first stars stripped electrons from hydrogen atoms. By this time in the universe, this ionized hydrogen was not dense enough to be opaque. It was transparent to the light from the first stars. Reionization continued with star formation in the first low-luminosity galaxies and with radiation from the first supermassive black holes inside the first quasars. Reionization was completed by about 900 million to 1 billion years after the Big Bang (**Figure 23.9**).

Only within the past decade have astronomers been detecting light from objects in the first billion years of the universe: Those objects have redshifts greater than 6. (Table 23.1 shows how the times in the history of the universe compare with observed redshifts.) Many astronomers were surprised by the identification of galaxies, quasars, and gamma-ray bursts (GRBs) at such high redshifts because those objects had been thought to have not formed until after the universe was at least a billion years old. The GRBs, for example, are extremely luminous and result from the explosive deaths of massive stars (see Chapter 18). For GRBs to be detected at $z = 8$, massive stars must have existed that had already died by 650 million years after the Big Bang. Similarly, the detection of quasars at $z = 7$ indicates that supermassive black holes must have formed in less than 750 million years after the Big Bang. The most distant galaxies observed have redshifts of 10–11, corresponding to 400 million to 500 million years after the

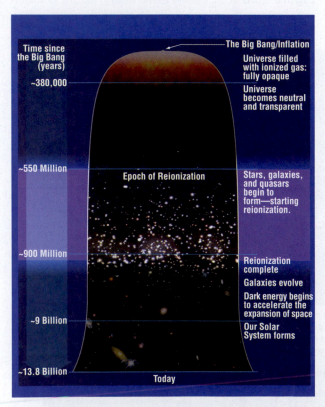

Figure 23.9 This time line illustrates the formation of the first stars and galaxies and the epoch of reionization.

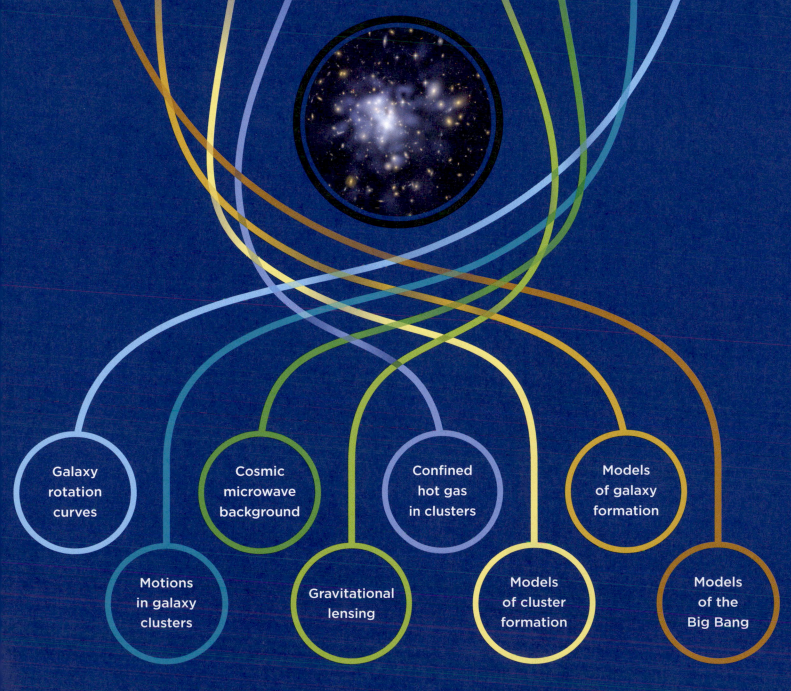

Multiple Streams of Evidence

Roughly 85 percent of the mass in the universe is an unknown form or composition called dark matter. So many observations, models, and experiments provide supporting evidence for dark matter that scientists have concluded that it exists.

Galaxy
rotation
curves

Motions
in galaxy
clusters

Cosmic
microwave
background

Gravitational
lensing

Confined
hot gas
in clusters

Models
of cluster
formation

Models
of galaxy
formation

Models
of the
Big Bang

Scientists do not have the luxury of ignoring evidence that does not fit current theories. They must see "being wrong" as an opportunity to learn and as a challenge to try harder.

what if . . .

What if you had time on a large telescope to search for first-generation stars? Where and how would you look for them?

Big Bang. The study of those highest-redshift objects and what they tell astronomers about the early universe is one of the most dynamic topics in astronomy today. New telescopes and new instruments are regularly detecting objects with higher and higher redshifts. By the time you read this book, the highest known redshifts will probably be even higher.

The First Stars

Astronomers use what has been learned in stellar evolution theory to study galaxy evolution and cosmology. The very first stars must have formed from the elements created in Big Bang nucleosynthesis: hydrogen, helium, and a very small amount of lithium. Observational astronomers look for stars that have only those elements, but so far none has been detected. Old stars in the halo of the Milky Way contain very low abundances of massive elements, but not as low as zero. Astronomers use computer simulations that combine data from those old stars with the conditions in the early universe to determine the history of star formation before those old stars formed.

The formation of the very first stars was different from the processes we discussed in Chapter 15. Because massive elements did not exist, the universe had no dust or molecular clouds filled with cold, dense gas for the stars to form in. Instead, those first stars formed inside dark matter minihalos, which were about 0.5 million to 1.0 million solar masses (M_{Sun}) and 100 parsecs (pc) across. Those minihalos formed a few hundred million years after the Big Bang ($z \approx 20$ to 30). Primordial gas clouds within those minihalos contained neutral hydrogen, and over time some small amounts of molecular hydrogen (H_2) formed. Radiation emitted by molecular hydrogen cooled the gas, and as it cooled, the gas pressure dropped and the gas collapsed to the center of the minihalo. A protostar grew in the gas cloud, accreting more gas to become a star. The illustration in **Figure 23.10** shows how such stars may have formed. Some theoretical models and computer simulations predict that those first stars were likely to be hot and massive. Estimates of the masses of the first stars range from 10 to more than 100 M_{Sun} for single stars and 10–40 M_{Sun} for double stars. Those stars had high luminosity, peaking in the ultraviolet, which ionized the gas near the star. Those stars were singles, doubles, or small multiples. Large star clusters probably didn't form.

Stars with those high masses have very short lifetimes (see Table 16.1 and Chapter 17), burning hydrogen in their cores for 10 million years or less. Today, massive stars use the carbon-nitrogen-oxygen (CNO) cycle for more efficient hydrogen burning, but carbon, nitrogen, and oxygen were unavailable to those first stars. The first stars ended their brief lives in supernova explosions, scattering some heavy elements into nearby space. If the core of such a star had rapid rotation during the supernova explosion, it might have emitted a GRB of extremely high luminosity. Astronomers have observed a few GRBs from as far as redshift 11.

Some of the first stars were massive enough to have become black holes after exploding. Black holes with companions can become energetic X-ray binary systems as mass falls onto the accretion disk of the black hole as the companion evolves. Because those stars were all so massive, the stars in a binary (or multiple) system would probably become black holes, which could then merge. (Recall from Chapter 18 that gravitational waves emitted during black-hole mergers have been detected.) Some theorists think that those merged black holes might have become the seeds for the supermassive black holes found in galaxies, but other models suggest that those stellar black holes would take too long to have built up to a mass of 1 million to 1 billion M_{Sun}.

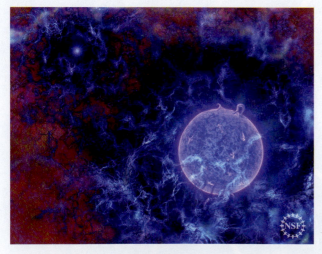

Figure 23.10 This supercomputer simulation shows the formation of the first stars from primordial gas in a dark matter minihalo a few hundred million years after the Big Bang. Here, two massive stars are forming a few hundred astronomical units apart. The brighter purple regions are denser than the darker purple regions.

Reading Astronomy News

Astronomers Spot Farthest Galaxy Known in the Universe

The farthest known galaxy is a moving target, as astronomers around the world compete to push the limits of observation even further into the past.

Maunakea, Hawai'i – An international team of astronomers using W. M. Keck Observatory have spectroscopic confirmation of the most distant astrophysical object known to date.

The researchers, led by Professor Linhua Jiang at the Kavli Institute for Astronomy and Astrophysics at Peking University, obtained near-infrared spectra with the Multi-Object Spectrograph for Infrared Exploration (MOSFIRE) on the Keck I telescope and successfully measured the distance of a very faint galaxy located 13.4 billion light-years away (redshift of z = 10.957).

Named GN-z11, the galaxy was generally believed to be at a redshift greater than 10, probably closer to 11, based on existing data from NASA's Hubble Space Telescope. But its exact redshift remained unclear, until now.

The results of the study, which are based on observations made under the time exchange program between Keck Observatory and Subaru Telescope on Maunakea, are published in the December 14, 2020 issue of the journal *Nature Astronomy*.

During their observations at Keck Observatory, the team also serendipitously detected a bright burst coming from the galaxy. After performing a comprehensive analysis, the team ruled out the possibility that the flash was from any known sources such as man-made satellites or moving objects in the solar system and determined it may have been produced by a gamma-ray burst.

A paper regarding this possible bright ultraviolet flash from GN-z11 is also published in the December 14, 2020 issue of *Nature Astronomy*.

Both studies are important to understanding the formation of stars and galaxies in the very early universe.

QUESTIONS

1. What is the redshift of this most-distant known galaxy, from December of 2020?
2. Using Table 23.1, estimate the age of the universe at the time the light left this galaxy.
3. This galaxy serendipitously underwent a gamma-ray burst (GRB) at the time of observation. What is the oldest possible age of the massive star that produced the GRB?
4. Search the Internet to find the farthest known galaxy at the time you read this. Has this record held, or has a more distant galaxy been discovered since 2020?

Source: https://keckobservatory.org/farthest-galaxy/?fbclid=IwAR2FBTBs2tCxyN555DklS3IA2_ZqO06YFVJqAMp1ws5oTm0x-hIdI6VYb4A.

Once a few stars had died, carbon, oxygen, and other elements mixed in and cooled nearby star-forming gas clouds. Some of those elements condensed into dust grains, which further cooled the clouds. Star formation in these clouds was similar to star formation in today's cold molecular clouds. Those "second-generation stars" had very low abundances of massive elements but measurably more than in the first stars. Because they formed in a cooler environment than the first stars, those second-generation stars could be lower in mass. Any stars less massive than 0.8–0.9 M_{Sun} that formed at that time have such long lifetimes that they are still burning hydrogen on the main sequence today. Those stars are not very luminous, but a few of them have been found in the halo of the Milky Way. Those halo stars have very small amounts of massive elements, but their spectra show small amounts of many elements in the periodic table—including uranium. Astronomers study those small second-generation stars because they offer clues about the nature of the first stars and the conditions of the very young Milky Way.

From a theoretical perspective, the minihalos just discussed did not form galaxies. After the first stars died, the energy from supernovae, GRBs, or X-ray binaries either may have heated any remaining gas in the minihalo too much for further star formation or such gas may have escaped the minihalo because of its

unanswered questions

How do supermassive black holes form? The detection of quasars at very high redshifts indicates that supermassive black holes formed early in the Dark Ages, but exactly how or how much they may have contributed to reionization is currently unknown.

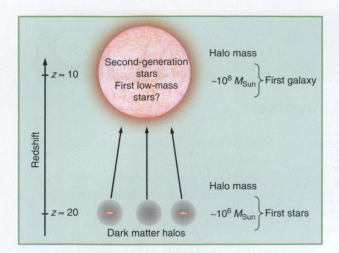

Figure 23.11 In "bottom up" (hierarchical) growth, the first stars formed in minihalos and then came together to form larger halos, which became the first small dwarf galaxies.

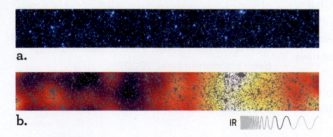

Figure 23.12 **a.** A standard infrared image from the Spitzer Space Telescope shows stars and some galaxies in this strip of sky. **b.** Here, the nearby stars and galaxies have been subtracted out (gray) and the remaining glow enhanced, showing some structure from the time when the earliest stars and galaxies were forming.

unanswered questions

How can astrophysicists better understand dark matter? Will a way be developed to detect dark matter by means other than gravity? If dark matter is a particle, is it stable against decay? Experiments in particle accelerators on the ground and observations of Local Group halos in space have put some limits on the type of dark matter particles that might be out there, but an answer is still not apparent.

relatively low gravitational pull. Thus, one generation of short-lived massive stars may have been produced from the minihalos, but they may not have been able to hold gas to form a stable structure like a galaxy.

The First Galaxies

The first galaxies were made up of the first systems of stars gravitationally bound in a dark matter halo. Those stars may have been first-generation stars or chemically enriched second-generation stars. The properties of the first galaxies were shaped by the first stars: their radiation, their production of some massive elements, and the black holes resulting from the deaths of those stars. The masses of those galaxies are thought to have been about 100 million M_{Sun}, and they were built up hierarchically from the merging of minihalos (**Figure 23.11**).

One piece of evidence for that theory comes from infrared observations (**Figure 23.12**). Figure 23.12a includes the usual nearby stars and galaxies, but when those are all subtracted, a glow remains. That remaining structure (Figure 23.12b) probably arose from the first stars and galaxies several hundred million years after the Big Bang.

Another piece of evidence comes from the discoveries of galaxies and quasars at higher and higher redshifts, so we see them when they were very young. Those observations constrain the time line by indicating how soon the first galaxies—and first supermassive black holes—formed after the Big Bang. The peak of the spectrum of those early galaxies has been cosmologically redshifted into the infrared (**Working It Out 23.2**). The near-infrared instruments installed on the Hubble Space Telescope (HST) and the infrared instruments on the Spitzer Space Telescope and the Herschel Space Observatory have provided images of those very young, highly redshifted objects (**Figure 23.13**). The images of the highest-redshift objects are small and faint, without the detail seen in the pictures of closer galaxies. Astronomers are excited by those images because just detecting those objects contributes to an understanding of when and how galaxies formed.

The first galaxies are thought to have formed by about $z \approx 15$ to 11, or by several hundred million years after the Big Bang. For the second generation of stars, the physics of formation was more complicated than for the first, so astronomers need to account for heavy elements and dust from the first generation mixed into the halo, magnetic fields, and turbulence. The heavier elements carbon, oxygen, and iron cooled the gas, which then collapsed to the center of that larger dark matter halo, probably to a disk, and stars formed in the dark matter halo. Those highest-redshift—that is, youngest—galaxies appear to be small (1/20 the size of the Milky Way), which adds support for the bottom-up models of galaxy formation. Only in the past decade have astronomers identified galaxies at $z = 7$ to 11, so as yet there are fewer confirmed galaxies from that time than older galaxies at $z = 2$ to 6.

The Local Group has small, faint dwarf galaxies orbiting the Milky Way (see Chapter 20). Streams of material from the dwarfs are falling onto the Milky Way, indicating that the Milky Way is still accreting mass. Recently, scientists have discovered many more of those dwarf galaxies. About a few dozen of the faint dwarf galaxies are called **ultrafaint dwarf galaxies** because they are dim—only 1000–100,000 times the Sun's luminosity. They contain mostly old, faint stars with small amounts of heavy elements. One such galaxy is Segue 1. It contains only about a thousand stars and fewer heavy elements than any other observed galaxy, suggesting it may not have had much more star formation after its first stars formed (unlike the additional bursts of star formation that occurred in regular dwarf galaxies). The ultrafaint dwarfs may have contributed to building the Milky Way's halo, and

Observing High-Redshift Objects

New instruments and new telescopes constructed in the past few years enable astronomers to detect galaxies at higher and higher redshifts. At high redshifts, the observed wavelengths of radiation are very different from the wavelengths emitted (see Working It Out 21.2). The equation is as follows:

$$1 + z = \frac{\lambda_{observed}}{\lambda_{emitted}}$$

where z is the redshift. For $z = 1$, the observed wavelength is twice that emitted; for $z = 2$, it is 3 times that emitted; and so on. For high-redshift galaxies with $z = 8$ to 11, the observed wavelengths are 9–12 times that emitted.

Neutral hydrogen in gas clouds along the line of sight to a distant galaxy scatters light at a wavelength of 121.6 nanometers (nm). As a result of that scattering by many clouds, a galaxy's spectrum is observed to be brighter at emitted wavelengths longer than 121.6 nm and is fainter, or "drops out," at emitted wavelengths shorter than 121.6 nm. That dropout is noticeable even in a weak spectrum from a very faint galaxy and even if other spectral details cannot be discerned. For a nearby galaxy at $z = 0$, that dropout occurs in the far-ultraviolet, but for distant galaxies the dropout is redshifted.

For example, at $z = 9$:

$$1 + 9 = \frac{\lambda_{observed}}{121.6\,nm}$$

Solving for the observed wavelength of the dropout: $\lambda_{observed} = 10 \times 121.6\,nm = 1216\,nm = 1.2\,microns\,(\mu m)$.

Galaxies at the highest redshifts are detected in observations at wavelengths longer than 1.2 μm in the near-infrared, but they do not show up in optical images. The infrared James Webb Space Telescope (the successor to HST) should observe many high-redshift galaxies.

they may be the oldest galaxies around that have been involved in galaxy mergers. Ultrafaint dwarfs may even be the fossil remains of the first galaxies or of the first minihalos. Because stars in those galaxies move too fast to be bound by the luminous matter, those very old galaxies are more massive than their luminosity suggests. Dark matter dominates even these small galaxies.

Parallels between Galaxy Formation and Star Formation

Both star formation (see Chapter 15) and galaxy formation involve the gravitational collapse of vast clouds to form denser, more concentrated structures. To help you see some similarities and differences, this section compares galaxy and star formation after the first generation.

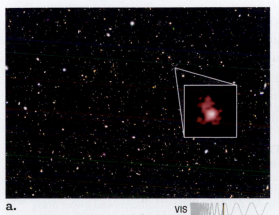
a.

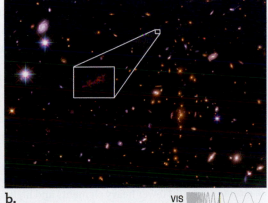
b.

Figure 23.13 a. This image from the Hubble Space Telescope shows a high-redshift galaxy, GN-z11 (at the X; enlarged in the inset). Spectroscopy indicates the galaxy is at redshift $z = 11.1$, seen as it was 13.4 billion years ago, when the universe was only 400 million years old. **b.** Hubble and Spitzer image of a $z = 10$ galaxy that has been gravitationally lensed. This galaxy is estimated to be only 2500 light-years in diameter.

Gravitational Instability   In both star and galaxy formation, the collapse begins with a gravitational instability. Regions only slightly denser than their surroundings are pulled together by their own self-gravity. As the matter in those regions becomes more compact, gravity becomes stronger, and the collapse process snowballs. One key difference between galaxy and star formation is that for a galaxy to form, the dark matter clump must collapse rapidly enough to counteract the overall expansion of the universe itself.

Fragmentation   The order of fragmentation differs between star and galaxy formation. In molecular clouds, first large regions begin to collapse, and then they fragment further to form individual stars. In contrast to that "top down" star formation process, galaxy formation is "bottom up": Smaller structures collapse first and then merge to form galaxies and, eventually, assemblages of galaxies.

Compression, Heating, and Thermal Support   As an interstellar molecular cloud collapses, its temperature increases, causing the pressure in the cloud to increase. The higher pressure would eventually be enough to prevent further collapse, except that the cloud core can radiate away thermal energy. That energy is the bright infrared radiation that enables astronomers to see star-forming cores. Compare that process with galaxy formation: As a dark matter clump collapses, the random velocities of its particles increase, and it too quickly reaches a point at which gravity is balanced with the random motions of the dark matter particles. Dark matter *cannot* radiate away energy, however, so once that balance is reached, the collapse of the dark matter is over. Only the normal matter within the cloud of dark matter can radiate away thermal energy and continue collapsing. That's why normal matter collapses to form galaxies, whereas dark matter remains in much larger dark matter halos.

As galaxies form, dark matter remains in extended halos. Dark matter may be the dominant form of matter in the universe, and it may determine the structure of galaxies, but dark matter can never collapse enough to play a role in the processes that shape stars, planets, or the interstellar medium.

Angular Momentum and the Formation of Disks   Conservation of angular momentum is responsible for the formation of disk galaxies, just as it is responsible for the formation of the accretion disks around young stars and for the flatness of both the Milky Way and the Solar System. The origin of the angular momentum is different, though. Whereas turbulent motions within star-forming molecular clouds produce the net angular momentum for stellar disks, gravitational interactions with nearby clumps are responsible for the angular momentum of the Milky Way.

Timescale   The time a star spends as a protostar is a small fraction of the time it will spend on the main sequence. The formation of a galaxy is a slower process, taking from redshift 10–20 down to redshift 1–2, or about half the current age of the galaxy.

The End Product   Once a stellar accretion disk forms, most of the matter moves inward and is collected into a star. In contrast, much of the matter in a spiral galaxy remains in the disk (see Chapter 19).

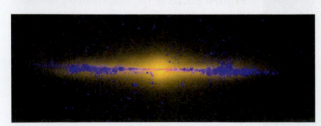

Figure 23.14 This supercomputer simulation of the formation of a Milky Way–sized spiral galaxy has reproduced the small bulge and big disk by using the Lambda-CDM model. At the bottom is an edge-on view of the galaxy. Blue colors indicate recent star formation, whereas older stars are redder.

CHECK YOUR UNDERSTANDING **23.3**

The first stars formed in the universe had _____ than the stars formed today.
(a) more heavy elements and higher mass; (b) more heavy elements and lower mass; (c) fewer heavy elements and higher mass; (d) fewer heavy elements and lower mass

23.4 Galaxies Evolve

Galaxies continued to evolve hierarchically in the young universe, with smaller "protogalactic" fragments merging to form larger ones. Those early fragments and galaxies were closer together because the universe was smaller, and therefore mergers were more common. The universe now is $z + 1$ times larger than it was when the light was emitted from a galaxy at redshift z (see Chapter 21), with a volume $(1 + z)^3$ times larger than its volume then. Computer simulations indicate that small concentrations of normal matter within the dark matter would have clumped and collapsed under their own gravity as they radiated and cooled, forming clumps of normal matter that ranged from the size of globular clusters to the size of dwarf galaxies.

In a large spiral galaxy such as the Milky Way, faint dwarf spheroidal galaxies (with dark matter) and the oldest globular clusters (without much dark matter) may be leftover protogalactic fragments. The gas collapsed to form a rotating disk as it cooled. A supercomputer simulation that included dark matter, gravity, star formation, and supernova explosions reproduced a Milky Way–like galaxy with a large disk and a small bulge (**Figure 23.14**). Observationally, astronomers conduct "stellar archaeology" on the oldest parts of the Milky Way to better understand how the components are all assembled into a galaxy. For example, the oldest globular clusters may be 1 billion to 2 billion years older than the halo. Most stars in the Milky Way formed between 11 billion and 7 billion years ago. The disk and bulge formed at about that same time.

The Most Distant Galaxies

Figure 23.15 shows images of galaxies throughout the history of the universe. The galaxies observed in the very early universe, before about 11 billion years ago, are so faint that no structure is visible. Observations of galaxies at about 11 billion years ago have shown visible structure much less regular than that of galaxies today. Even at 4 billion years ago, galaxies were much more irregular. Early irregular galaxies are merging galaxies: When galaxies were closer together, more mergers

Figure 23.15 A comparison of the Hubble classification of galaxies today (spirals on the top, barred spirals on the bottom, elliptical galaxies in the middle) with galaxies throughout the history of the universe. More irregular galaxies existed in the past, indicating that spirals took some time to form.
Credit: NASA, ESA, M. Kornmesser, https://esahubble.org/images/heic1315a/. https://creativecommons.org/licenses/by/4.0/.

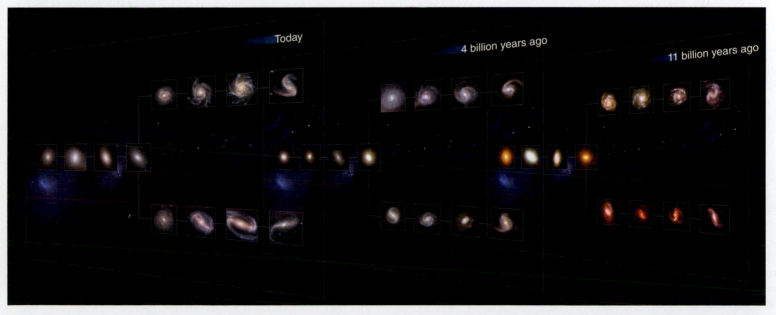

Figure 23.16 Chandra X-ray observations (red, orange, and yellow), combined with Hubble Space Telescope observations (blue and white), show that NGC 6240 has two black holes less than 1000 parsecs apart. The black holes at the center of the white features will probably merge in about 100 million years.

XRAY   VIS

were taking place. Most of today's galaxies conform to the Hubble classification, and only about 10 percent are irregular. Four billion years ago, however, more than half the galaxies were irregular, the number of elliptical and S0 galaxies was about the same, and many fewer spirals existed. That difference in galaxy types at different times suggests that spirals took time to form. Those later mergers probably produced spiral galaxies over time.

The bottom-up hierarchical merging also may have triggered the formation of the supermassive black holes at the centers of galaxies. The first supermassive black holes, which power the distant quasars seen at $z = 7$ with masses of $10^9\ M_{Sun}$, could have grown from the merging of minihalos with stellar black holes left after the first stars. Or they could have formed through the accretion of gas from the material between the galaxies during mergers of the first galaxies or through rapid collapse from hot, dense gas at the center of the first galaxies. In nearby galaxies, the mass of the supermassive black hole and the bulge properties are related, suggesting that the growth of the black hole and the bulge might have been linked when they were younger. Supermassive black holes could have grown even more massive from the mergers of large galaxies, too. **Figure 23.16** shows a nearby galaxy with two supermassive black holes about 900 pc (3000 light-years) apart that are merging.

The hierarchical merging and growth of the supermassive black holes also affected the rates of star formation in evolving galaxies. The tidal interactions between merging galaxies and the collisions between gas clouds probably triggered many regions of star formation throughout the combined system. Star formation in the universe increased sporadically, including a rapid increase in the 200 million years between $z = 10$ and 8. The graph in **Figure 23.17** shows that the star formation rate seems to have peaked around $z = 3$ (2.5 billion to 3 billion years after the Big Bang) before decreasing again to the current rate.

By observing galaxy mergers (**Figure 23.18a**) at different distances, astronomers can see how mergers differ at various times in the history of the universe. Large ellipticals are now thought to result from the merger of two or more spiral galaxies.

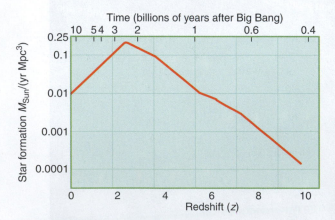

Figure 23.17 The rate of star formation per volume has changed in the universe, peaking at about 2 billion to 3 billion years after the Big Bang.

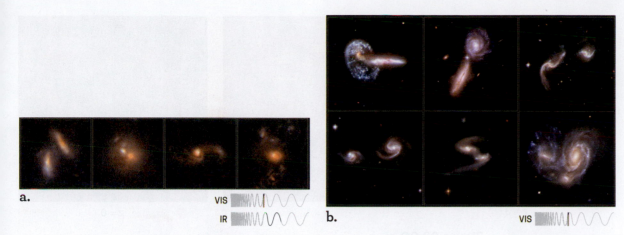

a. VIS ～～～～
IR ～～～～

b. VIS ～～～～

The dark matter halos of the galaxies merge, and the stars eventually settle down into the blob-like shape of an elliptical galaxy. Elliptical galaxies are more common in dense clusters, where mergers are likely to have been more frequent. Compare those young mergers with those of closer, older galaxies (**Figure 23.18b**—a computer simulation of such a merger is shown in Figure 6.31).

Just as galaxies merge, so do clusters of galaxies. **Figure 23.19a** shows the high-speed collision and merging of two galaxy clusters in the Bullet Cluster. Images in visible light show the individual galaxies. Ordinary matter, mostly hot gas, is seen in X-rays (shown in red), and the distribution of the total mass can be found from the gravitational lensing produced by the clusters. The ordinary matter slowed down in the collision, but the dark matter (shown in blue) did not. That separation serves as evidence for dark matter in galaxy clusters. The collision of four galaxy clusters is shown in **Figure 23.19b**. Clusters of galaxies also evolve hierarchically, growing from smaller structures to larger ones. As with galaxies themselves, younger, distant clusters are messier than older, nearby ones, yielding more evidence that the formation of structure in the universe was hierarchical.

Simulating Structure

Astronomers use the most powerful supercomputers available to simulate the universe. Those simulations start with billions of particles of dark matter and use the

Figure 23.18 a. These HST images show young, merging galaxies. From left to right, galaxies at 2.4 billion to 6.2 billion light-years from Earth merging at 11 billion to 7.5 billion years after the Big Bang. **b.** These tidally interacting galaxies in the more nearby universe show severe distortions, including stars and gas drawn into long tidal tails.

▶❚❚ **AstroTour:** Galaxy Interactions and Mergers

a. XRAY ～～～～
VIS ～～～～

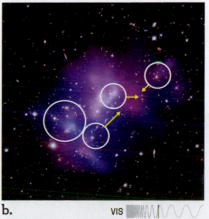

b. VIS ～～～～

Figure 23.19 a. The Bullet Cluster of galaxies, at a redshift of $z = 0.3$, represents a later stage in the merging of two giant clusters of galaxies. The smaller cluster on the right seems to have moved through the larger cluster like a bullet. **b.** Four galaxy clusters (circled) are merging in the direction of the yellow arrows. In the Chandra X-ray Observatory image, the cooler gas is magenta and the hotter gas is blue. This is one of the most complex galaxy clusters observed.

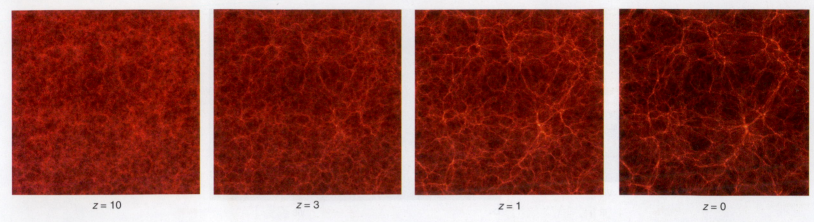

z = 10   z = 3   z = 1   z = 0

Figure 23.20 The Bolshoi supercomputer simulation of the evolution of dark matter shows how slight variations in density after inflation led to the formation of very large-scale structures from higher-density regions. These images show the growth of filaments and voids.

most recent observations of the CMB. The simulations model the formation and evolution of dark matter clumps and halos, filaments and voids, small and large galaxies, and galaxy groups and clusters. Those computations also simulate the flow of ordinary gas within those structures as stars form, and researchers use those simulations to create images of what the universe should have looked like at different times (and different redshifts). Those images are then compared with images of the actual universe. That comparison sets limits on the parameters of the universe: the amount of mass, for example, or the type of dark matter. If the inputs to the simulation are correct, the two sets of images will look very similar. If not, the two sets of images will look very different.

In one example, a Lambda-CDM simulation called the Bolshoi was run on NASA supercomputers. The simulation shows that slight variations in density after inflation led to higher-density regions that became the seeds for the growth of structure (**Figure 23.20**). During the first few billion years, dark matter fell together into structures comparable in size to today's clusters of galaxies. The spongelike filaments, walls, and voids became well defined later. Zooming in on some simulated filaments and voids shows a cluster of galaxies (**Figure 23.21**). The similarities between the results of the models and observations of large-scale structure are remarkable. **Figure 23.22** compares the simulated view with the observed slice of the universe from the Sloan Digital Sky Survey (SDSS). Only simulations with certain combinations of mass, CMB variations, types of dark matter, and values for the cosmological constant produce structure similar to what is actually observed. That result is very important. Models incorporate assumptions consistent with observational and theoretical knowledge of the early

Figure 23.21 Simulations enable astronomers to model structure at different-sized scales. Each image zooms in more to show smaller structures, down to the size of a cluster of galaxies. Dark matter halos are seen as bright blobs. The smallest blob in the last image could become a giant spiral galaxy like the Milky Way.

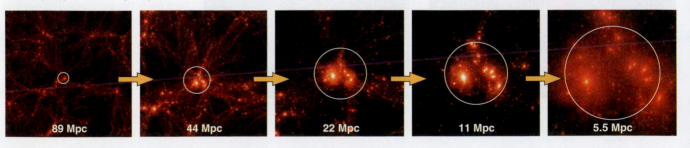

89 Mpc   44 Mpc   22 Mpc   11 Mpc   5.5 Mpc

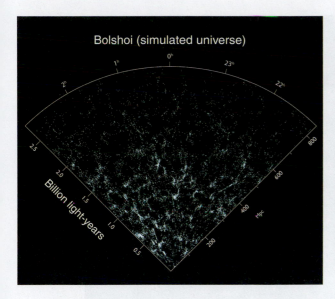

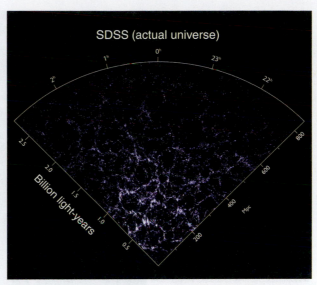

universe, and they predict the formation of large-scale structure similar to what is actually seen in today's universe.

The precision cosmology developed over the past two decades has given astronomers a detailed model of the universe in space and time, so that galaxy evolution can now be reliably sequenced. **Figure 23.23** neatly summarizes the galaxy formation process, in which smaller objects form first and merge into ever-larger structures, leading ultimately to the Hubble Ultra Deep Field 2014 image shown in **Figure 23.24**.

Figure 23.22 Comparing the large-scale structure of dark matter halos produced by the Bolshoi simulation (left) to the distribution of galaxies observed in the Sloan Digital Sky Survey (SDSS; right) shows that the simulated universe has properties similar to the actual universe.

Predictions about the Deep Future

What does the future hold for the universe and its structure? Scientists use well-established physics and our current cosmological understanding to speculate how the existing structures in the universe will evolve over a very long time. Any speculation might change with newer discoveries.

Figure 23.23 A schematic view of how structure formed in the universe, from smaller systems to larger ones.
Credit: NASA, ESA, and A. Feild (STScI), https://esahubble.org/images/heic0805c/. https://creativecommons.org/licenses/by/4.0/.

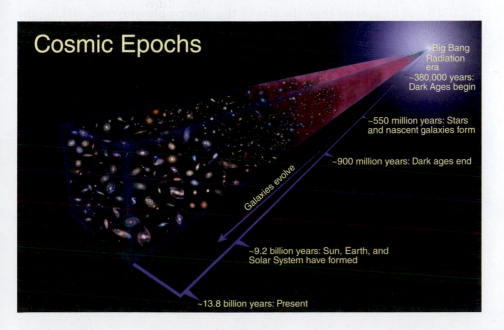

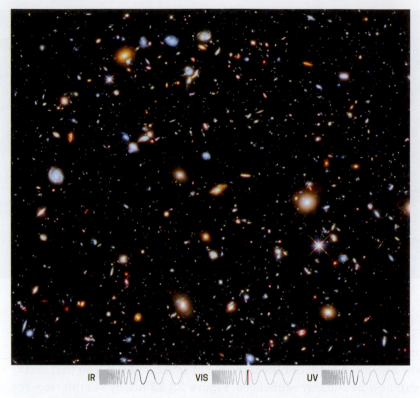

Figure 23.24 The Hubble Ultra Deep Field 2014 image shows about 10,000 galaxies in a very small area of the sky in the Southern Hemisphere. This image is a composite of ultraviolet, visible, and infrared light captured over a period of ten years.

In the distant future, studying galaxies and cosmology will be much more difficult, if not impossible. In 90 billion years or so, the only visible galaxy will be the one that resulted from the merger of the Milky Way and Andromeda (see Chapter 20). Not only will star formation have essentially ceased and most stars have burned out, but also, because of the acceleration of the expansion of the universe, the other galaxies will be too far away to be detectable from here. If exponential expansion continues, in a trillion years, even the wavelength of the CMB will be greater than the size of the observable universe, making the CMB invisible too. In 100 trillion (10^{14}) years from now (about 10,000 times as long as the current age of the universe), the last molecular cloud will collapse to form stars, and a mere 10 trillion years later the least massive of those stars will evolve to form white dwarfs.

At that time, most of the normal matter in the universe will be locked up in degenerate stellar objects: brown dwarfs, white dwarfs, and neutron stars. Unless a Big Rip (where everything is torn apart by expansion) occurs, every condensed object, including white dwarfs and neutron stars, will ultimately decay via quantum effects. Eventually, the only significant remaining concentrations of mass will be black holes, ranging from small ones with the masses of single stars to supermassive black holes that grew to the size of galaxy clusters. Those black holes will slowly evaporate into elementary particles through the emission of Hawking radiation (see Chapter 18). A black hole with a mass of a few solar masses will evaporate into elementary particles in 10^{65} years, and galaxy-sized black holes will evaporate in about 10^{98} years. By the time the universe reaches an age of 10^{100} years, even the largest black holes will be gone. A universe vastly larger than our current one will contain little but photons with colossal wavelengths, neutrinos, electrons, positrons, and other waste products of black-hole evaporation. The universe will continue to expand forever—into the long, cold, dark night of eternity. Ultimately, the universe will be structureless and empty except for any residual uniform dark energy.

CHECK YOUR UNDERSTANDING 23.4

We expect the kinds of galaxies that we see at a redshift of $z = 4$ to be: (a) much like what we see today; (b) smaller and much more irregular looking than today; (c) far more numerous but with more spiral galaxies; (d) larger versions of what we see today.

what if . . .

What if you ran a computer simulation of the universe, including galaxy formation and the stars within galaxies? How would you need to change this simulation if you compared it to an infrared survey of galaxies rather than an optical survey of galaxies?

Origins: We Are the 4 or 5 Percent

The very first stars initiated post–Big Bang nucleosynthesis, which chemically enriched the universe with elements heavier than lithium, including the elements necessary to form planets and life on Earth. The first stars and galaxies "lit up" the universe and brought it out of the Dark Ages, thanks to the glow of their ordinary matter. But that ordinary matter is not the primary constituent of the universe. In their study of the largest structures in the universe—the galaxies and groups and clusters of galaxies—astronomers have concluded that those objects are dominated by dark matter, not the stuff of stars and planets.

Several types of independent observations suggest that dark matter accounts for about 25 percent of the universe (see Chapters 19–22). Dark matter dominated over normal matter when the first stars and galaxies formed. Dark matter dominated the evolution of the galaxies as they went through mergers to become the larger systems seen today. But despite decades of study, astronomers don't know exactly what dark matter is.

As the universe evolved, dark energy became more important to the structure of the universe. When it was younger and galaxies were forming, the universe was dominated by matter. The expansion of the universe was slowing because of the pull of gravity from all that matter. About 5 billion to 7 billion years ago, though, dark energy began to dominate over matter in the universe, and the expansion of the universe accelerated. Dark energy now makes up about 70 percent of the mass and energy of the universe—nearly 3 times as much as the dark matter—and astronomers don't know exactly what it is, either.

Ordinary matter is only about 5 percent of the mass and energy of the present-day universe. Most of what astronomers have studied since people first looked at the sky is that 5 percent. The parts of the local universe important to life on Earth—the Sun, its planets and their moons, and the local environment in the Milky Way—are composed of that 5 percent. The matter that constitutes us is a surprisingly small component of the universe.

SUMMARY

Galaxies are not distributed uniformly. Instead, they are clumped into groups, clusters, superclusters, and walls. Galaxies develop because of gravitational instabilities in the presence of cold dark matter. The first stars formed in minihalos of dark matter, whereas the first galaxies formed later, in larger dark matter halos. Over time, smaller galaxy fragments merged to form larger galaxies. Mergers still happen today. Dark matter and dark energy are responsible for the formation of structure and the future of the universe. Ordinary matter, which makes up stars, planets, and us, is only about 20 percent of the total mass in the universe and, once dark energy is included, only 4 or 5 percent of all the universe. Dark energy is playing an increasingly dominant role in the universe and will shape its future.

1 Describe the distribution of galaxies in the universe. Galaxies are hierarchically gathered into groups, clusters, superclusters, and larger structures. The walls surround voids in which very few galaxies are present. Dark matter dominates the mass of galaxy groups and clusters.

2 Explain how the large-scale structure of today's universe evolved from the structure that began to form shortly after the Big Bang. Structure formed in the universe as slight variations grew in the density of the dark matter emerging from the Big Bang. Those "seeds" then collapsed under the force of gravity, pulling in normal matter as well. Structure formed from the bottom up.

3 Describe the formation of the first stars and the first galaxies. The first stars formed in dark matter minihalos rather than in clouds of dust and gas. Because of the absence of heavy elements in those dark matter halos, star formation and evolution were different from that in the nearby universe. Gravitationally bound groupings of stars formed, which were the earliest galaxies. Radiation from the first stars and their supernovae affected the growth of the first galaxies. Radiation from the first stars, galaxies, and black holes ended the Dark Ages.

4 Relate the observations of galaxies at different redshifts to the evolution of the large-scale structure of the universe. Larger structures form from smaller structures. The earliest galaxies merged to become larger galaxies, which in turn accumulated into clusters. The clusters grew hierarchically, through mergers, to become larger clusters, super-clusters, and walls. Distant young galaxies look very different from the nearby galaxies in the present-day universe: They are smaller, fainter, and more likely to be merging.

QUESTIONS AND PROBLEMS

TEST YOUR UNDERSTANDING

1. Place the following in order of size, from smallest to largest.
   - **a.** a galaxy
   - **b.** star clusters
   - **c.** the Local Group
   - **d.** a wall
   - **e.** Virgo Cluster
   - **f.** Laniakea
   - **g.** a star

2. The dominant force in the formation of galaxies is
   - **a.** gravity.
   - **b.** angular momentum.
   - **c.** the electromagnetic force.
   - **d.** the strong nuclear force.

3. Larger galaxies form from the merging of small protogalaxies. That process is similar to the formation of
   - **a.** stars.
   - **b.** planets.
   - **c.** molecular clouds.
   - **d.** asteroids.

4. Which of the following is a characteristic difference between cold and hot dark matter? (Choose all that apply.)
   - **a.** temperature
   - **b.** ability to emit radiation
   - **c.** the way they clump under the influence of gravity
   - **d.** mass density

5. Gravitational lenses can be used to find
   - **a.** dwarf galaxies near the Milky Way.
   - **b.** dust and gas in the voids.
   - **c.** the masses of galaxy clusters.
   - **d.** the structure of Laniakea.

6. If dark energy is constant, the universe in the far distant future
   - **a.** will be cold and dark.
   - **b.** will be bright and hot.
   - **c.** will collapse and re-form.
   - **d.** will be the same as it is now, on large scales.

7. What is the primary difference between galaxy groups and galaxy clusters?
   - **a.** how tightly they are bound by gravity
   - **b.** the size of the largest galaxy
   - **c.** the total mass of the galaxies
   - **d.** dark matter does not exist in galaxy groups

8. Once the redshift of a galaxy has been found, its _____ is also known.
   - **a.** mass
   - **b.** velocity
   - **c.** distance
   - **d.** both b and c

9. Galaxy formation is similar to star formation because both
   - **a.** are the result of gravitational instabilities.
   - **b.** are dominated by the influence of dark matter.
   - **c.** end with the release of energy through fusion.
   - **d.** result in the formation of a disk.

10. Dark matter clumps stop collapsing because
    - **a.** angular momentum must be conserved.
    - **b.** they are not affected by normal gravity.
    - **c.** fusion begins, and radiation pressure stops the collapse.
    - **d.** the particles are moving too fast to collapse any further.

11. Giant elliptical galaxies come from
    - **a.** the gravitational collapse of clouds of normal and dark matter.
    - **b.** the collision of smaller elliptical galaxies.
    - **c.** the fragmentation of large clouds of normal and dark matter.
    - **d.** the merging of two or more spiral galaxies.

12. Which of the following statements describe the Dark Ages of the universe? (Choose all that apply.)
    - **a.** The first stars began forming during the Dark Ages.
    - **b.** The end of the Dark Ages coincided with reionization.
    - **c.** The Dark Ages lasted from 200 million to 600 million years after the Big Bang.
    - **d.** During the Dark Ages, photons could travel freely through the universe.

13. All stars that have been observed have elements heavier than lithium. What does that imply about the first stars?
    a. They must have died before galaxies were fully formed.
    b. The first stars did not form until after galaxies formed.
    c. The first stars must have had very low masses.
    d. The first stars must have been enriched in heavy elements.

14. Reionization of the neutral gas in the universe occurred because of
    a. the decay of dark matter particles.
    b. the emission of neutrinos by the first stars that formed.
    c. the release of jets of charged particles from supermassive black holes.
    d. the radiation from the first stars, supernovae, and black holes that formed.

15. Place the following in increasing order of size.
    a. the fraction of the universe that is stars, planets, dust, and gas
    b. the fraction of the universe that is dark energy
    c. the fraction of the universe that is dark matter

THINKING ABOUT THE CONCEPTS

16. ★ WHAT AN ASTRONOMER SEES   Count the number of galaxies that you can see in the inset image in Figure 23.6. From this number, estimate the total number of galaxies visible in Figure 23.6 to the nearest power of 10. (Does the image contain 10 galaxies? 100? 1000? And so on.) This image spans only about an arcminute (1/30 of the diameter of the full Moon). Take a moment to consider that every piece of the sky looks like this, and write a sentence or two reacting to your conclusion about the number of galaxies in the observable universe. ⊙

17. Suppose you could view the early universe when galaxies were first forming. How would it be different from today's universe?

18. Are voids likely to be filled with dark matter? Why or why not?

19. Imagine that the universe has galaxies composed mostly of dark matter, with relatively few stars or other luminous normal matter. If that were true, how might you learn of the existence of such galaxies?

20. How are star formation and galaxy formation similar? How do they differ?

21. What is the origin of large-scale structure?

22. Why is dark matter essential to galaxy formation?

23. Why does the current model of large-scale structure require dark matter?

24. What is the difference between a galaxy cluster and a supercluster? Is our galaxy part of either? How do we know?

25. The theory of cosmology assumes that on large scales, the structure in the universe is uniform no matter where you look. Maps of structure, like the ones shown in Figure 23.3, support that assumption. Does the presence of large masses such as Laniakea violate that principle? Explain your answer. ⊙

26. What are some observational signs that dark matter exists? Explain why that evidence challenged earlier theories and forced astronomers to change their minds about the existence of matter they could not see.

27. Using the current model of galaxy formation, describe how galaxies should appear as you look further back in time. Are the features you described observed?

28. As clumps containing cold dark matter and normal matter collapse, they heat up. When a clump collapses to about half its maximum size, the increased thermal motion of particles tends to inhibit further collapse. Whereas normal matter can overcome that effect and continue to collapse, dark matter cannot. Explain the reason for that difference.

29. Describe structure formation in the universe, starting at recombination and ending today.

30. Why do scientists think that gravity, and not the other fundamental forces, is responsible for large-scale structure?

APPLYING THE CONCEPTS

31. Suppose a dwarf galaxy is orbiting a giant elliptical galaxy at the center of a cluster at a distance of 4 Mpc and a speed of 800 km/s. Estimate the core mass of the cluster. ●–■–●
    a. Make a prediction: Review Working It Out 23.1. Do you expect the core mass of a cluster of galaxies to be approximately billions, trillions, tens of trillions, or hundreds of trillions of solar masses?
    b. Calculate: Find the core mass of the cluster.
    c. Check your work: Verify that your answer has units of solar masses, and compare your answer to your prediction.

32. The Hα line of hydrogen is emitted with a wavelength of 656.28 nm. What is the observed wavelength of the Hα line from a galaxy with a redshift of 8? ●–■–●
    a. Make a prediction: Will the observed wavelength be larger or smaller than the emitted wavelength? By just a few tenths, or by a significant amount?
    b. Calculate: Follow Working It Out 23.2 to find the observed wavelength.
    c. Check your work: Verify that your answer has units of nm, and compare your answer to your prediction.

33. Suppose that you have independently measured the core mass of a cluster to be $7 \times 10^{14}\ M_{Sun}$. You observe a small galaxy orbiting at a speed of 700 km/s. How far is the small galaxy from the cluster core, in Mpc?

34. The Hα line of hydrogen is emitted with a wavelength of 656.28 nm. Suppose that you observe a galaxy in which the Hα line has been shifted to an observed wavelength of 4987.73 nm. What is the redshift of the galaxy?

35. Suppose that you observe a very strong absorption line in a galaxy at a wavelength of 4080.775 nm. From observing other spectral lines, the redshift of the galaxy is known to be 8.3. What is the rest wavelength of this absorption line? Use the Internet to find out which element emitted this line.

36. Figure 23.3a shows the redshifts and velocities for many galaxies. Find the average recession velocity for the galaxies in the wall indicated by the line labeled "Walls." ⊙

37. If 300 million neutrinos fill each cubic meter of space, and if neutrinos account for only 5 percent of the mass density (including dark matter) of the universe, estimate the mass of a neutrino.

38. What are the approximate masses of (a) an average group of galaxies, (b) an average cluster, and (c) an average supercluster?

39. The lifetime of a black hole varies in direct proportion with the cube of the black hole's mass. How much longer does a supermassive black hole of 3 million M_{Sun} take to decay than a stellar black hole of 3 M_{Sun}?

40. Knowing what elliptical galaxies are made of, estimate how old they must be. Knowing that ellipticals form via mergers of spirals, and knowing when galaxies first formed, estimate how long the merging events that formed the elliptical galaxies seen today took to complete.

41. The Bullet Cluster image in Figure 23.19a shows the collision of two galaxy clusters. Estimate the number of galaxies you can see in each cluster. ⊙

42. The initial fluctuations leading to large-scale structure probably arose from quantum fluctuations in the early universe. How would the universe look different today if those fluctuations had been 10 times bigger? What about 10 times smaller?

43. Is the early universe on the left or the right in the graph in Figure 23.17? (Alternatively, you could wonder whether "now" was on the left or right in the graph.) Compare the star formation rate today with the star formation rate at the peak. How much more star formation occurred during the peak than occurs now? ⊙

44. Figure 23.22 shows real data in the right panel and simulated data in the left panel. Those two panels are not in exact agreement. Do those differences indicate a significant problem in the simulation's ability to represent reality? Why or why not? ⊙

45. Compare galaxies with redshifts of z = 0.5, 4, and 8. About how old was the universe when the light was emitted from each galaxy? At what spectral wavelengths would you see the "dropout" of the spectrum? Can you observe those dropouts from the ground, or do you need a telescope in space?

EXPLORATION   The Story of a Proton

Now that you have surveyed the current astronomical under-standing of the universe, you are prepared to put the pieces together to make a story of how you came to be sitting in your chair and reading these pages. Taking a moment to work your way backward through the book is valuable: start at the Big Bang and review all the intervening steps that had to occur, back to the beginning of the book, which began with looking at the sky.

1 In the Big Bang, how did a proton form?

_____

_____

_____

_____

2 How might that proton have become part of one of the first stars?

_____

_____

_____

_____

3 Suppose that proton later became part of a carbon atom in a 4-M_{Sun} star. Through what type of nebula would it have passed before returning to the interstellar medium?

_____

_____

_____

_____

4 Suppose that carbon atom then became part of the molecular-cloud core forming the Sun and the Solar System. What two physical processes dominated the core's collapse as the Solar System formed and that carbon atom became part of a planet?

_____

_____

_____

_____

5 Beginning with the Big Bang, create a time line that traces the full history of a proton that becomes a part of the nucleus of a carbon atom in you.

_____

_____

_____

_____

_____

Life

When astronomers look for worlds that might harbor life, they look for water, as you will learn here in Chapter 24. Life as we know it depends on water because water is a terrific solvent; it can carry chemicals, minerals, and nutrients around in an organism. It has this property because it is a polar molecule—the hydrogen end of the molecule is positively charged, while the oxygen end is negatively charged. Water, therefore, is attracted to lots of different molecules. It can be attracted, in fact, to anything that carries a charge. Rub a plastic object, like a pen, a ruler, or an inflated balloon, all over your dry hair, so that the plastic object becomes electrically charged. Turn on the water from a faucet so that a thin stream falls from the tap. Take a moment to predict what will happen when you bring the charged plastic object close to the water. Then slowly bring the plastic object close to the stream of water. Sketch the result.

EXPERIMENT SETUP

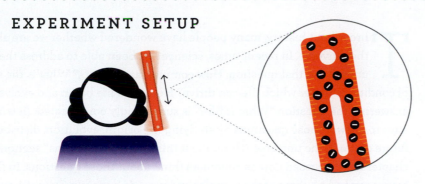

Rub a plastic object, like a pen, a ruler, or an inflated balloon, all over your hair, separating charges, so that the plastic object becomes electrically charged.

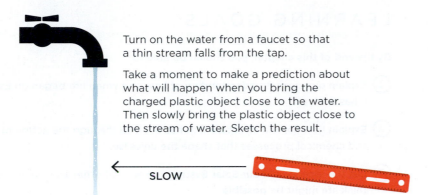

Turn on the water from a faucet so that a thin stream falls from the tap.

Take a moment to make a prediction about what will happen when you bring the charged plastic object close to the water. Then slowly bring the plastic object close to the stream of water. Sketch the result.

SLOW

PREDICTION

I predict that the stream of water will:

- ☐ **bend away from the object.**
- ☐ **bend toward the object.**
- ☐ **fall straight down.**
- ☐ **yank the object out of my hand.**

SKETCH OF RESULTS (in progress)

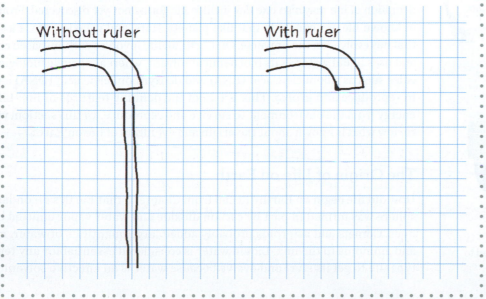

Without ruler

With ruler

24

Throughout history, many people have wondered whether we are alone in the universe. In recent times, science has been able to address the issues underlying that question: How common are planets? What is the range of conditions under which life can thrive? How does life begin and evolve? Even answering the question "What is life?" is surprisingly complicated. **Astrobiology** aims to answer those questions by studying the origin, evolution, distribution, and future of life in the universe. Throughout this book, the "Origins" section in each chapter has discussed how astronomers think about those questions. In this chapter, we expand on some of those topics and provide a more systematic overview of how scientists think about life in the universe and how they search for signs of it.

LEARNING GOALS

By the end of this chapter, you should be able to:

(1) Explain our current understanding of how and when life began on Earth and how it has evolved.

(2) Explain how life is a structure that has evolved through the action of the physical and chemical processes that shape the universe.

(3) List the locations in our Solar System and around other stars where astronomers think life might be possible.

(4) Describe some methods used to search for intelligent extraterrestrial life.

24.1 Life Evolves on Earth

What is **life**? Many scientists suggest that no single definition of life would encompass all the forms of life that may exist in the universe. To date, we have discovered only a single example of life: that found here on Earth. But even comparing varied organisms within that single example leads to complications in the definition of life. Viruses, for example, meet some criteria for life but not others. Life on Earth may be very different from life found in other places in the universe, and a complete definition of life may also one day include life-forms not yet discovered. From studies of known life, we conclude that like planets, stars, and galaxies, life is a structure that has evolved in the universe. On Earth, all life involves carbon-based chemistry and uses liquid water as its biochemical solvent, whereas specific biological molecules such as ribonucleic acid (RNA) and deoxyribonucleic acid (DNA) enable life to reproduce and evolve. This is one of the defining features of life: Life draws energy from the environment to survive and reproduce. In this section, we briefly review what is known about the origin and evolution of life on Earth.

The Origin of Life on Earth

How did life begin on Earth? Recall from Chapter 7 that Earth's secondary atmosphere was formed in part by carbon dioxide and water vapor emitted by volcanoes. Comets and asteroids probably added large quantities of water, methane, and ammonia to the mix. Liquid water is considered essential for any terrestrial-type life to get its start and evolve because water is an effective solvent that can

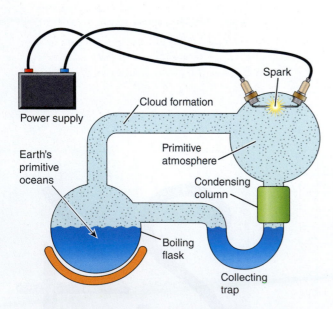

Figure 24.1 The Urey-Miller experiment was designed to simulate conditions in an early-Earth atmosphere.

move other atoms and molecules around, making them more accessible to cells. Early Earth had abundant sources of energy, such as lightning and ultraviolet solar radiation, that fragmented those simple molecules. Those fragments later reassembled into molecules of greater mass and complexity. Some of those were organic molecules—that is, molecules that contained hydrogen and carbon. Rain carried the heavier molecules out of the atmosphere into Earth's oceans, forming a primordial soup.

In 1952, chemists Harold Urey (1893–1981) and Stanley Miller (1930–2007) attempted to create something similar to those early-Earth conditions. Using equipment illustrated in **Figure 24.1**, they placed water in a sterilized laboratory jar to represent the ocean and then added methane, ammonia, and hydrogen as a primitive atmosphere; electric sparks simulated lightning as a source of energy. Within a week, the Urey-Miller experiment yielded molecules associated with life—namely, amino acids and components of nucleic acids. Proteins, the structural molecules of life, are made of 20 amino acids. Eleven of those acids were synthesized in the Urey-Miller experiment. Nucleic acids are the precursors of RNA and DNA.

Additional sealed samples from that old experiment were examined 50 years later. In those samples, hydrogen sulfide had been added to the "primitive atmosphere." When the samples were analyzed, 23 amino acids were found. That finding suggests that hydrogen sulfide, which would have come from volcanic plumes in the early Earth, was important. More recent experiments with carbon dioxide and nitrogen as the primitive atmosphere have produced results similar to those of Urey and Miller. A plausible atmospheric composition with an energy source can produce significant quantities of amino acids and other substances important to life.

From laboratory experiments such as those, scientists have developed various models to explain how life might have begun in an early-Earth environment. However, the details of how those precursor molecules evolved into the molecules of life are not yet clear. Some biologists think life began in the ocean depths, where volcanic vents provided the localized heat needed to create the highly organized molecules responsible for biochemistry (**Figure 24.2**). Other researchers think that life originated in tide pools, where lightning and ultraviolet radiation supplied the energy (**Figure 24.3**). Some researchers think life may have evolved in shallow lakes further inland. In either case, short strands of molecules that could replicate themselves may have formed first, later evolving into RNA, and finally into DNA, the huge molecule that serves as the biological "blueprint" for self-replicating organisms.

A few scientists have suggested that life on Earth may have been "seeded" from space in the form of microorganisms brought here by meteoroids or comets. This idea is called "panspermia." However, while quite long and complex carbon chains have been observed in nebulae, no scientific evidence exists to support this much more speculative hypothesis about *life* beginning in star-forming nebulae or in the outer reaches of the Solar System. Furthermore, seeding might explain how life came to Earth, but it does not satisfactorily answer the question: How did life begin?

When Life Began

If life did indeed get its start in Earth's oceans, when did it happen? Recall from Chapter 7 that Earth was bombarded by Solar System debris for several hundred

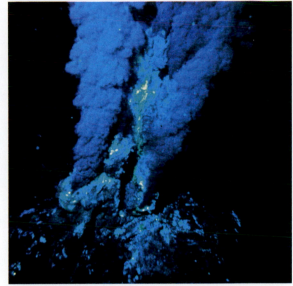

a.

b.

Figure 24.2 **a.** Life on Earth may have arisen near ocean hydrothermal vents such as this one. Similar environments might exist elsewhere in the Solar System. **b.** Living organisms around hydrothermal vents, such as the giant tube worms shown here, rely on hydrothermal rather than solar energy for their survival.

Figure 24.3 Life may have begun in tide pools, where lightning and ultraviolet light supply the energy for chemical processes.

Figure 24.4 These modern-day stromatolites are growing in colonies along an Australian shore.

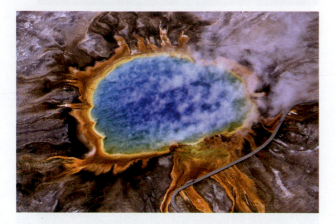

Figure 24.5 These thermophiles in the Grand Prismatic Spring in Yellowstone National Park live in temperatures of 70°C. The colors result from different amounts of chlorophyll.

million years after it formed roughly 4.6 billion years ago. Those conditions might have been too harsh for life to form and evolve on Earth. Once the bombardment abated and oceans formed, the opportunities for life to begin greatly improved. Terrestrial life seems to have quickly taken advantage of that more favorable environment. In the Nuvvuagittuq belt in Quebec, Canada, scientists recently found what they think are fossilized microorganisms that are 3.77 billion to 4.28 billion years old and seem similar to microorganisms from modern hydrothermal vents. Scientists debate whether carbonized material in Greenland rocks dating back 3.65 billion to 3.85 billion years provides indirect evidence of early life. Stronger and more direct evidence for early life appears in the form of fossilized **stromatolites** (masses of simple microorganisms) that date back about 3.5 billion years. Fossilized stromatolites have been found in western Australia and southern Africa, and living examples still exist today (**Figure 24.4**). An analysis of eleven 3.465-billion-year-old microfossils from western Australia found five species—two of which carried out primitive photosynthesis and the other three produced or consumed methane. That evidence suggests that the earliest life formed less than a billion years after the Solar System formed and within a few hundred million years of the end of the late heavy-bombardment period.

All life on Earth shares a similar genetic code that originated from a common ancestor. Close comparison of DNA of different species enables biologists to trace backward to the time when different types of life first appeared on Earth and to identify the species from which those life-forms evolved. The earliest organisms were **extremophiles**—life-forms that not just survive but actually thrive under extreme environmental conditions. Extremophiles include organisms such as thermophiles, which flourish in water temperatures as high as 120°C that occur near deep-ocean hydrothermal vents, such as the one shown in Figure 24.2a. Other extremophiles thrive in conditions of extraordinary cold, low oxygen, salinity, pressure, dryness, acidity, or alkalinity. Scientists today study extremophiles in boiling-hot sulfur springs in Yellowstone National Park, in salt crystals beneath the Atacama Desert in Chile, at the bottoms of glaciers, in ice fields in the Arctic, and in other extreme environments (**Figure 24.5**).

Among the early life-forms was an ancestral form of **cyanobacteria**, single-celled organisms otherwise known as **blue-green algae**. Those microorganisms form extensive sheets on the surface of bodies of water, as shown in **Figure 24.6a**.

a.

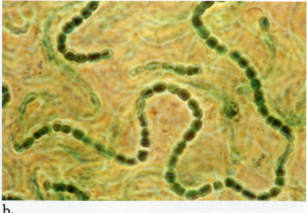

b.

Figure 24.6 a. Cyanobacteria today form sheets on lakes and other bodies of water. **b.** Under a microscope, the individual microorganisms are visible.

Under a microscope, it turns out that those sheets are colonies of individual microorganisms, as seen in **Figure 24.6b**. Cyanobacteria **photosynthesize**, using sunlight and carbon dioxide as food and generating oxygen as a waste product. Initially, the highly reactive oxygen that cyanobacteria produced was quickly removed from Earth's atmosphere by oxidation, or rusting, of surface minerals. Once most of the exposed minerals were oxidized and could no longer absorb oxygen, atmospheric levels of oxygen began to rise. Oxygenation of Earth's atmosphere and oceans began about 2 billion years ago, and the current level was reached only about 250 million years ago, as shown in **Figure 24.7**. Without cyanobacteria and other photosynthesizing organisms, Earth's atmosphere would be as oxygen-free as the atmospheres of Venus and Mars.

Biologists comparing DNA sequences find that terrestrial life is divided into two types: prokaryotes and eukaryotes. Prokaryotes, which include bacteria and archaea, are simple organisms that consist of free-floating DNA inside a cell wall; as shown in **Figure 24.8a**, they lack both cell structure and a nucleus. Eukaryotes (**Figure 24.8b**), which form the cells in animals, plants, and fungi, have a more complex form of DNA contained within the cell's membrane-enclosed nucleus. The first eukaryote fossils date from about 2 billion years ago, coincident with the rise of free oxygen in the oceans and atmosphere, although the first multicellular eukaryotes did not appear until a billion years later.

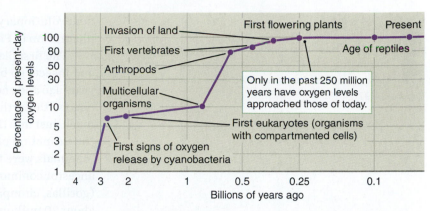

Figure 24.7 The amount of oxygen in Earth's atmosphere has built up as a result of photosynthesis by cyanobacteria and plant life on the planet.

Increasing Complexity

Scientists have used DNA sequencing to establish what is known as the "phylogenetic tree of life," shown in **Figure 24.9**. That complex tree describes the evolutionary interconnectivity of all species of Bacteria, Archaea, and Eukarya and has revealed some interesting relationships. For example, Archaea were initially thought to be the same as Bacteria, but genetic studies show they diverged long ago, and the Archaea have genes and metabolic pathways more similar to those of Eukarya than to those of Bacteria. On the macroscopic scale, the phylogenetic tree places animals closest to fungi, which branched off the evolutionary tree after slime molds and plants.

Living creatures in Earth's oceans remained much the same—a mixture of single-celled and relatively primitive multicellular organisms—for more than 3 billion years after terrestrial life appeared. Between 540 million and 500 million years ago, the number and diversity of biological species increased spectacularly. Biologists call that event the **Cambrian explosion**. The trigger of that sudden surge in biodiversity remains unknown, but possibilities include rising oxygen levels, an increase in genetic complexity, major climate change, or some combination of those factors. The "Snowball Earth" hypothesis suggests that before the Cambrian explosion, Earth was in a period of extreme cold between about 750 million and 550 million years ago and was covered almost entirely by ice. During that period of extreme cold, predatory animals died out, making it easier for new species to adapt and thrive. Another possibility is that the marked increase in atmospheric oxygen (O_2) would have been accompanied by a corresponding increase in stratospheric ozone (O_3), which shields Earth's surface from deadly solar ultraviolet radiation. With that protective ozone layer in place, life was free to leave the oceans and move to land. *Tiktaalik*, a fish with limblike fins and ribs, was an animal in a

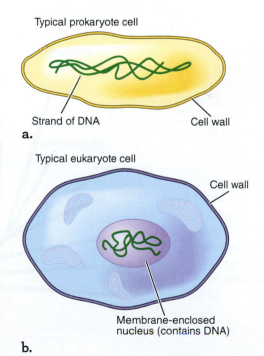

Figure 24.8 **a.** A simple prokaryote cell contains little more than the cell's genetic material. **b.** A eukaryote cell contains several membrane-enclosed structures, including a nucleus that houses the cell's genetic material.

what if . . .

What if life formed in both the ocean depths and in tide pools? Would you expect all life on Earth to be as interconnected as is shown in Figure 24.9?

midevolutionary step of leaving the water for dry land, as shown in the artist's illustration in **Figure 24.10**.

The first plants appeared on land about 475 million years ago. Large forests and insects go back 360 million years. The age of dinosaurs began 230 million years ago and ended abruptly 65 million years ago, when an asteroid or comet collided with Earth (see the "Origins" feature in Chapter 8). The collision threw so much dust into the atmosphere that the sunlight was dimmed for months, causing the extinction of more than 70 percent of all existing plant and animal species. Mammals were the big winners in the aftermath. Primates evolved from the last ancestor common with other mammals about 70 million years ago. The great apes (gorillas, chimpanzees, bonobos, and orangutans) split off from the lesser apes about 20 million years ago (Figure 24.9, inset). DNA tests show that humans and chimpanzees share about 98 percent of their DNA, indicating that they evolved from a common ancestor about 6 million years ago. By comparison, all humans share 99.9 percent of their DNA. The earliest human ancestors appeared a few million years ago, and the first civilizations occurred a mere 10,000 years ago. Present-day industrial society, barely more than two centuries old, is but a moment in the history of life on Earth.

Humans are here today because of a series of events that occurred throughout the history of the universe. Some of those events are common in the universe, such as the formation of heavy elements in earlier generations of stars and the formation of planets. Other events in Earth's history may have been less likely to happen elsewhere, such as the formation of a planet with life-supporting conditions like Earth or the development of self-replicating molecules that led to Earth's earliest life. A few events stand out, such as major extinctions that allowed the evolution of mammalian life and, ultimately, human beings.

Figure 24.9 This simplified version of the phylogenetic tree has been constructed from analysis of the DNA strands of different life-forms. Humans are included in the "Animals" twig on the Eukarya branch. The primate branch, which includes humans, is shown in the inset on the bottom. By tracing those common ancestors through DNA and other means, scientists can reconstruct the evolutionary history of a species.

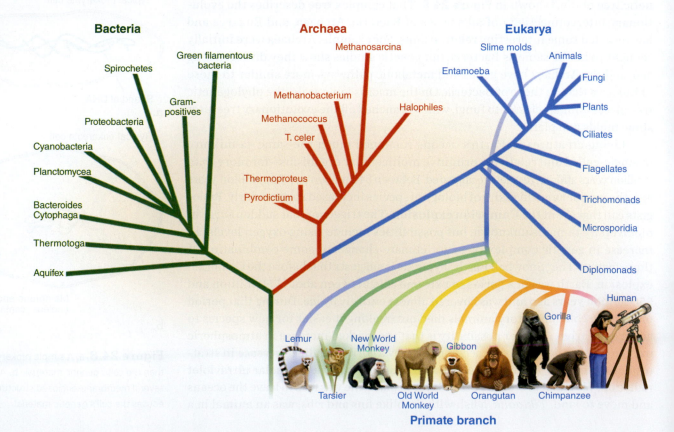

working it out 24.1

Exponential Growth

Self-replication is an example of exponential growth. The number of times a sample will double, n, for exponential growth is determined by the ratio of the original and final amounts:

$$\frac{P_F}{P_O} = 2^n$$

Assume a hypothetical self-replicating molecule makes one copy of itself each minute, and each copy in turn copies itself each minute. How many molecules will exist after an hour? Here, the number of generations is given by $n = 60$, because there are 60 minutes in an hour:

$$\frac{P_F}{P_O} = 2^{60} = 1.2 \times 10^{18}$$

Thus, a billion billion of those molecules will exist after 1 hour.

Now suppose a mutation occurs once every 50,000 times that a molecule reproduces itself, and one of every 200,000 mutations turns out to be beneficial. After 100 generations, how many molecules with those beneficial mutations might exist? That equation is similar to the previous equation, but here $n = 100$:

$$\frac{P_F}{P_O} = 2^{100} = 1.3 \times 10^{30}$$

The total number of molecules is 1.3×10^{30}. The number of mutations is that number divided by 50,000, or 2.6×10^{25} mutated molecules. The number of beneficially mutated molecules is that number divided by 200,000, or 1.3×10^{20} molecules. So, 100 million trillion (10^{20}) mutations will occur that, by chance, might improve the survivability of the original molecule. Because that number does not count earlier beneficial changes that themselves replicated, the total number of molecules with beneficial changes will be even larger!

Evolution as a Mechanism of Change

Imagine that just once during the first few hundred million years after Earth formed, a single molecule formed somewhere in Earth's oceans. That molecule had a very special property: chemical reactions between that molecule and other molecules in the surrounding water caused the molecule to make a copy of itself. The molecule became *self-replicating*. Chemical reactions then produced copies of each of those two molecules, making four molecules. Four molecules became eight, eight became 16, 16 became 32, and so on. By the time the original molecule had copied itself just 100 times, more than a million trillion trillion (10^{30}) of those molecules existed. That is about 100 million times more of those molecules than the number of stars in the observable universe. The molecules of DNA that make up the chromosomes in the nuclei of the cells of all advanced life today are direct descendants of those early self-replicating molecules that flourished in the oceans of the young Earth.

Over time, not all replications are exact. The likelihood that a copying variation will occur while a molecule is replicating increases significantly with the number of copies being made. For DNA, which contains the genetic code for an entire organism, a change in the genetic code is called a **mutation**. Sometimes a mutation has no effect. Other times, a mutation can prevent an organism from flourishing. In still other cases, a mutation can make an organism better suited to its environment. Organisms with those advantageous mutations will survive to reproduce successfully. Even if mutations are rare and only a small fraction turn out to be beneficial, after just 100 generations trillions of mutations occur that, by luck, might improve on the original (**Working It Out 24.1**). **Heredity**—the ability of one generation to pass on its genetic code to future generations—allow beneficial mutations to persist and be incorporated into a species' genetic code.

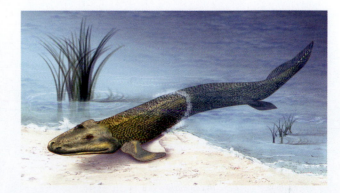

Figure 24.10 This illustration is an artist's reconstruction from a fossil of *Tiktaalik* found in the Canadian Arctic.

Figure 24.11 Fossils, such as this *Parasaurolophus* ("near crested lizard"), record the history of the evolution of life on Earth. This 10-meter-long plant-eating dinosaur lived in North America about 75 million years ago.

As the organisms of the early Earth continued to interact with their surroundings and make copies of themselves, mutations caused them to diversify into many species. Sometimes the resources they needed to reproduce became scarce. In the face of that scarcity, varieties that were more successful reproducers became more numerous. Competition for resources, predation by one species on another, and cooperation between organisms became important to the survival of different varieties. Some varieties were more successful and reproduced to become more numerous, whereas less successful varieties became less and less common. That process, in which better-adapted organisms reproduce and thrive, whereas less well-adapted organisms become extinct, is called **natural selection**.

Life has existed on Earth for about 4 billion years, which is a very long time—long enough for the combined effects of heredity and natural selection to shape the descendants of that early self-copying molecule into a huge variety of complex, competitive, successful structures. Geological processes on Earth have preserved a fossil record of the history of some of those structures (**Figure 24.11**). Among those descendants are human "structures" that can think about their own existence and unravel the mysteries of the stars.

CHECK YOUR UNDERSTANDING **24.1**

Extremophiles are organisms that: (a) are extremely reactive; (b) are extremely rare; (c) have an extreme quality, such as mass or size; (d) live in extreme conditions.

Answers to Check Your Understanding questions are in the back of the book.

24.2 Life Involves Complex Chemical Processes

The evolution of life on Earth cannot be separated from the narrative of astronomy: it is one of many examples of the emergence of structure in an evolving universe (see the **Process of Science Figure**). The emergence of that structure then leads to the following question: Has life arisen elsewhere? Unlike the study of planets, stars, and galaxies, only one known case exists for the study of life—Earth—and scientists do not know how much of that example can be generalized to other places. To explore that question, we need to take a closer look at the processes that have led to life on Earth. In this section, we explore the chemical and physical properties of life on Earth.

The infant universe was composed basically of hydrogen and helium and very little else. After 9 billion years of stellar nucleosynthesis, all the heavier chemical elements essential to life were present and available in the molecular cloud that gave birth to the Solar System. Those heavier elements were formed by nuclear fusion in the cores of earlier generations of stars and were then dispersed into space. At times, that dispersal was passive. For example, low-mass stars such as the Sun, when they become puffed-up, dying red giants, may shed their extended atmospheres, sending some newly created carbon into space. Other dispersals were more violent. High-mass stars produce even heavier elements through nucleosynthesis in their cores—up to and including iron. But some of the trace elements essential to biology on Earth are even more massive than iron. They are produced within a matter of minutes during

All of Science Is Interconnected

More than most other subjects, astrobiology relies on concepts from a number of scientific fields—showing that all of science is interconnected.

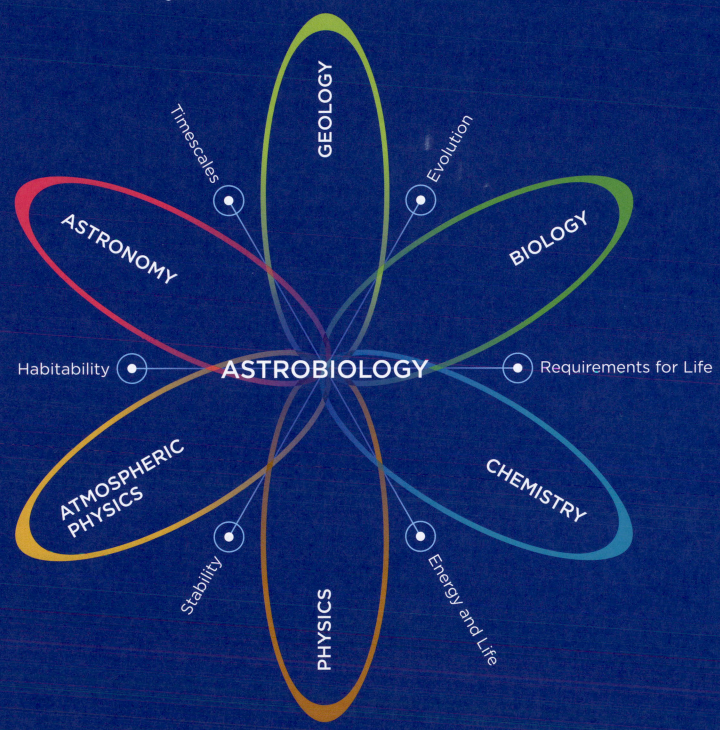

No science stands alone. All are connected. Interdisciplinary fields of study like astrobiology provide opportunities for new tests for theories of many fields.

what if . . .

What if we discovered self-replicating life on another planet, where the life is based on a molecule other than DNA? Would we expect life to be evolving on such a planet?

the violent supernova explosions that mark the death of high-mass stars and then are thrown into the chemical mix found in molecular clouds. Or they are produced in the collisions of neutron stars, the dense remains of prior supernova explosions (**Figure 24.12**).

All known living organisms on Earth are composed of a more or less common suite of complex chemicals. Approximately two-thirds of the atoms in the human body are hydrogen (H); about one-fourth are oxygen (O); a tenth are carbon (C); and a few hundredths are nitrogen (N). Carbon, nitrogen, and oxygen are the three most abundant products of stellar nucleosynthesis after helium (see the "astronomer's periodic table" in Figure 5.17). The several dozen remaining atomic elements in the human body make up only 0.2 percent of the total. All known living creatures are assemblages of molecules composed almost entirely of those four elements, sometimes called CHON (carbon, hydrogen, oxygen, nitrogen), along with small amounts of phosphorus and sulfur. Some of those molecules, such as RNA, DNA, and proteins, are enormous. A small piece of DNA, which is responsible for genetic codes, is illustrated in **Figure 24.13**. DNA is made up entirely of only five atomic elements—CHON and phosphorus—but the DNA in each cell of the human body is composed of combinations of *tens of billions* of atoms of those same five elements. Proteins, the huge molecules responsible for the structure and function of living organisms, are long chains of smaller molecules called amino acids. Terrestrial life uses 20 specific amino acids, which also consist of no more than five atomic elements—CHON plus sulfur.

Although life primarily uses the half-dozen atoms of CHON plus sulfur and phosphorus, some other elements, present in smaller amounts, are essential to the chemical processes that living organisms carry out. Those elements include sodium, chlorine, potassium, calcium, magnesium, iron, manganese, and iodine. Trace elements such as copper, zinc, selenium, and cobalt also play a crucial role in biochemistry but are needed in only tiny amounts.

Carbon, which can bond to four other atoms, forms the backbone of the DNA molecule shown in Figure 24.13. That is why carbon is so important to life on Earth. Forms of extraterrestrial life could exist that are also carbon-based but have chemistries different from that of life on Earth. For example, countless varieties of amino acids exist in addition to the 20 used by terrestrial life. Most other atoms are more limited than carbon in the number of bonds they can make, but silicon, like carbon, can bond to four other atoms, so many combinations are possible. As a potential life-enabling atom, silicon has both advantages and disadvantages in comparison with carbon. Silicon-based molecules remain stable at much higher temperatures than carbon-based molecules, perhaps enabling silicon-based life to thrive in high-temperature environments, such as on planets that orbit close to their parent star. But silicon is also a larger and more massive atom than carbon: It cannot form molecules as complex as those based on carbon. Any silicon-based life probably would be simpler than life-forms here on Earth, but it might exist in high-temperature niches somewhere within the universe. Although carbon's unique properties make it

Figure 24.12 This version of the periodic table shows the astronomical origin of the naturally occurring elements (up to uranium) in the Solar System.

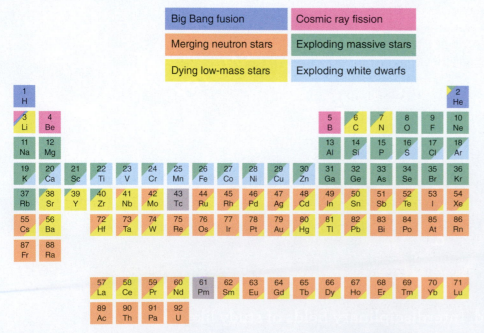

readily adaptable to the chemistry of life on Earth, other types of life might be found elsewhere.

CHECK YOUR UNDERSTANDING 24.2

Carbon is a favorable base for life because: (a) it can bond to many other atoms in long chains; (b) it is nonreactive; (c) it forms weak bonds that can be readily reorganized as needed; (d) it is organic.

24.3 Where Do Astronomers Look for Life?

One approach to the scientific search for extraterrestrial life is to use robotic spacecraft to explore the planets and moons of the Solar System (see Chapter 6). Spacecraft have visited all the planets and some moons and sent back at least some information about the conditions on those worlds. Another approach is to use telescopes to detect planets outside our Solar System (see Chapter 7). Telescopes on the ground and in space have detected several thousand planets orbiting other stars. In this section, we survey the locations where life might be found, both in our own Solar System and around other stars.

Life within Our Solar System

Scientists start the search for evidence of extraterrestrial life here in our own Solar System. Early conjectures about life in our Solar System seem naïve, considering what we now know. Two centuries ago, the eminent astronomer Sir William Herschel, discoverer of Uranus, proclaimed, "We need not hesitate to admit that the Sun is richly stored with inhabitants." In 1877, astronomer Giovanni Schiaparelli (1835–1910) observed what appeared to be linear features on Mars and dubbed them *canali* ("channels" in Italian). Another observer of Mars, Percival Lowell (1855–1916), misinterpreted Schiaparelli's *canali* as "canals," suggesting that they were constructed by intelligent beings.

Because Mercury and the Moon lacked atmospheres, astronomers determined that those worlds were not conducive to life. The giant planets and their moons were thought to be too remote and too cold to sustain life. By the 1960s it was understood that the surface of Venus was too hot to permit life. At the same time, astronomers using ground-based telescopes had discovered that Mars has a thin atmosphere, water ice, and carbon dioxide ice. During the 1960s, the United States and the Soviet Union sent reconnaissance spacecraft to the Moon, Venus, and Mars, but the instruments on those spacecraft probed the physical and geological properties of those astronomical bodies instead of searching for life. Serious efforts to look for signs of life—past or present—require more advanced spacecraft with specialized instrumentation.

In the mid-1970s, two American *Viking* spacecraft were sent to Mars with detachable landers containing a suite of instruments designed to find evidence of a terrestrial type of life. When the *Viking* landers failed to find convincing evidence of life on Mars, hopes faded for finding life on any other body orbiting the Sun. Since then, however, further exploration of the Solar System has generated renewed optimism. A better understanding of the history of Mars indicates the planet's climate has changed. Mars was once wetter and warmer than it is today.

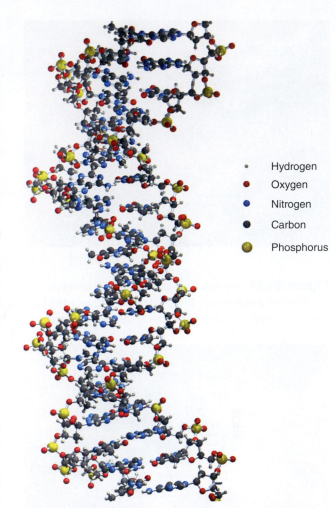

- Hydrogen
- Oxygen
- Nitrogen
- Carbon
- Phosphorus

Figure 24.13 DNA, the heritable molecule that forms the basis for life on Earth, contains the atoms of only five elements. Even so, those atoms are combined in billions of ways, giving rise to the diversity of life on Earth.

VIS

Figure 24.14 The Mars *Curiosity* rover detected evidence that Mars had a watery past. The rounded gravel surrounding the bedrock suggests that an ancient, flowing stream once existed.

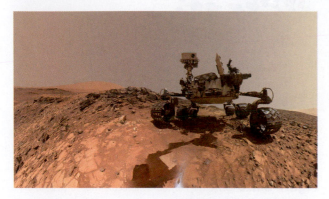

Figure 24.15 A "selfie" of the *Curiosity* Mars rover at the site where it drilled into some rocks and identified organic compounds.

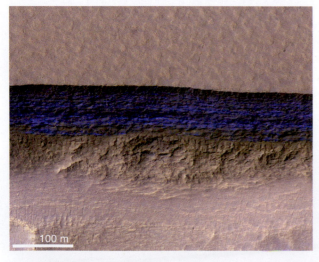

100 m

Figure 24.16 This image from the *MRO* shows 80 meters of exposed water ice on a steep slope. The color is exaggerated. Similar steep slopes with ice at other midlatitude locations also were seen.

In the 1990s, Mars missions began to map the planet's surface from the ground and from space. In 2008, NASA's *Phoenix* spacecraft landed at a far-northern latitude, inside the planet's arctic circle, where specialized instruments dug into and analyzed the Martian water-ice permafrost. *Phoenix* found that the Martian arctic soil has a chemistry similar to that of the Antarctic dry valleys on Earth, where life exists deep below the surface at the ice-soil boundary. Minerals that form in water, such as calcium carbonate, have been detected. That finding suggests that oceans existed in the past on Mars. However, *Phoenix* found no evidence of life.

The *Curiosity* rover (originally known as the *Mars Science Laboratory*) landed in Gale Crater on Mars in 2012. That rover studies the rocks and soil of Mars to acquire data for a better understanding of the history of the planet's climate and geology. Shortly after landing, *Curiosity* found evidence that a stream of liquid water had once flowed in the crater. The rover observed rounded, gravelly pebbles stuck together, which have been interpreted as coming from a stream that varied at times from ankle-deep to hip-deep, and moved at about 1 meter per second (**Figure 24.14**). The rover also found sedimentary rocks containing clay, which suggest the presence of a freshwater lake bed at one time. Later observations found that the surface soil contained up to 2 percent water by weight—or about 1 quart per cubic foot of Martian dirt. Those conditions are too dry to support Earth-like plant life, which permanently wilts in soil that is less than about 10 percent water by weight.

Curiosity also detected evidence of organic molecules in 3-billion-year-old rocks located in that crater (**Figure 24.15**). Those organic molecules could have come from space on asteroids and comets, from geological processes on Mars, or from ancient life. *Curiosity* also detected seasonal variation in the amount of methane gas within the crater; methane can come from water-rock chemistry or from life. Thus, a water lake inside Gale Crater could have been hospitable to life billions of years ago, but astrobiologists do not know whether any life existed.

The *Mars Atmosphere and Volatile EvolutioN* (*MAVEN*) mission, which arrived at Mars in September 2014, is studying the upper atmosphere to learn more about the escape of carbon dioxide, hydrogen, and nitrogen from the planet's atmosphere and how losing those gases affected surface pressure and the existence of liquid water. Recently the *Mars Reconnaissance Orbiter* (*MRO*) found water ice on eight steep, eroded slopes located at midlatitudes on Mars (**Figure 24.16**). *Mars Express* may have found evidence of a lake of liquid water under the south polar ice cap. Future experiments will look for liquid water—and fossil or living microorganisms—below the Martian surface. Eventually other missions will return samples of Martian rock and soil to Earth for more advanced analysis.

NASA's instrumented robotic spacecraft reached the outer Solar System starting in the 1970s, and many astrobiologists were surprised by the findings. Although the outer planets themselves did not appear to be habitats for life, some of their moons became objects of special interest. Jupiter's moon Europa is covered with a layer of water ice that appears to overlie a great ocean of briny liquid water (**Figure 24.17**). The water remains liquid because of high pressure and tidal heating by Jupiter. Impacts by comet nuclei may have added a mix of organic material, another essential ingredient for life. Once thought to be a frozen, inhospitable world, Europa is now a candidate for biological exploration. Recently, scientists using the Hubble Space Telescope observed water geysers taller than Mount Everest erupting from the icy surface of Europa. Ejected material from those geysers may make searching for life on Europa possible without drilling through the ice. NASA's *Europa Clipper* mission, scheduled for launch in the 2020s, will fly by Europa a few dozen times.

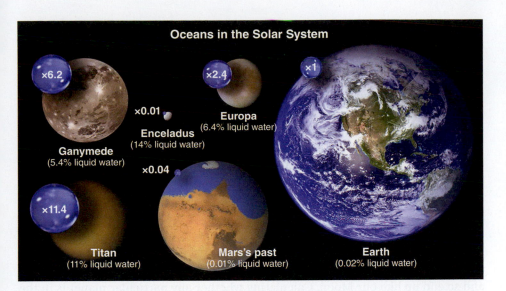

Figure 24.17 The total amount of liquid water (blue spheres) on Jupiter's moons and Saturn's moon Titan compared with the amount on Earth. All figures are drawn to scale and assume average ocean depths of 4 km (Earth), 100 km (Europa), and 200 km (Titan).

Jupiter's moons Ganymede and Calisto also have icy surfaces and possibly subsurface salty oceans.

Saturn's moon Titan has an atmosphere rich in organic chemicals, many of which are thought to be precursor molecules of a type that existed on Earth before life appeared here. A probe from the *Cassini* spacecraft in orbit around Saturn descended through Titan's atmosphere and found additional evidence for a variety of molecules that might be necessary for life, as well as liquid lakes of methane on the surface and probably a liquid-water ocean under the surface. The *Cassini* spacecraft also detected water-ice crystals spouting from cryovolcanoes (which erupt ice crystals instead of rocks) near the south pole of Saturn's tiny moon Enceladus. Liquid water must lie beneath its icy surface, so Enceladus is also a possible habitat of extremophile life—perhaps life similar to that found near hydrothermal vents deep within Earth's oceans.

The discovery of life on even one Solar System body beyond Earth would be exciting. If scientists discover that life arose independently *twice* in the same planetary system, that finding could suggest that the spontaneous appearance of life is not rare at all.

CHECK YOUR UNDERSTANDING 24.3a

Which of the following Solar System objects is a good candidate for future searches for life? (Choose all that apply.) (a) Mars; (b) Jupiter's moon Europa; (c) Saturn's moon Titan; (d) Uranus

Habitable Zones

Recall from Chapter 7 that thousands of confirmed and thousands of candidate exoplanets exist within the Milky Way Galaxy. To decide which planets to focus on for further study, astronomers are narrowing the possibilities by searching for planets with environments conducive to the formation and evolution of life, as we understand it, while eliminating clearly unsuitable planets. Astronomers consider issues such as each planet's orbit, inferred temperature, distance from its star, and location in the galaxy. One criterion astrobiologists look for is stability of planetary systems. As noted in Chapters 2, 3, and 4, astronomers think about the effects of a planet's rotation and orbit. Planets in stable systems have nearly circular orbits that preserve relatively uniform climatological environments. Planets in very elliptical

what if . . .

What if we find an exoplanet with a very small magnetic field and large amounts of greenhouse gases in its atmosphere? How would each of those factors increase or decrease the habitable zone distance for that exoplanet?

▶▶ **Interactive Simulation:** Habitable Zone

Figure 24.18 The habitable zone changes with the mass and temperature of a star. Habitable zones around hot, high-mass stars are larger and more distant than the zones around cooler, lower-mass stars.

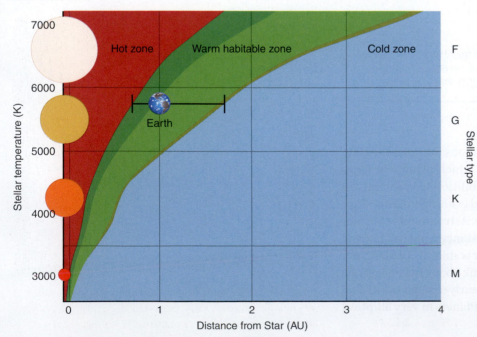

orbits or planets with a large axial tilt can experience more intense temperature swings that could be detrimental to the survival of life. A stable temperature that maintains the existence of water in a liquid state might be important. We know that liquid water was essential for life on Earth to form and evolve. We don't know whether liquid water is an absolute requirement for life elsewhere, but it's a reasonable starting assumption.

In the "Origins" sections of Chapters 7 and 13, we discussed the idea of the habitable zone, the location of a planet relative to its parent star that provides a range of temperatures in which liquid water can exist on the surface. On planets too close to their parent stars, water would exist only as a vapor—if at all. On planets too far from their stars, water would be permanently frozen as ice. Planet size is another consideration: Large gas giants retain most of their light gases during formation and have no solid surface. Small planets may be rocky or a mix of water, rock, and ice. Measuring the mass and radius enables scientists to estimate the density. Very small planets may not have enough surface gravity to retain their atmospheric gases and so end up like our Moon. Calculating whether any particular planet is in the habitable zone is complicated. Recall from Chapter 7 that even if a planet is located in the habitable zone, that means only that liquid water *could* exist on the surface: It does not mean that astronomers have confirmed the presence of liquid water or that the planet has inhabitants.

In our own Solar System, Venus, which orbits at 0.7 times Earth's distance from the Sun, has become an inferno because of its runaway greenhouse effect. Any liquid water that might once have existed on Venus has long since evaporated and been lost to space. Mars orbits about 1.5 times farther from the Sun than the orbit of Earth, and the water that we see on Mars today is nearly always frozen. But the orbit of Mars is more elliptical and variable than Earth's, giving Mars a greater variety of climate, including long-term cycles that might occasionally permit liquid water to exist. Most astrobiologists put the habitable zone of our Solar System at about 0.9–1.4 astronomical units (AU), which includes Earth but just misses Venus and Mars. However, that range ignores the possibility of liquid water under ice, as occurs on the moons of the outer Solar System.

Astronomers also must think about the type of star they are observing in their search for planets that could have liquid water. Stars less massive than the Sun and thus cooler will have narrower habitable zones. A planet in the habitable zone of a cool star is close in to its star. As a result, the planet is more likely to be tidally locked to the star so that no day/night cycle exists. Stars more massive than the Sun are hotter and will have a larger and more distant habitable zone (**Figure 24.18**). However, massive stars have shorter main-sequence lifetimes and might not last long enough for life to evolve in the first place. For example, a star of $3\,M_{\text{Sun}}$ has a lifetime of only a few hundred million years. Here on Earth, a billion years was long enough for bacterial life to form and cover the planet, but insufficient for anything more advanced to evolve. It took 3.5 billion years of evolution on Earth to reach the period of the Cambrian explosion. Even though evolution might happen at a different pace elsewhere, stellar lifetime is still a sufficiently strong consideration that astronomers focus their efforts on stars with longer lifetimes—specifically, stars of $0.6–1.4\,M_{\text{Sun}}$, which

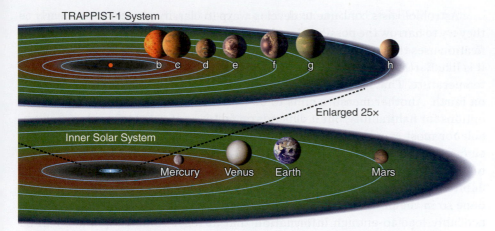

Figure 24.19 The Trappist-1 system. The Trappist planets are very close to their star: their distances are more comparable to those between Jupiter and its large moons than to those between the Sun and the terrestrial planets. (The star is not to scale.)

corresponds to spectral types F, G, K, and M. Most stars in our galaxy are low-mass M main-sequence stars, and many of those seem to have small, rocky planets. One example is the Trappist-1 system of seven planets in close orbits to their cool M dwarf star (**Figure 24.19**). Some of those planets could be tidally locked (see the Chapter 4 "Origins") to their star. Two or three of the planets are in the star's habitable zone.

Another factor to consider is a planet's atmosphere. A planet's ability to keep an atmosphere depends on the planet's mass and radius (and therefore its escape velocity) and its temperature. Very small planets may not have enough surface gravity to retain their atmospheric gases. In the inner Solar System, the Moon and Mercury were too small to keep any atmosphere. Mars lost most of its atmosphere, but the larger Earth and Venus kept thick atmospheres. Another important consideration is the greenhouse effect, which traps heat underneath an atmosphere and raises the temperature on a planetary surface. That has happened on Venus, Earth, and Mars, each of which has a higher surface temperature because of its atmosphere. The thickness and chemical content of the atmosphere affect the strength of the greenhouse effect, so that, for example, Venus is much hotter than its distance from the Sun would suggest because of its thick atmosphere of carbon dioxide (see Figure 5.25 and the Chapter 5 "Origins"). Some of those exoplanets might be more like Venus than like Earth if they have atmospheres filled with greenhouse gases. The total amount of atmosphere affects the atmospheric pressure at the surface, which, along with the temperature, determines whether water (or other molecules) can exist in a liquid state on the surface. The current thin atmosphere of Mars does not permit standing liquid water on its surface.

In the next few years, with new telescopes, many more observations will identify atmospheres on exoplanets. Water vapor, hydrogen, and carbon monoxide have been found on a few exoplanets already. In particular, discovering oxygen in the atmosphere of an exoplanetary atmosphere would be exciting, but not definitive. The oxygen in Earth's atmosphere makes it stand out from the rest of the planets and moons in the Solar System, and we know that most terrestrial oxygen was created by photosynthetic life. However, oxygen can also come from the breakup of water molecules, so the presence of oxygen alone doesn't necessarily mean life exists there. Aside from oxygen, astrobiologists are thinking about other possible **biosignatures**—chemical compounds that would show up in the spectra of an exoplanet atmosphere that would suggest life may be present (**Figure 24.20**).

Figure 24.20 Spectroscopy of rocky exoplanet atmospheres may pick up many chemicals important for life, including some that are biosignatures, thought to be produced overwhelmingly by life.

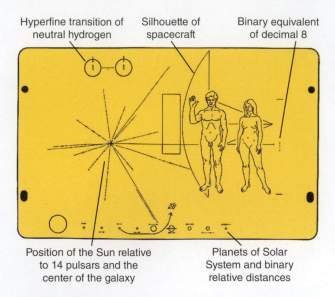

Hyperfine transition of neutral hydrogen

Silhouette of spacecraft

Binary equivalent of decimal 8

Position of the Sun relative to 14 pulsars and the center of the galaxy

Planets of Solar System and binary relative distances

Figure 24.21 This plaque is carried by the *Pioneer 11* probe, which launched in 1973 and will eventually leave the Solar System to travel in interstellar space.

Figure 24.22 This message was beamed toward the star cluster M13 in 1974. Reading right to left, this binary-encoded message contains the numbers 1–10, hydrogen and carbon atoms, some interesting molecules, DNA, a human figure and its size, the basics of the Solar System, and a depiction of the now-defunct Arecibo telescope.

Astrobiologists continue to develop ways to classify planets more clearly as they try to narrow the possibilities about where life might exist. One such classification uses the currently available data on an exoplanet to estimate how much it is like Earth. Factors include the radius, density, escape velocity, and surface temperature. That is an Earth-centric approach based on the experience of life on Earth. Another measure aims to be less Earth-centric and to broaden the options for habitability, but it depends on factors not yet measured or measurable for most exoplanets. That measure depends on whether the planet has a surface on which organisms can grow, as well as the right kind of chemistry, a source of energy, and the ability to hold a liquid solvent. Saturn's moon Titan or Jupiter's moon Europa might satisfy those conditions, and Mars might have done so in the past. Over the next decade, improvements in observations will probably lead to enough information that at least some exoplanets can be classified in that way. As observations of exoplanets become more complete, astrobiologists will undoubtedly develop new classification schemes that are more accurate and informative.

Astronomers also consider the *galactic habitable zone*—the idea that the Milky Way Galaxy may have some locations where planets might have a higher probability of hosting life. Stars situated too far from the galactic center may be without enough heavy elements—such as oxygen, silicon (silicates), iron, and nickel—in their protoplanetary disks to form rocky planets like Earth. Conversely, regions too close to the galactic center experience less star formation and therefore fewer opportunities to gather heavy elements into planetary environments. Stars too close to the galactic center may be affected by the high-energy radiation environment (X-rays and gamma rays from supermassive black holes or gamma-ray bursts), which can damage RNA and DNA. Stars that migrate within the galaxy and change their distance from the galactic center may move in and out of any galactic habitable zone.

CHECK YOUR UNDERSTANDING 24.3b

The habitable zone around a star depends most on the star's: (a) mass and age; (b) radius and distance; (c) age and radius; (d) color and distance; (e) luminosity and velocity.

24.4 Scientists Are Searching for Signs of Intelligent Life

Are we alone? Scientists approach that question from many directions. Biologists consider the origin and evolution of life and the definition of intelligence. Astronomers send messages and search for alien signals in the vast array of astronomical data. In this section, we describe the search for intelligent life and how scientists think about the probability of finding it.

Sending Messages

During the 1970s, messages were sent from Earth to space. The *Pioneer 11* spacecraft, which will probably spend eternity drifting through interstellar space, carries the plaque shown in **Figure 24.21**. It pictures humans and the location of Earth for any future interstellar traveler who might happen to find it and understand its content. Another message to the cosmos accompanied the two *Voyager* spacecraft on the "Golden Records"—identical phonograph records that contained greetings from planet Earth in 60 languages, samples of music, animal sounds, and a message from then-President Jimmy Carter. Some politicians were concerned that scientists were dangerously advertising our location in the galaxy, even though radio signals had already been broadcast into space for nearly 80 years. Some philosophers also worried that those messages contained anthropomorphic assumptions about aliens being enough like us to decode the messages. However, sending messages on spacecraft is an inefficient way to make contact with extraterrestrial life. The probability that an alien species will actually find any of those messages is very, very small.

In 1974, astronomers used the 300-meter-wide dish of the Arecibo radio telescope to beam a message in binary code (**Figure 24.22**) toward the star cluster M13, located 25,000 light-years away. That distance is far enough that by the time the message arrives, the core of M13 will have moved, and the radio signal will not actually arrive there. However, the intention of the experiment was to show that such a message could be sent, not to make contact, because 50,000 years would pass before any reply could come back. In 2008, a radio telescope in Ukraine sent a message to the exoplanet Gliese 581c. The message was composed of 501 digitized images and text messages selected by users on a social networking site and will arrive at Gliese 581c in 2029.

The Drake Equation

The first serious effort to quantify the probability of the existence of intelligent extraterrestrial life was made by astronomer Frank Drake in 1960. He developed the **Drake equation**, which estimates the likely number (N) of intelligent civilizations currently existing in the Milky Way Galaxy. The Drake equation is different from the other equations in this book because the values for many of the variables are uncertain. However, it is a useful way to categorize some of the factors that relate to the conditions that must be met for a civilization to exist. The equation is discussed further in **Working It Out 24.2.**

As illustrated in **Figure 24.23**, the conclusions we draw using the Drake equation depend a great deal on the assumptions we make and therefore on the numbers used in the equation. For the most pessimistic estimates, the Drake equation sets the number of technological civilizations in our galaxy at about 1, in which case we are the *only* technological civilization in the Milky Way at this time. Such a universe could still be full of intelligent life. With 100 billion galaxies in the observable universe, even those pessimistic assumptions mean that 100 billion technological civilizations could be out there somewhere. However, if the nearest neighbors are in another galaxy, they are *very* far away—millions of parsecs, on average.

At the other extreme—with the most optimistic numbers, which assume that intelligent life arises and survives everywhere it gets the chance—the Milky Way alone could have tens of millions of technological civilizations! If so, the nearest neighbors may be "only" 40 or 50 light-years away.

If humans did meet a technologically advanced civilization, what would it be like? The Drake equation suggests that neighbors nearby are highly unlikely, unless civilizations typically live for many thousands or even millions of years (see Working It Out 24.2). If so, any civilization we encountered would probably have been around for much longer than we have.

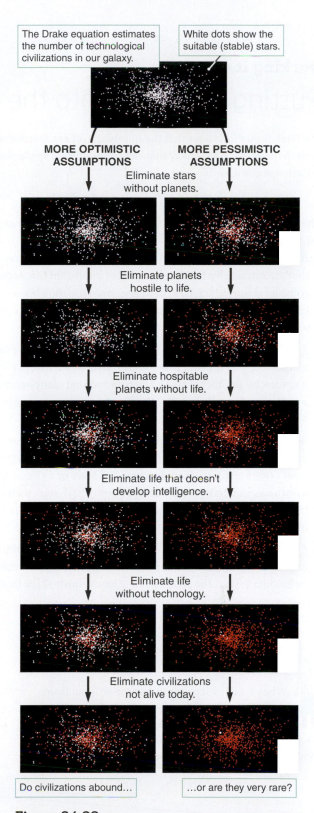

Figure 24.23 The two columns show estimates of the existence in the Milky Way Galaxy of intelligent, communicative civilizations as made using the Drake equation. White dots represent stars with possible civilizations. Notice how widely the estimates vary, given optimistic and pessimistic assumptions about the seven factors in the equation.

working it out 24.2

Putting Numbers into the Drake Equation

The Drake equation states that the number, N, of extraterrestrial civilizations in the Milky Way Galaxy that can communicate by electromagnetic radiation is given by

$$N = R^* \times f_p \times n_e \times f_l \times f_i \times f_c \times L$$

where the factors on the right-hand side of the equation are explained as follows:

1. R^* is the number of stars that form in the Milky Way Galaxy each year that are suitable for the development of intelligent life. Astronomers consider those to be F, G, K, or M spectral-type stars because their lifetimes are long enough. That is about five to seven stars per year, so $R^* = 5$ to 7.

2. f_p is the fraction of stars that form planetary systems. The discoveries of exoplanets over the past two decades have shown that planets form as a natural by-product of star formation and that many—perhaps most—stars have planets. For that factor, astronomers assume that f_p is between 0.5 and 1.

3. n_e is the number of planets and moons in each planetary system with an environment suitable for life. In the Solar System, that number is at least 1 (for Earth), but it could be more if Mars or an outer-planet moon or two has suitability for life. Only recently have stars with multiple planets been discovered, so astronomers are just starting to get data on that factor. Generally, they estimate that n_e is less than 3.

4. f_l is the fraction of suitable planets and moons on which life actually arises. Remember that just a single self-replicating molecule may be enough to start the process. Some biochemists think that life *will* develop if the right chemical and environmental conditions are present, but others disagree. Values of f_l range from 100 percent (life always develops) to 1 percent (life is more rare). Astronomers use a range of 0.01–1 for f_l.

5. f_i is the fraction of those planets harboring life that eventually develop intelligent life. Intelligence is certainly the kind of survival trait that might often be strongly favored by natural selection. Yet on Earth, tool-building intelligence took about 4 billion years—nearly half the expected lifetime of the Sun—to evolve. The correct value for f_i might be close to 0.01 or it might be closer to 1. No one knows.

6. f_c is the fraction of intelligent life-forms that develop technologically advanced civilizations—that is, civilizations that send communications into space. With only one example of a technological civilization to work with, f_c also is unknown. Astronomers estimate f_c to be between 0.1 and 1.

7. L is the number of years that technologically advanced civilizations exist. That factor is certainly difficult to estimate because it depends on the long-term stability of those civilizations. On Earth, the longest-lived civilizations have existed for, at most, thousands of years. Those civilizations, however, were not at the level of technology that allows interstellar communication—thus far, the first "technologically advanced civilization" on Earth is less than 100 years old. We do not know whether life intelligent enough to manipulate its environment technologically can maintain its planetary resources for any extended length of time. Astronomers usually put a value of L between 1000 years and 1 million years in their estimates.

what if . . .

What if SETI succeeds in finding a signal from a distant civilization—one that clearly must consist of technologically advanced beings? What (if anything) should we say to them, and who should have the authority to say it?

Technologically Advanced Civilizations

During lunch with colleagues, the physicist Enrico Fermi (1901–1954), a firm believer in extraterrestrial life, is reported to have asked, "If the universe is teeming with aliens . . . where is everybody?" Fermi's question—first posed in 1950 and sometimes called the *Fermi paradox*—remains unanswered. If intelligent life-forms are common but interstellar travel is difficult or impossible, would the aliens send out messages—perhaps by electromagnetic waves instead? And if they did, why haven't astronomers detected their signals?

Drake used what was then astronomy's most powerful radio telescope to listen for signals from intelligent life around two nearby stars, but he found nothing unusual. His original project has grown over the years into a much more elaborate program called the Search for Extraterrestrial Intelligence, or **SETI**. Scientists from around the world have thought carefully about what strategies might be useful for finding life in the universe. Most of those endeavors use radio telescopes to listen for signals from space that bear an unambiguous signature of an intelligent source. Some have focused on significant parts of the spectrum, assuming that a civilization will broadcast on a channel that astronomers throughout the galaxy should find interesting—for example, the 21-centimeter (21-cm) line from hydrogen gas. More recent searches have used advances in technology to record as broad a range of radio signals from space as possible. Analysts search those databases for regular signals that might be intelligent in origin.

The SETI Institute's Allen Telescope Array (ATA) received much of its initial financing from Microsoft cofounder Paul Allen. The ATA consists of a "farm" of small, inexpensive radio dishes like those used to capture signals from orbiting television-broadcasting satellites. One key project of the ATA is to observe the planets discovered by the Kepler Mission. Each dish has a diameter of 6.1 meters, but all the telescopes working together have a total signal-receiving area greater than that of a 100-meter radio telescope. Just as your brain can sort out sounds coming from different directions, that array of radio telescopes can determine the direction a signal is coming from, allowing it to listen to many stars at the same time. Over several years' time, astronomers using the ATA are expected to survey as many as a million stars, hoping to find a civilization that has sent a signal toward Earth.

As stated earlier, finding even one nearby civilization in the Milky Way Galaxy— that is, a *second* technological civilization in Earth's small corner of the universe—will make scientists optimistic that the universe as a whole is teeming with intelligent life. The likelihood of SETI's success is difficult to predict, but its potential payoff is enormous. Few discoveries would have a more profound impact than the certain knowledge that we on Earth are not alone.

Science fiction is filled with tales of humans who leave Earth to "seek out new life and new civilizations." Unfortunately, those scenarios are not scientifically realistic. The distances to the stars and their planets are enormous: To explore a significant sample of stars would require extending the physical search over tens or hundreds of light-years. Special relativity limits how fast one can travel. The speed of light is the limit, and even at that rate reaching the *nearest* star would take more than 4 years. The relativistic effect of time dilation means that time would pass slower for astronauts traveling at very high speeds, and they would return to Earth younger than if they had stayed at home. For example, suppose astronauts visited a star 15 light-years distant. Even if they traveled at speeds close to the speed of light, by the time they returned to Earth, 30 years would have passed at home.

Some science fiction writers get around that problem by invoking "warp speed" or "hyperdrive," which enables travel faster than the speed of light, or by using wormholes as shortcuts across the galaxy—but absolutely no evidence exists that any of those options are possible. And most of those imaginative stories ignore the vast number of other complications that accompany human space travel. Humans are just beginning to learn how to live in space even for short periods. Much work remains to be done before we can realistically contemplate even voyages within the Solar System.

Some people claim that aliens have already visited Earth: Tabloid newspapers, books, and websites are filled with tales of sightings of unidentified flying objects (UFOs), government conspiracies and cover-ups, alleged alien abductions, and UFO religious cults. However, none of those reports meets the basic standards of science. They are not falsifiable—they lack verifiable evidence and repeatability—so we must conclude that no scientific evidence currently exists for any alien visitations.

unanswered questions

Will humans spread life into space? Some scientists have suggested that seeding from Earth may have already happened as Earth microorganisms scattered into space after giant impacts. Humans also may have unintentionally sent microorganisms to space aboard our spacecraft. More intentional methods of seeding include sending microorganisms from Earth to other planets or moons to try to jump-start evolution, "terraforming" Mars or a moon to change conditions on it to make it more habitable for humanity, or sending humans in spaceships to colonize the galaxy.

Reading Astronomy News

When Reporting News about Aliens, Caution Is Advised

Elizabeth Howell, airspacemag.com

Astronomers struggle with how to handle announcements (or nonannouncements) about extraterrestrial contact.

Speculation about extraterrestrials seems to be everywhere these days. Last week it was "Tabby's Star" (more officially known as KIC 8462852), whose mysterious dimming and brightening, according to the latest analysis, is likely due to dust blocking different wavelengths of light rather than "alien megastructures." Before that came reports of an interstellar asteroid—not a spacecraft—entering our solar system and a UFO monitoring program conducted by the Department of Defense.

The attention given to such stories has some scientists worried, especially as social media amplifies claims of alien contact over other, more prosaic explanations.

"Currently, most SETI-related news seems to be interfering with conventional scientific discoveries, stealing the limelight—without following basic rules of science," wrote Dutch exoplanet researcher Ignas Snellen of Leiden Observatory, on a Facebook exoplanets discussion group for professional astronomers.

Although he has "great respect for SETI scientists," Leiden wrote, "there is no place for alien civilizations in a scientific discussion on new astrophysical phenomena, in the same way as there is no place for divine intervention as a possible solution. One may view it as harmless fun, but I see parallels in athletes taking banned substances. It may lead to short-term fame and medals, but in the long run it harms the sport. Same for astronomy: we should be very careful not to be ridiculed. I really hope we can stop mentioning SETI for every unexplained phenomenon."

Such worries aren't new. Nearly 20 years ago researchers in the field came up with the Rio Scale to guide them in reporting the significance of any candidate SETI signal. Modeled after the Richter Scale for earthquakes, it assigns a value between 0—no significance—and 10—extraordinary significance—to any detected signal. The method was first proposed by Ivan Almar and Jill Tarter at the 51st International Astronautical Congress in Rio de Janeiro, in 2000. Although the International Academy of Astronautics' SETI Permanent Committee adopted the scale two years later, it hasn't seen widespread use.

Meanwhile, the news business has changed drastically since 2000. Today much of the public doesn't get its information from TV and newspapers, but from Facebook updates and Twitter posts, which move—and change—at a much faster pace. In recognition of the altered media landscape, a new paper submitted to the *International Journal of Astrobiology* proposes streamlining the Rio Scale and having scientists take a short quiz to answer a few questions about their discovery. Journalists and other news providers could use the answers in their stories.

"You can go through the steps and ask yourself, do I believe the instrumentation is working properly? Do I believe I am looking at it objectively? Do I have enough people who looked at the signal with different instruments? Has there been a lot of scientific discussion? Have there been alternative explanations? Is it a hoax?" says the paper's lead author, Duncan Forgan, a postdoctoral scholar at the University of St. Andrews' School of Physics and Astronomy in Scotland, whose work includes problems related to astrobiology and SETI.

Forgan says that although it can be difficult to convey uncertainty about a particular result to the public, the revised Rio Scale would make that job easier. Some scientific speculation can be "a bit wibbly-wobbly," he says, "and those wibbly-wobbly bits end up in the press." He agrees that spurious SETI claims can sometimes distract from more legitimate science.

In recognition of the sensitivity around alien signal detection, SETI has a voluntary list of protocols to follow when something interesting is found. The first principle urges researchers to "verify that the most plausible explanation for the evidence is the existence of extraterrestrial intelligence, rather than some other natural phenomenon or anthropogenic phenomenon, before making any public announcement."

Morris Jones, an Australian space observer with both scientific and journalistic training, says we should label "fringe SETI" claims for what they are. "The media is under pressure to deliver attention-grabbing news, but it's hard to expect them to judge fringe SETI as spurious when it comes from reputable institutions and qualified researchers. The best way to reduce these reports is to stop the production of questionable scientific papers in the first place."

QUESTIONS

1. Why are astrobiologists concerned about media coverage of SETI?

2. How do the suggestions for journalists fit with what you have learned about the process of science?

3. Think of an article you have seen recently on evidence for alien life. Was it overly speculative?

4. How do you think people on Earth would react to an actual discovery of alien life?

Source: https://www.airspacemag.com/daily-planet/when-reporting-news-about-aliens-caution-advised-180967777/.

CHECK YOUR UNDERSTANDING **24.4**

The Drake equation enables astronomers to: (a) calculate precisely the number of alien civilizations; (b) organize their thoughts about probabilities for life; (c) locate the stars they should study to find life; (d) find new kinds of life.

Origins: The Fate of Life on Earth

Astronomers have used their understanding of physics and cosmology to look back through time and watch as structure formed throughout the universe and to look forward to the future of our Sun, our galaxy, and the ultimate fate of the universe. About 5 billion years from now, the Sun will end its long period of relative stability. The Sun will expand to become a red giant star, swelling to hundreds of times larger than it is now and thousands of times more luminous. The giant planets, orbiting outside the extended red giant atmosphere, will probably survive. But at least some of the planets of the inner Solar System will not. Just as an artificial satellite is slowed by drag in Earth's tenuous outer atmosphere and eventually falls to the ground, so, too, will a planet caught in the Sun's atmosphere be engulfed by the expanding Sun. If that happens to Earth, no trace of this planet will remain other than a slight increase in the amount of massive elements in the Sun's atmosphere.

Another possibility is that the red giant Sun will lose mass in a powerful wind, its gravitational pull on the planets will weaken, and the orbits of both the inner and outer planets will spiral outward. If Earth moves out far enough, it may survive as a seared cinder, orbiting the small, hot, white dwarf star that the Sun will become. Barely larger than Earth and with its nuclear fuel exhausted, the white dwarf Sun will slowly cool, eventually becoming a cold sphere of densely packed carbon, orbited by what remains of its planets. The ultimate outcome for Earth—consumed in the Sun or left behind as a cold, burned rock orbiting a long-dead white dwarf—is not yet known.

Alternatively, the threat to the future of Earth can come from within, as humanity itself changes conditions on Earth that are necessary for our habitability (for example, by releasing chemicals into the atmosphere that alter the ozone layer, or are greenhouse gases that alter the trapping of radiation).

In either case, however, Earth's status as a garden spot in the habitable zone will be at an end. If the Sun does not expand too far or Mars also migrates outward, Mars could become the habitable planet in the Solar System, at least for a while. As the dying Sun loses more and more of its atmosphere in a stellar wind, Earth's atoms might be expelled back into the reaches of interstellar space from which they came, perhaps to be recycled into new generations of stars, planets, and even life itself.

But even before the Sun's change into a red giant star, the Sun's luminosity will begin to rise. As solar luminosity increases, so will temperatures on all the planets. The inner edge of the Sun's habitable zone will slowly move out past the orbit of Earth. Eventually, Earth's temperatures will climb so high that all animal and plant life will perish. Even the extremophiles that inhabit the oceanic depths will die as the oceans themselves boil away. Models of the Sun's evolution are still not precise enough for astronomers to predict with certainty when that fatal event will occur, but the end of all terrestrial life may be 1 billion to 4 billion years away. In addition, the Milky Way Galaxy will collide with the Andromeda Galaxy in 4 billion to 5 billion years' time. Galaxies are mostly empty space, so the Sun is not likely to collide with another star, but one effect of the collision is that our Solar System may be gravitationally flung to a different part of our galaxy.

unanswered questions

Will humans themselves spread into space? At some point, humans must leave planet Earth if the species is to survive. But space is a dangerous place, and many of the problems that humans encounter in space have not yet been solved. Those problems are as diverse as purely physical issues, such as the loss of bone density in low gravity, and the societal problems that occur when a few people are confined together for long periods. We do not yet know whether humanity can overcome those problems and journey to other planets.

Figure 24.24 ★ **WHAT AN ASTRONOMER SEES** This image from the *Cassini* spacecraft shows Earth as seen from Saturn. The pale dot in the lower right, indicated by the arrow, is Earth. For this last "What an Astronomer Sees" image, we give you the words of Carl Sagan, describing what he saw in a similar image of Earth from a distant perspective:

"We succeeded in taking that picture, and, if you look at it, you see a dot. That's here. That's home. That's us. On it, everyone you ever heard of, every human being who ever lived, lived out their lives.

The aggregate of all our joys and sufferings, thousands of confident religions, ideologies and economic doctrines, every hunter and forager, every hero and coward, every creator and destroyer of civilizations, every king and peasant, every young couple in love, every hopeful child, every mother and father, every inventor and explorer, every teacher of morals, every corrupt politician, every superstar, every supreme leader, every saint and sinner in the history of our species, lived there—on a mote of dust, suspended in a sunbeam.

The Earth is a very small stage in a vast cosmic arena. Think of the rivers of blood spilled by all those generals and emperors so that in glory and in triumph they could become the momentary masters of a fraction of a dot. Think of the endless cruelties visited by the inhabitants of one corner of the dot on scarcely distinguishable inhabitants of some other corner of the dot. How frequent their misunderstandings, how eager they are to kill one another, how fervent their hatreds. Our posturings, our imagined self-importance, the delusion that we have some privileged position in the universe, are challenged by this point of pale light ...

To my mind, there is perhaps no better demonstration of the folly of human conceits than this distant image of our tiny world. To me, it underscores our responsibility to deal more kindly and compassionately with one another and to preserve and cherish that pale blue dot, the only home we've ever known."

— *Carl Sagan, speech at Cornell University, October 13, 1994*

However, whether the descendants of today's humanity will even be around a billion years from now is far from certain. Some threats to life come from beyond Earth. For the rest of the Sun's life, the terrestrial planets, including Earth, will continue to be bombarded by asteroids and comets. Perhaps a hundred or more of those impacts will involve kilometer-sized objects, capable of causing the kind of devastation that eradicated the dinosaurs. Those impacts may create new surface scars but they will have little effect on the integrity of Earth itself. Earth's geological record is filled with such events, and each time they happen, life manages to recover and reorganize. But we do not know if human life or civilization could survive.

Humans might protect themselves from the fate of the dinosaurs, but in the long run humanity will either leave this world or die out. Planetary systems are common to other stars, and many other Earth-like planets may well exist throughout the Milky Way Galaxy. Colonizing other planets is currently the stuff of science fiction, but if the descendants of modern-day humans are ultimately to survive the death of the home planet, off-Earth colonization must at some point become science fact.

By studying astronomy, you have learned where you come from. You have learned about the self-correcting nature of science—how it continually adapts to new information to give us the ability to make better and better predictions about the behavior of the physical world. That predictive ability makes science an extremely powerful tool. No other species in the history of Earth has been able to understand its position, predict what will happen next, and therefore adapt its behavior to seek the best possible future for its members.

Figure 24.24 shows Earth as seen by the *Cassini* spacecraft, looking near Saturn's rings. That tiny dot, which is Earth, is the only place in the entire universe where we know that life exists. Compare the size of that dot with the size of the universe. Compare the history of life on Earth with the history of the universe. Compare Earth's future with the fate of the universe. Astronomy is humbling. We occupy a tiny part of space and time. Yet we are unique, as far as we know. Think for a moment about what that means to you. That may be the most important lesson the universe has to offer.

SUMMARY

Astrobiology seeks to answer the question, Are we alone? All the sciences have a part to play in answering that question, from astronomy to zoology. Theories of how life began and evolved on Earth help limit the number of targets astronomers study in their search for life elsewhere. Even before the Sun expands and evolves to a red giant star, its luminosity will increase enough to alter the location of its habitable zone. When that happens, Earth may no longer be the planet at the right location for maintaining liquid water.

(1) **Explain our current understanding of how and when life began on Earth and how it has evolved.** Life on Earth is a form of complex carbon-based chemistry, made possible by self-replicating molecules. Life probably formed in Earth's oceans and then evolved chemically from simple molecules into self-replicating organisms through a combination of mutation and heredity.

(2) **Explain how life is a structure that has evolved through the action of the physical and chemical processes that shape the universe.** All terrestrial life is composed primarily of only six elements: carbon, hydrogen, oxygen, nitrogen, sulfur, and phosphorus. However, life-forms very different from those on Earth, including those based on silicon chemistry, cannot be ruled out.

(3) **List the locations in our Solar System and around other stars where astronomers think life might be possible.** Within the Solar System, Mars and some moons of Jupiter and Saturn are the most promising candidates for life. A habitable zone around a star is a location within which the temperature of a planet will support liquid water on its surface. Thus, astronomers look for exoplanets that orbit in habitable zones surrounding solar-type stars. A habitable zone can be extended by special circumstances, such as tidal heating from a giant planet or atmospheres containing greenhouse gases.

(4) **Describe some methods used to search for intelligent extraterrestrial life.** The Drake equation includes factors that astronomers consider when thinking about the possibility of life in the universe. Astronomers use radio telescopes to search for signals from extraterrestrial life, particularly in astronomically important regions of the spectrum. In recent times, that effort has been expanded to a search through as broad a region of the radio spectrum as possible. No intelligent extraterrestrial life has yet been detected.

QUESTIONS AND PROBLEMS

TEST YOUR UNDERSTANDING

1. The study of life and the study of astronomy are connected because (select all that apply)
   a. life may be commonplace in the universe.
   b. studying other planets may help explain why life exists on Earth.
   c. explorations of extreme environments on Earth suggest where to look for life elsewhere.
   d. life is a structure that evolved through physical processes, and life on Earth may not be unique.
   e. life elsewhere is most likely to be found by astronomers.

2. Scientists look for water to indicate places where life might exist because
   a. water is a common molecule in interstellar space.
   b. life on Earth depends on it.
   c. no other molecules are solvents.
   d. the spectrum of water is very complex.

3. The Urey-Miller experiment produced _____ in a laboratory jar.
   a. life
   b. RNA and DNA
   c. amino acids
   d. proteins

4. A mutation is
   a. always a deadly change to DNA.
   b. always a beneficial change to DNA.
   c. a change to DNA that is sometimes beneficial and sometimes not.
   d. a change to DNA that cannot be inherited.

5. Scientists think that terrestrial life probably originated in Earth's oceans, rather than on land, because (select all that apply)
   a. all the chemical pieces were in the ocean.
   b. energy was available in the ocean.
   c. the earliest evidence for life on Earth is from fossils of ocean-dwelling organisms.
   d. the deepest parts of the ocean have hydrothermal vents.

6. Any system with the processes of heredity, mutation, and natural selection will, over time (choose all that apply),
   a. change.
   b. become larger.
   c. become more complex.
   d. develop intelligence.

7. The fact that no alien civilizations have yet been detected indicates that
   a. they are not there.
   b. they are rare.
   c. Earth is in a "blackout," and they are not talking to us.
   d. we don't know enough yet to draw any conclusions.

8. During the Cambrian explosion
   a. the dinosaurs were killed.
   b. all the carbon that is now here on Earth was produced.
   c. biodiversity increased significantly.
   d. a lot of carbon dioxide was released into the atmosphere.

9. The habitable zone is the place around a star where
   a. life has been found.
   b. atmospheres can contain oxygen.
   c. liquid water has been confirmed.
   d. liquid water can exist on the surface of a planet.

10. The difference between a prokaryote and a eukaryote is that prokaryotes
    a. have no DNA.
    b. have no cell wall.
    c. have no nucleus.
    d. do not exist today.

11. A thermophile is an organism that lives in extremely _____ environments.
    a. salty
    b. hot
    c. cold
    d. dry

12. The search for life elsewhere in the Solar System is carried out primarily by
    a. astronauts.
    b. robotic spacecraft.
    c. astronomers using optical telescopes.
    d. astronomers using radio telescopes.

13. Astronomers think that intelligent life is more likely to be found around stars of types F, G, K, and M because
    a. those stars are hot enough to have planets and moons with liquid water.
    b. those stars are cool enough to have planets and moons with liquid water.
    c. those stars live long enough for life to begin and evolve.
    d. those stars produce no ultraviolet radiation or X-rays.

14. Life first appeared on Earth
    a. billions of years ago.
    b. millions of years ago.
    c. hundreds of thousands of years ago.
    d. thousands of years ago.

15. In the phrase "theory of evolution," the word *theory* means that evolution
    a. is an idea that can't be tested scientifically.
    b. is an educated guess to explain natural phenomena.
    c. probably doesn't happen anymore.
    d. is a tested and corroborated scientific explanation of natural phenomena.

THINKING ABOUT THE CONCEPTS

16. How does the evolution of life on Earth depend on RNA and DNA?

17. How do scientists think that amino acids first formed on Earth?

18. Today, most organisms on Earth enjoy relatively moderate climates and temperatures. Compare that environment with some of the conditions in which early life developed.

19. The Process of Science Figure represents many areas of science that all inform the science of astrobiology. Choose any two of those, and explain how one depends on the other. ⊙

20. How was Earth's carbon dioxide atmosphere changed into today's oxygen-rich atmosphere? How long did that transformation take?

21. What was the Cambrian explosion, and what might have caused it?

22. Why did plants and forests appear in large numbers before large animals did?

23. What is a habitable zone? What defines its boundaries?

24. Which general conditions were needed on Earth for life to arise?

25. Is evolution under way on Earth today? If so, how might humans continue to evolve?

26. Where did all the atoms in your body come from?

27. The two *Viking* spacecraft found no convincing evidence of life on Mars when they visited that planet in the late 1970s, nor did the *Phoenix* lander when it examined the martian soil in 2009. Do those results imply that life never existed on the planet? Why or why not?

28. Some scientists believe that humans may be the only advanced life in the galaxy today. If so, which factors in the Drake equation must be extremely small?

29. ★ **WHAT AN ASTRONOMER SEES** The image in Figure 24.24 shows Earth from a quite distant perspective. Take a moment to think back to the beginning of this course, and reflect on what you have learned. How has your concept of your place in the universe changed since then? How has your understanding of this image (and others like it) changed as you've learned about the universe beyond Earth? ⊛

30. Why is life on Earth as we know it likely to end long before the Sun runs out of nuclear fuel?

APPLYING THE CONCEPTS

31. Suppose that an organism grows in number in the exponential fashion of Working It Out 24.1. How many times more of the organism will there be after 3 doubling times? ●─●─●

    a. Make a prediction: Do you expect this number to be more or less than 1? What approximate power of 10 do you expect: 10, 100, 1000, or more?

    b. Calculate: Use your calculator and follow Working It Out 24.1 to find the answer.

    c. Check your work: Compare your answer to your prediction, to the numbers in Working It Out 24.1, and to the process of multiplying $2 \times 2 \times 2$ in your calculator. This will help you verify that you know how to work your calculator correctly to raise 2 to a power.

32. Use the most optimistic numbers in the Drake equation for all the terms on the right, except L. What, then, does L have to be equal to in order for a million civilizations to exist in the Milky Way? ●─●─●

    a. Make a prediction: Most of the "optimistic" values for terms in the Drake equation are around 1. Do you expect L to be a small number, like a century, or a large number, like millions of years?

    b. Calculate: Determine what L must be, under these conditions, for there to be a million civilizations in the Milky Way.

    c. Check your work: Compare your number to your prediction, and work the problem the other way (plug in your value for L, and solve for N) in order to check your work.

33. Suppose that an organism replicates itself each second. If you start with a single specimen, what will the final population be after 10 seconds?

34. Suppose that an organism replicates itself each second. After how many seconds will the population increase by a factor of 1024?

35. The "rule of 70" states that you can approximate the doubling time for exponential growth by dividing 70 by the rate of increase. So, if a population increases by 7 percent per year, the doubling time is 10 years ($70/7 = 10$). Suppose Earth's human population continues to grow by 1 percent annually. What is the doubling time? How much time will pass before Earth has 4 times as many humans?

36. Several missions are searching for planets in the habitable zones of stars. Explain which factors in the Drake equation are affected by that search and how the final number N will be affected if these missions find that most stars have planets in their habitable zones.

37. The white blocks at the top of Figure 24.22 represent the numbers 1–10, in order from right to left. Each representation of a number uses a maximum of four rows, where the bottom row of the set of four is a placeholder, and the top three rows of the four represent the number. Explain the "rule" for the kind of counting shown here. (For example, how do three white blocks represent the number 7?) ⊛

38. As discussed in Working It Out 24.1, the doubling time for exponential growth is given by $P_F/P_O = 2^n$. Assume a self-replicating molecule that makes one copy of itself each second. Make a graph of the number of molecules versus time for the first 60 seconds after the molecule begins replicating.

39. The doubling time for *Escherichia coli* is 20 minutes, and you start getting sick when just 10 bacteria enter your system. How many bacteria are in your body after 12 hours?

40. If the chance that a given molecule will mutate is 1 in 100,000, how many generations are needed before, on average, at least one mutation has occurred?

41. Study the Drake equation in Working It Out 24.2. Make your own most optimistic and most pessimistic assumptions for each variable in the equation. What values do you find for N?

42. Trace (or photocopy) the graph in Figure 5.25, and then add horizontal lines for the temperatures at which water freezes and boils. Which planets in the Solar System have measured temperatures that fall within those lines? Which planets have predicted temperatures (based on the equilibrium model) that fall within those lines? What does that tell you about assumptions about the habitable zone? ⊛

43. Figure 5.25 shows that Mercury's measured range of temperatures overlaps the temperatures at which water is a liquid. Is Mercury in the habitable zone? Why or why not? ⊛

44. Suppose astronomers announce the discovery of a new planet around a star with a mass equal to the Sun's. The planet has an orbital period of 87 days. Is that planet in the habitable zone for a Sun-like star?

45. As noted in Section 24.1, some scientists suspect the early Earth was "seeded" with primitive life stored in comets and meteoroids. Now that you know when and how our Solar System, galaxy, and universe formed, what time line is required for such seeding to be possible?

EXPLORATION   Exploring the Habitable Zone

digital.wwnorton.com/astro7

The habitable zone depends on the properties of both the star and the planet's orbit. This can be a little bit tricky to calculate for main-sequence stars, because the temperature, luminosity, and radius of the star are all connected. Open the Habitable Zone simulation, and study the habitable zone diagram in the large window for a moment. Click on "Legend," and identify the inner and outer radius of the conservative habitable zone and the optimistic habitable zone.

1   Which is wider: the conservative habitable zone or the optimistic habitable zone?

_____

_____

2   What color is the scale bar that shows the radius of the outer edge of the optimistic habitable zone?

_____

_____

Set the "Star Properties" to be equal to those for the Sun, and set the "Planet Orbit Properties" to be equal to those for Earth. It is good review for you to try to remember those, and look them up if you cannot! Answer the following questions.

3   For a star like the Sun, what is the inner radius of the optimistic habitable zone? How does this compare to the radius of Earth's orbit?

_____

_____

4   For a star like the Sun, what is the outer radius of the optimistic habitable zone? How does this compare to the radius of Earth's orbit?

_____

_____

5   Use the slider to change the eccentricity of the planet's orbit. How eccentric can the orbit of a planet with a semimajor axis of 1 AU be and still have the planet always within the *optimistic* habitable zone?

_____

_____

6   How eccentric can the orbit of a planet with a semimajor axis of 1 AU be and still have the planet always within the *conservative* habitable zone?

_____

_____

7   Double the luminosity of the star, keeping the temperature the same. What happened to the habitable zone? Did it become narrower or wider than the habitable zone around the Sun? Did it move closer or farther from the star than the habitable zone of the Sun? Why?

_____

_____

8   Reset the luminosity to the luminosity of the Sun, and double the temperature. What happened to the habitable zone? Why?

_____

_____

9   Double-check that the luminosity of the star is equal to the luminosity of the Sun, and set the temperature to half the temperature of the Sun. What happened to the habitable zone? Why?

_____

_____

10   Now set the star's luminosity to half the luminosity of the Sun, and set the temperature to half the temperature of the Sun. What happened to the habitable zone? Why?

_____

_____

Mathematics helps scientists understand the patterns they see and then communicate that understanding to others. Appendix 1 presents some tools that will be useful in our study of astronomy.

Powers of 10 and Scientific Notation

Astronomy is a science of both the very large and the very small. The mass of an electron, for example, is

0.0000000000000000000000000000009109 kilograms (kg)

whereas the distance to a galaxy far, far away might be about

100,000,000,000,000,000,000,000,000 meters

A quick glance at those two numbers shows why astronomers need a more convenient way to express numbers.

Our number system is based on powers of 10. Going to the left of the decimal place,

$$10 = 10 \times 1$$
$$100 = 10 \times 10 \times 1$$
$$1000 = 10 \times 10 \times 10 \times 1$$

and so on. Going to the right of the decimal place,

$$0.1 = \frac{1}{10} \times 1$$

$$0.01 = \frac{1}{10} \times \frac{1}{10} \times 1$$

$$0.001 = \frac{1}{10} \times \frac{1}{10} \times \frac{1}{10} \times 1$$

and so on. In other words, each place to the right or left of the decimal place in a number represents a power of 10. For example, 1 million can be written as

$$1 \text{ million} = 1,000,000 = 1 \times 10 \times 10 \times 10 \times 10 \times 10 \times 10$$

That is, 1 million is "1 multiplied by six factors of 10." **Scientific notation** combines those factors of 10 in convenient shorthand. Rather than all being written out, the six factors of 10 are expressed using an exponent:

$$1 \text{ million} = 1 \times 10^6$$

which also means "1 multiplied by six factors of 10."

Moving to the right of the decimal place, each step *removes* a power of 10 from the number. One-millionth can be written as

$$1 \text{ millionth} = 1 \times \frac{1}{10} \times \frac{1}{10} \times \frac{1}{10} \times \frac{1}{10} \times \frac{1}{10} \times \frac{1}{10}$$

That divides by powers of 10, so that number can be written by use of a negative exponent

$$\frac{1}{10} = 10^{-1}$$

as

$$1 \text{ millionth} = 1 \times 10^{-1} \times 10^{-1} \times 10^{-1} \times 10^{-1} \times 10^{-1} \times 10^{-1}$$
$$= 1 \times 10^{-6}$$

Let's return to our earlier examples. The mass of an electron is 9.109×10^{-31} kg, and the distant galaxy is located 1×10^{26} meters away. Scientific notation is a much more convenient way of writing those values. Notice that *the exponent in scientific notation gives you a feel for the size of a number at a glance.* The exponent of 10 in the electron mass is −31, which quickly indicates that it is a very small number. The exponent of 10 in the distance to a remote galaxy, +26, quickly indicates that it is a very large number. The exponent is often called the *order of magnitude* of a number. When you see a number written in scientific notation while reading *21st Century Astronomy* (or elsewhere), just remember to look at the exponent to understand the size of the number.

Scientific notation also is convenient because it makes multiplying and dividing numbers easier. For example, 2 billion multiplied by eight-thousandths can be written as

$$2,000,000,000 \times 0.008$$

but writing those two numbers in scientific notation is:

$$(2 \times 10^9) \times (8 \times 10^{-3})$$

We can regroup those expressions in the following form:

$$(2 \times 8) \times (10^9 \times 10^{-3})$$

The first part of the problem is just $2 \times 8 = 16$. The more interesting part of the problem is the multiplication in the right-hand parentheses. The first number, 10^9, is just shorthand for $10 \times 10 \times 10 \ldots$ nine times. That is, it represents nine factors of 10. The second number stands for three factors of 1/10—or removing three factors of 10 if you prefer to think of it that way. Altogether, that makes $9 - 3 = 6$ factors of 10. In other words,

$$10^9 \times 10^{-3} = 10^{(9-3)} = 10^6$$

Putting the problem together, we get

$$(2 \times 10^9) \times (8 \times 10^{-3}) = (2 \times 8) \times (10^9 \times 10^{-3})$$
$$= 16 \times 10^6$$

By convention, when a number is written in scientific notation, only one digit is placed to the left of the decimal point. Here, though, we have two. However, 16 is 1.6×10, so we can add that additional factor of 10 to the exponent at right, making the final answer

$$1.6 \times 10^7$$

Dividing is just the inverse of multiplication. Dividing by 10^3 means removing three factors of 10 from a number. Using the previous number,

$$(1.6 \times 10^7) \div (2 \times 10^3) = (1.6 \div 2) \times (10^7 \div 10^3)$$
$$= 0.8 \times 10^{(7-3)}$$
$$= 0.8 \times 10^4$$

This time we have only a zero to the left of the decimal point. To get the number into proper form, we can substitute 8×10^{-1} for 0.8, giving

$$0.8 \times 10^4 = (8 \times 10^{-1}) \times 10^4 = 8 \times 10^3$$

Adding and subtracting numbers in scientific notation is harder because all numbers must be written as values multiplied by the *same* power of 10 before they can be added or subtracted. Therefore, you will need to use a calculator that has scientific notation. Most scientific calculators have a button that says EXP or EE. Those abbreviations mean "times 10 to the ___." So for 4×10^{12}, you would type [4] [EXP][1][2] or [4][EE][1][2] into your calculator. Usually, that number shows up in the window on your calculator either just as you see it written in this book or as a 4 with a smaller 12 all the way over in the right side of the window.

Significant Figures

In the previous example, we actually broke some rules in the interest of explaining how powers of 10 are treated in scientific notation. The rules we broke involve the *precision* of the numbers. When expressing quantities in science, it is extremely important to know not only the value of a number but also how precise that value is.

The most complete way to keep track of the precision of numbers is to actually write down the uncertainty in the number. For example, suppose you know that the distance to a store (call it d) is between 0.8 and 1.2 kilometers (km); you can then write

$$d = 1.0 \pm 0.2 \text{ km}$$

where the symbol "$\pm$" is pronounced "plus or minus." Here, d is between $1.0 - 0.2 = 0.8$ km and $1.0 + 0.2 = 1.2$ km. That is an unambiguous statement about the limitations on knowing the value of d, but carrying along the formal errors with every number written would be cumbersome at best. Instead, you keep track of the approximate precision of a number by using *significant figures*.

The convention for significant figures is this: Assume the written number has been rounded from a number that had one additional digit to the right of the decimal point. If a quantity d, which might represent the distance to the store, is "1.", then d is close to 1. It is probably not as small as "0.", and it is probably not as large as "2.". If instead it is written as

$$d = 1.0$$

then d is probably not 0.9 and is probably not 1.1. It is roughly 1.0 to the nearest tenth. The greater the number of significant figures, the more precisely the number is being specified. For example, 1.00000 is not the same number as 1.00. The first number, 1.00000, represents a value that is probably not as small as 0.99999 and probably not as large as 1.00001. The second number, 1.00, represents a value that is probably not as small as 0.99 nor as large as 1.01. The number 1.00000 is much more precise than the number 1.00.

In mathematical operations, significant figures are important. For example, consider $2.0 \times 1.6 = 3.2$. It does *not* equal 3.20000000000. *The product of two numbers cannot be known to any greater precision than the numbers themselves.* In general, when you multiply and divide, the answer should have the same number of significant figures as the less precise of the numbers being multiplied or divided. In other words, $2.0 \times 1.602583475 = 3.2$. Because all you know is that the first factor is probably closer to 2.0 than to 1.9 or 2.1, all you know about the product is that it is between about 3.0 and 3.4. It is 3.2. It is not 3.205166950 (*even if that is the answer on your calculator*). The rest of the digits to the right of 3.2 just do not mean anything.

When two numbers are added or subtracted, if one number has a significant figure with a particular place value but another number does not, their sum or difference cannot have a significant figure in that place value. For example,

$$
\begin{array}{r}
1045. \\
+1.34567 \\
\hline
1046.
\end{array}
$$

The answer is "1046." *not* "1046.34567." Again, the extra digits to the right of the decimal place have no meaning because "1045." is not known to that precision.

What is the precision of the number 1,000,000? As it is written, the answer is unclear. Are all those zeros really significant, or are they merely placeholders? If the number is written in scientific notation, however, you never have to wonder. Instead of 1,000,000, you write 1.0×10^6 for a number that is known to the nearest hundred thousand or so, or you write 1.00000×10^6 for a number that is known to the nearest 10.

So the earlier example would have been more correct if written as

$$(2.0 \times 10^9) \times (8.0 \times 10^{-3}) = 1.6 \times 10^7$$

Algebra

Mathematics has many branches. The branch that focuses on the relationships between quantities is called **algebra**. Basically, algebra begins by using symbols to represent quantities.

For example, you could write the distance you travel in a day as d. As it stands, d has no value. It might be 10,000 miles. It might be 30 feet. It does, however, have **units**—here, the units of distance. The average speed at which you travel is equal to the distance you travel divided by the time you take. By using the symbol v to represent your average speed and the symbol t to represent the time you take, instead of writing out "Your average speed is equal to the distance you travel divided by the time you take" you can write

$$v = \frac{d}{t}$$

The meaning of that algebraic expression is the same as the sentence quoted before it, but it is much more concise. As it stands, v, d, and t still have no specific values. No numbers are assigned to them yet. However, the expression indicates what the relationship between those numbers will be when you look at a specific example. For example, if you go 500 km ($d = 500$ km) in 10 hours ($t = 10$ hours), that expression tells you that your average speed is

$$v = \frac{d}{t} = \frac{500 \text{ km}}{10 \text{ h}} = 50 \text{ km/h}$$

Notice that the units in this expression act exactly like the numerical values. Dividing the two shows that the units of v are kilometers divided by hours, or km/h (pronounced "kilometers per hour").

We introduced algebra as shorthand for expressing relations between quantities, but it is far more powerful than that. Algebra provides rules for manipulating the symbols used to represent quantities. We begin with a bit of notation for *powers* and *roots*. Similar to powers of 10, raising a quantity to a power means multiplying the quantity by itself some number of times. For example, if S is a symbol for something (anything), S^2 (pronounced "S-squared" or "S to the second power") means $S \times S$, and S^3 (pronounced "S-cubed" or "S to the third power") means $S \times S \times S$. Suppose S represents the length of the side of a square. The area of the square is given by

$$\text{Area} = S \times S = S^2$$

If $S = 3$ meters (m), the area of the square is

$$S^2 = 3 \text{ m} \times 3 \text{ m} = 9 \text{ m}^2$$

(pronounced "9 square meters"). That is why raising a quantity to the second power is called *squaring* the quantity. We could have done the same thing for the sides of a cube and found that the volume of the cube is

$$\text{Volume} = S \times S \times S = S^3$$

If $S = 3$ meters, the volume of the cube is

$$S^3 = 3 \text{ m} \times 3 \text{ m} \times 3 \text{ m} = 27 \text{ m}^3$$

(pronounced "27 cubic meters"). That is why raising a quantity to the third power is called *cubing* the quantity.

Roots are the reverse of that process. The square root of a number is the value that, when squared, gives the original quantity. The square root of 4 is 2, which means that $2 \times 2 = 4$. The square root of 9 is 3, which means that $3 \times 3 = 9$. Similarly, the cube root of a quantity is the value that, when cubed, gives the original quantity. The cube root of 8 is 2, which means that $2 \times 2 \times 2 = 8$. Roots are written with the symbol $\sqrt{}$. For example, the square root of 9 is written as

$$\sqrt{9} = 3$$

and the cube root of 8 is written as

$$\sqrt[3]{8} = 2$$

If the volume of a cube is $V = S^3$, then

$$S = \sqrt[3]{V} = \sqrt[3]{S^3}$$

Roots can also be written as powers. Powers and roots behave like the exponents of 10 in our discussion of scientific notation. (The exponents used in scientific notation are just powers of 10.) For example, if a, n, and m are all algebraic quantities, then

$$a^n \times a^m = a^{n+m} \quad \text{and} \quad \frac{a^n}{a^m} = a^{n-m}$$

(The square root of a can also be written $a^{1/2}$ and the cube root of a can be written $a^{1/3}$.)

The rules of arithmetic can be applied to the symbolic quantities of algebra. As long as the rules of algebra are applied properly, the relationships among symbols arrived at through algebraic manipulation remain true for the physical quantities that those symbols represent.

Here we summarize a few algebraic rules and relationships. Here, a, b, c, m, n, r, x, and y are all algebraic quantities.

Associative rule:

$$a \times b \times c = (a \times b) \times c = a \times (b \times c)$$

Commutative rule:

$$a \times b = b \times a$$

Distributive rule:

$$a \times (b + c) = (a \times b) + (a \times c)$$

Cross-multiplication:

$$\text{If } \frac{a}{b} = \frac{c}{d}, \text{ then } ad = bc$$

Working with exponents:

$$\frac{1}{a^n} = a^{-n} \quad a^n a^m = a^{n+m}$$

$$\frac{a^n}{a^m} = a^{n-m} \quad (a^n)^m = a^{n \times m} \quad \left(\frac{a}{b}\right)^n = \frac{a^n}{b^n}$$

Equation of a line with slope m and y-intercept b:

$$y = mx + b$$

Equation of a circle with radius r centered at $x = 0$, $y = 0$:

$$x^2 + y^2 = r^2$$

Angles and Distances

The farther away something is, the smaller it appears. That is common sense and everyday experience. Because astronomers cannot walk up to the object they are studying and measure it with a meterstick, their knowledge about the sizes of things usually depends on relating the size of an object, its distance, and the angle it covers in the sky.

One way to measure angles is to use a unit called the **radian**. As shown in **Figure A1.1a**, the size of an angle in radians is the length of the arc subtending the angle, divided by the radius of the circle. In the figure, the angle $x = S/r$ radians.

Because the circumference of a circle is 2π multiplied by the radius ($C = 2\pi r$), a complete circle has an angular measure of $(2\pi r)/r = 2\pi$ radians. In more conventional angular measure, a complete circle is $360°$, so

$$360° = 2\pi \text{ radians}$$

or

$$1 \text{ radian} = \frac{360°}{2\pi} = 57.2958°$$

Often, seconds of arc (**arcseconds**) are used to measure angles for stars and galaxies. A degree is divided into 60 minutes of arc (**arcminutes**), each of which is divided into 60 seconds of arc—so 3600 seconds of arc are in a degree. Therefore,

$$3600 \frac{\text{arcseconds}}{\text{degree}} \times 57.2958 \frac{\text{degree}}{\text{radian}} = 206{,}265 \frac{\text{arcseconds}}{\text{radian}}$$

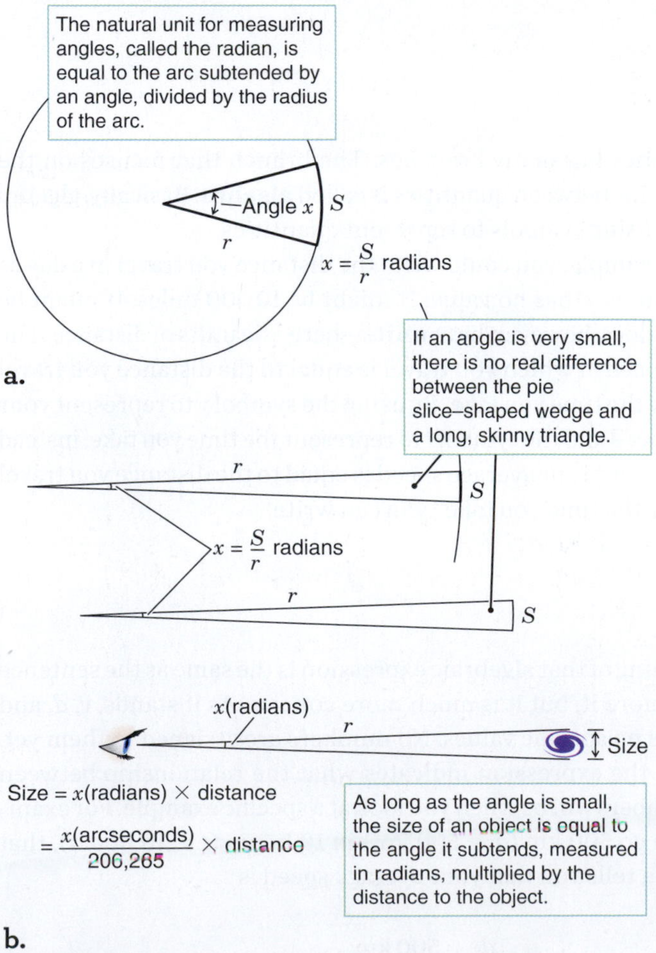

The natural unit for measuring angles, called the radian, is equal to the arc subtended by an angle, divided by the radius of the arc.

—Angle x S

r

$$x = \frac{S}{r} \text{ radians}$$

a.

If an angle is very small, there is no real difference between the pie slice–shaped wedge and a long, skinny triangle.

r S

$$x = \frac{S}{r} \text{ radians}$$

r S

x(radians) r Size

$$\text{Size} = x(\text{radians}) \times \text{distance}$$

$$= \frac{x(\text{arcseconds})}{206{,}265} \times \text{distance}$$

As long as the angle is small, the size of an object is equal to the angle it subtends, measured in radians, multiplied by the distance to the object.

b.

Figure A1.1 Measuring angles.

If the angle is small enough (which it usually is in astronomy), very little difference exists between the pie slice just described and a long skinny triangle with a short side of length S, as **Figure A1.1b** illustrates. So, if you know the distance d to an object and you can measure the angular size x of the object, the size of the object is given by

$$S = x \text{ (in radians)} \times d = \frac{x \text{ (in degrees)}}{57.2958 \text{ degrees/radian}} \times d$$

$$S = \frac{x \text{ (in arcseconds)}}{206,265 \text{ arcseconds/radian}} \times d$$

That is how astronomers relate an object's angular size, distance, and physical size.

Circles and Spheres

To round out those mathematical tools, here are a few useful formulas for circles and spheres. The circle or sphere in each case has a radius r.

$$\text{Circumference}_{\text{circle}} = 2\pi r$$
$$\text{Area}_{\text{circle}} = \pi r^2$$
$$\text{Surface area}_{\text{sphere}} = 4\pi r^2$$
$$\text{Volume}_{\text{sphere}} = \frac{4}{3}\pi r^3$$

Working with Proportionalities

Most of the mathematics in *21st Century Astronomy* involves **proportionalities**—statements about how one physical quantity changes when another quantity changes. We began a discussion of proportionality in Working It Out 1.1; here, we offer a few examples of working with proportionalities.

To use a statement of proportionality to compare two objects, begin by turning the proportionality into a ratio. For example, the price of a bag of apples is **proportional** to the weight of the bag:

$$\text{Price} \propto \text{Weight}$$

Here, the symbol $\propto$ is pronounced "is proportional to." That means the ratio of the prices of two bags of apples is equal to the ratio of the weights of the two bags:

$$\text{Price} \propto \text{Weight means}$$

$$\frac{\text{Price of A}}{\text{Price of B}} = \frac{\text{Weight of A}}{\text{Weight of B}}$$

Let's work a specific example. Suppose bag A weighs 2 pounds and bag B weighs 1 pound. That means bag A will cost twice as much as bag B. We can turn that proportionality into the following equation:

$$\frac{\text{Price of A}}{\text{Price of B}} = \frac{\text{Weight of A}}{\text{Weight of B}} = \frac{2 \text{ lb}}{1 \text{ lb}} = 2$$

In other words, the price of bag A is 2 times the price of bag B. The price per pound is an example of a **constant of proportionality**.

Now let's work another, more complicated example. In Chapter 13, we discuss how the luminosity, brightness, and distance of stars are related. The luminosity of a star—the total energy that the star radiates each second—is proportional to the star's brightness multiplied by the square of its distance:

$$\text{Luminosity} \propto \text{Brightness} \times \text{Distance}^2$$

We can turn that proportionality into a ratio for two stars, A and B:

$$\frac{\text{Luminosity of A}}{\text{Luminosity of B}} = \frac{\text{Brightness of A}}{\text{Brightness of B}} \times \left(\frac{\text{Distance of A}}{\text{Distance of B}}\right)^2$$

If we use the symbols L, b, and d to represent luminosity, brightness, and distance, respectively, that equation becomes

$$\frac{L_A}{L_B} = \frac{b_A}{b_B} \times \left(\frac{d_A}{d_B}\right)^2$$

As an example, suppose that star A appears twice as bright in the sky as star B, but star A is located 10 times as far away as star B. How do the luminosities of the two stars compare? We know that

$$\text{Luminosity} \propto \text{Brightness} \times \text{Distance}^2$$

we write

$$\frac{\text{Luminosity of A}}{\text{Luminosity of B}} = \frac{\text{Brightness of A}}{\text{Brightness of B}} \times \left(\frac{\text{Distance of A}}{\text{Distance of B}}\right)^2$$

$$\frac{\text{Luminosity of A}}{\text{Luminosity of B}} = \frac{2}{1} \times \left(\frac{10}{1}\right)^2 = 200$$

In other words, star A is 200 times as luminous as star B.

Two quantities may also be inversely proportional, such that making one of them smaller makes the other larger. For example, when you are driving to another town, if you drive twice as fast, getting there takes half the time. The travel time is inversely proportional to the travel speed. We write that relationship as

$$\text{Time} \propto \frac{1}{\text{Speed}}$$

Proportionalities are used to compare one object with another. Constants of proportionality are used to calculate actual values. In *21st Century Astronomy*, the proportionality is usually what is important.

APPENDIX 2

PHYSICAL CONSTANTS AND UNITS

Fundamental Physical Constants

Constant	Symbol	Value
Speed of light in a vacuum	c	2.99792×10^8 m/s
Universal gravitational constant	G	6.6743×10^{-11} m³/(kg s²) 6.6743×10^{-20} km³/(kg s²)
Planck's constant	h	6.62607×10^{-34} J-s
Boltzmann constant	k	1.38065×10^{-23} J/K
Stefan-Boltzmann constant	σ	5.67037×10^{-8} W/(m² K⁴)
Mass of electron	m_e	9.10938×10^{-31} kg
Mass of proton	m_p	1.67262×10^{-27} kg
Mass of neutron	m_n	1.67493×10^{-27} kg
Electric charge of electron or proton	e	1.60218×10^{-19} C

Source: Data from the Particle Data Group (http://pdg.lbl.gov).

Unit Prefixes

Prefix*	Name	Factor†
n	nano-	10^{-9}
μ	micro-	10^{-6}
m	milli-	10^{-3}
k	kilo-	10^{3}
M	mega-	10^{6}
G	giga-	10^{9}
T	tera-	10^{12}

*When appended to a unit, these prefixes change the size of the unit by the factor (†) given. For example, 1 km is 10^3 meters.

Units and Values

Quantity	Fundamental Unit	Values
Length	meters (m)	radius of Sun (R_{Sun}) = 6.957×10^8 m astronomical unit (AU) = 1.49598×10^{11} m 1 AU = 149,598,000 km light-year (ly) = 9.4607×10^{15} m 1 ly = 6.324×10^4 AU 1 parsec (pc) = 3.262 ly = 3.0857×10^{16} m 1 m = 3.281 feet
Volume	cubic meters (m³)	$1\,m^3$ = 1000 liters = 264.2 gallons
Mass	kilograms (kg)	1 kg = 1000 grams mass of Earth (M_{Earth}) = 5.9724×10^{24} kg mass of Sun (M_{Sun}) = 1.9885×10^{30} kg
Time	seconds (s)	1 hour (h) = 60 minutes (min) = 3600 s solar day (noon to noon) = 86,400 s sidereal day (Earth rotation period) = 86,164.2 s tropical year (equinox to equinox) = 365.24219 days = 3.15569×10^7 s sidereal year (Earth orbital period) = 365.25636 days = 3.15581×10^7 s
Speed	meters/second (m/s)	1 m/s = 2.237 miles/h 1 km/s = 1000 m/s = 3600 km/h c = 2.99792×10^8 m/s = 299,792 km/s
Acceleration	meters/second² (m/s²)	g = gravitational acceleration on Earth = 9.81 m/s²
Energy	joules (J)	1 J = 1 kg m²/s² 1 megaton = 4.18×10^{15} J
Power	watts (W)	1 W = 1 J/s solar luminosity (L_{Sun}) = 3.828×10^{26} W
Force	newtons (N)	1 N = 1 kg m/s² 1 pound (lb) = 4.448 N 1 N = 0.22481 lb
Pressure	newtons/meter² (N/m²)	atmospheric pressure at sea level = 1.013×10^5 N/m² = 1.013 bar
Temperature	kelvins (K)	absolute zero = 0 K = −273.15°C = −459.67°F

Source: Data from the Particle Data Group (http://pdg.lbl.gov).

PERIODIC TABLE OF THE ELEMENTS

Key:
- 1 — Atomic number
- H — Symbol
- Hydrogen — Name
- 1.00794 — Average atomic mass

Legend:
- Metals
- Metalloids
- Nonmetals

1 / 1A																	18 / 8A
1 **H** Hydrogen 1.00794	2 / 2A											13 / 3A	14 / 4A	15 / 5A	16 / 6A	17 / 7A	2 **He** Helium 4.002602
3 **Li** Lithium 6.941	4 **Be** Beryllium 9.012182											5 **B** Boron 10.811	6 **C** Carbon 12.0107	7 **N** Nitrogen 14.0067	8 **O** Oxygen 15.9994	9 **F** Fluorine 18.9984032	10 **Ne** Neon 20.1797
11 **Na** Sodium 22.98976928	12 **Mg** Magnesium 24.3050	3 / 3B	4 / 4B	5 / 5B	6 / 6B	7 / 7B	8 / 8B	9 / 8B	10 / 8B	11 / 1B	12 / 2B	13 **Al** Aluminum 26.9815386	14 **Si** Silicon 28.0855	15 **P** Phosphorus 30.973762	16 **S** Sulfur 32.065	17 **Cl** Chlorine 35.453	18 **Ar** Argon 39.948
19 **K** Potassium 39.0983	20 **Ca** Calcium 40.078	21 **Sc** Scandium 44.955912	22 **Ti** Titanium 47.867	23 **V** Vanadium 50.9415	24 **Cr** Chromium 51.9961	25 **Mn** Manganese 54.938045	26 **Fe** Iron 55.845	27 **Co** Cobalt 58.933195	28 **Ni** Nickel 58.6934	29 **Cu** Copper 63.546	30 **Zn** Zinc 65.38	31 **Ga** Gallium 69.723	32 **Ge** Germanium 72.64	33 **As** Arsenic 74.92160	34 **Se** Selenium 78.96	35 **Br** Bromine 79.904	36 **Kr** Krypton 83.798
37 **Rb** Rubidium 85.4678	38 **Sr** Strontium 87.62	39 **Y** Yttrium 88.90585	40 **Zr** Zirconium 91.224	41 **Nb** Niobium 92.90638	42 **Mo** Molybdenum 95.96	43 **Tc** Technetium [98]	44 **Ru** Ruthenium 101.07	45 **Rh** Rhodium 102.90550	46 **Pd** Palladium 106.42	47 **Ag** Silver 107.8682	48 **Cd** Cadmium 112.411	49 **In** Indium 114.818	50 **Sn** Tin 118.710	51 **Sb** Antimony 121.760	52 **Te** Tellurium 127.60	53 **I** Iodine 126.90447	54 **Xe** Xenon 131.293
55 **Cs** Cesium 132.9054519	56 **Ba** Barium 137.327	57 **La** Lanthanum 138.90547	72 **Hf** Hafnium 178.49	73 **Ta** Tantalum 180.94788	74 **W** Tungsten 183.84	75 **Re** Rhenium 186.207	76 **Os** Osmium 190.23	77 **Ir** Iridium 192.217	78 **Pt** Platinum 195.084	79 **Au** Gold 196.966569	80 **Hg** Mercury 200.59	81 **Tl** Thallium 204.3833	82 **Pb** Lead 207.2	83 **Bi** Bismuth 208.98040	84 **Po** Polonium [209]	85 **At** Astatine [210]	86 **Rn** Radon [222]
87 **Fr** Francium [223]	88 **Ra** Radium [226]	89 **Ac** Actinium [227]	104 **Rf** Rutherfordium [261]	105 **Db** Dubnium [262]	106 **Sg** Seaborgium [266]	107 **Bh** Bohrium [264]	108 **Hs** Hassium [277]	109 **Mt** Meitnerium [268]	110 **Ds** Darmstadtium [271]	111 **Rg** Roentgenium [272]	112 **Cn** Copernicium [285]	113 **Uut** Ununtrium [284]	114 **Fl** Flerovium [289]	115 **Uup** Ununpentium [288]	116 **Lv** Livermorium [292]	117 **Uus** Ununseptium [294]	118 **Uuo** Ununoctium [294]

6 Lanthanides

58 **Ce** Cerium 140.116	59 **Pr** Praseodymium 140.90765	60 **Nd** Neodymium 144.242	61 **Pm** Promethium [145]	62 **Sm** Samarium 150.36	63 **Eu** Europium 151.964	64 **Gd** Gadolinium 157.25	65 **Tb** Terbium 158.92535	66 **Dy** Dysprosium 162.500	67 **Ho** Holmium 164.93032	68 **Er** Erbium 167.259	69 **Tm** Thulium 168.93421	70 **Yb** Ytterbium 173.05	71 **Lu** Lutetium 174.967

7 Actinides

90 **Th** Thorium 232.03806	91 **Pa** Protactinium 231.03588	92 **U** Uranium 238.02891	93 **Np** Neptunium [237]	94 **Pu** Plutonium [244]	95 **Am** Americium [243]	96 **Cm** Curium [247]	97 **Bk** Berkelium [247]	98 **Cf** Californium [251]	99 **Es** Einsteinium [252]	100 **Fm** Fermium [257]	101 **Md** Mendelevium [258]	102 **No** Nobelium [259]	103 **Lr** Lawrencium [262]

We have used the U.S. system as well as the system recommended by the International Union of Pure and Applied Chemistry (IUPAC) to label the groups in this periodic table. The system used in the United States includes a letter and a number (1A, 2A, 3B, 4B, etc.), which is close to the system developed by Mendeleev. The IUPAC system uses numbers 1–18 and has been recommended by the American Chemical Society (ACS). While we show both numbering systems here, we use the IUPAC system exclusively in the book. Elements with atomic numbers higher than 112 have been reported but not yet fully authenticated.

PROPERTIES OF PLANETS, DWARF PLANETS, AND MOONS

Physical Data for Planets and Dwarf Planets

Planet	EQUATORIAL RADIUS		MASS		Average Density (relative to water*)	Rotation Period (days)	Tilt of Rotation Axis (degrees, relative to orbit)	Equatorial Surface Gravity (relative to Earth[†])	Escape Velocity (km/s)	Average Surface Temperature (K)[§]
	(km)	(R/R_{Earth})	(kg)	(M/M_{Earth})						
Mercury	2440	0.383	3.30×10^{23}	0.055	5.427	58.79	0.03	0.378	4.3	340 (100, 700)
Venus	6052	0.950	4.87×10^{24}	0.815	5.243	243.02[‡]	177.4	0.905	10.36	737
Earth	6378	1.000	5.97×10^{24}	1.000	5.514	1.000	23.44	1.000	11.19	288 (185, 331)
Mars	3396	0.532	6.42×10^{23}	0.107	3.933	1.027	25.19	0.379	5.03	210 (120, 293)
Ceres	473	0.074	9.39×10^{20}	0.0002	2.09	0.378	4	0.26	0.5	168
Jupiter	71,492	11.209	1.90×10^{27}	317.8	1.326	0.4135	3.13	2.530	59.5	165
Saturn	60,268	9.449	5.68×10^{26}	95.16	0.687	0.444	26.73	1.065	35.5	134
Uranus	25,559	4.007	8.68×10^{25}	14.54	1.271	0.7183[‡]	97.77	0.905	21.3	76
Neptune	24,764	3.883	1.02×10^{26}	17.15	1.638	0.6713	28.32	1.14	23.5	72
Pluto	1188	0.186	1.30×10^{22}	0.0022	1.854	6.387[‡]	122.53	0.063	1.21	31
Haumea	~800	0.13	4.0×10^{21}	0.0007	~2	0.163	?	0.025	0.71	<50
Makemake	715	0.11	3.1×10^{21}	0.0005	~2	0.95	3	~ 0.5	~0.9	~ 30
Eris	1163	0.182	1.65×10^{22}	0.0028	2.4	1.08	?	0.084	1.38	42

*The density of water is 1000 kg/m³.
[†]The surface gravity of Earth is 9.81 m/s².
[‡]Venus, Uranus, and Pluto rotate opposite to the directions of their orbits. Their north poles are south of their orbital planes.
[§]Values in parentheses give extremes of recorded temperatures.

Orbital Data for Planets and Dwarf Planets

| Planet | MEAN DISTANCE FROM SUN (A*) | | Orbital Period (P) (sidereal years) | Eccentricity | Inclination (degrees, relative to ecliptic) | Average Speed (km/s) |
	(10⁶ km)	(AU)				
Mercury	57.9	0.387	0.241	0.2056	7.0	47.36
Venus	108.2	0.723	0.615	0.0067	3.395	35.02
Earth	149.6	1.000	1.000	0.0167	0.000	29.78
Mars	227.9	1.524	1.881	0.0935	1.851	24.07
Ceres	414.0	2.768	4.603	0.076	10.59	17.90
Jupiter	778.57	5.204	11.862	0.0489	1.30	13.06
Saturn	1433.53	9.582	29.457	0.0565	2.485	9.68
Uranus	2872.46	19.20	84.011	0.0457	0.772	6.80
Neptune	4495.06	30.048	164.79	0.0113	1.769	5.43
Pluto	5869.656	39.237	247.94	0.2444	17.16	4.67
Haumea	6450.1	43.116	284.12	0.196	28.21	4.53
Makemake	6796.2	45.43	309.1	0.161	28.98	4.42
Eris	10,152	67.86	558	0.436	44.0	3.43

*A is the semimajor axis of the planet's elliptical orbit.

Properties of Selected Moons*

Planet	Moon	ORBITAL PROPERTIES		PHYSICAL PROPERTIES		Relative Density (g/cm³) (water = 1.00)
		P (days)	A (10³ km)	R (km)	M (10²⁰ kg)	
Earth (1 moon)	Moon	27.32	384.4	1738.1	734.6	3.34
Mars (2 moons)	Phobos	0.319	9.38	11.27	0.0001	1.88
	Deimos	1.26	23.46	6.2	0.00002	1.48
Jupiter (79 known moons)	Metis	0.29	128	21.5	0.0004	0.86
	Amalthea	0.50	181.4	83.5	0.021	0.86
	Io	1.77	421.7	1822	893	3.53
	Europa	3.55	670.9	1561	480	3.01
	Ganymede	7.15	1070	2634	1482	1.94
	Callisto	16.69	1883	2410	1076	1.83
	Himalia	248.3	11,389	70	0.042	1.6
	Pasiphae	722[†]	23,209	19	0.0030	2.6
	Callirrhoe	787[†]	24,583	4.8	0.00001	2.6
Saturn (82 known moons)	Pan	0.58	133.58	14.1	0.00005	0.42
	Prometheus	0.61	139.38	43.1	0.0016	0.48
	Pandora	0.63	141.70	40.7	0.0014	0.49
	Mimas	0.94	185.54	198	0.37	1.15
	Enceladus	1.37	237.95	252	1.08	1.6
	Tethys	1.89	294.62	531	6.18	0.98
	Dione	2.74	377.40	561	11.0	1.48
	Rhea	4.52	527.12	764	23.1	1.24
	Titan	15.95	1222	2575	1346	1.88
	Hyperion	21.28	1481	135	0.056	0.54
	Iapetus	79.33	3561	735	18.1	1.09
	Phoebe	550[†]	12,960	107	0.08	1.64
	Paaliaq	687	15,200	13	0.0007	2.3
Uranus (27 known moons)	Cordelia	0.34	49.8	20	0.0004	1.3
	Miranda	1.41	129.4	236	0.64	1.20
	Ariel	2.52	191.0	579	12.5	1.59
	Umbriel	4.14	266.0	585	12.8	1.39
	Titania	8.71	435.9	788	34.0	1.71
	Oberon	13.46	583.5	761	30.8	1.63
	Setebos	2225	17,418	24	0.0008	1.3

(continued)

Properties of Selected Moons*

(continued)

Planet	Moon	ORBITAL PROPERTIES		PHYSICAL PROPERTIES		Relative Density (g/cm³) (water = 1.00)
		P (days)	A (10^3 km)	R (km)	M (10^{20} kg)	
Neptune (14 known moons)	Naiad	0.29	48.2	30	0.002	1.3
	Larissa	0.55	73.5	97	0.04	1.2
	Proteus	1.12	117.6	210	0.44	1.3
	Triton	5.88[†]	354.8	1353	214	2.06
	Nereid	360.11	5513.9	170	0.3	1.5
Pluto (5 moons)	Charon	6.39	19.60	606	15.9	1.70
Haumea (2 moons)	Namaka	18.3	25.66	85	0.018	~1
	Hi'iaka	49	49.88	160	0.179	~1
Eris	Dysnomia	15.8	37.3	350	1.43	0.8

*Innermost, outermost, largest, and/or a few other moons for each planet.
[†]Irregular moon (has retrograde orbit).

APPENDIX 5

SPACE MISSIONS

Selected Recent and Current Solar System Missions

Spacecraft	Sponsoring Nation(s)*	Destination	Launch Year	Type	Status (mid-2021)
Voyager 1 and *2*	USA	Jupiter, Saturn, Uranus (2), Neptune (2)	1977	Flyby	Actively exploring outer edge of Solar System
Galileo	USA	Jupiter	1989	Orbiter/probe	Ended 2003
Ulysses	USA, Europe	Sun	1990	Solar polar orbiter	Ended 2008
SOHO	USA, Europe	Sun	1995	Orbiter	Active
Mars Global Surveyor	USA	Mars	1996	Orbiter	Ended 2006
Cassini-Huygens	USA, Europe, Italy	Saturn, Titan	1997	Saturn orbiter, Titan probe/lander	Ended 2017
Stardust	USA	Comets	1999	Sample return/flyby	Ended 2011
Mars Odyssey	USA	Mars	2001	Orbiter	Active
Mars Exploration Rover	USA	Mars	2003	Two landers	Ended 2019
Hayabusa	Japan	Asteroid	2003	Sample return	Ended 2010
Mars Express	Europe	Mars	2003	Orbiter	Active
Messenger	USA	Mercury (2011)	2004	Orbiter	Ended 2015
Rosetta	Europe	Comet 67P/Churyumov–Gerasimenko (2014)		Orbiter and lander	Ended 2016
Venus Express	Europe	Venus	2005	Orbiter	Ended 2014
Mars Reconnaissance Orbiter (*MRO*)	USA	Mars	2005	Orbiter	Active
Deep Impact/EPOXI	USA	Comet Hartley (2010)	2005	Impactor/flyby	Ended 2010
STEREO	USA	Sun	2006	Two orbiters	Active
New Horizons	USA	Pluto (2015)	2006	Flyby	Active
Chang'e 1	China	Moon	2007	Orbiter	Ended 2009
Kayuga	Japan	Moon	2007	Orbiter	Ended 2009
Themis	USA	Moon, solar wind	2007	Multiple Orbiters	Active
Dawn	USA	Vesta (2011), Ceres (2015)	2007	Orbiter	Ended 2018
Chandrayaan	India	Moon	2008	Orbiter/impactor	Ended 2009
Lunar Reconnaissance Orbiter (*LRO*)	USA	Moon	2009	Orbiter	Active
Lunar Crater Observation and Sensing Satellite (*LCROSS*)	USA	Moon	2009	Impactor	Ended 2009

(continued)

Selected Recent and Current Solar System Missions

(continued)

Spacecraft	Sponsoring Nation(s)*	Destination	Launch Year	Type	Status (mid-2021)
Chang'e 2	China	Moon	2010	Orbiter	Ended 2011
Akatsuki	Japan	Venus (2015)	2010	Orbiter	Active
Juno	USA	Jupiter (2016)	2011	Orbiter	Active
Gravity Recovery and Interior Laboratory (GRAIL)	USA	Moon	2011	Two orbiters	Ended 2012
Mars Science Laboratory (Curiosity rover)	USA	Mars	2011	Lander	Active
Mars Atmosphere and Volatile EvolutioN (MAVEN) mission	USA	Mars	2013	Orbiter	Active
Chang'e 3	China	Moon	2013	Lander	On lunar surface
Mangalyaan	India	Mars	2013	Orbiter	Active
Hayabasu 2	Japan	Asteroid 162173 Ryugu	2014	Orbiter, scheduled sample return	Delivered sample, on extended mission
DSCOVR	USA	Sun	2015	Orbiter	Active
Osiris-REx	USA	Asteroid 101955 Bennu	2016	Orbiter, scheduled sample return	Returning
ExoMars Trace Gas Orbiter (TGO)	Europe, Russia	Mars (2016)	2016	Orbiter (lander crashed)	Active
InSight	USA	Mars	2018	Lander	Active
BepiColombo	Europe, Japan	Mercury	2018	Orbiter	En route
Parker Solar Probe	USA	Sun	2018	Orbiter	Active
Solar Orbiter	Europe	Sun	2020	Orbiter	En route
Emirates Mars Mission	United Arab Emirates	Mars	2020	Orbiter	Active
Tianwen-1	China	Mars	2020	Orbiter	Active
Zhurong	China	Mars	2020	Rover	Active
Perseverance	USA	Mars	2020	Rover	Active
Ingenuity	USA	Mars	2020	Helicopter	Active

*Countries are represented by the following agencies: China = CNSA (China National Space Administration); Europe = ESA (European Space Agency); India = ISRO (Indian Space Research Organisation); Italy = Italian Space Agency; Japan = JAXA (Japan Aerospace Exploration Agency); Russia = ROSCOSMOS (State Corporation for Space Activities); USA = NASA (National Aeronautics and Space Administration).

APPENDIX 6

BRIGHTEST STARS

The 25 Brightest Stars in the Sky

Name	Common Name	Distance (ly)	Spectral Type	Relative Visual Luminosity* (Sun = 1.000)	Apparent Visual Magnitude	Absolute Visual Magnitude
Sun	Sun	1.58×10^{-5}	G2V	1.000	−26.74	4.83
Alpha Canis Majoris	Sirius	8.60	A1V	22.9	−1.46	1.43
Alpha Carinae	Canopus	309	F0II	14,900	−0.72	−5.60
Alpha¹ Centauri	Rigil Kentaurus A	4.36	G2V	1.51	−0.01	4.38
Alpha² Centauri	Rigil Kentaurus B	4.36	K1V	0.44	1.33	5.71
Alpha Bootis	Arcturus	37	K1.5III	113	−0.04	−0.30
Alpha Lyrae	Vega	25	A0Va	49.2	0.03	0.60
Alpha Aurigae	Capella	43	G5IIIe+G0III	137	0.08	−0.51
Beta Orionis	Rigel	860	B8Iab	54,000	0.12	−7.0
Alpha Canis Minoris	Procyon	11.5	F5IV-V	7.73	0.34	2.61
Alpha Eridani	Achernar	140	B3Vpe	1030	0.46	−2.70
Beta Centauri	Hadar	392	B1III	7180	0.61	−4.81
Alpha Orionis	Betelgeuse	570	M2Iab	13,600	0.7	−5.5
Alpha Aquilae	Altair	16.7	A7V	11.1	0.77	2.22
Alpha Crucis	Acrux	325	B0.5IV+B1V	3100	1.3	−3.9
Alpha Tauri	Aldebaran	67	K5III	163	0.85	−0.70
Alpha Scorpii	Antares	550	M1.5Ib	16,300	0.96	−5.7
Alpha Virginis	Spica	250	B1IV+B4V	1920	1.04	−3.38
Beta Geminorum	Pollux	34	K0III	32.2	1.14	1.06
Alpha Piscis	Fomalhaut	25	A3V	17.4	1.16	1.73
Beta Crucis	Mimosa	280	B0.5III	1980	1.25	−3.41
Alpha Cygni	Deneb	1425	A2Ia	58,600	1.25	−7.09
Alpha Leonis	Regulus	79	B7V	146	1.35	−0.58
Epsilon Canis Majoris	Adhara	405	B2II	3400	1.50	−4.0
Alpha Gemini	Castor	51	A1V+A5Vm	49	1.58	0.61
Gamma Crucis	Gacrux	88	M3.5III	138	1.63	−0.52

Sources: Data from Jim Kaler's *STARS* page (http://stars.astro.illinois.edu/sow/bright.html); SIMBAD Astronomical Database (http://simbad.u-strasbg.fr/simbad).
*Luminosity in this table refers only to radiation in "visual" light.

APPENDIX 7

OBSERVING THE SKY

This appendix gives you enough information to make sense of a star chart or list of astronomical objects and find a few objects in the sky.

Celestial Coordinates

In Chapter 2, we discuss the **celestial sphere**—the imaginary sphere with Earth at its center on which celestial objects appear to lie. Several coordinate systems are used to specify the positions of objects on the celestial sphere. The simplest of these is the *altitude-azimuth coordinate system*. The altitude-azimuth coordinate system is based on the "map" direction to an object (the object's azimuth, with north = 0°, east = 90°, south = 180°, and west = 270°) combined with how high the object is above the horizon (the object's altitude, with the horizon at 0° and the zenith at 90°). For example, an object 10° above the eastern horizon has an altitude of 10° and an azimuth of 90°. An object 45° above the horizon in the southwest is at altitude 45°, azimuth 225°.

The altitude-azimuth coordinate system is the simplest way to tell someone where in the sky to look at the moment, but it is not a good coordinate system for cataloging the positions of objects. The altitude and azimuth of an object are different for each observer, depending on the observer's position on Earth, and they are constantly changing as Earth rotates on its axis. To specify the direction to an object in a way that is the same for everyone requires a coordinate system that is fixed relative to the celestial sphere. The most common such coordinates are called *celestial coordinates*.

Celestial coordinates are illustrated in **Figure A7.1**. Celestial coordinates are much like the traditional system of latitude and longitude used on the surface of Earth. On Earth, latitude specifies how far you are from Earth's equator, as discussed in Chapter 2. If you are on Earth's equator, your latitude is 0°. If you are at Earth's North Pole, your latitude is 90° north. If you are at Earth's South Pole, your latitude is 90° south.

The latitude-like coordinate on the celestial sphere is called **declination**, often signified with the lowercase Greek letter δ (delta). The celestial equator has δ = 0°. The north celestial pole has δ = +90°. The south celestial pole has δ = −90°. (See Chapter 2 if you need to refresh your memory about the celestial equator or celestial poles.) Declination is usually expressed in degrees, minutes of arc, and seconds of arc. For example, Sirius, the brightest star in the sky,

has δ = −16°42′58″ meaning that it is located not quite 17° south of the celestial equator.

On Earth, east–west position is specified by longitude. Lines of constant longitude run north–south from one pole to the other. Unlike latitude, for which the equator provides a natural place to call "zero," longitude has no natural starting point, so one was invented. By arbitrary convention, the Royal Observatory in Greenwich, England, is defined to lie at a longitude of 0°. On the celestial sphere, the longitude-like coordinate is called **right ascension**, often signified with the lowercase Greek letter α (alpha). Unlike longitude, right ascension *does* have a natural starting point on the celestial sphere: the vernal equinox, or the point at which the ecliptic crosses the celestial equator with the Sun moving from the southern sky into the northern sky. The (Northern Hemisphere) vernal equinox defines the line of right ascension at which δ = 0°. The (Northern Hemisphere) autumnal equinox, located on the opposite side of the sky, is at δ = 180°.

Normally, right ascension is measured in units of time rather than degrees. Earth takes 24 hours (of sidereal time) to rotate on its axis, so the celestial sphere is divided into 24 hours of right ascension, with each hour of right ascension corresponding to 15°. Hours of right ascension are then subdivided into minutes and seconds of time. Right ascension increases going to the east. The right ascension of Sirius, for example, is α = 06^{h}45^{m}08.9^s, meaning that Sirius is about 101° (that is, 06^{h}45^m) east of the vernal equinox. Time is a natural unit for measuring right ascension because time naturally tracks the motion of objects due to Earth's rotation on its axis. If stars on the meridian at a certain time have α = 06^h, then an hour later the stars on the meridian will have α = 07^h, and an hour after that they will have α = 08^h. The *local sidereal time*, or *star time*, at your location right now is equal to the right ascension of the stars on your meridian at the moment. Because of Earth's motion around the Sun, a sidereal day is about 4 minutes shorter than a solar day, so local sidereal time constantly gains on solar time. At midnight on September 22, the local sidereal time is 0^h. By midnight on December 21, local sidereal time has advanced to 06^h. On March 20, local sidereal time at midnight is 12^h. And at midnight on June 20, local sidereal time is 18^h.

Putting all that together, right ascension and declination provide a convenient way to specify the location of any object on the celestial sphere. Sirius is located at α = 06^{h}45^{m}08.9^s, δ = −16°42′58″ which means that at midnight on December 21 (local sidereal

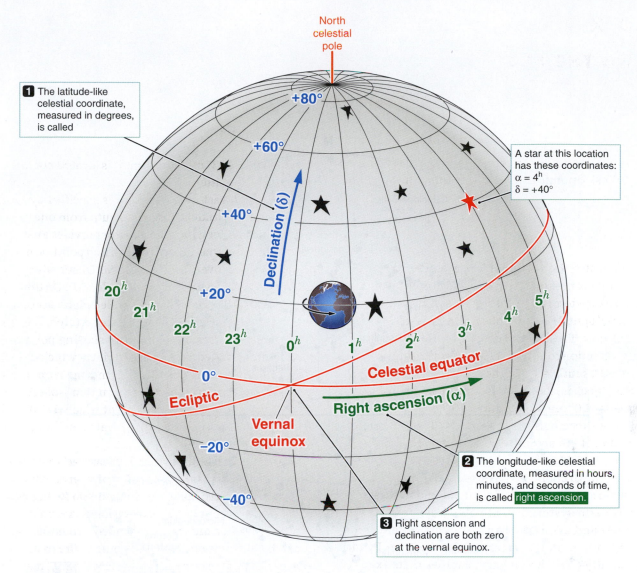

Figure A7.1 Celestial coordinates.

1 The latitude-like celestial coordinate, measured in degrees, is called

A star at this location has these coordinates:
$\alpha = 4^h$
$\delta = +40°$

North celestial pole

Declination (δ)

Celestial equator

Right ascension (α)

Ecliptic

Vernal equinox

2 The longitude-like celestial coordinate, measured in hours, minutes, and seconds of time, is called right ascension.

3 Right ascension and declination are both zero at the vernal equinox.

time = 06^h), you will find Sirius about 45^m east of the meridian, not quite 17° south of the celestial equator.

Just one final caveat remains. As we discussed in Chapter 2, the directions of the celestial equator, celestial poles, and vernal equinox are constantly changing as Earth's axis wobbles like the axis of a spinning top. In Chapter 2, we called that 26,000-year wobble the **precession of the equinoxes**, meaning that the location of the equinoxes is slowly advancing along the ecliptic. So when we specify the celestial coordinates of an object, we need to specify the date at which the positions of the vernal equinox and celestial poles were measured. By convention, coordinates are usually referred to with the position of the vernal equinox on January 1, 2000. A complete, formal specification of the coordinates of Sirius would then be $\alpha(2000) = 06^h45^m08.9^s$, $\delta(2000) = -16°42'58''$, where the "2000" in parentheses refers to the equinox of the coordinates.

Constellations and Names

Although it is certainly possible to specify any location on the surface of Earth exactly by giving its latitude and longitude, using a more descriptive address is usually more convenient. We might say, for example, that one of the coauthors of this book works near latitude 37° north, longitude 122° west; but it would probably mean a lot more to you if we said that George Blumenthal works in Santa Cruz, California.

Just as the surface of Earth is divided into nations and states with shared boundaries, the celestial sphere is divided into 88 **constellations**, the names of which are often used to refer to objects within their boundaries (see the star charts in **Figure A7.2**). The brightest stars within the boundaries of a constellation are named using a Greek letter combined with the name of the constellation.

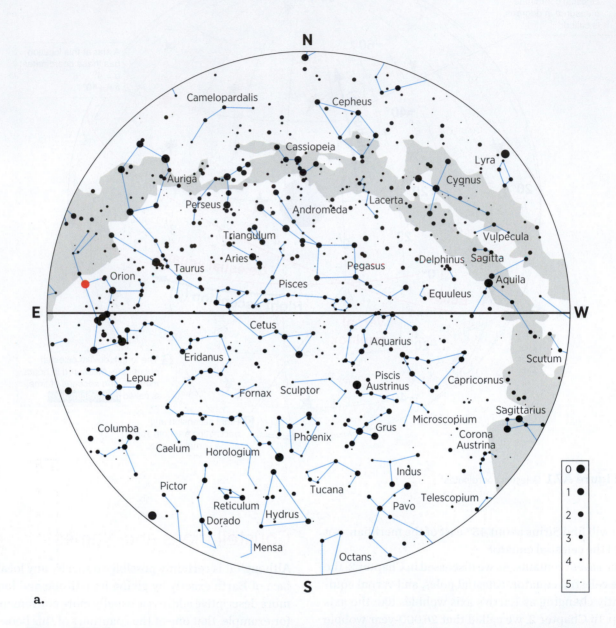

Figure A7.2A The night sky on the northern autumnal equinox, approximately September 21 each year. The celestial equator runs across the middle of the map. A red dot indicates a star that is noticeably red in the sky. The shaded area is the Milky Way.

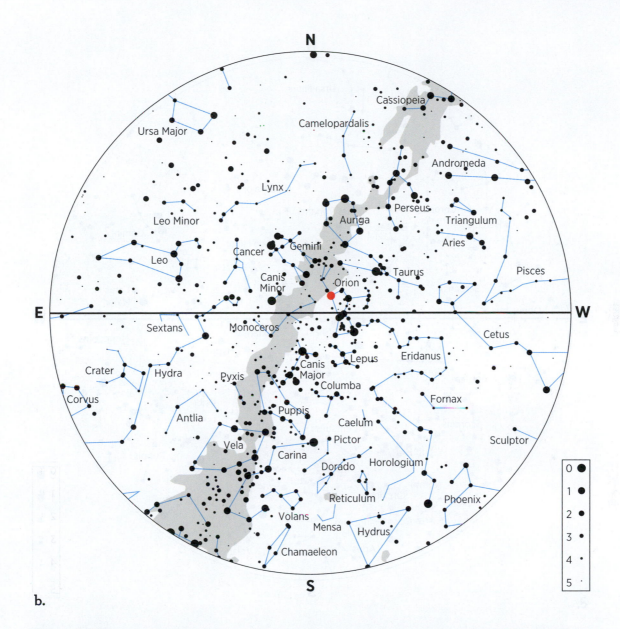

Figure A7.2B The night sky on the northern winter solstice, approximately December 21 each year. The celestial equator runs across the middle of the map. A red dot indicates a star that is noticeably red in the sky. The shaded area is the Milky Way.

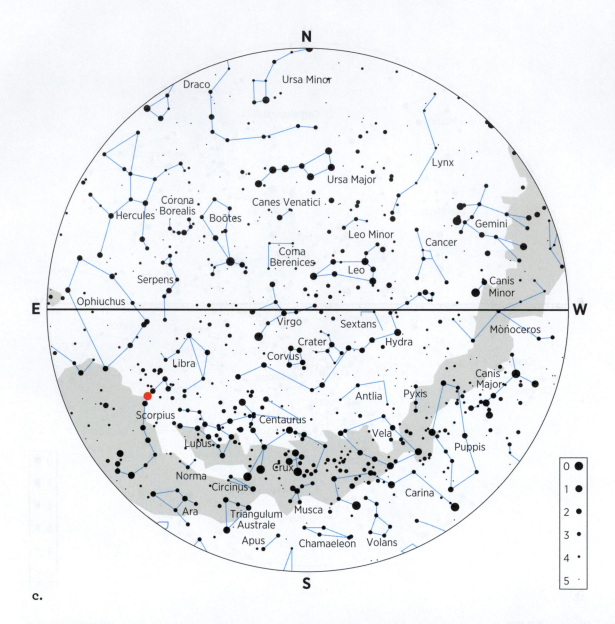

c.

Figure A7.2C The night sky on the northern vernal equinox, approximately March 21 each year. The celestial equator runs across the middle of the map. A red dot indicates a star that is noticeably red in the sky. The shaded area is the Milky Way.

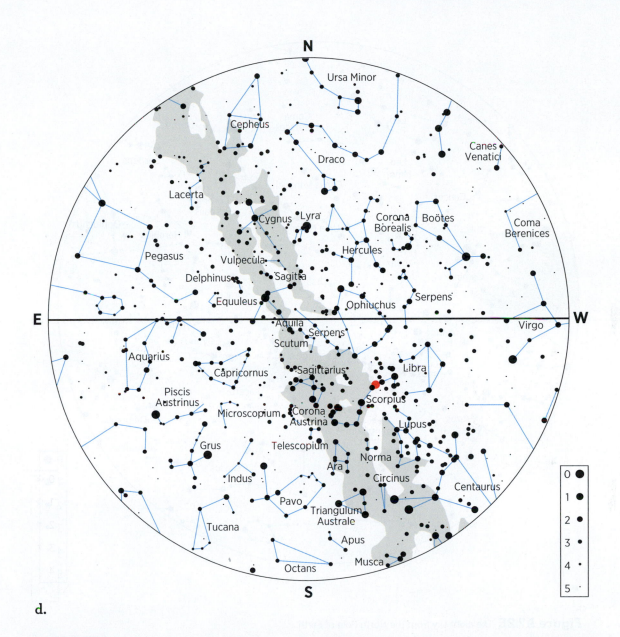

d.

Figure A7.2D The night sky on the northern summer solstice, approximately June 21 each year. The celestial equator runs across the middle of the map. A red dot indicates a star that is noticeably red in the sky. The shaded area is the Milky Way.

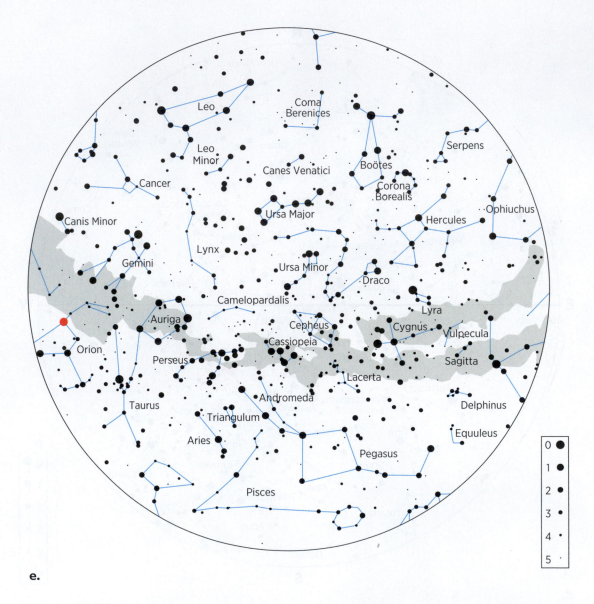

e.

Figure A7.2E The night sky from the North Pole of Earth.

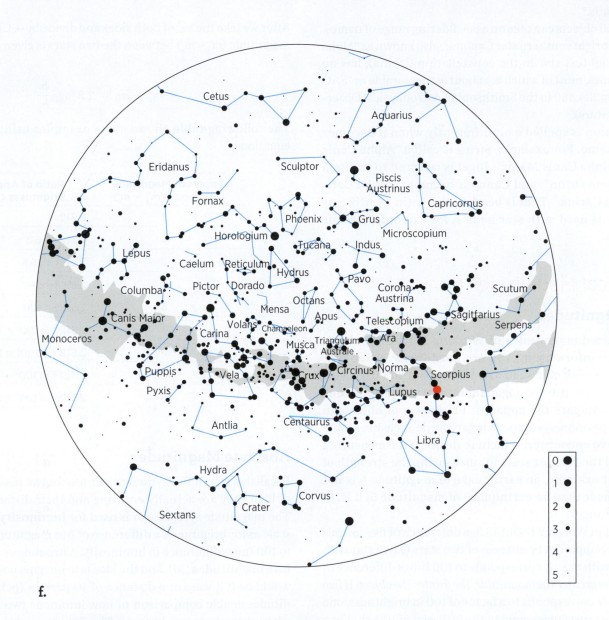

Figure A7.2F The night sky from the South Pole of Earth.

f.

For example, the star Sirius is the brightest star in the constellation Canis Major (literally, the "great dog"), so it is called "Alpha Canis Majoris." The bright red star in the northeastern corner of the constellation Orion is called "Alpha Orionis," also known as Betelgeuse. Rigel, the bright blue star in the southwest corner of Orion, is also called "Beta Orionis."

Astronomical objects can take on a bewildering range of names. For example, the bright southern star Canopus, also known as "Alpha Carinae" (the brightest star in the constellation Carina), has no fewer than 34 names, most of which are about as memorable as "SAO 234480" (number 234,480 in the Smithsonian Astrophysical Observatory catalog of stars).

A constellation is spelled a bit differently when it becomes part of a star's name. For example, Sirius is called "Alpha Canis Majoris," not "Alpha Canis Major"; Rigel is referred to as "Beta Orionis," not "Beta Orion"; and Canopus becomes "Alpha Carinae," not "Alpha Carina." That is because the Latin genitive, or possessive, case is used with star names; for example, *Orionis* means "of Orion."

Astronomical Magnitudes

Apparent Magnitudes

We first introduced magnitudes in Working It Out 13.2; here, we provide more information. You are most likely to see that system if you take a lab course in astronomy or if you use a star catalog. Astronomers use the logarithmic system of **apparent magnitudes** to compare the apparent brightness of objects in the sky. Other common systems of logarithmic measurements that you may have encountered include decibels for measuring sound levels and the Richter scale for measuring the strength of earthquakes. For example, an earthquake of magnitude 6 is not just a little stronger than an earthquake of magnitude 5; it is, in fact, 10 times stronger.

As discussed in Working It Out 13.2, a difference of five magnitudes between the apparent brightness of two stars (say, a star with $m = 6$ and a star with $m = 1$) corresponds to 100 times difference in brightness, and *the greater the magnitude, the fainter the object*. If five steps in magnitude corresponds to a factor of 100 in brightness, one step in magnitude must correspond to the fifth root of 100; that is, a factor of $100^{1/5}$ = approximately 2.512 in brightness ($100^{1/5} \times 100^{1/5} \times 100^{1/5} \times 100^{1/5} \times 100^{1/5} = 100$).

If star 1 has a brightness of b_1 and star 2 has a brightness of b_2, the ratio of the brightness of the stars is given by

$$\frac{b_1}{b_2} = (2.512)^{m_2 - m_1} = 100^{\frac{(m_2 - m_1)}{5}}$$

We can put that into the more common base 10 by noting that $100 = 10^2$, so the expression becomes

$$\frac{b_1}{b_2} = 10^{2 \times \frac{(m_2 - m_1)}{5}} = 10^{0.4(m_2 - m_1)}$$

After we take the log of both sides and divide by 0.4, the difference in magnitude ($m_2 - m_1$) between the two stars is given by

$$m_2 - m_1 = 2.5 \log_{10} \frac{b_1}{b_2}$$

The following table shows some examples using the preceding equations.

Apparent Magnitude Difference ($m_2 - m_1$)	Ratio of Apparent Brightness (b_1/b_2)
1	2.512
2	$2.512^2 = 6.3$
3	$2.512^3 = 15.8$
4	$2.512^4 = 39.8$
5	$2.512^5 = 100$
10	$2.512^{10} = 100^2 = 10,000$
15	$2.512^{15} = 100^3 = 1,000,000$
20	$2.512^{20} = 100^4 = 10^8$
25	$2.512^{25} = 100^5 = 10^{10}$

Absolute Magnitudes

Recall that stars differ in their brightness for two reasons: the amount of light they are actually emitting and their distance from Earth. The magnitude system also is used for **luminosity**, with the same scale as for brightness: a difference of five magnitudes corresponds to 100 times difference in luminosity. Astronomers call those **absolute magnitudes** (M), and the idea is to imagine how bright the star would be if it were at a distance of 10 parsecs (pc). Absolute magnitudes enable comparison of how luminous two stars really are, without the factor of distance. The Sun is very bright because it is so close (apparent visual magnitude = −27), but if the Sun were at a distance of 10 pc, its magnitude would be only about 5.[1] Thus, the absolute magnitude of the Sun is $M = 5$. Recall that the luminosity of a star is usually expressed by comparing it with the luminosity of the Sun. As with apparent magnitudes, higher magnitude numbers correspond to lower luminosity. Thus, a star 1/100 as luminous as the Sun will be 5 absolute magnitudes fainter, or $M = 10$. A star 10,000

[1]The apparent and absolute magnitudes of the Sun are −26.74 and +4.83, respectively. We use +5 for the Sun's absolute magnitude as an approximation.

times more luminous than the Sun will be 10 absolute magnitudes brighter, or $M = -5$.

Absolute magnitudes and luminosities follow the same equations as those we provided already, using L instead of b and M instead of m:

$$\frac{L_1}{L_2} = 10^{2 \times \frac{M_{(2)} - M_{(1)}}{5}} = 10^{0.4(M_{(2)} - M_{(1)})}$$

and

$$M_{(2)} - M_{(1)} = 2.5 \log_{10} \frac{L_1}{L_2}$$

Most often, astronomers think about the luminosity of a star in comparison with the luminosity of the Sun. Here $L_1 = L_{star}$ and $L_2 = L_{Sun}$. The following table compares luminosity (where $L_{Sun} = 1$) with absolute magnitude of a star.

L_{Star}/L_{Sun}	M
1,000,000	−10
10,000	−5
100	0
1	5
1/100	10
1/10,000	15

Distance Modulus

The difference between the apparent magnitude and the absolute magnitude depends on the star's distance. By definition, a star at a distance of exactly 10 pc will have an apparent magnitude equal to its absolute magnitude. Astronomers can always measure the brightness of a star and thus its apparent magnitude and can estimate the luminosity of a star and thus its absolute magnitude by using the Hertzsprung-Russell (H-R) diagram. That is how the distances to most stars are found.

Using the preceding equations and the definition of absolute magnitude, we can get to the following relatively simple expression:

$$m - M = 5 \log_{10} d - 5$$

where distance d is in parsecs.

We can rewrite that equation to solve for distance as follows:

$$d = 10^{\frac{(m - M_{abs} + 5)}{5}}$$

The following table shows how the difference between an object's apparent and absolute magnitudes leads to its distance in parsecs.

$m - M$	Distance (pc)
−3	2.5
−2	4.0
−1	6.3
0	10
1	16
2	25
3	40
4	63
5	100
10	1000
15	10,000
20	100,000

Although the system of astronomical magnitudes is convenient in many ways—which is why astronomers continue to use it—it can also be confusing to new students. Just remember three things and you will probably get by:

1. The greater the magnitude, the fainter the object.

2. One magnitude *smaller* means about two and a half times *brighter*.

3. The brightest stars in the sky have magnitudes of less than 1, and the faintest stars visible to the naked eye on a dark night have magnitudes of about 6.

A final note: Astronomers sometimes use "colors" based on the ratio of the brightness of a star as seen in two parts of the spectrum. The "b_B/b_V color," for example, is the ratio of the brightness of a star seen through a blue filter, divided by the brightness of a star seen through a yellow-green (visual) filter. Normally, astronomers instead discuss the "$B - V$ color" of a star, which is equal to the difference between a star's blue magnitude and its visual magnitude. We can use the previous expression for a magnitude difference to write

$$B - V \text{ color} = m_B - m_V = -2.5 \log_{10} (b_B/b_V)$$

Thus, a star with a b_B/b_V color of 1.0 has a $B - V$ color of 0.0, and a star with a b_B/b_V color of 1.4 has a $B - V$ color of −0.37. Notice that, as with magnitudes, $B - V$ colors are "backward": the bluer a star, the greater its b_B/b_V color but the less its $B - V$ color.

APPENDIX 8

UNIFORM CIRCULAR MOTION AND CIRCULAR ORBITS

Uniform Circular Motion

In Chapter 4 (see Section 4.2 and Figure 4.7), we discuss the motion of an object moving in a circle at a constant speed. That motion, called **uniform circular motion**, is the result of the fact that centripetal force always acts toward the center of the circle. The key question when thinking about uniform circular motion is, How hard does something have to pull to keep the object moving in a circle? Part of the answer to that question is obvious: the more massive an object is, the harder it will be to keep it moving on its circular path. According to Newton's second law of motion, $F = ma$, or here, the centripetal force equals the mass multiplied by the centripetal acceleration. The larger the mass, the greater the force required to keep it moving in its circle.

The centripetal force needed to keep an object moving in constant circular motion also depends on two other quantities: the speed of the object and the size of the circle. The faster an object is moving, the more rapidly it has to change direction to stay on a circle of a given size. The second quantity that influences the needed acceleration is the radius of the circle. The smaller the circle, the greater the pull needed to keep it on track. You can understand that relationship by looking at the motion. A small circle requires a continuous "hard" turn, whereas a larger circle requires a more gentle change in direction. It takes more force to keep an object moving faster in a smaller circle than it does to keep the same object moving more slowly in a larger circle. (To get a better feel for how that works, think about the difference between riding in a car taking a tight curve at high speed and a car moving slowly around a gentle curve.)

To arrive at the circular velocity and other results discussed in Chapter 4, those intuitive ideas about uniform circular motion are turned into a quantitative expression of exactly how much centripetal acceleration is needed to keep an object moving in a circle with radius r at speed v. **Figure A8.1** shows a ball moving around a circle of radius r at a constant speed v at two times. The centripetal acceleration keeping the ball on the circle is a. Remember that the acceleration is always directed toward the center of the circle, whereas the velocity of the ball is always perpendicular to the acceleration. The ball's velocity and its acceleration are always at right angles to each other. As the object moves around the circle,

the direction of motion and the direction of the acceleration change together in lockstep.

Figure A8.1 contains two triangles. Triangle 1 shows the velocity (speed and direction) at both times. The arrow labeled "Δv" connecting the heads of the two velocity arrows shows how much the velocity changed between time 1 (t_1) and time 2 (t_2). That change is the effect of the centripetal acceleration. If you imagine that points 1 and 2 are very close together—so close that the direction of the centripetal acceleration does not change by much between the two—the centripetal acceleration equals the change in the velocity divided by the time between the two, $\Delta t = t_2 - t_1$. So, $\Delta v = a\Delta t$.

Triangle 2 shows something similar. Here, the arrow labeled "Δr" indicates the change in the position of the ball between time 1 and time 2. Again, if you imagine that the time between the two points is very short, Δr is equal to the velocity multiplied by the time, or $\Delta r = v\Delta t$.

The line between the center of the circle and the ball is always perpendicular to the velocity of the ball. So if the direction of the ball's velocity changes by an angle θ, the direction of the line between the ball and the center of the circle must also change by the same angle α. In other words, triangles 1 and 2 are "similar triangles." They have

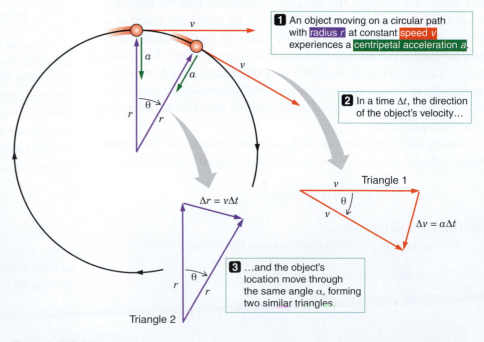

1 An object moving on a circular path with radius r at constant speed v experiences a centripetal acceleration a.

2 In a time Δt, the direction of the object's velocity…

Triangle 1

$\Delta v = a\Delta t$

$\Delta r = v\Delta t$

3 …and the object's location move through the same angle α, forming two similar triangles.

Triangle 2

Figure A8.1 Similar triangles are used to find the centripetal force needed to keep an object moving at a constant speed on a circular path.

the same *shape*. If the triangles are the same shape, the ratio of two sides of triangle 1 must equal the ratio of the two corresponding sides of triangle 2. Then,

$$\frac{a\Delta t}{v} = \frac{v\Delta t}{r}$$

If we divide by Δt on both sides of the equation and then cross-multiply, we obtain

$$ar = v^2$$

which, after we divide both sides of the equation by r, becomes

$$a_{\text{centripetal}} = \frac{v^2}{r}$$

The subscript "centripetal" is added to a to signify that this is the centripetal acceleration needed to keep the object moving in a circle of radius r at speed v. The centripetal force required to keep an object of mass m moving on such a circle is then

$$F_{\text{centripetal}} = ma_{\text{centripetal}} = \frac{mv^2}{r}$$

Circular Orbits

For an object moving in a circular orbit, no string is available to hold the ball on its circular path. Instead, that force is provided by **gravity**.

Think about an object with mass m in orbit about a much larger object with mass M. The orbit is circular, and the distance between the two objects is given by r. The force needed to keep the smaller object moving at speed v in a circle with radius r is given by the previous expression for $F_{\text{centripetal}}$. The force actually provided by gravity (see Chapter 4) is

$$F_{\text{grav}} = G\frac{Mm}{r^2}$$

If gravity is responsible for holding the mass in its circular motion, it should be true that $F_{\text{grav}} = F_{\text{centripetal}}$. That is, if mass m is moving in a circle under the force of gravity, the force provided by gravity must equal the centripetal force needed to explain that circular motion. Setting the two expressions for $F_{\text{centripetal}}$ and F_{grav} equal to each other gives

$$\frac{mv^2}{r} = G\frac{Mm}{r^2}$$

All that remains is a bit of algebra. Dividing by m on both sides of the equation and multiplying both sides by r gives

$$v^2 = G\frac{M}{r}$$

Taking the square root of both sides then yields the desired result:

$$v_{\text{circ}} = \sqrt{\frac{GM}{r}}$$

That is the **circular velocity** we presented in Chapter 4. It is the velocity at which an object in a circular orbit must be moving. If the object were not moving at that velocity, gravity would not be providing the needed centripetal force, and the object would not move in a circle.

A

aberration of starlight A star's apparent displacement in position due to the finite speed of light and Earth's orbital motion around the Sun. (Ch. 18)

absolute magnitude A measure of the intrinsic brightness, or luminosity, of a celestial object, generally a star. Specifically, the apparent magnitude an object would have if it were located at a standard distance of 10 parsecs (pc). Compare *apparent magnitude*. (Ch. 13)

absolute zero The temperature at which thermal motions cease. The lowest possible temperature. Zero on the Kelvin temperature scale. (Ch. 5)

absorption The process by which an atom captures energy from a passing photon. Compare *emission*. (Ch. 5)

absorption line A minimum in the intensity of a spectrum that is due to the absorption of electromagnetic radiation at a specific wavelength determined by the energy levels of an atom or molecule. Compare *emission line*. (Ch. 5)

acceleration (*a*) The rate at which the speed and/or direction of an object's motion is changing. (Ch. 3)

accretion disk A flat, rotating disk of gas and dust surrounding an object, such as a young stellar object, a forming planet, a collapsed star in a binary system, or a black hole. (Ch. 7)

achondrite A stony meteorite that does not contain chondrules. Compare *chondrite*. (Ch. 12)

active comet A comet nucleus that approaches close enough to the Sun to show signs of activity, such as the production of a coma and tail. (Ch. 12)

active galactic nucleus (AGN) A highly luminous, compact galactic nucleus whose luminosity may exceed that of the rest of the galaxy. (Ch. 19)

active region An area of the Sun's chromosphere anchoring bursts of intense magnetic activity. (Ch. 14)

adaptive optics Electro-optical systems that largely compensate for image distortion caused by Earth's atmosphere. (Ch. 6)

AGB See *asymptotic giant branch*. (Ch. 16)

AGN See *active galactic nucleus*. (Ch. 19)

albedo The fraction of electromagnetic radiation striking a surface that is reflected by that surface. (Ch. 5)

algebra A branch of mathematics in which letters represent numeric variables. (Appendix 1)

alpha particle A ^{4}He nucleus, consisting of two protons and two neutrons. Alpha particles are given off in the type of radioactive decay referred to as *alpha decay*. (Ch. 16)

altitude The location of an object above the horizon, measured by the angle formed between an imaginary line from an observer to the object and a second line from the observer to the point on the horizon directly below the object. (Ch. 2)

Amors A group of asteroids whose orbits cross the orbit of Mars but not the orbit of Earth. Compare *Apollos* and *Atens*. (Ch. 12)

amplitude In a wave, the maximum deviation from its undisturbed or relaxed position. For example, in a water wave the amplitude is the vertical distance from the wave's crest to the undisturbed water level. (Ch. 5)

angular momentum In a rotating or revolving system, a conserved property whose value depends on the velocity and distribution of the system's mass. (Ch. 7)

angular resolution The ability of an imaging device such as a telescope (or the eye) to separate two objects that appear close together. (Ch. 6)

annular solar eclipse The type of solar eclipse that occurs when the apparent diameter of the Moon is less than that of the Sun, leaving a visible ring of light ("annulus") surrounding the dark disk of the Moon. Compare *partial solar eclipse* and *total solar eclipse*. (Ch. 2)

Antarctic Circle The circle on Earth with latitude 66.5° south, marking the northern limit where at least one day per year is in 24-hour daylight. Compare *Arctic Circle*. (Ch. 2)

anthropic principle The idea that this universe (or this bubble in the universe) must have physical properties that allow intelligent life to develop. (Ch. 22)

anthropogenic climate change The release of greenhouse gases into the atmosphere from human activities, such as the burning of fossil fuels. (Ch. 9)

anticyclonic motion The rotation of a weather system resulting from the Coriolis effect as air moves outward from a region of high atmospheric pressure. Compare *cyclonic motion*. (Ch. 9)

antimatter Matter made up of antiparticles. (Ch. 14, 22)

antiparticle An elementary particle of antimatter identical in mass but opposite in charge and all other properties to its corresponding ordinary matter particle. (Ch. 22)

aperture The clear diameter of a telescope's objective lens or primary mirror. (Ch. 6)

aphelion (pl. aphelia) The point in a solar orbit that is farthest from the Sun. Compare *perihelion*. (Ch. 12)

Apollos A group of asteroids whose orbits cross the orbits of both Earth and Mars. Compare *Amors* and *Atens*. (Ch. 12)

apparent daily motion As seen from Earth's surface, the path along which each object seems to move across the sky. (Ch. 2)

apparent magnitude A measure of the apparent brightness of a celestial object, generally a star. Compare *absolute magnitude*. (Ch. 13)

arcminute (arcmin) A minute of arc ('), a unit used to measure angles. An arcminute is 1/60 of a degree of arc. (Ch. 6, 13)

arcsecond (arcsec) A second of arc ("), a unit used to measure very small angles. An arcsecond is 1/60 of an arcminute, or 1/3,600 of a degree of arc. (Ch. 6, 13)

Arctic Circle The circle on Earth with latitude 66.5° north, marking the southern limit where at least one day per year is in 24-hour daylight. Compare *Antarctic Circle*. (Ch. 2)

asteroid Also called *minor planet*. A primitive rocky or metallic body (planetesimal) that has survived planetary accretion. Asteroids are parent bodies of meteoroids. (Ch. 12)

asteroid belt The region between the orbits of Mars and Jupiter that contains most of the asteroids in our Solar System. (Ch. 12)

astrobiology An interdisciplinary science combining astronomy, biology, chemistry, geology, and physics to study life in the cosmos. (Ch. 1, 24)

astrology The belief that the positions and aspects of stars and planets influence human affairs and characteristics, as well as terrestrial events. (Ch. 1)

astronomical seeing A measurement of the degree to which Earth's atmosphere degrades the resolution of a telescope's view of astronomical objects. (Ch. 6)

astronomical unit (AU) The average distance from the Sun to Earth: approximately 150 million kilometers (km). (Ch. 3)

astronomy The scientific study of planets, stars, galaxies, and the universe as a whole. (Ch. 1)

astrophysics The application of physical laws to the understanding of planets, stars, galaxies, and the universe as a whole.

asymptotic giant branch (AGB) The path on the H-R diagram that goes from the horizontal branch toward higher luminosities and lower temperatures, asymptotically approaching and then rising above the red giant branch. (Ch. 16)

Atens A group of asteroids whose orbits cross the orbit of Earth but not the orbit of Mars. Compare *Amors* and *Apollos*. (Ch. 12)

atmosphere The gravitationally bound outer gaseous envelope surrounding a planet, moon, or star. (Ch. 9)

atmospheric greenhouse effect A warming of planetary surfaces produced by atmospheric gases that transmit optical solar radiation but partially trap infrared radiation. Compare *greenhouse effect*. (Ch. 9)

atmospheric probe An instrumented package designed to provide on-site measurements of the chemical and/or physical properties of a planetary atmosphere. (Ch. 6)

atmospheric window A region of the electromagnetic spectrum in which radiation can penetrate a planet's atmosphere. (Ch. 6)

atom The smallest unit of a chemical element that retains the properties of that element. Each atom is composed of a nucleus (neutrons and protons) surrounded by a cloud of electrons.

AU See *astronomical unit*. (Ch. 3)

aurora Emission in the upper atmosphere of a planet from atoms that have been excited by collisions with energetic particles from the planet's magnetosphere. (Ch. 9)

autumnal equinox 1. One of two points where the Sun crosses the celestial equator. 2. The day on which the Sun appears at that location, marking the first day of autumn (about September 22 in the Northern Hemisphere and March 20 in the Southern Hemisphere). Compare *vernal equinox*. See also *summer solstice* and *winter solstice*. (Ch. 2)

axion A hypothetical elementary particle first proposed to explain certain properties of the neutron and now considered a candidate for cold dark matter. (Ch. 23)

B

backlighting Illumination from behind a subject as seen by an observer. Fine material such as human hair and dust in planetary rings stands out best when viewed under backlighting conditions. (Ch. 11)

bar A unit of pressure. One bar is equivalent to 10^5 newtons per square meter—approximately equal to Earth's atmospheric pressure at sea level. (Ch. 9)

barred spiral (BS) galaxy A spiral galaxy with a bulge having an elongated, barlike shape. Compare *elliptical galaxy*, *irregular galaxy*, *S0 galaxy*, and *spiral galaxy*. (Ch. 19)

basalt Gray to black volcanic rock, rich in iron and magnesium. (Ch. 8)

beta decay 1. The decay of a neutron into a proton by emission of an electron (beta ray) and an antineutrino. 2. The decay of a proton into a neutron by emission of a positron and a neutrino.

Big Bang The event that occurred 13.8 billion years ago that marks the beginning of time and the universe. (Ch. 1, 21)

Big Bang nucleosynthesis The formation of low-mass nuclei (H, He, Li, Be) during the first few minutes after the Big Bang. (Ch. 21)

Big Crunch A hypothetical cosmic future in which the expansion of the universe reverses and the universe collapses onto itself. (Ch. 22)

Big Rip A hypothetical cosmic future in which all matter in the universe, from stars to subatomic particles, is progressively torn apart by expansion of the universe. (Ch. 22)

binary star A system in which two stars are in gravitationally bound orbits about their common center of mass. (Ch. 13)

binding energy The minimum energy required to separate an atomic nucleus into its component protons and neutrons. (Ch. 17)

biosignature Chemical compounds detectable in the spectra of an exoplanet's atmosphere that might indicate the presence of life. (Ch. 24)

biosphere The global sum of all living organisms on Earth (or any planet or moon). Compare *hydrosphere* and *lithosphere*.

bipolar outflow Material streaming away in opposite directions from both sides of the accretion disk of a young star. (Ch. 15)

black hole An object so dense that its escape velocity exceeds the speed of light; a *singularity* in spacetime. (Ch. 18)

blackbody An object that absorbs and can reemit all electromagnetic energy it receives. (Ch. 5)

blackbody spectrum See *Planck spectrum*. (Ch. 5)

blue-green algae See *cyanobacteria*. (Ch. 24)

blueshift The Doppler shift toward shorter (bluer) wavelengths of light from an approaching object. Compare *redshift*. (Ch. 5)

Bohr model A model of the atom, proposed by Niels Bohr in 1913, in which a small positively charged nucleus is surrounded by orbiting electrons, similar to a miniature solar system. (Ch. 5)

bound orbit An orbit in which an object is gravitationally bound to the body it is orbiting. A bound orbit's velocity is less than the escape velocity. Compare *unbound orbit*. (Ch. 4)

bow shock The boundary at which the speed of the solar wind abruptly drops from supersonic to subsonic in its approach to a planet's magnetosphere; the boundary between the region dominated by the solar wind and the region dominated by a planet's magnetosphere.

brightness The apparent intensity of light from a luminous object. Brightness depends on both the *luminosity* of a source and its distance. Units at the detector: watts per square meter (W/m^2).

brown dwarf A "failed star" without enough mass to fuse hydrogen in its core. An object whose mass is intermediate between that of the least massive stars and that of supermassive planets. (Ch. 7, 15, 16)

bulge The central region of a spiral galaxy that is similar in appearance to a small elliptical galaxy. (Ch. 19)

C

C See *Celsius*.

C-type asteroid An asteroid made of material that has remained mostly unmodified since the formation of the Solar System; the most primitive type of asteroid. Compare *M-type asteroid* and *S-type asteroid*. (Ch. 12)

caldera The summit crater of a volcano.

Cambrian explosion The spectacular rise in the number and diversity of biological species that occurred between 540 million and 500 million years ago. (Ch. 24)

carbon-nitrogen-oxygen (CNO) cycle One way in which hydrogen is converted to helium (hydrogen fusion in the interiors of main-sequence stars. Compare *proton-proton chain*. (Ch. 17)

carbon star A cool red giant or asymptotic giant branch star that has an excess of carbon in its atmosphere.

carbonaceous chondrite A primitive stony meteorite that contains chondrules and is rich in carbon and volatile materials. (Ch. 12)

Cassini Division The largest gap in Saturn's rings, discovered by Jean-Dominique Cassini in 1675. (Ch. 11)

catalyst An atomic and molecular structure that permits or encourages chemical and nuclear reactions but does not change its own chemical or nuclear properties. (Ch. 9)

CCD See *charge-coupled device*. (Ch. 6)

celestial equator The imaginary great circle that is the projection of Earth's equator onto the celestial sphere. (Ch. 2)

celestial sphere An imaginary sphere with celestial objects on its inner surface and Earth at its center. The celestial sphere has no physical existence but is a convenient tool for picturing the directions in which celestial objects are seen from Earth's surface. (Ch. 2)

Celsius (C) Also called *centigrade scale*. The arbitrary temperature scale, formulated by Anders Celsius (1701–1744), that defines 0°C as the freezing point of water and 100°C as the boiling point of water at sea level. Unit: degrees Celsius (°C). Compare *Fahrenheit* and *Kelvin temperature scale*.

center of mass 1. The weighted average location of all the mass in a system of objects. The point in any isolated system that moves according to Newton's first law of motion. 2. In a binary star system, the point between the two stars that is the focus of both their elliptical orbits. (Ch. 4, 13)

centigrade scale See *Celsius*.

centripetal force A force directed toward the center of curvature of an object's curved path. (Ch. 4)

Cepheid variable An evolved high-mass star with an atmosphere that is pulsating, leading to variability in the star's luminosity and color. (Ch. 17)

Chandrasekhar limit The upper limit on the mass of an object supported by electron degeneracy pressure; approximately 1.4 solar masses (M_{Sun}). (Ch. 16)

chaotic Behavior in complex systems in which a small change in the initial state of a system can lead to a large change in the final state of the system. (Ch. 7)

charge-coupled device (CCD) A common type of solid-state detector of electromagnetic radiation that transforms the intensity of light directly into electric signals. (Ch. 6)

charge destruction Electrons are absorbed by protons in atomic nuclei, forming neutrons and releasing neutrinos. (Ch. 17)

chondrite A stony meteorite that contains chondrules. Compare *achondrite*. (Ch. 12)

chondrule A small, crystallized, spherical inclusion of rapidly cooled molten droplets found inside some meteorites. (Ch. 12)

chromatic aberration A detrimental property of a lens in which rays of different wavelengths are brought to different focal distances from the lens. (Ch. 6)

chromosphere The region of the Sun's atmosphere located between the *photosphere* and the *corona*. (Ch. 14)

circular velocity The orbital velocity needed to keep an object moving in a circular orbit. (Ch. 4)

circumpolar Describing the part of the sky, near either celestial pole, that can always be seen above the horizon from a specific location on Earth. (Ch. 2)

climate The state of an atmosphere averaged over an extended time. Compare *weather*. (Ch. 9)

closed universe A finite universe with a curved spatial structure such that the sum of the angles of a triangle always exceeds 180 degrees. Compare *flat universe* and *open universe*. (Ch. 22)

CMB See *cosmic microwave background radiation*. (Ch. 6, 21)

CNO cycle See *carbon-nitrogen-oxygen cycle*. (Ch. 17)

cold dark matter Particles of dark matter that move slowly enough to be gravitationally bound even in the smallest galaxies. Compare *hot dark matter*. (Ch. 23)

color index The color of a celestial object, generally a star, based on the ratio of its brightness in blue light to its brightness in "visual" (yellow-green) light. The difference between an object's blue (B) magnitude and visual (V) magnitude, $B - V$. (Ch. 13, Appendix 7)

coma (pl. **comae)** The nearly spherical cloud of gas and dust surrounding the nucleus of an active comet. (Ch. 12)

comet A complex object consisting of a small, solid, icy nucleus; an atmospheric halo; and a tail of gas and dust. (Ch. 7)

comet nucleus A primitive planetesimal composed of ices and refractory materials that has survived planetary accretion. The "heart" of a comet, containing nearly the entire mass of the comet. A "dirty snowball." (Ch. 7)

comparative planetology The study of planets by comparing their chemical and physical properties. (Ch. 8)

composite volcano A large, cone-shaped volcano formed by viscous, pasty lava flows alternating with pyroclastic (explosively generated) rock deposits. Compare *shield volcano*. (Ch. 8)

compound lens A lens made up of two or more elements with different refractive indices, the purpose of which is to minimize chromatic aberration. (Ch. 6)

concave mirror A telescope mirror with a surface that curves inward toward the incoming light.

conduction The transfer of energy in which the thermal energy of particles is transferred to adjacent particles by collisions or other interactions. Conduction is the most important way that thermal energy is transported in solid matter. Compare *convection*. (Ch. 14)

conservation law A physical law stating that the amount of a particular physical quantity (such as energy or angular momentum) of an isolated system does not change.

conservation of angular momentum The physical law stating that the amount of angular momentum of an isolated system does not change. (Ch. 7)

conservation of energy The physical law stating that the amount of energy of an isolated, closed system does not change. (Ch. 7)

constant of proportionality The number by which one quantity is multiplied to get another number.

constellation An imaginary image formed by patterns of stars; any of 88 defined areas on the celestial sphere that astronomers use to locate celestial objects. (Ch. 2)

continental drift The slow motion (centimeters per year) of Earth's continents relative to one another and to Earth's mantle. See also *plate tectonics*. (Ch. 8)

continuous radiation Electromagnetic radiation with intensity that varies smoothly over a wide range of wavelengths.

continuous spectrum A spectrum containing all wavelengths, without specific spectral lines. (Ch. 5)

convection The transport of thermal energy from the lower (hotter) to the higher (cooler) layers of a fluid by motions within the fluid driven by variations in buoyancy. Compare *conduction*. (Ch. 8, 14)

convective zone A region within a star where energy is transported outward by convection. Compare *radiative zone*. (Ch. 14)

core 1. The innermost region of a planetary interior. Compare *crust* and *mantle*. 2. The innermost part of a star. (Ch. 8, 14)

core accretion–gas capture A process for forming giant planets, in which large amounts of surrounding hydrogen and helium gas are gravitationally captured onto a massive rocky core. (Ch. 7)

Coriolis effect The apparent displacement of objects in a direction perpendicular to their true motion as viewed from a rotating frame of reference. On a rotating planet, that effect arises from different latitudes' rotating at different speeds. (Ch. 2)

corona The hot, outermost part of the Sun's atmosphere. Compare *chromosphere* and *photosphere*. (Ch. 14)

coronal hole A low-density region in the solar corona containing "open" magnetic-field lines along which coronal material is free to stream into interplanetary space. (Ch. 14)

coronal mass ejection (CME) An eruption on the Sun that ejects hot gas and energetic particles at much higher speeds than are typical in the solar wind. (Ch. 14)

cosmic microwave background radiation (CMB) Also called simply *cosmic background radiation*. Isotropic microwave radiation from every direction in the sky having a 2.73-kelvin (K) blackbody spectrum. The CMB is residual radiation from the Big Bang. (Ch. 6, 21)

cosmic ray A very fast-moving particle (usually protons or another atomic nucleus) that originated in outer space; cosmic rays fill the disk of the Milky Way. (Ch. 18, 20)

cosmological constant A constant, introduced into general relativity by Einstein, that characterizes an extra, repulsive force in the universe due to the vacuum of space itself. (Ch. 22)

cosmological principle The (testable) assumption that the same physical laws that apply here and now also apply everywhere and at all times and that the universe has no special locations or directions. (Ch. 1)

cosmological redshift The redshift that results from the expansion of the universe rather than from the motions of galaxies or gravity. Compare *gravitational redshift*.

cosmology The study of the large-scale structure and evolution of the universe as a whole. (Ch. 21)

crescent Any phase of the Moon, Mercury, or Venus in which the object appears less than half illuminated by the Sun. Compare *gibbous*.

Cretaceous-Paleogene (K-Pg) boundary The boundary between the Cretaceous and Paleogene periods in Earth's history. That boundary corresponds to the time of the impact of an asteroid or comet and the extinction of the dinosaurs. (Ch. 8)

critical density The value of mass density of the universe that, ignoring any cosmological constant, can just barely halt expansion of the universe. (Ch. 22)

crust The relatively thin, outermost, hard layer of a planet, which is chemically distinct from the interior. Compare *core* and *mantle*. (Ch. 8)

cryovolcanism Low-temperature volcanism in which the magmas are composed of molten ices rather than rocky material. (Ch. 11)

cyanobacteria Also called *blue-green algae*. Single-celled organisms that created oxygen in Earth's atmosphere by photosynthesizing carbon dioxide and releasing oxygen as a waste product. (Ch. 24)

cyclonic motion The rotation of a weather system resulting from the Coriolis effect as air moves toward a region of low atmospheric pressure. Compare *anticyclonic motion*. (Ch. 9)

D

Dark Ages The epoch in the history of the universe during which no visible "light" came from astronomical objects. (Ch. 23)

dark energy A form of energy that permeates all space (including the vacuum), producing a repulsive force that accelerates the expansion of the universe. (Ch. 22)

dark matter Matter in galaxies that does not emit or absorb electromagnetic radiation. Dark matter is thought to constitute most of the mass in the universe. Compare *luminous matter*. (Ch. 19)

dark matter halo The centrally condensed, greatly extended dark matter component of a galaxy that accounts for up to 95 percent of the galaxy's mass. (Ch. 19)

daughter product An element resulting from radioactive decay of a more massive *parent element*. (Ch. 8)

decay 1. The process of a radioactive nucleus changing into its daughter product. 2. The process of an atom or molecule dropping from a higher energy state to a lower energy state. 3. The process of a satellite's orbit losing energy. (Ch. 5)

declination A measure, analogous to *latitude*, that tells you the angular distance of a celestial body north or south of the celestial equator (from 0° to ±90°). Compare *right ascension*. (Ch. 2)

degeneracy pressure Pressure exerted by closely packed electrons in the collapsing core of a star. Such pressure pushes outward against gravity and keeps the star from collapsing further. (Ch. 16)

density The measure of an object's mass per unit of volume. Possible units include kilograms per cubic meter (kg/m^3). (Ch. 4, 7)

deuterium An isotope of hydrogren; the nucleus contains one proton and one neutron. (Ch. 14)

differential rotation Rotation of different parts of a system at different rates. (Ch. 14)

differentiation The process by which materials of higher density sink toward the center of a molten or fluid planetary interior. (Ch. 8)

diffraction The spreading of a wave after it passes through an opening or beyond the edge of an object. (Ch. 6)

diffraction grating An optical component with many narrow parallel lines that separate the wavelengths of light to produce a spectrum. (Ch. 6)

diffraction limit The limit of a telescope's angular resolution caused by diffraction. (Ch. 6)

diffuse ring A sparsely populated planetary ring spread out both horizontally and vertically.

dispersion The separation of rays of light into their component wavelengths. (Ch. 6)

distance ladder A sequence of techniques for measuring cosmic distances: each method is calibrated using the results from other methods that have been applied to closer objects. (Ch. 19)

Doppler effect The change in wavelength of sound or light as a result of the relative motion of the source toward or away from the observer. (Ch. 5)

Doppler shift The amount by which the Doppler effect shifts the wavelength of light. (Ch. 5)

Drake equation A prescription for estimating the number of intelligent civilizations existing in the Milky Way Galaxy. (Ch. 24)

dust devil A small tornado-like column of air containing dust or sand. (Ch. 9)

dust tail A type of comet tail consisting of dust particles pushed away from the comet's head by radiation pressure from the Sun. Compare *ion tail*. (Ch. 12)

dwarf galaxy A small galaxy with a luminosity ranging from 1 million to 1 billion solar luminosities (L_{Sun}). Compare *giant galaxy*. (Ch. 19)

dwarf planet A body with characteristics similar to those of a planet except that it has not cleared smaller bodies from the neighboring regions around its orbit. Compare *planet* (definition 2). (Ch. 12)

dynamic equilibrium A state in which a system is constantly changing but its configuration remains the same because one source of change is exactly balanced by another source of change. Compare *static equilibrium*.

dynamo theory A theory postulating that Earth's magnetic field (and those of other planets) is generated from a rotating and electrically conducting liquid core. (Ch. 8)

E

eccentricity (*e*) The ratio of the distance between the two foci of an ellipse to the length of its major axis, which measures how noncircular the ellipse is. (Ch. 3)

eclipse 1. The total or partial obscuration of one celestial body by another. 2. The total or partial obscuration of light from one celestial body as it passes through the shadow of another celestial body. (Ch. 2)

eclipse season Any time during the year when the Moon's line of nodes points towards the Sun and eclipses can occur. (Ch. 2)

eclipsing binary A binary system in which the orbital plane is oriented such that the two stars appear to pass in front of each other as seen from Earth. Compare *spectroscopic binary* and *visual binary*. (Ch. 13)

ecliptic 1. The apparent annual path of the Sun against the background of stars. 2. The projection of Earth's orbital plane onto the celestial sphere. (Ch. 2)

ecliptic plane The plane of Earth's orbit around the Sun. The ecliptic is the projection of that plane onto the celestial sphere. (Ch. 2)

effective temperature The temperature at which a blackbody, such as a star, appears to radiate. (Ch. 14)

Einstein ring Light bent by gravitational lensing into a ring. (Ch. 18)

ejecta 1. Material thrown outward by the impact of an asteroid or comet on a planetary surface, leaving a crater behind. 2. Material thrown outward by a stellar explosion. (Ch. 8)

electric field A field that can exert a force on a charged object, whether at rest or moving. Compare *magnetic field*. (Ch. 5)

electric force The force exerted on electrically charged particles such as protons and electrons, arising from their electric charges. Compare *magnetic force*. See also *electromagnetic force*. (Ch. 5)

electromagnetic force The force, including both electric and magnetic forces, that acts on electrically charged particles. One of four fundamental forces of nature, along with the *strong nuclear force*, *weak nuclear force*, and *gravity* (definition 1). The force is mediated by the exchange of photons. (Ch. 22)

electromagnetic radiation A traveling disturbance in the electric and magnetic fields caused by accelerating electric charges. In quantum mechanics, a stream of photons. Light.

electromagnetic spectrum The spectrum made up of all possible frequencies or wavelengths of electromagnetic radiation, ranging from gamma rays through radio waves and including the portion our eyes can use. (Ch. 5)

electromagnetic wave A wave consisting of oscillations in the electric-field strength and the magnetic-field strength. (Ch. 5)

electron (e⁻) A subatomic particle having a negative electric charge of 1.6×10^{-19} coulomb (C), a rest mass of 9.1×10^{-31} kilogram (kg), and a rest energy of 8×10^{-14} joule (J). The antiparticle of the *positron*. Compare *proton* and *neutron*. (Ch. 5)

electron-degenerate Describing matter, compressed to the point at which electron density reaches the limit imposed by the rules of quantum mechanics. (Ch. 16)

electroweak theory The quantum theory that combines descriptions of both the electromagnetic force and the weak nuclear force. (Ch. 22)

element One of 92 naturally occurring substances (such as hydrogen, oxygen, and uranium) and more than 20 human-made ones (such as plutonium). Each element is chemically defined by the specific number of protons in the nuclei of its atoms. (Ch. 5)

elementary particle One of the basic building blocks of nature that is not known to have substructure, such as the *electron* or the *quark*. (Ch. 23)

ellipse A conic section produced by the intersection of a plane with a cone when the plane is passed through the cone at an angle to the axis other than 0° or 90°. The shape that results when you attach the two ends of a piece of string to a piece of paper, stretch the string tight with the tip of a pencil, and then draw around those two points while keeping the string taut. (Ch. 3)

elliptical galaxy A galaxy of Hubble type "E" class, with a circular to elliptical outline on the sky, and containing almost no disk and a population of old stars. Compare *barred spiral galaxy*, *irregular galaxy*, *S0 galaxy*, and *spiral galaxy*. (Ch. 19)

emission The production of a photon when an atom decays to a lower energy state. Compare *absorption*. (Ch. 5)

emission line A peak in the intensity of a spectrum that is due to the emission of electromagnetic radiation at a specific wavelength determined by the energy levels of an atom or molecule. Compare *absorption line*. (Ch. 5)

empirical science Descriptive scientific investigation based primarily on observations and experimental data rather than on theoretical inference. (Ch. 3)

energy The conserved quantity that gives objects and systems the ability to do work. Possible units include joules (J). (Ch. 5)

energy transport The transfer of energy from one location to another. In stars, energy transport is carried out mainly by radiation or convection. (Ch. 14)

entropy A measure of the disorder of a system related to the number of ways a system can be rearranged without its appearance being affected.

equator The imaginary great circle on the surface of a body midway between its poles that divides the body into northern and southern hemispheres. The equatorial plane passes through the center of the body and is perpendicular to its rotation axis. Compare *meridian*. (Ch. 2)

equilibrium The state of an object in which physical processes balance each other so that its properties or conditions remain constant.

equinox Literally, "equal night." 1. One of two positions on the ecliptic where it intersects the celestial equator. 2. Either of the two times of year (the *autumnal equinox* and *vernal equinox*) when the Sun is at one of these two positions. At this time, night and day are of the same length everywhere on Earth. Compare *solstice*.

equivalence principle The principle stating that no difference exists between a frame of reference freely floating through space and one freely falling within a gravitational field. (Ch. 18)

erosion The degradation of a planet's surface topography by the mechanical action of wind, water, or living organisms. (Ch. 8)

escape velocity The minimum velocity needed for an object to achieve a parabolic trajectory and thus permanently leave the gravitational grasp of another mass. (Ch. 4)

eternal inflation The idea that a universe might inflate forever. In such a universe, quantum effects could randomly cause regions to slow their expansion, eventually stop inflating, and experience an explosion resembling the Big Bang. (Ch. 22)

event Something that happens at a particular location in spacetime. (Ch. 18)

event horizon The effective "surface" of a black hole. Nothing inside that surface—not even light—can escape from a black hole. (Ch. 18)

evolutionary track The path that a star follows across the H-R diagram as it evolves through its lifetime. (Ch. 15)

excited state Any energy level of a system or part of a system, such as an atom, molecule, or particle, that is higher than its ground state. Compare *ground state*. (Ch. 5)

exoplanet A planet orbiting a star other than the Sun. (Ch. 7)

exosphere A very thin atmosphere or layer of atmosphere, where the molecules are bound by gravity to the moon or planet but their density is too low to behave like a gas of colliding particles. (Ch. 9)

extrasolar planet See *exoplanet*. (Ch. 7)

extremophile A life-form that thrives under extreme environmental conditions. (Ch. 24)

eyepiece A lens that is closest to the eye in a telescope. Changing the eyepiece will change the magnification of the image in the telescope. (Ch. 6)

F

F See *Fahrenheit*.

Fahrenheit (F) The arbitrary temperature scale, formulated by Daniel Gabriel Fahrenheit (1686–1736), that defines 32°F as the melting point of water and 212°F as the boiling point of water at sea level. Unit: degrees Fahrenheit (°F). Compare *Celsius* and *Kelvin temperature scale*.

falsified A hypothesis shown to be false. (Ch. 1)

fault A fracture in the crust of a planet or moon along which blocks of material can slide. (Ch. 8)

filter An instrument element that transmits a limited wavelength range of electromagnetic radiation. For the optical range, such elements are typically made of different kinds of glass and take on the hue of the light they transmit. (Ch. 13)

first quarter Moon The phase of the Moon in which only the western half of the Moon, as viewed from Earth, is illuminated by the Sun. It occurs about a week after a new Moon. Compare *third quarter Moon*. See also *full Moon* and *new Moon*. (Ch. 2)

fissure A fracture in the planetary lithosphere from which magma emerges. (Ch. 8)

flat universe An infinite universe whose spatial structure obeys Euclidean geometry, such that the sum of the angles of a triangle always equals 180 degrees. Compare *closed universe* and *open universe*. (Ch. 22)

flatness problem The surprising result that the sum of Ω_m plus Ω_Λ is extremely close to 1 in the present-day universe; equivalent to saying that it is surprising the universe is so close to being exactly flat.

flux (*f*) The total amount of energy passing through each square meter of a surface each second. Unit: watts per square meter (W/m²). (Ch. 5)

flux tube A strong magnetic field contained within a tubelike structure. Flux tubes are found in the solar atmosphere and connecting the space between Jupiter and its moon Io. (Ch. 10)

flyby A spacecraft that first approaches and then continues flying past a planet or moon. Flybys can visit multiple objects, but they remain near their targets only briefly. Compare *orbiter*. (Ch. 6)

focal length The optical distance between a telescope's objective lens or primary mirror and the plane (called the focal plane) on which the light from a distant object is focused. (Ch. 6)

focal plane The plane, perpendicular to the optical axis of a lens or mirror, on which an image is formed. (Ch. 6)

focus (pl. foci) 1. One of two points that define an ellipse. 2. A point in the focal plane of a telescope. (Ch. 3)

force (*F*) A push or a pull on an object. (Ch. 3)

frame of reference A coordinate system within which an observer measures positions and motions. (Ch. 2)

free fall The motion of an object when the only force acting on it is gravity. (Ch. 4)

frequency (*f*) The number of times per second that a periodic process occurs. Unit: hertz (Hz), or cycles per second (1/s). (Ch. 5)

full Moon The phase of the Moon in which the near side of the Moon, as viewed from Earth, is fully illuminated by the Sun. It occurs about two weeks after a *new Moon*. See also *first quarter Moon* and *third quarter Moon*. (Ch. 2)

G

galaxy A gravitationally bound system that consists of stars and star clusters, gas, dust, and dark matter; typically greater than 1,000 light-years across and recognizable as a discrete, single object. (Ch. 19)

galaxy cluster A large, gravitationally bound collection of galaxies containing hundreds to thousands of members; typically 3–5 megaparsecs (Mpc) across. Compare *galaxy group* and *supercluster*. (Ch. 23)

galaxy group A small, gravitationally bound collection of galaxies containing from several to a hundred members; typically 1–2 megaparsecs (Mpc) across. Compare *galaxy cluster* and *supercluster*. (Ch. 20)

gamma ray Also called *gamma radiation*. Electromagnetic radiation with higher frequency, higher photon energy, and shorter wavelength than all other types of electromagnetic radiation. (Ch. 5)

gamma-ray burst (GRB) A brief, intense burst of gamma rays from a distant energetic explosion. (Ch. 18)

gas giant A giant planet formed mostly of hydrogen and helium. In the Solar System, Jupiter and Saturn are the gas giants. Compare *ice giant*. (Ch. 10)

general relativistic time dilation The verified prediction that time passes more slowly in a gravitational field than in the absence of a gravitational field. Compare *time dilation*. (Ch. 18)

general relativity See *general theory of relativity*. (Ch. 18)

general theory of relativity Sometimes referred to as simply *general relativity*. Einstein's theory explaining gravity as the distortion of spacetime by massive objects, such that particles travel on the shortest path between two events in spacetime. This theory deals with all types of motion. Compare *special theory of relativity*. (Ch. 18)

geocentric model A historical cosmological model with Earth at its center, and all the other objects in the universe in orbit around Earth. Compare *heliocentric model*. (Ch. 3)

geodesic The path an object will follow through spacetime in the absence of external forces. (Ch. 18)

giant molecular cloud An interstellar cloud composed primarily of molecular gas and dust and having hundreds of thousands of solar masses. (Ch. 15)

giant planet Also called *Jovian planet*. One of the largest planets in the Solar System (Saturn, Jupiter, Uranus, or Neptune), typically 10 times the size and many times the mass of any *terrestrial planet* and lacking a solid surface. (Ch. 7)

gibbous Any phase of the Moon, Mercury, or Venus in which the object appears more than half illuminated by the Sun. Compare *crescent*.

global circulation The overall, planetwide circulation pattern of a planet's atmosphere. (Ch. 9)

globular cluster A spherically symmetric, highly condensed group of stars, containing tens of thousands to a million members. Compare *open cluster*. (Ch. 17)

gluon The particle that carries (or, equivalently, mediates) interactions due to the strong nuclear force. (Ch. 22)

grand unified theory (GUT) A unified quantum theory that combines the strong nuclear, weak nuclear, and electromagnetic forces but does not include gravity. (Ch. 22)

granite Rock that is cooled from magma and is relatively rich in silicon and oxygen. (Ch. 8)

grating An optical surface containing many narrow, closely and equally spaced parallel grooves or slits that spectrally disperse reflected or transmitted light. (Ch. 6)

gravitational lens A massive object that gravitationally focuses the light of a more distant object to produce multiple brighter, magnified, possibly distorted images.

gravitational lensing The bending of light by gravity. (Ch. 7, 18)

gravitational potential energy The stored energy in an object that is due solely to its position within a gravitational field. (Ch. 7)

gravitational redshift The shifting to longer wavelengths of radiation from an object deep within a gravitational well. Compare *cosmological redshift*. (Ch. 18)

gravitational wave A wave in the faric of spacetime emitted by accelerating masses. (Ch. 6, 18)

gravity 1. The mutually attractive force between massive objects. One of four fundamental forces of nature, along with the *electromagnetic force*, the *strong nuclear force*, and the *weak nuclear force*. 2. An effect arising from the bending of spacetime by massive objects. (Ch. 4)

GRB See *gamma-ray burst*. (Ch. 18)

Great Red Spot The giant, oval, brick red anticyclone seen in Jupiter's southern hemisphere. (Ch. 10)

greenhouse effect The solar heating of air in an enclosed space, such as a closed building or car, resulting primarily from the inability of the hot air to escape. Compare *atmospheric greenhouse effect*. (Ch. 9)

greenhouse gas One of a group of atmospheric gases such as carbon dioxide that are transparent to visible radiation but absorb infrared radiation. (Ch. 9)

greenhouse molecule A molecule such as water vapor or carbon dioxide that transmits visible radiation but absorbs infrared radiation.

Gregorian calendar The modern calendar. A modification of the Julian calendar decreed by Pope Gregory XIII in 1582. By then, the less accurate Julian calendar had developed an error of 10 days over the 13 centuries since its inception. (Ch. 2)

ground state The lowest possible energy state for a system or part of a system, such as an atom, molecule, or particle. Compare *excited state*. (Ch. 5)

GUT See *grand unified theory*. (Ch. 22)

H

H II region A region of interstellar gas that has been ionized by ultraviolet radiation from nearby hot, massive stars. (Ch. 15)

H-R diagram The Hertzsprung-Russell diagram, a plot of the luminosities versus the surface temperatures of stars. The evolving properties of stars are plotted as tracks across the H-R diagram. (Ch. 13)

habitable zone The distance from its star at which a planet must be located to have a temperature suitable for water to exist in a liquid state. (Ch. 7, 13)

Hadley circulation A simplified, and therefore uncommon, atmospheric global circulation that carries thermal energy directly from the equator to the polar regions of a planet. (Ch. 9)

half-life The time that half a sample of a particular radioactive parent element takes to decay into a daughter product. (Ch. 8)

halo The spherically symmetric, low-density distribution of stars and dark matter that defines the outermost regions of a galaxy.

harmonic law See *Kepler's third law*. (Ch. 3)

Hawking radiation Radiation from a black hole. (Ch. 18)

Hayashi track The path that a protostar follows on the H-R diagram as it contracts toward the main sequence. (Ch. 15)

head The part of a comet that includes both the nucleus and the inner part of the coma. (Ch. 12)

heavy element Also called *massive element*. Any element more massive than helium. (Ch. 10, 13)

heliocentric model A model of the Solar System, with the Sun at its center, and the planets, including Earth, in orbit around the Sun. Compare *geocentric model*. (Ch. 3)

helioseismology The use of solar oscillations to study the interior of the Sun. (Ch. 14)

heliosphere A region surrounding the Solar System in which the solar wind blows against the interstellar medium and clears out an area like the inside of a bubble. The heliosphere protects the Solar System from cosmic rays. (Ch. 14)

helium capture A helium nucleus is captured by another nucleus during nucleosynthesis. (Ch. 17)

helium flash The runaway explosive fusion of helium in the degenerate helium core of a red giant star. (Ch. 16)

Herbig-Haro (HH) object A glowing, rapidly moving knot of gas and dust that is excited by bipolar outflows in very young stars. (Ch. 15)

heredity The process by which one generation passes on its characteristics to future generations. (Ch. 24)

hertz (Hz) A unit of frequency equivalent to cycles per second. (Ch. 5)

Hertzsprung-Russell diagram See *H-R diagram*. (Ch. 13)

HH object See *Herbig-Haro object*. (Ch. 15)

hierarchical clustering The "bottom up" process of forming large-scale structure. Small-scale structure first produces groups of galaxies, which in turn form clusters, which then form superclusters. (Ch. 23)

high-mass star A star with a main-sequence mass of greater than about 8 solar masses (M_{Sun}). Compare *low-mass star* and *medium-mass star*. (Ch. 16)

homogeneous In cosmology, describing a universe in which observers in any location would observe the same properties. Compare *isotropic*. (Ch. 21)

horizon The boundary that separates the sky from the ground. (Ch. 2)

horizon problem The puzzling observation that the cosmic background radiation is so uniform in all directions, even though widely separated regions should have been "over the horizon" from each other in the early universe. (Ch. 22)

horizontal branch A region on the H-R diagram defined by stars fusing helium to carbon in a stable core. (Ch. 16)

hot dark matter Particles of dark matter that move so fast that gravity cannot confine them to the volume occupied by a galaxy's normal luminous matter. Compare *cold dark matter*. (Ch. 23)

hot Jupiter A large, Jupiter-type extrasolar planet located very close to its parent star. (Ch. 7)

hot spot A place where hot plumes of mantle material rise near the surface of a planet. (Ch. 8)

Hubble constant (H_0) The constant of proportionality relating the recession velocities of galaxies to their distances. Compare *Hubble time*. (Ch. 19)

Hubble flow The motion of galaxies as a result of the expanding universe. (Ch. 21)

Hubble's law The law stating that the speed at which a galaxy is moving away from Earth is proportional to the distance of that galaxy. (Ch. 19)

Hubble time An estimate of the age of the universe from the inverse of the *Hubble constant*, $1/H_0$. (Ch. 21)

hurricane A large tropical cyclonic system circulating counterclockwise in the Northern Hemisphere and clockwise in the Southern Hemisphere. Hurricanes can extend outward from their center to more than 600 kilometers (km) and generate winds in excess of 300 kilometers per hour (km/h). (Ch. 9)

hydrogen fusion The release of energy from the nuclear fusion of four hydrogen atoms into a single helium atom. (Ch. 14)

hydrogen shell fusion The fusion of hydrogen in a shell surrounding a stellar core that may be either degenerate or fusing more massive elements. (Ch. 16)

hydrosphere The portion of Earth that is largely liquid water. Compare *biosphere* and *lithosphere*. (Ch. 8)

hydrostatic equilibrium The condition in which the weight bearing down at a particular point within an object is balanced by the pressure within the object. (Ch. 8, 14)

hypernova (pl. **hypernovae**) A very energetic supernova from a very high-mass star.

hypothesis A well-considered idea, based on scientific principles and knowledge, that leads to testable predictions. Compare *theory*. (Ch. 1)

Hz See *hertz*. (Ch. 5)

I

ice The solid form of a volatile material; sometimes the *volatile material* itself, in any form. (Ch. 7)

ice giant A giant planet formed mostly of the liquid form of volatile substances (ices). In the Solar System, Uranus and Neptune are the ice giants. Compare *gas giant*. (Ch. 10)

impact crater The scar of the impact left on a solid planetary or moon surface by collision with another object. Compare *secondary crater*. (Ch. 8)

impact cratering The process in which solid planetary objects collide with each other, leaving distinctive scars. (Ch. 8)

index of refraction (*n*) The ratio of the speed of light in a vacuum (*c*) to the speed of light in an optical medium (*v*). (Ch. 6)

inert gas A gaseous element that combines with other elements only under conditions of extreme temperature and pressure. Examples include helium, neon, and argon.

inertia The tendency for objects to retain their state of motion. (Ch. 3)

inertial frame of reference 1. A frame of reference moving in a straight line at constant speed, that is, not accelerating. 2. In general relativity, a frame of reference falling freely in a gravitational field. (Ch. 3)

inertial reference frame See *inertial frame of reference*. (Ch. 18)

inferior planet A Solar System planet that orbits closer to the Sun than Earth does. Compare *superior planet*. (Ch. 3)

inflation An extremely brief phase of ultra-rapid expansion of the very early universe. After inflation, the standard Big Bang models of expansion apply. (Ch. 22)

infrared (IR) radiation Electromagnetic radiation with frequencies, photon energies, and wavelengths between those of visible light and microwaves. (Ch. 5)

instability strip A region of the H-R diagram containing stars that pulsate with a periodic variation in luminosity. (Ch. 17)

integration time The time interval during which photons are collected and added up in a detecting device. (Ch. 6)

intensity Of light, the amount of radiant energy emitted per second per unit area. Units for electromagnetic radiation: watts per square meter (W/m²).

intercloud gas A low-density region of the interstellar medium that fills the space between interstellar clouds. (Ch. 15)

interfere Typically pertaining to light, two sets of waves mutually interact to either amplify or reduce each other. See also *interference*. (Ch. 6)

interference The interaction of two sets of waves producing high and low intensity, depending on whether their amplitudes reinforce (*constructive interference*) or cancel (*destructive interference*). See also *interfere*.

interferometer Linked optical or radio telescopes whose overall separation determines the angular resolution of the system. (Ch. 6)

interferometric array An interferometer made up of several telescopes arranged in an array. (Ch. 6)

interstellar cloud A discrete, high-density region of the interstellar medium made up mostly of atomic or molecular hydrogen and dust. (Ch. 15)

interstellar dust Small particles or grains (0.01–10 microns [μm] in diameter) of matter, primarily carbon and silicates, distributed throughout interstellar space. (Ch. 15)

interstellar extinction The dimming of visible and ultraviolet light by interstellar dust. (Ch. 15)

interstellar gas The tenuous gas, far less dense than air, composing 99 percent of the matter in the interstellar medium. (Ch. 15)

interstellar medium The gas and dust that fill the space between the stars within a galaxy. (Ch. 14, 15)

inverse square law The rule stating that a quantity or effect diminishes with the square of the distance from the source. (Ch. 4)

ion An atom or molecule that has lost or gained one or more electrons. (Ch. 15)

ionize see *ionization*. (Ch. 9, 15)

ionization The process by which electrons are stripped free from an atom or molecule, resulting in free electrons and a positively charged atom or molecule. (Ch. 9, 15)

ionosphere A layer high in Earth's atmosphere in which most atoms are ionized by solar radiation. (Ch. 9)

ion tail A type of comet tail consisting of ionized gas. Particles in the ion tail are pushed directly away from the comet's head in the antisolar direction at high speeds by the solar wind. Compare *dust tail*. (Ch. 12)

IR Infrared. See *infrared radiation*. (Ch. 5)

iron meteorite A metallic meteorite composed mostly of iron-nickel alloys. Compare *stony-iron meteorite* and *stony meteorite*. (Ch. 12)

irregular galaxy A galaxy without regular or symmetric appearance. Compare *barred spiral galaxy*, *elliptical galaxy*, *S0 galaxy*, and *spiral galaxy*. (Ch. 19)

irregular moon A moon that has been captured by a planet instead of having formed along with that planet. Some irregular moons revolve in a direction opposite to the rotation of the planet, and many are in distant, unstable orbits. Compare *regular moon*. (Ch. 11)

isotope A forms of the same chemical element that has the same number of protons but a different number of neutrons. (Ch. 5, 8)

isotropic In cosmology, having the same appearance to an observer in all directions. Compare *homogeneous*. (Ch. 21)

J

J See *joule*.

jansky (Jy) The basic unit of flux density. Unit: watts per square meter per hertz (W/m²/Hz). (Ch. 6)

jet 1. A stream of gas and dust ejected from a comet nucleus by solar heating. 2. A stream of material that moves away from a protostar or active galactic nucleus at hundreds of kilometers per second. (Ch. 15)

joule (J) A unit of energy or work. 1 J = 1 newton meter.

Jovian planet See *giant planet*.

Jy See *jansky*. (Ch. 6)

K

K See *kelvin*.

K-Pg boundary See *Cretaceous-Paleogene boundary*. (Ch. 8)

KBO See *Kuiper Belt object*. (Ch. 12)

kelvin (K) The basic unit of the Kelvin scale of temperature. (Ch. 5)

Kelvin temperature scale The temperature scale, formulated by William Thomson, better known as Lord Kelvin (1824–1907), that uses Celsius-sized degrees but defines 0 K as absolute zero instead of as the melting point of water. Unit: kelvins (K). Compare *Celsius* and *Fahrenheit*. (Ch. 5)

Kepler's first law A rule of planetary motion that Johannes Kepler inferred, stating that planets move in elliptical orbits with the Sun at one focus. (Ch. 3)

Kepler's laws The three rules of planetary motion that Johannes Kepler inferred from data collected by Tycho Brahe. (Ch. 3)

Kepler's second law Also called *law of equal areas*. A rule of planetary motion that Johannes Kepler inferred, stating that a line drawn from the Sun to a planet sweeps out equal areas in equal times as the planet orbits the Sun. (Ch. 3)

Kepler's third law Also called *harmonic law*. A rule of planetary motion that Johannes Kepler inferred, describing the relationship between the period of a planet's orbit and its distance from the Sun. The law states that the square of the period of a planet's orbit, measured in years, is equal to the cube of the semimajor axis of the planet's orbit, measured in astronomical units: $(P_{\text{years}})^2 = (A_{\text{AU}})^3$. (Ch. 3)

kiloparsec A unit of distance equal to 1,000 parsecs, or 3,260 light-years.

kinetic energy (E_{K}) The energy of an object resulting from its motions. $E_{\text{K}} = \frac{1}{2}mv^2$ Possible units include joules (J). (Ch. 5, 7)

Kirkwood gap A gap in the main asteroid belt related to orbital resonances with Jupiter. (Ch. 12)

Kuiper Belt A disk-shaped population of comet nuclei extending from Neptune's orbit to perhaps several thousand astronomical units (AU) from the Sun. The highly populated innermost part of the Kuiper Belt has an outer edge approximately 50 AU from the Sun. (Ch. 12)

Kuiper Belt object (KBO) Also called *trans-Neptunian object*. An icy planetesimal (comet nucleus) that orbits within the Kuiper Belt beyond the orbit of Neptune. (Ch. 12)

L

Lambda-CDM The standard model of the Big Bang universe in which most of the energy density of the universe is dark energy (similar to Einstein's cosmological constant), and most of the mass in the universe is cold dark matter. (Ch. 23)

lander An instrumented spacecraft designed to land on a planet or moon. Compare *rover*. (Ch. 6)

large-scale structure Observable aggregates on the largest scales in the universe, including galaxy groups, clusters, and superclusters. (Ch. 23)

latitude The angular distance north (+) or south (–) from the equatorial plane of a nearly spherical body. Compare *longitude*. (Ch. 2)

lava Molten rock flowing out of a volcano during an eruption; also the rock that solidifies and cools from that liquid. (Ch. 8)

law of equal areas See *Kepler's second law*. (Ch. 3)

law of gravitation See *universal law of gravitation*. (Ch. 4)

leap year A year that contains 366 days. Leap years occur every 4 years when the year is divisible by 4, correcting for the accumulated excess time in a normal year, which is approximately 365 1/4 days long. (Ch. 2)

length contraction The relativistic compression of moving objects in the direction of their motion. (Ch. 18)

Leonids A November meteor shower associated with the dust debris left by comet Tempel-Tuttle. (Ch. 12)

life A biochemical process in which living organisms can reproduce, evolve, and sustain themselves by drawing energy from their environment. All terrestrial life involves carbon-based chemistry, assisted by the self-replicating molecules ribonucleic acid (RNA) and deoxyribonucleic acid (DNA). (Ch. 24)

light All electromagnetic radiation, which composes the entire electromagnetic spectrum.

light-year (ly) The distance that light travels in 1 year—about 9.5 trillion kilometers (km). (Ch. 1)

limb The outer edge of the visible disk of a planet, moon, or the Sun. (Ch. 14)

limb darkening The darker appearance caused by increased atmospheric absorption near the limb of a planet or star. (Ch. 14)

limestone A common sedimentary rock composed of calcium carbonate. (Ch. 9)

line of nodes 1. A line defined by the intersection of two orbital planes. 2. The line defined by the intersection of Earth's equatorial plane and the plane of the ecliptic. (Ch. 2)

lithosphere The solid, brittle part of Earth (or any planet or moon), including the crust and the upper part of the mantle. Compare *biosphere* and *hydrosphere*. (Ch. 8)

lithospheric plate A separate piece of Earth's lithosphere that can move independently. See also *continental drift* and *plate tectonics*. (Ch. 8)

Local Group The group of galaxies that includes the Milky Way and Andromeda galaxies. (Ch. 1, 20)

long-period comet A comet with an orbital period of greater than 200 years. Compare *short-period comet*. (Ch. 12)

longitude The angular distance east (+) or west (−) from the prime meridian at Greenwich, England. Compare *latitude*. (Ch. 2)

longitudinal wave A wave that oscillates parallel to the direction of the wave's propagation. Compare *transverse wave*. (Ch. 8)

look-back time The amount of time that the light from an astronomical object has taken to reach Earth. (Ch. 21)

low-mass star A star with a main-sequence mass of less than about 3 solar masses (M_{Sun}). Compare *high-mass star* and *medium-mass star*. (Ch. 16)

luminosity The total amount of light emitted by an object. Unit: watts (W). Compare *brightness*. (Ch. 5)

luminosity class A spectral classification based on stellar size, ranging from supergiants at the large end to white dwarfs at the small end. (Ch. 13)

luminosity-temperature-radius relationship A relationship among those three properties of stars indicating that if any two are known, the third can be calculated. (Ch. 13)

luminous matter Also called *normal matter*. Matter in galaxies—including stars, gas, and dust—that emits electromagnetic radiation. Compare *dark matter*. (Ch. 19)

lunar eclipse An eclipse that occurs when the Moon is partially or entirely in Earth's shadow. Compare *solar eclipse*. (Ch. 2)

lunar tide A tide on Earth caused by the differential gravitational pull of the Moon. Compare *solar tide*. (Ch. 4)

lunisolar calendar Calendar created by the Babylonians in which a month began with the first sighting of the lunar crescent, and a 13th month was added when needed to catch up to the solar year. (Ch. 2)

ly See *light-year*. (Ch. 1)

M

μm See *micron*. (Ch. 5)

M-type asteroid An asteroid made of material that was once part of the metallic core of a larger, differentiated body that has since broken into pieces; made mostly of iron and nickel. Compare *C-type asteroid* and *S-type asteroid*. (Ch. 12)

MaCHO Short for *massive compact halo object*. MaCHOs include brown dwarfs, white dwarfs, and black holes and are candidates for dark matter. Compare *WIMP*. (Ch. 19)

magma Molten rock, often containing dissolved gases and solid minerals.

magnetic field A field that can exert a force on a moving electric charge. Compare *electric field*. (Ch. 5)

magnetic force The force exerted on electrically charged particles such as protons and electrons, arising from their motion. Compare *electric force*. See also *electromagnetic force*. (Ch. 5)

magnetosphere The region surrounding a planet that is filled with relatively intense magnetic fields and plasmas. (Ch. 8)

magnitude A system used by astronomers to describe the brightness or luminosity of stars. The brighter the star, the lower its magnitude. (Ch. 13)

major axis The long axis of an ellipse. (Ch. 3)

main asteroid belt See *asteroid belt*. (Ch. 12)

main sequence The strip on the H-R diagram where most stars are found. Main-sequence stars are fusing hydrogen to helium in their cores. (Ch. 13)

main-sequence lifetime The amount of time a star spends on the main sequence, fusing hydrogen into helium in its core. (Ch. 16)

main-sequence turnoff The location on the main sequence of an H-R diagram made from a population of stars of the same age (such as a star cluster) where stars are just evolving off the main sequence. That location is determined by the age of the population of stars. (Ch. 17)

mantle The solid portion of a rocky planet that lies between the *crust* and the *core*. (Ch. 8)

mare (pl. **maria**) A dark region on the Moon composed of basaltic lava flows. (Ch. 8)

mass 1. Inertial mass: the property of matter that determines its resistance to changes in motion. Compare *weight*. 2. Gravitational mass: the property of matter defined by its attractive force on other objects. According to general relativity, the two are equivalent. (Ch. 3, 4)

mass-luminosity relationship An empirical relationship between the luminosity (L) and mass (M) of main-sequence stars; for example, $L \propto M^{3.5}$. (Ch. 13, 16)

mass transfer The transfer of mass from one member of a binary star system to its companion. Mass transfer occurs when one of the stars evolves to the point that it overfills its Roche lobe, so that its outer layers are pulled toward its binary companion. (Ch. 16)

massive element Also called *heavy element*. Any element more massive than helium. (Ch. 13)

matter 1. Objects made of particles that have mass, such as protons, neutrons, and electrons. 2. Anything that occupies space and has mass. (Ch. 5)

Maunder Minimum The period from 1645 to 1715 during which very few sunspots were observed. (Ch. 14)

medium The substance that a wave, such as light, travels through—for example, air or glass. Compare *vacuum*. (Ch. 5)

medium-mass star A star with a main-sequence mass between 3 and 8 solar masses (M_{Sun}). Compare *high-mass star* and *low-mass star*. (Ch. 16)

megaparsec (Mpc) A unit of distance equal to 1 million parsecs, or 3.26 million light-years. (Ch. 19)

meridian The imaginary arc in the sky running from the horizon at due north through the zenith to the horizon at due south. The meridian divides the observer's sky into eastern and western hemispheres. Compare *equator*. (Ch. 2)

mesosphere The layer of Earth's atmosphere immediately above the stratosphere, extending from an altitude of 50 kilometers (km) to about 90 km. Compare *troposphere*, *stratosphere*, and *thermosphere*. (Ch. 9)

meteor A meteoroid that enters and burns up in a planetary atmosphere, often leaving an incandescent trail. Compare *meteorite* and *meteoroid*. (Ch. 8)

meteor shower A larger-than-normal display of meteors, occurring when Earth passes through the orbit of a disintegrating comet, sweeping up its debris. Compare *sporadic meteor*. (Ch. 12)

meteorite A piece of rock or other fragment of material (a meteoroid) that survives to reach a planet's surface. Compare *meteor* and *meteoroid*. (Ch. 7, 8)

meteoroid A small cometary or asteroidal fragment, ranging in size from 100 microns (μm) to 100 meters. When entering a planetary atmosphere, the meteoroid creates a *meteor*. Compare *meteor* and *meteorite*; also *planetesimal* and *zodiacal dust*. (Ch. 8)

micrometer (μm) See *micron*. (Ch. 5)

micron (μm) One-millionth (10^{-6}) of a meter; a unit of length used for the wavelength of infrared light. (Ch. 5)

microwave radiation Electromagnetic radiation with frequencies, photon energies, and wavelengths between those of infrared radiation and radio waves. (Ch. 5)

Milky Way Galaxy The galaxy in which the Sun and Solar System reside. (Ch. 1)

minor axis The short axis of an ellipse, perpendicular to the major axis. (Ch. 3)

minor planet See *asteroid*.

model A simplified mathematical or conceptual representation of a physical system used to carry out calculations or predictions. (Ch. 1)

molecular cloud An interstellar cloud composed primarily of molecular hydrogen. (Ch. 15)

molecular-cloud core A dense clump within a molecular cloud that forms as the cloud collapses and fragments. Protostars form from molecular-cloud cores. (Ch. 15)

molecule Generally, the smallest particle of a substance that retains its chemical properties and is composed of two or more atoms. (Ch. 5)

momentum The product of the mass and velocity of a particle. Possible units include kilograms times meters per second (kg m/s).

moon A less massive satellite orbiting a more massive object. Moons are found around planets, dwarf planets, asteroids, and Kuiper Belt objects. The term is usually capitalized when referring to Earth's Moon. (Ch. 7)

Mpc See *megaparsec*. (Ch. 19)

multiverse A collection of parallel universes that together make up all that exists. (Ch. 22)

mutation In biology, an imperfect reproduction of self-replicating material. (Ch. 24)

N

N See *newton*. (Ch. 3)

nadir The point on the celestial sphere located directly below an observer, opposite the *zenith*. (Ch. 2)

nanometer (nm) One-billionth (10^{-9}) of a meter; a unit of length used for the wavelength of visible light. (Ch. 5)

natural selection The process by which forms of structure, ranging from molecules to whole organisms, that are best adapted to their environment become more common than less well-adapted forms. (Ch. 24)

NCP See *north celestial pole*. (Ch. 2)

neap tide An especially weak tide that occurs around the time of the first or third quarter Moon, when the gravitational forces of the Moon and the Sun on Earth are at right angles to each other. Compare *spring tide*. (Ch. 4)

near-Earth asteroid An asteroid whose orbit brings it close to the orbit of Earth. See also *near-Earth object*. (Ch. 12)

near-Earth object (NEO) An asteroid, comet, or large meteoroid whose orbit intersects Earth's orbit. (Ch. 12)

nebula (pl. **nebulae**) A cloud of interstellar gas and dust, either illuminated by stars (for bright nebulae) or seen in silhouette against a brighter background (for dark nebulae). (Ch. 7)

nebular hypothesis The first plausible theory of the formation of the Solar System, proposed by Immanuel Kant in 1755, which stated that the Solar System formed from the collapse of an interstellar cloud of rotating gas. (Ch. 7)

NEO See *near-Earth object*. (Ch. 12)

neutrino A very low-mass, electrically neutral particle emitted during beta decay. Neutrinos interact with matter only very feebly and so can penetrate great quantities of matter. (Ch. 6, 14)

neutron capture The process in which an atomic nucleus forms a heavier nucleus after colliding with a neutron. (Ch. 17)

neutrino cooling The process in which thermal energy is carried out of the center of a star by neutrinos rather than by electromagnetic radiation or convection. (Ch. 17)

neutron A subatomic particle having no net electric charge and a rest mass and rest energy nearly equal to that of the proton. Compare *electron* and *proton*. (Ch. 5)

neutron star The neutron-degenerate stellar core left behind by a Type II supernova. (Ch. 17)

new Moon The phase of the Moon in which the Moon is between Earth and the Sun; from Earth, we see only the side of the Moon not being illuminated by the Sun. Compare *full Moon*. See also *first quarter Moon* and *third quarter Moon*. (Ch. 2)

newton (N) The force required to accelerate a 1-kilogram (kg) mass at a rate of 1 meter per second per second (m/s²). Unit: kilograms multiplied by meters per second squared (kg m/s²). (Ch. 3)

Newton's first law of motion The law, formulated by Isaac Newton, stating that an object will remain at rest or will continue moving along a straight line at a constant speed until an unbalanced force acts on it. (Ch. 3)

Newton's laws The three physical laws of motion that Isaac Newton formulated.

Newton's second law of motion The law that Isaac Newton formulated, stating that if an unbalanced force acts on a body, the body will accelerate in proportion to the unbalanced force and in inverse proportion to the object's mass: $a = F/m$. The acceleration will be in the direction of the unbalanced force. (Ch. 3)

Newton's third law of motion The law that Isaac Newton formulated, stating that for every force an equal force in the opposite direction exists. (Ch. 3)

nm See *nanometer*. (Ch. 5)

normal matter See *luminous matter*. (Ch. 19)

north celestial pole (NCP) The northward projection of Earth's rotation axis onto the celestial sphere. Compare *south celestial pole*. (Ch. 2)

North Pole The location in the Northern Hemisphere where Earth's rotation axis intersects Earth's surface. Compare *South Pole*. (Ch. 2)

nova (pl. novae) A stellar explosion that results from runaway nuclear fusion in a layer of material on the surface of a white dwarf in a binary system. (Ch. 16)

nuclear burning See *nuclear fusion*.

nuclear fusion The combination of two less massive atomic nuclei into a single more massive atomic nucleus. (Ch. 14)

nucleosynthesis The formation of more massive atomic nuclei from less massive nuclei, either in the Big Bang (Big Bang nucleosynthesis) or in the interiors of stars (stellar nucleosynthesis). (Ch. 17)

nucleus (pl. nuclei) 1. The dense, central part of an atom. 2. The central core of a galaxy, comet, or other diffuse object. (Ch. 5)

 O

objective lens The primary optical element in a telescope or camera that produces an image of an object. (Ch. 6)

oblate Pertaining to a sphere, flattened or squashed along the polar axis. (Ch. 10)

obliquity The inclination of a celestial body's equator to its orbital plane.

observable universe The part of the universe from which light has had time to reach us since shortly after the Big Bang. (Ch. 21)

observational uncertainty The fact that real measurements are never perfect; all observations are uncertain by some amount.

Occam's razor The principle that the simplest hypothesis is the most likely, named after William of Occam (circa 1285–1349), the medieval English cleric to whom the idea is attributed. (Ch. 1)

Oort Cloud A spherical distribution of comet nuclei stretching from beyond the Kuiper Belt to more than 50,000 astronomical units (AU) from the Sun. (Ch. 12)

opacity A measure of how effectively a material blocks the radiation going through it. (Ch. 14)

open cluster A loosely bound group of a few dozen to a few thousand stars that formed together in the disk of a spiral galaxy. Compare *globular cluster*. (Ch. 17)

open universe An infinite universe with a negatively curved spatial structure (much like the surface of a saddle) such that the sum of the angles of a triangle is always less than 180 degrees. Compare *closed universe* and *flat universe*. (Ch. 22)

orbit The path taken by one object moving around another object under the influence of their mutual gravitational or electric attraction. (Ch. 4)

orbital resonance A situation in which the orbital periods of two objects are related by a ratio of small integers. (Ch. 11)

orbiter A spacecraft placed in orbit around a planet or moon. Compare *flyby*. (Ch. 6)

organic Containing the element carbon. (Ch. 7)

 P

P wave See *primary wave*. (Ch. 8)

pair production The creation of a particle-antiparticle pair from a source of electromagnetic energy. (Ch. 22)

paleoclimatology The study of changes in Earth's climate throughout its history. (Ch. 9)

parallax Also called *parallactic angle*. The displacement in the apparent position of a nearby star caused by the changing location of Earth in its orbit. (Ch. 13)

parent element A radioactive element that decays to form more-stable *daughter products*. (Ch. 8)

parsec (pc) Short for *parallax second*. The distance to a star with a parallax of 1 arcsecond (arcsec) using a base of 1 astronomical unit (AU). One parsec is approximately 3.26 light-years. (Ch. 13)

partial lunar eclipse An eclipse that occurs when the Moon is partially in Earth's shadow. (Ch. 2)

partial solar eclipse The type of eclipse that occurs when Earth passes through the penumbra of the Moon's shadow, so that the Moon blocks only a portion of the Sun's disk. Compare *annular solar eclipse* and *total solar eclipse*. (Ch. 2)

pc See *parsec*. (Ch. 13)

peculiar velocity The motion of a galaxy relative to the overall expansion of the universe. (Ch. 23)

penumbra (pl. penumbrae) 1. The outer part of a shadow, where the source of light is only partially blocked. Compare *umbra* (definition 1). 2. The region surrounding the umbra of a sunspot. The penumbra is cooler and darker than the surrounding surface of the Sun but is not as cool or dark as the umbra. Compare *umbra* (definition 2). (Ch. 2, 14)

penumbral lunar eclipse A lunar eclipse in which the Moon passes through the penumbra of Earth's shadow. Compare *total lunar eclipse*. (Ch. 2)

perihelion (pl. perihelia) The point in a solar orbit that is closest to the Sun. Compare *aphelion*. (Ch. 12)

period The time that a regularly repetitive process takes to complete one cycle. (Ch. 17)

period-luminosity relationship The relationship between the period of variability of a pulsating variable star, such as a Cepheid or RR Lyrae variable, and the luminosity of the star. Longer-period pulsating variable stars are more luminous than shorter-period ones. (Ch. 17)

Perseids A prominent August meteor shower associated with the dust debris left by comet Swift-Tuttle. (Ch. 12)

phase One of the various appearances of the sunlit surface of the Moon or a planet caused by the change in viewing location of Earth relative to both the Sun and the object. Examples include crescent phase and gibbous phase. (Ch. 2)

photino A hypothetical subatomic particle related to the photon. One of the candidates for cold dark matter. (Ch. 23)

photochemical Resulting from light acting on chemical systems.

photodisintegration An atomic nucleus absorbs an energetic gamma ray and emits some particles. (Ch. 17)

photodissociation The breaking apart of molecules into smaller fragments or individual atoms by the action of photons. Compare *recombination* (definition 1). (Ch. 11)

photoelectric effect The emission of electrons from a substance illuminated by electromagnetic radiation greater than a certain critical frequency.

photon Also called *quantum of light*. A discrete unit or particle of electromagnetic radiation. The energy of a photon is equal to Planck's constant (*h*) multiplied by the frequency (*f*) of its electromagnetic radiation: $E_{photon} = h \times f$. The photon is the carrier of the electromagnetic force. (Ch. 5)

photosphere The apparent surface of the Sun as seen in visible light. Compare *chromosphere* and *corona*. (Ch. 14)

photosynthesis The process by which green plants use sunlight to create food from water and carbon dioxide. (Ch. 9)

photosynthesize The process by which plants and algae absorb sunlight, water, and carbon dioxide, and release oxygen. (Ch. 24)

physical law A broad statement that predicts a particular aspect of how the physical universe behaves and that is supported by many empirical tests. See also *theory*.

pixel The smallest picture element in a digital image array. (Ch. 6)

Planck era The early time, just after the Big Bang, for which the universe as a whole must be described with quantum mechanics. (Ch. 22)

Planck spectrum Also called *blackbody spectrum*. The spectrum of electromagnetic energy emitted by a blackbody per unit area per second, which is determined only by the temperature of the object. (Ch. 5)

Planck's constant (*h*) The constant of proportionality between the energy and the frequency of a photon. This constant defines how much energy a single photon of a given frequency or wavelength has. Value: 6.63×10^{-34} joule-second.

planet 1. A large body that orbits the Sun or other star that shines only by light reflected from the Sun or star. 2. In the Solar System, a body that orbits the Sun, has enough mass for self-gravity to overcome rigid body forces so that it assumes a spherical shape, and has cleared smaller bodies from the neighborhood around its orbit. Compare *dwarf planet*. (Ch. 7)

planet migration The theory that a planet can move to a location away from where it formed, through gravitational interactions with other bodies or loss of orbital energy from interaction with gas in the protoplanetary disk. (Ch. 7)

planetary nebula The expanding shell of material ejected by a dying asymptotic giant branch star. A planetary nebula glows from fluorescence caused by intense ultraviolet light coming from the hot, stellar remnant at its center. (Ch. 16)

planetary system A system of planets and other smaller objects in orbit around a star. (Ch. 7)

planetesimal A primitive body of rock and ice, 100 meters or more in diameter, that combines with others to form a planet. Compare *meteoroid* and *zodiacal dust*. (Ch. 7)

plasma A gas composed largely of charged particles but that also may include some neutral atoms. (Ch. 10, 14)

plate tectonics The geological theory concerning the motions of lithospheric plates, which in turn serves as the theoretical basis for *continental drift*. (Ch. 8)

positron A positively charged subatomic particle; the antiparticle of the *electron*. (Ch. 14)

power Energy per unit time. Possible units include watts (W) and joules per second (J/s). (Ch. 15)

precession of the equinoxes The slow change in orientation between the ecliptic plane and the celestial equator caused by the wobbling of Earth's axis. (Ch. 2)

pressure Force per unit area. Possible units include newtons per square meter (N/m^2) and bars.

primary atmosphere An atmosphere, composed mostly of hydrogen and helium, that forms at the same time as its host planet. Compare *secondary atmosphere*. (Ch. 7)

primary mirror The principal optical mirror in a reflecting telescope. The primary mirror determines the telescope's light-gathering power and resolution. Compare *secondary mirror*. (Ch. 6)

primary wave Also called *P wave*. A longitudinal seismic wave, in which the oscillations involve compression and decompression parallel to the direction of travel. Compare *secondary wave*. (Ch. 8)

principle A general idea or sense about the universe that guides us in constructing new scientific theories. Principles can be testable theories. (Ch. 1)

prograde motion 1. Rotational or orbital motion of a moon that is in the same direction as the planet it orbits. 2. The counterclockwise orbital motion of Solar System objects as seen from above Earth's orbital plane. Compare *retrograde motion*. (Ch. 3)

prominence An archlike projection above the solar photosphere often associated with a sunspot. (Ch. 14)

proportional See *proportionality*. (Appendix 1)

proportionality A relationship between two things whose ratio is a constant. (Appendix 1)

proton (*p* or *p*⁺) A subatomic particle having a positive electric charge of 1.6×10^{-19} coulomb (C), a rest mass of 1.67×10^{-27} kilogram (kg), and a rest energy of 1.5×10^{-10} joule (J). Compare *electron* and *neutron*. (Ch. 5)

proton-proton chain One way in which hydrogen fusion can take place. That path is the most important for hydrogen fusion in low-mass stars such as the Sun. Compare *carbon-nitrogen-oxygen cycle*. (Ch. 14)

protoplanetary disk The remains of the accretion disk around a young star from which a planetary system may form. Sometimes called *circumstellar disk*. (Ch. 7)

protostar A young stellar object that derives its luminosity from converting gravitational energy to thermal energy rather than from nuclear reactions in its core. (Ch. 7, 15)

pulsar A rapidly rotating neutron star that beams radiation into space in two searchlight-like beams. To a distant observer, the star appears to flash on and off. (Ch. 17)

pulsating variable star A variable star that undergoes periodic radial pulsations. (Ch. 17)

Q

QCD See *quantum chromodynamics*. (Ch. 22)

QED See *quantum electrodynamics*. (Ch. 22)

quantized Existing as discrete, irreducible units.

quantum chromodynamics (QCD) The quantum theory describing the strong nuclear force and its mediation by gluons. Compare *quantum electrodynamics*. (Ch. 22)

quantum efficiency The likelihood that a particular photon falling on a detector will actually produce a response in the detector. (Ch. 6)

quantum electrodynamics (QED) The quantum theory describing the electromagnetic force and its mediation by photons. Compare *quantum chromodynamics*. (Ch. 22)

quantum mechanics The branch of physics that deals with the quantized and probabilistic behavior of atoms and subatomic particles. (Ch. 5)

quantum of light See *photon*. (Ch. 5)

quark The building block of protons and neutrons. (Ch. 22)

quasar Short for *quasi-stellar radio source*. The most luminous of the active galactic nuclei, seen only at great distances from the Milky Way. (Ch. 19)

R

radial velocity (v_r) The component of velocity directed toward or away from the observer. (Ch. 5)

radian The angle at the center of a circle subtended by an arc equal to the length of the circle's radius; 2π radians equals 360°, and 1 radian equals approximately 57.3°. (Appendix 1)

radiant The point in the sky from which the meteors in a meteor shower appear to come. (Ch. 12)

radiation Waves or particles of energy traveling through space or a medium. (Ch. 14)

radiation belt A toroidal ring of high-energy particles surrounding a planet. (Ch. 9)

radiative transfer The transport of energy from one location to another by electromagnetic radiation.

radiative zone A region within a star where energy is transported outward by radiation. Compare *convective zone*. (Ch. 14)

radio galaxy A type of elliptical galaxy that has an active galactic nucleus at its center and very strong emission (10^{35} to 10^{38} watts [W]) in the radio part of the electromagnetic spectrum. Compare *Seyfert galaxy*. (Ch. 19)

radio telescope An instrument for detecting and measuring radio frequency emissions from celestial sources. (Ch. 6)

radio wave Electromagnetic radiation in the extreme long-wavelength region of the spectrum, beyond the region of microwaves. (Ch. 5)

radioisotope A radioactive element. (Ch. 8)

radiometric dating Use of the radioactive decay of elements to measure the ages of materials such as minerals. (Ch. 8)

ratio The relationship in quantity or size between two or more things.

ray 1. A beam of electromagnetic radiation. 2. A bright streak emanating from a young impact crater.

recombination 1. The combining of ions and electrons to form neutral atoms. Compare *photodissociation*. 2. An event early in the evolution of the universe in which hydrogen and helium nuclei combined with electrons to form neutral atoms. The removal of electrons caused the universe to become transparent to electromagnetic radiation. (Ch. 11, 21)

red giant A low-mass star that has evolved beyond the main sequence and is now fusing hydrogen in a shell surrounding a degenerate helium core. (Ch. 16)

red giant branch A region on the H-R diagram defined by low-mass stars evolving from the main sequence toward the horizontal branch. (Ch. 16)

reddening The effect by which stars and other objects, when viewed through interstellar dust, appear redder than they actually are. Reddening occurs because blue light is more strongly absorbed and scattered than red light. (Ch. 15)

redshift The Doppler shift toward longer (redder) wavelengths of light from an approaching object. Compare *blueshift*. (Ch. 5)

reflecting telescope A telescope that uses mirrors to collect and focus incoming electromagnetic radiation to form an image in their focal planes. The size of a reflecting telescope is defined by the diameter of the primary mirror. Compare *refracting telescope*. (Ch. 6)

reflection The redirection of a beam of light that strikes, but does not cross, the surface between two media having different refractive indices. If the surface is flat and smooth, the angle of incidence equals the angle of reflection. Compare *refraction*. (Ch. 6)

refracting telescope A telescope that uses objective lenses to collect and focus incoming electromagnetic radiation to form an image. Compare *reflecting telescope*. (Ch. 6)

refraction The redirection or bending of a beam of light when it crosses the boundary between two media having different refractive indices. Compare *reflection*. (Ch. 6)

refractory material Material that remains solid at high temperatures. Compare *volatile material*. (Ch. 7)

regular moon A moon that formed together with the planet it orbits. Compare *irregular moon*. (Ch. 11)

reionization A period after the Dark Ages during which objects formed that radiated enough energy to ionize neutral hydrogen, at redshift $6 < z < 20$. (Ch. 23)

relative humidity The amount of water vapor held by a volume of air at a given temperature compared (stated as a percentage) to the total amount of water that could be held by the same volume of air at the same temperature. (Ch. 9)

relative motion The difference in motion between two individual frames of reference. (Ch. 2)

relativistic Describing systems that travel at nearly the speed of light or are located near very strong gravitational fields. (Ch. 18)

relativistic beaming The effect created when material moving at nearly the speed of light beams the radiation it emits in the direction of its motion. (Ch. 19)

relativistic speed A speed high enough that special relativity, rather than Newtonian physics, is needed to describe the motion. Speeds greater than about 10% the speed of light are relativistic. (Ch. 18)

remote sensing The use of images, spectra, radar, or other techniques to measure the properties of an object from a distance.

resolution The ability of a telescope to separate two point sources of light. Resolution is determined by the telescope's aperture and the wavelength of light it receives. (Ch. 6)

rest wavelength The wavelength of light that is seen coming from an object at rest with respect to the observer. (Ch. 5)

retrograde motion 1. Rotation or orbital motion of a moon that is in the opposite direction to the rotation of the planet it orbits. 2. The clockwise orbital motion of Solar System objects as seen from above Earth's orbital plane. Compare *prograde motion*. 3. Apparent retrograde motion is a motion of the planets with respect to the "fixed stars," in which the planets appear to move westward for a time before resuming their normal eastward motion. (Ch. 3)

revolve Motion of one object in orbit around another. (Ch. 2)

right ascension A measure, analogous to *longitude*, that tells you the angular distance of a celestial body eastward along the celestial equator from the vernal equinox. Compare *declination*. (Ch. 2)

ring An aggregation of small particles orbiting a planet or star. The rings of the four giant planets of the Solar System are composed variously of silicates, organic materials, and ices. (Ch. 11)

ring arc A discontinuous, higher-density region within an otherwise continuous, narrow ring. (Ch. 11)

ringlet A narrowly confined concentration of ring particles. (Ch. 11)

Roche limit The distance at which a planet's tidal forces exceed the self-gravity of a smaller object—such as a moon, asteroid, or comet—causing the object to break apart. (Ch. 4)

Roche lobes The hourglass-shaped or figure eight–shaped volume of space surrounding two stars, which constrains material that is gravitationally bound by one or the other. (Ch. 16)

rotation curve A plot showing how the orbital velocity of stars and gas in a galaxy changes with radial distance from the galaxy's center. (Ch. 19)

rover A remotely controlled instrumented vehicle designed to move and explore the surface of a terrestrial planet or moon. Compare *lander*. (Ch. 6)

RR Lyrae variable A variable giant star whose regularly timed pulsations are good predictors of its luminosity. RR Lyrae variables are used in measuring the distance to globular clusters. (Ch. 17)

S

S-type asteroid An asteroid made of material that was once part of the outer layer of a larger, differentiated body that has since broken into pieces. Compare *C-type asteroid* and *M-type asteroid*. (Ch. 12)

S wave See *secondary wave*. (Ch. 8)

S0 galaxy A galaxy with a bulge and a disk-like spiral, but smooth in appearance like ellipticals. Compare *barred spiral galaxy*, *elliptical galaxy*, *irregular galaxy*, and *spiral galaxy*. (Ch. 19)

satellite An object in orbit around a more massive body, for example, a moon of any planet or an artificial satellite. (Ch. 4)

scale factor (R_U) A dimensionless number proportional to the distance between two points in space. The scale factor increases as the universe expands. (Ch. 21)

scattering The random change in the direction of travel of photons, caused by their interactions with molecules or dust particles. (Ch. 9)

Schwarzschild radius The distance from the center of a nonrotating, spherical black hole at which the escape velocity equals the speed of light. (Ch. 18)

scientific method The formal procedure—including hypothesis, prediction, and experiment or observation—used to test (that is, to attempt to falsify) the validity of scientific hypotheses and theories. (Ch. 1)

scientific notation The standard expression of numbers with one digit (which can be zero) to the left of the decimal point and multiplied by 10 to the exponent required to give the number its correct value. Example: $2.99 \times 10^8 = 299,000,000$. (1, Appendix 1)

SCP See *south celestial pole*. (Ch. 2)

second law of thermodynamics The law stating that the entropy or disorder of an isolated system always increases as the system evolves.

secondary atmosphere An atmosphere that forms—as a result of volcanism, comet impacts, or another process—sometime after its host planet has formed. Compare *primary atmosphere*. (Ch. 7)

secondary crater A crater formed from ejecta thrown from an *impact crater*. (Ch. 8)

secondary mirror A small mirror placed on the optical axis of a reflecting telescope that returns the beam back through a small hole in the *primary mirror*, thereby shortening the mechanical length of the telescope. (Ch. 6)

secondary wave Also called *S wave*. A transverse seismic wave, which involves the sideways motion of material. Compare *primary wave*. (Ch. 8)

sediment Solid material such as rock or sand that has been affected by erosion and weather, and transported by water, wind, or ice. (Ch. 8)

seismic wave A vibration due to an earthquake, a large explosion, or an impact on the surface that travels through a planet's interior. (Ch. 8)

seismometer An instrument that measures the amplitude and frequency of seismic waves. (Ch. 8)

self-gravity The gravitational attraction among all parts of the same object. (Ch. 4)

semimajor axis Half of the longer axis of an ellipse. (Ch. 3)

SETI The Search for Extraterrestrial Intelligence project, which uses advanced technology combined with radio telescopes to search for evidence of intelligent life elsewhere in the universe. (Ch. 24)

Seyfert galaxy A type of spiral galaxy with an active galactic nucleus at its center; first discovered in 1943 by Carl Seyfert. Compare *radio galaxy*. (Ch. 19)

shepherd moon A moon that orbits close to rings and gravitationally confines the orbits of the ring particles. (Ch. 11)

shield volcano A volcano formed by very fluid lava flowing from a single source and spreading out from that source. Compare *composite volcano*. (Ch. 8)

short gamma-ray burst Bursts of gamma rays lasting a few milliseconds to a few minutes, thought to originate in neutron star mergers. (Ch. 17)

short-period comet A comet with an orbital period of less than 200 years. Compare *long-period comet*. (Ch. 12)

sidereal day Earth's period of rotation with respect to the stars—about 23 hours 56 minutes—which is the time Earth takes to make one rotation and face the same star on the meridian. It differs from the *solar day* because of Earth's motion around the Sun. (Ch. 2)

sidereal period An object's orbital or rotational period measured with respect to the stars. Compare *synodic period*. (Ch. 2)

silicate One of the family of minerals composed of silicon and oxygen in combination with other elements. (Ch. 7)

singularity The point at which a mathematical expression or equation becomes meaningless, such as a fraction whose denominator approaches zero See also *black hole*. (Ch. 18)

solar abundance The relative amount of an element detected in the atmosphere of the Sun, expressed as the ratio of the number of atoms of that element to the number of hydrogen atoms. (Ch. 5)

solar day The day in common use—24 hours, which is Earth's period of rotation that brings the Sun back to the same local meridian where the rotation started. Compare *sidereal day*. (Ch. 2)

solar eclipse An eclipse that occurs when the Moon partially or entirely blocks the Sun. Compare *lunar eclipse*. (Ch. 2)

solar flare An explosion on the Sun's surface associated with a complex sunspot group and a strong magnetic field. (Ch. 14)

solar maximum (pl. **maxima**) The time, occurring about every 11 years, when the Sun is at its peak activity, meaning that sunspot activity and related phenomena (such as prominences, flares, and coronal mass ejections) are at their peak. (Ch. 14)

solar neutrino problem The historical observation that only about a third as many neutrinos as predicted by theory seemed to be coming from the Sun. (Ch. 14)

Solar System The gravitationally bound system made up of the Sun, planets, dwarf planets, moons, asteroids, comets, and Kuiper Belt objects, along with their associated gas and dust. (Ch. 1)

solar tide A tide on Earth caused by the differential gravitational pull of the Sun. Compare *lunar tide*. (Ch. 4)

solar wind The stream of charged particles emitted by the Sun and that flows at high speeds through interplanetary space. (Ch. 9)

solstice Literally, "Sun standing still." 1. One of the two most northerly and southerly points on the ecliptic. 2. Either of the two times of year (the *summer solstice* and *winter solstice*) when the Sun is at one of these two positions. Compare *equinox*.

south celestial pole (SCP) The southward projection of Earth's rotation axis onto the celestial sphere. Compare *north celestial pole*. (Ch. 2)

South Pole The location in the Southern Hemisphere where Earth's rotation axis intersects Earth's surface. Compare *North Pole*. (Ch. 2)

spacetime A concept that combines space and time into a four-dimensional continuum with three spatial dimensions plus one dimension of time. (Ch. 18)

special relativity See *special theory of relativity*. (Ch. 18)

special theory of relativity Sometimes referred to as simply *special relativity*. Einstein's theory explaining how the fact that the speed of light is a constant affects nonaccelerating frames of reference. Compare *general theory of relativity*. (Ch. 18)

spectral type A classification system for stars that is based on the presence and relative strength of absorption lines in their spectra. Spectral type is related to the surface temperature of a star. (Ch. 13)

spectrograph Also called *spectrometer*. A device that spreads out the light from an object into its component wavelengths. (Ch. 6)

spectrometer See *spectrograph*. (Ch. 6)

spectroscopic binary A binary star system whose existence and properties are revealed to astronomers only by the Doppler shift of its spectral lines. Most spectroscopic binaries are close pairs. Compare *eclipsing binary* and *visual binary*. (Ch. 13)

spectroscopic parallax Use of the spectroscopically determined luminosity and the observed brightness of a star to determine the star's distance. (Ch. 13)

spectroscopy The study of an object's electromagnetic radiation in terms of its component wavelengths. (Ch. 6)

spectrum (pl. **spectra**) Waves sorted by wavelength. See also *electromagnetic spectrum*. (Ch. 5)

speed The rate of change of an object's position with time, without regard to the direction of movement. Possible units include meters per second (m/s) and kilometers per hour (km/h). Compare *velocity*. (Ch. 3)

spherically symmetric Describing an object whose properties depend only on distance from the object's center, so that the object has the same form viewed from any direction. (Ch. 4)

spin-orbit resonance A relationship between the orbital and rotation periods of an object such that the ratio of their periods can be expressed with simple integers.

spiral density wave A stable, spiral-shaped change in the local gravity of a galactic disk that can be produced by periodic gravitational kicks from neighboring galaxies or from nonspherical bulges and bars in spiral galaxies. (Ch. 20)

spiral galaxy A galaxy of Hubble type "S" class, with a discernible disk in which large spiral patterns exist. Compare *barred spiral galaxy*, *elliptical galaxy*, *irregular galaxy*, and *S0 galaxy*. (Ch. 19)

spoke One of several narrow radial features seen occasionally in Saturn's B Ring. Spokes appear dark in backscattered light and bright in forward, scattering light, indicating that they are composed of tiny particles. Their origin is not well understood. (Ch. 11)

spreading center A zone from which two tectonic plates diverge. (Ch. 8)

spring tide An especially strong tide that occurs around the time of a new or full Moon, when lunar tides and solar tides reinforce each other. Compare *neap tide*. (Ch. 4)

sputtering Collision of fast-moving ions with a surface or an atmosphere that causes the loss of material from the surface or atmosphere. (Ch. 9)

stable equilibrium An equilibrium state in which the system returns to its former condition after a small disturbance. Compare *unstable equilibrium*.

standard candle An object whose luminosity either is known or can be predicted in a distance-independent way, so its brightness can be used to determine its distance via the inverse square law of radiation. (Ch. 19)

standard model The theory of particle physics that combines electroweak theory with quantum chromodynamics to describe the structure of known forms of matter. (Ch. 22)

star A luminous ball of gas held together by gravity. A normal star is powered by nuclear reactions in its interior.

star cluster A group of stars that all formed at the same time and in the same general location. (Ch. 15)

static equilibrium A state in which the forces within a system are all in balance so that the system does not change. Compare *dynamic equilibrium*.

Stefan-Boltzmann constant (σ) The constant of proportionality that relates the flux emitted by an object to the fourth power of its absolute temperature. Value: 5.67×10^{-8} W/(m^2 K^4) (W = watts, m = meters, K = kelvins). (Ch. 5)

Stefan-Boltzmann law The law, formulated by Josef Stefan and Ludwig Boltzmann, stating that the amount of electromagnetic energy emitted from the surface of a body (flux), summed over the energies of all photons of all wavelengths emitted, is proportional to the fourth power of the temperature of the body: $\mathcal{F} = \sigma T^4$. (Ch. 5)

stellar evolution The stages of development that a star goes through over its life cycle. Astronomers observe many stars at the same time to build a picture of how stars change. (Ch. 16)

stellar-mass loss The loss of mass from the outermost parts of a star's atmosphere during the star's evolution. (Ch. 16)

stellar occultation An event in which a planet or other Solar System body moves between the observer and a star, eclipsing the light emitted by that star. (Ch. 10)

stellar population A group of stars with similar ages, chemical compositions, and dynamic properties. (Ch. 17)

stereoscopic vision The way an animal's brain combines the different information from its two eyes to perceive the distances to objects around it. (Ch. 13)

stony-iron meteorite A meteorite consisting of a mixture of silicate minerals and iron-nickel alloys. Compare *iron meteorite* and *stony meteorite*. (Ch. 12)

stony meteorite A meteorite composed primarily of silicate minerals, similar to those found on Earth. Compare *iron meteorite* and *stony-iron meteorite*. (Ch. 12)

stratosphere The atmospheric layer immediately above the troposphere. On Earth, it extends upward to an altitude of 50 kilometers (km). Compare *troposphere*, *mesosphere*, and *thermosphere*. (Ch. 9)

string theory See *superstring theory*. (Ch. 22)

stromatolite A structure created by living or fossilized cyanobacteria. (Ch. 24)

strong nuclear force The attractive short-range force between protons and neutrons that holds atomic nuclei together. One of the four fundamental forces of nature, along with the *electromagnetic force*, the *weak nuclear force*, and *gravity* (definition 1). The force is mediated by the exchange of gluons. (Ch. 14, 22)

subduction zone A region where two tectonic plates converge, with one plate sliding under the other and being drawn downward into the interior. (Ch. 8)

subgiant A giant star that is smaller and lower in luminosity than normal giant stars of the same spectral type. Subgiants evolve to become giants. (Ch. 16)

subgiant branch A region of the H-R diagram defined by stars that have left the main sequence but have not yet reached the red giant branch. (Ch. 16)

sublimation The process by which a solid becomes a gas without first becoming a liquid. (Ch. 12)

summer solstice 1. One of two points where the Sun is at its greatest distance from the celestial equator. 2. The day on which the Sun appears at that location, marking the first day of summer (about June 20 in the Northern Hemisphere and December 21 in the Southern Hemisphere). Compare *winter solstice*. See also *autumnal equinox* and *vernal equinox*. (Ch. 2)

Sun The star at the center of the Solar System. (Ch. 1)

sungrazer A comet whose perihelion is within a few solar diameters of the surface of the Sun. (Ch. 12)

sunspot A cooler, transitory region on the solar surface produced when loops of magnetic flux break through the surface of the Sun. (Ch. 14)

sunspot cycle The approximate 11-year cycle during which sunspot activity increases and then decreases. This is one-half of a full 22-year cycle, in which the magnetic polarity of the Sun first reverses and then returns to its original configuration. (Ch. 14)

supercluster A large conglomeration of galaxy clusters and galaxy groups, typically more than 100–300 megaparsecs (Mpc) in size and containing tens of thousands to hundreds of thousands of galaxies. Compare *galaxy cluster* and *galaxy group*. (Ch. 1, 23)

super-Earth An extrasolar planet with about 2–10 times the mass of Earth. (Ch. 7)

superior planet A Solar System planet that orbits farther from the Sun than Earth does. Compare *inferior planet*. (Ch. 3)

supermassive black hole A black hole of 1,000 solar masses (M_{Sun}) or more that resides in the center of a galaxy and whose gravity powers active galactic nuclei. (Ch. 19)

supernova (pl. **supernovae**) A stellar explosion resulting in the release of tremendous amounts of energy, including the high-speed ejection of matter into the interstellar medium. See also *Type Ia supernova* and *Type II supernova*.

supernova remnant The material ejected from the outer layers of a star after a supernova explosion. (Ch. 16)

supersonic Moving within a medium at a speed faster than the speed of sound in that medium. Compare *subsonic*.

superstring theory The theory that conceives of particles as strings in 10 dimensions of space and time; the current contender for a theory of everything. (Ch. 22)

surface brightness The amount of electromagnetic radiation emitted or reflected per unit area. (Ch. 6, 19)

surface wave A seismic wave that travels on the surface of a planet or moon. (Ch. 8)

surface gravity The acceleration (downward) due to gravity measured at the surface of a planet or star. (Ch. 9)

symmetry In theoretical physics, the properties of physical laws that remain constant when certain things change, such as the symmetry between matter and antimatter even though their charges may be different.

synchronous rotation The case that occurs when a body's rotation period equals its orbital period around another body. A special type of spin-orbit resonance. (Ch. 2)

synchrotron radiation Radiation from electrons moving at close to the speed of light as they spiral in a strong magnetic field, so named because that kind of radiation was first identified on Earth in particle accelerators called synchrotrons. (Ch. 10)

synodic period An object's orbital or rotational period measured with respect to the Sun. Compare *sidereal period*. (Ch. 2)

T

T Tauri star A young stellar object that has dispersed enough of the material surrounding it to be seen in visible light. (Ch. 15)

tail A stream of gas and dust swept away from the coma of a comet by the solar wind and by radiation pressure from the Sun. See also *ion tail* and *dust tail*. (Ch. 12)

tectonism Deformation of the lithosphere of a planet. (Ch. 8)

telescope The basic tool of astronomers. Working over the entire range from gamma rays to radio waves, astronomical telescopes collect and concentrate electromagnetic radiation from celestial objects. (Ch. 6)

temperature A measure of the average kinetic energy of the atoms or molecules in a gas, solid, or liquid. (Ch. 5)

terrestrial planet An Earth-like planet, made of rock and metal and having a solid surface. In the Solar System, the terrestrial planets are Mercury, Venus, Earth, and Mars. Compare *giant planet*. (Ch. 7)

theoretical model A description of the properties of a particular object or system in terms of known physical laws or theories; often, a computer calculation of predicted properties based on such a description.

theory A well-developed idea or group of ideas that are tied solidly to known physical laws and make testable predictions about the world. A very well-tested theory may be called a *physical law*, or simply a *fact*. Compare *hypothesis*. (Ch. 1)

theory of everything (TOE) A theory that unifies all four fundamental forces of nature: strong nuclear, weak nuclear, electromagnetic, and gravitational forces. (Ch. 22)

thermal conduction See *conduction*. (Ch. 14)

thermal energy The energy that resides in the random motion of atoms, molecules, and particles, by which we measure their temperature. (Ch. 5, 7)

thermal equilibrium The state in which the rate of thermal-energy emission by an object is equal to the rate of thermal-energy absorption. (Ch. 5)

thermal motion The random motion of atoms, molecules, and particles that gives rise to thermal radiation. (Ch. 5)

thermal radiation Electromagnetic radiation resulting from the random motion of the charged particles in every substance. (Ch. 5)

thermosphere The layer of Earth's atmosphere at altitudes greater than 90 kilometers (km), above the mesosphere. Near its top, at an altitude of 600 km, the temperature can reach 1000 K. Compare *troposphere*, *stratosphere*, and *mesosphere*. (Ch. 9)

thick disk Extends above and below the thin disk of a spiral galaxy, and contains the older population of disk stars, distinguishable by lower abundances of massive elements. (Ch. 20)

thin disk Central layer of the disk of a spiral galaxy. The youngest stars are concentrated in the thin disk because that is where the molecular clouds and gas are most concentrated. (Ch. 20)

third quarter Moon The phase of the Moon in which only the eastern half of the Moon, as viewed from Earth, is illuminated by the Sun. It occurs about one week after the full Moon. Compare *first quarter Moon*. See also *full Moon* and *new Moon*. (Ch. 2)

tidal bulge Distortion of a body resulting from tidal stresses. (Ch. 4)

tidal force A force caused by the change in the strength of gravity across an object. (Ch. 4)

tidal locking Synchronous rotation of an object caused by internal friction as the object rotates through its tidal bulge. (Ch. 4)

tide On Earth, the rise and fall of the oceans as Earth rotates through a tidal bulge caused by the Moon and the Sun. See also *lunar tide*, *neap tide*, *solar tide*, and *spring tide*. (Ch. 4)

time dilation The relativistic "stretching" of time. Compare *general relativistic time dilation*. (Ch. 18)

TOE See *theory of everything*. (Ch. 22)

topography The study of the natural physical features of the surface of a planet or moon. (Ch. 8)

tornado A violent rotating column of air, typically 75 meters across with 200-kilometer-per-hour (km/h) winds. Some tornadoes can be more than 3 km across, and winds up to 500 km/h have been observed. (Ch. 9)

torus (pl. **tori**) A three-dimensional, doughnut-shaped ring. (Ch. 10, 19)

total lunar eclipse A lunar eclipse in which the Moon passes through the umbra of Earth's shadow. Compare *penumbral lunar eclipse*. (Ch. 2)

total solar eclipse The type of eclipse that occurs when Earth passes through the umbra of the Moon's shadow, so that the Moon blocks the disk of the Sun. Compare *annular solar eclipse* and *partial solar eclipse*. (Ch. 2)

transit method A method of detecting extrasolar planets by measuring the decrease in light from a star as its orbiting planet passes in front of the star as viewed from Earth. (Ch. 7)

trans-Neptunian object See *Kuiper Belt object*.

transverse wave A wave that oscillates perpendicular to the direction of the wave's propagation. Compare *longitudinal wave*. (Ch. 8)

triple-alpha process The nuclear fusion reaction that combines three ^{4}He nuclei (alpha particles) into a single nucleus of carbon. (Ch. 16)

Trojans A group of asteroids orbiting in the L_4 and L_5 Lagrangian points of Jupiter's orbit. (Ch. 12)

tropical year The time between two crossings of the vernal equinox. Because of the precession of the equinoxes, a tropical year is slightly shorter than the time Earth takes to orbit once around the Sun. Compare *year*. (Ch. 2)

Tropics The region on Earth between latitudes 23.5° south and 23.5° north, where the Sun appears directly overhead twice during the year. (Ch. 2)

tropopause The top of a planet's troposphere. (Ch. 9)

troposphere The convection-dominated layer of a planet's atmosphere. On Earth, the atmospheric region closest to the ground within which most weather phenomena take place. Compare *stratosphere, mesosphere*, and *thermosphere*. (Ch. 9)

turbulence The random motion of blobs of gas within a larger cloud of gas.

Type Ia supernova A supernova explosion with a calibrated peak luminosity that occurs as a result of runaway carbon fusion in a white dwarf star that accretes mass from a companion and approaches the Chandrasekhar mass limit of 1.4 M_{Sun}. (Ch. 16)

Type II supernova A supernova explosion in which the degenerate core of an evolved massive star suddenly collapses and rebounds. (Ch. 17)

U

ultrafaint dwarf galaxy A dim dwarf galaxy with only 1,000–100,000 times the Sun's luminosity. Ultrafaint dwarf galaxies differ from globular clusters in that they are composed of large amounts of dark matter. (Ch. 23)

ultraviolet (UV) radiation Electromagnetic radiation having frequencies and photon energies greater than those of visible light but less than those of X-rays and having wavelengths shorter than those of visible light but longer than those of X-rays. (Ch. 5)

umbra (pl. **umbrae**) 1. The darkest part of a shadow, where the source of light is blocked. Compare *penumbra* (definition 1). 2. The darkest, innermost part of a sunspot. Compare *penumbra* (definition 2). (Ch. 2, 14)

unbound orbit An orbit in which an object is no longer gravitationally bound to the body it was orbiting. An unbound orbit's velocity is greater than the escape velocity. Compare *bound orbit*. (Ch. 4)

uncertainty principle The physical limitation that the product of the position and the momentum of a particle cannot be smaller than a well-defined value, Planck's constant (h). (Ch. 22)

unified model of AGN A model in which many types of activity in the nuclei of galaxies are all explained by accretion of matter around a supermassive black hole. (Ch. 19)

uniform circular motion Motion in a circular path at a constant speed. (Ch. 4)

unit A fundamental quantity of measurement. The meter is an example of a metric unit; the foot is an example of an English unit. (Appendix 1)

universal gravitational constant (G) The constant of proportionality in the universal law of gravitation. Value: 6.67×10^{-11} meters cubed per kilogram second squared [m^3/kg s^2 = N m^2/kg^2]. (Ch. 4)

universal law of gravitation The law, formulated by Isaac Newton, stating that the gravitational force between any two objects is proportional to the product of their masses and inversely proportional to the square of the distance between them: $F_{grav} = G \times \dfrac{m_1 \times m_2}{r^2}$. (Ch. 4)

universe 1. All of space and everything contained therein. 2. Our own universe in a collection of parallel universes that together make up all that exists.

unstable equilibrium An equilibrium state in which a small disturbance will cause a system to move away from equilibrium. Compare *stable equilibrium*.

UV Ultraviolet. See *ultraviolet radiation*. (Ch. 5)

V

vacuum A region of space devoid of matter. In quantum mechanics and general relativity, however, even a perfect vacuum has physical properties. (Ch. 5, 22)

variable star A star with varying luminosity. Many periodic variables are found within the instability strip on the H-R diagram. (Ch. 17)

velocity (*v*) The rate and direction of change of an object's position with time. Possible units include meters per second (m/s) and kilometers per hour (km/h). Compare *speed*. (Ch. 3)

vernal equinox 1. One of two points where the Sun crosses the celestial equator. 2. The day on which the Sun appears at that location, marking the first day of spring (about March 20 in the Northern Hemisphere and September 22 in the Southern Hemisphere). Compare *autumnal equinox*. See also *summer solstice* and *winter solstice*. (Ch. 2)

very low-mass star A star with mass between 0.08 and 0.5 M_{Sun}. (Ch. 16)

virtual particle A particle that, according to quantum mechanics, comes into existence only momentarily. According to theory, fundamental forces are mediated by the exchange of virtual particles. (Ch. 18)

visual binary A binary system in which the two stars can be seen individually from Earth. Compare *eclipsing binary* and *spectroscopic binary*. (Ch. 13)

void A region in space containing little or no matter. Examples include regions in cosmological space that are largely empty of galaxies. (Ch. 23)

volatile material Generally called *ice* in its solid form. Material that remains gaseous at moderate temperature. Compare *refractory material*. (Ch. 7)

volcanism A form of geological activity on a planet or moon in which molten rock (magma) erupts onto the surface. (Ch. 8)

W

W See *watt*.

waning Describing the changing phases of the Moon as it becomes less fully illuminated between full Moon and new Moon as seen from Earth. Compare *waxing*. (Ch. 2)

waning crescent Moon The phases of the Moon between third quarter and new Moon. (Ch. 2)

waning gibbous Moon The phases of the moon between full Moon and third quarter. (Ch. 2)

water cycle The flow of water on, above, and through Earth's surface. (Ch. 9)

watt (W) A measure of *power*. Unit: joules per second (J/s).

wave A disturbance moving along a surface or passing through a space or a medium. (Ch. 5)

wavefront The imaginary surface of an electromagnetic wave, either plane or spherical, oriented perpendicular to the direction of travel.

wavelength (λ) The distance on a wave between two adjacent points having identical characteristics. The distance a wave travels in one period. Possible units include meters (m). (Ch. 5)

waxing Describing the changing phases of the Moon as it becomes more fully illuminated between new Moon and full Moon as seen from Earth. Compare *waning*. (Ch. 2)

waxing crescent Moon The phases of the Moon between new and first quarter. (Ch. 2)

waxing gibbous Moon The phases of the Moon between first quarter and full. (Ch. 2)

weak nuclear force The force underlying some forms of radioactivity and certain interactions between subatomic particles. It is responsible for radioactive beta decay and for the initial proton-proton interactions that lead to nuclear fusion in the Sun and other stars. One of the four fundamental forces of nature, along with the *electromagnetic force*, the *strong nuclear force*, and *gravity* (definition 1). The force is mediated by the exchange of *W* and *Z* particles. (Ch. 22)

weather The state of an atmosphere at any given time and place. Compare *climate*. (Ch. 9)

weight The gravitational force acting on an object, that is, the force equal to the mass of an object multiplied by the local acceleration due to gravity. In general relativity, the force equal to the mass of an object multiplied by the acceleration of the frame of reference in which the object is observed. Compare *mass*. (Ch. 4)

white dwarf The stellar remnant left at the end of the evolution of a low-mass star. A typical white dwarf has a mass of 0.6 solar mass (M_{Sun}) and a size about equal to that of Earth; it is made of non-fusing, electron-degenerate carbon. (Ch. 16)

Wien's law A law, named for Wilhelm Wien, stating that location of the peak wavelength in the electromagnetic spectrum of an object is inversely proportional to the temperature of the object. (Ch. 5)

WIMP Short for *weakly interacting massive particle*. A hypothetical massive particle that interacts through the weak nuclear force and gravity but not with electromagnetic radiation. WIMPs are candidates for dark matter. Compare *MaCHO*. (Ch. 19)

winter solstice 1. One of two points where the Sun is at its greatest distance from the celestial equator. 2. The day on which the Sun appears at that location, marking the first day of winter (about December 21 in the Northern Hemisphere and June 20 in the Southern Hemisphere). Compare *summer solstice*. See also *autumnal equinox* and *vernal equinox*. (Ch. 2)

X

X-ray Electromagnetic radiation having frequencies and photon energies greater than those of ultraviolet (UV) light but less than those of gamma rays and having wavelengths shorter than those of UV light but longer than those of gamma rays. (Ch. 5)

X-ray binary A binary system in which mass from an evolving star spills over onto a collapsed companion, such as a neutron star or black hole. The material falling in is heated to such high temperatures that it glows brightly in X-rays. (Ch. 17)

Y

year The time Earth takes to make one revolution around the Sun. A solar year is measured from equinox to equinox. A sidereal year, Earth's true orbital period, is measured relative to the stars. Compare *tropical year*. (Ch. 2)

Z

zenith The point on the celestial sphere located directly overhead of an observer. Compare *nadir*. (Ch. 2)

zero-age main sequence The strip on the H-R diagram plotting where stars of all masses in a cluster begin their lives.

zodiac The 12 constellations lying along the plane of the ecliptic. (Ch. 2)

zodiacal dust Particles of cometary and asteroidal debris smaller than 100 microns (μm) that orbit the inner Solar System close to the plane of the ecliptic. Compare *meteoroid* and *planetesimal*. (Ch. 12)

zodiacal light A band of light in the night sky caused by sunlight reflected by zodiacal dust. (Ch. 12)

zonal wind The east–west component of a wind. (Ch. 9)

These pages contain selected answers for all 24 chapters in the complete volume of *21st Century Astronomy, Seventh Edition*. The Solar System Edition does not include Chapters 15–23. The Stars and Galaxies Edition does not include Chapters 8–12.

Chapter 1

CHECK YOUR UNDERSTANDING

1.1. d, a, e, c, f, b
1.2. b
1.3. c

WHAT IF . . .

p. 4. If our Solar System were located in the Andromeda Galaxy our cosmic address would be the same except for replacing Milky Way with Andromeda.

p. 9. A theory that is not likely to be falsifiable within many human lifetimes can still be a scientific theory if our knowledge and technology need those lifetimes to falsify it.

p. 11. If there were a region in the universe where the laws of physics were different than those on Earth, we would call into question the cosmological principle and begin a search for other regions that might also be different.

p. 17. If Earth formed a mere 4 billion years after the Big Bang, it would probably not have the abundances of elements that are formed in stars that it has today.

TEST YOUR UNDERSTANDING

3. b
5. d
9. d
12. c
15. b, d, a, c, e

THINKING ABOUT THE CONCEPTS

16. Figure 1.1 places each object in reference to its location: Earth in the Solar System, Solar System in the Milky Way, Milky Way in the Local Group, Local Group in the Virgo Supercluster, and our supercluster in Laniakea. Figure 1.3 directly relates difference sizes of these objects to an equivalent time—the longer the time, the larger the object. The sizes of the distance steps in Figure 1.1 are greater than those of Figure 1.3. Figure 1.1 is simply showing us our cosmic address.

20. 2.5 million years

23. A scientific theory has a more defined, stringent definition than the everyday use of "theory." A scientific theory puts together a well-developed idea that must agree with known physical laws and accounts for all of the relevant data leading to its development. It must make predictions that are testable and must stand up to those tests. Everyday usage is more fluent, meaning "theory" can be an idea or applied to something tossed aside as unimportant.

30. Our composition is the same as what remains after the death of a star. The current theory of how stars are born, live, and die—and observations of stellar remnants—support this statement.

APPLYING THE CONCEPTS

32. (a) Hours are units of time. (b) The units would be km^2/h, which are not units of time. (c) Division of speed by distance gives $\frac{km/h}{km} = \frac{1}{h}$, which is not units of time. (d) Dividing the distance, km, by the speed, km/h, will give units of time. (e) $\frac{100\,km}{30\,km/h} = 3.3\,h$.

45. (a) 9.42×10^8 km. (b) 1.1×10^5 km/hr. (c) 2.6×10^6 km/dy.

Chapter 2

CHECK YOUR UNDERSTANDING

2.1a. b
2.1b. d
2.2. d
2.3. full moon; third quarter
2.4. longer
2.5. b

WHAT IF . . .

p. 30. If Earth did not rotate with respect to the stars (it is not in a synchronous orbit), then a day would last a year. Any given star would take half a year to rise and set. The seasons would be more extreme.

p. 37. With a slightly larger tilt of its axis, the seasons on Mars would have larger variations. This is because the tilt toward and away from the Sun is greater.

p. 43. If the Moon were a cube, it would still go through phases, but the shape of the Moon would depend on the orientation of the cube.

p. 45. The dates of the "wandering" holidays would not change.

p. 51. There would be more eclipses. With the two moons being 120° apart in their orbit, it would take about 10 days after the first eclipse for the second moon to pass through the node. The eclipses would occur in the same eclipse season.

TEST YOUR UNDERSTANDING

3. b
6. e
10. a
14. b
15. a

THINKING ABOUT THE CONCEPTS

18. From Figure 2.13 we can see that the constellation Gemini is in the sky during the daytime.

25. (a) Earth would always remain in the same position in the sky. (b) The phases of Earth would be opposite the current phase of the Moon as seen from Earth.

29. We can determine that Earth is curved, no matter what the location or time of the observations; that is, it is curved "all the way around," making it a sphere.

APPLYING THE CONCEPTS

37. 50° for both the altitude of Polaris and the latitude

42. (a) 66.5° N. (b) summer solstice.

Chapter 3

CHECK YOUR UNDERSTANDING

3.1a. While the stars always remained in the same location, the planets moved against those stars.
3.1b. c
3.2. c, b, d, a
3.3. d
3.4. b

WHAT IF . . .

p. 62. Because the astronauts would be orbiting Earth faster than the Moon, they would observe retrograde motion of the Moon.

p. 68. The Northern Hemisphere winters would be much warmer and the Southern Hemisphere summers much hotter. When Earth was at its farthest, the Northern Hemisphere summers would be much colder and Southern Hemisphere winters cooler for a much longer period of time.

p. 71. Given $A = 2$ AU, $P = 4$ years, $P^2 = A^3$; $4^2 \neq 2^3$.

p. 73. One might hypothesize that the relationship is different for moons orbiting planets versus planets orbiting the Sun. Kepler applied his third law to the planets.

TEST YOUR UNDERSTANDING

3. a
8. d
12. c
15. c

THINKING ABOUT THE CONCEPTS

16. Mars is larger at that time because it is closer to Earth. The changing size of Venus is due to its changing distance from Earth.

22. The Moon's orbital eccentricity of 0.05 is greater than Earth's, which is 0.017. Since Earth's orbit is very close to a circle, whether there is a total

eclipse or an annular eclipse of the Sun will depend on the Moon's distance from Earth.

27. We wear seat belts to protect us from acceleration. Due to our inertia, we resist a change in our state and if not held by a seat belt, we would react to the force being applied.

APPLYING THE CONCEPTS

33. (a) Because it is closer to the Sun, Venus's sidereal orbital period will be slightly shorter than Earth's. (b) $P = 0.62$ yr. (c) The answer agrees with the prediction. (d) $P = 1.9$ yr.

37. It is approximately three times larger at the thin crescent versus the gibbous phase. The ratio of distances gives $\dfrac{1.72\,\text{AU}}{0.28\,\text{AU}} = 6$.

Chapter 4

CHECK YOUR UNDERSTANDING

4.1a. d
4.1b. c, a
4.2. c, a, b
4.3. d, b, a, c
4.4. a

WHAT IF . . .

p. 86. If the gravitational force did not depend on an object's mass, then we could assume the force would be the same for all objects. More massive objects would still have more inertia, would resist acceleration, and thus would fall more slowly.

p. 89. Assuming you did not reach terminal velocity, you would free-fall like an astronaut in orbit. Astronauts train for weightless conditions in deep pools.

p. 100. As the Moon's orbital radius shrank to half, the tidal forces would increase by a factor of eight. If it shrank to one-third, the tidal forces would increase by a factor of 27. Life on Earth's surface would need to move to higher ground and ocean life would have to deal with very strong currents during the extremes of the changing tides.

p. 102. If the Moon were orbiting as close as two Earth radii, then the probability that the two worlds would be tidally locked (like Pluto and Charon) would greatly increase.

TEST YOUR UNDERSTANDING

3. c
8. a
12. d
14. d

THINKING ABOUT THE CONCEPTS

20. It would be unbound.
25. Because Earth rotates from west to east faster than the Moon orbits, the high tide gets dragged away from the location of the Moon.
30. The tidal tail shows the trajectory of the galaxy on the right and indicates that that galaxy has

entered the Roche limit of the other galaxy and the two will eventually collide.

APPLYING THE CONCEPTS

38. $\dfrac{F_{\text{ground}}}{F_{\text{space}}} = 1.11$

39. 7698 m/s, or a little over 17,000 mph

Chapter 5

CHECK YOUR UNDERSTANDING

5.1a. e, c, b, d, a
5.1b. c
5.2. Each element has a unique pattern of spectral lines.
5.3. d
5.4. d, b, a, c, e
5.5. c

WHAT IF . . .

p. 112. Assuming the medium was the same all through space, the speed of light would not change.

p. 114. If we could see at infrared wavelengths, we would be able to tell the temperatures of objects and see them in the dark. Our lives would not change if we had X-ray vision because very few objects on Earth emit that radiation and our atmosphere blocks X-rays from space.

p. 126. The Sun is almost all hydrogen and helium. There would not be enough carbon, nitrogen, or oxygen for life to evolve.

p. 127. If an ambulance exceeded the speed of sound, we would see it go by first and then hear it.

p. 132. If a planet emitted less radiation as it heated up, and if its star kept heating it, then eventually the planet would reach very high temperatures and never reach equilibrium.

p. 135. The length of time Earth would take to reach equilibrium would depend on its radius and other factors such as ocean and mantle depth.

TEST YOUR UNDERSTANDING

2. b
5. b
11. c
13. d

THINKING ABOUT THE CONCEPTS

16. Ultraviolet radiation runs from about 50 to 350 nm. The radio part of the spectrum is very broad, running from around 10 cm to over 1 km in wavelength. When combining multiple parts of the spectrum, astronomers gain a great deal more information about an object.

21. Astronomers find the same elements across the entire universe—locally as well as in the farthest objects observed.

28. The brightnesses of stars do not reveal their distances because stars vary greatly in their luminosities.

APPLYING THE CONCEPTS

36. 10,000 km/s. Away from you.
38. 0.00005 times or 0.005 percent as bright

Chapter 6

CHECK YOUR UNDERSTANDING

6.1a. a, c
6.1b. a
6.2. c
6.3. d
6.4. c
6.5. a, b, d

WHAT IF . . .

p. 146. One possibility is that our eyes would capture more light.

p. 156. With a quantum efficiency of 100% and an integration time of 1 ms, we would be able to see in the dark and distinguish separate events that happen far faster than we can distinguish now.

p. 159. With almost no atmosphere, astronomers on this planet would see celestial objects far more sharply and would not have to worry about atmospheric blurring.

p. 165. A primary goal would be to find out if there is life on any planets in those planetary systems.

p. 166. We might conclude that mergers of black holes and neutron stars were more rare than expected.

TEST YOUR UNDERSTANDING

1. c
7. a
10. b
13. d
14. b

THINKING ABOUT THE CONCEPTS

19. Information gathered from observations at all wavelengths tells us so much more about the objects we observe.

23. An adaptive optic system will observe a laser-generated star and use sophisticated computer programs to control a deformable mirror to keep the image steady. Figure 6.14 reveals the fine details that become apparent with the use of adaptive optics.

29. Neutrino detectors are located deep inside mines. Neutrinos are neutral and interact very weakly with matter, and the only way to detect one is to watch the result of it impacting another atom.

APPLYING THE CONCEPTS

36. (a) 14 arcsec. (b) This limit is 5 to 10 times smaller than the stated resolution. (c) The diffraction limit gives the theoretical limit. The lens of an eye and the placement of the rods and cones on the retina are far from the perfection needed for such high resolution.

43. (a) 2.1×10^{10} km. (b) 19 hr. (c) 5.2×10^{-4}, or 0.05 percent.

Chapter 7

CHECK YOUR UNDERSTANDING

7.1. a, b, d
7.2. a
7.3. b
7.4. a
7.5. c

WHAT IF . . .

p. 178. A planetary system with planets orbiting in oblique (slanted to each other), separate planes could still support the nebular hypothesis. We would expect there to be multiple collisions between planets during formation. Perhaps collisions would be the cause for the different orbital planes. For example, Pluto's orbit.

p. 186. The dwarf planets far from the Sun have no atmospheres and their compositions are mostly solid ices. So far, none has a mass as great as Earth. With the paucity of planet-forming material far from its central star, the planet probably formed close to its star and was gravitationally flung out by interactions with one or more massive planets.

p. 194. One would conclude that solar systems are quite rare; fewer than 10 million of them in our galaxy. We would not yet conclude that there are no other planetary systems in the Milky Way. More stars may have planets, but the planets would be of very low mass and/or very far from their central stars; in other words, not similar to those that we have discovered with current technology.

TEST YOUR UNDERSTANDING

2. c
7. a
8. a
12. c
15. d

THINKING ABOUT THE CONCEPTS

16. What an Astronomer Sees, Figure 7.8.

19. A protoplanetary disk is the spinning disk of gas and dust out of which planets form around a central star. The inner part will be hotter because it is closer to the central star and because the inner material gained energy when it moved in from the outer regions.

28. Stars are very bright and far away, and the very faint reflected light from an exoplanet would come from a position extremely close to its star. Thus, it is difficult to mask out the light of the star and still see close enough to it to see the reflected light of an exoplanet.

30. The Kepler satellite used the planetary transit method and the monitoring of hundreds of millions of stars in the constellation of Cygnus to search for exoplanets. By "Earth-like" we mean planets that are terrestrial and have roughly the same mass as Earth, have thick protective atmospheres, and orbit in the habitable zones of their stars.

APPLYING THE CONCEPTS

36. $30 \text{ m/s} \times 2.23694 \text{ mph/1 m/s} = 67$ mph. Earth's orbital speed is 1000 times that.

38. $$\frac{L_{\text{Venus}}}{L_{\text{Earth}}} = \frac{m_{\text{Venus}}}{m_{\text{Earth}}} \times \left(\frac{R_{\text{Venus}}}{R_{\text{Earth}}}\right)^2 \times \frac{P_{\text{Earth}}}{P_{\text{Venus}}}$$

$$= 0.815 \times 0.950^2 \times \frac{1}{243} = 0.003$$

42. $\triangle \lambda = 500 \text{ nm} \dfrac{0.09 \text{ m/s}}{3 \times 10^8 \text{ m/s}} = 1.5 \times 10^{-7}$ nm

Chapter 8

CHECK YOUR UNDERSTANDING

8.1. b
8.2. d
8.3a. d
8.3b. c
8.4. c
8.5. a, c, d
8.6. a, c, d, e

WHAT IF . . .

p. 207. We would expect equal or greater surface cratering.

p. 212. Yes, we would adjust for the beginning ratio.

p. 216. We would want the poles to migrate over time and not leave Earth unprotected from charged particles from space.

p. 219. There would be just one extremely large Hawai'ian island.

p. 225. Observing a volcanic eruption on an Earth-like planet might lead to our concluding that the planet has a hot interior and is still cooling.

p. 228. If large quantities of liquid water on Mars obeyed the water cycle, we might observe the erosion of Olympus Mons and features in Valles Marineris.

TEST YOUR UNDERSTANDING

1. b
5. d
10. c
14. c

THINKING ABOUT THE CONCEPTS

22. We can detect a shadow zone for secondary earthquake waves.

25. Mars has the largest volcanoes with Olympus Mons standing 27 km high and others that reach over 14 km. Mauna Kea on Earth reaches over 9 km high from its base.

27. Mercury and the Moon have many more craters of all sizes than Earth, Venus, and Mars.

30. Given that the region surrounding the large crater has very few craters, we can state that the surface is relatively young.

APPLYING THE CONCEPTS

36. (a) Up until about 3 billion years ago, the cratering rate fell off abruptly. Since then, the rate has fallen gradually. (b) It is hard to quantify without a y-axis scale, but today's rate is very close to zero. (c) The drop in the cratering rate indicates that almost all of the leftover material from the formation of the Solar System has been incorporated into the larger worlds.

45. 1.2×10^8, or 120 million years

Chapter 9

CHECK YOUR UNDERSTANDING

9.1. a, b, d
9.2. c
9.3a. c, a, e, b, d
9.3b. a
9.4. d, c, b. Mercury's variations are due to rotation, not revolution.
9.5. a, b, c

WHAT IF . . .

p. 243. Sublimation of CO_2 and H_2O from polar caps and impacts from asteroids and comets would help restore the atmosphere.

p. 247. The secondary atmospheres would be composed of outgassing from volcanoes and any volatiles from asteroid impacts.

p. 249. We would look for oxygen, O_2, because life is the primary producer of this molecule and it is short-lived in an atmosphere.

p. 253. Without the protection of Earth's magnetic field, astronauts would be bombarded with charged particles.

p. 258. With Mars's rotation period being similar to Earth's, we would expect the zonal winds to be more like Earth's.

p. 263. A few changes might include rising sea levels and more frequent and violent storms. Some areas on Earth will be so hot as to be inhabitable.

TEST YOUR UNDERSTANDING

3. c
6. b
7. b
11. c
14. d

THINKING ABOUT THE CONCEPTS

18. The secondary atmospheres were formed by cometary impacts and outgassing from volcanoes.
25. Venus is surrounded by a thick opaque cloud of gas.
26. The three terrestrial planets with atmospheres all have surface temperatures that are much higher than can be explained through the balance of energy received by the Sun and given off by the heated surface.

APPLYING THE CONCEPTS

37. From Figure 9.7, we see that we would have to go at least 40 km above Earth's surface to experience the same atmospheric pressure as the surface of Mars.
45. (a) 3×10^{15} kg. (b) 0.01. (c) 4×10^{38} molecules/yr. (d) Because it is the number of molecules that matters, not the percentage.

Chapter 10

CHECK YOUR UNDERSTANDING

10.1. a
10.2. b
10.3. b
10.4. d
10.5. a, d, b, c

WHAT IF . . .

p. 280. As with seasons on Uranus, Earth's poles would have 3 months of more direct Sun and 3 months of darkness. Weather during each season would be more extreme.

p. 284. Calculations show that Jupiter has enough gravity to hold on to its atmosphere of hydrogen and helium should its orbit shrink to that of Venus. However, it might get "puffed out" and become larger.

p. 290. The smaller shrinking rate would produce less thermal energy than is produced now. The shrinking rate would increase if Jupiter had more mass.

p. 294. Searching for magnetospheres would be a good way to detect other large planets due to the large sizes of these magnetic fields and the radiation they emit.

p. 296. Earth might have been either ejected from the Solar System or sent into a highly elliptical orbit. Either way, life would probably not develop.

TEST YOUR UNDERSTANDING

3. b
6. b
9. d
15. d

THINKING ABOUT THE CONCEPTS

21. Giant planets rotate very rapidly, and rapidly rotating spheres tend to expand along their equator and contract along their poles. This would make these planets oblate, or "squished."

22. Colorful regions on Jupiter are those for which we see deeper into its atmosphere and effectively see "smog"; that is, chemical pollutants that have caused the gases and ices to take on reddish colors. Methane gas, which is abundant in the atmospheres of Uranus and Neptune, strongly absorbs the longer red wavelengths of visible light but reflects the shorter blue and green wavelengths.

28. Jupiter moved from right to left in the image, reflecting its orbital prograde motion of west to east.

APPLYING THE CONCEPTS

36. Figure 10.1a shows the relative sizes of the planets and Figure 10.1b shows the relative angular diameters when viewed from Earth.

42. (a) 1320 Earths. (b) $p = \dfrac{M}{V} = \dfrac{318 \text{ Earths}}{1{,}320 \text{ Earths}} = 0.24$.

Chapter 11

CHECK YOUR UNDERSTANDING

11.1. a, b, c
11.2a. b, c, d
11.2b. a, d, b, c
11.3. a
11.4. b

WHAT IF . . .

p. 306. Possible answer: The sixth moon orbits either inside all other moons or it orbits outside them.

p. 314. Answers will vary. Possible moons are those thought to have salty oceans beneath their icy crusts.

p. 320. The resulting ring would fall along the Moon's orbital path. The ring could persist for a length of time as the Moon orbits outside Earth's Roche limit. We would see light reflected from the volcanic dust particles emitted.

TEST YOUR UNDERSTANDING

2. b
7. c
11. a
15. c

THINKING ABOUT THE CONCEPTS

19. Ultraviolet photons from the Sun photodissociate the methane molecules, producing organic molecules such as ethane.

24. Planetary ring material is likely produced by the breakup of moons, asteroids, or comets that fall inside a planet's Roche limit. It is also produced as nearby moons shed material through volcanic processes or impact events.

25. Gaps in Saturn's rings are created by small moons that clear out a lane of material as they orbit within a ring. Gaps are also created by orbital resonances of shepherd moons.

APPLYING THE CONCEPTS

37. (a) Both scales are linear. (b) Figure 11.2b covers 225 times as much area.
45. (a) A more massive planet will have a higher escape velocity. (b) The escape velocity is less. (c) Find the planet's radius.

Chapter 12

CHECK YOUR UNDERSTANDING

12.1. because they orbit the Sun and not another planet
12.2. c
12.3a. c
12.3b. c
12.4. a, b, c, e
12.5. d

WHAT IF . . .

p. 336. round moons, additional Kuiper Belt objects

p. 344. lots of dust, tenuous surface conditions, low gravity

p. 349. There would be two tails. Both comets would orbit together.

p. 355. Stony meteorites are hard to distinguish from regular rock. You could use a magnet to test for stony irons and irons.

p. 358. Instigate a slow, steady change in the asteroid's orbit.

TEST YOUR UNDERSTANDING

4. a
9. b
12. b
15. d

THINKING ABOUT THE CONCEPTS

19. An old surface would have had time to become heavily cratered.
22. Asteroids that cross Earth's orbit are potentially hazardous.
27. Comet tails have extremely low densities.

APPLYING THE CONCEPTS

36. 340 km/hr.
43. (a) 29 times. (b) 9×10^{12} kg. (c) 0.04 or 4 percent.

Chapter 13

CHECK YOUR UNDERSTANDING

13.1. b
13.2. d
13.3. a, b, d
13.4. b, d

WHAT IF . . .

p. 370. We would expect stars at very different distances, implying a range of luminosities.

p. 376. The stars have different compositions, but similar temperature.

p. 381. You could still get velocities, and Kepler's laws apply.

p. 385. as a linear relationship resembling Figure 13.18b, since higher masses mean higher temperatures

TEST YOUR UNDERSTANDING

1. b
5. b
8. a
10. d
15. a

THINKING ABOUT THE CONCEPTS

21. The original spectral sequence was based on the strengths of the hydrogen lines in the spectra and was developed before we understood atoms. When that knowledge came, we realized that O and B stars had weak hydrogen lines due to ionization, and they were moved ahead of A stars. The temperature sequence runs from the very hottest stars to the very coolest.

24. That third star would be Proxima Centauri. Since it is at the same distance as A and B, it would be dimmer with a much lower luminosity.

26. On Mars, we will be able to measure the distances to about 1.5 times as far. On Jupiter, we would get accurate parallaxes to about 5 times farther away. If we were on Venus, we would measure accurate parallaxes only for stars closer than about 70 percent of what we currently measure.

APPLYING THE CONCEPTS

36. Star B has ½ the parallax as star A. The third star would have ¼ the parallax. The third star has ½ the parallax as star B.

38. (a) Almost 10^6 (1 million) times as luminous. (b) About 10^{-4} (0.0001) times as luminous.

Chapter 14

CHECK YOUR UNDERSTANDING

14.1a. c
14.1b. b
14.2. c
14.3. b
14.4. d

WHAT IF . . .

p. 403. There would be no variation between night and day.

p. 407. Either the opacity or the method of energy transport or both could be different.

p. 411. The photon energy would be much higher.

p. 417. There would be fewer sunspots.

TEST YOUR UNDERSTANDING

1. a
7. c
11. b, c, d, f, g, e, a
15. c

THINKING ABOUT THE CONCEPTS

20. Fission requires heavy elements to exist in the core, such as uranium with atomic number 92. The Sun is made up almost exclusively of light elements; mainly H and He.

24. The winds from solar flares contain huge amounts of charged particles that can disrupt or destroy electronics on satellites and orbiting telescopes. The periods of increased solar activity cause Earth's upper atmosphere to expand slightly, increasing the drag on orbiting satellites and causing their orbits to decay. Unless the satellites are boosted back up again, this effect can cause them to crash to the ground.

28. Data from sunspots give us the Sun's rotation period, the direction of rotation, and the fact that it experiences differential rotation.

APPLYING THE CONCEPTS

37. Figure 14.19 shows (b) the location and number of sunspots versus time and (c) the strength and position of magnetic fields versus time. If we are to argue that spots result from strong magnetic fields, then they should appear in roughly the same regions over time. Inspection of the figures shows that the butterfly shape seen in panel (b) is reproduced in panel (c), confirming that the two effects are correlated.

42. 3.75×10^5 s ~ 4.3 dy

Chapter 15

CHECK YOUR UNDERSTANDING

15.1. a, d
15.2. c
15.3. a
15.4. b

WHAT IF . . .

p. 433. The dust in the cloud would be "blown" away by the high-energy photons.

p. 439. O stars that form fast and first tend to clear out material from the star-forming region and thus prevent some stars from forming.

p. 445. Jupiter would be much brighter as a brown dwarf, especially at infrared wavelengths.

p. 448. That it was still hidden by lots of dust.

TEST YOUR UNDERSTANDING

4. c
8. c
11. a
12. c
15. a

THINKING ABOUT THE CONCEPTS

17. The regions where stars are forming are usually shrouded by a lot of dust that blocks visible light. Infrared wavelengths, however, are not blocked so astronomers can observe details otherwise hidden.

22. Low-density regions have very few collisions that lead to the cooling of the gas, unlike high-density regions, where collisions are frequent.

26. The core of the star eventually starts fusing hydrogen to helium, which provides the pressure needed to halt the gravitational collapse.

27. Any regions that contain a lot of molecular hydrogen, H_2, would be seen at radio wavelengths. Visible wavelengths may show the dust disk surrounding the star, blocking the star's light, and maybe regions where the jets are creating shock waves in the interstellar medium.

APPLYING THE CONCEPTS

36. Reddening by the interstellar dust will make stars appear redder than they are by reducing the amount of blue light observed.

41. $\dfrac{v_e}{v_p} = 43$

Chapter 16

CHECK YOUR UNDERSTANDING

16.1. a
16.2. b
16.3. a, c
16.4. a
16.5. d

WHAT IF . . .

p. 461. The stars would live a lot longer on the main sequence.

p. 462. The orbit of the Moon would not change, but the acceleration due to gravity on Earth's surface would be 100^2 or 10,000 times greater.

p. 467. The core would keep contracting until the temperature and density were high enough to start helium-to-carbon fusion.

p. 473. The inner planets might be engulfed. Mass loss from the Sun might mean the size of planetary orbits would increase.

p. 475. The planets survived the evolution of their star and are probably all rocky.

p. 476. Each star would evolve according to its mass.

p. 478. It would depend on how much mass it accreted from its companion and what its current mass is. With a companion having 0.8 times Sun's mass, there would probably be no worry about it going supernova.

TEST YOUR UNDERSTANDING

2. a, e, d, c, f, b, g
8. d
10. c
14. a
15. c

THINKING ABOUT THE CONCEPTS

20. For having the same brightness, the derived distances to Type Ia supernovae would increase if it were found that they were more luminous than previously thought.

24. The horizontal branch star has a smaller radius.

27. It is cooling while staying the same size.

30. Low-mass stars spend enough time on the main sequence to increase the possibility that complex life has formed.

APPLYING THE CONCEPTS

41. (a) 1.5. (b) 45.

45. 112,000 m = 112 km

Chapter 17

CHECK YOUR UNDERSTANDING

17.1. b

17.2. b

17.3. d

17.4. a

WHAT IF . . .

p. 492. The white dwarf must have been more massive originally as it evolved faster. The currently variable star must have gained mass from its companion.

p. 497. neutrinos and then the visible light

p. 506. The gas may be totally dispersed. The neutron star would be rotating more slowly.

p. 510. There were many more generations of stars that formed and died that enriched the interstellar medium from which this star formed.

TEST YOUR UNDERSTANDING

2. e, a, f, b, c, d

4. d

7. e

12. a

14. b

THINKING ABOUT THE CONCEPTS

16. The bumps most noticeably change in brightness. It doesn't appear that the ring changed size.

18. High-mass stars begin fusing helium to carbon in their core when the temperatures and pressures get high enough. There is no helium flash.

21. We just need to measure the period of the variability to get the luminosity of the star through the observed period-luminosity relation.

28. There is a relationship between the mass of a main-sequence star and how long it stays on the main sequence.

APPLYING THE CONCEPTS

38. The bottom image of Figure 17.4 is not to scale.

$$\frac{0.01\,R_{Sun}}{1,000\,R_{Sun}} = 0.00001$$

42. (a) 3.8×10^{23} kg/min. (b) 5.2 (Moon masses).

Chapter 18

CHECK YOUR UNDERSTANDING

18.1. d

18.2. b

18.3. d

18.4. c

WHAT IF . . .

p. 520. Converting the speed of light to meters per second: 200 km/hr = 0.056 km/s = 56 m/s. The length of a soccer field is ~110 m. It would take ~2 s for a goalie to see what was happening at the other goal.

p. 524. Using $E = mc^2$ and solving for m: $m = E/c^2$, leading to the conclusion that a change in energy of the containers results in a change in its mass, but one that is extremely small.

p. 528. The ball would remain in orbit and would thus be accelerated by Earth's gravity. If the space station is an inertial frame, then so is the ball.

p. 538. Detection could occur through the black hole's gravitational influence on other stars or through gravitational lensing of starlight.

TEST YOUR UNDERSTANDING

1. c, a, b

6. d

8. b

11. c

15. c, d

THINKING ABOUT THE CONCEPTS

16. There is no difference between the shifts in wavelengths due to gravity or radial motion.

19. Twin A could never return before twin B was born, because time continues to pass along normally for twin B.

23. The astronaut is moving with the ship and is in the same reference frame, thus he will see no length contraction. An outside viewer who is *not* moving at that high speed would observe length contraction.

27. The color of the star would change to that of longer and longer wavelengths of light, say from yellow to orange to red to infrared and to radio.

APPLYING THE CONCEPTS

40. We might not get the *exact* number, but γ does not deviate significantly from 1 until the speed is about 0.33 to 0.5c. The Lorentz factor goes to infinity as the speed of an object approaches the speed of light.

44. (a) Working It Out 18.1 tells us for that spaceship, $v = 0.99c$, $\gamma = 7.09$. It will take 25.25 years (25/0.99)

to reach the alien civilization from both Earth's and the aliens' viewpoints. A spaceship traveling at 0.99c would take 25 ly/7.09 = 3.53 years to get there. (b) A return will happen 50.5 years later for the people on Earth.

Chapter 19

CHECK YOUR UNDERSTANDING

19.1a. d

19.1b. c

19.2. c

19.3. a

19.4. c

WHAT IF . . .

p. 555. If two giant spiral galaxies merged perpendicularly, then a giant elliptical galaxy might form with stars orbiting in all directions. If they merged edge on, and the disks were rotating in the same direction, then maybe most stars would eventually make up an even larger giant spiral.

p. 556. The distances to galaxies would all be greater.

p. 563. WIMPS would be streaming from all directions due to the random motions of halo objects. Since WIMPS would be detected by their mass, sometimes most of the particles would be coming at 250 km/s and 6 months later most of the particles would be coming at 190 km/s.

p. 568. Even before the supermassive black holes powering the AGNs merged, we would expect to observe gravitational waves.

TEST YOUR UNDERSTANDING

2. c

5. a

10. d

11. c

15. b

THINKING ABOUT THE CONCEPTS

16. Between 2 spirals: 7. Between 2 ellipticals: 0. Between an elliptical and a spiral: 1

19. Observing different kinds of standard candles gives one more confidence in the results.

25. Quasars appear to be associated with mergers and interactions between galaxies. Because they are observed at great distance, and thus back to an earlier time, these mergers must have been much more prevalent in the past when the universe was younger and the distances between galaxies smaller.

29. A nearby rung is needed to calibrate a more distant one; there must be overlap so that we can reliably use the more distant rungs.

APPLYING THE CONCEPTS

41. Because the object varies within an 83-minute period, it can be no larger than 83 light-minutes

across. Light travel time for 1 AU is 8.3 light-minutes, implying the object is 10 AU across.

45. (a) $E = \triangle mc^2 = (5.97 \times 10^{24}\,\text{kg}) \times 0.1 \times (3 \times 10^8\,\text{m/s})^2$
$= 5.4 \times 10^{40}\,\text{J}$. (b) $\dfrac{5.4 \times 10^{40}\,\text{J}}{3.85 \times 10^{26}\,\text{J}} = 1.4 \times 10^{14}$

Chapter 20

CHECK YOUR UNDERSTANDING

20.1. The Milky Way has lots of star formation and dust. Observations over a wide range of wavelengths show spiral structure and the motion of stars orbiting the center on a disk.

20.2. The bulge, halo, and thick disk contain mostly old stars whereas the thin disk contains young stars.

20.3. c

20.4. The gravitational force between the Andromeda and Milky Way galaxies is strong enough to overcome other motions.

WHAT IF . . .

p. 583. There would not be a lot of dust and gas in these arms as no new star formation is occurring.

p. 585. The star would be low mass (a red dwarf) and crossing the disk in its halo orbit. We would expect this star to be a high-velocity star.

p. 593. If the Large Magellanic Cloud were to merge with our galaxy and came close enough to the center to merge with the black hole there, the black hole may then become an active galactic nucleus.

p. 597. With a much more crowded cluster of galaxies, the Milky Way might come close enough to merge with one or more of the other galaxies. Stars with planets would be affected, and if the mergers occurred early enough, and Earth were ejected from solar system, life may not have had time enough for humans to appear.

TEST YOUR UNDERSTANDING

2. a
7. b
8. b
13. a, b, c

THINKING ABOUT THE CONCEPTS

16. We observe many spiral galaxies and note that there are central bright areas surrounded by relatively flat disks. These disks contain dust and hot, bright stars. Since we observe many spiral galaxies with a range of viewing angles, we can assume that the Milky Way would be similar if, for example, it were viewed face on. If we lived in an elliptical galaxy, we would not see any dust or formation of massive stars. The stars we do see would be distributed spherically around us.

21. Halo stars are distinguished from disk stars by their low abundances of elements heavier than helium and their relatively fast orbits that trace back to the halo.

23. Lots and lots of dust between Earth and the center of the galaxy is opaque at visible wavelengths but transparent at X-ray, IR, and radio wavelengths.

26. The Large and Small Magellanic clouds are dwarf satellite galaxies that have interacted many times with the Milky Way, exchanging gas and dust and possibly stars.

APPLYING THE CONCEPTS

40. (a) 480 Myr. (b) 27 orbits.

45. Ratio: $\dfrac{158\,\text{kpc}^3}{296\,\text{kpc}^3} = 0.53$.

Chapter 21

CHECK YOUR UNDERSTANDING

21.1. b, a
21.2. d
21.3a. c
21.3b. b
21.4. a

WHAT IF . . .

p. 607. We would still not live in a special place because anyone anywhere else would see the same patterns that we do.

p. 612. Still, nothing special would be implied because the same relationship would be observed from a distant galaxy looking back at the Milky Way.

p. 615. If the universe were contracting, the light from other galaxies would be blueshifted.

p. 618. A dark cloud having a temperature of 1.5K would have a very high density.

TEST YOUR UNDERSTANDING

3. c
5. d
10. c
11. d
13. d

THINKING ABOUT THE CONCEPTS

16. The features in each view are in the same areas at the same relative scales. What has improved is the resolution, something with which astronomers are very familiar.

18. The galaxies are fixed in space, and it is that space that is expanding.

22. The expansion of the universe applies to the spaces among galaxy clusters. Gravity is in control for the Sun (as it is for all stars) and the Milky Way (as it is for all galaxies).

29. As it expanded, the universe became too cool to fuse any elements heavier than hydrogen to helium.

APPLYING THE CONCEPTS

38. (a) theory. (b) observations. (c) Within the ranges given, the theory and observations match.

41. (a) 4100 Mpc. (b) The age of the universe times the scale factor gives an age of about 2 billion years. Table 23.1 gives an age between 900 million ($z = 6$) and 1.2 billion ($z = 5$) years. (c) $R_U = 0.15$ The universe has expanded by about 7 times.

Chapter 22

CHECK YOUR UNDERSTANDING

22.1. c
22.2. c
22.3. The flatness and horizon problems.
22.4. d
22.5. a

WHAT IF . . .

p. 631. With $\Omega > 1$, the universe is closed and astronomers far in the future would observe the universe contracting and the light from galaxies blueshifted.

p. 632. All the galaxies would be much farther apart.

p. 641. The four forces would still break, but the ripples in the cosmic microwave background would not be smoothed out. There would be no flatness or horizon problem.

p. 645. Yes. Since an interaction of dark matter particles and antiparticles today would be observable at some distance away.

TEST YOUR UNDERSTANDING

2. c
5. a, f, j, b, e, g, d, i, h, c
10. b
14. b
15. b

THINKING ABOUT THE CONCEPTS

18. Dark energy is vacuum energy—that is, a repulsive force that results from the energy present in totally empty space.

19. Gravity is in control for the Milky Way (and all galaxies and galaxy clusters), the Solar System (and all planetary systems), and for all planets.

23. The early universe was hot and dense, which are situations that are replicated within the realm of high-energy particle physics. At these high temperatures and densities, all sorts of exotic forms of sub-atomic matter can exist that may have had a substantial influence on how the early universe evolved.

28. Grand unified theories (GUTs) attempt to unite the electroweak and strong nuclear forces. Theories of everything (TOEs) attempt to unify suitable GUTs with gravity.

APPLYING THE CONCEPTS

34. Each gamma ray has an energy: $E = 8.2 \times 10^{-14}\,\text{J}$, and 2 gamma rays are emitted.

40. (a) The time axis is approximately logarithmic. (b) Density has dropped by 125 orders of magnitude. (c) Temperature has dropped by about 31 orders of magnitude.

Chapter 23

CHECK YOUR UNDERSTANDING

23.1. c, b, d, a
23.2. b
23.3. c
23.4. b

WHAT IF . . .

p. 659. With dark matter filling the voids, the total mass of the universe would increase, thereby increasing the density, perhaps exceeding the critical level.

p. 666. Without the gravity from dark matter, it would take longer for structure to form. More time would be needed for intelligent life to appear.

p. 668. We should look at "blank" regions of the sky at infrared wavelengths.

p. 678. Infrared wavelengths would be used to observe galaxies that are far away, at higher redshifts, and thus younger than those seen at visible wavelengths. We would need to pause the simulation to match the age of the universe corresponding to the redshifts.

TEST YOUR UNDERSTANDING

3. b
6. a
8. d
9. a
14. d

THINKING ABOUT THE CONCEPTS

17. If we could see the early universe when galaxies were first forming, it would be smaller and more active. There was a large amount of hot gas but little dust. Smaller, irregularly shaped protogalaxies dominated the visible universe. Galaxies were merging, making irregular galaxies a common sight.

23. Inhomogeneities in the density of normal matter were never strong enough to cause the structure seen today. There had to be another source of gravity.

27. As we look back in time, we should see more spiral galaxies than elliptical galaxies. As we look back even farther in distance and time, we should see galaxies that have not yet formed spiral arms, and galaxies should have a lot of massive stars in them because they have just formed and are not yet old enough to have exploded. This is what is observed.

30. The strong and weak nuclear forces are too short ranged to work over cosmic distances. Electromagnetic forces work only for charged particles, and most matter in the universe is neutral. This leaves only gravity.

APPLYING THE CONCEPTS

34. $z = 6.6$

43. Early universe is on the right, at a redshift of 10. The star formation rate now is about 0.01 M_{Sun}/yr Mpc³ whereas the star formation rate at its peak, some 2.5 billion years after the Big Bang, was approximately 0.2 M_{Sun}/yr Mpc³. This is 20 times higher than today's rate.

Chapter 24

CHECK YOUR UNDERSTANDING

24.1. d
24.2. a
24.3a. a, b, c
24.3b. a
24.4. b

WHAT IF . . .

p. 689. The environments of the ocean depths and tide pools are very different and thus we would expect that perhaps two different strands of life would evolve.

p. 694. Yes. Even a small change in a basic molecule of self-replicating life may lead to evolution if that change is beneficial and can be passed on.

p. 699. Even a weak magnetic field would offer some protection against cosmic rays, and the greenhouse gases would keep the planet warmer if it is at the outer edge of the star's habitable zone, thus increasing the size of the zone.

p. 702. These are good questions posed by Carl Sagan with many, many answers.

TEST YOUR UNDERSTANDING

4. c
6. a, c
9. d
13. c
15. d

THINKING ABOUT THE CONCEPTS

20. Cyanobacteria slowly broke CO_2 molecules apart through photosynthesis to form oxygen. This required about 2.25 billion years to reach the oxygen levels we have today.

26. The hydrogen and most of the helium in our bodies are from the Big Bang. All the other atoms are from the winds produced by stars as they ascended the giant branches and supernovae.

30. Main-sequence stars slowly become more luminous over time, pushing the inner edge of the habitable zone out past Earth. In a billion years or so, the increase in the Sun's luminosity will cause the oceans to evaporate and scorch all of Earth.

APPLYING THE CONCEPTS

39. 6.87×10^{11} or 687 billion bacteria.

43. Mercury is not in the habitable zone. Its range of temperatures is due to its lack of an atmosphere and the differences between its "day" and its "night."

These pages contain credits for all 24 chapters in the complete volume of *21st Century Astronomy*, Seventh Edition. The Solar System Edition does not include Chapters 15–23. The Stars and Galaxies Edition does not include Chapters 8–12.

Photos

FRONT MATTER

Page ix: Left_Coast_Photographer/Istockphoto/Getty Images Plus; **p. xi:** All Canada Photos/Alamy Stock Photo; **p. xiv:** SDO/AIA; **p. xvi:** NG Images/Alamy Stock Photo; **p. xxi:** Anne DeMarinis Design for W. W. Norton & Company; **p. xxii:** NASA, ESA, and the Hubble 20th Anniversary Team (STScI); **p. xxiv both:** W. W. Norton & Company; **p. xxxiv author Palen:** Courtesy Stacy Palen; **p. xxxiv author Blumenthal:** Courtesy George Blumenthal.

CHAPTER 1

Pages 2–3 composite: Anne DeMarinis Design for W. W. Norton & Company; **p. 3 trees:** Andrew Bret Wallis/Getty Images; **p. 6 moon:** NASA; **p. 6 Saturn:** NASA, ESA, J. Clarke (Boston University, USA), and Z. Levay (STScI), https://esahubble.org/products/calendars/cal200604/, https://creativecommons.org/licenses/by/4.0/; **p. 7 sea growth:** Chase Studio/Science Source; **p. 7 Earth:** NASA; **p. 13:** GL Archive/Alamy Stock Photo; **p. 15 tree:** Don Hammond/Design Pics/Science Source; **p. 17:** ESA/Hubble & NASA, https://esahubble.org/images/potw1208a/, https://creativecommons.org/licenses/by/4.0/.

CHAPTER 2

Pages 22–23 composite: Anne DeMarinis Design for W. W. Norton & Company; **p. 22:** Deanna Truesdale/EyeEm/Getty Images; **p. 23 moon day 14:** Anne DeMarinis Design for W. W. Norton & Co.; **p. 23 moon day 5:** Anthony Qualkinbush/EyeEm/Getty Images; **p. 24 top:** robertharding/Alamy Stock Photo; **p. 24 bottom:** Daniela Constantinescu/Shutterstock; **p. 28 left:** Pekka Parviainen/Science Source; **p. 28 right:** D. Nunuk/Science Source; **p. 39:** Dr. Juerg Alean/Science Source; **p. 44:** akg images/Rabatti – Dominige; **p. 47 top:** © Laura Kay; **p. 47 center:** HEINZ-PETER BADER/REUTERS/Newscom; **p. 47 bottom:** Laura Kay and Lisa Rand; **p. 51 left:** knickohr/Getty Images; **p. 51 right:** Courtesy © Wang Letian.

CHAPTER 3

Pages 58–59 composite: Anne DeMarinis Design for W. W. Norton & Company; **p. 58 lightbulb:** Jakub Gojda/Alamy Stock Photo; **p. 58 paper balls:** Anne DeMarinis Design for W. W. Norton & Co.; **p. 60:** Image Select/Art Resource, NY; **p. 61 top:** © Tunc Tezel; **p. 61 bottom:** GL Archive/Alamy Stock Photo; **p. 62:** Bettmann/Getty Images; **p. 66:** Art Collection 2/Alamy Stock Photo; **p. 67 top:** Science History Images/Alamy Stock Photo; **p. 67 bottom:** North Wind Picture Archives/Alamy Stock Photo; **p. 69 Copernicus:** Science History Images/Alamy Stock Photo; **p. 69 Tycho:** Pictorial Press Ltd/Alamy Stock Photo; **p. 69 Kepler:** GL Archive/Alamy Stock Photo; **p. 69 Newton:** Heinz-Dieter Falkenstein/age fotostock/Superstock; **p. 71:** FineArt/Alamy Stock Photo; **p. 72 top:** GRANGER; **p. 72 bottom:** David Hajnal/Shutterstock; **p. 73:** Lebrecht Music & Arts/Alamy Stock Photo.

CHAPTER 4

Pages 82–83 composite: Anne DeMarinis for W. W. Norton & Company; **p. 92 Earth:** NASA Johnson Space Center; **p. 92 moon:** NASA/JPL/USGS; **p. 92 lunar subsatellite:** JSC/NASA; **p. 92 David Scott:** JSC/NASA; **p. 92 Galileo:** ITAR-TASS News Agency/Alamy Stock Photo; **p. 99 left:** Eric Carr/Alamy Stock Photo; **p. 99 right:** Sandi Culifer/Shutterstock; **p. 103:** NASA, Holland Ford (JHU), the ACS Science Team and ESA, https://esahubble.org/images/heic0206b/, https://creativecommons.org/licenses/by/4.0/.

CHAPTER 5

Pages 108–109 composite: Anne DeMarinis for W. W. Norton & Company; **p. 108 pencils:** Julladit Portfolio/Shutterstock; **p. 108 single slit diffraction:** Ted Kinsman/Science Source; **p. 109 tape:** fStop Images GmbH/Alamy Stock Photo; **p. 116 Romer:** Paul Fearn/Alamy Stock Photo; **p. 116 Bradley:** Heinz-Dieter Falkenstein/age fotostock/Superstock; **p. 116 Maxwell:** Iberfoto/Superstock; **p. 116 Einstein:** Underwood Photo Archives/Superstock; **p. 124:** ESO, https://www.eso.org/public/images/eso1413a/, https://creativecommons.org/licenses/by/4.0/; **p. 137 Mercury:** NASA/Johns Hopkins University Applied Physics Laboratory/Carnegie Institution of Washington; **p. 137 Venus:** NASA/JPL-Caltech/ASU; **p. 137 Earth:** NASA; **p. 137 Mars:** NASA//Hubble Heritage Team (STScI/AURA); **p. 137 Jupiter, Saturn, Uranus, Neptune & Pluto:** NASA/NSSDC/GSFC.

CHAPTER 6

Pages 142–143 composite: Anne DeMarinis for W. W. Norton & Company; **p. 142 water and glass:** Richard Sharrocks/Getty Images; **p. 145:** ASU Physics Instructional Resource Team. © 2009 Arizona State Board of Regents. Used with permission; **p. 146:** Richard Dreiser, Yerkes Observatory; **p. 148 light:** ASU Physics Instructional Resource Team. © 2009 Arizona State Board of Regents. Used with permission; **p. 148 telescope:** Jim Sugar/Getty Images; **p. 150 Keck reflectors:** Enrico Sacchetti/Science Source; **p. 150 angular resolution, both:** The Space Telescope Science Institute; **p. 152 both:** Nick Law (Caltech) and Craig Mackay (Univ. of Cambridge); **p. 153 top:** ESO/L. Calçada, https://www.eso.org/public/images/eso1440e/?lang=no, https://creativecommons.org/licenses/by/4.0/; **p. 153 bottom:** NASA, Earth Observatory; **p. 154:** Courtesy, Victoria Girgis/Lowell Observatory; **p. 155 top:** Hulton-Deutsch Collection/CORBIS/Corbis via Getty Images; **p. 155 bottom:** Jean-Charles Cuillandre (CFHT)/NASA; **p. 157 Fermi:** NASA E/PO, Sonoma State University, Aurore Simonnet; **p. 157 Chandra:** NASA (Illustration: NASA/CXC/NGST); **p. 157 HST:** NASA; **p. 157 Keck:** Left_Coast_Photographer/Istockphoto/Getty Images Plus; **p. 157 Spitzer:** NASA/JPL-Caltech/R. Hurt (IPAC); **p. 157 JCMT:** POLARBEAR Consortium, UC Berkeley; **p. 157 EVLA:** NRAO/AUI/NSF; **p. 157 Green Bank:** ANDREW CABALLERO-REYNOLDS/AFP via Getty Images; **p. 157 FAST:** © Liu Xu/Xinhua/Alamy Live News/Alamy Stock photo; **p. 157 Green Bank:** ANDREW CABALLERO-REYNOLDS/AFP via Getty Images; **p. 157 FAST:** Liu Xu/Xinhua/Alamy Live News/Alamy Stock photo; **p. 158 top:** Photo by Dave Finley, courtesy National Radio Astronomy Observatory and Associated Universities, Inc.; **p. 158 bottom:** Markus Thomenius/Alamy Stock Photo; **p. 159 top:** NASA; **p. 159 bottom:** NASA/Tom Tschida; **p. 161:** NASA; **p. 164 top:** NASA/JPL-Caltech/MSSS; **p. 164 bottom:** NASA/GSF and NASA's Scientific Visualization Studio; **p. 165 photograph:** Maximilien Brice, © 2005-2021 CERN, http://cds.cern.ch/record/910381, https://creativecommons.org/licenses/by/4.0/; **p. 166:** Yuya Makino, IceCube/National Science Foundation; **p. 167 Weber:** Volker Steger/Science Source; **p. 167 LIGO:** Courtesy Caltech/MIT/LIGO Laboratory; **p. 167 LISA:** NASA; **p. 168 all:** Patrik Jonsson, Greg Novak & Joel Primack, UC Santa Cruz, 2008.

CHAPTER 7

Pages 174–175 composite: Anne DeMarinis for W. W. Norton & Company; **p. 177 left:** NASA/Hubble/STScI; **p. 177 right:** ALMA (NRAO/ESO/NAOJ); C. Brogan, B. Saxton (NRAO/AUI/NSF)/Science Source; **p. 178:** NASA; **p. 180:** John Schults/REUTERS/Newscom; **p. 183:** NASA, ESA, and the Hubble 20th Anniversary Team (STScI); **p. 188:** NASA/Johns Hopkins University Applied Physics Laboratory/Carnegie Institution of Washington; **p. 193 top:** ESO/A.-M. Lagrange et al., https://www.eso.org/public/images/eso0842b/, https://creativecommons.org/licenses/by/4.0/; **p. 193 bottom:** NASA, ESA and P. Kalas (University of California, Berkeley and Seti Institute); **p. 197:** NASA.

CHAPTER 8

Pages 202–203 composite: Anne DeMarinis for W. W. Norton & Co.; **pp. 202–203 moon, background, and inset:** NASA/GSFC/Arizona State University; **p. 205:** NASA/GSFC/Arizona State University; **p. 206 sequence:** RGB Ventures/SuperStock/Alamy Stock Photo; **p. 207 top:** Kit Leong/Shutterstock; **p. 207 bottom:** NASA/Goddard/MIT/Brown; **p. 208:** European Space Agency/Science Source; **p. 209:** NASA's Scientific Visualization Studio; **p. 212 both:** NASA; **p. 216:** Spencer Grant/Science Source; **p. 218:** Francois Gohier/Science Source; **p. 222 top four:** NASA; **p. 222 bottom left:** NASA/USGS; **p. 222 bottom right:** European Space Agency/DLR/FU Berlin/G. Neukum/Science Source; **p. 223:** NASA/JPL; **p. 224 Mauna Kea:** WorldSurlo/Shutterstock; **p. 224 Mt. Fuji:** Jukurae/Shutterstock; **p. 224 rock:** NASA/JSC; **p. 225 top:** NASA/JSC/Arizona State University; **p. 225 bottom:** NASA/Johns Hopkins University Applied Physics Laboratory/Carnegie Institution of Washington; **p. 226 top:** NASA/JPL/Malin Space Science Systems; **p. 226 center and bottom:** NASA/JPL-Caltech/ESA; **p. 227:** NASA/JPL-Caltech/Univ of Arizona; **p. 228 top:** NASA/JPL/University of Arizona; **p. 228 bottom:** NASA/JPL-Caltech/MSSS; **p. 229 top:** NASA/JPL-Caltech/University of Arizona/Texas A&M University; **p. 229 bottom:** NASA/JPL-Caltech/UA/USGS; **p. 230 both:** NASA; **p. 231:** NASA; **p. 233:** Joe Tucciarone/Science Source.

CHAPTER 9

Pages 238–239 composite: Anne DeMarinis Design for W. W. Norton & Company; **p. 245:** NASA's Goddard Space Flight Center; **p. 249:** Seth White; **p. 253 left:** NASA; **p. 253 right:** All Canada Photos/Alamy Stock Photo; **p. 256:** NASA/JPL-Caltech; **p. 258 top:** Russian Academy of Sciences/Ted Stryk; **p. 258 bottom:** NASA/JPL; **p. 259 top:** NASA/JPL/USGS; **p. 259 bottom:** NASA, James Bell (Cornell Univ.), Michael Wolff (Space Science Inst.), and The Hubble Heritage Team (STScI/AURA); **p. 260:** NASA/JPL-Caltech/University of Arizona; **p. 267 Venus:** NASA/JPL-Caltech; **p. 267 Earth:** NASA, Johnson Space Center; **p. 267 Mars:** Phil James (Univ. Toledo), Todd Clancy (Space Science Inst., Boulder, CO), Steve Lee (Univ. Colorado), and NASA/ESA, https://esahubble.org/images/opo9715c/, https://creativecommons.org/licenses/by/4.0/.

CHAPTER 10

Pages 272–273 composite: Anne DeMarinis Design for W. W. Norton & Company; **pp. 272–273 background spread:** Jordana Meilleur/Alamy Stock Photo; **p. 276 Jupiter:** NASA, ESA, STScI, A. Simon (Goddard Space Flight Center), M.H. Wong (University of California, Berkeley), and the OPAL team; **p. 276 Saturn:** NASA/JPL/Space Science Institute; **p. 276 Uranus:** NASA/JPL-Caltech; **p. 276 Neptune:** NASA/JPL; **p. 277 Jupiter:** NASA, ESA, STScI, A. Simon (Goddard Space Flight Center), M.H. Wong (University of California, Berkeley), and the OPAL team; **p. 277 Saturn:** NASA/JPL/Space Science Institute; **p. 277 Uranus:** NASA/JPL-Caltech; **p. 277 Neptune:** NASA/JPL; **p. 278 Uranus:** NASA/JPL-Caltech; **p. 278 Neptune:** NASA/JPL; **p. 279 top:** NASA/JPL/Space Science Institute; **p. 279 bottom:** NASA/JPL; **p. 280 top:** NASA, ESA, and the Hubble Heritage Team (STScI/AURA); **p. 280 bottom:** NASA/JPL-Caltech/SwRI/MSSS/Gerald Eichstädt/© Seán Doran; **p. 281:** Enhanced image by Jason Major based on images provided courtesy of NASA/JPL-Caltech/SwRI/MSSS; **p. 282 top:** NASA/JPL-Caltech/SwRI/ASI/INAF/JIRAM; **p. 282 bottom left:** NASA/JPL-Caltech/Space Science Institute; **p. 282 bottom right:** NASA/JPL-Caltech/SSI; **p. 283 Uranus:** Lawrence Sromovsky, University of Wisconsin-Madison/W.W. Keck Observatory; **p. 283 Neptune:** NASA, ESA, and M.H. Wong and J. Tollefson (UC Berkeley); **p. 283 Neptune spot:** NASA/JPL; **p. 287:** NASA/JPL; **p. 288:** NASA, ESA, L. Sromovsky and P. Fry (University of Wisconsin), H. Hammel (Space Science Institute), and K. Rages (SETI Institute); **p. 293 water over rocks:** Sigur/Shutterstock; **p. 293 Jupiter:** NASA/ESA, https://esahubble.org/images/heic1613a/, https://creativecommons.org/licenses/by/4.0/; **p. 293 inset top:** John Clarke (University of Michigan), and NASA/ESA, https://esahubble.org/images/opo9804b/, https://

This index contains page references for all 24 chapters in the complete volume of 21st Century Astronomy, Seventh Edition. The Solar System Edition does not include Chapters 15–23 (pp. 426–683). The Stars and Galaxies Edition does not include Chapters 8–12 (pp. 202–363).